ADVANCES IN CHEMICAL PHYSICS

VOLUME XLIX

Advances in
CHEMICAL PHYSICS

EDITED BY

I. PRIGOGINE

University of Brussels
Brussels, Belgium
and
University of Texas
Austin, Texas

AND

STUART A. RICE

Department of Chemistry
and
The James Franck Institute
The University of Chicago
Chicago, Illinois

VOLUME XLIX

1807 1982
175 YEARS OF PUBLISHING

AN INTERSCIENCE® PUBLICATION

JOHN WILEY & SONS

NEW YORK · CHICHESTER · BRISBANE · TORONTO · SINGAPORE

Library of Congress Catalog Card Number: 58-9935

ISBN 0-471-09361-0
Printed in the United States of America

10 9 8 7 6 5 4 3 2 1

CONTRIBUTORS TO VOLUME XLIX

MICHAEL BAER, Soreq Nuclear Research Center, Yavne, Israel

P. BORCKMANS, Service de Chimie-Physique II, Université Libre de Bruxelles, Brussels, Belgium

JEREMY K. BURDETT, Department of Chemistry, The University of Chicago, Chicago, Illinois

H. T. DAVIS, Department of Chemical Engineering and Materials Science and Department of Chemistry, University of Minnesota, Minneapolis, Minnesota

G. DEWEL, Service de Chimie-Physique II, Université Libre de Bruxelles, Brussels, Belgium

W. EBERHARDT, Department of Physics, University of Pennsylvania, Philadelphia, Pennsylvania

GRAHAM R. FLEMING, Department of Chemistry and James Franck Institute, University of Chicago, Chicago, Illinois

L. LATHOUWERS, Dienst Teoretische en Wiskundige Natuurkunde, University of Antwerp, Antwerp, Belgium

E. W. PLUMMER, Department of Physics, University of Pennsylvania, Philadelphia, Pennsylvania

L. E. SCRIVEN, Department of Chemical Engineering and Materials Science, University of Minnesota, Minneapolis, Minnesota

S. A. SOLIN, Department of Physics, Michigan State University, East Lansing, Michigan

P. VAN LEUVEN, Dienst Teoretische en Wiskundige Natuurkunde, University of Antwerp, Antwerp, Belgium

D. WALGRAEF, Service de Chimie-Physique II, Université Libre de Bruxelles, Brussels, Belgium

INTRODUCTION

Few of us can any longer keep up with the flood of scientific literature, even in specialized subfields. Any attempt to do more, and be broadly educated with respect to a large domain of science, has the appearance of tilting at windmills. Yet the synthesis of ideas drawn from different subjects into new, powerful, general concepts is as valuable as ever, and the desire to remain educated persists in all scientists. This series, *Advances in Chemical Physics*, is devoted to helping the reader obtain general information about a wide variety of topics in chemical physics, which field we interpret very broadly. Our intent is to have experts present comprehensive analyses of subjects of interest and to encourage the expression of individual points of view. We hope that this approach to the presentation of an overview of a subject will both stimulate new research and serve as a personalized learning text for beginners in a field.

ILYA PRIGOGINE
STUART A. RICE

CONTENTS

ADVANCES IN CHEMICAL PHYSICS

VOLUME XLIX

APPLICATIONS OF CONTINUOUSLY OPERATING, SYNCHRONOUSLY MODE-LOCKED LASERS

GRAHAM R. FLEMING

Department of Chemistry and James Franck Institute
The University of Chicago
Chicago, Illinois

CONTENTS

I. INTRODUCTION

The ability to produce wavelength-tunable, ultrashort light pulses at very high repetition rates has significantly extended the scope and reliability of picosecond spectroscopy. Important applications of such light sources are also being found in other areas of spectroscopy, for example, in the tour de force of Heritage, Levine, et al[1, 2] of measuring stimulated Raman spectra of monolayers without any surface enhancement, and in the two-photon Doppler free measurements of Hänsch and co-workers on the sodium $3s$-$4d$ transition.[3]

This chapter describes the basic physical principles involved in synchronously pumped, mode-locked dye lasers and the operating characteristics of the most common type of laser—the actively mode-locked argon or krypton ion–pumped dye laser combination. Methods of application to time-resolved spectroscopic studies are then described. The chapter concludes with discussions of the applications of synchronously pumped lasers to vibrational spectroscopy and to high-resolution spectroscopy.

II. SYNCHRONOUSLY PUMPED LASERS

A synchronously pumped laser is one in which the cavity length is set equal to (or as a submultiple of) the interpulse spacing of a pump laser. In this way the cavity gain is modulated at the round-trip frequency and mode locking results. The initial synchronously pumped lasers were dye lasers pumped by high-power, mode-locked ruby [4] or Nd-glass[5] lasers, and their output thus consisted of a burst of ultrashort pulses following the pulse train of the pump laser. Synchronous pumping also enables a truly continuous train of ultrashort pulses to be generated, provided the pump laser operates continuously. The most common high-power continuous lasers are the argon and krypton ion lasers, and the finding that these lasers are readily actively mode locked to provide a stable, continuous train of pulses of 100 to 200 psec duration has led to a good deal of interest in the use of these lasers as a synchronous pump source. The synchronously pumped laser has generally been an organic dye laser, although more recently F-center lasers have also been used.[6] The synchronously pumped organic dye laser retains the tunability of the normal continuous-wave (cw) dye laser, and is capable of producing pulses of <1 psec.

A typical experimental setup is shown in the lower portion of Fig. 1, with an actively mode-locked argon laser pumping a cw dye laser with its cavity length extended to match the pump laser and thus achieve synchronous pumping. Mode locking of the ion laser is achieved by an acousto-optic modulator placed close to the rear mirror. About one watt of radio frequency (rf) power is applied through a transducer to a quartz prism, and the

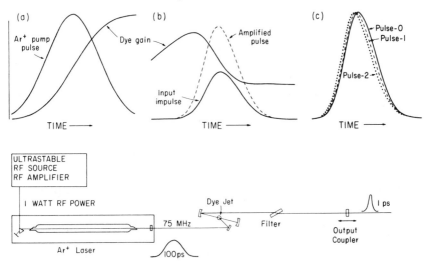

Fig. 1.   Optical system and pulse-shortening mechanism in synchronously pumped dye lasers.
(*a*) Argon laser pumping pulse and dye gain as a function of time. (*b*) Qualitative representation of the dye gain and pulse shape of input (solid curve) and output (dashed curve) dye laser pulses. The circulating dye laser pulse has arrived late in the gain medium and the amplified pulse envelope has its peak advanced (dye laser cavity slightly too short). (*c*) The input and output pulses normalized to the same peak height, for two round trips: solid curve, initial pulse; dashed curve, first round trip; dotted curve, second round trip.

frequency of the rf source is set so that light passing through the prism is diffracted at the cavity round-trip frequency. This periodic loss results in locking of the phase of the longitudinal laser modes, and a train of pulses with period ω_M^{-1} ($\sim 2L/c$), where ω_M is the modulation frequency, results.[7] All the major argon and krypton lines have been mode locked; and with a stable rf source, pulse widths are typically 100 to 150 psec. Pulses as short as 50 psec have been reported with a mode-locked krypton laser.[8] Average powers may be as high as 1.5 W for the stronger lines.

Adoption of a standard cw dye laser for synchronous pumping simply requires extension of the cavity, and provision of a sensitive length adjustment on the output mirror. When the dye laser cavity length is correctly set, the pulses emerging may be as much as 100 times shorter than the pumping ion laser pulses. The next section gives a brief qualitative description of the pulse-shortening mechanism at work in the dye laser.

The repetition rate of the dye laser is typically 75 to 80 MHz, and this enables the use of sophisticated signal-averaging detection techniques, giving very precise data, while the low pulse energy (~ 1 nJ) allows investigators to avoid the problems of nonlinear behavior and sample damage, which plagued much earlier picosecond spectroscopy. The very high repetition rate can lead to problems of its own, however. Sample heating and the building up of steady-state concentrations of transient species are possible problems: these can be overcome by the use of flowing sample cells and pulse repetition rate reduction by electro-optic or acousto-optic techniques. Hesselink and Wiersma[9] have exploited the build-up of steady-state transient populations in their observation of photon echoes from an accumulated grating in the electronic ground state.

A. Pulse-Shortening Mechanism

Qualitatively the pulse-shortening mechanism at work in the dye pulse laser results from the increasing gain on the rising edge of the dye pulse, followed by rapid depletion of the gain (gain saturation) at the peak of the dye pulse. These two factors produce greater amplification of the center of the pulse compared with the wings, thus produce pulse shortening. Figure 1 represents a qualitative attempt to depict the pulse-shortening process.

In the absence of a circulating dye pulse, the gain in the dye medium will rise as the convolution of the argon pump pulse with the dye response function (Fig. 1a). Since the excited-state lifetime of most laser dyes is long compared with the pumping pulse, once the gain has reached its maximum value it will decay only very slowly. Figure 1b shows the sudden depletion in gain when a circulating dye laser pulse arrives in the jet stream. The pulse shortening is produced when the increasing gain on the rising edge is followed by rapid depletion of the gain (gain saturation). In the example in Fig. 1b the dye pulse has arrived a little late in the gain profile. The result is to advance the peak of the amplified pulse. This effect is clearly seen when the input and output pulses from the amplifying medium are compared normalized to the same height (Fig. 1c). If the pulse were to arrive too soon in the dye medium, the maximum would be retarded. An extensive discussion of this type of phenomenon has been given by Icsevgi and Lamb.[10] The equilibrium situation then is that the interval between dye pulses emerging from the dye laser is equal to the interval between the argon pulses, even if the dye laser length is not exactly equal to (the inverse) of this frequency. This is an important point because it places specific stability requirements on the rf source driving the argon laser, since, in turn, the argon pulse repetition rate is precisely equal to (twice) the rf source frequency. If there is an optimum dye laser length for minimum dye pulse duration, then any jitter in the rf source frequency will perturb the dye laser operation and produce longer pulses. An rf source stability of at least 1 part per million is required for generation of pulses of less than 5 psec. Amplitude variations in the argon pulses are also equivalent to a timing jitter, and so high pulse-to-pulse amplitude stability of the argon laser is required.

B. Pulse Duration and Structure Measurements

One of the most useful methods for measuring pulse lengths is the zero-background, second-harmonic autocorrelation technique.[11] The technique (Fig. 2) is simple and convenient to operate. Two replica pulses are produced by a beam splitter. One pulse traverses a fixed and the other a variable optical path. The two pulse trains are then brought parallel (but not collinear) and focused to a common spot in a thin crystal of, for example, lithium iodate. The intensity of ultraviolet light generated along a line bisecting the two input beams is then measured as a function of delay (τ) between the pulses. The signal generated by this method is proportional to the autocorrelation of the pulse intensities:

$$G(\tau)=\left\langle \int_{-\infty}^{\infty} \langle I(t)\rangle_{\phi}\langle I(t+\tau)\rangle_{\phi}dt \right\rangle_{N} \qquad (1)$$

where the time fluctuations at the spectral frequency are averaged in the inner brackets, and the outer brackets indicate an average over a large number of pulses.

The influence of dye laser length is illustrated in Fig. 3, where the laser-tuning element was a three-plate birefringent filter. When the cavity length is optimal, the autocorrelation trace is smooth and has negligible intensity between pulses. For cavities too long, the trace becomes broader and develops structure. For cavities that are too short, structure again develops but now, because of the finite duration of the argon pumping pulse, a second pulse appears. Referring again to Fig. 1, if the dye pulse arrives early, the gain will increase again after the passage of the dye pulse and it is possible that threshold will be exceeded a second time, allowing a second pulse to circulate. This will not occur for a pulse arriving late (cavity too long), since there will be insufficient pump pulse remaining to build up the gain back to threshold.

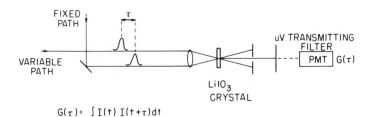

Fig. 2.   Optical arrangement for zero-background autocorrelation measurement.

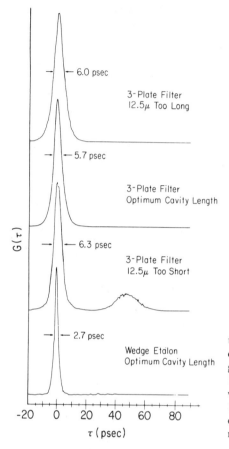

Fig. 3. Dye laser output pulses as a function of dye laser cavity length. Top three curves, tuning element: three-plate birefringent filter, 45% T output coupler 1·2 W pump, 120 mW output. Bottom curve, interference wedge tuning element 80% T output coupler, 1·2 W pump, 20 mW output. The marked durations are the measured full width half-maxima of $G(\tau)(\Delta\tau)$.

The shortest pulse obtainable depends on several factors: (a) the bandwidth of the intracavity filter, (b) the gain in the cavity, and (c) pump pulse duration. Table I summarizes our own and published data. The pulse durations shown were obtained by dividing $\Delta\tau$ by 1.41 to 2 to obtain "the pulse duration." This method ignores the influence of coherence, and below we show how to analyze $G(\tau)$ to obtain the true envelope autocorrelation.

In the noise burst model of Pike and Hersher,[12] in which the pulse is treated as a burst of bandwidth limit noise, $G(\tau)$ is decomposed into the product of two autocorrelations: one for the pulse envelope and one for the bandwidth-limited substructure. Thus

$$G(\tau)=G_P(\tau)(1+G_N(\tau))\qquad(2)$$

TABLE I

Reported[a] Pulse Durations for Synchronously Pumped Dye Lasers

Intracavity filter	Reported[a] pulsewidth (psec)	References	Average power (mW)	References	Dyes used	References
Three-plate BRF	4–10	14,24,26,32,111,113	85–270	14,26,111,113	Rhodamine 6G	14,19,24,26,32,111–115
Two-plate BRF	1.7–4.5	26,32,111,113	60–85	111,113	Rhodamine B	26
					Fluorescein	26
					Oxazine[d]	113
One-plate BRF	0.8–1.1	19,112,[b]113	16–45	19,113	DOTC,HITC[d]	27
					Stilbene 3	25
Wedge-shaped interference filter	0.7–1.9	14,19,21,25,113,114	20–85	14,113		
Prism	0.6–1.4	112,[b]115[c]	20	112		
Dye stream only	0.7–0.9	14,112[b]	15	14		

[a]All the authors assume a transform-limited pulse and divide the FWHM of $G(\tau)$ by 1.41, 1.8, or 2.0, depending on the assumed pulse shape. This ignores the influence of coherence[12] for nontransform-limited pulses and probably leads to significant underestimates in some cases.[13,14]
[b]Tandem pumping.
[c]Hybrid passive-active system.
[d]Krypton pump.

where $G_p(\tau)$ is the autocorrelation of the pulse envelope and $G_N(\tau)$ is a Gaussian function resulting from the noise bandwidth.

By combining detailed fits of (2) with measurements of the laser spectrum (obtained with a 1 m spectrograph/SIT Vidicon combination), the influence of coherence in the measured autocorrelation traces can be reliably determined.[13] The influence of cavity detuning on pulse envelope can then be obtained without the distortion imposed by coherence, and without confining observations to the "region of good mode locking." In fact the influence of coherence is particularly insidious at close to optimum cavity length, since (1) the spectral width of the laser changes very rapidly in this region (see Fig. 4) and (2) very smooth autocorrelation traces can be obtained when the pulse envelope–bandwidth product is two to three times the transform limit.[14] Once the coherence and envelope widths can be extracted reliably from autocorrelation measurements, it should be possible to deconvolute rise times containing both the coherent coupling contribution,[15-17] for which the appropriate time scale is determined by the coherence width, and the contribution from the molecular response with the pulse envelope.

Figure 4 shows the spectral full width at half-maximum (FWHM) and the FWHM of $G_N(\tau)(\Delta\tau_N)$ obtained by fitting autocorrelations to

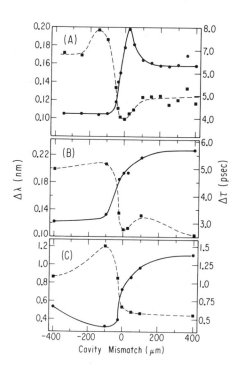

Fig. 4. Spectral width (solid curves) and substructure width $\Delta\tau_N$ (dashed curves) versus cavity length mismatch for (a) 70% output coupler and birefringent filter, (b) 55% output coupler and birefringent filter, and (c) 55% output coupler and wedge etalon. Note the change in the vertical scale in Fig. 4c.

(13). Gaussian fits were excellent to both the spectrum and the coherence spike and the product $\Delta t_N \Delta \nu = 0.43 \pm 0.06$ within experimental error for all cavity lengths. For a Gaussian spectrum this product should be 0.441. Figure 5 shows the dependence of the pulse envelope autocorrelation width ($\Delta \tau_p$) on cavity length. Also shown in Fig. 5a is the FWHM of the full autocorrelation trace $G(\tau)$ over the region where $G(\tau)$ appears as a smooth function. This function is clearly a shallower function of cavity length than $\Delta \tau_p$ and also leads to a considerable underestimate of the actual pulse duration. The variation in pulse envelope is qualitatively quite similar to the calculations of Kim et al.[18] Generally our envelope width does not increase as rapidly for short cavities as in the calculations of Kim et al.[13] A second difference is that the minimum envelope duration occurs for cavities in exact synchrony with the pump laser,[14] rather than for slightly longer cavities.[18]

The excellent fit of autocorrelations obtained with close-to-optimum cavity lengths to functions of the form $\exp(-a|x|)$ has led several authors [19-21] to hypothesize that their pulses are bandwidth-limited, single-sided exponentials (see Fig. 6). This conclusion is not supported by our spectral data;[13] we do not observe the required Lorentzian spectral profile for a single- or double-sided exponential profile. Also the time-bandwidth product for an

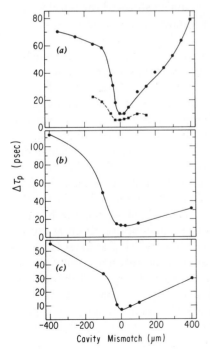

Fig. 5.   Pulse envelope width $\Delta \tau_p$ versus cavity length mismatch for (a) 70% output coupler and birefringent filter, 55% output coupler and birefringent filter, and (c) 55% output coupler and wedge etalon. For comparison, the FWHM of the autocorrelation $\Delta \tau$ is included in (a).

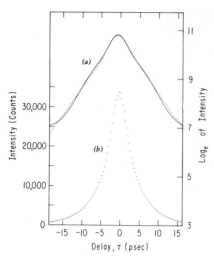

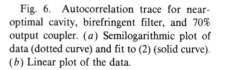

Fig. 6. Autocorrelation trace for near-optimal cavity, birefringent filter, and 70% output coupler. (*a*) Semilogarithmic plot of data (dotted curve) and fit to (2) (solid curve). (*b*) Linear plot of the data.

exponential pulse is almost one order of magnitude smaller than that observed for our pulse if we assume exponential shape. Synchronously pumped dye lasers do not give bandwidth-limited, single-sided exponentials, and by assuming that they do, many authors have underestimated the pulse widths by about a factor of 3. For example, a symmetric exponential $G(\tau)$ with FWHM of 2 psec would correspond to a Δt_p of about 2.7 rather than 1 psec.

How then does the exponential shape of $G(\tau)$ for a perfectly matched cavity arise? Autocorrelations very closely resembling the experimental exponential shape are given by (2) with G_1 and G_2 both Gaussian and Δt_p in the range two to three times Δt_N.[14] The envelope shapes obtained through (2) for closely matched cavities, although much shorter than for mismatched cavities, are essentially the same shape at all cavity lengths and are roughly Gaussian. Van Stryland[22] has pointed out the importance of remembering that (1) contains an ensemble average over more than 10^7 pulses for each data point. Some pulses are likely to be longer than others, for example, those occurring shortly after lasing has been interrupted by a bubble in the dye jet. It is possible to generate almost any shape of autocorrelation by summing the appropriate distribution of Gaussian functions of differing widths. Van Stryland[22] obtains a symmetric exponential by using a rather large distribution. He ignores, however, the presence of the coherence spike, the inclusion of which obviates the necessity for such a large distribution. As pointed out above, the sum of two Gaussian functions comes very close to the mark, and a large distribution of pulse widths is not required to explain our results. A distribution of exponential or Lorentzian pulses does not reproduce our data. We conclude, then, that the pulses are Gaussian or skewed Gaussian.

We have also looked at the autocorrelation function obtained from the optical Kerr effect in CS_2.[23] This is a third-order autocorrelation

$$G^3(\tau) = \left\langle \int_\infty^\infty \langle I(t) \rangle_\phi^2 \langle I(t-\tau) \rangle_\phi \right\rangle_N \tag{3}$$

convoluted with the rotational correlation function of CS_2. The $G^3(\tau)$ is sensitive to pulse asymmetry, but our measured traces are symmetric, indicating that the pulses are skewed only slightly, if at all.

C. Pulse-to-Pulse Reproducibility

The technique shown in Fig. 2 measures the ensemble average of the autocorrelation function, since different pulses give rise to different parts of $G(\tau)$ and very many (10^9 to 10^{10}) pulses contribute to a single measurement. The pulse-to-pulse reproducibility is of considerable significance for studies of molecular population risetimes and coherence phenomena. We have studied this reproducibility by measuring cross-correlations of the form $\int I_n(t) \times I_{n-m}(t+\tau)dt$, where n labels an individual pulse and m the number of round trips separating the pair, as a function of m.[14] Our results are illustrated in Fig. 7. When the dye laser cavity is set for optimum pulse length with no discernible structure or satellite pulses in the autocorrelation trace, the cross-correlation of a pulse with its near neighbors $(n, n-m)$ is indistinguishable from the autocorrelation (n, n) function. We studied $m=0$ to 6 and found identical results. The results for $m=0$ and 6 (Figs. 7d and 7e) are very reassuring and strongly imply that there are no rapid pulse-to-pulse variations in shape or duration.

Perhaps more revealing are the results presented in Figs. 7a to 7c, where the dye laser cavity length is incorrectly set and partial mode locking results. Autocorrelation traces such as the curve in Fig. 7c, consisting of a broad base with a sharp central spike, are characteristic of a noise burst.[12] In other words, $I(t)$ is not a single smooth pulse but has considerable random amplitude structure. In this case the duration of the burst of noise is related to the FWHM of the broad base. The $G(\tau)$ function has this shape because since the noise is random, only when the pulses are exactly superimposed ($\tau=0$) do the noise spikes exactly overlap; at all other delay times there is considerable cancellation. If the noise is nearly random and closely approaches zero amplitude between maxima, the spike and base will have heights in the ratio $2:1$. The contributions from G_N and G_p (equation 2) can be easily observed in Figs. 7a to 7c, where there is large (500 μm) cavity mismatch.

Figures 7a and 7b show the results of cross-correlation measurements $(n, n-6)$ for dye laser cavities either too short or too long.[14] Here the noise spike marches to one side of the broad base, the direction depending on whether the dye cavity is too long or too short, and the distance depending

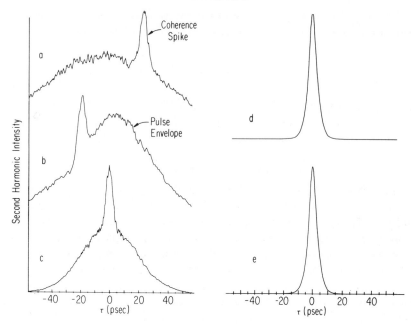

Fig. 7. Autocorrelation (n, n) and cross-correlation $(n, n-m)$ function measurements. (a)–(c) Cavity length mismatched; (d),(e) cavity length optimized. (a) $n, n-6$, cavity length 500 μm too short; (b) $n, n-6$, cavity length 500 μm too long; (c) n, n, cavity length 500 μm too long; (d) $n, n-6$, cavity length optimum; (e) n, n, cavity length optimum.

linearly on m. The center of the broad base remains in the same position in all cases.

Figure 8 depicts our explanation. Recalling the discussion of Fig. 1, if the circulating dye pulse arrives late in the gain profile, the rising edge of the pulse receives more amplification than the trailing edge and the pulse peak is advanced. The simulations in Fig. 8 shows that the "new" part of the pulse has "new" noise unrelated to noise on the same part of the pulse on its previous round trip. The "old" noise on the "old" part of the pulse is, however, replicated. But the pulse shape has changed, and the old noise no longer occurs on precisely the same part of the pulse profile as on the pulse from the previous round trip. Thus the noise spikes add in phase for a value of τ different from zero. This argument predicts a linear dependence of spike displacement on both m and cavity mismatch (in micrometers). Both linear dependences are observed experimentally (Fig. 9).

Perhaps the main significance of this experiment is that it allows experimental determination of the dye laser cavity length corresponding to exact match with the argon laser. The discussion above indicates that this will

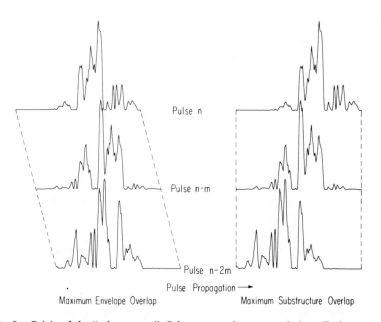

Pulse n

Pulse n−m

Pulse n−2m

Pulse Propagation →

Maximum Envelope Overlap   Maximum Substructure Overlap

Fig. 8.   Origin of the "coherence spike" for auto- and cross-correlations. Each trace represents the same sample of random noise shaped with a Gaussian envelope. The envelopes are of equal width but are progressively displaced in the series $n, n-m, n-2m$. The cases shown correspond to maximum envelope overlap and maximum coherence of the noise (maximum substructure overlap).

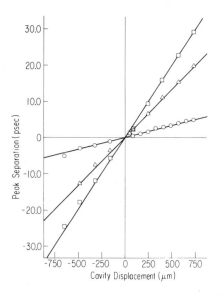

Fig. 9.   Plot of displacement of coherence spike (in picoseconds) from the center of the pulse envelope against cavity mismatch (in micrometers): squares, $m=6$; triangles, $m=4$; circles, $m=1$. The solid lines are calculations according to our model.

13

occur when the cross-correlation trace becomes precisely symmetrical. We conclude, in contrast to theoretical predictions of optical pulse duration for dye laser slightly short[24] or slightly long,[18] that the shortest pulses are formed at exact cavity match.

III. DETECTION SYSTEMS

Results obtained from the application of synchronously pumped lasers to a range of relaxation processes are described in Section IV. This section briefly reviews the basic experimental techniques used in these applications.

A very wide range of wavelengths (400–1000 nm for the fundamental[25-27] and 265–350 nm for the second harmonic[26]) have been obtained from synchronously pumped dye lasers, and this range will certainly be extended in the near future both through new dyes and through frequency-mixing techniques (e.g., with ion laser lines[28]). Economou et al.[29] have reported the generation of tunable vacuum ultraviolet radiation near 1700 Å by resonantly enhanced four-wave mixing in strontium vapor. Thus a single ion–dye laser combination can provide almost all the excitation wavelengths one could wish for and is an ideal source for fluorescence spectroscopy. On the other hand, time-resolved absorption spectroscopy in general also requires an independent range of monitoring wavelengths to record the spectra of transient species. There are a number of solutions to this problem of varying complexity and generality. All the techniques described here use the pulses themselves to provide the time resolution — they are all variants of the pump-probe principle where the sample is excited by a strong pump pulse and the response of the sample to a probe pulse measured as a function of time delay between pump and probe. The optical setup is very similar to that in Fig. 2, where the frequency doubling crystal is replaced with sample cell and the intensity of the variable path length beam monitored as a function of delay time (τ).

A. Ground-State Recovery

The return of excited molecules to the ground state can be followed by the decrease in transmission of a weak probe pulse through the sample as a function of time delay after the arrival of the strong pump pulse. In this special case the pump and probe pulses may have the same wavelength, that is, they may originate from the same dye laser. A significant advantage of the high repetition rate, synchronously pumped source in any kind of absorption measurement is that lock-in amplifier detection may be used. If the pump beam only is chopped, then very small modulation depths may be detected on the probe beam, since the lock-in rejects the large dc component in the probe beam. A variant on the basic technique is to use the second harmonic for excitation and the fundamental as the probe.[29] This technique, however, has its dangers because with single-wavelength probing the effects of electronic relaxation between states and spectral relaxation within a single state cannot be disentangled. A straightforward example is the case of intersystem crossing where the initially formed triplet state has a different spectrum from the relaxed triplet[30, 31] and single-wavelength probing will not provide accurate intersystem crossing rates. Moreover, if two replica pulses derived by beam-splitting are used for pump and probe, the coherence between the two pulses distorts the observed signal around zero delay time.[16, 17] We discuss this point in more detail in Section IV.

B.   Double Dye Laser Technique

A partial solution to the problem of providing complete wavelength coverage for the probing pulses is to synchronously pump two different dye lasers with some ion laser.[32, 33] Interlaser jitters as low as 5 psec have been reported by Heritage and co-workers.[34] To obtain minimum jitter, it is necessary to match the gain in the two lasers and to have high pulse-to-pulse amplitude stability in the ion laser. With current mode locker design, the ion laser is mode locked on a single line; thus the range of wavelengths obtainable in this technique is limited to the dyes that can be pumped by the same ion laser line. Mixing dyes and using energy transfer will extend this range somewhat. Using a non-dispersive mode-locking element (e.g., a rhomb), it is possible to mode lock ion lasers on "all lines" or a group of lines, and although longer pulses are expected (since ion laser cavity length will not be perfectly matched for all wavelengths), this may also provide a means of pumping a wider range of dyes. The double wavelength technique has been used to study the surface Raman effect,[2, 34-36] photon echoes,[33, 37] and Raman line shapes in a pulsed CARS (Coherent Anti-Stokes Raman Scattering) experiment.[38]

C.   Amplification and Continuum Generation

The generation of picosecond white light continua by self-phase modulation in a variety of liquids (e.g., D_2O, CCl_4, phosphoric acid) has become a common technique in experiments involving mode-locked solid-state lasers.[39] The extension of this technique to cw mode-locked dye lasers has been pioneered by Shank and Ippen,[40] using a passively mode-locked dye laser. The principle is identical for a synchronously pumped dye laser and has been applied by Martin et al.[41] An amplified Q-switched Nd-YAG laser is frequency doubled and pumps three stages of dye laser amplification, giving a total gain of 10^6. The picosecond continuum is then generated by focusing the intense picosecond pulses into D_2O. The major disadvantage of this technique, aside from its cost and complexity, is that the repetition rate is lowered to about 10 Hz. On the other hand, this still represents a high repetition rate when compared with a Nd-glass laser system ($\sim 10^{-2}$ Hz)! Copper vapor lasers are capable of much higher (>1 kHz) repetition rates, but these are not yet standard laboratory lasers.

D. Read-In–Read-Out Technique

Read-in–read-out provides a wide spectral coverage while maintaining the high repetition rates of the cw dye lasers. Again, the technique was developed by Ippen and Shank,[40, 42] and it has been applied to synchronously pumped dye lasers by Gillbro and Sundstrom.[43] The essential feature of the technique is that the probing pulse is obtained from an independent dye laser that produces pulses long enough to be essentially flat on a picosecond time scale. This means that a small amount of jitter will not affect the probe intensity, and the transient information is read out of the probe pulse by a second picosecond pulse (derived from the first by beam splitting and therefore with zero jitter) in a sum frequency technique. Again, lock-in amplifier detection is used so that only the desired information is recorded. Any wavelength that can be obtained from an ion laser-pumped dye laser can be used for probing, and this technique should assume considerable importance in the near future. The experimental arrangement is shown in Fig. 10.

E. The Coherent Coupling Phenomenon

Pump-probe experiments with both pulses derived from the same pulse show an unexpected peaking or sharp spike at zero time delay.[10] This spike

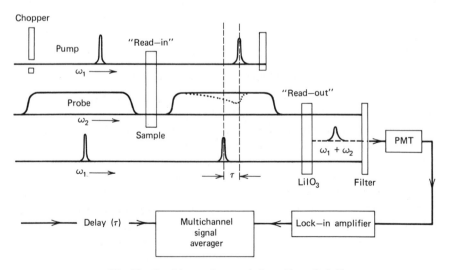

Fig. 10.   Read-in–read-out technique. From Ref. 40.

arises from a transient grating created in the sample by the pump and probe beams. The beams intersect at an angle θ to the propagation direction and, as a result of pathlength differences across the focal spot, there are regions of constructive and destructive interference across the sample. These give rise to regions of excited and unexcited molecules, that is, a grating of light and dark strips in a bleachable sample. The grating spacing is

$$d = \frac{\lambda}{2 \sin \theta}$$

where λ is the laser wavelength. The first-order diffraction from this grating is for one beam to be diffracted into the other, and the sharp spike results from the diffraction of some pump photons into the probe beam direction.

A brief summary of the phenomenon has been given by Ippen and Shank,[15] and an extensive discussion has recently been provided by von Jena and Lessing.[16, 17] These authors considered the influence of pulse shape and polarization, population kinetics, and orientational relaxation.[16] The major conclusion from both Ippen and Shank and von Jena and Lessing is that in a standard ground-state recovery experiment with identical pump and probe pulses, the coherent coupling artifact exactly doubles the measured signal at zero delay time. In the anisotropic absorption experiment, where the signal is proportional to the square of population differences, coherent coupling quadruples the signal at $\tau = 0$.[17] As noted earlier, the duration of the coherent coupling interaction is controlled by the coherence length of the pulse, which is only equal to the pulse duration for a transform-limited pulse. In a related paper von Jena and Lessing[17] consider the influence of a phase grating, generated by thermally induced refractive index modulation, in addition to the amplitude grating described above. However, they conclude that only the amplitude grating is important in transient absorption measurements. By using a third pulse, the dynamics of the induced grating can be probed by recording the diffracted intensity as a function of time delay. The grating decays through (1) excited-state relaxation, (2) orientation relaxation, and (3) energy transfer. Since this technique has inherent spatial resolution, it holds great promise for the study of diffusion. To date studies have been made of the rotational diffusion of rhodamine 6G[44] and of electronic energy migration in pentacene crystals.[45] Although the coherent coupling phenomenon should occur with any laser system, it is not commonly observed with mode-locked Nd-glass lasers, presumably because the coherence length is generally much shorter than the pulse duration in these lasers.

F. Emission Spectroscopy

Detection of emission profiles with picosecond resolution presents a problem slightly different from that of measurements of time-resolved absorption profiles. A method must be found for measuring the fluorescence intensity $I_t(t)$ as a function of time. Further information can be obtained by measuring the emission spectrum $I_f(\omega)$ at fixed times.

G.   Fluorescence Up-Conversion Technique

Mahr and co-workers[46, 47] developed the up-conversion technique, which enables direct measurement of fluorescence decay profiles and time-resolved spectra with a time resolution limited by the pulse duration. The technique (Fig. 11) is a development of the pump-probe methods described in the preceding section. Fluorescence excited by a one-picosecond pulse is mixed in a lithium iodate crystal with a second picosecond pulse and the sum frequency $\omega_f + \omega_L$ detected as a function of time delay. Provided the laser pulse duration is less than that of the fluorescence, the recorded signal displays $I_f(t)$, with excellent time resolution, good dynamic range ($10^2 - 10^3$), and good signal-to-noise ratio. The phase-matching condition in the up-converting crystal limits the portion of the fluorescence spectrum up-converted for a particular ω_L. If ω_L is scanned, at fixed delay time, time-resolved spectra can be built up.[48]

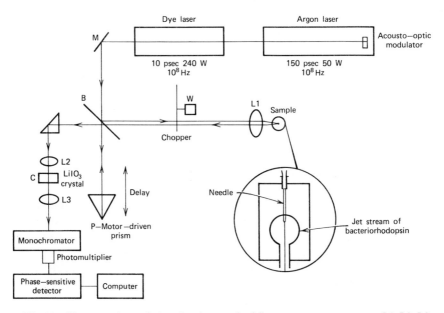

Fig. 11.   Up-conversion technique for time-resolved fluorescence measurements. L1, L2, L3, lenses; M, mirror; B, beamsplitter; C, frequency doubling crystal; W, mechanical chopper. From Ref. 47.

H. Synchronously Scanning Streak Cameras

The deflection ramp of an image converter streak camera can be swept in synchrony with the rf source driving the mode locker.[49] Successive streak images are then superimposed, and the sensitivity and precision of the streak camera are significantly enhanced. A number of applications to photochemical systems have been described by Sibbett and co-workers.[50, 51]

I. Time-Correlated, Single-Photon Counting

Ware and co-workers[52, 53] have given an excellent description of the principles of the time-correlated, single-photon counting technique for spark lamp excitation, and this section describes only the improvements we have made using the synchronously pumped dye laser as an excitation source.[54–57] Our experimental setup is shown in Fig. 12. The recorded decay profile $f(t)$ is in general the convolution of the molecular response to delta function excitation $g(t)$ and the instrument response function $i(t)$.

$$f(t) = \int_0^t i(t-t')g(t')\,dt' \qquad (4)$$

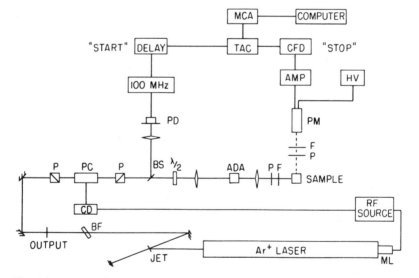

Fig. 12. Experimental arrangement for time-correlated, single-photon counting method for fluorescence decay measurements: $\lambda/2$, halfwave plate; P, polarizer; ADA, frequency doubling crystal; PM, photomultiplier; AMP, rf amplifier; CFD, constant fraction discriminator; 100 MHz, discriminator; TAC, time-to-amplitude converter; MCA, multichannel analyzer, ML, mode locker; BF, birefringent filter; PC, electro-optic modulator, CD, countdown; PD, photodiode; BS, beamsplitter; F, filter; HV, high voltage power supply.

Here $i(t)$ contains contributions from the finite duration of the excitation pulse and the time resolution of the detection system. The time resolution of the photon-counting apparatus, measured in terms of the shortest lifetime that can be deconvoluted reliably, is determined by the width, shape, and reproducibility of the instrument response function. When using ultrashort (<10 psec) excitation pulses, the measured instrument response function, which is recorded by scattering the laser from a dilute suspension of milk or talc, is due only to timing jitter in the electronic apparatus and transit time variations in the detectors. For Gaussian broadening, the measured width of the instrument response function τ_M is

$$\tau_M = \left(\sum_i \tau_i^2 \right)^{1/2}$$

where the τ_i are the widths of jitters due to the individual components. The major sources of timing jitter are the constant fraction discriminator and the photomultiplier tube. It is very difficult to decouple these two, since it is the failure of the constant-fraction discriminator to deal with the pulse height variation of the anode pulses from the photomultiplier tube that causes the

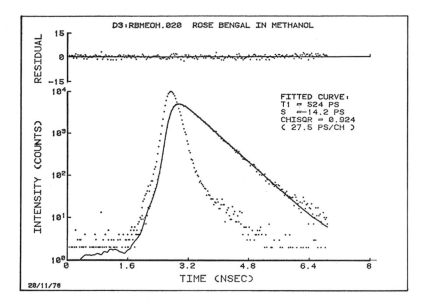

Fig. 13.   Fluorescence decay of rose bengal in methanol. The solid line represents the convolution of an exponential decay with $\tau = 524$ psec with the measured instrument response function, while the upper curve shows the weighted residuals for the fit ($\chi_r^2 = 0.96$).

discriminator jitter. With a small pulse height distribution, the constant fraction discriminator exhibits very small jitter.

Figure 13 gives a typical example of our system's performance. The narrow dotted curve is the instrument function; the wider the fluorescence decay of rose bengal in methanol. The solid line is the best-fit, single exponential, convoluted with the instrument response function. The value of 535 ± 15 psec is obtained from results in three laboratories,[58-60] the first two using the photon-counting technique and the third using an Nd-glass–streak camera system. This leads us to suggest that rose bengal in methanol provides a useful standard for subnanosecond fluorescence lifetimes in the visible region.

Experimental details of improvements in time-correlated photon counting have also been provided recently by Koester and Dowben,[61, 62] Koester,[63] Spears et al.,[64] and Harris et al.[65] With the development of microchannel plate photomultipliers and improved discriminators, it does not seem unreasonable to expect instrument functions of 50 to 100 psec in the next few years. This would imply a time resolution of 10 to 20 psec with deconvolution.

IV. APPLICATIONS TO TIME-RESOLVED SPECTROSCOPY

A. Vibrational Studies in Solids and Liquids

Synchronously pumped lasers have been applied to the study of dephasing and energy relaxation of excited vibrational levels in a number of systems.

1. Mixed Crystals

Hesselink and Wiersma have conducted an elegant and extensive study of dephasing processes in pentacene dissolved in naphthalene and p-terphenyl mixed crystals.[9, 33, 37, 66] Their experimental setup is shown in Fig. 14. Each of two dye lasers synchronously pumped by a single argon laser is amplified twice by nitrogen laser, pumped-dye amplifiers. The jitter between the two dye lasers (determined by cross-correlation measurements) was about 14 psec. For the observation of the two-pulse photon echo, the excitation dye laser is tuned to the electronic origin of pentacene, while the probe dye laser is set for maximum output. The excitation pulse is split and, after a delay, is recombined to give the desired two pulses with intensity ratios of 1 : 4. Both pump beams are focused on the sample, which is inside a temperature-variable cryostat. After passing through the sample, the pump pulses and the echo are combined with the probe pulse via a 50–50 beam splitter.

The collinear beams are now focused in an ADP (Ammonium Dihydrogen Phosphate) crystal oriented for phase matching at the sum frequency of the two dye lasers. The probe pulse is delayed to coincide with the echo, and

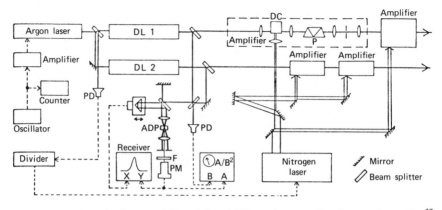

Fig. 14. Schematic diagram of Hesselink and Wiersma's system for photon echo studies.[37] Also shown is the autocorrelator for measuring the autocorrelations and cross-correlations. Solid lines are laser beams; dashed lines indicate electrical connections. PD, photodiode; PM, photomultiplier; F, Schott UG11 filter; DC, dye cell; P, prism, DL, dye laser; ADP, frequency mixing crystal.

the sum frequency echo + probe is detected through a monochromator with a photomultiplier as a function of time delay of the probe laser pulse. Figure 15 shows the cross-correlation of the pump and probe pulses and of the pump pulse and two-photon echo. The echo signal is clearly much wider than the laser pulse cross-correlation. The entire inhomogeneous line is coherently excited, since the laser width and absorption line width are approximately equal. The photon echo width[37] is 33 psec, whereas the calculated width from the 0.85 cm^{-1} inhomogeneous width of the absorption line is 24.5 psec. Hesselink and Wiersma attribute this difference to differences in the inhomogeneous width in the volume of the sample excited in the echo experiment as compared with the entire crystal volume observed in the absorption spectrum. Since the full inhomogeneous line is excited, the observed relaxation rates represent ensemble averages over the distribution of absorbing sites.

In a related experiment Hesselink and Wiersma also recorded three pulse-stimulated echoes[37] and accumulated three-pulse echoes.[9] The pulse cycles and phase-matching conditions for the various types of photon echo are shown in Fig. 16. The temperature dependence of the dephasing of the origin in pentacene was investigated by two-pulse echo and accumulated three-pulse echo techniques. Figure 17 shows the dependence of the pure dephasing time T_2^* on inverse temperature. Below 10°K exponential behavior is observed with the data well fit by

$$T_2^*(T) = T_2^*(\infty)\exp\frac{\Delta E}{kT}$$

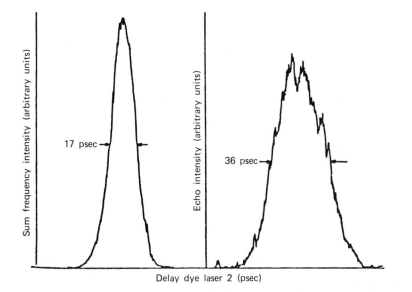

Fig. 15. *Left*: cross-correlation between the excitation pulse and the probe pulse; *right*: cross-correlation between the two-pulse photon echo and the probe pulse. The photon echo was generated in the 0–0 band of pentacene band in naphthalene at 1.6°K with an excitation pulse separation of 116 psec.[37]

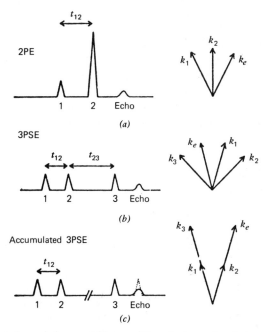

Fig. 16. Pulse cycles and phase-matching conditions for two-pulse photon echo (2PE) (top), three-pulse photon echo (3PE) (middle), and accumulated three-pulse echo (bottom).[37]

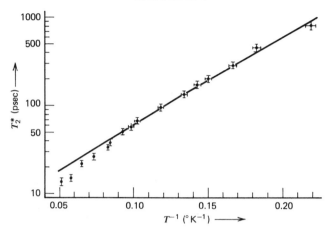

Fig. 17. Pure dephasing time T_2^* of the electronic origin of pentacene in naphthalene as a function of the inverse temperature from Ref. 37. Results are from accumulated three-pulse echo experiments.[9, 37] The solid line is a calculated line; see text and Ref. 37 for details. Deviation from exponential behavior above 12°K.

with $\Delta E = 16 \pm 1$ cm^{-1} and $T_2^*(\infty) = 6.5 \pm 1.5$ psec. Deviation from simple exponential behavior is clear from 10 to 20°K, the highest temperature measured. Analysis of these data in combination with spectroscopic measurements leads to the conclusion that the optical dephasing both in the electronic origin and in several vibronic levels is induced by pseudolocal phonon scattering in both ground and excited states. The four-level scheme used by Hesselink and Wiersma[37] is shown in Fig. 18. The solid line in Fig. 17 is a fit using this scheme and yields $\omega_{13} = 18$ cm^{-1} ($\tau_3 = 3.5$ psec) and $\omega_{24} = 13.8$ cm^{-1} ($\tau_4 = 11$ psec) for the system pentacene in naphthalene. The local phonon is suggested to be on in-plane librational mode. For the system

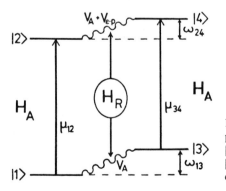

Fig. 18. The four-level scheme used by Hesselink and Wiersma[37] to describe the dephasing of pentacene in naphthalene: $|1\rangle$ and $|2\rangle$ are the ground and excited electronic state; $|3\rangle$ and $|4\rangle$ are the electronic states plus a local phonon.

pentacene in p-terphenyl the photon echo measurements indicate the existence of a librational mode of $\simeq 30$ cm^{-1} with $\simeq 1.5$ psec lifetime.

2. Semiconductors

Von der Linde and co-workers have recently described measurements of the decay of LO (Longitudinal Optical) phonons in gallium arsenide (GaAs) generated during the interaction of photoexcited hot electrons and holes with the lattice.[67] The experimental technique is a variant of the pump-probe technique in which the material is excited by two successive pulses with the same intensity and frequency, but orthogonal polarizations. The anti-Stokes Raman signal from the second pulse is measured through a polarizer oriented to remove the Raman signal from the first pulse. The observed signal consists of a constant background from the second pulse and a time delay–dependent signal from the excitations created by the first pulse. The electron and hole energy loss rate in GaAs is very rapid (4×10^{11} eV/sec[68]) and a very rapid rise in population of LO phonons is expected. Figure 19 shows

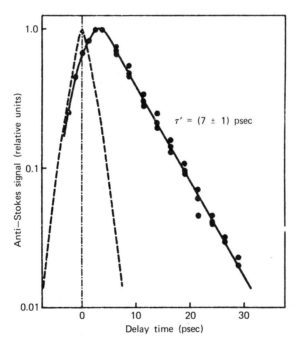

Fig. 19. Semilog representation of the anti-Stokes signal showing the decay of the LO-phonon population in photoexcited GaAs at 77°K.[67] The dashed curve is the measured autocorrelation function of the pulses. The solid curve is calculated with a decay constant of 7 ± 1 psec. The dash-dot line marks zero time delay.

the observed signal with the expected rapid rise (~ 6 psec) consistent with a formation time of about 2 psec convoluted with the probe pulse. The anti-Stokes signal decays exponentially with $\tau' = 7 \pm 1$ psec at 77°K, where τ' is the incoherent phonon lifetime; τ' differs from τ, the phonon lifetime proper, in the same sense that population relaxation and dephasing times differ in molecular vibrations and liquids. From line-width measurements a value of $\tau = 6.3 \pm 0.7$ was obtained, thus $\tau' \simeq \tau$ within experimental error, indicating that intrabranch LO phonon scattering is not important in GaAs, although more precise measurements of τ with coherent Raman scattering may reveal a small difference between τ' and τ.[67]

3. Liquids

Although a great deal of research on dephasing and vibrational relaxation has been done in liquids with mode-locked Nd-glass lasers, to date rather little work has been accomplished with synchronously pumped lasers. The major reason for this is the inability of dye lasers to excite single quanta of vibrations directly (although F-center lasers should have this capacity[6]), and unamplified the dye lasers have insufficient intensity for the stimulated

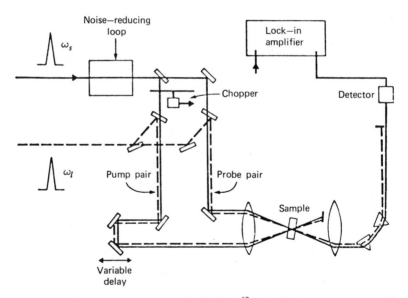

Fig. 20. Experimental arrangement used by Heritage[69] for dephasing measurements with cw mode-locked lasers (ω_p and ω_s). A synchronized collinear pump pair and a separate synchronized collinear probe pair cross in the sample volume. The gain (loss) in the probe Stokes beam is measured at the chopping frequency. Coincidence of each pair of pulse trains is maintained with two additional delay lines (not shown).

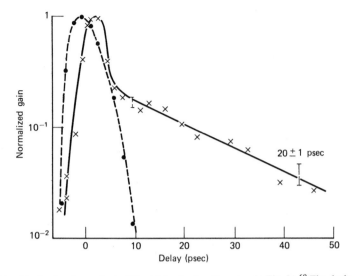

Fig. 21.   Dephasing dynamics in CS_2 obtained using the setup in Fig. 21.[69] The dashed curve is the cross-correlation of the two pulse trains and determines zero time. The solid line curve, indicating the four-pulse gain, is drawn through the experimental points for the time-resolved dephasing experiment. The delay peak gain and two decay regions corresponding to the aniso-tropic components are evident.

Raman excitation used with solid-state lasers. Heritage has described a pump-probe double dye laser technique based on Raman gain for dephasing studies.[69] Here the excitation problem is overcome by using two pulses whose wavelengths differ by the frequency of the desired vibration. The gain (or loss) of one of the applied frequencies is then measured. The experimental arrangement is shown in Fig. 20. The pump Stokes pulse is chopped, and gain in the probe Stokes pulse is measured with a lock-in amplifier as a function of time delay between the pump and probe pairs. Results with the four-pulse technique were obtained for ν_1 (656.5 cm^{-1}) of liquid CS_2 (Fig. 21). There was a rapid initial loss of gain due to molecular reorientation followed by an exponential decay with time constant 20 ± 1 psec, in excellent agreement with the dephasing time of 21 psec obtained from the spontaneous Raman line width.

A double dye laser technique has also been used by Kamga and Sceats[70] in their pulse-sequenced CARS method. The basis of their technique is that the nonresonant contribution to the CARS signal can be eliminated by interposing a time delay between the sequence of events leading to the excitation of real vibrational states and the subsequent events leading to the generation of the anti-Stokes radiation. In the normal situation the

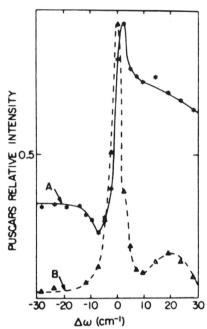

PUSCARS RELATIVE INTENSITY

0.5

A

B

-30 -20 -10   0   10   20   30

$\Delta\omega$ (cm^{-1})

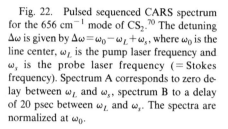

Fig. 22. Pulsed sequenced CARS spectrum for the 656 cm^{-1} mode of CS_2.[70] The detuning $\Delta\omega$ is given by $\Delta\omega = \omega_0 - \omega_L + \omega_s$, where ω_0 is the line center, ω_L is the pump laser frequency and ω_s is the probe laser frequency ($= $ Stokes frequency). Spectrum A corresponds to zero delay between ω_L and ω_s, spectrum B to a delay of 20 psec between ω_L and ω_s. The spectra are normalized at ω_0.

nonresonant (background) signal interferes with the resonant signal (since both signals are coherently generated), and a distorted line shape is produced. However, if the time delay described above is of the order of the vibrational dephasing time for the intermediate virtual state, a resonant line shape is restored. The results for delay times of 0 and 20 psec for the 656 cm^{-1} mode of CS_2 are shown in Fig. 22. A number of applications of the technique are discussed by Kamga and Sceats.[70]

B. Orientational Relaxation in Liquids

The development of picosecond spectroscopic techniques has prompted considerable interest in direct measurements of orientational correlation functions of medium-sized molecules in liquids.[71-74] The basic principle of the experimental methods used is that an intense, polarized, ultrashort pulse of light disturbs the equilibrium distribution of molecular orientations (burns a hole in the orientational distribution). The relaxation of the perturbed distribution is monitored either by absorption of a second polarized pulse or by time-dependent fluorescence depolarization. Since the initial work, a number of groups have made measurements of rotational motion of medium-sized dye molecules in solutions of low to medium viscosity.[75-81]

Thus far the results have been discussed in terms of a hydrodynamic description of the motion, where solvent structure is explicitly neglected. In the hydrodynamic description of rotational diffusion, two limits exist relating to the boundary condition between the moving solute molecule and its surrounding solvent molecules. The stick boundary condition[78] assumes coherence between the motion of the test particle and its nearest-neighbor solvent molecules. By contrast, the slip boundary condition[78, 82] assumes zero frictional forces between the motion of the test particle and its nearest-neighbor solvent molecules (no coefficient of tangential stress). Here the resistance to motion for a nonspherical solute molecule arises because for the solute molecule to move, solvent molecules must be pushed out of the way. For translational diffusion the difference between the two limits is small ($6\pi\eta r$ vs. $4\pi\eta r$); but for rotational motion very large differences exist between slipping and sticking boundary conditions, particularly for motions around axes where little solvent displacement occurs. The slip boundary condition has been found to give good agreement for rotational diffusion of small molecules is noninteracting solvents,[82, 83] although it has recently been suggested[84] that this agreement may be fortuitous. For large molecules the stick boundary condition is appropriate, and it has been demonstrated by Zwanzig[86] that for a large rough cylinder ($R \sim 500$ Å) the two boundary conditions are identical. This work follows that of Richardson, who showed that slip on a rough surface is asymptotically the same as stick on the equivalent smooth surface.[87] The physical reason for this is that solvent molecules are "caught up" in the indentations of the rough surface and are carried along without the necessity for strong attractive intermolecular forces. Directional intermolecular forces (e.g., hydrogen bonds) should also give the effect of microscopic roughness.[78]

Using a single-pulse, Nd-glass laser–streak camera method to determine time-dependent fluorescence depolarization, we have found three cases for the form of the diffusion coefficient for a series of medium-sized molecules in alcohol and aqueous solutions: slip, stick, and superstick. In the latter case numerical agreement with experiments can be obtained only by using a significantly larger molecular volume than is obtained from models or from Van der Waals increments. Agreement with experimental values is obtained if the oblate base molecule is padded out with solvent to form a sphere of roughly the largest molecular dimension.[74] In the case of the normal stick or slip boundary conditions, agreement is found when the true molecular shape is approximated by an ellipsoid— no solvent added. These three forms of the boundary condition can be rationalized in terms of the strength of the solvent-solute interaction.[78] The idea suggested in Ref. 78 is that the two extremes represent cases where solvent-solute bonds (H-bonds) are breaking and reforming on times scales much less than or much longer than the time scale of the rotational motion. The strongest hydrogen bonds are formed with molecules containing negatively charged groups, while by contrast the neutral molecule BBOT rotates in essentially the same time in both ethanol and cyclohexane (which have similar viscosities but very different H-bonding capabilities). An important by-product of this work is that the shape of the normal and photoisomer forms of the important mode-locking dye DODCI has been obtained.[77]

The importance of hydrogen bond interactions in determining molecular motion has been confirmed by the work of Spears and Cramer[76] and von Jena and Lessing,[79] both groups using continuously mode-locked lasers. For the fluorescein derivatives (dianions), where strong H-bonds are expected, the rotation time is significantly longer (two to three times) in alcohol solution than in a very polar but non-H-bond donor solvent of comparable viscosity. Von Jena and Lessing have also suggested the influence of internal flexibility in determining the observed boundary condition. For methyl red, which contains an ionized carboxylic acid function, these authors observed a result compatible with a slip boundary condition[79] and attributed this to internal flexibility. They also suggested that this may explain our previous result with BBOT, rather than simply the expected weaker H-bonding, as suggested in Ref. 78.

Von Jena and Lessing concluded that the carboxylic acid group is rather poor in H-bonding ability,[79] a conclusion in line with our results on the acid and base forms of rhodamine B,[87] where the rotation times of the COO$^-$ and COOH forms were identical within experimental error.

To date, rather surprisingly, all the directly measured rotational correlation functions have been well described by a single exponential. In other words, the motion observed has been isotropic. A Raman study of liquid benzene,[88] on the other hand, concluded that the motion is highly anisotropic, with the spinning motion (around the C_6 axis) two to three times faster than the tumbling motion (around the C_2 axis).

In a series of papers Hynes, Kapral, and Weinberg[84, 89] have developed a theory for diffusional motion, taking account of microscopic boundary layer effects. They write the rotational diffusion coefficient as the sum of a Stokes-Einstein-Debye term D_S, and an Enskog diffusion constant, D_E. The Enskog term contains a slip coefficient β, which is both mass (through the moment of inertia) and density dependent. The model calculations of Hynes et al.[89] for rough spheres and Tanabe's[88] Raman study of benzene (pure liquid) imply that the contribution from the Enskog diffusion constant is significantly greater than the hydrodynamic constant D_S. This was the case even for the tumbling motion in benzene. It would be especially interesting to apply this theory to solutes and solvents of increasing size and determine the relative hydrodynamic contribution.

The most chemically satisfying description of molecular motion would be one in which the dynamics was derived directly from the intermolecular potentials. Peralta-Fabi and Zwanzig[91] have made a first step in this direction, and it is hoped that further developments will lead to a true molecular theory in which the effects of attractive forces will arise in a natural way, rather than being added on to the slip hydrodynamic theory.

C.   Anisotropic Absorption

Shank and Ippen[73] have described a novel variant of the conventional methods of measuring orientational relaxation. In their technique (Fig. 23), linearly polarized pump and probe pulses have an angle of 45° between their polarization axes. The intensity of the probe pulse is monitored as a function of time delay through a polarizer crossed with the input probe polarizer. Differential absorption (dichroism) of the probe components parallel with and perpendicular to the pump polarization produces a rotation of the polarization of the probe beam. This rotated linearly, polarized probe is partially transmitted through the analyzer and detected. In this ideal case, and given the small signal limit, the measured signal is

$$T(t) = \text{const}\left[r(t)K(t)\right]^2 \qquad (5)$$

where $r(t) = 2/5 \langle P_2(e(o) \cdot e(t)) \rangle$ is the rotational correlation function, and $K(t) = N_{\parallel}(t) + 2N_{\perp}(t)$ is the excited state decay function. A typical experimental result is shown in Fig. 24.

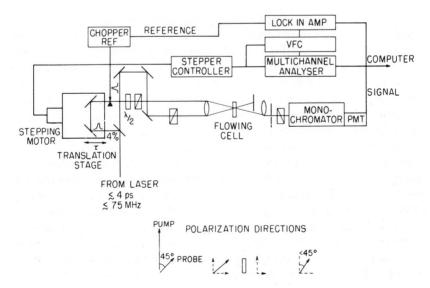

Fig. 23.   Schematic of the experimental arrangement for anisotropic absorption measurements.

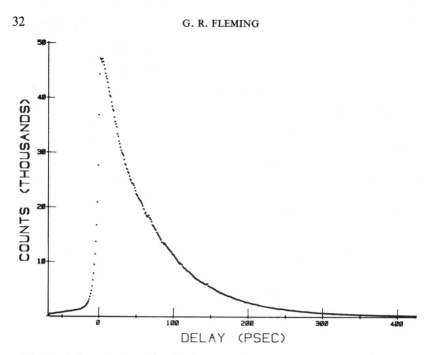

Fig. 24. Anisotropic absorption signal from the dye DODCI in ethanol (10^{-4} M at 22°C) measured with an extinction ratio of 2×10^{-7}. The fitted single exponential decay time is 68 psec, giving a rotation time of 154 psec in this solvent.

The technique is very sensitive and should give more accurate correlation times than the conventional techniques involving subtraction of two large signals. However, since there are a number of experimental complications with the technique, it should be used with caution. We have found[91] that the measured decay time is very sensitive to any external birefringence in the optics (e.g., focusing and collimating lenses). Values as much as twice the correct value are readily obtained unless high-quality, strain-free lenses and polarizers are used. An extinction ratio (I_{trans}/I_0) of $\sim 10^{-7}$ is required before the equation above for $T(t)$ can be used with confidence.

In addition to transient dichroism, transient birefringence caused by the change in the anisotropy of the polarizability will also give rise to a signal in this experiment. We have found that some samples (e.g., oxazine 725 excited at 590 nm) give signals arising almost entirely from birefringence. In this case the initial height of the signal gives the change in the anisotropic polarizability between the ground and excited singlet states. A full analysis of these effects will be given in a forthcoming publication in which the influence of the sample and various optical components is handled with a Jones matrix for-

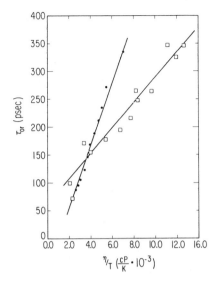

Fig. 25.   Plot of rotational reorientation time (τ_{or}) against the function η/T for DODCI.[92] Circles, data points and least-squares fit for DODCI in ethanol over the temperature range -10–50°C; squares, data points and least-squares line for DODCI in a series of solvents at 22°C. Solvents used were water, methanol, ethanol, 2-propanal, and a series of propanal-water mixtures.

malism.[91] When the signal arises from dichroism, the experimental results are much less sensitive to imperfect optics.

We have also used the anisotropic absorption technique to study the influence of temperature on the rotational reorientation time of DODCI in ethanol. Figure 25 plots τ_{or} versus η/T for the temperature range -8 to 40°C for ethanol and also τ_{or} versus η/T for a series of solvents at room temperature. The slope of the plot is much greater for the temperature study than for the solvent study, and we are continuing to investigate this effect.[92]

D.   Photochemistry and Photophysics

1.   Gas-Phase Studies

The idea of combining synchronously pumped dye lasers with supersonic molecular beam studies is an extremely exciting prospect. To date, however, results have been obtained only for thermal samples. Langelaar and co-workers have used the single-photon counting technique to study energy relaxation in S-tetrazine vapor.[93] They observed a small decrease in fluorescence lifetime with increasing rotational energy for both the zero-point level and several vibronic levels. The excitation bandwidth (60 pm) was such that a group of rotational levels were excited for each excitation wavelength. McDonald and Rice[94] used a similar technique to measure the single vibronic fluorescence lifetimes of pyrazine vapor, which were too short for conventional photon-counting studies. The decay is nonexponential, and the short component ($\tau \sim 100$ psec) is pressure independent. Pyrazine appears to be an intermediate case molecule, where the initially prepared state dephases very rapidly into the quasistationary states, which can then collisionally relax to triplet levels.[94] Spears and Hoffland have also used single-photon counting to study rapid radiationless process in CF_3NO.[95] Evidence was found for a rapid photochemical process competing with internal conversion for shorter excitation wavelengths.[95]

2. Solution Studies

The influence of solvent on the dynamics of excited states is very poorly understood, and there is a great need for systematic studies. The effects are often very large—for example, the fluorescence probe molecule 1, 8-anilino-naphthalene sulfonate[57, 96] has a fluorescence lifetime of about 8 nsec in ethanol with the emission maximum at 472 nm, whereas in water the life-time is about 250 psec with the emission maximum at 515 nm. Solvent viscosity is also capable of strongly influencing excited-state decay as in stilbene,[51, 97] the triphenylmethane dyes,[98–100] and the cyanine dyes.[101–103] The nonexponential decays observed by Hirsch and Mahr[104] and by Ippen et al.[105] for the triphenylmethane dyes emphasize the importance of time-resolved studies to unravel these phenomena.

E. Photobiology

The improved time resolution possible with synchronously pumped lasers has proved very attractive to groups working in the photobiology area. Processes with characteristic time scales in the 50 psec to 5 nsec region seem to be remarkably common, and applications to light-harvesting systems in green plants have been described.[54, 106] Research has also been performed on bacteriorhodopsin,[47] tryptophan, peptides, and proteins,[55, 56, 62] and the torsional dynamics of DNA.[107] The experimental method most commonly used is that of time-correlated, single-photon counting, although the bacterio-rhodopsin study of Hirsch et al.[47] by the up-conversion technique constitutes a notable exception.

V. APPLICATIONS TO TIME-INDEPENDENT SPECTROSCOPY

A. Surface Raman Spectroscopy with Synchronously Pumped Dye Lasers

In a remarkable series of papers Heritage and co-workers and Levine and co-workers[1, 2, 34–46] described the application of picosecond synchronously pumped dye lasers to the Raman spectroscopy of surfaces. The sensitivity of the technique is such that ordinary (unenhanced) Raman spectra can be obtained from monolayers of materials deposited on a surface. For example, the Raman spectrum of a monolayer of p-nitrobenzoic acid on an Al_2O_3 surface was recorded with 1 cm^{-1} resolution by Heritage.[2]

The experimental technique involves two temporally synchronized pulse trains from a pair of dye lasers synchronously pumped by the same ion laser. When the two pulse train wavelengths (ω_l and ω_s, respectively) are tuned such that $\omega_l - \omega_s = \omega_r$, where ω_r is the frequency of a Raman active mode gain is observed at the Stokes frequency ω_s. Since this is a coherent technique, the fluorescent background from the bulk material is eliminated, and much higher sensitivity is achieved than in ordinary incoherent Raman scattering. Of the several possible coherent Raman techniques [e.g., CARS and RIKES (Raman-induced Kerr effect)], the stimulated Raman gain technique was found to be the best for surface studies.[1] The calculated Raman gain for a monolayer of benzene under typical experimental conditions is about 10^{-8}: using the techniques described below, Levine and Bethea have achieved (at the shot noise limit) a minimum detectable gain of 2×10^{-9} for a 30 mW dye laser probe using a 1 sec integration time. Thus a wide range of surface Raman studies now seems to be possible.

Detection of such a minute gain obviously requires extremely stable lasers, with amplitude fluctuations of less than 10^{-8} in a bandwidth corresponding to a 1 sec integration time. Levine and Bethea[35, 36] have described in detail the modifications of their laser and detection equipment required to achieve this performance. Amplitude noise in the dye lasers arises mainly from the jet stream and is much worse in the kilohertz region than in the megahertz region.[35, 36] The Raman gain signal is detected (see Fig. 26) by means of a silicon detector with a lock-in amplifier; thus chopping the pump beam at 10 MHz with an electro-optic modulator results in a substantial decrease in noise over that obtained by 2 kHz mechanical chopping. Amplitude stability in the probe laser was increased by over an order of magnitude by insertion of a Fabry-Perot etalon. With 20 psec pulses from both lasers, a signal-to-noise ratio of 200,000 was achieved for bulk benzene with a 1 sec time constant, a focused pump intensity of 20 MW/cm^2; and a probe average power of 20 mW.

Heritage[2] and Heritage and Allara[116] have described the application of such a laser system to the observation of Raman spectra of a monolayer of chemisorbed p-nitrobenzoic acid (PNBA) on aluminum oxide. The dielectric substrate does not absorb the laser light, and a simple transmission geometry was employed for detection of the gain in the Stokes beam. The two synchronized synchronously pumped dye lasers were turned to about 580 and 640 nm. The gain in the Stokes (red) beam was then measured as the pump (yellow) laser wavelength was scanned. The signal obtained by subtraction of a substrate blank signal from that obtained with the PNBA monolayer present is shown in Fig. 27. The arrowed features at 1610,

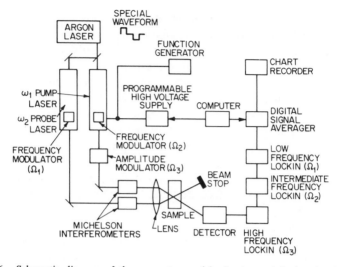

Fig. 26. Schematic diagram of the apparatus used by Levine and Bethea for frequency-modulated, stimulated Raman gain spectroscopy.[36] The special FM wave form used to reduce the background appears at the top.

1600, and 1580 cm^{-1} are assigned as Raman peaks. The 1610 cm^{-1} mode is assigned to the PNBA ring stretch, the 1600 and 1580 lines are suggested to result from photochemical and/or thermal reactions induced by the focused beams. The spectral resolution is about 6 cm^{-1} and the signal-to-noise ratio about 10.[116]

A further problem arises when the substrate absorbs at the laser wavelength: the thermal background due to reflectivity change produced by substrate heating is typically 10^4 times larger than the desired monolayer Raman signal, and thus would completely obscure it. Levine and Bethea's solution is to frequency modulate the pump laser, thus shifting the pump laser in and out of resonance at the modulation frequency. Thus the Raman signal is modulated but the thermal background is not. Rapid frequency tuning of dye lasers can be achieved with electro-optic tuners,[36, 108] however, high-frequency (10 MHz) frequency modulation introduces significant amplitude modulation on top of the desired frequency modulation, in the dye laser output. Levine and Bethea frequency modulated at low frequency Ω_L and then amplitude modulated the resulting beam at high frequency Ω_H to produce a carrier wave allowing high-frequency detection of the doubly modulated signal. As shown in Fig. 26, two lock-in amplifiers are required to detect the desired signal. Frequency modulation is at 1 kHz and amplitude modulation at 10 MHz. The carrier wave at Ω_H and the sidebands at

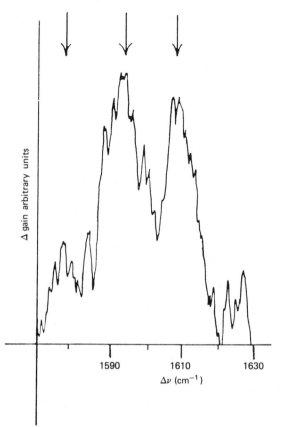

Fig. 27. Raman gain spectrum detected by Heritage and Allara[116] of a monolayer of *p*-nitrobenzoic acid (PNBA) on a thin film of alumina. The figure is obtained by subtracting a structureless background signal obtained from an alumina blank from the spectrum obtained with the PNBA monolayer adsorbate present. Three principal spectral features are indicated by arrows. See text for assignments.[116]

$\Omega_H \pm \Omega_L$ are passed through a high-frequency lock-in to extract the thermal background free signal at Ω_L, which is then detected by a low-frequency lock-in amplifier. The frequency-modulated pump beam generates amplitude modulation in the probe through the Raman interaction, which is being switched on and off at both Ω_L and Ω_H. The Stokes signal is thus

$$I_s(t) = I_s \sin \Omega_L t \sin \Omega_H t$$

which has sidebands at $\Omega_H \pm \Omega_L$. A high-frequency lock-in amplifier was modified so that its output frequency response was sufficiently high to pass signals at Ω_L. This resulting low-frequency signal, detected by a standard low-frequency lock-in, is the usual way of giving a final dc signal proportional to the original frequency modulation (FM) at Ω_L. Since there is no heating at Ω_L, the thermal background is eliminated.

A special FM waveform (shown in Fig. 26) is used to eliminate any (laser) frequency-dependent thermal background. The background rejection was tested by measuring the heating signal produced by a number of metal mirrors. The background signal dropped 10^3 to 10^4 times when the FM was turned on.

A similar technique has also been reported recently by Heritage and Bergman[117] in which the "carbonate" Raman spectrum is observed on a roughened silver surface despite the strong absorption at the roughened silver surface. The ability to study Raman spectra of surface monolayers on substrates used in heterogeneous catalysis is now demonstrably possible, and a wealth of new information should appear over the next few years.

B. High-Resolution Spectroscopy

Rather surprisingly, perhaps, picosecond pulses from synchronously pumped dye lasers have also been used for high-resolution, Doppler-free spectroscopy.[3, 109, 110] The physical basis for the use of picosecond pulses for high-resolution spectroscopy becomes apparent when their underlying structure is considered.

In the frequency domain the dye laser pulse consists of an (in-phase) sum of cavity modes whose spacing is determined exactly by the laser repetition

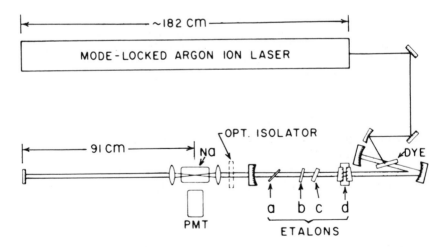

Fig. 28.   Schematic diagram of the laser setup for high-resolution, two-photon spectroscopy using a synchronously pumped dye laser.[2]

rate and whose envelope is determined by the spectrum of a single pulse. As described below, several of the techniques of Doppler-free spectroscopy have been used, but the resolution in each case is directly related to the line width of each mode making up the pulse. Couillard et al.[110] locked the frequency of a single cavity mode to an external reference cavity; the mode-locking process then transfers this stability over all the modes making up the pulse. The resultant width of each mode was less than 500 kHz.

Eckstein et al.[3] and Ferguson et al.[109] have demonstrated, respectively, two-photon spectroscopy and polarization spectroscopy with a synchronously pumped dye laser. In their initial experiments Eckstein et al.[3] recorded the Doppler-free two-photon spectrum of the sodium $3s$–$4d$ transition. They note that since the intermode spacing is controlled by the rf source

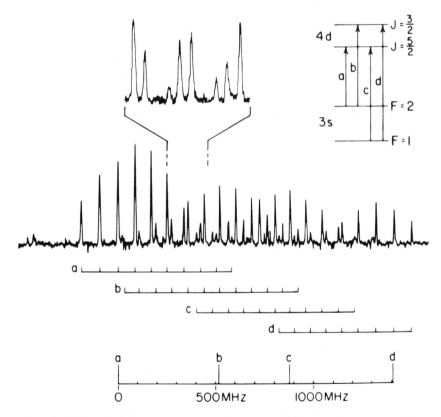

Fig. 29.   Multipulse two-photon spectrum of the sodium $3s$-$4d$ transition (center) with expanded portion shown above, recorded with pulses of 500 psec duration.[2] The entire comb of laser modes has been scanned in frequency with the help of piezotranslators. Each of the four line components a to d appears as a comb of narrow fringes, identified by the markers below. The fringe spacing equals half the pulse repetition rate.

driving the argon laser mode locker, which can be stable to 1 part in 10^8, a precise frequency calibration scale is provided. It is thus possible to apply accurate electronic frequency-counting techniques to the measurement of large line separations, up to high multiples of the pulse repetition rate, where direct modulation or beat frequency techniques would be difficult or impossible to apply. Using this calibration they determined a new value of the sodium $4d$ fine-structure splitting as 1028.5 ± 0.4 MHz. The experimental setup is shown in Fig. 28. The output of the dye laser, after passing through an optical isolator, was focused into the sample, recollimated by a lens, and

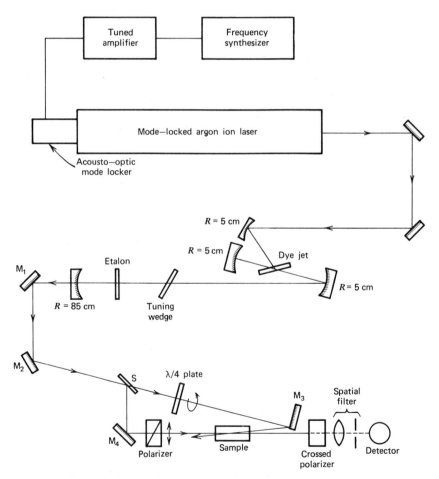

Fig. 30.   Schematic diagram of the synchronously pumped dye laser–polarization spectrometer combination of Ferguson et al.[109]

reflected back by a plane mirror. The distance between the focal spot and the mirror was adjusted to equal the dye laser cavity length. Thus each pulse meets its counterpropagating predecessor, thus forming a pulsed standing wave at the focus. Resonant two-photon absorption is possible whenever the sum of two mode frequencies coincides with an atomic transition frequency. If the atoms are in a region where the two counterpropagating pulses form a standing wave field, the excitation is nearly free of first-order Doppler broadening. When the frequency of the laser modes is scanned and the $3s$–$4d$ sodium transition monitored by observing the cascade decay of the $4p$ state to the $3s$ state, a fringelike spectrum is obtained (Fig. 29), each line component giving a comb of resonances, separated by half the laser intermode frequency spacing, as each laser mode in turn comes into resonance. (As the frequencies are scanned, adjacent modes will be in resonance when the frequency is half that giving resonance for a single mode, thus resonance occurs twice per mode spacing.) The fringe spacing is thus directly relatable to the frequency of the rf source driving the mode locker. In other words, the spectrum is self-calibrating and the spacing between two transitions (giving two interleaved combs) can be determined directly, provided the splitting is known to the order of the fringe spacing. Eckstein et al.[3] suggest that the technique holds particular promise for two-photon, Doppler-free studies of the hydrogen $1S$-$2S$ transition using a synchronously pumped blue dye laser. The authors calculate a signal enhancement of more than 10^6 for a 5 psec synchronously pumped laser as compared with a single-frequency dye laser.

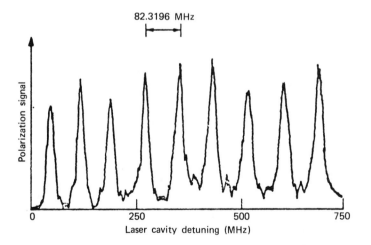

Fig. 31.   Resonances in the polarization signal for the transition from ^{20}Ne $1s_5$ to $2p_2$, obtained as the laser cavity in Fig. 29 was scanned.[109] The spacing between the resonances is half the laser repetition rate.

Ferguson et al.[109] present a theoretical description of polarization spectroscopy with a synchronously pumped dye laser. The principle is identical with conventional saturation spectroscopy, the saturating and probe modes interact with the same atoms twice as the laser wavelength is scanned by one mode spacing. Polarization spectroscopy serves to reduce the background signal arising from those modes outside the Doppler width of the transition. The experimental setup is shown in Fig. 30. As with the two-photon technique discussed above, if several transitions are present the signal consists of interleaved combs of resonances. In the example demonstrated by Ferguson et al.,[109] the transition from ^{20}Ne $1s_5$ to $2p_2$ is an isolated resonance free of hyperfine structure, and the polarization spectrum consists of a series of resonances separated by one-half the intermode spacing (82.3196 MHz) (Fig. 31). The line width is 16 MHz, which is due to the natural line width of the transition (8 MHz) the laser frequency jitter (8 MHz) and the nonzero crossing angle of the probe and saturating beam, which does not completely cancel the Doppler broadening.

V. CONCLUDING REMARKS

Although the properties of continuous synchronously pumped dye lasers now seem to be quite well understood, their applications to physical problems are just beginning. This chapter has surveyed the possibilities, with the intention of whetting the appetite for the rich range of phenomena that will be studied with these lasers over the next few years.

Acknowledgments

This work was supported by the Camille and Henry Dreyfus Foundation, the Louis Block Fund of the University of Chicago, and in part by National Science Foundation grant CHE-8009216. Thanks go to my students and colleagues, particularly Dan McDonald, Dave Waldeck, and Jan Rossell. Thanks are also due to the many authors who sent preprints and reprints of their work.

References

1. B. F. Levine, C. V. Shank, and J. P. Heritage, *IEEE J. Quantum Electron. Appl.*, **QE15**, 1418 (1979).
2. J. P. Heritage, in *Picosecond Phenomena*, Vol. 2, C. V. Shank, Ed., Springer, New York, 1980.
3. J. N. Eckstein, A. I. Ferguson, and T. W. Hänsch, *Phys. Rev. Lett.*, **40**, 847 (1978).
4. D. J. Bradley and A. J. F. Durrant, *Phys. Lett.*, **27A**, 73 (1968); D. J. Bradley, A. J. F. Durrant, G. M. Gale, P. D. Moore, and P. D. Smith, *IEEE J. Quantum Electron. Appl.*, **QE4**, 707 (1968).
5. W. H. Glenn, M. J. Brienza, and A. J. DeMaria, *Appl. Phys. Lett.*, **12**, 54 (1968); B. H. Soffer and J. W. Linn, *J. Appl. Phys.*, **39**, 5859 (1968).
6. L. Isganitis, M. G. Sceats, and K. R. German, *Opt. Lett.*, **5**, 7 (1980).
7. A. Yariv, *J. Appl. Phys.*, **36**, 388 (1965).

8. L. L. Steinmetz, J. H. Richardson, and B. W. Wallin, *Appl. Phys. Lett.*, **33**, 163 (1978).
9. W. H. Hesselink and D. A. Wiersma, *Phys. Rev. Lett.*, **43**, 1991 (1979).
10. A. Icsevgi and W. E. Lamb, Jr., *Phys. Rev.*, **185**, 517 (1969).
11. E. P. Ippen and C. V. Shank, *Appl. Phys. Lett.*, **27**, 488 (1975).
12. H. A. Pike and M. Hercher, *J. Appl. Phys.*, **41**, 4562 (1970).
13. D. B. McDonald, J. L. Rossel, and G. R. Fleming, *IEEE J. of Quantum Electron. Appl.*, **QE17**, 1134 (1981).
14. D. B. McDonald, D. Waldeck, and G. R. Fleming, *Opt. Commun.* **34**, 127 (1980).
15. E. P. Ippen and C. V. Shank, in *Ultrashort Light Pulses* S. L. Shapiro, Ed., Springer, Berlin, 1977.
16. A. von Jena and H. Lessing, *Appl. Phys.*, **19**, 131 (1979).
17. A. von Jena and H. Lessing, *Opt. Quant. Electron.*, **11**, 419 (1979).
18. D. M. Kim, J. Kuhl, R. Lambrich, and D. von der Linde, *Opt. Commun.*, **27**, 123 (1978).
19. R. K. Jain and C. P. Ausschnitt, *Opt. Lett.*, **2**, 117 (1978).
20. C. P. Ausschnitt and R. K. Jain, *Appl. Phys. Lett.*, **32**, 727 (1978).
21. K. L. Sala, G. A. Kenney-Wallace, and G. E. Hall, *IEEE J. Quantum Electron. Appl.*, **QE16**, 990 (1980).
22. E. W. Van Stryland, *Opt. Commun.*, **31**, 93 (1979).
23. E. P. Ippen and C. V. Shank, *Appl. Phys. Lett.*, **26**, 92 (1975).
24. N. J. Frigo, R. Daley, and H. Mahr, *IEEE J. Quantum Electron. Appl.*, **QE13**, 101 (1977).
25. J. N. Eckstein, A. J. Ferguson, T. W. Hänsch, C. A. Minard, and C. K. Chan, *Opt. Commun.*, **27**, 466 (1978).
26. J. de Vries, D. Bebelaar, and J. Langelaar, *Opt. Commun.*, **18** (1976).
27. J. Kuhl, R. Lambrich, and D. von der Linde, *Appl. Phys. Lett.*, **31**, 657 (1977).
28. S. Blit, E. G. Weaver, T. A. Rabson, and F. K. Tittel, *Appl. Opt.*, **17**, 721 (1978).
29. C. V. Shank, E. P. Ippen, and O. Teschke, *Chem. Phys. Lett.*, **45**, 291 (1977).
30. G. R. Fleming, O. L. J. Gijzeman, and S. H. Lin, *J. Chem. Soc., Faraday Trans. II*, **70**, 1074 (1974).
31. B. I. Greene, R. M. Hochstrasser, and R. B. Weisman, *J. Chem. Phys.*, **70**, 1247 (1979).
32. R. K. Jain and J. P. Heritage, *Appl. Phys. Lett.*, **32**, 41 (1978).
33. W. H. Hesselink and D. A. Wiersma, *Chem. Phys. Lett.*, **56**, 227 (1978).
34. J. P. Heritage, J. G. Bergman, A. Pinczuk, and J. M. Worlock, *Chem. Phys. Lett.*, **67**, 229 (1979).
35. B. F. Levine and C. G. Bethea, *IEEE J. Quantum Electron. Appl.*, **QE16**, 85 (1980).
36. B. F. Levine and C. G. Bethea, *Appl. Phys. Lett.*, **36**, 245 (1980).
37. W. H. Hesselink and D. A. Wiersma, *J. Chem. Phys.*, **73**, 648 (1980).
38. F. M. Kamga and M. G. Sceats, *Opt. Lett.*, **5**, 126 (1980).
39. S. C. Pyke and M. W. Windsor, in *Chemical Experimentation Under Extreme Conditions*, B. W. Rossitor, Ed., Interscience, New York, 1978.
40. C. V. Shank and E. P. Ippen, in *Picosecond Phenomena*, Vol. I, C. V. Shank, E. P. Ippen, and S. L. Shapiro, Eds., Springer, New York, 1978, p. 103.
41. J. L. Martin, R. Astier, A. Antonetti, C. A. Minard, and A. Orszag *C.R. Hebd. Seances Acad. Sci. Ser. B*, **289**, 45 (1979).
42. J. M. Wesenfeld and E. P. Ippen, *Chem. Phys. Lett.*, **67**, 213 (1979).
43. T. Gillbro and V. Sündstrom, *Chem. Phys. Lett.*, **74**, 188 (1980).
44. D. W. Phillion, D. J. Kuizenga, and A. E. Siegman, *Appl. Phys. Lett.*, **27**, 85 (1975).
45. D. D. Dlott, M. D. Fayer, J. Salcedo, and A. E. Siegman, in *Picosecond Phenomena*, Vol. I, C. V. Shank, E. P. Ippen, and S. L. Shapiro, Eds., Springer, New York, 1978, p. 240.
46. H. Mahr and M. D. Hirsch, *Opt. Commun.*, **13**, 96 (1975).
47. M. D. Hirsch, M. A. Marcus, A. Lewis, H. Mahr, and N. Frigo, *Biophys. J.*, **16**, 1399 (1976).

48. T. Daly and H. Mahr, *Solid State Commun.*, **25**, 323 (1978).
49. M. C. Adams, W. Sibbett, and D. J. Bradley, *Opt. Commun.*, **26**, 273 (1978).
50. J. R. Taylor, M. C. Adams, and W. Sibbett, *Appl. Phys. Lett.*, **35**, 590 (1979).
51. J. R. Taylor, M. C. Adams, and W. Sibbett, *J. Photochem.*, **12**, 127 (1980).
52. W. R. Ware, in *Creation and Detection of the Excited State*, Vol. 1A, A. A. Lamola, Ed., Dekker, New York, 1971.
53. C. Lewis, W. R. Ware, L. J. Doemeny, and T. L. Nemzek, *Rev. Sci. Instrum.*, **44**, 107 (1973).
54. G. S. Beddard, G. R. Fleming, G. Porter, G. F. W. Searle, and J. A. Synowiec, *Biochim. Biophys. Acta*, **545**, 165 (1979).
55. G. S. Beddard, G. R. Fleming, G. Porter, and R. J. Robbins, *Phil. Trans. R. Soc. Lond.* **A298**, 321 (1980).
56. R. J. Robbins, G. R. Fleming, G. S. Beddard, G. W. Robinson, P. J. Thistlethwaite, and G. J. Woolfe, *J. Am. Chem. Soc.*, **102** 6271 (1980).
57. P. J. Sadkowski and G. R. Fleming, *Chem. Phys.*, **54**, 79 (1980).
58. G. S. Beddard, R. J. Robbins, and G. R. Fleming, unpublished results.
59. L. E. Cramer and K. G. Spears, *J. Am. Chem. Soc.*, **100**, 221 (1978).
60. R. J. Robbins, thesis, Melbourne, 1980.
61. V. J. Koester and R. M. Dowben, *Rev. Sci. Instrum.*, **49**, 1186 (1978).
62. V. J. Koester and R. M. Dowben, *Biophys. J.*, **24**, 245 (1978).
63. V. J. Koester, *Anal. Chem.*, **51**, 459 (1979).
64. K. G. Spears, L. E. Cramer, and L. D. Hoffland, *Rev. Sci. Instrum.*, **49**, 255 (1978).
65. J. M. Harris, R. W. Chrisman, F. E. Lytle, and R. S. Tobias, *Anal. Chem.*, **48**, 1937 (1976).
66. W. H. Hesselink and D. A. Wiersma, *Chem. Phys. Lett.*, **65**, 300 (1979).
67. D. von der Linde, J. Kuhl, and H. Klingenberg, *Phys. Rev. Lett.*, **44**, 1505 (1980).
68. D. M. Auston, S. McAfee, C. V. Shank, E. P. Ippen, and O. Teschke, *Solid State Electron.*, **21**, 147 (1978).
69. J. P. Heritage, *Appl. Phys. Lett.*, **34**, 470 (1979).
70. F. M. Kamga and M. G. Sceats, *Opt. Lett.*, **5**, 126 (1980).
71. T. J. Chuang and K. E. Eisenthal, *Chem. Phys. Lett.*, **11**, 368 (1971).
72. H. E. Lessing, A. von Jena, and M. Reichart, *Chem. Phys. Lett.*, **36**, 517 (1975).
73. C. V. Shank and E. P. Ippen, *Appl. Phys. Lett.*, **26**, 62 (1975).
74. G. R. Fleming, J. M. Morris, and G. W. Robinson, *Chem. Phys.*, **17**, 91 (1976).
75. G. Porter, P. J. Sadkowski, and C. J. Tredwell, *Chem. Phys. Lett.*, **49**, 416 (1977).
76. K. G. Spears and L. E. Cramer, *Chem. Phys.*, **30**, 1 (1978).
77. G. R. Fleming, A. E. W. Knight, J. M. Morris, R. J. Robbins, and G. W. Robinson, *Chem. Phys. Lett.*, **49**, 1 (1977).
78. G. R. Fleming, A. E. W. Knight, J. M. Morris, R. J. Robbins, and G. W. Robinson, *Chem. Phys. Lett.*, **51**, 399 (1977).
79. A. von Jena and H. E. Lessing, *Chem. Phys.*, **60**, 245 (1979).
80. D. P. Millar, R. Shah, and A. H. Zewail, *Chem. Phys. Lett.*, **66**, 435 (1979).
81. H. J. Eichler, U. Klein, and D. Longhans, *Chem. Phys. Lett.*, **67**, 21 (1979).
82. C. M. Hu and R. Zwanzig, *J. Chem. Phys.*, **60**, 4354 (1974).
83. D. R. Bauer, J. I. Brauman, and R. Pecora, *J. Am. Chem. Soc.*, **96**, 6840 (1974).
84. J. T. Hynes, R. Kapral, and M. Weinberg, *J. Chem. Phys.*, **70**, 1456 (1979).
85. R. Zwanzig, *J. Chem. Phys.*, **68**, 4325 (1978).
86. S. Richardson, *J. Fluid Mech.*, **59**, 707 (1973).
87. P. J. Sadkowski and G. R. Fleming, *Chem. Phys. Lett.*, **57**, 526 (1978).
88. K. Tanabe, *Chem. Phys.*, **31**, 319 (1978).
89. J. T. Hynes, *Annu. Rev. Phys. Chem.*, **28**, 301 (1977).
90. R. Peralta-Fabi and R. Zwanzig, *J. Chem. Phys.*, **70**, 504 (1979).

91. D. Waldeck, A. J. Cross, D. B. McDonald, and G. R. Fleming, *J. Chem. Phys.*, **74** 3381 (1981).
92. D. Waldeck and G. R. Fleming, *J. Phys. Chem.*, in press (1981).
93. J. Langelaar, M. Leeuw, D. Bebelaar, and R. P. H. Rettschnick, in *Picosecond Phenomena*, Vol. II, C. V. Shank, Ed., Springer, New York, 1980.
94. D. B. McDonald, S. A. Rice, and G. R. Fleming, *Chem. Phys.*, in press (1981).
95. K. G. Spears and L. Hoffland, *J. Chem. Phys.*, **66**, 1755 (1977).
96. G. R. Fleming, A. E. W. Knight, J. M. Morris, R. J. S. Morrison, and G. W. Robinson, *J. Am. Chem. Soc.*, **99**, 4306 (1977).
97. D. Gegion, K. A. Muszkat, and E. Fischer, *J. Am. Chem. Soc.*, **90**, 12 (1968).
98. T. Forster and G. Hoffman, *Z. Phys. Chem. NF*, **75**, 63 (1971).
99. W. Yu, F. Pellegrino, M. Grant, and R. R. Alfano, *J. Chem. Phys.*, **67**, 1766 (1977).
100. D. A. Cremers and M. W. Windsor, *Chem. Phys. Lett.*, **71**, 27 (1980).
101. A. T. Eske and K. Razi Naqvi, *Chem. Phys. Lett.*, **63**, 128 (1979).
102. J. C. Mialocq, P. Goujon, and M. Arvis, *J. Chim. Phys.*, **76**, 1067 (1979).
103. J. Jaraudias, *J. Photochem.*, **13**, 35 (1980).
104. M. D. Hirsch and H. Mahr, *Chem. Phys. Lett.*, **60**, 299 (1979).
105. E. P. Ippen, C. V. Shank, and A. Bergman, *Chem. Phys. Lett.*, **38**, 611 (1976).
106. G. S. Beddard, G. R. Fleming, G. Porter, and J. A. Synowiec, *Biochem. Soc. Trans.*, **6**, 1385 (1978).
107. D. P. Millar, R. J. Robbins, and A. H. Zewail, *Proc. Natl. Acad. Sci.* (U.S.A.), **77**, 5593 (1980).
108. C. G. Bethea and B. F. Levine, *IEEE J. Quantum Electron. Appl.*, **QE15**, 547 (1979).
109. A. I. Ferguson, J. N. Eckstein, and T. W. Hänsch, *Appl. Phys.*, **18**, 257 (1979).
110. B. Couilland, A. Ducasse, L. Sarger, and D. Boscher, *Appl. Phys. Lett.*, **36**, 1 (1980).
111. G. R. Fleming and G. S. Beddard, *Opt. Laser Technol.* **10**, 217 (1978).
112. J. P. Heritage and R. K. Jain, *Appl. Phys. Lett.*, **32**, 101 (1978).
113. J. Kuhl, H. Klingenberger, and D. von der Linde, *Appl. Phys.*, **18**, 279 (1979).
114. R. H. Johnson, *IEEE J. Quant Electron. Appl.*, **QE15**, 84 (1979).
115. J. P. Ryan, L. S. Goldberg, and D. J. Bradley, *Opt. Commun.*, **27**, 127 (1978).
116. J. P. Heritage and D. L. Allara, *Chem. Phys. Lett.*, **74**, 507 (1980).
117. J. P. Heritage and J. G. Bergman, *Opt. Commun.*, **35**, 373 (1980).

PREDICTIONS OF THE
STRUCTURE OF COMPLEX SOLIDS

JEREMY K. BURDETT*

Department of Chemistry
University of Chicago
Chicago, Illinois

CONTENTS

I. Introduction 48
   A. Comparisons Between Molecules and Solids 48
   B. Description of Solid-State Structures 50
II. Structural Models of Solids 53
   A. Radius Ratio Rules 53
   B. Pearson Diagrams 59
   C. The Scheme of Phillips and Van Vechten 63
   D. Pseudopotential Structural Maps 65
III. Structural Enumeration 65
   A. The Philosophy 65
   B. Pólya's Theorem 68
   C. Edge-Sharing Octahedra 70
IV. Theory of the Electronic Structure of Solids 72
   A. Molecules and Solids 72
   B. Fragment-Within-the-Solid Orbitals 75
V. Techniques of Molecular Orbital Analysis 79
   A. Perturbation Theory 79
   B. Instability of Occupied High-Energy Orbitals 80
   C. Site Preferences and Bond Lengths 83
   D. The Fragment Formalism 84
VI. Some Illustrative Examples 85
   A. The Structure of Red PbO 85
   B. Defect Diamond Structures 89
   C. The Cuprite (Cu_2O) Structure 93
   D. Structures Derived from the Breakup of the Rock Salt Structure 95
   E. Structures Derived from Puckered Sheets 99
VII. A Linking Thread? 107
Acknowledgments 110
References 110

*Fellow of the Alfred P. Sloan Foundation and Camille and Henry Dreyfus Teacher-Scholar.

I. INTRODUCTION

A. Comparisons Between Molecules and Solids

The rich panoply of the structural solid state presents a variety of architectural arrangements that have held the attention of physicists, chemists, and materials scientists for over three-quarters of a century. It is, however, a truism that the gigantic strides made in the past two decades in appreciating the electronic structure of molecules at a basic and global level have not been matched by analogous advances in extended solid-state arrays. To the molecular chemist, able to rationalize structures and reaction pathways using molecular orbital theory supported by symmetry, overlap, and perturbation theoretic arguments,[1-5] current ideas concerning the structure and bonding in solids often appear naive and rather archaic. Current work in the molecular field involves detailed molecular orbital studies on hypothetical (and sometimes rather bizarre) molecules,[6] to examine their geometrical and electronic structures: such is the confidence of the molecular scientist. By way of contrast, there are virtually no comparable explorations in the solid-state field at present. Historically a similar situation occurred in the area of spectroscopy. It was not until the 1950s that the electronic spectrum of solid germanium was understood at a basic level, whereas a comparable level of appreciation concerning the spectrum of benzene vapor was achieved during the 1930s after the ideas of molecular orbital theory had developed some roots.

The reasons for these differences are not hard to find, of course. Solids are molecules of infinite extent and, although crystalline materials are characterized by a translational three-dimensional periodicity, they are undoubtedly more "complex" than molecules. In addition solids are more dense and, since this implies that the interaction of an atom with second nearest neighbors may often be important, there are extra problems in understanding the structure of solids compared to molecules, where nearest-neighbor interactions give a good picture. As a practical point, while the electronic structure of a molecule is understandable (in principle at least) from the results of a single molecular orbital calculation, that of the solid requires calculation of the level structure at a large number of different points within the Brillouin zone. Our comments of course only apply to the lack of a detailed quantum-mechanical theory to predict and understand structural arrangements. There is a high level of understanding concerning the electrical, optical, and thermal properties of solids in the realm of solid-state physics. It is in the area, perhaps best described as crystal chemistry, that a major advance is overdue.

Continuing our molecule-solid comparison, it is interesting to delve a little more deeply into the different philosophies that solid-state and molecular structural chemists use in plying their trade. Among other things an organometallic chemist might be interested in (a) the conformation of various organic units attached to a metal; (b) whether the metal ligand distances are unusually short or long by comparison with other species that may differ by having a different number of d electrons, or a metal from a different row of the periodic table; (c) the type of atoms (π, σ, and electronegativity properties) that occupy symmetry-inequivalent sites in the structure (site preferences), and (d) the difference in bond lengths not set by symmetry requirements. (See, e.g., recent work of the Hoffmann school.[7-8]) The structural organic chemist is particularly interested in the conformations of molecules and the (perhaps rather small) variations in bond distances induced by substituent and structural changes in the molecule, and how collections of structural results may give clues to likely reaction pathways, following the ideas of Bürgi and Dunitz.[9] The polyhedral chemist is again interested in site preferences (e.g., of boron and carbon in a carborane), in bond lengths, and naturally in the shape of the polyhedron itself.[10]

Both inorganic and polyhedral chemists are interested in structural variations induced by the presence of different numbers of electrons (d electrons or the number of skeletal electron pairs) in the species concerned. Since most organometallic complexes have a total of 18 valence electrons associated with the metal, we do not need to look much further than the 18-electron (or effective atomic number) rule[7] to understand the coordination number of a particular system. In transition metal complexes with hard ligands or main group complexes, the structural chemist is more interested in the configurational nuances of a given series (e.g., the sulfuranes SX_4) than in trying to understand the relative stabilities of four- and six-coordination. For transition metal complexes there is a simple theoretical approach based on the crystal field[11] or molecular orbital stabilization energy[12] that allows comment on octahedral versus tetrahedral coordination and has been long used in crystal chemistry[11, 13, 14] to rationalize the occurrence of normal and inverse spinels $A^{II}B_2^{III}O_4$ where A or B or both are transition metals.

Generally, however, apart from normal valence considerations and the use of qualitative steric arguments, the molecular chemist does not tackle the coordination number problem. The reason is simple. It is much easier theoretically to probe the angular geometry and variations in bond lengths of a complex than to calculate numerically the energetics of ligand attachment. In the former, perturbation theory, symmetry, and overlap play a role. In the latter, numerical results are much more important. Progress in determining energy profiles for chemical reactions of molecules, especially in the area of transition metal systems, has been rather slow because of this.

On the other hand, the major question crystal chemists ask is why a particular structure, defined by the details of the coordination polyhedra and topology, is adopted for a species of given stoichiometry. Structural alternatives very often involve polyhedra with different coordination numbers. For binary AX systems there are over 20 distinct structures known. (Some of these are not represented by more than a handful of examples, however.) It is also true to say that there is a greater variety of coordination polyhedra exhibited by the extended solid state than by the structures of molecules. In many cases the only differences between two structures occur between second nearest neighbors or at an even longer range. Note, for example, the zinc sulfide polytypes wurtzite and sphalerite and more complex polytypes of both SiC^{15} and $CdCl_2{}^{16}$ which may repeat every 1000 Å or so. The energy differences between such structures is clearly rather small. Neither of these two areas, the coordination number problem and second nearest neighbor structural variations, is well served by qualitative arguments; they are areas which molecular chemists usually avoid. Tackling the solid state structural problem in the most obvious fashion requires an immense task of electronic structural calculation. By using pseudopotentials[17] this capability is probably now within our grasp, but let us halt a minute and ask whether this is really our aim. If we had infinite computing capacity and could actually calculate accurately the energies of all structural alternatives, triumphantly report that the observed structure was the one calculated to be of lowest energy, how much improvement in understanding would we have generated? Certainly the results of our quality quantum calculations would be very useful to those who wish to calculate optical and electrical properties of the solid, and the importance of such efforts in this area should not be understated. But the number of results that could be transferred to other systems to build a global structure view might well be small. By analogy with the historical reception of such calculations on molecules, the more exact the calculation, the less understandable the results are to the nonspecialist. It is those theories that are often rather drastic simplifications of the quantum-mechanical truth that have been of greatest use to the organic and inorganic chemist. In the realm of the one-electron approximation, witness the success of simple Hückel theory in organic chemistry[18, 19] and the developments induced in organic and inorganic areas by the use of the extended Hückel theory,[20] such as the Woodward-Hoffmann rules.[21] Such striking advances have overshadowed the solid progress made in "quality" calculations. Simple theories whose workings are transparent to the average chemist, in which a structural result has an overlap or symmetry-based explanation, are much more readily appreciated than those whose result is buried in the quantum-molecular numerology.

B. Description of Solid-State Structures

Even before theoretical questions are asked about the electronic ideas behind the crystal structure, the crystal chemist has two other problems to face. How is the structure geometrically described, and how does it fit into an overall classification scheme? Present ideas largely focus on the nature of the coordination polyhedra present and their linkage into three-dimensional nets.[22, 23] The molecular chemist rarely faces these problems. An immediate pigeonhole is provided because of the coordination number of the complex, and a secondary classification as regards its geometry usually immediately follows. Ambiguities in the description of the coordination geometry may be readily tackled by using the ideas of Muetterties and Guggenberger[24] or of Porai-Koshits and Aslanov,[25] to resolve whether, for example, an eight-coordinate system is best described as a square antiprism or a bicapped trigonal prism. Even before the development of Wade's rules,[26] the rich area of polyhedral molecules had succumbed to a topological approach. The organic chemist has long regarded molecules as being assembled from smaller fragments, both from a synthetic and a descriptive viewpoint. In the solid state the cuprite structure (Cu_2O) has been described (equally correctly) as being made up of two interpenetrating β-cristobalite nets[23] (Fig. 1), as a defect derivative superstructure of cubic diamond (itself constructed by filling half the tetrahedral holes in a cubic close-packed array) as shown in Fig. 2, or simply as the CsCl structure with a tetrahedron inscribed in the cube (Fig. 3).

The description of the solid-state structure is so extremely important because often the way the structural problem is tackled theoretically leans very heavily on it. For example, the early ideas of Barlow[27] in which he envisaged the packing together of billiard ball-like atoms or ions in solids encouraged the use of mechanical models (radius ratio rules) to view structural preferences in solids, as we will see later.

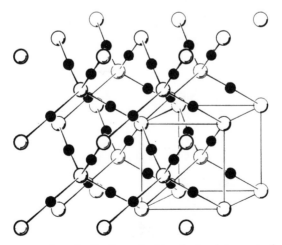

Fig. 1.   The structure of cuprite Cu_2O, emphasizing the two interpenetrating β-cristobalite frameworks.

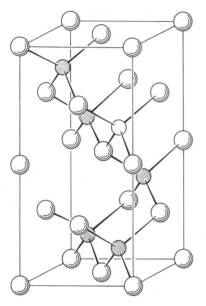

Fig. 2. Cuprite viewed as a double defect diamond derivative superstructure. This is the chalcopyrite structure ($CuFeS_2$) with the iron atoms removed.

The importance of structural classification and description has long been recognized. Perhaps the most recent progress in this field has come by way of the concepts of chemical twinning[28] and crystallographic shear,[29] where complex structures may be readily built up by successive twinning or shear operations on the unit cell. In addition the description of several structures in terms of rod packings,[30] where the geometrical arrangement is built up from units that are sometimes very large, simplifies some of the conceptual descriptive problems enormously.

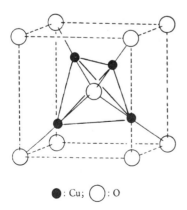

● : Cu; ○ : O

Fig. 3. The unit cell of cuprite showing a stuffed tetrahedron within a primitive cube or, alternatively, a tetrahedron inscribed within a body-centered-cubic lattice.

This rather lengthy introduction has indicated the nature of the problem we face in being able to predict and therefore understand the structures of solids. The problems we meet differ in many ways from those encountered in molecules.[31] Below we describe several methods that for relatively simple systems (mainly the AB binaries) are rather good at separating one structural type from another. These are methods that lead, by way of a two-dimensional plot of judiciously chosen variables for A and B, to regions where only one structural type is found. They are not predictive methods in the absolute sense, since the boundaries defining each region are not determined by any analytic, theoretically determined function but are drawn in by hand afterward. We also show how progress can be made in understanding structural preferences within a related series of structures by first using the techniques of modern combinatorial analysis[32] to enumerate all possibilities and then investigating energetic differences by means of electronic structure calculations. The focal point of the discussion is to show how the techniques of analysis used to such effect in the molecular area[12] (symmetry, overlap, and perturbation theory), are powerful tools in the construction of theories of structure and bonding in solids. Throughout, our interest lies with nonmetallic systems that fall into the traditional bonding categories of ionic, covalent, and Van der Waals solids.

II. STRUCTURAL MODELS OF SOLIDS

A. Radius Ratio Rules

As mentioned, Barlow's ideas concerning the packing of spherical atoms or ions together to build up an extended three-dimensional array encouraged the use of mechanical models based on rigid atoms or ions to predict structures. The alkali halides and alkaline earth oxides might be considered ideal candidates to include in such a treatment. They could be expected to be reasonably well represented as A^+X^- systems (throughout this chapter we use A and X to represent electropositive and electronegative atoms, respectively), and the total electrostatic energy of an array of oppositely charged billiard balls of different sizes is readily computed as a function of their radius ratio $\rho(=r_+/r_-)$. Whenever the anion and cation are of approximately the same size, the preferred structure is one wherein each ion has the largest number of oppositely charged neighbors. This occurs in the 8 : 8-coordinate CsCl structure, where the dimensions of the unit cell are set by the anion-cation contact along the body diagonal of the cube. As ρ decreases from unity, the stage is reached when the eight-coordinating anions touch. For smaller ρ values this would imply that the cation is able to rattle in the cavity formed by the anions. This gives a larger cation-anion distance than would be set by the rigid ion radii and clearly close anion-anion contacts, both of which are energetically unfavorable. Geometrical considerations show that for $\rho < 0.732$ the rock salt (6:6) structure becomes more stable. Here the anion and cation can remain in contact and the anion-anion contact is eliminated by replacing cubal eight-coordination of the ions by octahedral six-coordination. As ρ decreases further to 0.414, these six anions come in contact and the tetrahedral four-coordinated zinc sulfide structures [wurtzite and sphalerite (zincblende)] become of lower energy. The variation in the total energy of the three structures as a function of ρ is shown in Fig. 4. Figure 5 shows how badly this rigid-ion model fares in predicting crystal structures, even in these most favorable cases. The model implies that some of the lithium halides will have the four-coordinate ZnS structures, but none in fact do. It also overestimates the number of eight-coordinate species. In the alkaline earth chalcogenide series, analogous results are found. There are no oxides with the CsCl structure, for example. No improvement is found if Pauling radii[34] are replaced by those of Shannon and Prewitt,[35] for example; and in spite of these failings it is surprising to find these rules still said to have predictive use in the crystal chemical folklore. In *qualitative* terms they do work of course just as similar steric ideas concerning the coordination numbers of molecular species work in an analogous fashion. First-row atoms Li through F are rarely more than four-coordinate, second-row atoms are frequently six-coordinate in *molecular* examples. The larger actinide and lanthanide ions are very often eight-coordinate.

Improvements in the scheme could be made in two areas. First the bond length or radii sum of the ions is actually dependent on the coordination number.[35] This affects cations more than anions and leads to an extra degree of freedom in interpreting the predictions of the rules. As it turns out this does not help very much. Second, the ions can be made deformable. Where accurate electron density maps[36] are available for simple systems, a field receiving increasing study, it is clear that the electron distributions are

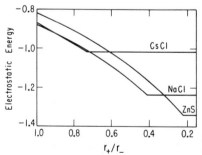

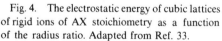

Fig. 4.   The electrostatic energy of cubic lattices of rigid ions of AX stoichiometry as a function of the radius ratio. Adapted from Ref. 33.

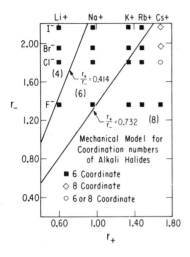

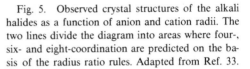

Fig. 5. Observed crystal structures of the alkali halides as a function of anion and cation radii. The two lines divide the diagram into areas where four-, six- and eight-coordination are predicted on the basis of the radius ratio rules. Adapted from Ref. 33.

far from spherical in these systems. Also, for example, as Schwartz has shown[37] for MgO, the radial extent of the electron distribution function for the "anion" is large enough to partially envelop the "cation" such that assessment of "ion charges" is a nontrivial matter. Given this result, it is of course also difficult to calculate an "ionic radius" for the species or even to decide what the term really means. In spite of such ambiguities, there are several sets of results that allow considerable structural sorting using the radii as parameters. The results of one very impressive study[33] are shown in Fig. 6, where A_2BO_4 structures are cleanly separated, although the quantitative reasons behind the placement of the boundary lines between the various

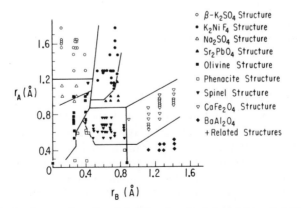

Fig. 6. Structural sorting of A_2BO_4 minerals. Adapted from Ref. 33.

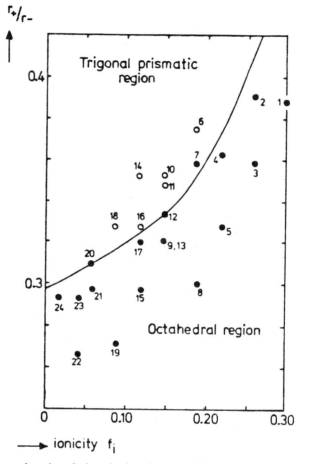

Fig. 7.   Structural sorting of trigonal prismatic (open circles) and octahedral (solid circles) coordination. Pauling's definition of ionicity is used as the ordinate. From *Crystal Structure and Chemical Bonding in Inorganic Chemistry*, C. J. M. Rooymans and A. Rabenau, Eds., North-Holland Publishing Co., Amsterdam, 1975, with permission.

structures are far from clear. Figure 7 shows sorting of trigonal prismatic (MoS_2 structure) and octahedral ($CdHal_2$ structures) coordination in AX_2 species.[38] Clearly as the X atom gets larger compared to the central atom, trigonal prismatic coordination will be disfavored because of nonbonded repulsions, but the dependence of the trigonal prismatic octahedral border on the electronegativity difference is a feature not readily understood.

A rather different type of rigid-sphere model has recently been applied to solid-state structures. This is the Bartell-Glidewell approach[39, 40] to (usually

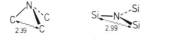

1

subtle) features of molecular geometry, by seeing how repulsions between nonbonded atoms might influence the structure. The extra degree of freedom allowed by including these effects is usefully seen for SiO_2. An ionic model would regard the system as being made up of large O^{2-} ions and small Si^{4+} ions; the nonbonded model allows a contribution from the converse, namely, large Si atoms surrounded by small O atoms. In the molecular case while $N(CH_3)_3$ has a pyramidal NC_3 skeleton, as would be predicted by VSEPR,[41] Walsh's rules,[3] and allied approaches to the structure problem,

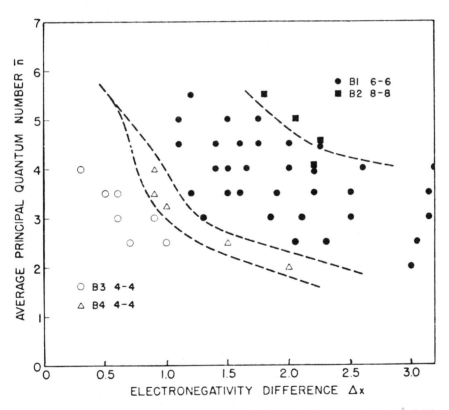

Fig. 8. Pearson diagram for nontransition metal AX species (A: group A cation). Solid circles, rock salt; open circles, sphalerite; triangles, wurtzite; squares; CsCl structure. From W. B. Pearson, *Crystal Chemistry and Physics of Metals and Alloys*, Wiley, New York, 1972, with permission.

$N(SiH_3)_3$ is planar. It is Glidewell's hypothesis[40] that the nonbonded repulsions between the large Si atoms is smaller at the planar geometry (**1**) (Si—N—Si angle = 120°) than at the pyramidal structure (Si—N—Si angle = 90°). A similar approach has been used[42] in the solid state to rationalize the observation that many Si—O—Si angles in silicates are considerably greater than tetrahedral. (In coesite[43] there is good crystallographic evidence for a linear Si—O—Si unit.) As we discuss later, this structural flexibility has an important bearing on silicate structures. An alternative explanation exists[12] for these structural features. This is in terms of superior π-bonding between central atom p and ligand σ orbitals at the linear or planar structure, which would give rise to a structural effect very similar to that expected from nonbonded repulsions. Recent *ab initio* calculations[44] on some isoelectronic ylid structures focused on such π-bonding arguments as the dominant feature influencing the stereochemistry. Theoretically it is difficult to numerically sort out the two effects.

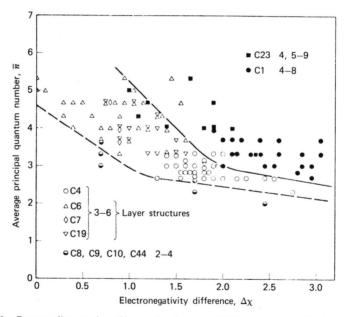

Fig. 9. Pearson diagram for AX_2 species. Squares, $PbCl_2$ structure; solid circles, fluorite structure; open circles, rutile structure; half-solid circles, GeS_2 and silica structures; right triangles, CdI_2 structure; inverted triangles, $CdCl_2$ structure; diamonds, MoS_2 structure. From W. B. Pearson, *Crystal Chemistry and Physics of Metals and Alloys*, Wiley, New York, 1972, with permission.

B. Pearson Diagrams

Mooser and Pearson found[45, 46] that if the average principal quantum number ($\bar{n}$) of the valence shell of an A_nX_m system was plotted against the AX electronegativity difference ($\Delta\chi$), quite a striking separation of structures with different coordination numbers could be achieved. Figures 8 to 12 show some of the sorting found. If a modified ordinate $r_-/r_+ \cdot \Delta\chi$ is used, the separation is a little better. The usual interpretation of such a diagram is that as $\Delta\chi$ increases, the bonding becomes less directional because there is a larger ionic contribution to the bonding. As $\bar{n}$ increases, a decrease in directional effects is also found because of a rather ill-defined process called metallization. Directional bonding is usually associated with four-coordinate structures, perhaps via a historical preoccupation with the chemistry of carbon-containing molecules, and its description in terms of the sp^3 hybrids of valence-bond theory. It is interesting to examine how these diagrams are understandable from the perspective of ideas based on covalent molecular orbital theory.[47]

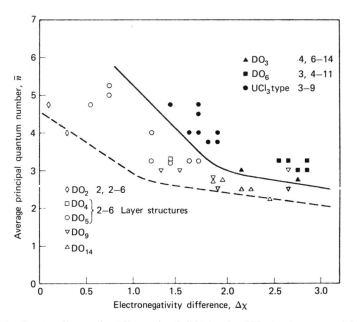

Fig. 10. Pearson diagram for AX_3 species. Solid triangles, BiF_3 structure; open right triangles, AF_3 structure; inverted triangles, ReO_3 structure; solid squares, LaF_3 structure; open squares, $CrCl_3$ structure; solid circles, UCl_3 structure; open circles, BiI_3 structure; diamonds, $CoAs_3$ structure. From W. B. Pearson, *Crystal Chemistry and Physics of Metals and Alloys*, Wiley, New York, 1972, with permission.

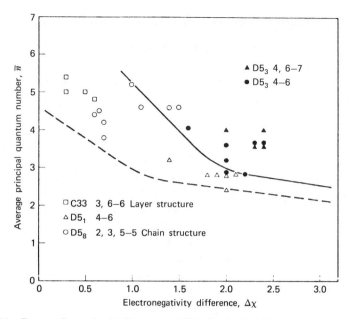

Fig. 11. Pearson digram for M_2X_3 species. Solid triangles, La_2O_3 structure; open triangles, α-Al_2O_3 structure; solid circles, Mn_2O_3 structure; open circles, Sb_2S_3 structure. From W. B. Pearson, *Crystal Chemistry and Physics of Metals and Alloys*, Wiley, New York, 1972, with permission.

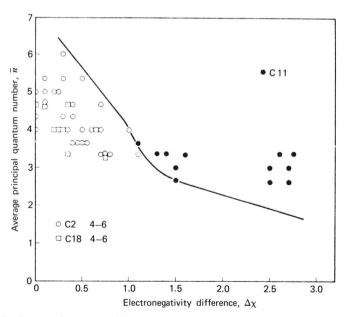

Fig. 12. Pearson diagram for AX_2 species containing X_2 pairs. Solid circles, CaC_2 structure; open circles, pyrite structure; squares, marcasite structure. From W. B. Pearson, *Crystal Chemistry and Physics of Metals and Alloys*, Wiley, New York, 1972, with permission.

2

Structure **2** shows the three coordination numbers found in the zinc-blende (sphalerite) or wurtzite, rock salt, and CsCl structures. Figure 13 shows molecular orbital diagrams for the fragments AX_4^{3-}, AX_6^{5-} and AX_8^{7-}, which perhaps can give us qualitative clues to the electronic factors influencing the coordination number. Using this approach we can see that the bonding is just as "directional" in all three structures. An important feature of the latter two structures is the presence of occupied central atom–ligand nonbonding orbitals that are ligand-ligand antibonding and destabilized with respect to the orbitals of the free ligand. These give rise to non-bonded repulsions between the ligands. In addition in all three structures some of the orbitals, stabilized by central atom–ligand interactions contain some ligand-ligand antibonding character.

Within this simple molecular orbital framework we can identify three factors that can influence the stabilities of the arrangements **2**. (a) On moving from four- to six- to eight-coordination, the A—X distance increases. This is the dependence of "ionic radius" on coordination number mentioned above. Concurrently the A—X overlap integrals are reduced and the stabilization of the bonding orbitals decreases accordingly. (b) The increase in coordination number is reflected in a decrease in X $\cdots$ X nonbonded distances, partially offset by the increase in A—X distance. Nonbonded repulsions of the type described above become more important as the coordination number increases. (c) Because of the increase in coordination number, the total stabilization energy of the AX_y^{1-y} system increases. While the stabilization energy per close contact decreases (and the AX distance increases), the increasing number of such contacts results in increasing stabilization for increasing y. A plot of the preferred structure as a function of the Slater exponent of the X atom is shown in Fig. 14, which is remarkably similar to the Pearson diagram of Fig. 8. As the electronegativity of X increases with respect to that of A, the ligand orbitals become more contracted, leading to a decrease in the molecular orbital-based nonbonded repulsions that make the higher coordination numbers unstable. The balance of all three factors on this model controls the coordination number of the lowest energy structure. It is interesting that a very crude covalent model also gives a similar variation of coordination with $\Delta\chi$, although a much better calculational method would have to be used to stand a chance of being able to predict the structure of a given system.

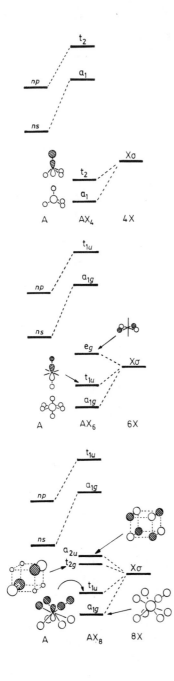

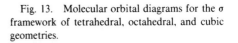

Fig. 13. Molecular orbital diagrams for the σ framework of tetrahedral, octahedral, and cubic geometries.

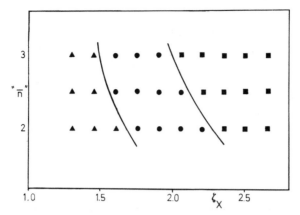

Fig. 14.   Calculated preferred geometry for an AX_y^{1-y} unit as a function of $\bar{n}$ and the Slater exponent ζ_X of the orbitals on the X atom. Squares, eight-, circles, six-, triangles, four-coordination.

C.   The Scheme of Phillips and Van Vechten

Phillips and Van Vechten devised[48, 49] an ingenious scheme with which to examine the structures of the octet binary system $A^N B^{8-N}$ that derived from earlier research into the physical properties of these systems. First the static polarizability (ε) of the solid is written in terms of an average energy gap E_g as

$$\varepsilon = 1 + \frac{\left(h\omega_p\right)^2}{E_g^2} \tag{1}$$

where ω_p is the plasma frequency.

This describes a polarizability that increases as the energy gap between filled and unfilled levels decreases, as in the perhaps more familiar expression for atoms and molecules. The key step of the scheme is to divide E_g into ionic (C) and covalent (E_h) contributions in the following way:

$$E_g^2 = E_h^2 + C^2 \tag{2}$$

where E_h depends on the distance between the AB bonded atoms in the crystal and C is given explicitly by

$$C = 1.5\left[\frac{Z_A e^2}{r_A} - \frac{Z_B e^2}{r_B}\right] \exp\left(-\frac{1}{2}k_s[r_A + r_B]\right) \tag{3}$$

In (3) $r_{A,B}$ are the covalent radii of the atoms A, B, $Z_{A,B}$ are the ion core charges, and k_s^{-1} is the Thomas-Fermi screening radius for a free-electron gas; C and E_h are effectively obtained for a given system by knowledge of the interatomic AB distance and the static dielectric constant of the crystal. A two-dimensional plot of these functions leads to a striking resolution of the structures adopted by these systems (Fig. 15). If the fractional ionic character is defined as

$$F_i = \frac{C^2}{\left(E_h^2 + C^2\right)} = \frac{C^2}{E_g^2} \tag{4}$$

then a line of slope $F_i = 0.785$ separates the fourfold- from the sixfold-coordinated solids. Thus, as in the Pearson diagrams, increasing ionic character leads to higher coordination numbers. Why the dividing line appears at this particular value of F_i is of course not easily extracted directly from the model. It is interesting to note, however, that Cohen[17] finds that the bond charge goes to zero near $F_i \sim 0.8$, hence it has been suggested that the stability of some of these more ionic solid-state systems is determined by electrostatic forces. The cohesive energies of these systems may also be matched by a simple function that depends on the bond length and the ionicity. Parenthetically we note that the method has been used to devise a dielectric scale of electronegativity, interesting in itself because the Pearson diagram ordinate was an electronegativity difference. A similar division of structural types may be achieved[50] by using the splitting of peaks in the X-ray Photoelectron Spectra of these systems.

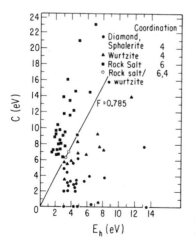

Fig. 15. Structural sorting of eight-electron AB systems using Phillips and Van Vechten's approach. Adapted from Ref. 33.

D. Pseudopotential Structural Maps

In several studies, various authors have achieved a very impressive structural sorting of AB compounds. The structural indices used here are two sums and differences of atomic "radii":

$$R_\pi = |r_p^A - r_s^A| + |r_p^B - r_s^B| \qquad (5)$$

$$R_\sigma = |(r_p^A + r_s^A)| - |(r_p^B + r_s^B)| \qquad (6)$$

$r_{s,p}$ are the crossing points of the pseudopotential $V_{eff}^l(r)$ for the $l=0(s)$ and $1(p)$ valence orbitals on A and B. This is the point where $V_{eff}^l(r)=0$. In the study of St. John and Bloch[51] these were obtained empirically from the spectra of nontransition elements, in Zunger's study[53] by actual calculation of the pseudopotentials for both main group and transition metal ions. In the treatments by St. John and Bloch[51] and Phillips and Chelikowsky[53] the absolute value signs of (5) and (6) are not included. (The semiempirical radii used by these authors always show $r_p - r_s > 0$ in contrast to the first-principles radii of Zunger.[52]) Zunger's results (Fig. 16) show an excellent separation of several different structure types. (The sorting would be even better if the artificial restriction of linear boundaries were relaxed.) Very interestingly, the structures of several intermetallics are resolved. One very important point concerning these and other systems that contain transition metals is that the result is obtained without any explicit consideration of the d orbitals or the electrons in them. Consideration[54] of the lattice energies of the transition metal halides or the heats of hydration as a function of the number of d electrons shows that the d-orbital contribution to the stabilization energy is small, perhaps a maximum 20% at the left-hand side of the periodic table, so that this result is not too surprising. In assessing the angular geometries of transition metal complexes we have devised a scheme[55] that indicates that the observed geometry is a balance between s, p, and d-orbital forces.

The radii r_l are related to the ionization potential from the relevant orbital, and hence are linked to an orbital electronegativity for the atom concerned. In this way there is connection, albeit a difficult one to explore in a concrete fashion, with the approaches to the structural problem of Phillips and Van Vechten and of Pearson. We return to this relationship at the end of the chapter.

III. STRUCTURAL ENUMERATION

A. The Philosophy

Section II indicated that absolute structure prediction on chemical grounds is a goal that has not yet been achieved, even for simple systems, although structural sorting is really quite good. Therefore the rest of the chapter in-

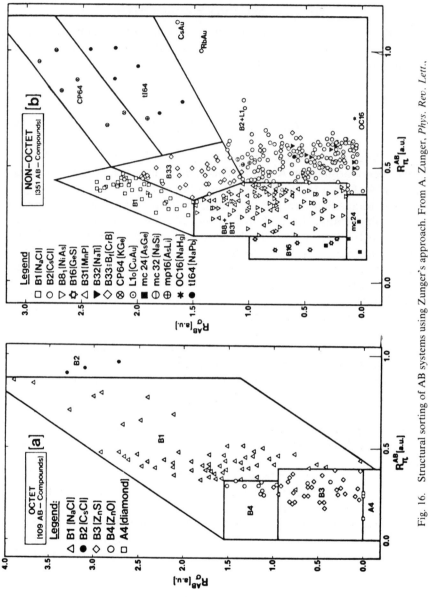

Fig. 16. Structural sorting of AB systems using Zunger's approach. From A. Zunger, *Phys. Rev. Lett.*, **44**, 582 (1980), with permission.

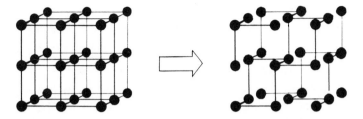

Fig. 17. A derivation of the arsenic layer structure by the selective fission of some of the linkages of the rock salt structure.

vestigates the possibility of making progress by examining the factors influencing the stabilities of different structures derived from a parent in different ways. This is clearly a small subset of the structural problem, but by investigating sufficient numbers of these sets a global chemical picture may perhaps be built up.

The arsenic structure of Fig. 17 may be derived[22] from that of rock salt by breaking a specific set of the linkages between adjacent atoms. Figure 18 shows a similar way to regard the chain structure of selenium. In this case one subset of the structural problem contains all those arrangements derived from rock salt in this way. In a later section we ask why this particular pattern is favored over other possibilities. Here we investigate the ways of enumerating all possible bond-breaking routes. As an indicator of the nontrivial nature of the problem, there are 24 linkages contained in the unit cell of rock salt. The number of different patterns in general will be around 2^{24}, since each linkage is either broken or unbroken. Since some patterns are related to one another by the symmetry of the lattice, this figure may be reduced, approximately by division by the number of symmetry elements of the space group. Even this smaller figure of around 44,000 hardly encourages hand enumeration. Fortunately the present state of combinatorial analysis[31, 56–60]

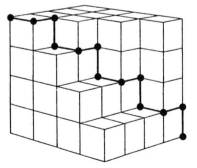

Fig. 18. A way of viewing the structure of elemental selenium by selectively breaking the linkages of the rock salt structure. (For clarity, only the selenium chain atoms are shown.)

allows considerable headway to be made to solve the problem, although there are still areas in which the necessary group theoretical techniques are largely undeveloped.

B. Pólya's Theorem

Let us start with a near-trivial molecular problem, one that is readily enumerable by hand. How do we count the number of isomers MA_nB_{6-n} ($n=0$ –6)for an octahedrally coordinated transition metal complex? Put another way, the combinatorial problem is to ask how many different ways there are of coloring the vertices of the octahedron in two colors (corresponding to the presence of an A or a B atom). Since there are two colors and six vertices there are a total of $2^6 = 4096$ possibilities, but how many are actually different? Because of the high symmetry of the octahedron, this number actually reduces to a mere 10. In fact if we do not really care which is A or B but are interested only in the patterns, there are only six possibilities. The general problem, the number of distinct colorings of a set D permuted by a group G is treated by algebra developed by Burnside, Pólya,[56] and De Bruijn.[57, 58, 61] Of these Pólya's enumeration theorem is by far the best known.[56]

If D and R are finite sets, G is a finite group permuting D and $\Omega = \Omega(G; R^D)$ contains one representative of every pattern of colorings of the set D with colors from the set R, then the number of patterns is given by

$$|\Omega| = Z(G; |R|, |R| \cdots |R|) \tag{7}$$

where Z is the cycle index of G and is defined as follows. Every permutation may be written uniquely as a product of disjoint cycles. So for an operation $g \in G$ we may simply define the permutational properties of g as the product of disjoint cycles $\pi_1 \pi_2 \cdots \pi_n$ where every element of D appears in exactly one π_i. The cycle π_i is a cycle of length m_i or just an m_i-cycle if it contains m_i elements of D. The number $c_j(g)$ of j-cycles in the permutation is then readily obtained. The cycle structure of $g \in G$ is defined to be the monomial

$$\prod_{j=1}^{|D|} x_j^{c_j(g)}$$

where the x_j are dummy variables. The cycle index of G is simply defined as the average

$$Z(G; x_1, x_2 \cdots x_{|D|}) = |G|^{-1} \sum_{g \in G} \prod_{j=1}^{|D|} x_j^{c_j(g)} \tag{8}$$

In evaluating Z for a given problem a simplification arises, since a weighted sum need only be performed, using one representative from each class contained in G weighted by the number of operations in each class. For the trivial problem described at the beginning of this section the permutations associated with the operations, one from each class, contained in $G=O$, the octahedral group, are shown in Table I. The set D is the collection of vertices ($|D|=6$) and the set R the collection of colors ($|R|=2$). From these the cycle structures are derived and the cycle index becomes

$$Z(O; x_1, x_2 \cdots x_6) = \tfrac{1}{24}\left(x_1^6 + 8x_3^2 + 6x_2^3 + 6x_1^2x_4 + 3x_1^2x_2^2\right)$$

Since we are interested in the number of patterns using two colors, each x_i is put equal to $|R|=2$, hence $Z=10$. Use of the full point symmetry $G=O_h$ leads to the same result, but below we describe an example in which this is not the case.

A weighted form of Pólya's theorem also exists[58] to enable the number of MA_nB_{6-n} variants to be extracted. If ω is a weight function on R, then the weighted sum of all the patterns $\phi \in \Omega$ is given by

$$\sum_{\phi \in \Omega} \omega(\phi) = Z\left(G; \sum_{r \in R} \omega(r), \sum_{r \in R} \omega(r)^2, \ldots, \sum_{r \in R} \omega(r)^{|D|}\right) \qquad (9)$$

In our present case if the colors A, B ($\in R$) are given the weights a, b, then the cycle index becomes $Z = a^6 + a^5b + 2a^4b^2 + 2a^3b^3 + 2a^2b^4 + ab^5 + b^6$ obtained simply from the previous example by the substitution $x_i = a^i + b^i$. This directly informs us by way of the weighted form of the theorem that

TABLE I
Permutations for Point Group $G=O$

Operation (g)	E	$8C_3$	$6C_2'$	$6C_4$	$3C_2(\equiv C_4^2)$
Cycle structure	(1)(2)(3)(4)(5)(6)	(134)(256)	(12)(34)(56)	(1)(2)(3456)	(1)(2)(35)(46)
$\prod x_j^{c_j(g)}$	x_1^6	x_3^2	x_2^3	$x_1^2x_4$	$x_1^2x_2^2$

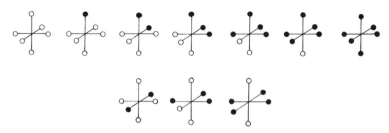

Fig. 19. The 10 distinct ways of coloring the vertices of the octahedron in two colors.

there is but one arrangement MA_6 from the coefficient of the a^6 term (obvious), one MA_5B isomer from the a^5b term, two MA_4B_2 isomers from the a^4b^2 term (actually cis and trans), two MA_3B_3 isomers (actually mer and fac) from the a^3b^3 term, and so on, as shown in Fig. 19.

Suppose we did not really care which of the two colors is A or B, as might be the case if we were interested in rearrangement processes of MA_6 species. A generalization of Pólya's theorem due to de Bruijn allows[57] ready enumeration in this case. In the present case we simply evaluate two cycle indices and take their average, using quite a simple procedure. The result is simply six, as may be derived from Fig. 19. A more recent advance by White[60] allows sorting of the different patterns into space groups. We refer the reader to some elegant papers by McLarnan, which put these combinatorial problems in perspective.[58, 59]

C. Edge-Sharing Octahedra

We now apply the foregoing ideas to a question of crystallographic interest that as it turns out, will eventually allow us to answer the question posed by the arsenic structure (Fig. 17), although it may not be clear to the reader at present. How many different ways are there of linking edge-sharing octahedra around a central octahedron, as in the polymolybdate and polytungstate systems of Fig. 20? In the language of the preceding section, how many different ways are there of coloring the edges of an octahedron in two colors? Here one color represents a shared edge and the other an unshared edge of the central octahedron. From Pólya's theorem

$$Z(O_h; x_1, x_2, \ldots, x_{12}) = \tfrac{1}{48}\left(x_1^{12} + 3x_1^4x_2^4 + 12x_1^2x_2^5 + 4x_2^6 + 8x_3^4 + 12x_4^3 + 8x_6^2\right)$$

so that the total number of possibilities found by setting each $x_i = 2$ is 144. Evaluation of the cycle index in the point group O, a subgroup of O_h, leads to a different result:

$$Z(O; x_1, x_2, \ldots, x_{12}) = \tfrac{1}{24}\left(x_1^{12} + 8x_3^4 + 6x_1^2x_2^5 + 6x_4^3 + 3x_2^6\right),$$
$$\text{so } Z(O; 2, 2, \ldots, 2) = 218$$

$W_4O_{16}{}^{8-}$

Fig. 20.   An example of a polytungstate that is constructed of edge-sharing octahedra.

The difference lies in the number of chiral pairs, that is, arrangements that may not be converted into each other by rotation of the molecule but are so related by an $\bar{n}$ operation. The analysis by way of the point group O_h includes such operations and eliminates one partner of each pair. Its subgroup O lacks such symmetry operations and counts separately the two such arrangements. From these two results we may readily conclude that there are a total of 144 different structures, of which 74 are chiral and 70 are achiral.

Although this particular problem of edge-shared octahedra was solved by Moore[62] without the use of combinatorial techniques, but by systematic enumeration, the use of Pólya's theorem is very desirable in making sure that all possibilities are included. It is tempting at this state to ask how many of

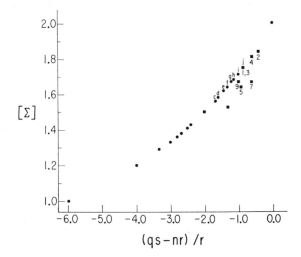

Fig. 21.   Plot of $\langle\Sigma\rangle$, the mean electrostatic valence balance per O^{2-} ion, versus the average negative charge per octahedral unit. Squares show the clusters that are actually observed. Some of the points represent several of the 144 possibilities. Adapted from Ref. 62.

the 144 structures are actually found experimentally, and why. First there are two rules that describe the empirical results. (a) No octahedron has more than two unshared vertices. (This is Lipscomb's rule.) (b) The observed clusters have the highest point symmetry (i.e., highest degeneracy) among their geometrical and topological isomers. Moore[62] has rationalized their behavior by using a simple electrostatic model (Fig. 21) that plots the mean electrostatic valence balance per oxygen anion against the average negative charge per octahedral unit. Each oxide anion is undersaturated in the sense of Pauling's rules,[27] a feature that is directly related to the charge on the cluster $((qs-nr)/r$ of Fig. 21). Below a value of $\langle \Sigma \rangle$ of about 1.5 this is so severe that such a species, if it existed, would polymerize further.

IV. THEORY OF THE ELECTRONIC STRUCTURE OF SOLIDS

A. Molecules and Solids

Our treatment differs slightly from that usually encountered but emphasizes the similarities in electronic structure calculations between molecules and solids. A more complete physical description may be found in many places elsewhere.[63-65]

If we wished to evaluate the molecular orbital level energies for an approximately square planar molecule (e.g., S_2N_2, **3**) we might proceed as follows.[66] If G is the symmetry group of the molecule, then using a basis set of atomic orbitals $\{\phi_j\}$ having the property that $g\{\phi_j\}=\{\phi_j\}$ for all symmetry operations $g \in G$ we may construct symmetry orbitals ψ_i that transform as the ith irreducible representation of G in a well-established way. The character of g in the ith representation is $\chi_i(g)$ and so

$$\psi_i = \frac{1}{M} \sum_{g \in G} \chi_i(g) g \phi_p \tag{10}$$

Here ϕ_p is any member of $\{\phi_j\}$; M is a normalization constant. For the two sets of s orbitals located on the sulfur and nitrogen atoms, respectively (**4**), in $S_2N_2(D_{2h})$ we may simply write

$$\psi_a = \psi_{a_{1g}}(s,S) = \frac{1}{\sqrt{2}}(\phi_1 + \phi_3)$$

$$\psi_b = \psi_{b_{2u}}(s,S) = \frac{1}{\sqrt{2}}(\phi_1 - \phi_3)$$

$$\psi_c = \psi_{a_{1g}}(s,N) = \frac{1}{\sqrt{2}}(\phi_2 + \phi_4)$$

$$\psi_d = \phi_{b_{3u}}(s,N) = \frac{1}{\sqrt{2}}(\phi_2 - \phi_4)$$

$$\tag{11}$$

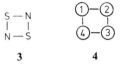

3   4

The normalization constant assumes zero overlap between the basis s orbitals. The energies of the b_{2u} and b_{3u} levels may be obtained by simple substitution into the Schrödinger wave equation, but a secular determinant (12)

$$\begin{vmatrix} H_{aa}-E & H_{ac}-S_{ac}E \\ H_{ac}-S_{ac}E & H_{cc}-E \end{vmatrix} = 0 \qquad (12)$$

needs to be solved to allow for the interaction between ψ_a and ψ_c, both of a_{1g} symmetry. In the one-electron approximation H_{aa}, H_{cc}, and H_{ac} contain Coulomb and resonance integrals; S_{ac} is the overlap integral between ψ_a and ψ_c.

In crystalline solids, characterized by translational symmetry, the set of translation vectors taking one lattice point to another form an infinite group for our infinite "molecule," with of course an infinite set of irreducible representations. In this case the character is given by $e^{i\mathbf{k}\cdot\mathbf{r}}$ where $\mathbf{r}$ is the translation vector linking one lattice point to the next (usually represented as $\Sigma l_i\mathbf{a}_i$ where $\mathbf{a}_i$ ($i=1-3$) are the primitive translation vectors of the Bravais lattice, l_i are integers, and $\mathbf{k}$ is the wave vector). There are as many irreducible representations as there are values of $\mathbf{k}$ (i.e., an infinite number), and the sets of energy levels for values of $\pi/a < k < -\pi/a$ (for a one-dimensional system) map out the electronic structure of the first Brillouin zone. The linear combination of basis orbitals generated by the translation group is by analogy with (10) given by the Bloch sum

$$\psi_\mu(k) = \frac{1}{\sqrt{N}} \sum_j e^{i\mathbf{k}\cdot\mathbf{r}_j}\phi_j \qquad (13)$$

where N is the (in principle infinite) number of basis orbitals included in the sum. In the evaluation of the energy, however, N drops out, since there will always be N interactions to be included when matrix elements involving such sums are evaluated. For the trivial case of a chain of $1s$-orbital-bearing atoms (5), the energy as a function $\mathbf{k}$ is readily derived as (14):

$$E(k) = H_{ss} + 2H_{ss'}\cos 2\pi\alpha \qquad (14)$$

5

where for the first Brillouin zone, α takes values from $-\frac{1}{2}$ to $\frac{1}{2}$; H_{ss} is the Coulomb integral for an electron in an isolated atom $1s$ orbital, and $H_{ss'}$ is the resonance integral between two adjacent $1s$ orbitals. Clearly because of time-reversal symmetry $E(\mathbf{k}) = E(-\mathbf{k})$, and thus only the positive half of the zone needs to be considered. In general of course the zone will be a three-dimensional one. When the basis set is not so simple—when the linear chain contains valence p orbitals, for example [as perhaps in $(SN)_x$ polymer]—then a secular determinant needs to be solved at each value of $\mathbf{k}$ just as in the molecular case above:

$$|H_{\mu\nu}(\mathbf{k}) - S_{\mu\nu}(\mathbf{k})E| = 0 \qquad (15)$$

In principle then we need to evaluate the energy of the solid-state structure by summing the integrals of the energy of each occupied level over $\mathbf{k}$-space, a process in general of great complexity. The problem is somewhat simplified, but not much, since the symmetry of the Bravais lattice imposes some restrictions[67] on the behavior of $E(\mathbf{k})$. In the Gilat-Raubenheimer[68] method of electronic spectral synthesis the Brillouin zone is divided into blocks of equal volume and a theoretical spectrum is constructed by evaluating the contributions from each block. A good approximation to the total energy (or charge distribution, etc.) however may be achieved by using the special points method of Baldareschi[69] and of Chadi and Cohen.[70, 71] By expanding the energy $E(\mathbf{k})$ of a given band in terms of a Fourier series in the Brillouin zone, that point (or collection of points) in $\mathbf{k}$-space may be found which gives the best value of the average energy of this k-dependent function. As an indicator of the accuracy of the method, the energy of solid silicon changed by 0.004 eV per atom on moving from 2 to 10-special points.[72] For systems with high dispersion we need to take a larger number of points to get accurate results, and problems sometimes do arise when cusps are found where two bands touch.

The methods available for obtaining $E(\mathbf{k})$ numerically are in many ways similar to those used for molecules.[65] With accurate pseudopotentials it is possible to generate band structures for simple solids (the eight-electron binary AX systems have been very well studied), which by existing measures of appraisal are very good indeed. Quantitative studies are also able to differentiate energetically between polytypes in some cases, for example, the cubic diamond and its structural alternatives. At a simpler level Harrison[64] has pioneered the use of parameterized values of the Coulomb and resonance integrals involved by way of an ingenious link with free-electron theory in solids. Just as there are various ways, of differing degrees of sophistication and accuracy, to evaluate the energies of the molecular orbitals of molecules, so there are similar options open for the study of solids. There is the important practical proviso that for a molecule, a single calculation is sufficient for a given geometry; for a solid, however, several calculations may be needed at various points in **k**-space to reach a similar level of appreciation of the electronic structure.

B.   Fragment-Within-the-Solid Orbitals

After the band structure of a system has been obtained, we face the conceptual problem of determining what it means in terms of localized bonding pictures. One approach is to look at the orbitals of a fragment removed from the lattice with suitable simulation of its electronic environment. The crudest method, which saturates the "loose ends" of the fragment with electrons, has been used extensively in studies of the photoelectron spectra of solids.[73] We used this "molecular cluster" approach in Section II.B. A second approach, applicable really only to tetrahedrally coordinated "valence compounds," ties off the loose ends with hydrogen atoms. Thus the bonding in silicates, especially the question of the energetics involved in bending the Si–O–Si unit, has been tackled[74, 75] by performing calculations at various levels of sophistication on the molecule disiloxane (**6**). The rather soft bending mode calculated for this molecule fits rather well with the observed wide range of angles found in silicates. The flexibility of this unit allows the construction of a huge number of silicate structures, although no rules have yet been formulated to help understand their natural or synthetic existence. On the other hand, the structural chemistry of SiS_2, where the Si–S–Si angle is much more rigid, is extraordinarily limited in comparison to that of SiO_2.

The third technique simulates the environment exactly by tying the ends of the fragment to give the orbitals of the fragment-within-the-solid[76] or the orbitals of the small periodic cluster, as it has also been called.[77–79]

With reference to **7**, which shows a one-dimensional chain, this may be done in a particularly simple way. If we choose as a repeating unit the set of atoms 2 through 5, then in the solid-state environment atom 2 feels the influence of atom 1 and atom 5 feels the influence of atom 6. Neither atom 1 nor atom 6 is included in the fragment we have chosen. However we may readily calculate the orbital overlap integrals between these atom pairs and the corresponding off-diagonal elements of the secular determinant of, for example, an extended Hückel molecular orbital (EHMO) calculation. Addition of the off-diagonal terms between atoms 5 and 6 to those between 2 and

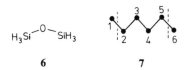

$$H_3Si \overset{O}{\diagup} \diagdown SiH_3$$

6   **7**

5 (small because of the large distance between them) and completion of the molecular orbital calculation leads to the orbitals of the "fragment-within-the-solid." Atom 2 feels an interaction of just the right magnitude expected from atom 1, and atom 5 feels a similar interaction with atom 6. As an example we derive the "fragment-within-the-solid" orbitals for simple linear polyenes (**8, 9**). Figure 22 shows the derivation of the orbital diagram of **8** within the polymer from that of the C_4H_4 fragment of the same geometry, and its distortion to the lower energy form **9**. With the aid of the atomic orbital composition of the orbitals of the isolated C_4H_4 fragment before its ends are tied together, the generation of the diagram for **8** is simple. The only orbitals that change in energy significantly when the ends of the unit are tied together are those with sizable coefficients of orbitals located on the end atoms, which point in the correct direction to interact with the next atom in the chain . If the coefficients of these frontier orbitals at each end of the chain are in phase, a stabilization results; if of opposite sign, a destabilization. Figure 22 clearly shows a σ-bonding and a σ-antibonding interaction associated with the orbitals labeled σ and σ^*. These are the orbitals responsible for linking a fragment of **8** to its surroundings in the polymer. They cross over in energy on moving from the isolated fragment to the polymer orbitals. Thus the nature of the highest occupied molecular orbital (HOMO) is different in the isolated fragment and in the fragment-within-the-solid.

All four π orbitals of the fragment change in energy on moving to the polymer; the highest energy orbital is destabilized and the lowest energy orbital is stabilized. The middle two orbitals of the fragment become degenerate in the polymer structure. The π-orbital structure of this unit is readily seen to be isomorphous with that of cyclobutadiene. Here then is an explanation for the distortion of **8** to the experimentally observed structure **9**. Just as this small organic molecule in its lowest singlet state is predicted to distort away[80] from the D_{4h} geometry to relieve the asymmetric occupation of degenerate orbitals by way of a pseudo-Jahn-Teller process, so the regular linear polymer is predicted to distort to a structure containing alternating single and double bonds. This is the Peierls's distortion[81] in one-dimensional systems at its simplest. The degeneracy is clearly imposed in our model by the translational symmetry of the system. Further, more detailed studies on the structural instability or otherwise of such half-filled levels have recently been published[82, 83] by Whangbo, who goes beyond our simple one-electron picture. Interestingly, with two extra electrons per

8   **9**

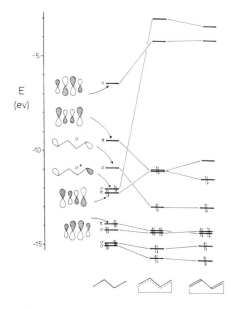

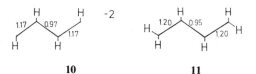

Fig. 22.   The derivation of the fragment-within-the-solid orbitals of a polyene starting from those of the isolated C_4H_4 fragment showing the effect of a distortion leading to bond alternation.

fragment [$(NH)_n$ or S_n] the symmetric structure **8** is predicted to be most stable just as $C_4H_4^{2-}$ is predicted to be square planar and actually found in this geometry in, for example, $Fe(CO)_3(C_4H_4)$, formally written as $Fe^{II}(CO)_3(C_4H_4)^{2-}$. This nonalternating bond arrangement is found experimentally in the structures of fibrous sulfur and of elemental selenium and tellurium (although here the atoms are arranged in spirals and do not all lie in the same plane).

The occupied orbitals of the isolated fragment do then differ in nature from the fragment-within-the-solid, as was to be expected. We noted above that within the σ manifold two orbitals crossed over in energy on tying together the ends of the unit. Use of the molecular orbitals of the isolated fragment to describe the polymeric material is then not a good one. Structure **10** shows the population analysis for a $C_4H_4^{2-}$ unit where the ends of the fragment have been saturated with electrons. Even though all C—C bond lengths were put equal in the calculation, the bond overlap populations alternate along the chain. A similar result occurs (**11**) if the ends of the unit are saturated with hydrogen atoms. Equal bond overlap populations are

10   **11**

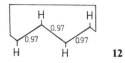

found (**12**) by using the present method, since both the σ and π bonding requirements of the terminal atoms are satisfied.

What then do the orbitals of the fragment-within-the-solid actually represent? For the π manifold of the polyene chain, the connection between these
orbitals and the band structure of the infinite solid is easy to see. As we will
show, the present approach produces a subset of the crystal orbitals and may
be regarded as an approximant to a band structure calculation. With reference to Fig. 23, the π orbitals of the C_4H_4 unit of **8** before its ends are tied
together are those of butadiene with orbital energies according to simple
Hückel theory[18] of $\varepsilon = \alpha \pm \beta(\sqrt{5} \pm 1)/2$. As the length of the chain increases,
the energy of the highest π orbital approaches $\alpha - 2\beta$ and the energy of the
lowest π orbital approaches $\varepsilon = \alpha + 2\beta$. (The general equation for the jth-level
energy of an n-membered chain is $\varepsilon_j + \alpha - 2\beta \cos[j\pi/(n+1)]$.) In between lie
a collection of other orbitals. The band structure calculation for the infinite
solid shows[84] two π bands, the lower stretching in energy from $\alpha + 2\beta$ to
α and the other from α to $\alpha - 2\beta$. In this case the molecular orbital structure
of the "fragment-within-the-solid," by applying the cyclic boundary condi-
tions, defines the energies of the top and bottom of the two π bands of the

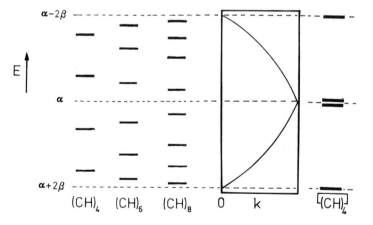

Fig. 23.   The π energy levels of $(CH)_n$ species using a Hückel model. The π orbitals of $(CH)_n$
fragments are shown as a function of the chain length, to be compared with the π-band structure of the polymer. The fragment-within-the-solid orbitals of a four-carbon atom fragment appear at right.

polymer. In general, depending on fragment choice, the orbitals will represent a particular subset of the crystal orbitals. Selection of just the repeat unit itself leads to the band structure at $k=0$, two fragments will give the orbital energies at the zone boundaries, and larger fragments give orbital energies that fall between the two.

A detailed description of the method has been given by Zunger,[77-79] who outlines some of the more subtle differences compared to a real band structure calculation. Messmer and co-workers[85, 86] have employed this method, simulating a solid surface by tying the edges together of an 18-atom graphite-based raft, to view chemisorption processes. For reasonably smooth energy bands in k-space the error involved is perhaps not too large, but it does mean that the energy of a system is evaluated using a weighting scheme different from that of the special points method.

In calculations on the polyenes (**8**) and related conformational isomers, significantly larger energy differences are calculated[76] between them than are obtained[84] by way of proper integration of $E_{total}(k)$. The numerical advantage of the technique is that the presence of a finite rather than an infinite collection of atoms means that the total energy of the system may be obtained by a finite summation over all occupied orbitals rather than by an integration of this sort. Its most important advantage as far as we are concerned is that the use of a rather small number of atoms means that the reasons for energy changes on distortion—namely, how the overlap integrals change and molecular orbitals mix as the molecule is deformed—are readily extracted from the calculation. In this way we have a simple way to use the analytical techniques that have been found to be very useful in studies on the structures of molecules in the area of the solid state.

V. TECHNIQUES OF MOLECULAR ORBITAL ANALYSIS

A. Perturbation Theory

Having outlined the theoretical formalisms and philosophies that provide the differences between electronic structure calculations on molecules and solids, it will be useful to summarize the techniques that molecular chemists have used to great effect over the years in understanding structural aspects of molecules. Later we show how they may be used to understand aspects of the structures of solids. Of central importance is the use of perturbation theory applied to the orbitals ψ_j of molecules or molecular fragments, or the orbitals ϕ_j of interacting atoms.[1] Second-order perturbation theory allows us to understand how two such orbitals may mix together as the result of some perturbation V. Consider two interacting orbitals $\chi_{1,2}$ with unperturbed eigenvalues $E_{1,2}$ such that $|E_1| > |E_2|$. Two rules govern their interaction as a result of the perturbation. First the lower energy orbital is depressed in energy such that

$$E_1' = E_1 + \frac{|\langle 1|V|2\rangle|^2}{E_1 - E_2} \tag{16}$$

and the higher energy orbital is pushed to higher energy

$$E_2' = E_2 + \frac{|\langle 1|V|2\rangle|^2}{E_2 - E_1}$$

$$= E_2 - \frac{|\langle 1|V|2\rangle|^2}{E_1 - E_2} \tag{17}$$

Second, the higher energy orbital mixes into the lower energy one in a bonding way such that

$$\chi_1' = \chi_1 + \frac{\langle 1|V|2\rangle}{E_1 - E_2} \chi_2 \tag{18}$$

and the lower energy orbital mixes into the higher energy orbital in an antibonding way such that

$$\chi_2' = \chi_2 + \frac{\langle 1|V|2\rangle}{E_2 - E_1} \chi_1$$

$$= \chi_2 - \frac{\langle 1|V|2\rangle}{E_1 - E_2} \chi_1 \tag{19}$$

The off-diagonal element of the Hamiltonian matrix, $\langle 1|V|2\rangle$, is invariably of opposite sign to the overlap integral between the two functions $\chi_{1,2}$. (In our case we arbitrarily define $S_{12} > 0$.) These ideas are obvious and familiar ones that guide our synthesis of the molecular orbital diagrams, aided by group theoretical ideas, of molecules.[12] Development of (16) and (17) has led to the angular overlap model, an extremely useful way to view structural and electronic properties of transition metal and main group coordination compounds.[87] Here however we are most interested in the mixing of molecular orbitals of a molecular or solid-state geometrical configuration when the perturbation V arises by way of a change in the geometry of the species.

B. Instability of Occupied High-Energy Orbitals

A pervading theme in the analysis of the geometrical structures of molecules by molecular orbital techniques is that occupation of high-lying orbitals is energetically unfavorable.[5, 12] Invariably the system will distort so as to

lower the energy of the highest occupied molecular orbital (HOMO). The most effective stabilization of the HOMO is often as a result of that perturbation V, which mixes HOMO and LUMO (lowest unoccupied molecular orbital) together as described algebraically by (16) to (19). The closer in energy the two orbitals are in the undistorted structure, the larger the effect; hence the importance of HOMO and LUMO (via the energy denominator). But sometimes overlap forces, determined by the numerator, are more important than simple energy gap arguments. Such dominance[88] of the structural problem by the energetics of the HOMO is the basis of Walsh's scheme[3] for geometry rationalization and in the second-order Jahn-Teller approach of Bartell[89] and Pearson.[90, 91] We must bear in mind, however, that the energetic behavior of the lower occupied orbitals may cumulatively resist such changes.

As an example, which will be useful below, we view the distortion of the five valence electron pair SF_4 molecule from the regular tetrahedral geometry (observed for the four pair CF_4 species) to the (observed) C_{2v} butterfly structure (**13**). A σ-only molecular orbital diagram is shown in Fig. 24 for the tetrahedral structure. For CF_4 the orbitals are filled through the $1t_2$ set of levels. For SF_4 at this geometry two extra electrons occupy the high-energy $2a_1$ orbital and the system should distort. The perturbation V is simply $(\partial \mathcal{H}/\partial q)_0 q$, where q is a distortion coordinate. From (16) to (19) there is clearly a symmetry restriction on the type of distortion that will allow mixing between HOMO and LUMO.[89] This is simply $a_1 \in \Gamma_1 \times \Gamma_2 \times \Gamma_q$ where

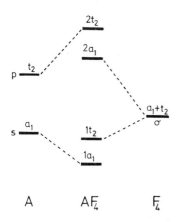

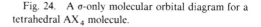

Fig. 24. A σ-only molecular orbital diagram for a tetrahedral AX_4 molecule.

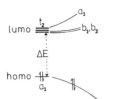

Fig. 25. Orbital behavior on distortion of a tetrahedral AX_4 molecule.

$\Gamma_{1,2,q}$ are the symmetry species of HOMO, LUMO, and distortion coordinate, respectively. From this relationship the distortion mode must be of symmetry t_2. Figure 25 shows how the HOMO is rapidly stabilized and LUMO destabilized as a result of bending by way of one component of the t_2 bending mode. The result is a dramatic increase in HOMO-LUMO gap. Note also that the high-energy orbital of the tetrahedral structure, antibonding between central atom and ligands, has become a lone pair in the distorted structure. This follows simply as in Fig. 26, as a result of HOMO-LUMO mixing after the style of (18) and (19), and appears to be a common feature of such approaches to the structure of molecules. Sometimes we might wish to view the generation of one structure from another as the result of a bond-breaking process. Here the high-energy electrons, initially in an orbital antibonding between the two atoms, become lone pairs at the ends of the severed bond.

We need to ask an obvious question. If SF_4 is unstable at the tetrahedral geometry as a result of effective mixing between $2a_1$ and $2t_2$ orbitals during a t_2 distortion, why is CF_4 stable at the tetrahedral structure when $1t_2$ and $2a_1$ could couple in an analogous way? The answer lies[89] in the size of the denominator in (16). As a rule, the stabilization of the HOMO on distortion will be large enough only to overwhelm the energy changes associated with the electrons in lower energy orbitals if the HOMO and LUMO are close in energy. Arguments based on the size of the HOMO-LUMO gap in isoelectronic series (e.g., NX_3 at the planar geometry) as a function of ligand (X) electronegativity have been used[89] to qualitatively order the size of the driving force away from the symmetric structure; hence the equilibrium bond angle in the distorted pyramidal geometry.

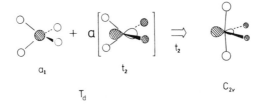

Fig. 26. Generation of the lone pair in SF_4 as a result of HOMO-LUMO mixing.

In the solid state similar ideas have been little used. However Heine and Weaire suggest[92] that the sometimes distorted structures of metals may be understood by noting that they are reached by moving the lattice vectors of a more symmetrical arrangement so that the band gaps increase, in an exactly analogous fashion to the changes in the HOMO-LUMO gap in our molecular examples.

C. Site Preferences and Bond Lengths

It is often very useful to be able to predict in a basic molecular structure containing symmetry-unrelated ligand positions (and symmetry-unrelated bond lengths) where electronegative ligands will reside in a partially substituted molecule to give the lowest energy structure, and which of these bond lengths in the parent will be longer.[12] (In molecular compounds, especially organometallic ones, we are also often interested in the site preferences of ligands that can behave as π acceptors or donors). SF_4 provides a useful example. The axial linkages (of a VSEPR trigonal bipyramid) are longer than their equatorial counterparts in the sulfuranes (14) and also attract [e.g., in $S(CH_3)_2F_2$] the more electronegative ligands (15). A population analysis of the molecular orbital results for an SF_4 species where all SF distances are maintained to be of equal length is shown in 16. It is clear that the axial linkages have a smaller bond overlap population than the equatorial ones and so in the real molecule should indeed be weaker, hence longer. Also the axial ligands carry a larger negative charge than the equatorial ones. This implies that the ligand with the deepest lying atomic energy levels (i.e., the most electronegative, following the ideas of Mulliken) will search out these axial sites because here will be the largest electronic stabilization. (There is in fact a very simple explanation for differences in overlap populations and charges in this system,[12] which we do not discuss here.) A general feature of such population analyses in molecules is that the lower the coordination number, the higher the atomic charge. This implies that electronegative atoms, given the choice, will search out situations of low coordination number. N_2O is therefore found as NNO and not NON, OF_2 as FOF and not OFF, and so on. When the atoms concerned have similar electronegativities, both structural alternatives may be found. OCl_2 is found both as ClOCl and as the (unstable) OClCl. We consider below some solid-state systems in which this general "molecular" rule does not apply.

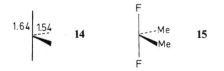

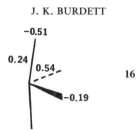

16

D. The Fragment Formalism

It was noted some years ago that several structural and chemical features of organometallic compounds could be understood[93] by the simple replacement of a fragment of an organic molecule by a specific unit containing a transition metal. For example, replacement of one CH_2 unit in cyclopropane (**17**) by $Fe(CO)_4$ leads to the $Fe(CO)_4C_2H_4$ species, usually regarded as coordination of ethylene by the $Fe(CO)_4$. Replacement of all three CH_2 units by $Os(CO)_4$ (Os is a congener of Fe) leads to $Os_3(CO)_{12}$ (**18**). Note that the replacement is not geometrically arbitrary but that there is a specific geometrical relationship between the $Fe(CO)_4$ or $Os(CO)_4$ configuration and that of the CH_2 it replaces. Hoffmann[7] and Mingos[5] have examined the problem theoretically and pointed out that the "frontier" orbitals of CH_2 and $Fe(CO)_4$ [and $Os(CO)_4$] are very similar (Fig. 27) in spatial extent and can be envisaged as holding the same number of electrons. In this sense they are "isolobal" and isoelectronic. These frontier orbitals by definition are those that are located in the vacant coordination sites of the unit, and are responsible for the dominant interaction of the fragment with its environment as in cyclopropane or $Os_3(CO)_{12}$. This observation provides us with an extremely powerful way to understand the structures of complex molecules. Instead of performing a single molecular orbital calculation on the species, the diagram can be assembled from smaller fragments whose orbital structure is simpler and readily appreciated. At each stage in the assembly process the energy changes associated with the relatively small number of frontier orbitals of each fragment can be readily followed to give vital clues to the electronic reasons for the adoption of a particular conformation or structure. We will see that the method gives us a similar capability of examining the structures of complex solids in an exactly analogous way.

             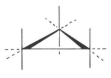

17   18

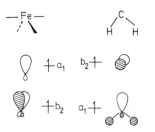

Fig. 27. The frontier orbitals of $Fe(CO)_4$ [$Os(CO)_4$] and CH_2.

VI. SOME ILLUSTRATIVE EXAMPLES

A. The Structure of Red PbO

We start with a very simple example that illustrates the immediate use of "molecular" techniques in the solid state. One way the red PbO structure may be conceptually derived[94] is to imagine it as a distorted version of the CsCl arrangement where the central atom of this structure is moved toward a face of the cube (Fig. 28). The CsCl structure itself places both atoms in cubal coordination. In the red form of PbO the lead atom is in a somewhat un-usual tetragonal pyramidal environment [found also in molecular complexes such as diethythiocarbamato Pb(II)], and the oxygen atom is tetrahedrally coordinated. First we need to look at the electronic structure of the un-distorted (CsCl) arrangement. Using the fragment-within-the-solid orbitals approach of Section IV.B, we may readily obtain the energy levels at the point $\Gamma(\mathbf{k}=0)$ in the Brillouin zone. Both A and X atoms of this AX structure are located at sites of O_h point symmetry, and the diagram of Fig. 29 readily re-sults. With eight valence electrons, all the four low-lying orbitals (mainly X located) are occupied, a large HOMO-LUMO gap results, and a stable structure is predicted. With 10 valence electrons, as in PbO, the high-energy $2a_{1g}$ orbital is occupied, and this geometry might now be expected to be un-stable toward distortion. In fact the form of the orbital diagram is very simi-lar indeed to that found for tetrahedral SF_4 in Fig. 24. With exactly equiva-lent reasoning, the eight-electron AX solid-state system should be stable (cf. CF_4), and the 10-electron system (cf. SF_4) unstable at this geometry. As may be seen in Fig. 30, HOMO-LUMO mixing to give a lone pair orbital on the A atom occurs in a similar way to our molecular example and is approach-able by an equivalent line of reasoning using the perturbation theory ap-proach of Sections V.A and V.B. Here the distortion process involves fission of four of the eight Pb—O linkages of the CsCl structure.

Chemical intuition led us to set up the problem with the lead atom ini-tially at the center of the cube, since eventually we expected a lone pair to evolve on distortion. Repeating our calculation, but with identical atomic orbital input parameters on the two atoms, leads to little change in the

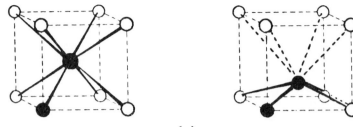

(a)

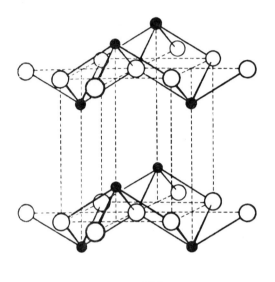

(b)

Fig. 28. (*a*) Derivation of the structure of red PbO from that of CsCl. (*b*) The structure of red PbO.

qualitative picture but does allow generation of charge densities on the atoms, which are now symmetry inequivalent. The result[94] (**19**) is that the tetrahedral site should preferentially be occupied by the more electronegative atom, as indeed is observed in the real structure.

We may observe very similar level behavior at other points in the Brillouin zone. Figure 31 shows the energetic behavior of the highest occupied bands on distortion. A detailed analysis is out of place here, but the important feature of note is that the valence band is stabilized by analogous sp mixing all along the symmetry lines (where s and p mix together before distortion because of the lower-than-octahedral point symmetry at points other than Γ) to the zone edges.

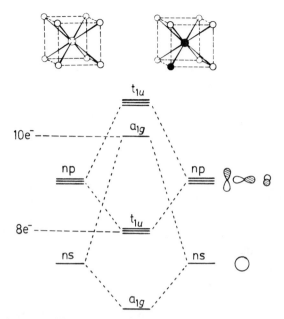

Fig. 29. The fragment-within-the-solid orbitals (c) of a two-atom repeat unit (b) of the CsCl structure (a), showing the effect of distortion on the red PbO structure.

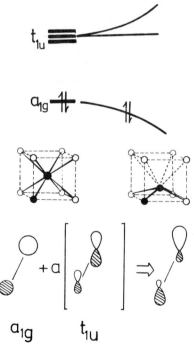

Fig. 30. Generation of the lone pair in the red PbO structure by HOMO-LUMO mixing.

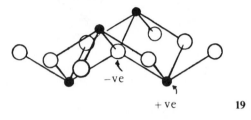

$-ve$

$+ve$　　　　　**19**

It is of some interest to examine why the alternate up-down distortion of Fig. 28*b* and **20** is observed rather than the two alternative arrangements of **21** and **22**. All three contain lead atoms in tetragonal pyramidal coordination but differ in the oxygen coordination. In **20**, **21**, and **22** the coordinations are tetrahedral, square planar, and tetragonal pyramidal respectively. The tetrahedral structure is preferred over the other two because this is the arrangement expected for an AX_4 moiety with four valence pairs of electrons; also it is the one that minimizes the nonbonded repulsions between the lead atoms.

Parenthetically we note that the CsCl structure, not a distorted variant, is observed as the stable room temperature–pressure polymorph for other 10-electron systems[95] including TlBr and TlCl. This feature is not restricted to the solid state. Seven valence pair, six-coordinate molecules, predicted to have distorted octahedral structures on the VSEPR scheme, are very often[12] regular octahedral, albeit sometimes only tenuously so. In binary AX systems the tendency for regular coordination geometries (NaCl or CsCl) increases as X or A become heavier. The observation has been categorized for some time under the title the "inert pair effect," the reluctance of the ns^2 electron pair to exert a full chemical and stereochemical influence. Recent years have allowed an explanation[96] for the effect based on relativistic corrections that become very significant for the heavy elements at the bottom

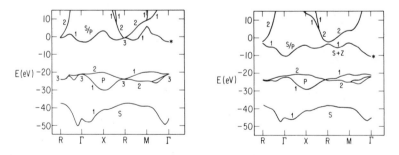

Fig. 31. Band structure of CsCl showing the stabilization of the valence band (asterisk) on distortion to the red PbO structure.

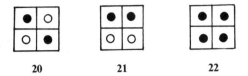

of the periodic table. The valence s orbital contracts much more than the corresponding p orbitals (effectively a larger Slater exponent), and this results in smaller overlap integrals of the s orbital with its environment. The numerator of (16) to (19) may then be considerably reduced, and the energetic changes on distortion of the HOMO may be rather small.

B. Defect Diamond Structures

A simple rule, the Grimm-Sommerfeld valence rule,[23, 97] controls the stoichiometry of structures based on tetrahedral coordination. Traditional bonding ideas in such systems, after the ideas of Pauling,[27] envisage four sp^3 hybrid orbitals that point toward the vertices of a tetrahedron. With a total of four electrons per atom, an infinite three-dimensional structure may be built up with simple two center–two electron ($2c-2e$) bonds between each pair of close atoms in the structure. If we write the formula of such a system as $0_a 1_b 2_c \cdots 7_h$ or in general as ΠN_n, where there are n atoms in the formula unit contributing N valence electrons to the bonding, the Grimm-Sommerfeld rule is simply

$$\frac{\Sigma n N_n}{\Sigma n} = 4 \qquad (20)$$

which tells us that the average number of electrons per tetrahedral site should be 4. Examples include diamond itself (4_1), zincblende ZnS ($2_1 6_1$), CuCl ($1_1 7_1$), and AlN($3_1 5_1$). Figure 32 shows two defect diamond derivatives that may be written in an exactly analogous way as $0_1 2_1 3_2 6_4$ where 0 represents a defect site. To emphasize this fact the structures are conventionally written as $\square$CdAl$_2$S$_4$ and $\square$CdIn$_2$Se$_4$, where $\square$ = vacancy. Parthé has enumerated[98] many of the possibilities corresponding to different values of a–h and a rather satisfying structural picture results. This simple bonding pattern and the high symmetry of the diamond structure and its derivatives have resulted in a tremendous amount of experimental and theoretical studies on these systems.[17, 99]

Our aim here is to show how the ejection of an atom in a (hypothetical) electron-rich diamond structure to give a defect arrangement consistent with the valence rule also leads to stabilization of high-energy orbitals of the initial structure, and to tie the result to analogous results found in molecular

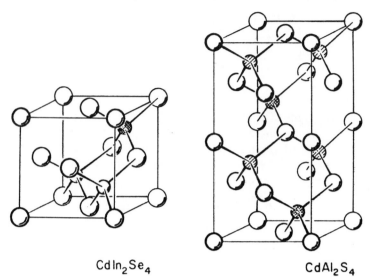

CdIn$_2$Se$_4$ CdAl$_2$S$_4$

Fig. 32. The defect structures of CdIn$_2$Se$_4$ and CdAl$_2$S$_4$. From W. B. Pearson, *Crystal Chemistry and Physics of Metals and Alloys*, Wiley, New York, 1972, with permission.

polyhedral species. Figure 33 shows a cell of the (cubic) diamond structure. It is not the smallest we could have chosen but it will be useful to use one of this size in what follows. Figure 34 shows the fragment-within-the solid orbitals of such a C$_8$ arrangement with a substantial HOMO-LUMO gap for the configuration reached by including four electrons per atom. (As an aside we note that the calculation on the fragment itself before the ends are tied together leads to no gap for this electronic configuration.) The energy changes that occur when one of the four coordinate atoms is removed from the structure are also shown. Some dramatic energy changes are observed when an atom is ejected from the lattice with a new substantial HOMO-LUMO gap developing at the C$_7^{4-}$ configuration. Thus the electron-rich C$_8^{8-}$ structure is considerably stabilized by the introduction of a single-atom defect. The forms of these new orbitals are interesting to examine. In C$_8^{8-}$ it-

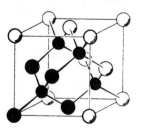

Fig. 33. The cubic diamond structure, show the repeat unit used in the calculations.

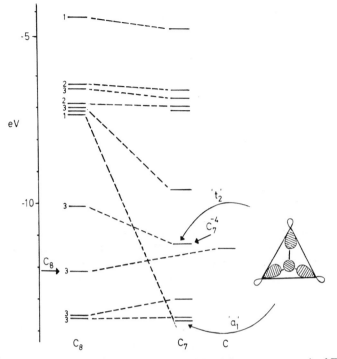

Fig. 34.   The fragment-within-the-solid orbitals of the eight-atom repeat unit of Fig. 33 and their correlation with the orbitals of a C_7 unit, where one atom has been ejected, and those of the freed C atom. The deeper lying, largely $2s$ orbitals are not shown. There are seven of these for C_8, six for C_7, and of course one for the isolated C atom. Orbital occupancies corresponding to C_8 and C_7 configurations are indicated with arrows. The small numbers by the side of the levels indicate their degeneracy.

self they are high-energy orbitals, antibonding between the carbon atoms. On distortion they become lone pair orbitals "coordinating" the vacant tetrahedral site. On our simple orbital model we can see that the local symmetry demands the formation of a_1 and t_2 symmetry-adapted combinations shown in Fig. 34. A similar result occurs if another atom is ejected. The new double defect structure is stable for the C_6^{8-} configuration. A population analysis for the C_7^{4-} structure is shown in **23**. Interestingly the atoms with the largest negative charge are those surrounding the vacancy. A similar result is found for C_6^{8-}. These then are the sites that should contain the most electronegative ligands, as indeed are found in the two examples of Fig. 32, and other more complex structures such as nowackiite ($\square Cu_6Zn_3As_4S_{12}$, $0_1 6_2 3_5 4_6 12$). As a result, systems that are stoichiometrically short of electronegative ligands might adopt some structure other than a tetrahedrally coordinated one, and we explore this suggestion later.

$$C_7^{-4}$$

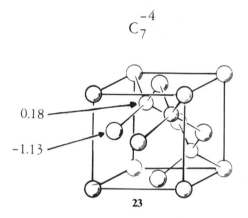

0.18

−1.13

23

The double defect structure is actually that found for cuprite (Cu_2O), which we discuss below. This is a species that does not satisfy the Grimm-Sommerfeld rule (only an average of two electrons per site), and here the site preferences are the opposite of those expected from our analysis with four electrons per site. It is an example of a general rule that inverse structures are often observed in AX_2 systems with a total of eight valence electrons (e.g., Cu_2O) compared to their 16-electron [e.g., $Cd(CN)_2$] analogs. A corresponding population analysis for the C_6^{8+} configuration (two electrons per site) indeed shows a reversal of the signs of the atomic populations.

It is pertinent to view[100] the results of this section in the broader light of the structures of polyhedral molecules, which have received considerable attention in recent years.[10, 26] The extraordinary collection of geometries adopted by these cage, basket, and ring molecules, exemplified by the boranes, carboranes, transition metal cluster compounds, and metallocarboranes, have recently succumbed to a molecular orbital-based set of electron-counting rules[26, 101, 102] developed primarily by Wade.[26] Figure 35 shows three simple boranes that in contrast to our solid-state systems, are electron deficient. In other words there are insufficient electrons to form normal $2c$—$2e$ bonds between each pair of atoms that are in close contact. Wade's rules are based on the general principle[103] that in an n-vertex deltahedron (e.g., the octahedron, trigonal, and pentagonal bipyramids, etc.) the number of skeletal bonding orbitals is always $n+1$, except, as we see in the next section, for the tetrahedron, there are $n+2$ skeletal bonding orbitals. By viewing the frontier orbitals of BH, CH, or transition metal carbonyl fragments we may compute the total number of electrons that will eventually contribute to the skeletal bonding.[26] If there are a total of $n+1$ skeletal electrons pairs and n vertices, all the skeletal bonding orbitals are filled and a stable configuration results. In Fig. 35 $B_6H_6^{2-}$ is an example described as a

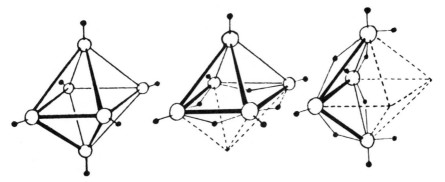

Fig. 35.   The structures of $B_6H_6^{2-}$ (left), B_5H_9 (middle), and B_4H_{10} (right), showing their geometrical resemblance to the parent octahedron.

closo-octahedron. Each BH unit contributes two electrons to skeletal bonding and with the -2 charge, seven bonding pairs result. If however there are n skeletal atoms and $n+2$ electrons, it turns out that the geometry adopted is based on the $n+1$ vertex polyhedron (the $n+2$ skeletal pairs fill all the bonding orbitals of this arrangement) but with one vertex missing. Figure 35 shows the structure of B_5H_9, a *nido*-octahedron. With $n+3$ pairs of electrons, two atoms are missed out of the $n+2$ vertex polyhedron. Figure 35 also shows B_4H_{10}, an *arachno*-octahedron. Similar rules apply to all the various types of polyhedral molecule mentioned above.

The similarities between the structural features of the molecular polyhedral examples of Fig. 35 and the defect solid-state structures of Figs. 32 and 33 are very striking.[100] Exactly analogous theoretical reasoning lies behind both. It is also interesting to note here that bond-breaking ideas similar to the ones described for the CsCl-to-PbO transition of the preceding section have also been applied to the structures of these polyhedral molecules.[102]

C.   The Cuprite (Cu₂O) Structure

The structure of cuprite supplies a simple example that illustrates [104] the use of the fragment formalism in relating a complex structure to a much simpler one. In Section VI.A we determined the fragment-within-the-solid orbitals for the CsCl structure. The central atom of this structure, located at a site of cubic symmetry, used a degenerate group of three orbitals (p) and a single nondegenerate (s) orbital to interact with its surroundings. We might ask whether there are any other structural units that are isolobal and isoelectronic that could replace this central atom in an exactly analogous way to the organometallic examples of Section V.D. One obvious example is a tetrahedron of atoms. The orbital structure (Fig. 36) consists of a set of

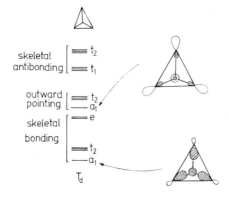

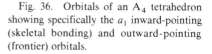

Fig. 36. Orbitals of an A_4 tetrahedron showing specifically the a_1 inward-pointing (skeletal bonding) and outward-pointing (frontier) orbitals.

skeletal bonding orbitals at low energy, a set of skeletal antibonding orbitals at high energy, and a set of approximately nonbonding orbitals in between. It is important to understand the spatial extent of this collection of levels. All the orbitals involved in skeletal interactions point either toward the interior of the tetrahedron or along its edges or faces—hence the label "skeletal." The nonbonding orbitals point outward from the tetrahedron. In the P_4 molecule with this geometry they are filled with electrons and are lone pairs. In $C_4 R_4$, tetrahedrane derivatives, they are involved in $2c-2e$ bonds between the skeletal carbon atoms and the R groups. Clearly in a naked tetrahedron with the right electronic configuration, these orbitals are the frontier orbitals of the structure and will be responsible for binding the unit to its environment. Importantly, their spatial extent and degeneracy are similar to those of an isolated atom, and the two units are isolobal. We have not been able to find any structures of this type, but a closely related arrangement, that of cuprite, which contains a tetrahedron with a central atom (picturesquely described by crystal chemists as a stuffed tetrahedron) is well known (Figs. 1 to 3). Each oxygen atom in the structure is four-coordinated by copper atoms and each of the latter are two coordinate. In contrast to the rule of thumb noted in Section V.C and also discussed above, the more electronegative atom has the higher coordination number.

Although the resulting fragment-within-the-solid orbital diagram (Fig. 37) is more complex than that for CsCl, the manner in which it has been assembled shows rather clearly the electronic forces involved in holding the solid together. A similar revealing picture of the electronic structure of this system may also be obtained by adding the orbital picture of the simple Cu_4 tetrahedron to that of a body-centered-cubic lattice in an analogous way. Although we do not discuss it here, a corresponding picture is obtained[104] if the band structures of the relevant components are added at points other than Γ in the Brillouin zone.

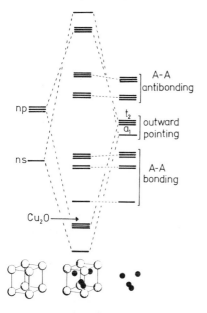

Fig. 37. The fragment-within-the-solid orbitals of the structure that results from replacing the central atom of the CsCl structure by a tetrahedron. (The diagram for the stuffed tetrahedron is similar.)

An alternative view of the cuprite structure is of two interpenetrating, but independent β-cristobalite (SiO_2)-type nets, where there are no formal linkages between the two frameworks. From Fig. 37 we see in fact that the skeletal bonding orbitals of the tetrahedron (responsible for Cu—Cu interactions) are empty. To first order at any rate, there are then no chemical forces between the two nets. However, there may indeed be a weak interaction between the atoms by way of d-s mixing as found for several Cu^I molecular examples. (Our simple considerations here have excluded the filled d shell from the orbital diagram for simplicity.) We might ask, however, for which electronic configuration these skeletal bonding orbitals are filled. Six extra pairs of electrons are needed per A_4O_2 unit compared to Cu_2O. This will be the case for a group IV element. For A—A interactions to be significant, the A atom needs to be much larger than oxygen. An obvious candidate is A = Pb, and Pb_2O in fact was claimed to have the cuprite structure,[105] but the existence of such a phase is now strongly in doubt.

D. Structures Derived from the Breakup of the Rock Salt Structure

In the rock salt structure (Fig. 17) the sodium and chlorine atoms are octahedrally coordinated and the fragment-within-the-solid orbitals will be qualitatively similar to the structure of Fig. 29. Analogously, occupation of high-energy orbitals should lead to structural instability, with the qualification concerning the inert pair effect noted above. With nine electrons per

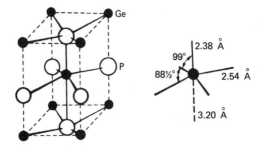

Fig. 38.   The structure of GeP.

formula unit, the GeP structure is found,[22] where one bond around each atom has been broken (Fig. 38) leaving both Ge and P in a square pyramidal environment. With 10 electrons per atom pair we may envisage the arsenic (antimony and bismuth) structure of Fig. 17 and black phosphorus structure of Fig. 39 as being derived from that of rock salt by the selective fission of three perpendicular linkages around each center. The resulting movement of the atoms leads to a trigonal pyramidal, three-coordinate arsenic or black phosphorus atoms with a lone pair pointing along the body diagonal of a "small cube" of the rock salt structure (24). This result is precisely analogous to the generation of the lone pair and tetragonal pyramidal coordination of the lead atom in PbO from the CsCl structure.

The question we would like to answer was posed above—namely, how many different ways are there of breaking the linkages of the rock salt structure? Of the very large number of possibilities, we need to extract those in which each atom retains three linkages approximately perpendicular to one another. We note that between any three adjacent atoms lying along [100] or symmetry-equivalent directions, broken and unbroken linkages alternate (Figs. 17 and 39.)

The combinatorial problem is set up as follows. How many ways are there of coloring these lines of atoms in the rock salt cube in two colors that represent A— — —A—A and A—A— — —A? Pólya's theorem is inappropriate here, since how one line is colored automatically restricts the

Fig. 39.   The structure of black phosphorus, emphasizing its relationship to the rock salt structure.

24

possibilities for coloring other lines if each atom is to remain three-coordinate. The combinatorial problem can in fact be treated by an elegant method due to McLarnan,[106] in actuality a more general form of de Bruijn's theorem. Quite a neat approach, however, uses a result generated in Section III.C, namely, the number of different ways edge-sharing octahedra may pack around a central octahedron. We first note that the bond-breaking pattern of the large 27-atom cube of Fig. 17 is determined by whether the edges of a single small eight-atom cube are broken or unbroken, since broken and unbroken bonds alternate along all lines of nearest-neighbor atoms. Second, the cube is the topological dual of the octahedron, so that the two-colorings of the edges of the octahedron (shared or unshared) enumerated above are directly equivalent to the two-colorings of the edges of the cube (broken or unbroken). Not all these two-colorings will give rise to different bond-breaking patterns of the large cube. By inspection only, a total of 36 will actually give rise to distinctly different, three-dimensional extended structures.

Of these 36 possibilities we have mentioned the empirical observation of the arsenic and black phosphorus (also found for GeS and SnS) layered arrangements. There are two other frameworks known,[22] those of $La_2Be_2O_5$ and $Hg_3S_2Cl_2$, which are slightly more complex in that they have "spacer" atoms between the rock salt lattice positions and also contain some interstitial atoms (Fig. 40). Claudetite, a form of As_2O_3, has a structure[22] very similar to that of arsenic itself. Each two-coordinate oxygen atom acts as a spacer between two arsenic atoms (**25**).

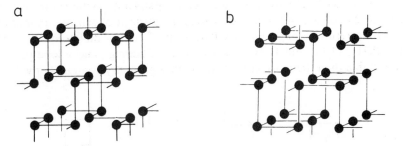

Fig. 40.   The simplified structures of (a) La_2BeO_5 and (b) $Hg_3S_2Cl_2$, emphasizing their relationship to the rock salt structure. (Spacer and interstitial atoms have been omitted for clarity.)

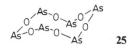

25

Band structure calculations on all these 36 possibilities show[107] that these four arrangements do occur at low energy, and are among the lowest energy group of seven structures. What of course is important to investigate, however, is what structural and electronic features are responsible for these energy differences. Each atom is simply trigonally three-coordinate on a local basis and so we must look for structural features that involve more than one atom. Structures **26–29** show some geometrical arrangements found in these structures, which provide a useful energetic classification of these different possibilities. If they are labeled c, t, g, and s respectively (cis, trans, gauche, and square) a set of indices w to z may be used to indicate the number of arrangements of each type in the structure using the notation $c^w t^x g^y s^z$. Because of the way the structures are constructed from the rock salt cube, $w + x + y = 12$ for all the permutations. Figure 41 shows a least-squares fit of the calculated band structure energy, using a sufficient number of special points to have reached good energetic convergence, against an equation of the form

$$E_{\text{band}} = E_0 + wE_c + xE_t + yE_g + zE_s \qquad (21)$$

where the E_i are the contributions to the energy from each of the units **26** to **29**.

The following values for the E_i are found: $E_c = -56.79$ eV, $E_t = -57.04$ eV, $E_g = -57.07$ eV, and $E_s = 0.99$ eV, and we are now in a position to discuss the reasons for the adoption of a particular structure. The orientations of the lone pairs generated on bond fission in the three arrangements **26** to **28** are given in **30** to **32**. Energetically the most unfavorable of these is that of the two lone pairs eclipsed, as might be expected on the basis of the VSEPR scheme.[41] The structure in which the pairs are oriented in the gauche fashion (**28**, **32**) are of lower energy than the trans arrangement, an initially surprising result perhaps, but one that is also found in molecules. In N_2H_4 (**33**), P_2H_4, O_2F_2, and O_2H_2 the structure containing gauche lone pairs is more stable than the more intuitively obvious geometry containing the trans

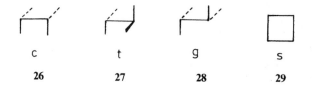

c   t   g   s

26   27   28   29

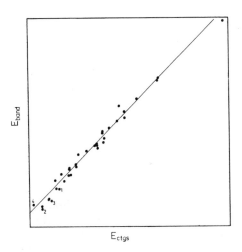

E_{band}

E_{ctgs}

Fig. 41. The band structure energies of the 36 structural possibilities, as a result of breaking three mutually perpendicular linkages around each lattice site of the rock salt structure, plotted against a calculated energy using (21). Numbers indicate observed structures: 1, arsenic, antimony, bismuth; 2, black phosphorus, GeS, SnS; 3, La_2BeO_5; 4, $Hg_3S_2Cl_2$. The highest energy structure is one that produces an isolated perfect cube of bonded atoms.

arrangement. This "gauche effect" has been studied by Wolfe,[108] who discussed the orbital details underlying the effect. The large destabilization associated with the presence of squares of bonded atoms in the structure reflects the inherent instability of such an arrangement from four sp^3 hybridized (in Pauling's language) atoms. None of the low-energy structures contains any such feature.

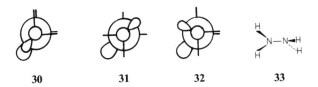

30   31   32   33

A very clean theoretical picture thus arises from this rather different way of viewing the occurrence and stabilities of solid-state structures. It will be interesting to see how applicable the approach is in general.

E.   Structures Derived from Puckered Sheets

A further example that illustrates the use of the fragment approach allows a prediction[109] of how puckered sheets of atoms [which we may visualize as arising through distortion (Fig. 42) of planar, graphite-like sheets] are linked together as a function of electronic configuration. The structure of GaSe may be viewed as being assembled from two such sheets (Fig. 43) by way of Ga—Ga linkages and the cubic (sphalerite) and hexagonal (wurtzite) structures of ZnS by way of the indefinite stacking of such sheets with Zn—S linkages. In the cubic structure each sheet is staggered with respect to the

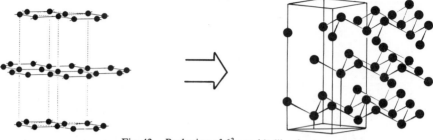

Fig. 42.   Puckering of 6^3 graphitelike sheets.

one beneath it, in the hexagonal form they are eclipsed. In covellite (Cu^IS) and klockmannite (Cu^ISe) pairs of sheets occur[22] in the structure linked by S—S or Se—Se interactions. In advance we note that these AX structures contain X—X linkages for seven electrons per AX unit, A—X linkages for eight electrons per AX unit, and A—A linkages for nine electrons per AX unit.

We can tackle the problem in several ways, perhaps assembling the structure using the fragment-within-the-solid orbitals or a more complete band structure, but in this case the most direct pedagogic approach is to take a puckered six-membered ring, tie off the loose ends with hydrogen atoms to mimic the intrasheet bonding using Gibb's method of Section IV.B, and see how pairs of such units interact. In Fig. 44, an orbital diagram for such an $A_3X_3H_6$ fragment, the X atom is more electronegative (i.e., its atomic energy levels lie deeper) than the A atom. Such a diagram would hold for the hypothetical puckering of the (planar) borazine molecule, $B_3N_3H_6$.

The frontier orbitals of the sheets are clear to see. They are sp^3-type hybrid orbitals pointing toward the vacant tetrahedral coordination site of each atom, and they are derived from the π and π^* orbitals of the planar unit. Because of the electronegativity difference between A and X, the "π"-type orbitals are largely X located and the "π^*"-type orbitals are largely A located. With seven electrons per AX formula unit, half of the block of "π" levels is occupied; with eight electrons these orbitals are full (as in $B_3N_3H_6$). With nine electrons per formula unit, half of the block of "π" levels is occupied in addition; with 10 electrons these orbitals are now full, as in the electronic configuration for elemental arsenic.

As the members of a pair of puckered sheets are brought together, simple overlap considerations allow generation of a qualitative molecular orbital picture.[109] The results are shown in Fig. 45. For AA-linked sheets the largest energy changes will be associated with the "π^*" block of orbitals, since they are largely A located. To benefit from this interaction all the orbitals through

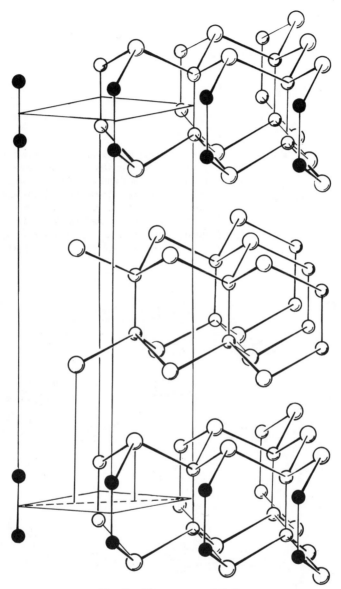

Fig. 43.   The structure of GaSe.

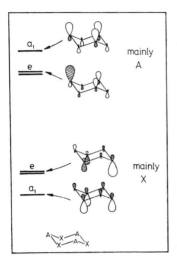

Fig. 44. The frontier orbitals of a puckered $A_3X_3H_6$ sheet. The low-energy trio of orbitals derive from the π levels of the planar molecule, the high-energy trio from the corresponding π^* levels. The deeper lying σ-bonding orbitals are not shown. For four electrons per AX unit, all these are filled in addition to the low-energy π-bonding trio of e and a_1 symmetry. From J. K. Burdett, *J. Am. Chem. Soc.*, **102**, 450 (1980), with permission.

a.

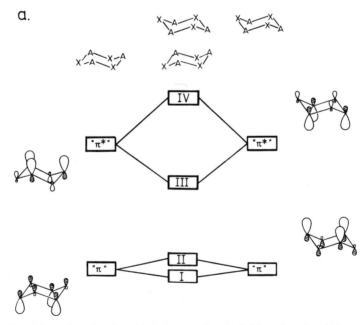

Fig. 45. Schematic molecular orbital diagrams for the linking of puckered six-membered $A_3X_3H_6$ sheets by way of (*a*) AA, (*b*) XX, and (*c*) AX linkages. The blocks of orbitals labeled "π" and "π^*" are the trios of orbitals shown in Fig. 44 derived from the corresponding π and π^* levels of the planar sheet. From J. K. Burdett, *J. Am. Chem. Soc.*, **102**, 450 (1980), with permission.

b.

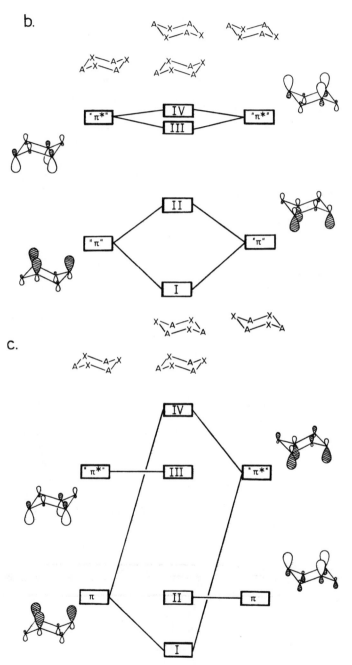

c.

Fig. 45. (*Continued*)

block III need to be filled, which occurs for the case of nine electrons per AX formula unit as found indeed in GaS. For XX-linked sheets the largest energy change is associated with the "π" block of orbitals, since these are largely X located. For the two sheets to be chemically bonded, only the block of three orbitals labeled I must be occupied. (Any more electrons would enter the block labeled II, antibonding between the sheets and destabilizing the structure.) This occurs for the case of seven electrons per AX formula unit, as found experimentally for covellite. For the AX-linked case the A located orbitals ("π^*") on one sheet interact with the X-located orbitals ("π") on the next, leaving the block of orbitals II ready to interact by way of A atom interactions with the next sheet. AX linkages should then be found when blocks I and II, both stabilized in an infinite structure, are full (i.e., for eight electrons per formula unit, as in ZnS, CuCl, AlN, etc.). For 10 electrons per formula unit, in whatever manner the sheets are linked, intersheet antibonding orbitals are filled and the paired structure will not be stable. Thus elemental arsenic, antimony, and bismuth exist as isolated single puckered sheets held together by Van der Waals forces.

Some of these site preferences can be viewed in other ways. For example, **34** shows a population analysis of the fragment-within-the-solid orbitals of the GaSe structure where all the atomic input parameters were put equal. Basically there are two different sites in this structure, one three- and one four-coordinate, and the population analysis strongly directs the electronegative S or Se atom to the three-coordinate position.

Similar arguments to these may also be used[109] to see how the cadmium halide (AX_2) structures are derived. Each of the layers of this structure may be assembled from a puckered AX sheet and an extra atom that "caps" one side of the sheet (Fig. 46). In principle the capping procedure may take place either on the A atom side or the X atom side (**35**). The former, the observed arrangement, gives rise to octahedrally coordinated A atoms and two equivalent sandwiching sets of trigonally pyramidally coordinated X atoms. To keep track of the electrons, we view the structure as being assembled from a puckered AX^+ layer (eight electrons per AX unit and isoelectronic with ZnS)

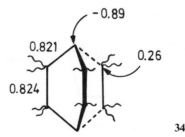

34

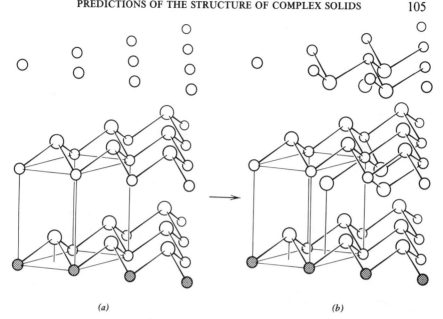

Fig. 46. Assembly of the $CdCl_2$ structure (b) by capping a puckered $CdCl^+$ sheet (a) with a Cl^- atom.

and a capping X^- atom with filled valence orbitals. Figure 47 shows why the observed arrangement is more stable, since capping on the X atom side results in strong destabilization of occupied orbitals. Again the electronegative atoms occupy the sites of low coordination number. A similar situation occurs with the molybdenite (MoS_2) structure, where the capping procedure results in trigonal prismatic coordination of A. For the eight-electron A_2X species (e.g., Cs_2O) the inverse structure is found; that is, the A^+ atoms cap the X atom side of the puckered AX^- sheet. Figure 47 also shows why this is the favored structure, using exactly analogous arguments. Just as in the cuprite structure of Section VI.C, the inverse structure is naturally found as the total number of valence electrons is changed from eight to 16, and vice versa. Structure **36** shows a charge distribution for the fragment-within-the-solid orbitals of the capped puckered sheet for the 16 electron system taken from a calculation in which all the atoms had the same input atomic orbital parameters. A distribution of opposite sign is found for the eight-electron case.

35

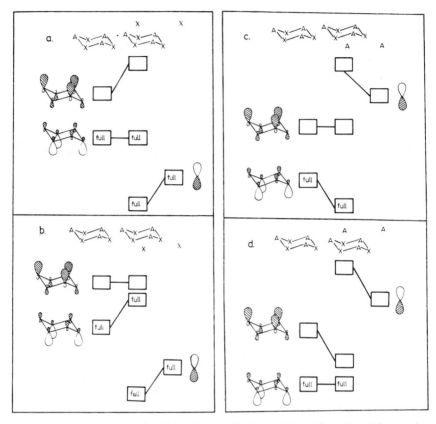

Fig. 47. Schematic molecular orbital diagrams showing capping of the puckered sheet to give the cadmium halide structures. The a_1 orbital from each block is pictorially represented. An electronegative atom X is capping a puckered AX sheet (a) on the A atom side and (b) on the X atom side. An atom of low electronegativity (A) is capping an AX sheet (c) on the X atom side and (d) on the A atom side. The electron occupancy represents in (a) and (b) a Cl^- ion capping a $CdCl^+$ sheet, for example, and in (c) and (d) a Cs^+ ion capping a CsO^- sheet. (b) and (d) are unfavorable arrangements, since either occupied orbitals are destabilized (b) or the best stabilization occurs with unoccupied orbitals (d). From J. K. Burdett, *J. Am. Chem. Soc.*, **102**, 450 (1980), with permission.

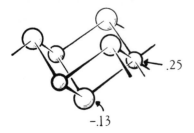

36

VII. A LINKING THREAD?

In this section we look further at the structural sorting methods and attempt to link them to the more detailed molecular orbital studies of the second part of the chapter. The coordination number problem of Section II could be phrased in terms of two-dimensional plots of AX electronegativity difference (or in its synthetic form by variation of Slater exponents) against the average value of the principal quantum number. The latter was suggested to be a measure of the degree of metallization, which traditionally weighs the occurrence of metallic rather than typically covalent structures. More specifically, as a measure of the "directionality" of the chemical bonding, it arises historically by way of the valence bond concept of the s-p promotion energy needed to form sp^3 hybrid orbitals. The structural maps use a similar philosophy. The orbital electronegativity is defined as

$$X_l = \frac{1}{r_l} \qquad (22)$$

and a total electronegativity for atom A as

$$X_A = a \sum_{l=0}^{2} X_l(A) + b \qquad (23)$$

with a, b chosen arbitrarily to fit Pauling's scale for the first-row elements. Now $X_A - X_B$ may be used as an ordinate in place of Pearson's ΔX (Pauling). As a measure of the extent of s versus p participation in bonding (the metallization concept), a weighted difference of orbital electronegativities may be used, namely $(X_0 - X_1)/X_0 [= (r_1 - r_0)/r_1]$. Such a plot gave [51] quite a good separation of structural types, but a better one was achieved, not by using weighted electronegativities in both ordinate and abscissa but by using their inverses, the r_l values as in (5) and (6). Zunger's r_l values have the advantage of linearly scaling with the inverse of the average multiplet energy of the level; thus they are directly related to the valence-state ionization potential used in molecular orbital theory. Parenthetically, associated with this comment and (22), Mulliken's electronegativity scale is defined as the arithmetic mean of ionization potential and electron affinity. An analysis in orbital terms allows an understanding of the features behind these diagrams, albeit in a qualitative fashion, and realizes a connection between the plots of Section II and the molecular orbital approaches of Section VI.

Consider two orbitals $\phi_{i,j}$ located on two different centers with Coulomb integrals (in the extended Hückel sense) of H_{ii} and H_{jj}. The perturbation result of (16) implies that their interaction energy will be given by

$$\Delta E = \frac{H_{ij}^2}{H_{ii} - H_{jj}} \qquad (24)$$

There are various ways to estimate H_{ij}. Following Mulliken we may write[12] $H_{ij} = \frac{1}{2}KS_{ij}(H_{ii}H_{jj})^{1/2}$, where S_{ij} is the overlap integral between $\phi_{i,j}$ and K is an adjustable constant. Thus

$$\Delta E \alpha \frac{S_{ij}^2 H_{ii} H_{jj}}{H_{ii} - H_{jj}} \tag{25}$$

or

$$(\Delta E)^{-1} \alpha \frac{1}{S_{ij}^2} \left(\frac{1}{H_{jj}} - \frac{1}{H_{ii}} \right) = \frac{1}{S_{ij}^2} (r_i - r_j) \tag{26}$$

The abscissas of the structural maps therefore measure the weighted sum of the s-s and p-p interactions between the two atoms A, B, and is an indicator of the covalent interaction between the two centers. As this decreases [with increasing $(\Delta E)^{-1}$], the coordination number increases as shown in the figures of Sections II.B and II.D. The effect is also seen in the Phillips–Van Vechten approach of Fig. 15. This is traditionally viewed in terms of an increase in ionic contributions, although as we showed in Section II.B, a purely covalent model gives a similar result. A completely ionic system is expected to have the largest coordination number possible (subject to the limitations imposed by repulsions between the ion cores themselves) since with the largest number of nearest neighbors, the Coulombic attractive forces will be strongest. The relative importance of covalent and ionic contributions to the energy difference between two structures for a given AX system is, however, not easily determined, and similar problems have been found for molecules.

The origin of the ordinate of the Bloch-Zunger approach is less obvious in such orbital language, but our discussion will link it intimately to the site preference problem described in Sections V.C and VI. The charge distribution ρ for a given structure (i), obtained from a calculation on an idealized system where the atomic orbital input parameters for each atom are set equal, can in general be broken down into contributions from atomic s and p orbitals as (27).

$$\rho = p_i \phi_s^2 + q_i \phi_p^2 \tag{27}$$

Approximately the energy of an AB system will be given by

$$E_i = p_i \left(H_{ss}^A + H_{ss}^B \right) + q_i \left(H_{pp}^A + H_{pp}^B \right) \tag{28}$$

Since $p_i + q_i = 4$, the total atomic density in octet systems, for example,

$$E_i = p_i \left(H_{ss}^A - H_{pp}^A + H_{ss}^B - H_{pp}^B \right) + 4 \left(H_{pp}^A + H_{pp}^B \right) \tag{29}$$

which leads to an energy difference between two structures as

$$\Delta E_{ij} = (p_i - p_j)\left(H_{ss}^A - H_{pp}^A + H_{ss}^B - H_{pp}^B \right) \qquad (30)$$

which contains in parentheses a function of a type similar to that found in the structural maps. In the case of tetrahedral and octahedral coordination, simple calculations show that $p_4 > p_6$, reflecting the greater importance of the s orbital in bonding interactions in the lower coordinate structures, an idea inherent in existing qualitative valence bond ideas. Figure 48 is a simple plot, clearly separating four- and six-coordinate, eight-electron binary structures of the elements of the first two rows using this function with the usual values of H_{ss} and H_{pp} employed in molecular orbital calculations.[12] One apparently has considerable flexibility in the choice of ordinate and abscissa in the construction of these plots, all of which result in good structural sorting.

The metallization concept is not one that is current in the molecular field, but we note a recent study by Hall[110] in which the s-p energy separation of the central atom orbitals was found to be of crucial importance in determining the *angular* geometry of simple molecules such as H_2O and NH_3. We have also pointed out[12] the vital importance of the s-p separation in determining the unsymmetrical structures of hypervalent molecules such as ClF_3. The inert pair effect noted in Section VI.A is probably another manifestation of the same result.

The structural maps actually resolve several different structure types, and it is our contention that such structural sorting arises via a similar charge effect described above specifically for four-versus six-coordination. In general, the difference in energy between two structures after the style of (28) may reflect differences in s and p densities as a result of both coordination

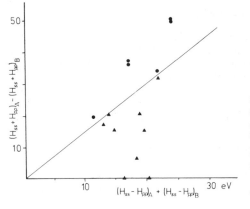

Fig. 48. A separation of four- and six-coordinate AX structures (and carbon and silicon) with a total of eight valence electrons using standard valence-state ionization potentials. Similar separations may be achieved by using $(\Delta H_{ss})^{-1} + (\Delta H_{pp})^{-1}$ as abscissa after the style of (26). Circles, six-coordination; triangles, four-coordination.

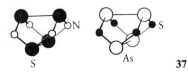

37

number and geometrical differences. Overall of course, the difference in the total energy between one geometrical configuration compared to another determines the structure adopted, but these charge ideas provide a useful theoretical construct. Often the approach allows rationalization of a result that is counter to general chemical intuition. Structure **37** shows an example from the molecular field. On simple valence grounds As and N might be expected to occupy three-coordinate sites in these cradle-shaped molecules and the sulfur atoms the two-coordinate positions. However calculation of the charge distribution shows a much larger electron density at the lower coordination number sites, which attracts the more electronegative nitrogen atom in S_4N_4 but the more electronegative sulfur atom in As_4S_4.

The charge distribution will naturally be a function of the total electronic configuration, and we note in this context that only the anion-rich (16 electron) AX_2 structures, for example, are plotted on the Pearson diagram of Fig. 9. For Cu_2O (cation rich, eight electrons) the point lies right in the middle of the diagram and is clearly unresolved from the other structures with different coordination numbers, as ought to be expected. In another system of this type in Section VI.E, where the site preferences are again dependent on the number of electrons, the molecular orbital assembly process of Fig. 47 showed rather clearly in orbital terms how the latent charge distribution of **36** controlled the geometry of the lowest energy structure. Ideas such as these may well be very useful in the future for dissecting in molecular orbital terms the underlying reasons for the adoption of one structure rather than another.

Acknowledgments

It is a pleasure to acknowledge the contributions made by my co-workers in this field, Peter Haaland, Jung-Hui Lin, Timothy J. McLarnan, and Guy L. Rosenthal. I also acknowledge the donors of the Petroleum Research Fund, administered by the American Chemical Society for their partial support of this research, and the Alfred P. Sloan and Camille and Henry Dreyfus foundations.

References

1. R. Hoffmann, *Acct. Chem. Res.*, **4**, 1 (1971).
2. M. J. S. Dewar and R. C. Dougherty, *The PMO Theory of Organic Chemistry*, Plenum Press, New York, 1975.
3. A. D. Walsh, *J. Chem. Soc.*, 2260, 2266, 2288, 2296, 2301, 2306 (1953).
4. M. J. S. Dewar, *The Molecular Orbital Theory of Organic Chemistry*, McGraw-Hill, New York, 1969.

5. D. M. P. Mingos, *Adv. Organomet. Chem.*, **15**, 1 (1977).
6. For example, K. Krogh-Jespersen, D. Cremer, D. Poppinger, J. Pople, P. von R. Schleyer, and J. Chandresekhar, *J. Am. Chem. Soc.*, **101** 4843 (1979)
7. M. Elian and R. Hoffmann, *Inorg. Chem.*, **14**, 1058 (1975).
8. R. Hoffmann, T. A. Albright, and D. L. Thorn, *Pure Appl. Chem.*, **50**, 1 (1978).
9. H.-B. Bürgi, J. D. Dunitz, and E. Sheffer, *J. Am. Chem. Soc.*, **95**, 5065 (1973).
10. E. L. Muetterties, *Boron Hydride Chemistry*, Academic, New York, 1975.
11. T. M. Dunn, D. S. McClure, and R. G. Pearson, *Some Aspects of Crystal Field Theory*, Harper & Row, New York, 1965.
12. J. K. Burdett, *Molecular Shapes*, Wiley, New York, 1980.
13. R. G. Burns, *Mineralogical Applications of Crystal Field Theory*, Cambridge University Press, New York, 1970.
14. J. D. Dunitz and L. E. Orgel, *Adv. Inorg. Chem. Radiochem.*, **2**, 1 (1960).
15. P. T. B. Schaffer, *Acta Crystallogr.*, Sect. B, **25**, 477 (1969).
16. R. S. Mitchell, *Z. Kristallogr.*, **108**, 296 (1956).
17. (a) For an optimistic statement in this regard, see J. C. Phillips, *Solid State Commun.*, **22**, 549 (1977); (b) M. L. Cohen, in Ref. 31.
18. A. Streitweiser, *Molecular Orbital Theory for Organic Chemists*, Wiley, New York, 1961.
19. E. Heilbronner and H. Bock, *The HMO Model and Its Application*, Wiley, New York, 1976.
20. R. Hoffmann and W. N. Lipscomb, *J. Chem. Phys.*, **36**, 2179, 3489 (1962).
21. R. B. Woodward and R. Hoffmann, *The Conservation of Orbital Symmetry*, Verlag Chemie, Weinheim, 1970.
22. A. F. Wells, *Structural Inorganic Chemistry*, 4th ed., Oxford University Press, New York, 1975.
23. W. B. Pearson, *Crystal Chemistry and Physics of Metals and Alloys*, Wiley, New York, 1972.
24. E. L. Muetterties and L. J. Guggenberger, *J. Am. Chem. Soc.*, **96**, 1798 (1974).
25. M. A. Porai-Koshits and L. A. Aslanov, *J. Struct. Chem.* (Engl. trans.), **13**, 244 (1972).
26. K. Wade, *Adv. Inorg. Chem. Radiochem.*, **18**, 1 (1976).
27. L. Pauling, *The Nature of the Chemical Bond*, Cornell University Press, Ithaca, N.Y., 1960.
28. E. Parthé, *Acta Crystallogr.*, Sect. B, **32**, 2813 (1980).
29. R. J. D. Tilley in "MTP Review of Science," *Inorganic Chemistry*, Vol. 10, L. E. J. Roberts, Ed., Series One 279 (1972), Series Two 74 (1975).
30. M. O'Keeffe and S. Andersson, *Acta Crystallogr.*, Sect. A, **33**, 914 (1977).
31. A. Navrotsky and M. O'Keeffe, Eds., *Structure and Bonding in Crystals*, Academic, New York, 1981.
32. T. J. McLarnan and P. B. Moore, in Ref. 31.
33. J. C. Phillips, in *Treatise on Solid State Chemistry*, Vol. 1, N. B. Hannay, Ed., Plenum, New York, 1974.
34. L. Pauling, *Proc. R. Soc. London*, Sec. A, **114** 181 (1972).
35. R. D. Shannon, *Acta Crystallogr*, Sect. A, **32**, 751 (1976).
36. P. Becker, Ed., *Electron and Magnetization Densities in Molecules and Crystals*, Plenum, New York, 1980.
37. J. Redinger and K. Schwarz, Zeits. f. Physik **B40**, 269 (1981).
38. C. Haas in *Crystal Structure and Chemical Bonding in Inorganic Chemistry*, C. J. M. Rooymans and A. Rabenau, Eds., North-Holland, New York, 1975.
39. L. S. Bartell, *J. Chem. Phys.*, **32**, 827 (1960).
40. C. Glidewell, *Inorg. Chim. Acta*, **12**, 219 (1975).
41. R. J. Gillespie, *Molecular Geometry*, Van Nostrand-Reinhold, London, 1972.
42. M. O'Keeffe, in Ref. 31.

43. G. V. Gibbs. C. T. Prewitt, and K. J. Baldwin, Z. *Kristallogr.*, **145**, 108 (1977).
44. T. A. Albright, P. Hoffman, and A. R. Rossi, *Naturforsch. B*, **35**, 34 (1980).
45. E. Mooser and W. B. Pearson, *Acta Crystallogr.*, **12**, 1015 (1959).
46. W. B. Pearson, *J. Phys. Chem. Solids*, **23**, 103 (1962).
47. J. K. Burdett and G. L. Rosenthal, *J. Solid State Chem.*, **33**, 173 (1980).
48. J. C. Phillips and J. A. Van Vechten, *Phys. Rev. B*, **2**, 2147 (1970).
49. J. C. Phillips, *Covalent Bonding in Crystals, Molecules and Polymers*, University of Chicago Press, Chicago, 1969.
50. S. P. Kowakzyk, L. Ley, F. R. McFeely, and D. A. Shirley, *J. Chem. Phys.*, **61**, 2850 (1974).
51. (a) J. St. John and A. N. Bloch, *Phys. Rev. Lett.*, **33**, 1095 (1974); (b) A. N. Bloch and G. Schatteman, in Ref. 31.
52. (a) A. Zunger, *Phys. Rev. Lett.*, **44**, 582 (1980); *Phys. Rev. B*, **22**, 5839 (1980); (b) see also E. S. Machlin and B. Loh, *Phys. Rev. Lett.*, **45**, 1642 (1980).
53. (a) J. R. Chelikowsky and J. C. Phillips, *Phys. Rev. B*, **17**, 2453 (1978); (b) E. S. Machlin, T. P. Chow, and J. C. Phillips, *Phys. Rev. Lett.*, **38**, 1292 (1977).
54. J. K. Burdett, *J. Chem. Soc., Dalton Trans.*, 1725 (1976).
55. J. K. Burdett, *Inorg. Chem.*, **14**, 375 (1975).
56. G. Pólya, *Acta Math.*, **68**, 145 (1937).
57. N. G. de Bruijn, *Niew. Arch. Wiskdë*, **19**, 89 (1971).
58. T. J. McLarnan, Z. *Kristallogr.*, **155** 227, 247, 269 (1981).
59. T. J. McLarnan, *J. Solid State Chem.*, **26** 235 (1978).
60. D. E. White, *Proc. Am. Math. Soc.*, **47**, 41 (1975).
61. D. H. Rouvray, *Chem. Soc. Rev.*, **3** 355 (1974).
62. P. B. Moore, *Neue J. Jahrb. Miner. Abh.*, **120**, 205 (1974).
63. A. J. Dekker, *Solid State Physics*, Prentice-Hall, Englewood Cliffs, N.J., 1957.
64. W. A. Harrison, *Electronic Structure and the Properties of Solids*, Freeman, San Francisco, 1980.
65. J. R. Reitz, *Solid State Phys.*, **1**, 1 (1955).
66. F. A. Cotton, *Group Theory and Its Applications*, 2nd ed., Wiley, New York, 1971.
67. H. Jones, *The Theory of the Brillouin Zone and Electronic States in Crystals*, North-Holland, New York, 1960.
68. L. J. Raubenheimer and G. Gilat, *Phys. Rev.*, **144**, 390 (1966).
69. A. Baldereschi, *Phys. Rev.*, **137**, 5212 (1973).
70. D. J. Chadi and M. L. Cohen, *Phys. Rev. B*, **8**, 5747 (1973).
71. D. J. Chadi and M. L. Cohen, *Phys. Rev. B*, **7**, 692 (1973).
72. D. J. Chadi and R. M. Morton, *Solid State Commun.*, **19**, 643 (1976).
73. J. A. Tossell and G. V. Gibbs, *Phys. Chem. Min.*, **2**, 21 (1977).
74. S. J. Louisnathan, R. J. Hill, and G. V. Gibbs, *Phys. Chem. Min.*, **1**, 53 (1977).
75. M. D. Newton, in Ref. 31.
76. J. K. Burdett, *J. Am. Chem. Soc.*, **102**, 5458 (1980).
77. A. Zunger, *J. Phys. C*, **7**, 76, 96 (1974).
78. A. Zunger, *J. Chem. Phys.*, **62** 1861 (1975); **63**, 1713, 4854 (1975).
79. A. Zunger, *Phys. Rev. B*, **17**, 625, 642 (1978).
80. M. J. S. Dewar and G. J. Gleicher, *J. Am. Chem. Soc.*, **87**, 3255 (1965).
81. R. E. Peierls, *Quantum Theory of Solids*, Oxford University Press, London, 1955, p. 108.
82. M.-H. Whangbo, *Inorg. Chem.*, **19**, 1728 (1980).
83. M.-H. Whangbo, *J. Chem. Phys.*, **70**, 4963 (1979).
84. M.-H. Whangbo, R. Hoffmann, and R. B. Woodward, *Proc. R. Soc. London, Sec. A*, **366**, 23 (1979).
85. A. J. Bennett, B. McCarroll, and R. P. Messmer, *Surf. Sci.*, **24**, 191 (1971).
86. A. J. Bennett, B. McCarroll, and R. P. Messmer, *Phys. Rev. B*, **3**, 1397 (1971).

87. J. K. Burdett, *Adv. Inorg. Chem. Radiochem.*, **21**, 113 (1978).
88. B. M. Gimarc, *Molecular Structure and Bonding*, Academic, New York, 1979.
89. L. S. Bartell, *J. Chem. Educ.*, **45**, 754 (1968).
90. R. G. Pearson, *J. Chem. Phys.*, **52**, 2167 (1970).
91. R. G. Pearson, *J. Am. Chem. Soc.*, **91**, 1252, 4947 (1969).
92. V. Heine and D. Weaire, *Solid State Phys.*, **24**, 249 (1970).
93. J. Halpern, in *Advances in Chemistry Series*, No. 70, American Chemical Society, Washington, D.C., 1968, p. 1.
94. J. K. Burdett and J.-H. Lin, Acta Cryst. (in press).
95. R. G. W. Wyckoff, *Crystal Structures*, Wiley, New York, 1963.
96. K. S. Pitzer, *Acct. Chem. Res.*, **12**, 271 (1979).
97. H. G. Grimm and A. Sommerfeld, *Z. Phys.*, **36**, 36 (1926).
98. E. Parthé, *Crystal Chemistry of Tetrahedral Structures*, Gordon & Breach, New York, 1964.
99. A. A. Levin, *Solid State Quantum Chemistry*, McGraw-Hill, New York, 1977.
100. J. K. Burdett, *Nature (London)*, **279**, 121 (1979).
101. R. E. Williams, *Adv. Inorg. Chem. Radiochem.*, **18**, 67 (1976).
102. D. M. P. Mingos, *Nature (London) Phys. Sci.*, **236**, 99 (1972).
103. A. J. Stone, *Inorg. Chem.*, **20**, 563 (1981).
104. J. K. Burdett and S. Lee, unpublished.
105. E. Zintl and G. Brauer, *Z. Phys. Chem.*, **20B**, 245 (1933).
106. J. K. Burdett and T. J. McLarnan, to be published.
107. J. K. Burdett, P. Haaland, and T. J. McLarnan, to be published.
108. S. Wolfe, *Acct. Chem. Res.*, **5**, 102 (1972).
109. J. K. Burdett, *J. Am. Chem. Soc.*, **102**, 450 (1980).
110. M. B. Hall, *Inorg. Chem.*, **17**, 2261 (1978).

GENERATOR COORDINATE THEORY OF NUCLEAR MOTION IN MOLECULES

L. LATHOUWERS* AND P. VAN LEUVEN

Dienst Teoretische en Wiskundige Natuurkunde
University of Antwerp (RUCA)
Antwerp, Belgium

CONTENTS

*Also at the Quantum Theory Project, University of Florida, Gainesville, Florida.

I. INTRODUCTION

The generator coordinate method (GCM) was introduced into nuclear physics by Hill and Wheeler in 1953.[1] It is now a well-known technique for describing collective motion in the nucleus, and a large number of applications appear each year in the nuclear physics literature. In 1977 it was suggested[2] that this method be applied to the totally different problem of molecular motion. Since then a number of papers have been published in which the GCM description of nuclear motion in molecules was developed until it came to represent a self-contained and coherent quantum-mechanical picture of molecules. This chapter describes and evaluates the main ideas of these studies.

Let us first discuss some of the reasons for the need for a new theoretical approach to molecular dynamics. The present accuracy of high-resolution spectroscopy requires a very precise phenomenological analysis of molecular spectra. This is usually carried out in terms of spectroscopic constants whose definition and physical interpretation is rooted in the adiabatic approximation. However, by its very assumptions the inherent accuracy of the whole adiabatic procedure is limited by nonadiabatic errors. Several examples exist,[3] however, of experimental spectra that have been fitted with a precision exceeding this limit. The quality of these fits, using relatively few spectroscopic parameters, is often impressive. This apparently contradictory situation calls for clarification. It has been argued[4] that the GCM provides a theoretical framework that could reestablish consistency between experimental and theoretical error bars.

From the purely theoretical point of view, a number of fundamental criticisms against the traditional description of molecules using "energy surfaces," "equilibrium shapes," and so on have been put forward.[5] It is asserted that the adiabatic approximation relies too heavily on semiclassical concepts that render its physical interpretation and quantum-mechanical foundations dubious. On the other hand, every practitioner of molecular physics would agree that these semiclassical concepts play an important role in an intuitive understanding of many of the mechanisms of molecular motion. It has been shown that in the GCM one can write down a wave function for the molecule in which electrons and nuclei are treated on the same footing, that is, both as quantal particles, but all semiclassical concepts are put into the "generator coordinates." These coordinates serve only as a mathematical auxiliary; hence no problems of physical interpretation arise.[5]

Although the complete Born-Huang method is in principle exact, it leads to a system of coupled differential equations that have some practical drawbacks. Part of the problems associated with this approach stem from the fact that nonadiabatic effects are built in through detailed information about different electronic states. The GCM, on the contrary, makes it possible to build in nonadiabatic correlation still using a *single* electronic state. In this respect it fits into the same class of approximations as the conventional clamped nuclei, and adiabatic approximation and can be considered to be the logical nonadiabatic extension of these approaches.

This chapter consists of several parts. Section II outlines the GCM as a representation in quantum mechanics in which a continuous basis is used. This part does not aim at a complete review of all technical aspects of GCM. Other excellent papers exist.[6] There have been reports of other attempts, involving electronic structure, to use GCM in atomic and molecular physics. We have not attempted to trace back these points of contact in the existing literature.

Section III situates the GCM as a stage in a hierarchy of approximations, all based on the separation of electronic and nuclear motion. The three stages considered are the clamped-nuclei approximation, the adiabatic approximation, and the GCM. At each level in this hierarchy an increased amount of information is transferred from the electronic problem to a nuclear motion equation. To compare these three methods, we have analyzed them in terms of the Born-Oppenheimer perturbation theory. From the general theory it follows that certain nonadiabatic corrections are contained in the GCM wave function. The analysis of an exactly solvable model system confirms this statement. A number of alternative approaches to nonadiabaticity as well as more extended GCM schemes are presented but not evaluated.

Section IV presents applications of GCM to diatomic molecules. It is self-contained and written in textbook style. A relatively large amount of space is devoted to the implications of rotational invariance. This is to show how the use of generator coordinates avoids the complications of moving axes. Our discussion of symmetry properties also illustrates how the symmetry of the molecule influences the structure of the kernels in the basic integral equation and helps to prepare generalizations necessary in the next section. A GCM version of the Dunham series is established, which explains its applicability to high-resolution spectroscopy.[4]

Section V describes schematically how the GCM can be developed in the polyatomic case. Again the problem of rotational invariance can be solved elegantly by the use of projection operator techniques. This formulation provides a basis for the construction of GCM term formulas for polyatomics. On this subject very little has appeared in the literature; therefore we limit ourselves to some speculations that may help to define future applications.

II. THE GENERATOR COORDINATE METHOD

A. The Generator Coordinate Variational Principle

Consider a function $\chi(x|\alpha)$, depending on the dynamical variables x of a many-body system and a real parameter α that can vary in the interval $[a, b]$. There are a variety of ways in which one can generate approximate eigenstates of the system from the family $\{\chi(\alpha)\}$ with α in $[a, b]$. Most of them can be formulated in terms of Ritz's variational principle.[7] A common approach is to minimize the energy associated with $\chi(x|\alpha)$, that is,

$$E(\alpha) = \frac{\langle \chi(\alpha)|H|\chi(\alpha)\rangle}{\langle \chi(\alpha)|\chi(\alpha)\rangle} \qquad (2.1)$$

where H is the many-body Hamiltonian. This gives an optimal value α_0 for which the function above is stationary:

$$\left[\frac{\partial E}{\partial \alpha}\right]_{\alpha=\alpha_0} = 0 \qquad (2.2)$$

The associated energy $E(\alpha_0)$ and wave function $\chi(x|\alpha_0)$ may or may not, depending on a combination of luck and skill in the choice of $\chi(x|\alpha)$, be good approximations to an exact energy level and wave function. A logical extension of this approach is to consider not one, but a superposition of a selected number of functions out of the set $\{\chi(\alpha)\}$. Assuming wave functions of the form

$$\Psi(x) = F_0\chi(x|\alpha_0) + F_1\chi(x|\alpha_1) + \cdots = \sum_i F_i\chi(x|\alpha_i) \qquad (2.3)$$

one can now apply Ritz's variational principle to the undetermined coefficients F_i. Minimization of

$$E(F_1, F_2, \ldots) = \frac{\Sigma\Sigma F_i^* H_{ij} F_j}{\Sigma\Sigma F_i^* \Delta_{ij} F_j} \qquad (2.4)$$

$$H_{ij} = \langle \chi(\alpha_i)|H|\chi(\alpha_j)\rangle = H_{ji}^*, \qquad \Delta_{ij} = \langle \chi(\alpha_i)|\chi(\alpha_j)\rangle = \Delta_{ji}^* \qquad (2.5)$$

where H_{ij} and Δ_{ij} are the Hamiltonian and overlap matrices, leads to the well-known secular equation for a nonorthogonal basis set[8]

$$\sum_j \left[H_{ij} - E\Delta_{ij}\right] F_j = 0 \qquad (2.6)$$

The eigenvalues of this equation are, in order, upper bounds to the ground state, the first excited state $\cdots$ of the Hamiltonian under consideration. One can now imagine going to the extreme by including all members of the family $\{\chi(\alpha)\}$ in a trial state. This is accomplished if the summation in (2.3) is replaced by the integration over the parameter domain $[a, b]$. The resulting trial state reads

$$\Psi(x) = \int_a^b F(\alpha)\chi(x|\alpha)\,d\alpha \qquad (2.7)$$

in which the function $F(\alpha)$ can be looked on as a continuously labeled set of superposition coefficients. The energy becomes a functional of $F(\alpha)$

$$E[F(\alpha)] = \frac{\iint F^*(\alpha)H(\alpha,\beta)F(\beta)\,d\alpha\,d\beta}{\iint F^*(\alpha)\Delta(\alpha,\beta)F(\beta)\,d\alpha\,d\beta} \qquad (2.8)$$

$$H(\alpha,\beta) = \langle\chi(\alpha)|H|\chi(\beta)\rangle = H^*(\beta,\alpha), \qquad \Delta(\alpha,\beta) = \langle\chi(\alpha)|\chi(\beta)\rangle = \Delta^*(\beta,\alpha) \qquad (2.9)$$

which is the continuous analog of (2.4) with the matrices H_{ij} and Δ_{ij} replaced by the integral kernels $H(\alpha,\beta)$ and $\Delta(\alpha,\beta)$. The functions $F(\alpha)$ that minimize (2.8) satisfy the following integral equation:

$$\int_a^b [H(\alpha,\beta) - E\Delta(\alpha,\beta)]F(\beta)\,d\beta = 0 \qquad (2.10)$$

The eigenvalues of this generalized secular equation again furnish upper bounds to the exact energy levels. Since one includes more and more elements out of $\{\chi(\alpha)\}$ as one goes from (2.2) to (2.10) by way of (2.6), the energy is gradually lowered. In the final GC stage one has incorporated all states $\chi(\alpha)$ such that one can say that the GCM *is the variationally optimal way to dispose of a parameter.*

Wave functions of the form (2.7) and the associated variational principle were first suggested by Wheeler in connection with the problem of collective motions in atomic nuclei.[9] The now common terminology, which for historical reasons we adopt here also, was introduced:

$\chi(x|\alpha)$	*Intrinsic state*	
α	*Generator coordinate*	
$F(\alpha)$	*Weight function*	
$\Delta(\alpha,\beta)$	*Overlap kernel*	
$H(\alpha,\beta)$	*Hamiltonian kernel*	(2.11)

A possible justification, which should not be taken too strictly, goes as follows. The $\chi(x|\alpha)$ assume the role of basis functions in the procedure above. The choice of the basis functions is always crucial for the success of a variational method. One cannot expect good results unless they are of "intrinsic importance" to the system under consideration. Hence the term *intrinsic states* for the $\chi(x|\alpha)$ is appropriate. As for the parameter α, one can argue that it appears in $\chi(x|\alpha)$ as an extra coordinate in addition to the true dynamical coordinates x. However, since α serves only to "generate" the trial functions through the integration (2.7) and no longer appears in the final wave functions, it is referred to as a *generator coordinate* (GC). Each intrinsic state $\chi(x|\alpha)$ has a "weight" $F(\alpha)$ in the superposition (2.7). On the interval $[a, b]$ the coefficients of the $\chi(x|\alpha)$ define a *weight function*. Finally $\Delta(\alpha, \beta)$ and $H(\alpha, \beta)$ are straightforward generalizations of the overlap matrix and the Hamiltonian matrix, respectively.

The entire variational scheme involving the choice of $\chi(x|\alpha)$, the computation of the Hamiltonian and overlap kernel and the solution of the integral equation (2.10) is called the *generator coordinate method* (*GCM*). Evidently, the GC integral equation (2.10) should be supplemented with the proper boundary condition according to whether one wants to describe bound states or scattering situations.

In principle any parameter in a many-body state can be considered to be a generator coordinate. However, as one usually concentrates on a specific part of the energy spectrum or a special kind of dynamics among the particles, the art of applying the GCM successfully lies in building in a GC into an intrinsic function in such a way that superpositions of the type (2.7) actually describe the desired features. In such attempts one may be guided by physical intuition, symmetry considerations, mathematical elegance, computational advantages, and other factors. Ideally the reason for preferring the GCM over any other approach should be a combination of these guidelines. The application to molecular systems will prove to be a typical example.

As a simple illustrative application of the GCM we have chosen the hydrogen atom problem (in atomic units)

$$H = -\frac{1}{2}\Delta_r - \frac{1}{r} \tag{2.12}$$

More precisely, we would like to describe the s-states of (2.12) in terms of scaled Gaussian intrinsic states:

$$\chi(r|\alpha) = e^{-\alpha r^2} \tag{2.13}$$

$$\Psi(r) = \int_0^\infty F(\alpha)e^{-ar^2}\,d\alpha \tag{2.14}$$

The Hamiltonian and overlap kernel are easily calculated:

$$\Delta(\alpha,\beta)=\left(\frac{\pi}{\alpha+\beta}\right)^{3/2} \tag{2.15}$$

$$H(\alpha,\beta)=3\left(\frac{\alpha\beta}{\alpha+\beta}\right)\left(\frac{\pi}{\alpha+\beta}\right)^{3/2}-\frac{2\pi}{\alpha+\beta} \tag{2.16}$$

Although a direct solution of the GC integral equation is not feasible for these kernels, the weight functions can be obtained indirectly by considering the following integral transformations between exponential and Gaussian functions:

$$e^{-\sigma r}=\frac{\sigma}{2\sqrt{\pi}}\int_0^\infty \alpha^{-3/2}e^{-\sigma^2/4\alpha}e^{-\alpha r^2}\,d\alpha \tag{2.17}$$

This expression has proved useful in electronic structure calculations for molecules as a transformation between Gaussian and Slater-type orbitals.[10] Here it solves the GC problem for the ground state, since (unnormalized)

$$\Psi_0(r)=e^{-r}=\int F_0(\alpha)e^{-\alpha r^2}\,d\alpha \tag{2.18}$$

$$F_0(\alpha)=\frac{1}{2\sqrt{\pi}}\alpha^{-3/2}e^{-1/4\alpha} \tag{2.19}$$

That is, we have found a weight function that gives the exact ground state of the hydrogen atom. Observing that

$$\frac{de^{-\sigma r}}{d\sigma^k}=(-)^k r^k e^{-\sigma r} \tag{2.20}$$

and that all s-states are polynomials in r times e^{-r}, it is clear also that these states can be represented exactly in the form (2.14). Indeed, one simply has to apply (2.20) to (2.17), take suitable linear combinations, and put $\sigma=1$.

An important concept to be added to the list (2.11) is that of the *generator coordinate subspace*, which is defined as follows. All square-integrable functions of the form (2.7) clearly form a linear metric space. By including the limit points of all Cauchy sequences with elements

$$\Psi_n(x)=\int F_n(\alpha)\chi(x|\alpha)\,d\alpha \tag{2.21}$$

one obtains a Hilbert space $\mathcal{H}_{GC}$ to which one refers as the GC subspace. A problem involved is that taking the limit $n \to \infty$ of (2.21) in general does not commute with the integration of the GC. Therefore, it may happen that elements of $\mathcal{K}_{GC}$ cannot be written as a GC integral (2.7). Attempts have been made to characterize $\mathcal{K}_{GC}$ in a more strict mathematical fashion, that is, by providing a basis for it. In particular one can prove[11] that it is always possible to select a countable subset of $\{\chi(\alpha)\}$ that spans $\mathcal{K}_{GC}$. This provides a justification for discretization techniques used to approximately solve the GC integral equation. Indeed by taking enough GC values in (2.3) one knows that the eigenvalues of the associated secular equations (2.6) will eventually converge to those of (2.10). It is clear, however, that the resulting basis for $\mathcal{K}_{GC}$ is neither unique nor orthogonal. A uniquely defined, complete, and orthogonal set of functions in $\mathcal{K}_{GC}$ can be obtained by applying a natural-state analysis to $\chi(x|\alpha)$. Within the so-called natural-state formalism,[12] one proves that under suitable normalization of $\chi(x|\alpha)$, there exists a norm convergent expansion

$$\chi(x|\alpha) = \sum_n \sqrt{\lambda_n}\, y_n(x) b_n^*(\alpha) \tag{2.22}$$

The sets $\{y_n(x)\}$ and $\{b_n(\alpha)\}$ are termed the coordinate and generator coordinate natural states. They are the eigenfunctions of the left- and right-iterated kernels associated with the intrinsic state, while the λ_n are the corresponding eigenvalues. It is easy to prove that the coordinate natural states $\{y_n(x)\}$ are orthogonal and complete in $\mathcal{K}_{GC}$. For further details on the natural-state formalism, see Ref. 12.

In conclusion it should be observed that there is no difficulty in extending the GCM to several, possibly complex, generator coordinates. The domain in which such a set of GCs is allowed to vary is then referred to as the space of GCs (not to be confused with the GC subspace).

B. Symmetry Properties and Generator Coordinates

The combined use of group theory and the GCM will prove to be of considerable importance in the GC approach to molecular dynamics. More precisely, we examine the problem of generating symmetry-adapted wave functions within the framework of the GCM and the use of group parameters of continuous groups as GCs. This section introduces the necessary concepts and notations to be used later.

If the Hamiltonian has a constant of motion, there exists a unitary operator S such that

$$[H, S] = 0 \quad \text{and} \quad S^+ = S^{-1} \tag{2.23}$$

where S^+ is the adjoint and S^{-1} the inverse of S. One can easily prove that an exact eigenstate of H is also an eigenfunction of S. The question of whether this property can be extended to approximate variational wave functions has been studied by Laskowski and Löwdin.[13] They showed that a sufficient condition is the stability of the variational subspace under the symmetry operator S. Stability here means that if $\Psi(x)$ belongs to the variational subspace, so does $S\Psi(x)$. The stability of GC subspaces has been examined by Brink and Weiguny.[14] In turn, they proved that $\mathcal{H}_{GC}$ *is stable under S provided the intrinsic states are closed with respect to S* in the sense that

$$S\chi(x|\alpha)=\chi(x|\sigma(\alpha)) \qquad (2.24)$$

This equation, known as the Brink-Weiguny condition, implies that conjugate to the symmetry operation S there exists a transformation σ in the space of GCs having the same effect as S. The stability of $\mathcal{H}_{GC}$ is easily verified, since using (2.24) we have

$$S\Psi(x)=\int F_\sigma(\alpha)\chi(x|\alpha)\,d\alpha \qquad (2.25)$$

$$F_\sigma(\alpha)=\frac{F\big(\sigma^{-1}(\alpha)\big)}{|J(\alpha)|} \qquad (2.26)$$

where $J(\alpha)$ is the Jacobian and $\sigma^{-1}(\alpha)$ the inverse of the transformation $\sigma(\alpha)=\alpha'$. *In applying the GCM it will thus be to our advantage to use intrinsic states that satisfy the Brink-Weiguny condition (2.24), since this guarantees that the resulting GC wave functions will be symmetry adapted.* One can easily deal with a finite or a continuous group of transformations by generalizing the demonstration above along the lines suggested in Refs. 13 and 14.

If H is rotationally invariant, we can consider the rotation group as a prototype continuous symmetry group. In this case the exact wave functions can be chosen as simultaneous eigenfunctions of J^2 and J_z with corresponding quantum numbers J and M, $\mathbf{J}$ being the total angular momentum of the system. The elements of the rotation group induce a set of unitary rotation operators $\mathcal{R}(\Omega)$ in Hilbert space such that

$$[H,\mathcal{R}(\Omega)]=0 \qquad (2.27)$$

$$\mathcal{R}(\Omega)=e^{-i\varphi J_z}e^{-i\theta J_y}e^{-i\gamma J_z} \qquad (2.28)$$

A parameterization in terms of Euler angles $\Omega\equiv(\varphi,\theta,\gamma)$ has been adopted, and (2.28) corresponds to the active point of view of the rotation operators. For a detailed derivation of (2.28) and clarification of some well-hidden errors in the standard texts on angular momentum, see Ref. 15.

For an arbitrary function $\chi(x)$ we have

$$\mathfrak{R}(\Omega)\chi(x)=\chi\big(A^{-1}(\Omega)x\big)\equiv\chi(x|\Omega) \qquad (2.29)$$

where $A(\Omega)$ stands for the coordinate transformation

$$A(\Omega)=\begin{pmatrix} \cos\gamma\cos\theta\cos\varphi-\sin\gamma\sin\varphi & -\sin\gamma\cos\theta\cos\varphi-\cos\gamma\sin\varphi & \sin\theta\cos\varphi \\ \cos\gamma\cos\theta\sin\varphi+\sin\gamma\cos\varphi & -\sin\gamma\cos\theta\sin\varphi+\cos\gamma\cos\varphi & \sin\theta\sin\varphi \\ -\cos\gamma\sin\theta & \sin\gamma\sin\theta & \cos\theta \end{pmatrix} \qquad (2.30)$$

The functions $\chi(x|\Omega)$ define an orientation in space through the Euler angles Ω. They can be considered to be intrinsic states if the Ω assume the role of generator coordinates. The resulting trial states are of the form

$$\Psi(x)=\int F(\Omega)\chi(x|\Omega)\,d\Omega \qquad (2.31)$$

The space generated in this way is obviously stable under rotations, since

$$\mathfrak{R}(\Omega')\chi(x|\Omega)=\mathfrak{R}(\Omega)\mathfrak{R}(\Omega')\chi(x)=\mathfrak{R}(\Omega'')\chi(x)=\chi(x|\Omega'') \qquad (2.32)$$

where Ω'' are the Euler angles of a rotation describing the combined effect of $\mathfrak{R}(\Omega)$ and $\mathfrak{R}(\Omega')$. Hence the variational principle applied to (2.31) will yield wave functions consistent with the rotational invariance of the Hamiltonian.

Because the GCs are the group parameters, we can use representation theory to give a more explicit form of the GC equations. Using the properties of the Wigner functions,[16]

$$D_{MK}^{J}(\Omega)=e^{-iM\varphi}d_{MK}^{J}(\theta)e^{-iK\gamma} \qquad (2.33)$$

one can show that *the weight function $F(\Omega)$ is a linear combination of elements of the same row M of a single irreducible representation J*, that is,

$$F^{JM}(\Omega)=\sum_{K}C_{K}^{J}D_{MK}^{J*}(\Omega) \qquad (2.34)$$

The coefficients in this expansion satisfy the secular equation

$$\sum_{L}\big[H_{KL}^{J}-E^{J}\Delta_{KL}^{J}\big]C_{L}^{J}=0 \qquad (2.35)$$

$$H_{KL}^{J}=\int D_{KL}^{J*}(\Omega)\langle\chi|H\mathfrak{R}(\Omega)|\chi\rangle\,d\Omega \qquad (2.36)$$

$$\Delta_{KL}^{J}=\int D_{KL}^{J*}(\Omega)\langle\chi|\mathfrak{R}(\Omega)|\chi\rangle\,d\Omega \qquad (2.37)$$

which is equivalent to the Wheeler equation associated with the trial state (2.31). In this way the degeneracy of the intrinsic states $\chi(x|\Omega)$ is lifted and a set of $2J+1$ levels is obtained for each value of J. The corresponding wave functions are of the form

$$\Psi^{JM}(x) = \sum_K C_K^J \int D_{MK}^{J*}(\Omega) \mathfrak{R}(\Omega) \chi(x) \, d\Omega \qquad (2.38)$$

In this expression one recognizes the angular momentum projection operators[17]

$$P_{MK}^J = \frac{2J+1}{8\pi^2} \int D_{MK}^{J*}(\Omega) \mathfrak{R}(\Omega) \, d\Omega \qquad (2.39)$$

Operating on an arbitrary function P_{MK}^J gives a J^2, J_z eigenstate with associated eigenvalues J, M. Absorbing the constant $(2J+1)/8\pi^2$ in the coefficients C_K^J, we can write (2.38) as

$$\Psi^{JM}(x) = \sum_K C_K^J P_{MK}^J \chi(x) \qquad (2.40)$$

We can therefore conclude that *using Euler angles as GC's is equivalent to angular momentum projection*. From the general properties of the P_{MK}^J

$$[H, P_{MK}^J] = 0 \quad \text{and} \quad (P_{MK}^J)^+ = P_{KM}^J \quad \text{and} \quad P_{MK}^J P_{KL}^J = P_{ML}^J \qquad (2.41)$$

one easily recovers the secular problem (2.35). It should be remarked that strictly speaking, only the operators P_{KK}^J are true projection operators. The P_{MK}^J with $M \neq K$ are not idempotent and are better known as "shift operators."

C. Generator Coordinate Perturbation Theory

If an exact solution of (2.10) is not possible but the Wheeler equation can be solved by minor modifications in the Hamiltonian and overlap kernels, a perturbation theory (PT) approach is called for. Let us denote the exact and unperturbed GC equations in the abbreviated forms:

$$[H - E\Delta]F = 0 \quad \text{and} \quad [H^{(0)} - E^{(0)}\Delta^{(0)}]F^{(0)} = 0 \qquad (2.42)$$

The perturbations by which the unperturbed kernels $H^{(0)}$ and $\Delta^{(0)}$ differ from the exact ones are

$$V = H - H^{(0)} \quad \text{and} \quad W = \Delta - \Delta^{(0)} \qquad (2.43)$$

These are referred to as dynamical and geometrical perturbations, respectively. The two problems (2.42) can be linked to each other by studying the equation

$$\left[H(\lambda)-E(\lambda)\Delta(\lambda)\right]F(\lambda)=0 \tag{2.44}$$

$$H(\lambda)=H^{(0)}+\lambda V \quad \text{and} \quad \Delta(\lambda)=\Delta^{(0)}+\lambda W \tag{2.45}$$

where a PT expansion parameter, or coupling strength, λ has been introduced. Classical Rayleigh-Schrödinger PT then proceeds by expanding $E(\lambda)$ and $F(\lambda)$ in power series:

$$E(\lambda)=E^{(0)}+\lambda E^{(1)}+\lambda^2 E^{(2)}+\cdots \tag{2.46}$$

$$F(\lambda)=F^{(0)}+\lambda F^{(1)}+\lambda^2 F^{(2)}+\cdots \tag{2.47}$$

To be of any use, these expansions must have a convergence radius of at least unity or be summable by using Padé approximants or related techniques. Substitution of the series above in (2.44) leads to a set of inhomogeneous integral equations for the weight function corrections

$$\left[H^{(0)}-E^{(0)}\Delta^{(0)}\right]F^{(k)}=\sum_{l=1}^{k}\left[E^{(l)}\Delta^{(0)}+E^{(l-1)}W\right]F^{(k-l)}-VF^{(k-1)} \tag{2.48}$$

It should be noticed that unlike the Wheeler integral equation, the equations (2.48) are of the classical type studied by Fredholm.[18] We assume the weight functions to be normalized such that

$$\left(F^{(0)}|\Delta^{(0)}|F^{(0)}\right)=\left(F^{(0)}|\Delta^{(0)}|F\right)=1 \tag{2.49}$$

where we have introduced the following notation for double integrals

$$(F|C|F)\equiv\int\int F^*(\alpha)C(\alpha,\beta)F(\beta)\,d\alpha\,d\beta \tag{2.50}$$

The formula for the energy corrections then readily follows by taking the scalar product to the left of (2.48) with $F^{(0)}$ and using (2.49):

$$E^{(k)}=\left(F^{(0)}|V|F^{(k-1)}\right)-\sum_{l=1}^{k}E^{(l-1)}\left(F^{(0)}|W|F^{(k-l)}\right) \tag{2.51}$$

In particular for $k=1$ one obtains

$$E^{(1)}=\left(F^{(0)}|V|F^{(0)}\right)-E^{(0)}\left(F^{(0)}|W|F^{(0)}\right) \tag{2.52}$$

This result shows that also in the case of a generalized eigenvalue problem such as (2.42), the first-order energy correction in PT is equal to the average of the perturbation with respect to the unperturbed state. In this case the total perturbation is $V - E^{(0)}W$. A more detailed, double PT using a different expansion parameter for V and W is given in Ref. 19. Such an approach is appropriate if V and W are of different orders of magnitude. One can also generalize a theorem by Wigner stating that the kth-order eigenfunction correction determines the energy up to and including order $2k+1$. Explicit formulas and examples are given in Ref. 19.

In classical Schrödinger PT the energy to first order furnishes an upper bound, since it can be written as an expectation value of the Hamiltonian. In the present case this property is lost because (2.52) is not a GC energy expectation value. However, one can still prove the following useful theorem: if the first-order energy correction is negative, the true GC eigenvalues satisfy the inequalities

$$E_{\text{exact}} \leq E < E^{(0)} \tag{2.53}$$

The left-hand side follows trivially from the variational principle, while on the other hand we have by assumption

$$E^{(0)} + E^{(1)} = E^{(0)} + \left(F^{(0)} | H | F^{(0)} \right) - E^{(0)} \left(F^{(0)} | \Delta | F^{(0)} \right) < E^{(0)} \tag{2.54}$$

Using the GC variational principle once more, it follows that

$$E^{(0)} > \frac{\left(F^{(0)} | H | F^{(0)} \right)}{\left(F^{(0)} | \Delta | F^{(0)} \right)} \geq E \tag{2.55}$$

which completes the proof of (2.53).

This theorem is important because although the unperturbed and perturbed energies are not directly related, one can still draw a conclusion about their relative magnitude on the basis of first-order PT. The application of (2.53) in connection with the adiabatic and GC approximations for molecules will make this clearer.

As an example of an exactly solvable GC equation, which often serves as a useful unperturbed problem, we consider the kernels:

$$\Delta^{(0)}(\alpha, \beta) = e^{-S(\alpha - \beta)^2} \tag{2.56}$$

$$K^{(0)}(\alpha, \beta) = \frac{H^{(0)}(\alpha, \beta)}{\Delta^{(0)}(\alpha, \beta)}$$

$$= E(\alpha_0) + \tfrac{1}{2} \left[B(\alpha - \alpha_0)^2 + 2A(\alpha - \alpha_0)(\beta - \alpha_0) + B(\beta - \alpha_0)^2 \right] \tag{2.57}$$

In many cases assuming a Gaussian behavior of the overlap kernel is not a bad approximation. Indeed, if the intrinsic states are normalized to one, the overlap $\langle \chi(\alpha)|\chi(\beta)\rangle$ will decrease as $\chi(\alpha)$ differs more and more from $\chi(\beta)$, that is, as $\alpha - \beta$ increases. The scale factor S then measures the rate at which the overlap function drops off. On the other hand, if $H(\alpha, \beta)/\Delta(\alpha, \beta)$ is a smooth function of the GCs, it seems appropriate to expand this ratio in a Taylor series around the energetic minimum $E(\alpha_0)$. Truncating the expansion at second order, one obtains (2.57). The combined use of the Gaussian overlap (2.56) and the quadratic expansion (2.57) is known as the *harmonic approximation* because the resulting energy spectrum is that of a harmonic oscillator

$$E_v^{(0)} = E(\alpha_0) - \frac{A}{4S} + \left(v + \frac{1}{2}\right)\Omega \qquad (2.58)$$

$$\Omega = \frac{1}{2S}(A^2 - B^2)^{1/2} \qquad (2.59)$$

This result can be derived directly from the Wheeler equation[9] or by converting the unperturbed problem in a differential equation.[14] The following approach, although somewhat indirect, may clarify the nature of the harmonic approximation.

Consider the harmonic oscillator Hamiltonian and a set of translated Gaussian intrinsic states:

$$H = -\frac{1}{2M}\frac{d^2}{dx^2} + \frac{M\Omega^2}{2}x^2 \qquad (2.60)$$

$$\chi(x|\alpha) = \left(\frac{4S}{\pi}\right)^{1/4} e^{-2S(x-\alpha)^2} \qquad (2.61)$$

It is known that the GCM will be exact from a theorem by Wiener[20] known as the "closure of translated functions," which in this case proves that the GC subspace is the full space of square-integrable functions. Computation of the overlap kernels yields (2.56), while for the Hamiltonian kernel one obtains

$$e^{-S(\alpha-\beta)^2}\left\{\frac{1}{2M}\left[2S - 4S^2(\alpha-\beta)^2\right] + \frac{M\Omega^2}{2}\left[\frac{1}{8S} + \left(\frac{\alpha+\beta}{2}\right)^2\right]\right\} \qquad (2.62)$$

Observing that $\alpha_0 = 0$, it is easily verified that (2.62) is of the form (2.57) if one makes the identifications

$$E(\alpha_0) = \frac{S}{M} + \frac{M\Omega^2}{16S} \tag{2.63}$$

$$A = \frac{M\Omega^2}{4} + \frac{4S^2}{M} \quad \text{and} \quad B = \frac{M\Omega^2}{4} - \frac{4S^2}{M} \tag{2.64}$$

Solving for Ω and M one obtains (2.59) and

$$M = \frac{8S^2}{A - B} \tag{2.65}$$

The weight functions corresponding to the oscillator eigenstates can be found by solving

$$\Psi_v^{osc}(x) = \int_{-\infty}^{+\infty} F_v^{(0)}(\alpha) e^{-2S(x-\alpha)^2} \, dx \tag{2.66}$$

for $F_v^{(0)}(\alpha)$ by Fourier transformation. For the ground state one obtains (unnormalized):

$$F_0^{(0)}(\alpha) = \exp\left[\frac{-\frac{1}{2}M\Omega\alpha^2}{1 - M\Omega/4S} \right] \tag{2.67}$$

The derivation of the excited state weight functions can be found in Ref. 12.

These results demonstrate that a many-body system whose GC kernels are well approximated by the harmonic ones (2.56) and (2.57) behaves effectively as a harmonic oscillator with mass M and frequency Ω. It is in fact in this spirit that the first application of the GCM was made. In his study of collective oscillations of the ^{16}O nucleus, Griffin[21] starts from a Slater determinant of single-particle states in a potential well. The GC is a deformation parameter describing the shape of the potential. Fitting the overlap to a Gaussian (2.56) and expanding the ratio H/Δ up to second order, a vibrational spectrum is obtained. In this way the complicated many-nucleon system can be identified with a simple harmonic oscillator with effective collective frequency Ω and an effective collective mass M.

The role of the harmonic approximation in GC-PT is now clear. *If part of the spectrum is expected to be quasivibrational and a GC related to the collective oscillations has been identified, the harmonic approximation is an appropriate zeroth-order problem.* Anharmonicities can then be taken into account by GC-PT.

III.   ADIABATIC AND NONADIABATIC APPROXIMATIONS
FOR MOLECULAR SYSTEMS

A.   Separation of Electronic and Nuclear Motion

The nonrelativistic Hamiltonian for a system of electrons and nuclei, with charges Z_i and masses M_i, is given by

$$H = - \sum_i \frac{1}{2M_i} \Delta_{\mathbf{R}_i} - \frac{1}{2m} \sum_i \Delta_{\mathbf{r}_i} + \sum_{i<j} \frac{Z_i Z_j}{|\mathbf{R}_i - \mathbf{R}_j|}$$

$$+ \sum_{i<j} \frac{1}{|\mathbf{r}_i - \mathbf{r}_j|} - \sum_{i,j} \frac{Z_j}{|\mathbf{r}_i - \mathbf{R}_j|} \qquad (3.1)$$

Here $\mathbf{r}_i$ and $\mathbf{R}_i$ are the electronic and nuclear position vectors in a laboratory, or space-fixed, reference frame. Molecules are distinguishable from other many-body systems (atoms, nuclei) because their dynamics involves two, essentially different types of particles: the light electrons and the heavy nuclei ($m \ll M_i$). Since all Coulomb potentials occurring in H are of the same order of magnitude, we can say, classically speaking, that the electronic motion is much faster than that of the nuclei. From a traditional point of view one therefore pictures a molecule as an electronic cloud, formed by the "swiftly" revolving electrons, in which the "sluggish" nuclei are embedded. During an appreciable amount of time the electrons will see the nuclei as practically fixed in space, that is, their quantum-mechanical state will be determined by the Coulombic force field created by point charges Z_i located at the instantaneous nuclear positions. As the nuclei move, the electronic cloud is dragged along and assumed not to lag behind. These arguments motivate what is commonly known as the separation of electronic and nuclear motion, which is expressed mathematically by writing the total Hamiltonian as the sum of the nuclear kinetic energy and the so-called electronic Hamiltonian:

$$H(r, R) = T(R) + H_{el}(r, R) \qquad (3.2)$$

According to the arguments above, the behavior of the electrons in the molecule is described by the eigenstates of $H_{el}(r, R)$ considered as an operator on electronic variables only

$$H_{el}(r, R) \varphi_n(r|R) = U_n(R) \varphi_n(r|R) \qquad (3.3)$$

The nuclear positions enter this electronic eigenvalue problem only as a parameterization of the geometry of the field of force felt by the electrons.

Consequently the eigenvalues and eigenfunctions $U_n(R)$ and $\varphi_n(r\,|\,R)$, labeled by the electronic quantum numbers n, depend parametrically on R. The fact that in $\varphi_n(r\,|\,R)$ the nuclear positions are not to be considered to be dynamical variables is emphasized by separating them from the electronic ones by a vertical bar.

Powerful methods have been developed for the solution of (3.3). Hartree-Fock theory, configuration interaction techniques, propagator methods, and other approaches have become disciplines by themselves. To use their results for the solution of the full Schrödinger equation, one must transfer electronic information to an equation governing the nuclear motion. This section considers several alternatives in supplementing the electronic eigenvalue problem, which is assumed to be solved, with a nuclear motion equation or set of equations. *We consider this approach to the solution of the molecular Schrödinger equation to be a methodological definition of electronic and nuclear motion separation.*

We express all quantities in atomic units but, for obvious reasons, we keep the notation m for the electron mass.

B.   The Born-Oppenheimer Perturbation Theory

The fundamental paper of M. Born and J. Oppenheimer,[22] published in 1927, is probably the most quoted reference in the molecular physics literature. It has had a tremendous impact on the development of the chemistry and physics of molecules and solids within the framework of quantum mechanics. The paper provides a clear quantum-mechanical foundation for concepts such as molecular structure, potential energy surfaces, force constants, and moments of inertia, which were introduced earlier on a purely classical or semiclassical basis.[23-25] From the mathematical point of view it is a prime example of (singular) Schrödinger perturbation theory.[26] In the following summary we closely follow Born's exposition in his book on lattice dynamics.[27] Afterwards we focus on another merit of Born-Oppenheimer perturbation theory (BO-PT), namely, its capability in the analysis and comparison of the commonly used approximation schemes for the computation of molecular energy levels.

To simplify the equations that follow, we rewrite the nuclear kinetic energy symbolically:

$$T = -\sum \frac{1}{2M} \frac{\partial^2}{\partial R^2} \qquad (3.4)$$

Because of the large mass of the nuclei, it is reasonable to assume that the nuclear kinetic energy operator is "small" compared to the electronic Hamiltonian. In the total Hamiltonian (3.2) we can therefore regard T as a perturbation of H_{el}. The question of how to introduce an expansion parameter in H was solved by Born and Oppenheimer, who argued that this quantity had to be some power of the ratio m/M_0, where M_0 is the average nuclear mass in the system. They found that the correct choice is

$$\kappa = \left(\frac{m}{M_0} \right)^{1/4}, \qquad M_0 = \frac{(\Sigma M)}{N} \qquad (3.5)$$

N being the number of nuclei in the system under consideration. In the literature little explanation for the particular power $\frac{1}{4}$ is given. In fact, one often gets the impression that this is some kind of "magic number." There is, however, no mystery about it. The value $\frac{1}{4}$ is simply the largest one possible such that on expansion of the total Hamiltonian, only integer powers of the expansion parameter appear. In most cases (3.5) leads to $\kappa \cong \frac{1}{10}$. This number is typical for a molecule containing some hydrogen atoms (e.g., CH_4). For light diatomics (e.g., H_2, D_2, hydrides) κ is somewhat bigger, whereas for compounds consisting solely of heavier atoms it may be smaller than $\frac{1}{10}$.

The derivation of a perturbation theory formalism for the full molecular Schrödinger equation requires the following basic assumption: *the nuclear motion is confined to a small vicinity of a certain preferred nuclear configuration, say R_0*. To observe more closely what happens around R_0, one can introduce the transformation, sometimes called "microscopic canonical transformation,"[26]

$$R = R_0 + \kappa u \qquad (3.6)$$

Indeed, since $\kappa \ll 1$, expressing the problem in terms of the scaled nuclear variables u allows a "close-up" study of the nuclear motion. In the following we shall see that consistency of the PT uniquely determines the primary configuration R_0.

We can now expand the electronic Hamiltonian, its eigenvalues, and its eigenfunctions around R_0 using (3.6). We obtain

$$U(R) = U(R_0 + \kappa u) = U^{(0)} + \kappa U^{(1)}(u) + \kappa^2 U^{(2)}(u) + \cdots \qquad (3.7)$$

$$\varphi(r|R) = \varphi(r|R_0 + \kappa u) = \varphi^{(0)}(r) + \kappa \varphi^{(1)}(r|u) + \kappa^2 \varphi^{(2)}(r|u) + \cdots \qquad (3.8)$$

$$H_{el}(r, R) = H_{el}(r|R_0 + \kappa u) = H_{el}^{(0)}(r) + \kappa H_{el}^{(1)}(r, u) + \kappa^2 H_{el}^{(2)}(r, u) + \cdots \qquad (3.9)$$

where all functions and operators are homogeneous, of the degree indicated by the superscripts, in u. Since we are concentrating on a particular electronic state, the electronic quantum numbers n are dropped. Inserting these expansions into the electronic eigenvalue problem and identifying κ-powers gives

$$\left[H_{el}^{(0)} - U^{(0)} \right] \varphi^{(0)} = 0 \tag{3.10}$$

$$\left[H_{el}^{(0)} - U^{(0)} \right] \varphi^{(1)} = - \left[H_{el}^{(1)} - U^{(1)} \right] \varphi^{(0)} \tag{3.11}$$

$$\left[H_{el}^{(0)} - U^{(0)} \right] \varphi^{(2)} = - \left[H_{el}^{(1)} - U^{(1)} \right] \varphi^{(1)} - \left[H_{el}^{(2)} - U^{(2)} \right] \varphi^{(0)} \tag{3.12}$$

$$\dots \quad = \quad \dots$$

As for the nuclear kinetic energy operator, one observes that since $\partial / \partial R = (1/\kappa)\partial/\partial u$,

$$T = - \frac{m}{M_0} \sum \frac{M_0}{M} \frac{1}{2m} \frac{\partial^2}{\partial R^2} \tag{3.13}$$

$$T = \kappa^2 T^{(2)} \quad \text{with} \quad T^{(2)} = - \frac{1}{2m} \sum \frac{M_0}{M} \frac{\partial^2}{\partial u^2} \tag{3.14}$$

Thus it seems that the nuclear kinetic energy contributes a single second-order term to the Hamiltonian. The situation is not quite as simple as this because we have neglected the possibility of translation and rotation of the system. For these motions one cannot assume that the nuclei are confined to the immediate neighborhood of R_0. If one introduces explicitly three translational coordinates, three Euler angles for the rotational motion, and $3N - 6$ variables describing the vibrational motion, it can be shown that translations can be separated off and that the proper expression for T is[22]

$$T = \kappa^2 T^{(2)} + \kappa^3 T^{(3)} + \kappa^4 T^{(4)} + \cdots \tag{3.15}$$

where $T^{(2)}$ corresponds to purely vibrational motion while $T^{(3)}$ and $T^{(4)}$ contain vibration-rotation couplings and rotational terms. We neglect the effect of rotational motion at present and postpone the problems of rotational invariance and vibration-rotation interaction for later sections. This does not, however, affect the conclusions that follow. It should also be observed that for the treatment of solids the extra terms in (3.15) are redundant.

Combining (3.9) and (3.14) we can now write the total Hamiltonian as

$$H = H_{el}^{(0)} + \kappa H_{el}^{(1)} + \kappa^2 \left[H_{el}^{(2)} + T^{(2)} \right] + \kappa^3 H_{el}^{(3)} + \cdots \tag{3.16}$$

It will turn out that when acting on the actual molecular wave functions, all operators in this series can be considered to be of the same order of magnitude. We now look for solutions of Schrödinger's equation of the form

$$E = E^{(0)} + \kappa E^{(1)} + \kappa^2 E^{(2)} + \cdots \tag{3.17}$$

$$\Psi(r, u) = \Psi^{(0)}(r, u) + \kappa \Psi^{(1)}(r, u) + \kappa^2 \Psi^{(2)}(r, u) + \cdots \tag{3.18}$$

A procedure similar to the one that led to (3.10) to (3.12) now gives

$$\left[H_{el}^{(0)} - E^{(0)} \right] \Psi^{(0)} = 0 \tag{3.19}$$

$$\left[H_{el}^{(0)} - E^{(0)} \right] \Psi^{(1)} = - \left[H_{el}^{(1)} - E^{(1)} \right] \Psi^{(0)} \tag{3.20}$$

$$\left[H_{el}^{(0)} - E^{(0)} \right] \Psi^{(2)} = - \left[H_{el}^{(1)} - E^{(1)} \right] \Psi^{(1)} - \left[H_{el}^{(2)} + T^{(2)} - E^{(2)} \right] \Psi^{(0)} \tag{3.21}$$

$$\cdots \quad = \quad \cdots$$

Comparison of (3.10) and (3.19) immediately identifies $E^{(0)}$ with $U^{(0)}$ and $\varphi^{(0)}(r) = \varphi(r|R_0)$ as a solution of (3.19). However, it is clear that since $H_{el}^{(0)}$ does not operate on u, one can always multiply $\varphi^{(0)}$ with an arbitrary function of u such that in general

$$\Psi^{(0)}(r, u) = \varphi^{(0)}(r) F^{(0)}(u) \tag{3.22}$$

It will be shown that $F^{(0)}(u)$ is uniquely determined by higher order equations.

For the first-order equation (3.20) to have a solution, the inhomogeneous part [right-hand side of (3.20)] must be orthogonal to the solution of the homogeneous equation. This solubility condition gives

$$\langle \varphi^{(0)} | H_{el}^{(1)} - E^{(1)} | \Psi^{(0)} \rangle_r = F^{(0)} \langle \varphi^{(0)} | H_{el}^{(1)} - E^{(1)} | \varphi^{(0)} \rangle_r = 0 \tag{3.23}$$

where the scalar products refer to integration over the electron coordinates only. On the other hand, multiplying (3.11) to the left with $\varphi^{(0)}(r)$ and integrating over r, one finds that

$$\langle \varphi^{(0)} | H_{el}^{(1)} - U^{(1)} | \varphi^{(0)} \rangle_r = - \langle \varphi^{(0)} | H_{el}^{(0)} - U^{(0)} | \varphi^{(1)} \rangle_r = 0 \tag{3.24}$$

Combining the last two series of equalities, we can conclude that

$$U^{(1)} = U^{(1)}(u) = E^{(1)} \tag{3.25}$$

Since $U^{(1)}(u)$ is a linear homogeneous function in u while $E^{(1)}$ is a constant, this is clearly impossible unless

$$U^{(1)}(u) = \sum_i \left[\frac{\partial U}{\partial R_i} \right]_{R_0} u_i = 0 \tag{3.26}$$

that is, the preferred nuclear configuration in the vicinity of which the nuclei are localized must correspond to an extremum on the potential energy surface $U(R)$

$$\left[\frac{\partial U}{\partial R_i} \right]_{R_0} = 0 \tag{3.27}$$

With this choice for R_0, which is the result of a consistency requirement for the PT, the first-order energy correction vanishes:

$$E_1 = 0 \tag{3.28}$$

Taking this into account, on comparison of (3.11) and (3.20) one sees that $\varphi^{(1)}(r, u)F^{(0)}(u)$ is a particular solution of (3.20) to which we can always add an arbitrary solution of the homogeneous problem. Therefore

$$\Psi^{(1)}(r, u) = \varphi^{(1)}(r, u)F^{(0)}(u) + \varphi^{(0)}(r)F^{(1)}(u) \tag{3.29}$$

where $F^{(1)}(u)$ also remains undetermined.

After manipulation of the second-order equations (3.11) and (3.20) (see Ref. 27), one is led to the solubility condition

$$\langle \varphi^{(0)} | T^{(2)} + U^{(2)} - E^{(2)} | \Psi^{(0)} \rangle_r = 0 \tag{3.30}$$

$$[T^{(2)} + U^{(2)} - E^{(2)}] F^{(0)} \langle \varphi^{(0)} | \varphi^{(0)} \rangle_r = 0 \tag{3.31}$$

which gives us the equation

$$[T^{(2)} + U^{(2)}(u) - E^{(2)}] F^{(0)}(u) = 0 \tag{3.32}$$

Multiplied to the left with κ^2, (3.32) describes the motion of the nuclei in the potential $\kappa^2 U^{(2)}(u)$. Since $U^{(2)}(u)$ is a homogeneous quadratic function in u, this motion will consist of harmonic vibrations about the configuration R_0. For these oscillations to be stable, $U^{(2)}(u)$ must be positive definite; that is, the extremum R_0 must be a local minimum of the electronic eigenvalue $U(R)$. After a transformation to normal coordinates (3.32) reduces to a sum of one-dimensional oscillators such that the second-order energy $\kappa^2 E^{(2)}$ is a sum of oscillator quanta. It is still rather easy to obtain the second-order term in the wave function. The result reads

$$\Psi^{(2)}(r, u) = \varphi^{(2)}(r, u)F^{(0)}(u) + \varphi^{(1)}(r, u)F^{(1)}(u) + \varphi^{(0)}(r)F^{(2)}(u)$$

(3.33)

If one wants to proceed to higher orders the procedure becomes rather tedious. We therefore quote some results that can be found in Ref. 27.

Mathematical manipulation of the two sets of equations following (3.10) and (3.18) shows that the wave function correction of order k can be written as

$$\Psi^{(k)}(r, u) = \sum_{l=0}^{k} \varphi^{(l)}(r, u)F^{(k-l)}(u) + \varepsilon^{(k)}(r, u)$$

(3.34)

where $k = 3, 4, \ldots$. Here the $\varepsilon^{(k)}(r, u)$ are functions that do not depend on the electronic state through terms of the form $\varphi^{(l)}F^{(k-l)}$. The equations satisfied by the $\varepsilon^{(k)}$ and $F^{(k)}$ are inhomogeneous differential equations, in particular, the one for $\varepsilon^{(3)}$ reads

$$\left[H_{\text{el}}^{(0)} - U^{(0)} \right] \varepsilon^{(3)} = \frac{1}{m} \sum \left(\frac{M_0}{M} \right) \frac{\partial \varphi^{(1)}}{\partial u} \frac{\partial F^{(0)}}{\partial u}$$

(3.35)

In the derivation of these results, it is assumed that the electronic states are real and normalized to one for all R, that is,

$$\langle \varphi(R)|\varphi(R)\rangle_r = 1$$

(3.36)

This convention also is respected in the following sections.

It should be stressed that the BO-PT scheme is valid only for quantum states in which the nuclei are confined near local minima of the electronic eigenvalue. The situation is illustrated schematically in Fig. 1. Levels near R_0 and R_0' can be described within the BO framework, provided the tunneling rate between the potential wells is small. The procedure necessarily breaks down for energy levels situated in the shaded region. We are thus restricting ourselves at present low-lying levels of semirigid molecules. Extension to highly excited states and nonrigid systems are included in some sections.

As far as actual computations are concerned, BO-PT is of limited value. To reproduce molecular energies to an accuracy of $\pm 10^{-k}$ a.u., for example, one must carry out the scheme to kth order if $\kappa \approx 10^{-1}$. This means that a moderate experimental accuracy of ± 1 cm$^{-1} \cong \pm 10^{-6}$ a.u. already requires sixth-order BO-PT. In view of the complexity of the PT equations

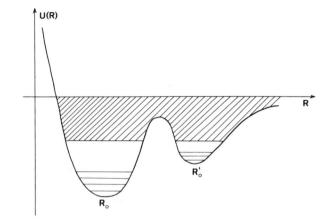

Fig. 1. Schematic illustration of energy ranges in within which BO-PT is valid.

at this order, especially when taking account of rotational and rotation-vibration features, this is not feasible. In fact, it is a general problem associated with PT that highly accurate results can be obtained in principle, but only at the cost of solving increasingly difficult equations. On the other hand, variational principles often lead to simpler equations but have the drawback that their results are not accompanied by definite error bars. *Both schemes may be complementary if the variational wave functions and energies can be shown to coincide to a certain order with those furnished by PT. In this case one may compute quantities variationally and decide on their accuracy by PT arguments.* This is how we use BO-PT in the analysis of variational approaches to the molecular Schrödinger equation.

C. The Adiabatic Approximation

If one neglects the terms $\varepsilon^{(k)}(r, u)$ the wave function reduces to

$$\varphi(r|R)\left[F^{(0)}(u) + \kappa F^{(1)}(u) + \kappa^2 F^{(2)}(u) + \cdots \right] \qquad (3.37)$$

This expression is the basis of what is commonly known as the "adiabatic approximation."[27] The term "adiabatic," used mostly in connection with problems in time-dependent PT, refers here to a decoupling of the electronic and nuclear motion. Indeed, in a wave function of the form (3.37) the motion of the electrons is affected by that of the nuclei solely through a change of the nuclear positions in $\varphi(r|R)$. In classical terms, this is frequently expressed by saying that the electrons adapt themselves instantaneously to a modification of the nuclear configuration. The actual, quantum-mechanical behavior of the nuclei is described by a function $F(R)$ consisting of the terms in square brackets in (3.37).

To optimally determine the nuclear motion, one can start from the ansatz

$$\Psi(r, R) \cong \varphi(r|R)F(R) \qquad (3.38)$$

and apply the variational principle, that is, minimize the energy expectation value

$$E = \frac{\langle \varphi F | H | \varphi F \rangle}{\langle \varphi F | \varphi F \rangle} \qquad (3.39)$$

with respect to $F(R)$. Assuming that the electronic state is normalized as in (3.36), this leads to the eigenvalue problem

$$\left[T + U(R) + \langle \varphi(R) | T | \varphi(R) \rangle_r \right] F_{AD}(R) = E_{AD} F_{AD}(R) \qquad (3.40)$$

where the adiabatic nuclear motion functions $F_{AD}(R)$ are square integrable in R. According to this equation the nuclei move in a potential

$$U_{AD}(R) = U(R) + \Delta U_{AD}(R) \qquad (3.41)$$

$$\Delta U_{AD}(R) = \langle \varphi(R) | T | \varphi(R) \rangle_r \qquad (3.42)$$

made up from the electronic eigenvalue $U(R)$ and the so-called adiabatic correction $\Delta U_{AD}(R)$. It should be observed that although the nuclear positions were considered to be parameters in the initial stage (3.3), they are again treated as dynamical variables in the variational principle above. It is this subtle switch that causes the appearance of the term $\Delta U_{AD}(R)$.

The procedure described above is a well-defined approximation scheme that leads to the results

$$E \cong E_{AD} \quad \text{and} \quad \Psi(r, R) \cong \Psi_{AD}(r, R) = \varphi(r, R)F_{AD}(R) \qquad (3.43)$$

How good is the adiabatic approximation? This question remains to be answered. This is most conveniently done by applying BO-PT, not only to the full Schrödinger equation, but to the adiabatic eigenvalue problem (3.40) also. It is readily verified that the adiabatic nuclear Hamiltonian can be expanded as follows:

$$H_{AD} = T + U(R) + \Delta U_{AD}(R) \qquad (3.44)$$

$$H_{AD} = U^{(0)} + \kappa^2 [T^{(2)} + U^{(2)}] + \kappa^3 U^{(3)} + \kappa^4 [U^{(4)} + \Delta U_{AD}^{(0)}] + \cdots \qquad (3.45)$$

The adiabatic correction term starts contributing at fourth order through the term

$$\Delta U_{AD}^{(0)} = -\frac{1}{2m} \Sigma \left(\frac{M_0}{M} \right) \left\langle \varphi^{(0)} \left| \frac{\partial^2}{\partial u^2} \right| \varphi^{(2)} \right\rangle_r \tag{3.46}$$

The adiabatic energy and nuclear motion function can now be written in the form

$$E_{AD} = E_{AD}^{(0)} + \kappa E_{AD}^{(1)} + \kappa^2 E_{AD}^{(2)} + \cdots \tag{3.47}$$

$$F_{AD}(R) = F_{AD}^{(0)}(u) + \kappa F_{AD}^{(1)}(u) + \kappa^2 F_{AD}^{(2)}(u) + \cdots \tag{3.48}$$

and substituted into the adiabatic eigenvalue problem (3.40). One immediately sees that $E_{AD}^{(0)} = U^{(0)}$ and $E_{AD}^{(1)} = 0$, while further identification of κ powers gives

$$[T^{(2)} + U^{(2)} - E^{(2)}] F_{AD}^{(0)} = 0 \tag{3.49}$$

$$[T^{(2)} + U^{(2)} - E^{(2)}] F_{AD}^{(1)} = -[U^{(3)} - E^{(3)}] F_{AD}^{(0)} \tag{3.50}$$

$$[T^{(2)} + U^{(2)} - E^{(2)}] F_{AD}^{(2)} = -[U^{(3)} - E^{(3)}] F_{AD}^{(1)} - [U^{(4)} + \Delta U_{AD}^{(0)} - E^{(4)}] F_{AD}^{(0)} \tag{3.51}$$

$$\cdots = \cdots$$

These equations are the ones obtained by Born and Huang[27] for the functions $F^{(0)}$, $F^{(1)}$, and $F^{(2)}$ in the original BO-PT. That $F^{(l)} = F_{AD}^{(l)}$ for $l = 0, 1, 2$ shows that the adiabatic approximation (i.e., the assumption of a product-type wave function and the application of the variational principle to the nuclear motion function) does not distort the low-order contributions in κ to the exact wave function. All terms up to and including second order coincide, such that we can state that

$$\Psi = \Psi_{AD} + O(\kappa^3) \tag{3.52}$$

Because the adiabatic approximation is variational, we have for the energy eigenvalues

$$E = E_{AD} + O(\kappa^6) \tag{3.53}$$

Similarly, as in the BO harmonic approximation (3.32), one can picture the nuclei in the adiabatic model as moving in an effective potential, namely, $U_{AD}(R)$. However, the presence of the nonproduct term $\varepsilon^{(3)}(r, u)$ in the exact wave function forces us to give up this simplifying feature if we aim at energies and eigenfunctions more accurate than the adiabatic ones. Thus the concept of a potential energy surface has its limitations. For example, "*it is not permissible to formally treat the dynamics of the nuclei on the basis of a potential function containing fifth or higher order terms in κ.*"[27]

According to the definition in Section III.A, the adiabatic approximation is indeed based on the separation of electronic and nuclear motion. The information transferred from the electronic eigenvalue problem to the adiabatic nuclear motion equation (3.40) consists of the potential energy surface $U(R)$ and the adiabatic correction term (3.42). Sometimes a further simplification is introduced by dropping $\Delta U_{AD}(R)$. Computationally this is motivated by the difficulty of evaluating the matrix element (3.42),[28] especially for high-quality electronic states that have a complicated dependence on R. Neglecting the adiabatic correction is sometimes referred to as the crude adiabatic or *clamped nuclei approximation.*[29] In principle, since $\Delta U_{AD}(R)$ is of order κ^4, the energies of the resulting nuclear motion equation may contain errors of the same order of magnitude. The situation is more favorable if one considers transition energies, since the κ^4 contribution of $\Delta U_{AD}(R)$ is the constant (3.46), which does not affect energy differences. It is also well known that the ground-state energy in the clamped-nuclei approximation is a lower bound to the exact value.[30]

D.   The Generator Coordinate Approximation

It is well known[6] that one of the qualities of the GCM is that many methods can be formulated as a GC procedure through a specific choice of intrinsic states. The adiabatic approximation is a typical example. Indeed, (3.43) can easily be rewritten as follows:

$$\Psi_{AD}(r, R) = \int F_{AD}(\alpha)\varphi(r|\alpha)\delta(R - \alpha)\, d\alpha = \varphi(r|R)F_{AD}(R) \quad (3.54)$$

This trivial identity shows that the adiabatic wave functions are in fact GC states in which the nuclear positions play the role of GCs, the adiabatic nuclear motion functions are the weight functions, and the intrinsic states are the so-called "fixed-nuclei" states[16]

$$\chi_\delta(r, R|\alpha) = \varphi(r|\alpha)\delta(R - \alpha) \quad (3.55)$$

These functions describe the electrons as moving in a Coulombic field determined by the configuration α at which the nuclei are strictly localized by the Dirac delta function $\delta(R - \alpha)$. As such, they clearly demonstrate that in the adiabatic approximation the electronic state is deformed progressively as the nuclei change positions. In the superposition each of the nuclear configurations is then assigned a variationally determined weight factor $F_{AD}(\alpha)$.

Considered as GC basis states, the functions (3.55) are subject to some criticism. First, they are not square integrable, and second, they violate the position-momentum uncertainty relations for the nuclear variables. *In the usual product form of the adiabatic wave functions these problems apparently do not show up. They are, however, disguised by considering the nuclei as classical point charges, when it comes to interpreting the electronic part of the wave functions and treating them as quantum-mechanical, finite mass particles only at the level of the nuclear motion equation.*

To remove the above-mentioned problems and at the same time to generate nonproduct-type molecular states, one may replace $\delta(R-\alpha)$ in (3.54) by a function having a small, but finite width W. A new GC trial function of the form

$$\Psi(r, R) \cong \int F(\alpha)\varphi(r|\alpha)\Phi\left(\frac{|R-\alpha|}{W}\right) d\alpha \qquad (3.56)$$

is then obtained. To easily recover the adiabatic trial states, we assume that Φ is chosen such that under the proper normalization condition we have*

$$\lim_{W\to 0} \Phi\left(\frac{|R-\alpha|}{W}\right) = \delta(R-\alpha) \qquad (3.57)$$

We leave the explicit form of Φ undetermined because it will later prove to be of secondary importance. The limit $W\to 0$ is referred to as the *adiabatic limit*, since in this process the GC trial state (3.56) changes into the adiabatic one (3.38).

It is clear that by modifying the Dirac delta function to a function with a nonzero width, the integrations in (3.56) can no longer be performed. The new GC trial state is not of the product form (3.38), since the electronic and nuclear variables are coupled through the α integration. The intrinsic states

$$\chi_\Phi(r, R|\alpha) = \varphi(r|\alpha)\Phi\left(\frac{|R-\alpha|}{W}\right) \qquad (3.58)$$

are no longer in contradiction to Heisenberg's principle. Indeed if $W\neq 0$, there is a small uncertainty, of order W, on the nuclear positions.

Application of the GC variational principle to (3.56) yields the integral equation

$$\int [T(\alpha, \beta) + H_{el}(\alpha, \beta)] F_{GC}(\beta) d\beta = E_{GC} \int \Delta(\alpha, \beta) F_{GC}(\beta) d\beta \qquad (3.59)$$

---

*We consider only one W-parameter in analogy with the use of a single mass ratio in BO-PT.

where the integral kernels are given by the matrix elements of the nuclear kinetic energy, the electronic Hamiltonian, and the unit operator with respect to basis states (3.58) at nuclear configurations α and β, that is,

$$T(\alpha, \beta) = \langle \chi_\Phi(\alpha) | T | \chi_\Phi(\beta) \rangle = \langle \varphi(\alpha) | \varphi(\beta) \rangle_r \langle \Phi(\alpha) | T | \Phi(\beta) \rangle_R$$
$$(3.60)$$

$$H_{el}(\alpha, \beta) = \langle \chi_\Phi(\alpha) | H_{el} | \chi_\Phi(\beta) \rangle = \langle \varphi(\alpha)\Phi(\alpha) | H_{el} | \varphi(\beta)\Phi(\beta) \rangle$$
$$(3.61)$$

$$\Delta(\alpha, \beta) = \langle \chi_\Phi(\alpha) | \chi_\Phi(\beta) \rangle = \langle \varphi(\alpha) | \varphi(\beta) \rangle_r \langle \Phi(\alpha) | \Phi(\beta) \rangle_R \qquad (3.62)$$

Equation 3.59 is the GC nuclear motion equation in which the superposition amplitudes $F_{GC}(\alpha)$ assume the role of nuclear motion functions. *As expected in a nonadiabatic theory, the dynamics of the nuclei is no longer governed by an effective potential function. The concept of a potential energy surface is here replaced by that of Wheeler integral kernels.*

The nonproduct-type wave functions associated with the GC eigenvalues E_{GC} are obtained by folding the (product-type) basis states (3.58) with the solutions F_{GC} of (3.59).

$$\Psi_{GC}(r, R) = \int F_{GC}(\alpha)\varphi(r|\alpha)\Phi\left(\frac{|R-\alpha|}{W}\right) d\alpha \qquad (3.63)$$

The analysis of these functions in terms of BO-PT is in two steps. First we observe that since we expect $\Phi(|R-\alpha|/W)$ to be close to $\delta(R-\alpha)$, the remainder of the integrand in (3.56) will vary slowly over the range where the nuclear intrinsic state is essentially different from zero (i.e., $R \cong \alpha$). This allows us to expand the product $\varphi(r|\alpha)F_{GC}(\alpha)$ in a Taylor series around $\alpha = R$ and to perform the GC integration term by term. Including terms up to second-order, one obtains the following approximate form of the GC wave function

$$\Psi_{GC}(r, R) \cong \varphi(r|R)F_{GC}(R) + \frac{W^2}{2} \sum \Phi^{(2)} \frac{\partial^2}{\partial R^2} \left[\varphi(r|R)F_{GC}(R)\right]$$
$$(3.64)$$

where $\Phi^{(2)}$ are the second moments of the nuclear intrinsic state defined by

$$\Phi_i^{(2)} = \int_{-\infty}^{+\infty} x_i^2 \Phi(|x|) \, dx \qquad (3.65)$$

Since for Φ to have a delta function limit, it is chosen as an even function, the odd moments vanish, explaining the absence of a linear term in (3.64).

This procedure shows that a GC wave function can be separated in an adiabatic, product-type state and a nonadiabatic correction. The latter consists of numerical factors, depending solely on the nuclear basis state, and nonproduct functions of r and R containing first- and second-order derivatives of the electronic state and the nuclear motion function. Several remarks concerning this nonadiabatic term should be made. First, it is consistent with the adiabatic limit, since it vanishes as $W \to 0$. However, as soon as $W \neq 0$ it can become important if the electronic states and/or the nuclear motion functions vary rapidly with the internuclear distances. This is known to happen in regions where potential energy surfaces approach each other and for highly excited vibration-rotation levels. From the perturbative treatment of coupled differential equations, these are well-known cases in which the adiabatic approximation breaks down.[31] In the absence of such a breakdown, the magnitude of the nonadiabatic term is dominated by the value of the width parameter W. Indeed, the only traces of the functional form of the nuclear intrinsic states in (3.64) are the second moments $\Phi^{(2)}$, which for reasonable choices of $\Phi(|R-\alpha|/W)$ (built from Gaussians, exponentials, etc.), are of the same order of magnitude.

To relate W, which thus far has been left undetermined, to the BO mass parameter, we proceed to the second step in the analysis of the GC wave function. In analogy to the expansion of the adiabatic nuclear motion function in κ, we assume the following form for the GC weight function:

$$F_{GC} = F_{GC}^{(0)} + \kappa F_{GC}^{(1)} + \kappa^2 F_{GC}^{(2)} + \cdots \qquad (3.66)$$

Using scaled nuclear coordinates and inserting κ-expansions of F_{GC} and the electronic state into (3.64), we can rewrite the nonadiabatic terms as

$$\frac{1}{2}\left(\frac{W}{\kappa}\right)^2 \sum \Phi^{(2)} \left[\varphi^{(0)} \frac{\partial^2 F_{GC}^{(0)}}{\partial u^2} + \kappa \left(\varphi^{(0)} \frac{\partial^2 F_{GC}^{(1)}}{\partial u^2} + \varphi^{(1)} \frac{\partial^2 F_{GC}^{(0)}}{\partial u^2} \right. \right.$$
$$\left. \left. + 2 \frac{\partial \varphi^{(1)}}{\partial u} \frac{\partial F_{GC}^{(0)}}{\partial u} \right) + O(\kappa^2) \right] \qquad (3.67)$$

The electronic functions in the first three terms in the square brackets have the proper dependence on u to be absorbed in the product-type part of Ψ_{GC} [first term in (3.64)]. Because $\partial \varphi_1/\partial u$ is no longer homogeneous in the u's, however, the last term is a truly nonadiabatic contribution. As such, combined with the factor $(W/\kappa)^2$, it should be of third order in κ. One verifies immediately that for the κ-dependence of the width parameter this implies that

$$W = O(\kappa^2) \qquad (3.68)$$

That is, the nonadiabaticity of the GC wave functions is of right order in κ, provided the width of the nuclear intrinsic state is of the order of the square of the Born-Oppenheimer perturbation parameter.

Clearly the conclusion reached, that is, (3.68), should not contradict the initial assumptions, namely, that we should check whether the product φF_{GC} is indeed slowly varying over a range of order W around $\alpha = R$. For the typical case $\kappa = 10^{-1}$ the nuclear intrinsic state has a width of approximately 10^{-2} a.u. As for the nuclear motion functions, one can associate with them widths W_F for which

$$W_F = (M_0 E_{\text{vib}})^{-1/2} = O(\kappa) \tag{3.69}$$

We therefore find that $W/W_F = O(\kappa)$ such that the F_{GC} are indeed smooth over the range where $\Phi(|R - \alpha|/W)$ is essentially nonzero. The same can be claimed for the electronic states, since even in a crossing region one rarely observes a drastic difference between $\varphi(R)$, and $\varphi(R + \Delta R)$ if $\Delta R \cong 10^{-2}$ a.u. Hence we can conclude that the analysis above is quite general.

Care must be taken with the foregoing prescription for W that one does not distort the lower order κ contributions to the wave function. Indeed if $W = O(\kappa^2)$, the first term in (3.67) will be of order κ^2. Comparing the exact, adiabatic, and generator coordinate states, it can be seen that if one puts

$$F_{GC}^{(0)} = F_{AD}^{(0)}, \qquad F_{GC}^{(1)} = F_{AD}^{(1)} \tag{3.70}$$

and

$$F_{GC}^{(2)} = F_{AD}^{(2)} - \frac{1}{2} \sum \Phi^{(2)} \frac{\partial^2 F_{AD}^{(0)}}{\partial u^2} \tag{3.71}$$

all terms, up to and including second-order ones, coincide. Equations 3.70 and 3.71 demonstrate that the GC nuclear motion function differs from the adiabatic one through second order in κ only:

$$F_{GC} = F_{AD} + O(\kappa^2) \tag{3.72}$$

We shall see later how one can take advantage of this similarity between the GC and adiabatic nuclear motion functions for the computation of molecular energy levels by way of GC perturbation theory.

Considered as methods based on the separation of electronic and nuclear motion, there are some basic differences between the GC and the adiabatic approximations. The correlation between electronic and nuclear variables, introduced into (3.63) by the wave packet-type structure of $\Phi(|R - \alpha|/W)$, results in an additional amount of electronic information transferred to the GC nuclear motion equation. Aside from the diagonal elements of the Hamiltonian kernels, equivalent to the adiabatic potential surface $U_{AD}(R)$, nondiagonal elements, which couple electronic states at different nuclear configurations, are taken into account. The increased sophistication of the GCM is also apparent from the expansion of the GC trial state around $R = \alpha$, the first terms of which are given by (3.64). Indeed, this shows that *not only $\varphi(r|R)$ but also its derivatives with respect to R enter the GC wave functions implicitly.*

As a measure of the correlation between electronic and nuclear variables in a wave function, the conditional probability

$$P_R(r) = \frac{|\Psi(r, R)|^2}{\langle \Psi(R)|\Psi(R)\rangle_r} \qquad (3.73)$$

has been suggested.[32] The $P_R(r)$ gives the probability distribution for the electrons if the nuclear configuration is known to be R. In the adiabatic case one easily computes

$$P_R^{AD}(r) = |\varphi(r|R)|^2 \qquad (3.74)$$

Thus the probability distribution of the electrons is that of the electronic state with the nuclei fixed at the positions R. The term $P_R^{AD}(r)$ is completely independent of the quantum-mechanical motion of the nuclei. This demonstrates, once more, the semiclassical nature of the nuclei in the adiabatic approximation. On the other hand, the GC probability $P_R^{GC}(r)$ contains the nuclear motion function F_{GC} explicitly. Therefore, the electronic distribution is directly influenced by the quantum-mechanical behavior of the nuclei.

How good is the GC approximation? In principle, this question could also be answered by applying the BO-PT to the GC eigenvalue problem. Because the exact (partial differential equation) and the GC (integral equation) eigenvalue problems are of a different nature, this analysis has not been carried out yet. It is, however, to be expected that the increased flexibility of the GC wave function allows a lowering of the energy compared to the adiabatic results. It is an open question whether this energy gain can be identified with a certain power of κ.

A partial answer to these questions can be given by applying GC perturbation theory. Indeed, the similarity between F_{GC} and F_{AD} may be exploited by considering the adiabatic approximation as a zeroth-order problem in GC perturbation theory. Since the adiabatic approximation was formulated as a GC procedure we can apply (2.52) to give

$$E_{GC} \cong E_{AD} + (F_{AD}|V|F_{AD}) - E_{AD}(F_{AD}|W|F_{AD}) \qquad (3.75)$$

in first-order PT. In this form, however, the proposed PT is of little use because the perturbations $V = H_{GC} - H_{AD}$ and $W = \Delta_{GC} - \Delta_{AD}$ are not known explicitly as a result of the delta function character of the adiabatic integral kernels. Fortunately, $E_{GC-PT}^{(1)}$ can be rewritten, using the GC kernels only, as

$$E_{GC-PT}^{(1)} = (F_{AD}|H_{GC}|F_{AD}) - E_{AD}(F_{AD}|\Delta_{GC}|F_{AD}) \qquad (3.76)$$

This enables us to compute the GC energies approximately, making use of the adiabatic results. Also if $E_{GC}^{(1)} < 0$, the theorem stated in Section II.C allows us to conclude that $E_{GC} < E_{AD}$. The importance of this result lies in the impossibility, since the adiabatic subspace is not contained in the GC one, of ever deciding on the relative magnitude of E_{AD} and E_{GC} by variational arguments alone.

E.   An Exactly Soluble Test Model

It is desirable that the adiabatic and generator coordinate approximations, described in the preceding sections, be tested on a simple model for which the exact solution is available. A problem that allows a rather detailed illustration of the ideas presented before is that of two coupled oscillators with unequal mass. The Hamiltonian is written as

$$H(r, R) = H_0(R) + h_0(r) + V(r, R) \qquad (3.77)$$

where

$$H_0 = -\frac{1}{2M}\frac{\partial^2}{\partial R^2} + \frac{k}{2}R^2 \qquad (3.78)$$

$$h_0 = -\frac{1}{2m}\frac{\partial^2}{\partial r^2} + \frac{k}{2}r^2 \qquad (3.79)$$

$$V = -\lambda k r R \qquad (3.80)$$

The coupling strength λ will be considered as a parameter ($|\lambda| \leq 1$). This Hamiltonian arises if one considers the quantum-mechanical behavior of a linear system of two point masses, m and M, and three weightless springs. The springs corresponding to the force constant k are connected to fixed points such that no translations or rotations enter. The variables r and R then refer to the displacements of the masses from their equilibrium positions.[33] The exact eigenvalues and eigenfunctions are obtained in a straightforward way by transforming H in two uncoupled oscillators with frequencies ω and Ω given by

$$\begin{aligned}\omega^2 \\ \Omega^2\end{aligned} = \frac{1}{2}(\omega_0^2 + \Omega_0^2) \pm \frac{1}{2}\left[(\omega_0^2 - \Omega_0^2)^2 + (2\omega_0\Omega_0\lambda)^2\right]^{1/2} \qquad \begin{aligned}(3.81a) \\ (3.81b)\end{aligned}$$

where $\omega_0 = (k/m)^{1/2}$ and $\Omega_0 = (k/M)^{1/2}$ are the frequencies of the uncoupled oscillators ($\lambda = 0$). The exact wave function, for the ground state, is of the form

$$\Psi(r,R) = \left(\frac{m\omega M\Omega}{\pi^2} \right)^{1/4} \exp\left[-\frac{1}{2}(ar^2 + brR + cR^2) \right] \qquad (3.82)$$

with

$$a = \tfrac{1}{2}m\left[(\omega + \Omega) + (\omega - \Omega)\cos 2\theta \right] \qquad (3.83)$$

$$b = (mM)^{1/2}(\omega - \Omega)\sin 2\theta \qquad (3.84)$$

$$c = \tfrac{1}{2}M\left[(\omega + \Omega) - (\omega - \Omega)\cos 2\theta \right] \qquad (3.85)$$

The angle θ is related to the coupling strength λ and the BO mass parameter $\kappa = (m/M)^{1/4}$ by

$$\tan 2\theta = \frac{2\lambda\kappa^2}{\kappa^4 - 1} \qquad (3.86)$$

If $m \ll M$, the present system can be considered to be a caricature of a diatomic molecule in which the Coulomb potentials have been replaced by harmonic oscillator interactions. It is easily seen that if $\kappa \ll 1$, one has $\omega_0 \cong \omega$, $\Omega_0 \cong \Omega$, and $\Omega \ll \omega$. The unperturbed oscillator frequencies ω_0 and Ω_0, and their exact counterparts ω and Ω, can be compared to the electronic and vibrational molecular energies, respectively. Similar arguments and a Hamiltonian analogous to (3.77) have been used by Kittel and Moshinsky[34] as a test of the clamped-nuclei approximation. Here we compare the exact results for the "vibrational band" built on the "ground electronic state" with those obtained in the various approximation schemes. We therefore consider the power series in κ of the exact quantities. The vibrational frequency Ω may be expanded as follows:

$$\Omega = \omega_0(1 - \lambda^2)^{1/2}\left[\kappa^2 - \tfrac{1}{2}\lambda^2\kappa^6 + O(\kappa^{10}) \right] \qquad (3.87)$$

The ground-state energy $E = \tfrac{1}{2}(\omega + \Omega)$ has the expansion

$$E = \tfrac{1}{2}\omega_0\left[1 + (1 - \lambda^2)^{1/2}\kappa^2 + \tfrac{1}{2}\lambda^2\kappa^4 - \tfrac{1}{2}\lambda^2(1 - \lambda^2)^{1/2}\kappa^6 + O(\kappa^8) \right] \qquad (3.88)$$

It is now illustrative to consider the clamped-nuclei and adiabatic approximations to this problem. Separating the electronic and nuclear motions (i.e., discarding the nuclear kinetic energy) leads to the "electronic Hamiltonian"

$$H_{el}(r, R) = -\frac{1}{2m}\frac{\partial^2}{\partial r^2} + \frac{k}{2}(r - \lambda R)^2 + \frac{k}{2}(1 - \lambda^2)R^2 \qquad (3.89)$$

The electronic eigenvalue problem is therefore that of a displaced harmonic oscillator. The electronic eigenvalue and eigenfunction for the ground electronic state are

$$U(R) = \frac{k}{2}(1 - \lambda^2)R^2 + \frac{\omega_0}{2} \qquad (3.90)$$

$$\varphi(r|R) = \left(\frac{m\omega_0}{\pi}\right)^{1/4} \exp\left[-\frac{1}{2}m\omega_0(r - \lambda R)^2\right] \qquad (3.91)$$

The adiabatic correction term is easily seen to be

$$\Delta U_{AD} = \tfrac{1}{4}\omega_0\lambda^2\kappa^4 \qquad (3.92)$$

Consistent with the analysis of Section III.C, this constant is of order κ^4. When added to (3.90), we see that the full adiabatic potential is quadratic in R such that the adiabatic nuclear motion equation also corresponds to a one-dimensional harmonic oscillator with frequency

$$\Omega_{AD} = \Omega_0(1 - \lambda^2)^{1/2} \qquad (3.93)$$

The ground-state vibrational wave function

$$F_{AD}(R) = \left(\frac{M\Omega_{AD}}{\pi}\right)^{1/4} \exp\left[-\frac{1}{2}M\Omega_{AD}R^2\right] \qquad (3.94)$$

multiplied by (3.91) gives the adiabatic ground state

$$\Psi_{AD}(r, R) = \left(\frac{m\omega_0 M\Omega_{AD}}{\pi^2}\right)^{1/4} \exp\left[-\frac{1}{2}(a_{AD}r^2 + b_{AD}rR + c_{AD}R^2)\right] \qquad (3.95)$$

where

$$a_{AD} = m\omega_0 \qquad (3.96)$$

$$b_{AD} = -2m\omega_0\lambda \qquad (3.97)$$

$$c_{AD} = m\omega_0(1 - \lambda^2)^{1/2}\left[\kappa^{-2} + \lambda^2(1 - \lambda^2)^{-1/2}\right] \qquad (3.98)$$

To compare with the exact result, the BO perturbation parameter must be introduced into the results above. Clearly the adiabatic frequency is

$$\Omega_{AD} = \omega_0 (1 - \lambda^2)^{1/2} \kappa^2 \qquad (3.99)$$

For the ground-state energy we find, first in the clamped nuclei approximation corresponding to the potential (3.90)

$$E_C = \tfrac{1}{2}(\omega_0 + \Omega_{AD}) = \tfrac{1}{2}\omega_0 \left[1 + (1 - \lambda^2)^{1/2} \kappa^2 \right] \qquad (3.100)$$

In the adiabatic approximation, corresponding to the potential (3.90) + (3.92), we get

$$E_{AD} = E_C + \Delta U_{AD} = \tfrac{1}{2}\omega_0 \left[1 + (1 - \lambda^2)^{1/2} \kappa^2 + \tfrac{1}{2}\lambda^2 \kappa^4 \right] \qquad (3.101)$$

We now turn to the discussion of the generator coordinate approximation. In view of its Gaussian character, we supplement the electronic ground state (3.91) with a Gaussian nuclear part to form the intrinsic state (normalized to one):

$$\chi(r, R | \alpha) = \left(\frac{2}{\pi w W} \right)^{1/2} \exp\left[-\left(\frac{r - \lambda \alpha}{w} \right)^2 \right] \exp\left[-\left(\frac{R - \alpha}{W} \right)^2 \right] \qquad (3.102)$$

where $w = (2/m\omega_0)^{1/2}$ and W is the (as yet undetermined) width parameter. With this choice of intrinsic state, the kernels can be calculated analytically. The overlap $\Delta(\alpha, \beta)$ is given by (2.56), while one obtains for the ratio $H(\alpha, \beta)/\Delta(\alpha, \beta)$ the expression

$$E_g + \frac{1}{2M_{GC}} \left[2S - 4S^2(\alpha - \beta)^2 \right] + \frac{M_{GC}\Omega_{GC}^2}{2} \left[\frac{1}{8S} + \left(\frac{\alpha + \beta}{2} \right)^2 \right] \qquad (3.103)$$

That is, the GC kernels for the present problem coincide exactly with the harmonic ones discussed in Section II.C. The scale factor S is given by

$$S = \frac{1}{2} \left[\left(\frac{1}{W} \right)^2 + \left(\frac{\lambda}{w} \right)^2 \right] \qquad (3.104)$$

Since we expect $W \ll w$, the value of S is largely determined by W. It is therefore useful to introduce the variable $\xi=(W/w)^2$. In terms of the function

$$\Xi(\xi)=\frac{\left(1+\lambda^2\xi\right)^2}{1+\lambda^2\xi^2\kappa^{-4}} \tag{3.105}$$

we can then write the effective GC mass and frequency as

$$M_{GC}=M\Xi(\xi) \tag{3.106}$$

$$\Omega_{GC}=\Omega_{AD}\left[\Xi(\xi)\right]^{-1/2} \tag{3.107}$$

Since the GC spectrum is given by (2.58), the latter is the GC approximation to the vibrational excitation energy (3.81b). Explicit calculation gives the constant E_g in (3.103):

$$E_g+\frac{S}{M_{GC}}=\frac{1}{2}\left(\frac{1}{mw^2}+\frac{1}{MW^2}\right)+\frac{1}{8}\left(m\omega_0^2w^2+M\Omega_0^2W^2\right) \tag{3.108}$$

The GC ground-state energy becomes, according to (2.58),

$$E_{GC}=\frac{1}{2}\omega_0\frac{1+\lambda^2\xi+\frac{1}{2}\lambda^2\left(\xi^2+\kappa^4\right)+\left(1-\lambda^2\right)^{1/2}\left(\kappa^4+\lambda^2\xi^2\right)^{1/2}}{1+\lambda^2\xi} \tag{3.109}$$

The corresponding ground-state weight function is (unnormalized)

$$F_{GC}(\alpha)=\exp\left[\frac{-\frac{1}{2}M_{GC}\Omega_{GC}\alpha^2}{1-\frac{M_{GC}\Omega_{GC}}{4S}}\right] \tag{3.110}$$

Expressions 3.107 and 3.109 provide an illustration of the adiabatic limit (3.57). Indeed, one verifies that since $W\to0$ implies $\xi\to0$, one has

$$\lim_{W\to0}E_{GC}=E_{AD}\quad\text{and}\quad\lim_{W\to0}\Omega_{GC}=\Omega_{AD} \tag{3.111}$$

It is worthwhile to observe that the adiabatic limit applies also to the effective mass

$$\lim_{W\to0}M_{GC}=M \tag{3.112}$$

The determination of ξ, hence of W, by minimizing E_{GC} leads to a quartic equation in ξ. If, since both ξ and κ are small, we expand E_{GC} in powers of ξ and κ^4 to second order and minimize, the optimum value obtained is $\xi = \kappa^4 + O(\kappa^6)$. On the other hand, $\xi = \kappa^4$ exactly minimizes Ω_{GC}. This demonstrates that the ξ-dependence of E_{GC} near its minimum is determined mainly by Ω_{GC}. With either choice for ξ we can conclude that the energetically optimal width parameters satisfy

$$W = O(\kappa^2) \tag{3.113}$$

a result obtained in Section III.D using general arguments. Because of obvious computational advantages, we will fix the width parameter by putting $W = w\kappa^2 = (2/M\omega_0)^{1/2}$. With this choice we can give the various GC quantities more explicitly. The vibrational frequency (3.107) then has the κ-expansion

$$\Omega_{GC} = \omega_0 (1 - \lambda^2)^{1/2} \left[\kappa^2 - \tfrac{1}{2}\lambda^2\kappa^6 + O(\kappa^{10}) \right] \tag{3.114}$$

while for the GC ground-state energy we get

$$E_{GC} = \tfrac{1}{2}\omega_0 \left[1 + (1 - \lambda^2)^{1/2}\kappa^2 + \tfrac{1}{2}\lambda^2\kappa^4 - \tfrac{1}{2}\lambda^2(1 - \lambda^2)^{1/2}\kappa^6 + O(\kappa^8) \right] \tag{3.115}$$

Integration of (3.110) with (3.102) can be done analytically and yields the GC ground-state wave function in the form

$$\Psi_{GC}(r, R) = N_{GC} \exp\left[-\tfrac{1}{2}\left(a_{GC}r^2 + b_{GC}rR + c_{GC}R^2 \right) \right] \tag{3.116}$$

where the scale factors to relevant order in κ are given by

$$a_{GC} = m\omega_0 \left[1 + O(\kappa^4) \right] \tag{3.117}$$

$$b_{GC} = -2m\omega_0\lambda \left[1 - (1 - \lambda^2)^{1/2}\kappa^2 + O(\kappa^4) \right] \tag{3.118}$$

$$c_{GC} = m\omega_0(1 - \lambda^2)^{1/2} \left[\kappa^{-2} + \lambda^2(1 - \lambda^2)^{-1/2} - \tfrac{3}{2}\lambda^2\kappa^2 + O(\kappa^4) \right] \tag{3.119}$$

Comparison of the κ-expansions for the ground-state energies and excitation frequencies with the exact results shows that the errors introduced by the various approximation schemes are those given in Table I.

TABLE I

Comparison of Exact and Approximate Energies and Excitation Frequencies
in the Coupled Oscillator Model

Approximation scheme	Energy	Excitation frequency
Clamped nuclei	$\Delta E_C = O(\kappa^4)$	$\Delta \Omega_C = O(\kappa^6)$
Adiabatic	$\Delta E_{AD} = O(\kappa^6)$	$\Delta \Omega_{AD} = O(\kappa^6)$
Generator coordinate	$\Delta E_{GC} = O(\kappa^8)$	$\Delta \Omega_{GC} = O(\kappa^{10})$

The results for the ground-state energy in the clamped-nuclei and adiabatic approximations confirm the theoretically predicted error bounds for these methods. The GC energy is seen to coincide with the exact expression (3.88) to eighth order in κ. As observed earlier, there is no direct proof of the generality of this result.

As for the excitation energies, the clamped-nuclei approximation works as well as the adiabatic one because the adiabatic correction term is canceled in making energy differences. Apparently, a similar phenomenon occurs in the GC approximation, where the excitation energy agrees with the exact one to a higher order in κ than the absolute energies. The precise nature of this effect (i.e., its explanation by BO analysis) is lacking.

The wave functions (3.82), (3.95), and (3.116) can consistently be compared through their κ-expansions if the scaled variable $u = R/\kappa$ is introduced. Because all functions are of the same exponential form, their similarity can be discussed by considering the differences of the scale factors a, b, and c, in the respective approximations, with the exact ones. The expansions of the exact scale factors in κ agree with the GC expressions (3.117) and (3.118) to the order listed. Therefore one easily checks that

$$\frac{\Psi_{AD}(r,u)}{\Psi(r,u)} = 1 - \frac{1}{2}\left[\Delta a_{AD} r^2 + \Delta b_{AD} ru + \Delta c_{AD} u^2\right] + \cdots \quad (3.120)$$

$$\frac{\Psi_{GC}(r,u)}{\Psi(r,u)} = 1 - \frac{1}{2}\left[\Delta a_{GC} r^2 + \Delta b_{GC} ru + \Delta c_{GC} u^2\right] + \cdots \quad (3.121)$$

where

$$\Delta a_{AD} = O(\kappa^4), \qquad \Delta a_{GC} = O(\kappa^4) \quad (3.122)$$

$$\Delta b_{AD} = O(\kappa^3), \qquad \Delta b_{GC} = O(\kappa^5) \quad (3.123)$$

$$\Delta c_{AD} = O(\kappa^4), \qquad \Delta c_{GC} = O(\kappa^6) \quad (3.124)$$

The adiabatic wave functions therefore agree to third order with the exact ones, a result consistent with the theoretical analysis in Section III.C. It is the deviation in the coefficient b_{AD} (i.e., the coupling between "electron" and "nuclear" coordinates) that causes the third-order error. In the GCM the proper κ^3-terms are included and the wave function differs from the exact one through fourth-order contributions only. In addition, we observe that it is now the pure r-dependence that is responsible for the error term. This seems logical, since the electronic behavior is not directly affected by the GC variational principle. This is illustrated by noting that Δa_{AD} and Δa_{GC} are of the same order of magnitude. On the other hand, the GC coupling terms and the purely "nuclear" factors show a marked improvement over the adiabatic ones. It is clear that this must be attributed to the additional flexibility of the GC trial state. Here again, however, we cannot now provide any arguments supporting the general validity of this result.

A comparison of the exponents in the adiabatic wave function (3.94) and the GC weight function (3.110) shows that

$$\frac{\tfrac{1}{2}M_{GC}\Omega_{GC}}{1-\dfrac{M_{GC}\Omega_{GC}}{4S}} = \frac{1}{2}M\Omega_{AD}\left(1+O(\kappa^2)\right) \qquad (3.125)$$

This implies that for the present model the functions F_{AD} and F_{GC} indeed differ through second-order terms in κ, as demonstrated at the end of Section III.D.

In summary, *the general statements made in Sections III.B, III.C, and III.D are all confirmed by the present model calculations.* This illustrates the observation made by Born and Oppenheimer that their results do not depend on the special (e.g., Coulomb) form of the interactions. Whether the result that the κ^6-term in the ground-state energy is rendered exactly by the GCM is model dependent, is an open problem. A similar question arises in connection with the excitation frequency, which agrees to order $O(\kappa^{10})$. That the adiabatic and clamped-nuclei energies are not better than expected shows that the model is unbiased in its predictions. This supports our belief in the relevance of the GCM as applied to more realistic systems.

F.   Interplay Between Theory and Experiment

This section compares theory and experiment at the level of energy eigenvalues. We have outlined in detail how to construct nuclear motion equations of various types that can provide approximate eigenenergies. Experimentally, however, one measures transitions between various levels (i.e., differences of energies). Theoretically computed and experimental transition frequencies are seldom compared (a notable exception is found in Ref.

35). Instead experimentalists rely on so-called term formulas to para-meterize their data. These are expressions that have a definite dependence on vibration-rotation quantum numbers but leave a number of constants undetermined. The latter can then be fitted to reproduce the observed spectral lines. Within the adiabatic approximation this is a well-established procedure.

The derivation of the adiabatic term formulas was pioneered by Dunham[36] (diatomic molecules), Howard and Wilson,[37] and Hill and Van Vleck[38] (polyatomic molecules). The basic principle is to expand the potential energy surface around the equilibrium configuration R_0 and separate the nuclear kinetic energy into vibrational, rotational, and vibration-rotation interaction parts (after separating off the center of mass motion). This technique, often referred to as the vib-rotor model, introduces a number of spectroscopic constants such as normal vibration frequencies, anharmonicity constants, moments of inertia, centrifugal distortion coefficients, and Coriolis coupling constants. All these are theoretically well defined and can, in principle, also be computed (for concise treatments see Refs. 33 and 39).

The comparison between theory and experiment then usually proceeds as illustrated schematically in Fig. 2. Clearly, getting to molecular constants or

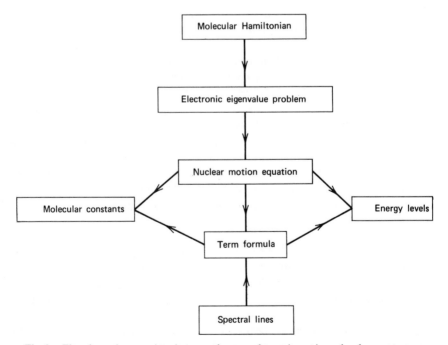

Fig. 2.   Flowchart of comparison between theory and experiment in molecular spectroscopy.

energy levels starting from the full Hamiltonian or the experimental spectrum always involves an approximation scheme through the corresponding nuclear motion equation. Comparison between theory and experiment can be carried out using either one of the methods described in the preceding sections, provided the term formulas are available. However, care should be taken to keep the diagram internally consistent. That is, *the accuracy of the scheme used to compute the molecular constants and energies must be the same as the one obtained by fitting differences of term formulas to observed spectral lines.* In general, this limits the number of spectroscopic constants available to the fitting procedure and for comparison between theory and experiment. Let us discuss this important observation in some more detail.

The BO analysis of the clamped-nuclei, adiabatic, and generator coordinate approximations and the results of the test calculation on the coupled oscillator model allow us to classify these methods into a *hierarchy of approximation schemes.* As demonstrated in Table II, each stage in the hierarchy corresponds a particular power of κ specifying at which order the corresponding approximate BO energy series differ from the exact ones.

It should be observed that the κ^8-dependence of ΔE_{GC} has not been proved explicitly: we have merely extrapolated the results for the model Hamiltonian (3.77). Also, since all methods considered use a single electronic state as input, the accuracy should increase with the amount of information transferred to the respective nuclear motion equations. Therefore, the place of the GCM in the hierarchy can be attributed to the incorporation by the GC trial state of the larger amount of electronic information.

The importance of the classification in Table II lies in its prescription of the approximation scheme that is appropriate for the analysis of experiments performed with a certain spectral resolution; that is, the internal consistency of the diagram (Fig. 2) requires the theoretical accuracy to be at least that of the experiment. In principle, therefore, adiabatic term formulas should be used only if the spectral resolution does not exceed 1 to 10^{-1} cm^{-1}. Through the use of laser technology and methods to avoid line broadening,

TABLE II

The Hierarchy of Molecular Approximation Schemes According to Their
Energetic Error Expressed in Terms of the BO Expansion Parameter

Approximation scheme	Errors on molecular energies[a]
Clamped nuclei	$\Delta E_C \cong \kappa^4 E_{el}$ (10–10^2 cm^{-1})
Adiabatic	$\Delta E_{AD} \cong \kappa^6 E_{el}$ (1–10^{-1} cm^{-1})
Generator coordinate	$\Delta E_{GC} \cong \kappa^8 E_{el}$ (10^{-2}–10^{-3} cm^{-1})

[a]Values in parentheses are based on $E_{el} \cong 1$ a.u. and $\kappa \cong 10^{-1}$.

experiments of considerably higher precision can now be performed.[40] However, since only adiabatic term formulas are available, these high-resolution spectra have so far been analyzed within the adiabatic scheme. This is a quite unsatisfactory state of affairs, since at that stage one can no longer decide whether the accuracy with which one reproduces the spectral lines is due to the workability of the adiabatic approximation or to the actual fitting procedure. The situation would be more favorable in the GC approximation, since the error on the energies is expected to be only 10^{-2} to 10^{-3} cm^{-1}, which is more on a par with today's experimental accuracy. We consider this to be an important motivation for the detailed study of the GC wave functions and nuclear motion equation. Sections IV and V give the present status of the method as applied to diatomic and polyatomic molecules.

G. Alternative Nonadiabatic Approaches

Section III.D presented the GC approximation as the third stage in the BO hierarchy of approximation schemes: clamped-nuclei approximation, adiabatic approximation, and generator coordinate approximation. Historically, however, nonadiabaticity has been introduced following a different line of thought. Two alternative approaches have been suggested: one is based on the so-called Born-Huang expansion of the wave function; the other relies on direct use of the variational principle, using basis sets for electrons and nuclei. Both differ essentially from the previously discussed methods in that they do not lead to a nuclear equation that contains information from a single electronic state. This makes them unsuitable for analysis by BO-PT. For the sake of completeness, we outline these methods briefly.

Born[27, 41] observed that since the set of electronic states $\{\varphi_n(r, R)\}$ form a complete set in r for each configuration R, the molecular wave function may be expanded as

$$\Psi(r, R) = \sum_n \varphi_n(r \mid R) F_n(R) \tag{3.126}$$

where the $F_n(R)$ are unknown "nuclear coefficients." In principle the sum in (3.126) includes an integral over the electronic continuum. Inserting (3.126) into Schrödinger's equation, multiplication to the left with $\varphi_m(r \mid R)$, and integration over the electron coordinates yields the following set of coupled differential equations for the $F_n(R)$:

$$\left[T + U_m^{\mathrm{AD}}(R) \right] F_m(R) = - \sum_{n \neq m} C_{mn}\left(R, \frac{\partial}{\partial R} \right) F_n(R) \tag{3.127}$$

$$C_{mn}\left(R, \frac{\partial}{\partial R} \right) = \langle \varphi_m(R) | T | \varphi_n(R) \rangle_r - \sum \frac{1}{M} \langle \varphi_m(R) \left| \frac{\partial}{\partial R} \right| \varphi_n(R) \rangle_r \frac{\partial}{\partial R} \tag{3.128}$$

These equations decouple, provided the $C_{mn}(R, \partial/\partial R)$ are small, that is, when the electronic states are slowly varying functions of R. In this case the Born-Huang series reduces to a single term and the approach becomes equivalent to the adiabatic approximation. On the other hand, it is well known that if two potential energy surfaces come close together, there is a rapid change in the structure of the electronic states in the neighborhood of the point of closest approach. The coupling operators may then become quite large,[42] and the wave functions will have a significant component along both electronic states. Since the strength of the coupling depends on the detailed form of the electronic eigenfunctions involved, its effect cannot be analysed systematically by BO-PT. Whether these large nonadiabaticities may be accounted for by the GCM through the nonproduct term in (3.64) remains to be investigated.

Since the Born-Huang approach is in principle exact, it has been used extensively for H_2^+ and its isotopic substitutions for which the full set of electronic states is available.[31, 43-45] The quasiexact results obtained in this way confirm the error estimates for the clamped nuclei and adiabatic approximations.

For systems of more than one electron, the application of the method is complicated because of the necessity of a number of accurate electronic eigenfunctions over a wide range of R values. In general these are not readily available. However, it should be stressed that even with the use of relatively poor electronic states, the coupled differential equations are the basis of microscopic descriptions of chemical reactions (for review, see Refs. 46 and 47). As such it is worthwhile to mention that by summing the GC wave function (3.63) over the electronic quantum numbers one obtains a GC version of the Born-Huang series:

$$\Psi(r, R) = \sum_n \int F_n(\alpha)\varphi_n(r|\alpha)\Phi(|R-\alpha|)\, d\alpha \qquad (3.129)$$

In view of Wiener's theorem[20] and the completeness of the sets $\{\varphi_n(r|\alpha)\}$ in electronic configuration space, (3.129) is an exact representation of the molecular wave function. The coupled differential equations are then replaced by a set of coupled integral equations

$$\int [H_{mm}(\alpha, \beta) - E\Delta_{mm}(\alpha, \beta)] F_m(\beta)\, d\beta$$

$$= -\sum_{n \neq m} \int [H_{mn}(\alpha, \beta) - E\Delta_{mn}(\alpha, \beta)] F_n(\beta)\, d\beta \qquad (3.130)$$

$$H_{mn}(\alpha, \beta) = \langle \varphi_m(\alpha)\Phi(\alpha)| H |\varphi_n(\beta)\Phi(\beta)\rangle \qquad (3.131)$$

$$\Delta_{mn}(\alpha, \beta) = \langle \varphi_m(\alpha)\Phi(\alpha)|\varphi_n(\beta)\Phi(\beta)\rangle \qquad (3.132)$$

On the basis of these equations one can construct a GC description of molecular scattering. Because of the attractive features of the GCM in connection with consequences of rotational invariance (see Sections IV.B and V.B), this may turn out to be of importance in molecular reaction dynamics. Results for atom-atom collisions are given in Ref. 48.

In the standard GC intrinsic states (3.63) the electronic states cusp at the average position of the nuclei. If one gives up this "perfect following" of the electronic cloud, still another GC wave function results. Introducing two sets of GCs: α_e for the electrons and α_n for the nuclei, one obtains the trial state:

$$\Psi(r, R) = \int G(\alpha_e, \alpha_n) \varphi(r|\alpha_e) \Phi(|R - \alpha_n|) \, d\alpha_e \, d\alpha_n \qquad (3.133)$$

From a physical point of view it can be said that in this way one actually allows the electrons to lag behind the nuclear motion. The GC subspace thus generated contains the one corresponding to (3.63). Indeed, imposing the perfect following of the electronic cloud by putting

$$G(\alpha_e, \alpha_n) = F(\alpha_n) \delta(\alpha_e - \alpha_n) \qquad (3.134)$$

gives the standard GC trial state as a special case of (3.133). However, contrary to the GC analog of the Born-Huang series, the present form (3.133) is not necessarily exact.

Whereas the Born-Huang series can still be considered to be based on the separation of electronic and nuclear motion, electronic information from several electronic states being incorporated, one may prefer to totally cast aside the splitting of the Hamiltonian into an electronic part and a nuclear part and apply the variational principle directly to the full operator H. Applications to three-particle systems can be done in this way with extreme accuracy. The theoretical transition frequencies for H_2^+ reported by Bishop and Chueng,[35] taking account of relativistic and radiative corrections, can compete with the experimental results of Wing.[49] These authors[50] have also given a very precise determination of the ground state of H_2. Although calculations of this type provide a test of the foundations of quantum mechanics as applied to molecules, they are of limited practical value. Because of basis set dependence, no systematic behavior in κ of the results can be predicted. Also, the direct variational method is restricted to very small systems, one- and two-electron diatomics, and is obviously incapable of generating phenomenological insight; that is, it does not yield term formulas. These drawbacks are illustrated in attempts by Thomas (in Ref. 51 and references therein) to compute vibration-rotation levels of polyatomic molecules using one-particle basis functions for electrons and nuclei.

In summary one can conclude that *nonadiabatic approaches, alternative to the standard GC approximation, do not lend themselves to an error analysis by BO-PT*. They are not size consistent, that is, the accuracy of their results for a particular molecule cannot simply be carried over to other molecular systems.

IV. GENERATOR COORDINATE THEORY OF DIATOMIC SYSTEMS

A. Notations and Definitions

The Hamiltonian for a diatomic molecule with nuclear charges and masses (Z_1, M_1) and (Z_2, M_2) can, after separating off the center-of-mass motion, be written as[31]

$$H = -\frac{1}{2\mu}\Delta_{\mathbf{R}} - \frac{1}{2}\sum_i \Delta_{\mathbf{r}_i} - \frac{1}{2M}\left(\sum_i \nabla_{\mathbf{r}_i}\right)^2 + \frac{Z_1 Z_2}{R}$$

$$- Z_1 \sum_i \frac{1}{|\mathbf{r}_i - \mathbf{R}_1|} - Z_2 \sum_i \frac{1}{|\mathbf{r}_i - \mathbf{R}_2|} + \sum_{i<j} \frac{1}{|\mathbf{r}_i - \mathbf{r}_j|} \qquad (4.1)$$

The origin from which the particle positions are measured is the center of mass of the nuclei, such that

$$\mathbf{R}_1 = -\frac{(M_2\mathbf{R})}{M} \quad \text{and} \quad \mathbf{R}_2 = \frac{(M_1\mathbf{R})}{M} \qquad (4.2)$$

where $\mathbf{R} = \mathbf{R}_2 - \mathbf{R}_1$ denotes the relative positions of the nuclei. Furthermore, the reduced nuclear mass $\mu = (M_1 M_2)/M$, where M is the total nuclear mass $M_1 + M_2$. Other alternatives for the origin of the reference frame are possible[52] (center of mass of all particles, geometrical center of the nuclei, etc.), however the choice above is convenient because of the absence of electronic-nuclear cross-derivatives. Therefore, the separation into an electronic Hamiltonian and nuclear kinetic energy operator is simple:

$$H(r, \mathbf{R}) = -\frac{1}{2\mu}\Delta_{\mathbf{R}} + H_{el}(r, \mathbf{R}) \qquad (4.3)$$

The total angular momentum about the center of mass of the nuclei also is a sum of a nuclear and electronic part:

$$\mathbf{J} = \mathbf{R} \times \mathbf{P} + \sum_i \mathbf{r}_i \times \mathbf{p}_i = \mathbf{L} + \mathbf{l} \qquad (4.4)$$

where $\mathbf{P}$ and $\mathbf{p}_i$ are the momenta conjugate to $\mathbf{R}$ and $\mathbf{r}_i$. The Hamiltonian (4.1) commutes with the components of $\mathbf{J}$ such that using the notations of Section II.C, the solutions of Schrödinger's equation may be labeled as follows:

$$H(r,\mathbf{R})\Psi^{JM}(r,\mathbf{R})=E^J\Psi^{JM}(r,\mathbf{R}) \qquad (4.5)$$

It should be repeated that our approach is purely nonrelativistic and that coupling effects between the particle spins and the orbital angular momenta (4.4) are neglected. In common spectroscopic terms this means we are considering Hund's case b, which is appropriate for light diatomic molecules.[53] For heavy systems the angular momentum algebra in the following sections has to be modified by working in the correct coupling scheme.

Since we will be concerned mainly with the GCM, we consider the electronic eigenvalue problem for $H_{\text{el}}(r,\boldsymbol{\alpha})=H_{\text{el}}(r,\mathbf{R}=\boldsymbol{\alpha})$. The components of the vector $\boldsymbol{\alpha}$ parameterize the possible relative positions of the nuclei and will later assume the role of GCs. However, let us first establish the relationship between electronic Hamiltonians corresponding to different orientations of the internuclear axis. It can easily be shown that the following operator equality holds:

$$H_{\text{el}}(\boldsymbol{\alpha})=\mathcal{R}_e(\omega_\alpha)H_{\text{el}}(\alpha\mathbf{e}_z)\mathcal{R}_e^{-1}(\omega_\alpha) \qquad (4.6)$$

where $\mathcal{R}_e(\omega_\alpha)\equiv\mathcal{R}_e(\varphi_\alpha,\theta_\alpha,0)$ is the active rotation operator in the electronic configuration space corresponding to the polar angles of $\boldsymbol{\alpha}$, that is, $\omega_\alpha\equiv(\varphi_\alpha,\theta_\alpha)$. Equation 4.6 states that $H_{\text{el}}(\boldsymbol{\alpha})$ and $H_{\text{el}}(\alpha\mathbf{e}_z)$ are unitary equivalent.[54] This implies that these operators have the same energy spectrum while their eigenfunctions transform into each other under the operation $\mathcal{R}_e(\omega_\alpha)$, that is,

$$U(\boldsymbol{\alpha})=U(\alpha\mathbf{e}_z)\equiv U(\alpha) \qquad (4.7)$$

$$\varphi(r|\boldsymbol{\alpha})=\mathcal{R}_e(\omega_\alpha)\varphi(r|\alpha\mathbf{e}_z)=\varphi\left(A^{-1}(\omega_\alpha)r|\alpha\mathbf{e}_z\right) \qquad (4.8)$$

where $A(\omega_\alpha)$ is the coordinate transformation (2.30) for $\Omega=\omega_\alpha$. The first result is well known and shows that the electronic eigenvalue for a diatomic is a potential energy curve. The second property states that rotating the force centers in the electronic state in one direction is equivalent to rotating all the electrons in the opposite direction. This will prove to be of crucial importance in the GC approximation.

Given (4.7) and (4.8) we write the electronic eigenvalue problem as follows:

$$H_{\text{el}}(r,\alpha)\varphi(r|\alpha)=U(\alpha)\varphi(r|\alpha) \qquad (4.9)$$

Considered as an operator on the electronic variables only, $H_{el}(\alpha)$ is clearly invariant under rotations around α and under reflections $S(\alpha)$ through planes containing α. The axial symmetry implies that the projection of the electronic angular momentum on the nuclear axis α is a conserved quantity. The electronic eigenstates will therefore be proportional to $e^{\pm i\Lambda\gamma_\alpha}$ where $\Lambda = 0, 1, 2, \ldots$, and γ_α is the angle for rotations around α. Under a reflection $S(\alpha)$ the electronic energy is unchanged; however the electronic angular momentum about α of an electronic eigenfunction changes sign. The eigenvalues $U_\Lambda(\alpha)$ corresponding to nonzero Λ are therefore doubly degenerate and the corresponding eigenfunctions $\varphi_\Lambda(r|\alpha)$ and $\varphi_{-\Lambda}(r|\alpha)$ transform into each other under $S(\alpha)$. One can form the functions

$$\varphi_\Lambda^\pm(r|\alpha) = \frac{\varphi_\Lambda(r|\alpha) \pm \varphi_{-\Lambda}(r|\alpha)}{\sqrt{2}} \qquad (4.10)$$

which are eigenfunctions of $S(\alpha)$ with corresponding eigenvalues ± 1. For $\Lambda = 0$ the electronic wave functions $\varphi_0(r|\alpha)$ can be multiplied only by a constant as a result of a reflection. Since $S^2(\alpha)$ is the identity, this constant is ± 1. We therefore distinguish between $\varphi_0^+(r|\alpha)$ and $\varphi_0^-(r|\alpha)$ states according to whether these functions remain unaltered or change sign in a reflection. The corresponding eigenvalues $U_0^\pm(\alpha)$ are nondegenerate. In common spectroscopic terminology,[55] one speaks of Σ^+ and Σ^- states if $\Lambda = 0$ and of (doubly degenerate) $\Pi, \Delta, \ldots$, states for $\Lambda = 1, 2, \ldots$.

B.   Rotational Invariance

As a consequence of the rotational invariance of (4.1), exact diatomic eigenstates can be chosen as simultaneous eigenfunctions of J^2 and J_z. It is therefore desirable that wave functions generated by approximation schemes share this fundamental property. This section seeks to verify whether the adiabatic and generator coordinate methods can fulfill this requirement.

According to (3.40), the adiabatic nuclear motion Hamiltonian is given by

$$H_{AD}(\mathbf{R}) = -\frac{1}{2\mu}\Delta_\mathbf{R} + U_\Lambda(R) - \langle\varphi_\Lambda(\mathbf{R})|\frac{1}{2\mu}\Delta_\mathbf{R}|\varphi_\Lambda(\mathbf{R})\rangle_r \qquad (4.11)$$

It is easy to prove, using (4.8), that the adiabatic correction term is a function of the relative distance between the nuclei only. Therefore a separation of radial and angular variables yields nuclear motion functions proportional to spherical harmonics $Y_{LM}(\omega_R)$, where $\omega_R \equiv (\varphi_R, \theta_R)$ are the polar angles of $\mathbf{R}$. These are eigenstates of L^2 and L_z. As a result, the total adiabatic wave functions are not eigenfunctions of J^2 and J_z. Put a different way, the electronic contribution to the total angular momentum has been disregarded. The adiabatic approximation in the form presented in Section III.C, common also to many textbooks,[16, 56] is not adapted to rotational invariance.

This unsatisfactory situation can be remedied by introducing a so-called body-fixed reference frame. Since the electronic cloud should follow the nuclei in their rotational motion, one argues[57] that the electronic states must be defined in a coordinate system rigidly attached to the nuclei. In this way the angular momentum of the electrons would automatically be accounted for. For diatomic molecules an obvious choice for the body-fixed unit vectors is obtained by choosing the body-fixed z-axis along the internuclear vector $\mathbf{R}$.[58] The body-fixed unit vectors $(\mathbf{e}_{x'}, \mathbf{e}_{y'}, \mathbf{e}_{z'})$ will therefore be linear combinations of the space-fixed ones $(\mathbf{e}_x, \mathbf{e}_y, \mathbf{e}_z)$, with sines and cosines of the angles (φ_R, θ_R) as coefficients. This implies that the electron coordinates r' in the body-fixed frame are implicit functions of the angles of $\mathbf{R}$, [i.e., $r' = r'(\omega_R)$]. Consequently, to carry out the adiabatic variational principle, the Hamiltonian (4.1), which is expressed in terms of space-fixed variables, has to be transformed to the body-fixed reference frame. The dependence of the body-fixed electron coordinates on the nuclear variables is, in the case of diatomic molecules, still simple enough to perform this transformation explicitly. The resulting adiabatic Hamiltonian reads[28]

$$H'_{\mathrm{AD}}(\mathbf{R}) = -\frac{1}{2\mu R^2}\frac{\partial}{\partial R}R^2\frac{\partial}{\partial R} + \frac{J^2(\Lambda) - \Lambda^2}{2\mu R^2} + U_\Lambda(R)$$

$$+ \frac{1}{2\mu R^2}\langle \varphi_\Lambda(R\mathbf{e}_{z'})|l_{x'}^2 + l_{y'}^2|\varphi_\Lambda(R\mathbf{e}_{z'})\rangle_{r'}$$

$$- \langle \varphi_\Lambda(R\mathbf{e}_{z'})\left|\frac{1}{2\mu}\frac{\partial^2}{\partial R^2}\right|\varphi_\Lambda(R\mathbf{e}_{z'})\rangle_{r'} \qquad (4.12)$$

where $l_{x'}$ and $l_{y'}$ are the components of $\mathbf{l}$ along $\mathbf{e}_{x'}$ and $\mathbf{e}_{y'}$. The term $J^2(\Lambda)$ is the total angular momentum squared:

$$J^2 = L^2 + 2i\frac{\cotan\theta_R}{\sin\theta_R}l_{z'}\frac{\partial}{\partial\varphi_R} - \frac{1}{\sin^2\theta_R}l_{z'}^2 \qquad (4.13)$$

in which $l_{z'}$ has been replaced by the electronic eigenvalue Λ. The operator $l_{z'}$ appears because there is no nuclear angular momentum about the body-fixed axis:

$$J_{z'} = \mathbf{J}\cdot\mathbf{e}_{z'} = (\mathbf{l} + \mathbf{L})\cdot\mathbf{e}_{z'} = \mathbf{l}\cdot\mathbf{e}_{z'} = l_{z'} \qquad (4.14)$$

Because of the separation of $J^2(\Lambda)$ in H'_{AD} and because in view of (4.14) the

$\varphi_\Lambda(r'|R\mathbf{e}_{z'})$ are eigenstates of $J_{z'}$, we can conclude that the body-fixed adiabatic states are simultaneous eigenfunctions of J^2 and J_z.

The intermediate mathematical manipulations necessary to arrive at this result are far from trivial. A detailed exposition of the foregoing is rarely included in standard textbooks and can be found only in the original papers[59] and in highly specialized articles. *It is rather embarassing for the adiabatic approximation that it seems so difficult to account for a basic conservation law such as rotational invariance.*

We now consider this problem in the GC approximation. The GC wave function for a diatomic molecule reads

$$\Psi(r,\mathbf{R}) = \int F(\boldsymbol{\alpha})\varphi_\Lambda(r|\boldsymbol{\alpha})\Phi(|\mathbf{R}-\boldsymbol{\alpha}|)\,d\boldsymbol{\alpha}$$

$$= \int F(\boldsymbol{\alpha})\chi_\Lambda(r,\mathbf{R}|\boldsymbol{\alpha})\,d\boldsymbol{\alpha} \qquad (4.15)$$

That is, the GCs are the components of a vector $\boldsymbol{\alpha}$, which is the counterpart of the internuclear vector $\mathbf{R}$. To prove that the GC variational principle, applied to (4.15), will yield angular momentum eigenstates, we know that it is sufficient to demonstrate the stability of the GC trial space under rotations. One readily verifies, using (4.8) that the electronic and nuclear parts of the intrinsic states satisfy the relations

$$\varphi_\Lambda(r|\boldsymbol{\alpha}) = \varphi_\Lambda(r|A(\omega_\alpha)\alpha\mathbf{e}_z) = \varphi_\Lambda\big(A^{-1}(\omega_\alpha)r|\alpha\mathbf{e}_z\big)$$

$$= \mathscr{R}_e(\omega_\alpha)\varphi_\Lambda(r|\alpha\mathbf{e}_z) \qquad (4.16)$$

$$\Phi(|\mathbf{R}-\boldsymbol{\alpha}|) = \Phi(|\mathbf{R}-A(\omega_\alpha)\alpha\mathbf{e}_z|) = \Phi\big(|A^{-1}(\omega_\alpha)\mathbf{R}-\alpha\mathbf{e}_z|\big)$$

$$= \mathscr{R}_n(\omega_\alpha)\Phi(|\mathbf{R}-\alpha\mathbf{e}_z|) \qquad (4.17)$$

where $\mathscr{R}_e(\omega_\alpha)$ and $\mathscr{R}_n(\omega_\alpha)$ refer to (active) rotation operators in the electron and nuclear space, respectively. For the full intrinsic states this implies that

$$\chi_\Lambda(r,\mathbf{R}|\boldsymbol{\alpha}) = \mathscr{R}_e(\omega_\alpha)\varphi_\Lambda(r|\alpha\mathbf{e}_z)\mathscr{R}_n(\omega_\alpha)\Phi(|\mathbf{R}-\alpha\mathbf{e}_z|)$$

$$= \mathscr{R}(\omega_\alpha)\chi_\Lambda(r,\mathbf{R}|\alpha\mathbf{e}_z) \qquad (4.18)$$

$$\mathscr{R}(\omega_\alpha) = \mathscr{R}_e(\omega_\alpha)\mathscr{R}_n(\omega_\alpha) \qquad (4.19)$$

that is, the Brink-Weiguny condition (2.24) is satisfied. This proves that the GC variational principle will automatically lead to angular momentum eigenstates.

One can demonstrate this more explicitly by actually determining part of the GC weight function from first principles. We observe that

$$J_z \chi_\Lambda(\alpha e_z) = (L_z + l_z)\varphi_\Lambda(\alpha e_z)\Phi(\alpha e_z) = \Lambda \chi_\Lambda(\alpha e_z) \qquad (4.20)$$

that is, the intrinsic states $\chi_\Lambda(\alpha e_z)$ are eigenstates of the total angular momentum component J_z with eigenvalue Λ. Using (4.18) and (4.20), we can put the GC diatomic states (4.15) in the form

$$\int F(\alpha)\mathcal{R}(\omega_\alpha)\chi_\Lambda(r, R|\alpha e_z)\,d\alpha$$

$$= \int F(\alpha)e^{+i\Lambda\gamma_\alpha}\mathcal{R}(\omega_\alpha)e^{-i\gamma_\alpha J_z}\chi_\Lambda(r, R|\alpha e_z)\,d\alpha$$

$$= \int F(\alpha)e^{+i\Lambda\gamma_\alpha}\mathcal{R}(\Omega_\alpha)\chi_\Lambda(r, R|\alpha e_z)\alpha^2\,d\alpha\,d\Omega_\alpha \qquad (4.21)$$

thus introducing the third, in this case redundant, Euler angle γ_α. In the last expression one starts to recognize the projection operator $P_{M\Lambda}^J$. Indeed we simply have to specify the angular part of the GC weight function as

$$F^{JM}(\alpha) = f^J(\alpha)e^{+iM\varphi_\alpha}d_{M\Lambda}^J(\theta_\alpha)$$

$$= f^J(\alpha)D_{M\Lambda}^{J*}(\varphi_\alpha, \theta_\alpha, 0) \equiv f^J(\alpha)D_{M\Lambda}^{J*}(\omega_\alpha) \qquad (4.22)$$

to obtain (up to a normalization constant)

$$\Psi^{JM}(r, R) = \int f^J(\alpha)D_{M\Lambda}^{J*}(\Omega_\alpha)\mathcal{R}(\Omega_\alpha)\chi_\Lambda(r, R|\alpha e_z)\alpha^2\,d\alpha\,d\Omega_\alpha$$

$$= \int f^J(\alpha)P_{M\Lambda}^J\chi_\Lambda(r, R|\alpha e_z)\alpha^2\,d\alpha \qquad (4.23)$$

that is, the GC wave functions are superpositions of projected intrinsic states $P_{M\Lambda}^J\chi_\Lambda(\alpha e_z)$ with radial weight functions $f^J(\alpha)$. In fact, what we have shown here is that the polar angles of the GC vector α have been used to project out good angular momentum quantum numbers. As a special case of the projection technique outlined in Section II.B, the analysis above illustrates that in this scheme there is no need for a body-fixed reference frame. Hence, no transformation of the Hamiltonian and use of the explicit expression of J^2 was necessary. It is safe to say that *implementing the consequences of rotational invariance in the GCM is considerably simpler than in the adiabatic approximation.*[60]

The nuclear motion equation corresponding to the projected form (4.23) of the GC wave function reads

$$\int \left[H_{\Lambda\Lambda}^{J}(\alpha,\beta) - E^{J}\Delta_{\Lambda\Lambda}^{J}(\alpha,\beta) \right] f^{J}(\beta)\beta^{2}\, d\beta = 0 \qquad (4.24)$$

and contains the so-called angular momentum projected kernels

$$H_{\Lambda\Lambda}^{J}(\alpha,\beta) = \langle \chi_{\Lambda}(\alpha\mathbf{e}_{z}) | H P_{\Lambda\Lambda}^{J} | \chi_{\Lambda}(\beta\mathbf{e}_{z}) \rangle$$

$$= \frac{2J+1}{2} \int d_{\Lambda\Lambda}^{J}(\theta) \langle \chi_{\Lambda}(\alpha\mathbf{e}_{z}) | H e^{-i\theta J_{y}} | \chi_{\Lambda}(\beta\mathbf{e}_{z}) \rangle\, d\cos\theta$$

$$= \frac{2J+1}{2} \int d_{\Lambda\Lambda}^{J}(\theta) \langle \chi_{\Lambda}(\alpha\mathbf{e}_{z}) | H | \chi_{\Lambda}(\beta\cos\theta\mathbf{e}_{z} + \beta\sin\theta\mathbf{e}_{x}) \rangle\, d\cos\theta$$

$$(4.25)$$

and analogous expressions for $\Delta_{\Lambda\Lambda}^{J}$ obtained from (4.25) by dropping H. The eigenvalues E_{v}^{J} of (4.24) represent vibration-rotation levels of the molecule in a state of total angular momentum J and "vibrational quantum number v."

C. Symmetry Labels of Molecular States

Besides being rotationally invariant, H remains unchanged under a simultaneous inversion of all particle position vectors; that is, the Hamiltonian (4.1) is parity invariant. The exact wave functions can therefore be labeled by an additional quantum number, $+$ or $-$, according to whether they do or do not change sign under inversion of all coordinates. In addition one should keep in mind that the general permutation symmetries imposed by the particle statistics should be respected; that is, for identical fermions (half-integer spin particles) the wave function must change sign if subjected to a two-particle permutation, while for bosons (integer spin particles) it must remain invariant. This is automatically accounted for as far as the electrons are concerned, if the electronic states are assumed to be properly antisymmetrized, but remains to be studied for the nuclei in case of homonuclear (identical nuclei) diatomics. This section studies the interplay between geometric symmetries (rotation, inversion) and particle statistics (fermions, bosons) for a given set of electronic quantum numbers.

The derivation of symmetry properties of energy levels within the adiabatic approximation is complicated by the use of a body-fixed reference frame. Indeed, the adiabatic wave functions are expressed in body-fixed variables, Euler angles, and a set of internal coordinates, while the symmetry operators are defined through their action on space-fixed variables. This necessitates a "translation" of the effect of the symmetry elements in "body-fixed language." Because of this, a number of rather obscure statements can be found in the literature. Some examples are: "the operation of inversion in a fixed system of coordinates is equivalent in the moving system to a reflection in a plane passing through the axis of the molecule,"[58] and "the transformation properties of a function of Euler angles under a sense-reversing operation is [sic] the same as that of the pure rotation obtained by multiplying the operation by the operator which inverts the body-fixed axes."[61] The situation is particularly complicated for the so-called sense-reversing operations (inversions and reflections), since in a passive point of view, they change the handedness of the axes system. This implies that their effect on a function of Euler angles cannot be described by a new set of angles, for these can represent only relative orientations.[61]

For these reasons we do not discuss the symmetry properties of diatomic levels in the adiabatic approximation. Instead we immediately proceed to the GC version, where the above-mentioned problems are not to be expected because of the use of the active point of view for symmetry operations and the absence of a body-fixed frame in the theory.

If an approximate wave function Ψ is not a parity eigenstate, one can construct the function $\Psi \pm I\Psi$, where I is the inversion operator of the space-fixed coordinates, which have even or odd parity according to whether one takes the $+$ or $-$ combination. We therefore compute

$$I\Psi^{JM}(r,\mathbf{R}) = \int f^J(\alpha) D_{M\Lambda}^{J*}(\omega_\alpha) \chi_\Lambda(-r, -\mathbf{R}|\alpha)\, d\alpha$$

$$= \int f^J(\alpha) D_{M\Lambda}^{J*}(\omega_\alpha) \varphi_\Lambda(-r|\alpha) \Phi(|\mathbf{R}+\alpha|)\, d\alpha$$

$$= \int f^J(\alpha) D_{M\Lambda}^{J*}(\bar\omega_\alpha) \varphi_\Lambda(-r|-\alpha) \Phi(|\mathbf{R}-\alpha|)\, d\alpha \quad (4.26)$$

where $\bar\omega_\alpha = (\pi + \varphi_\alpha, \pi - \theta_\alpha)$ are the polar angles of $-\alpha$. The Wigner D-functions have the property

$$D_{M\Lambda}^J(\bar\omega_\alpha) = (-)^J D_{M-\Lambda}^J(\omega_\alpha) \quad (4.27)$$

while using (4.8) and the conventions of Section II.C one can verify that

$$\varphi_\Lambda(-r|-\alpha) = \varphi_\Lambda\big(-A^{-1}(\bar\omega_\alpha) r|\alpha e_z\big) = \varphi_{-\Lambda}\big(A^{-1}(\omega_\alpha) r|\alpha e_z\big) = \varphi_{-\Lambda}(r|\alpha)$$

$$(4.28)$$

$$\varphi_0^\pm(-r|-\alpha) = \varphi_0^\pm\big(-A^{-1}(\bar\omega_\alpha) r|\alpha e_z\big) = \pm\varphi_0^\pm\big(A^{-1}(\omega_\alpha) r|\alpha e_z\big) = \pm\varphi_0^\pm(r|\alpha)$$

$$(4.29)$$

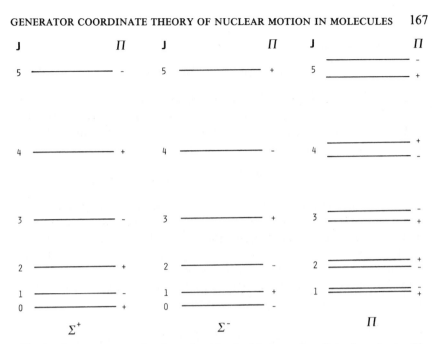

Fig. 3. Symmetry properties of rotational levels of heteronuclear diatomic molecules. The Λ-doubling is exaggerated.

for $\Lambda \neq 0$ and $\Lambda = 0$ electronic states, respectively. These results, which are nothing but the mathematics behind the foregoing quotation from Ref. 58, show that the angular momentum eigenstates (4.23) for Σ electronic states have parity $(-)^{J+s}$ where $s=0$ for Σ^+ and $s=1$ for Σ^-. For $\Lambda \neq 0$ electronic terms the partiy eigenfunctions are given by

$$\Psi^{JM\Pi}(r,\mathbf{R}) = \int f^J(\alpha) \left[P^J_{M\Lambda} \chi_\Lambda(r,R|\alpha) \pm P^J_{M-\Lambda} \chi_{-\Lambda}(r,\mathbf{R}|\alpha) \right] d\alpha$$

$$(4.30)$$

where $\Pi = (-)^{J+s}$ with $s=0$ for the $+$ combination and $s=1$ for the $-$ combination. The symmetry labels of the rotational levels are summarized in Fig. 3. The double degeneracy of the levels in the $\Lambda \neq 0$ case is lifted by an effect known as Λ-doubling. For this reason the positive and negative parity states are drawn slightly apart. A GC derivation of these splittings is given at the end of Section IV.D.

The discussion above is appropriate for heteronuclear diatomics only. In the homonuclear case the additional symmetries caused by the presence of two identical nuclei require special considerations. The electronic Hamiltonian is then invariant with respect to an inversion of the electron coordinates. Therefore electronic eigenstates are even or odd ("*gerade*" or "*ungerade*") under I_e:

$$I_e\varphi_\Lambda^{\Pi_e}(r|\alpha)=(-)^p\varphi_\Lambda^{\Pi_e}(r|\alpha) \tag{4.31}$$

That is, the electronic parity $\Pi_e=(-)^p$ where $p=0$ for gerade and $p=1$ for ungerade states. The total Hamiltonian is now invariant under a permutation of the nuclei. To adjust the GC wave functions to the implications of this symmetry, we introduce gerade and ungerade nuclear basis functions

$$\Phi^{\Pi_n}(\mathbf{R}|\alpha)=\Phi(|\mathbf{R}-\alpha|)+(-)^P\Phi(|\mathbf{R}+\alpha|) \tag{4.32}$$

These states have nuclear parity $\Pi_n=(-)^P$, where $P=0$ for gerade nuclear states and $P=1$ for ungerade. The use of such nuclear intrinsic states guarantees that after combination with a symmetric or antisymmetric spin function, the GC wave functions will change sign (remain unchanged) under a permutation of the fermion (boson) nuclei.

In view of the foregoing considerations, the GC wave functions for homonuclear diatomic molecules are of the form

$$\Psi^{JM\Pi}(r,\mathbf{R})=\int f^J(\alpha)\left[D_{M\Lambda}^{J*}(\omega_\alpha)\chi_\Lambda^\Pi(r,\mathbf{R}|\alpha)\pm D_{M-\Lambda}^{J*}(\omega_\alpha)\chi_{-\Lambda}^\Pi(r,\mathbf{R}|\alpha)\right]d\alpha \tag{4.33}$$

where

$$\chi_\Lambda^\Pi(r,\mathbf{R}|\alpha)=\varphi_\Lambda^{\Pi_e}(r|\alpha)\Phi^{\Pi_n}(\mathbf{R}|\alpha) \tag{4.34}$$

These states clearly have a definite parity, since

$$I\chi_\Lambda^\Pi(\alpha)=I_e\varphi_\Lambda^{\Pi_e}(\alpha)I_n\Phi^{\Pi_n}(\alpha)=(-)^{p+P}\chi_\Lambda^\Pi(\alpha) \tag{4.35}$$

To determine the interrelationship between the quantum numbers of various types (total parity, electronic and nuclear parity, total angular momentum, etc.), we evaluate the terms containing $\Phi(|\mathbf{R}+\alpha|)$ in (4.33). The following set of equalitites is readily verified using (4.27), (4.28), and (4.35):

$$(-)^P\int f^J(\alpha)\left[D_{M\Lambda}^{J*}(\omega_\alpha)\varphi_\Lambda^{\Pi_e}(r|\alpha)\pm D_{M-\Lambda}^{J*}(\omega_\alpha)\varphi_{-\Lambda}^{\Pi_e}(r|\alpha)\right]\Phi(|\mathbf{R}+\alpha|)\,d\alpha$$

$$=(-)^P\int f^J(\alpha)\left[D_{M\Lambda}^{J*}(\bar\omega_\alpha)\varphi_\Lambda^{\Pi_e}(r|-\alpha)\pm D_{M-\Lambda}^{J*}(\bar\omega_\alpha)\varphi_{-\Lambda}^{\Pi_e}(r|-\alpha)\right]\Phi(|\mathbf{R}-\alpha|)\,d\alpha$$

$$=(-)^{J+P}\int f^J(\alpha)\left[D_{M-\Lambda}^{J*}(\omega_\alpha)\varphi_\Lambda^{\Pi_e}(r|-\alpha)\pm D_{M\Lambda}^{J*}(\omega_\alpha)\varphi_{-\Lambda}^{\Pi_e}(r|-\alpha)\right]\Phi(|\mathbf{R}-\alpha|)\,d\alpha$$

$$=(-)^{J+P+p}\int f^J(\alpha)\left[D_{M-\Lambda}^{J*}(\omega_\alpha)\varphi_{-\Lambda}^{\Pi_e}(r|\alpha)\pm D_{M\Lambda}^{J*}(\omega_\alpha)\varphi_\Lambda^{\Pi_e}(r|\alpha)\right]\Phi(|\mathbf{R}-\alpha|)\,d\alpha$$

$$=(-)^{J+P+p+s}\int f^J(\alpha)\left[D_{M\Lambda}^{J*}(\omega_\alpha)\varphi_\Lambda^{\Pi_e}(r|\alpha)\pm D_{M-\Lambda}^{J*}(\omega_\alpha)\varphi_{-\Lambda}^{\Pi_e}(r|\alpha)\right]\Phi(|\mathbf{R}-\alpha|)\,d\alpha \tag{4.36}$$

This rather lengthy, but trivial, formularium proves that the $\Phi(|\mathbf{R}-\boldsymbol{\alpha}|)$ and $\Phi(|\mathbf{R}+\boldsymbol{\alpha}|)$ parts of the GC wave functions are equal up to a sign factor. This implies for the full GC wave function that

$$\Psi^{JM\Pi}(r,\mathbf{R})=(-)^{J+P+p+s}\Psi^{JM\Pi}(r,\mathbf{R}) \qquad (4.37)$$

a formula that proves to be valid also in case of Σ electronic states. The importance of (4.37) lies in the fact that it determines the compatibility of the total angular momentum, electronic parity, nuclear parity, and total parity, and the sign of the electronic term; that is, the relation between these quantum numbers has to be such that $J+P+p+s$ is an even integer. This condition summarizes all the symmetry properties of molecular levels, which are usually given in the form shown in Fig. 4. The derivation of these properties in the traditional way, by referring to the body-fixed frame, is tedious. The derivation from the GCM wave function, as given above, is concise and transparent[62]. The diagrams in Fig. 4 can be obtained as a trivial exercise. The compatibility condition between J, s, p, and P must be used in constructing the GCM intrinsic state; that is, for a given electronic state (p,s)

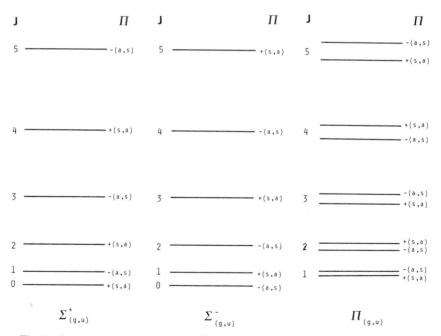

Fig. 4.   Symmetry properties of rotational levels of homonuclear diatomic molecules. The Λ-doubling is exaggerated. The labels s and a denote symmetry and antisymmetry, respectively, of the nuclear spin function.

and a given rotational state (J) we must choose the nuclear basis function with corresponding parity P. The nuclear spin function is then also determined from statistics.

D. Vibrations, Rotations, and Vibration-Rotation Interactions

The general features of diatomic spectra can be explained by a threefold structure of the molecular energy

$$E \cong E_{\text{el}} + E_{\text{vib}} + E_{\text{rot}} \qquad (4.38)$$

where E_{el} stands for the electronic energy and E_{vib} and E_{rot} for the vibrational and rotational energy, respectively, of the nuclei. Theoretically this result follows easily from low-order BO-PT, which also shows[22] that

$$E_{\text{vib}} \cong \kappa^2 E_{\text{el}} \qquad \text{and} \qquad E_{\text{rot}} \cong \kappa^4 E_{\text{el}} \qquad (4.39)$$

such that

$$E_{\text{el}} \gg E_{\text{vib}} \gg E_{\text{rot}} \qquad (4.40)$$

Within the adiabatic approximation one easily recovers the pattern (4.38). The radial nuclear motion equation arising from (4.12) contains the potential

$$U_{\text{AD}}(R) + \frac{J(J+1) - \Lambda^2}{2\mu R^2} \qquad (4.41)$$

which consists of the adiabatic part and a centrifugal term. Expanding $U_{\text{AD}}(R)$ around R_0, retaining terms of order κ^2, and replacing the centrifugal potential by its value at R_0, giving a κ^4 contribution, yields the approximate energy spectrum

$$E_v^J \cong U_{\text{AD}}(R_0) + (v + \tfrac{1}{2})\Omega_{\text{AD}} + \frac{J(J+1) - \Lambda^2}{2 I_{\text{AD}}} \qquad (4.42)$$

$$\Omega_{\text{AD}} = \left(\frac{[\partial^2 U_{\text{AD}} / \partial R^2]_{R_0}}{\mu} \right)^{1/2} \qquad \text{and} \qquad I_{\text{AD}} = \mu R_0^2 \qquad (4.43)$$

It should be observed that to derive this result, the variable $R - R_0$ was assumed to run from $-\infty$ to $+\infty$. This is justified only for states sufficiently localized around R_0, in accordance with the validity condition of BO-PT. Equation 4.42 identifies E_{el} with the binding energy of the electrons in the field of nuclei fixed at R_0, E_{vib} with the vibrational energy of a one-dimensional harmonic oscillator of frequency Ω_{AD}, and E_{rot} with a rigid-rotorlike energy containing the moment of inertia I_{AD}.

The parameterization of diatomic band spectra is usually carried out using an extended version of (4.42). It turns out that the position of spectral lines can be explained on the basis of the formula

$$E_v^J = \sum_{k,l} Y_{kl} \left[v + \tfrac{1}{2} \right]^k \left[J(J+1) - \Lambda^2 \right]^l \qquad (4.44)$$

This expression, known as the Dunham series,[36] states that the vibration-rotation levels E_v^J are "power series" in the respective rotational and vibrational quantum numbers. In most cases it is possible to fit, with high precision, thousands of spectral lines using some tens of spectroscopic constants Y_{kl}.[3] Dunham showed, using semiclassical methods, that

$$Y_{00} \cong U_{AD}(R_0), \qquad Y_{10} \cong \Omega_{AD}, \qquad Y_{01} \cong \frac{1}{2I_{AD}} \qquad (4.45)$$

and furthermore that the Y_{kl} could be written as expansions in the BO parameter κ.

In the adiabatic approximation (4.44) can be derived by considering the higher order terms in the expansion of (4.41) as perturbations to those yielding the separated form (4.42). However, as observed before, the adiabatic spectroscopic constants cannot be expected to fully agree with the experimental ones because of the limited accuracy of the adiabatic scheme. For some explicit calculations we refer to Refs. 63 and 64.

From the literature, one gets the impression that the approximate separation (4.42) and the form of the Dunham series are tied together with the adiabatic assumption, more specifically with the factorization of the adiabatic wave function as a product of an electronic, a vibrational, and a rotational part. We demonstrate that this is in fact not so by deriving the GC versions of (4.42) and (4.44). The first step in this process is the definition of a suitable unperturbed problem giving an approximate spectrum of the form (4.42). Second, a perturbation scheme has to be devised to incorporate higher order effects.

According to the results of Section IV.B, the GC energy of a state with total angular momentum J is given by the GC expectation value

$$E^J = \frac{(f|H^J|f)}{(f|\Delta^J|f)} \qquad (4.46)$$

In the projected kernels (4.25) we drop subscripts referring to the electronic state under consideration. The radial weight function $f(\alpha)$ is so far undetermined. An approximate evaluation of (4.46) requires a closer study of the structure of the GC kernels in the neighborhood of a suitable point of reference. Since the GC analog of the potential energy curve is the intrinsic energy

$$E(\alpha) = \frac{\langle \chi(\alpha \mathbf{e}_z)|H|\chi(\alpha \mathbf{e}_z)\rangle}{\langle \chi(\alpha \mathbf{e}_z)|\chi(\alpha \mathbf{e}_z)\rangle} \qquad (4.47)$$

we will examine the behavior of the integral operators around the point $\alpha_0 \mathbf{e}_z$, where α_0 is the value for which $E(\alpha)$ is minimal. Thus α_0 corresponds to R_0 in the adiabatic and BO methods.

The shape of the unprojected overlap

$$\Delta(\alpha, \beta, \theta) = \langle \chi(\alpha \mathbf{e}_z)|e^{-i\theta J_y}|\chi(\beta \mathbf{e}_z)\rangle$$

$$= \langle \varphi(\alpha \mathbf{e}_z)|e^{-i\theta l_y}|\varphi(\beta \mathbf{e}_z)\rangle_r \langle \Phi(\alpha \mathbf{e}_z)|e^{-i\theta L_y}|\Phi(\beta \mathbf{e}_z)\rangle_R$$

$$(4.48)$$

in α, β, θ space is determined mainly by the integral involving the nuclear basis functions. Indeed, because of the small width of the latter, the electronic overlap is practically constant over the range in which the nuclear one is essentially different from zero.

We now consider the intrinsic states as obtained by dilations and rotations from the one corresponding to the reference configuration $\alpha_0 \mathbf{e}_z$, illustrated schematically in Fig. 5. Because of the peaked structure of the nuclear intrinsic states, the overlap will rapidly decrease as $\chi(\alpha_0 \mathbf{e}_z)$ is dilated or rotated away from $\alpha_0 \mathbf{e}_z$.

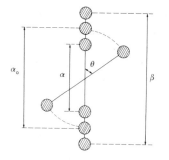

Fig. 5. Schematic illustration of the factorization property (4.49).

This suggests a factorization of the form

$$\langle \chi(\alpha e_z) | e^{-i\theta J_y} | \chi(\beta e_z) \rangle \cong \langle \chi(\alpha e_z) | \chi(\beta e_z) \rangle \langle \chi(\alpha_0 e_z) | e^{-i\theta J_y} | \chi(\alpha_0 e_z) \rangle$$
(4.49)

which holds, for normalized intrinsic states, to third order in the displacements $\alpha - \alpha_0$, $\beta - \alpha_0$, and θ. To prove this statement, one may use (4.20) and the fact that odd powers of J_y have vanishing matrix elements between eignefunctions of J_z corresponding to the same eigenvalue. Equation 4.48 shows that to a good approximation, the overlap of two intrinsic states is the product of overlaps of functions related to each other by a dilation along the z-axis and by a rotation around the y-axis. We should note that in the result (4.49) α_0 stands for any reference value, although we shall apply the formula with α_0 corresponding to the minimum of the intrinsic energy.

The right-hand side of (4.49) suggests a tractable, unperturbed form of the unprojected kernel

$$\Delta(\alpha, \beta, \theta) \cong \Delta_V^{(0)}(\alpha, \beta) \Delta_R^{(0)}(\theta) = \Delta^{(0)}(\alpha, \beta, \theta)$$
(4.50)

where it should be kept in mind that the kernels $\Delta_V^{(0)}(\alpha, \beta)$ and $\Delta_R^{(0)}(\theta)$ are sharply peaked around $\alpha = \beta = \alpha_0$ and $\theta = 0$. The subscripts V and R stand for "vibration" and "rotation"; that is, we have anticipated that dilating and rotating the intrinsic state $\chi(\alpha_0 e_z)$ will describe vibrational and rotational motion, respectively. In view of the results of Sections II.B and II.C, this seems logical.

The strong localization of the basis functions will have approximately the same effect on the unprojected Hamiltonian kernel that it had on $\Delta(\alpha, \beta, \theta)$. Indeed, there are no terms in (4.1) that can significantly couple intrinsic states $\chi(\alpha)$ and $\chi(\beta)$ if α and β refer to entirely different regions in space. We can therefore safely assume that

$$H(\alpha, \beta, \theta) \cong \Delta(\alpha, \beta, \theta) K(\alpha, \beta, \theta)$$
(4.51)

where $K(\alpha, \beta, \theta)$ is a slowly varying function of α, β, and θ. Expanding the smooth kernel $K(\alpha, \beta, \theta)$ in a Taylor series around $\alpha_0 e_z$, one obtains to second order

$$K(\alpha, \beta, \theta) \cong E(\alpha_0) + \tfrac{1}{2} \Big[B(\alpha - \alpha_0)^2 + 2A(\alpha - \alpha_0)(\beta - \alpha_0)$$

$$+ B(\beta - \alpha_0)^2 \Big] - \tfrac{1}{2} C\theta^2$$

$$= E(\alpha_0) + K_V^{(0)}(\alpha, \beta) + K_R^{(0)}(\theta)$$
(4.52)

where $K_V^{(0)}(\alpha, \beta)$ and $K_R^{(0)}(\theta)$ denote vibrational and rotational unperturbed kernels. Linear contributions in $\alpha - \alpha_0$ and $\beta - \alpha_0$ are absent because we are expanding around the energetic minimum α_0. Terms proportional to θ vanish because of the remark above on matrix elements of J_y. The constants A, B, and C are given by

$$A = \left\langle \frac{\partial \chi}{\partial \alpha}(\alpha_0 e_z) | [H - E(\alpha_0)] | \frac{\partial \chi}{\partial \alpha}(\alpha_0 e_z) \right\rangle \qquad (4.53)$$

$$B = \left\langle \chi(\alpha_0 e_z) | [H - E(\alpha_0)] | \frac{\partial^2 \chi}{\partial \alpha^2}(\alpha_0 e_z) \right\rangle \qquad (4.54)$$

$$C = \left\langle \chi(\alpha_0 e_z) | [H - E(\alpha_0)] J_y^2 | \chi(\alpha_0 e_z) \right\rangle \qquad (4.55)$$

Inserting (4.50) and (4.52) into the GC energy (4.46) yields

$$E^J \cong E(\alpha_0) + \frac{(f | \Delta_V^{(0)} K_V^{(0)} | f)}{(f | \Delta_V^{(0)} | f)} + \frac{\int d_{\Lambda\Lambda}^J(\theta) \Delta_R^{(0)}(\theta) K_R^{(0)}(\theta) \, d\cos\theta}{\int d_{\Lambda\Lambda}^J(\theta) \Delta_R^{(0)}(\theta) \, d\cos\theta}$$

$$(4.56)$$

We can compute the GC energy explicitly by combining (4.52) with a Gaussian overlap assumption for the unperturbed overlaps, that is,

$$\Delta_V^{(0)}(\alpha, \beta) = e^{-S_V(\alpha - \beta)^2} \qquad \Delta_R^{(0)}(\theta) = e^{-S_R \theta^2} \qquad (4.57)$$

$$S_V = \frac{1}{2} \left\langle \frac{\partial \chi}{\partial \alpha}(\alpha_0 e_z) | \frac{\partial \chi}{\partial \alpha}(\alpha_0 e_z) \right\rangle \quad S_R = \frac{1}{2} \left\langle \chi(\alpha_0 e_z) | J_y^2 | \chi(\alpha_0 e z) \right\rangle \quad (4.58)$$

and taking into account the form (4.25) of the projected kernels. Minimization of the second term in (4.56) with respect to the radial weight factor $f(\alpha)$ gives the vibrational eigenvalue problem associated with (2.57) if $\alpha - \alpha_0$ and $\beta - \alpha_0$ are assumed to run from $-\infty$ to $+\infty$. Since $S_V \cong \kappa^{-4} \gg 1$, this is justified. As for the third term, a similar argument, $S_R \cong \kappa^{-4} \gg 1$, allows us to consider the integrands for small angles θ only and to extend the integration range for θ from 0 to $+\infty$. The θ-integral can then be computed analytically[65] if one replaces $d_{\Lambda\Lambda}^J(\theta)$ by its asymptotic expansion around $\theta = 0$, that is,

$$d_{\Lambda\Lambda}^J(\theta) \underset{\theta \to 0}{\sim} 1 - \tfrac{1}{4}[J(J+1) - \Lambda^2]\theta^2 \qquad (4.59)$$

The resulting energy spectrum under the assumptions above reads

$$E_v^J \cong E(\alpha_0) - \frac{A}{4S_V} - \frac{C}{2S_R} + \left(v + \frac{1}{2}\right)\Omega_{GC} + \frac{J(J+1) - \Lambda^2}{2I_{GC}} \quad (4.60)$$

$$\Omega_{GC} = \frac{1}{2S_V}\sqrt{A^2 - B^2} \quad \text{and} \quad I_{GC} = \frac{4S_R^2}{C} \quad (4.61)$$

This approximate GC spectrum is of the separated form (4.38). There are some noticeable differences, however, with the adiabatic counterpart (4.42). The first three terms in the expression above represent the so-called "electronic energy." It is clear not only that this quantity is now dependent on the electrons but that also the nuclei contribute directly in $E(\alpha_0)$ and through the additional, energy-lowering terms $-A/4S_V$ and $-C/2S_R$. This phenomenon is also present in the adiabatic version of the Dunham series in which Y_{00}^{AD} turns out to be dependent on nuclear factors.[66] Also, the GC frequency Ω_{GC} and the GC moment of inertia I_{AD} are directly dependent on the electronic eigenfunction through the matrix elements (4.53 to 4.55) and (4.58). This illustrates that under the assumptions leading to the separated form of the energy, the GC still takes account of the fact that the electrons follow the nuclei in their vibrational and rotational motion. A more important new aspect is that the wave functions in the foregoing approximation are not of the adiabatic product form. Although the energies separate and the weight functions factorize as

$$F_v^{(0)}(\boldsymbol{\alpha}) = f_v^{(0)}(\alpha) D_{M\Lambda}^{J*}(\omega_\alpha) \quad (4.62)$$

the GC wave functions, obtained by integrating $F_v^{(0)}(\boldsymbol{\alpha})$ with the intrinsic states (4.18) are not of the adiabatic product type. This demonstrates clearly that the separation of the molecular energy in the form (4.38) is not directly coupled to the adiabatic ansatz.

Since the GC wave functions introduce couplings between the space-fixed coordinates of electrons and nuclei, we can give a simple explanation of the so-called Λ-doubling[67] mentioned earlier. Experimentally, one finds that the rotational energies corresponding to nonzero Λ are not doubly degenerate but are actually slightly split doublets of opposite total parity levels. The GC wave functions corresponding to these states are given by (4.30) and (4.33). It is easily seen that the difference of the corresponding energy expectation values, which by definition is the GC Λ-doubling splitting, can be approximated as

$$\Delta E^J = |E^{J+} - E^{J-}| \cong |(f|H^J_{\Lambda\Lambda}|f)(f|\Delta^J_{\Lambda-\Lambda}|f) - (f|H^J_{\Lambda-\Lambda}|f)| \tag{4.63}$$

where the kernels $H^J_{\Lambda-\Lambda}(\alpha, \beta)$ and $\Delta^J_{\Lambda-\Lambda}(\alpha, \beta)$ are of the form (4.25) with $d^J_{\Lambda\Lambda}(\theta)$ replaced by $d^J_{\Lambda-\Lambda}(\theta)$. The latter function has the following behavior near $\theta = 0$:

$$d^J_{\Lambda-\Lambda}(\theta) \underset{\theta \to 0}{\sim} \binom{J+\Lambda}{J-\Lambda}\theta^{2\Lambda} \tag{4.64}$$

Using arguments similar to those used before to perform the θ-integration, one obtains for the Λ-doubling energy, by retaining the lowest order θ-powers only,

$$\Delta E^J = C_\Lambda \binom{J+\Lambda}{J-\Lambda} \tag{4.65}$$

The constant C_Λ can be shown to be of order $\kappa^{8\Lambda}$ (see also Ref. 53). Therefore Λ-doubling is most important for small Λ and large J values. Equation 4.65 contains as special cases the results for Π and Δ electronic states derived in Ref. 58 by applying PT to the Born-Huang coupled differential equations. The calculation above is much easier and more general because the binomial coefficient in (4.65) specifies the J-dependence of the Λ-doubling energy for any given electronic state.

The derivation of correction terms to (4.52) can most conveniently be done by way of GC perturbation theory. To optimally study the interplay between vibration and rotation, we write the unprojected overlap in the form

$$\Delta(\alpha, \beta, \theta) = \Delta_V^{(0)}(\alpha, \beta)\Delta_R^{(0)}(\theta)[1 + W_V(\alpha, \beta) + W_R(\theta) + W_{VR}(\alpha, \beta, \theta)]$$
$$= \Delta^{(0)}(\alpha, \beta, \theta)[1 + W(\alpha, \beta, \theta)] \tag{4.66}$$

where a separated vibrational, rotational, and vibration-rotation perturbation kernel has been introduced. The Gaussian forms (4.57) do not have to be maintained and may be replaced by any type of kernel providing a good approximation to the matrix elements on the right-hand side of (4.49). Similarly we write (4.51) in the form

$$H(\alpha, \beta, \theta) = \Delta(\alpha, \beta, \theta)[E(\alpha_0) + K_V^{(0)}(\alpha, \beta) + K_R^{(0)}(\theta)$$
$$+ V_V(\alpha, \beta) + V_R(\theta) + V_{VR}(\alpha, \beta, \theta)]$$
$$= \Delta(\alpha, \beta, \theta)[K^{(0)}(\alpha, \beta, \theta) + V(\alpha, \beta, \theta)] \tag{4.67}$$

The vibration, rotation, and vibration-rotation perturbations in (4.67), for example, can be formed in grouping the higher order Taylor expansion terms following (4.52). However, here also there is freedom in choosing convenient unperturbed kernels. In particular, the commonly used vibrational potentials (Morse potential, Eckart potential, etc.[68]) generate suitable unperturbed vibrational kernels if the corresponding eigenvalue problems are put in a GC form following the procedure applied to the harmonic oscillator in Section II.C. Therefore the separated form of the GC energy (4.60) is not a consequence of any special functional form of the unperturbed kernels. For any of the choices above, the unperturbed energy can be written as

$$E^{(0)}(J,v)=E(\alpha_0)+E_V^{(0)}(v)+E_R^{(0)}(J) \qquad (4.68)$$

where $E_V^{(0)}(v)=(f_v^{(0)}|\Delta_V^{(0)}K_V^{(0)}|f_v^{(0)})$ and $E_R^{(0)}(J)$ is given by the last term in (4.60). We assume $f_v^{(0)}$ normalized with respect to $\Delta_V^{(0)}$ such that the PT formulas of Section II.C apply.

Combination of (4.66) and (4.67) leads to an alternative expression for the Hamiltonian kernel.

$$H(\alpha,\beta,\theta)=\Delta_V^{(0)}(\alpha,\beta)\Delta_R^{(0)}(\theta)\big[K^{(0)}(\alpha,\beta,\theta)+L_V(\alpha,\beta)$$
$$+L_R(\theta)+L_{VR}(\alpha,\beta,\theta)\big]$$
$$=\Delta^{(0)}(\alpha,\beta,\theta)\big[K^{(0)}(\alpha,\beta,\theta)+L(\alpha,\beta,\theta)\big] \qquad (4.69)$$

$$L_V=V_V+E(\alpha_0)W_V+K_V^{(0)}W_V \qquad (4.70)$$

$$L_R=V_R+E(\alpha_0)W_R+K_R^{(0)}W_R \qquad (4.71)$$

$$L_{VR}=V_{VR}+E(\alpha_0)W_{VR}+\big(K_R^{(0)}+V_R\big)W_V+\big(K_V^{(0)}+V_V\big)W_R \qquad (4.72)$$

Incorporating the effect of the L-perturbations by first-order GC-PT, one obtains

$$E_v^J\cong E^{(0)}(J,v)+\big(f_v^{(0)}|\Delta_V^{(0)}\langle L\rangle^J|f_v^0\big)$$
$$-E^{(0)}(J,v)\big(f_v^0|\Delta_V^{(0)}\langle W\rangle^J|f_v^{(0)}\big) \qquad (4.73)$$

where the following notation for the "rotational average" of an integral kernel $C(\alpha,\beta,\theta)$ has been introduced

$$\langle C\rangle^J=\frac{\int d_{\Lambda\Lambda}^J(\theta)C(\alpha,\beta,\theta)\Delta_R^{(0)}(\theta)\,d\cos\theta}{\int d_{\Lambda\Lambda}^J(\theta)\Delta_R^{(0)}(\theta)\,d\cos\theta} \qquad (4.74)$$

Inserting the explicit form of both $L(\alpha, \beta, \theta)$ and $W(\alpha, \beta, \theta)$ and adopting the notations

$$\langle C \rangle_v = \left(f_v^{(0)} | C | f_v^{(0)} \right) \quad \text{and} \quad \langle C \rangle_v^J = \left(f_v^{(0)} | \langle C \rangle^J | f_v^{(0)} \right) \qquad (4.75)$$

for "vibrational averages," we can write the first-order GC PT result (4.73) as

$$E_v^J \cong E(\alpha_0) + E_V^{(0)}(v) + E_R^{(0)}(J) + \Delta E_V(v) + \Delta E_R(J) + \Delta E_{VR}(J, v)$$
$$(4.76)$$

$$\Delta E_V(v) = \langle V_V \rangle_v + \langle K_V^{(0)} W_V \rangle_v - \langle K_V^0 \rangle_v \langle W_V \rangle_v \qquad (4.77)$$

$$\Delta E_R(J) = \langle V_R \rangle^J + \langle K_R^{(0)} W_R \rangle^J - \langle K_R^{(0)} \rangle^J \langle W_R \rangle^J \qquad (4.78)$$

$$\Delta E_{VR}(J, v) = \langle V_V \rangle_v \langle W_R \rangle^J + \langle V_R \rangle^J \langle W_V \rangle_v$$
$$+ \langle V_{VR} \rangle_v^J - \left(\langle K_R^{(0)} \rangle^J + \langle K_V^{(0)} \rangle_v \right) \langle W_{VR} \rangle_v^J \qquad (4.79)$$

The physical interpretation of the extra terms is obvious: $\Delta E_V(v)$ and $\Delta E_R(J)$ correct for deviations from the unperturbed vibrational and rotational energies $E_V^0(v)$ and $E_R^0(J)$, while $\Delta E_{VR}(J, v)$ describes the effects arising from the vibration-rotation coupling. The explicit derivation of these terms for Gaussian overlaps, quadratic kernels $K_R^{(0)}$ and $K_V^{(0)}$, and higher order Taylor expansions for the perturbations and the $d_{\Lambda\Lambda}^J(\theta)$ yields the GC Dunham series. The GC spectroscopic constants Y_{kl}^{GC} are given elsewhere.[69]

In the adiabatic approximation, centrifugal stretching and vibration-rotation coupling is described by considering the centrifugal potential as a perturbation of the potential energy curve.[70] On the other hand, in the GC energy (4.76) each of the constituents consists of averages of integral kernels. Pure rotational and pure vibrational effects correspond to $\langle \ \rangle^J$ and $\langle \ \rangle_v$ averages, respectively, while vibration-rotation terms contain double averages $\langle \ \rangle_v^J$ and products $\langle \ \rangle_v \langle \ \rangle^J$. Each type of energy therefore acquires more of an identity that is, however, dependent on the choice of the unperturbed kernels.

V.   GENERATOR COORDINATE THEORY OF POLYATOMIC SYSTEMS

A.   Notations and Definitions

The aim of Section V is to formulate the GC approximation for polyatomic molecules in analogy to the exposition for diatomic systems in Section IV. In doing so we emphasize the simplifying aspects, typical of diatomic molecules, that carry over to polyatomic ones and how the different topics compare as developed in the adiabatic or generator coordinate approximations.

We consider the center of mass of the nuclei to be the origin of the space-fixed frame in which the Hamiltonian (3.1) is defined. The generator coordinate vectors $\{\boldsymbol{\alpha}_i: i=1,2,\dots,N\}$ measured from this point satisfy the condition

$$\sum M\boldsymbol{\alpha}=0 \tag{5.1}$$

This equation assures us that the intrinsic states (3.58) will not describe pure translations of the system as a whole. The behavior under rotations of the $\chi(r,R|\alpha)$ can be established as follows. The unitary equivalence of diatomic electronic Hamiltonians corresponding to different nuclear orientations generalizes directly to the polyatomic case, that is,

$$\mathcal{R}_e(\Omega)H_{el}(\alpha)\mathcal{R}_e^{-1}(\Omega)=H_{el}(A(\Omega)\alpha) \tag{5.2}$$

which implies that

$$U(\alpha)=U(A(\Omega)\alpha) \quad \text{and} \quad \mathcal{R}_e(\Omega)\varphi(r|\alpha)=\varphi(r|A(\Omega)\alpha) \tag{5.3}$$

The first of these two equations merely states the well-known fact that the potential energy surfaces do not depend on the orientation of the nuclear skeleton, while the second relates polyatomic electron states, corresponding to differently oriented skeletons, to each other. As implied by the notation, the nuclear basis functions $\Phi(|R-\alpha|)$ are to be chosen such that

$$\Phi(|R-\alpha|)=\Phi(|\mathbf{R}_1-\boldsymbol{\alpha}_1|,|\mathbf{R}_2-\boldsymbol{\alpha}_2|,\dots,|\mathbf{R}_N-\boldsymbol{\alpha}_N|) \tag{5.4}$$

In our symbolic notation we then have

$$\mathcal{R}_n(\Omega)\Phi(|R-\alpha|)=\Phi(|R-A(\Omega)\alpha|) \tag{5.5}$$

Combination of (5.3) and (5.5) gives us the closure relation under rotations of the polyatomic intrinsic states

$$\mathcal{R}(\Omega)\chi(r,R|\alpha)=\chi(r,R|A(\Omega)\alpha) \tag{5.6}$$

To distinguish between configurations that are related to each other by a pure rotation only, we formally introduce three conditions, in addition to (5.1), such that the geometric figures defined by the vectors $\{\boldsymbol{\alpha}_i\}$ cannot be brought in coincidence by a rotation. The auxiliary restrictions on the GCs can be the ones given by Eckart[71] for nuclear variables or one can, for example, choose $\boldsymbol{\alpha}_1$ to lie along the z-axis and require $\boldsymbol{\alpha}_2$ to belong to the xz-plane.[72] However, in the following it will not be necessary to actually specify these three extra conditions. In any case, the remaining $3N-6$ independent GCs can be subject to a transformation to give a convenient set of internal generator coordinates $\mu=\mu(\alpha)$. In terms of these internal GCs we can rewrite the closure relations (5.6) as

$$\mathcal{R}(\Omega)\chi(\mu,0)=\chi(\mu,\Omega) \tag{5.7}$$

which is the polyatomic analog of (4.18).

B. Rotational Invariance and Inversion Symmetry

Since we have verified that the polyatomic intrinsic states are closed under rotations, the results of Section II.C readily apply. The weight function, expressed in the Euler angles Ω and the internal GCs μ, is therefore of the form

$$F^{JM}(\mu,\Omega)=\sum_K f_K^J(\mu)D_{MK}^{J*}(\Omega) \tag{5.8}$$

Substitution into the GC wave function (3.56) yields angular momentum eigenstates

$$\Psi^{JM}(r,R)=\sum_K \int f_K^J(\mu)D_{MK}^{J*}(\Omega)\chi(r,R|\mu,\Omega)\,d\mu\,d\Omega \tag{5.9}$$

Introducing the projection operators (2.39), these wave functions can be written as

$$\Psi^{JM}(r,R)=\sum_K \int f_K^J(\mu)P_{MK}^J\chi(r,R|\mu,0)\,d\mu \tag{5.10}$$

The essential difference, as compared to the diatomic case, is that the polyatomic intrinsic states are no longer eigenfunctions of J_z. Therefore (5.10) is a superposition of $2J+1$ components, each having angular momentum quantum numbers JM. The variational equations satisfied by the internal weight functions $f_K^J(\mu)$ form a set of coupled integral equations

$$\sum_L \int \left[H_{KL}^J(\mu,\nu)-E^J\Delta_{KL}^J(\mu,\nu)\right]f_L^J(\nu)\,d\nu=0 \tag{5.11}$$

$$H_{KL}^J(\mu,\nu)=\frac{2J+1}{8\pi^2}\int D_{KL}^{J*}(\Omega)\langle\chi(\mu)|H\mathcal{R}(\Omega)|\chi(\nu)\rangle\,d\Omega \tag{5.12}$$

and a similar expression for $\Delta^J_{KL}(\mu, \nu)$. For fixed total angular momentum J, the solutions $E_v^{J\tau}$ of these equations represent a multiplet of $2J+1$ rotational levels $(\tau = 1, 2, \ldots, 2J+1)$ for the molecule in a certain "vibrational state v."

The outline above demonstrates that *application of angular momentum projection techniques to polyatomic molecules does not present any formal difficulties.* Compared to the diatomic case, the only modification is that, because of the loss of axial symmetry, the angular momentum projected kernels (5.12) require integrations over three Euler angles instead of just one. The situation is considerably more complicated in the adiabatic approximation. If one wants to follow the procedure sketched in Section IV.B, a body-fixed frame must be defined explicitly (i.e., a choice must be made for the three auxiliary conditions that so far, has been arbitrary). Indeed, to account for rotational invariance in the adiabatic scheme, the Hamiltonian must be transformed to a body-fixed frame, which necessitates the definition of this axis system in terms of particle coordinates. Whereas in the diatomic case the choice was obvious and the transformation could be done explicitly, this becomes prohibitively complex for polyatomic molecules. To our knowledge no workable nuclear motion equations yielding adiabatic total angular momentum eigenstates have been derived. Traditionally one considers the primitive version (3.40) of the adiabatic nuclear motion problem as satisfactory. However, since the additional operators arising from a transformation to a body-fixed frame are then neglected, it is clear that this violates rotational invariance. Furthermore even at this level, the popular choice of the Eckart frame as the body-fixed axis system is not free of troublesome aspects, as demonstrated by Sutcliffe.[73]

We now turn to the investigation of implications of inversion symmetry, which will be performed in two alternative ways. First one can simply require the angular momentum eigenstates (5.9) to also have a certain parity, that is,

$$I\Psi^{JM}(r, R) = \sum_K \int f_K^J(\mu) D_{MK}^{J*}(\Omega)\chi(-r, -R|\mu, \Omega)\, d\mu\, d\Omega$$

$$= \sum_K \int f_K^J(\mu) D_{MK}^{J*}(\Omega)\chi(r, R|\bar{\mu}, \bar{\Omega})\, d\mu\, d\Omega$$

$$= \sum_K \int f_K^J(\bar{\mu}) D_{MK}^{J*}(\bar{\Omega})\chi(r, R|\mu, \Omega)\left|\frac{d\bar{\mu}}{d\mu}\right| d\mu\, d\Omega$$

$$= \sum_K \int f_K^J(\bar{\mu})(-)^{J+K} D_{M-K}^{J*}(\Omega)\chi(r, R|\mu, \Omega)\left|\frac{d\bar{\mu}}{d\mu}\right| d\mu\, d\Omega$$

$$= \Pi \sum_K \int f_K^J(\mu) D_{MK}^{J*}(\Omega)\chi(r, R|\mu, \Omega)\, d\mu\, d\Omega \qquad (5.13)$$

where $\bar{\mu}$ defines the mirror image of the n-pod described by the internal GCs μ and $\bar{\Omega}=(\pi+\varphi, \pi-\theta, \pi-\gamma)$. Comparing the last two lines in (5.13), we can conclude that

$$\Pi f^J_{-K}(\mu)=(-)^{J+K}\left|\frac{d\bar{\mu}}{d\mu}\right|f^J_K(\bar{\mu})\tag{5.14}$$

This condition fixes the internal weight functions at the inverted configurations $\bar{\mu}$ in terms of the ones at μ. It is similar to the conditions on internal wave functions derived by Wigner.[72] In fact (5.14) is a sort of "boundary condition" on the coupled Wheeler equations (5.11). However, it is not clear at this point what is practical implications are. We therefore also present the straightforward projection of good parity quantum numbers along the lines followed for diatomic systems; that is, we introduce the parity-projected intrinsic states

$$\chi^\Pi(\mu,\Omega)=\chi(\mu,\Omega)\pm I\chi(\mu,\Omega)=\chi(\mu,\Omega)\pm\chi(\bar{\mu},\bar{\Omega})\tag{5.15}$$

The GC wave functions using these basis functions are

$$\Psi^{JM\Pi}(r,R)=\sum_K\int f^J_K(\mu)D^{J*}_{MK}(\Omega)\left[\chi^\Pi(r,R|\mu,\Omega)\right]d\mu\,d\Omega$$
$$=\sum_K\int f^J_K(\mu)\left[P^J_{MK}\chi(r,R|\mu,0)\pm(-)^{J+K}P_{M-K}\chi(r,R|\bar{\mu},0)\right]d\mu\,d\Omega$$

$$\tag{5.16}$$

and obviously have the parity of $\chi^\Pi(\mu,\Omega)$. In this case there are no extra conditions on the weight functions, but a number of additional terms appear in the Wheeler integral kernels.

$$H^{J\Pi}_{KL}(\mu,\nu)=H^J_{KL}(\mu,\nu)\pm(-)^{J+K}H^J_{-KL}(\bar{\mu},\nu)\tag{5.17}$$
$$\pm(-)^{J+L}H^J_{K-L}(\mu,\bar{\nu})+(-)^{K+L}H^J_{-K-L}(\bar{\mu},\bar{\nu})\tag{5.18}$$

and similar expressions for the $\Delta^{J\Pi}_{KL}(\mu,\nu)$. Aside from the direct part $H^J_{KL}(\mu,\nu)$, kernels appear that correspond to matrix elements involving inverted configurations, accompanied by sign factors depending on total and azimuthal angular momentum quantum numbers. The discussion of the physical meaning of these terms is postponed until the end of Section V.D.

C. Vibrations, Rotations, and Vibration-Rotation Interaction

Although the separation of the molecular energy in the form (4.38) is sometimes not so pronounced in polyatomic spectra as in diatomic ones, it still proves an excellent starting point in the majority of cases.

Using what we have called the primitive version of the adiabatic approximation, Wilson and Howard[37] showed that the nuclear kinetic energy can be written as the sum of a vibrational and a rotational contribution and a vibration-rotation part, the so-called Coriolis energy. They arrived at this result by transforming T to the Eckart frame. The molecular energy splits in the described form if the rotation-vibration interactions are neglected. Expansion of the potential energy surface to second order in the nuclear displacements from the equilibrium configuration R_0 then yields a vibrational eigenvalue problem consisting of $3N-6$ coupled oscillators ($3N-5$ if the R_0 structure is linear). Introducing normal coordinates,[33] these oscillators decouple such that the vibrational energy is a sum of oscillator frequencies. If the intertia tensor of the molecule is assumed to be the one corresponding to the configuration R_0, the rotational portion of the nuclear kinetic energy can be identified with a rigid-rotor Hamiltonian

$$\frac{L_{x'}^2}{2I_1^{(0)}} + \frac{L_{y'}^2}{2I_2^{(0)}} + \frac{L_{z'}^2}{2I_3^{(0)}} \qquad (5.19)$$

where the $I_i^{(0)}$ are the principal moments of inertia and the $L_{x'}$, $L_{y'}$, and $L_{z'}$ the components of the nuclear angular momentum along the axes of the Eckart frame. The spectrum of (5.19) has been studied in detail.[39,75] Various refinements taking into account anharmonicities, centrifugal distortions, and Coriolis terms have been given, and a variety of term formulas for the analysis of the spectra are available. We give no details here and refer the reader to the excellent papers and books on the subject.[29,31,33,39,61,74,75] What is of primary interest here is whether a simple and well-founded approximate form of the GC kernels can be given such that the Wheeler coupled integral equations (5.11) yield the energy pattern just described. This would have the merit of making available a basis for the derivation of GC polyatomic term formulas. In view of the increased accuracy of the GCM, these formulas can be expected to perform better than their adiabatic counterparts. Also it would have the reassuring property that rotational invariance is properly accounted for. Since (5.19) described rotations of the nuclei only this is not the case in the adiabatic approximation.

The function $E(\mu)$ defined by the expectation value

$$E(\mu) = \frac{\langle \chi(\mu) | H | \chi(\mu) \rangle}{\langle \chi(\mu) | \chi(\mu) \rangle} \qquad (5.20)$$

where $\chi(\mu) \equiv \chi(\mu, 0)$, determines a $(3N-6)$-dimensional surface closely resembling the adiabatic potential energy surface. If the minimum of (5.20) occurs for the internal GC values μ_0, an equivalent set of GCs α_0 exists, for which $\mu_0 = \mu(\alpha_0)$. These define the GC nuclear equilibrium configuration and furnish the proper reference point for the study of the polyatomic Wheeler kernels.

The discussion in the diatomic case concerning the peakedness of the intrinsic states and its consequences on the behavior of the overlap and Hamiltonian kernels may be repeated for polyatomic molecules. We therefore assume that the unprojected kernel $\Delta(\mu, \nu, \Omega)$ factorizes as follows:

$$\Delta(\mu, \nu, \Omega) = \langle \chi(\mu) | \mathcal{R}(\Omega) | \chi(\nu) \rangle \cong \langle \chi(\mu) | \chi(\nu) \rangle \langle \chi(\alpha_0) | \mathcal{R}(\Omega) | \chi(\alpha_0) \rangle \tag{5.21}$$

It should be observed that since $\chi(\mu)$ is no longer an eigenfunction of J_z, this approximation is not as well founded as the corresponding one (4.49) for the diatomic overlap. As for the ratio $K(\mu, \nu, \Omega) = H(\mu, \nu, \Omega)/\Delta(\mu, \nu, \Omega)$, a convenient separated form turns out to be

$$\begin{aligned}
K(\mu, \nu, \Omega) &= \frac{\langle \chi(\mu) | H\mathcal{R}(\Omega) | \chi(\nu) \rangle}{\langle \chi(\mu) | \mathcal{R}(\Omega) | \chi(\nu) \rangle} \\
&\cong E(\alpha_0) + \frac{\langle \chi(\mu) | [H - E(\alpha_0)] | \chi(\nu) \rangle}{\langle \chi(\mu) | \chi(\nu) \rangle} \\
&\quad + \frac{\langle \chi(\alpha_0) | [H - E(\alpha_0)] \mathcal{R}(\Omega) | \chi(\alpha_0) \rangle}{\langle \chi(\alpha_0) | \mathcal{R}(\Omega) | \chi(\alpha_0) \rangle} \\
&= E(\alpha_0) + K_V^{(0)}(\mu, \nu) + K_R^{(0)}(\Omega) \tag{5.22}
\end{aligned}$$

A first check shows that this expression results in an equality if no deformation ($\chi(\mu) = \chi(\nu) = \chi(\alpha_0)$) or no rotation ($\Omega = 0$) of the equilibrium structure is involved. In fact, (5.22) is nothing but the brute force generalization of the diatomic Taylor expansion (4.52), where higher order powers in pure vibrational and pure rotational GCs are incorporated but all mixed terms are neglected. The assumptions above are therefore analogous to discarding the Coriolis operators in the adiabatic case.

Considering (5.21) and (5.22) as the proper approximate forms of the unprojected kernels and writing the unperturbed weight functions as a product,

$$F^{(0)}(\mu, \Omega) = f_V^{(0)}(\mu) f_R^{(0)}(\Omega) \tag{5.23}$$

the Wheeler equations decouple into a vibrational and a rotational eigenvalue problem. The resulting zeroth-order vibrational equation has the form

$$\int [K_V^{(0)}(\mu, \nu) - E_V^{(0)}] \Delta_V^{(0)}(\mu, \nu) f_V^{(0)}(\nu) \, d\nu = 0 \tag{5.24}$$

The unperturbed kernel $K_V^{(0)}(\mu, \nu)$ is defined through (5.22) and of course $\Delta_V^{(0)}(\mu, \nu) = \langle \chi(\mu) | \chi(\nu) \rangle$. In this way molecular vibrations are described by distortions of the intrinsic state $\chi(\alpha_0)$ corresponding to internal GCs μ only. A multidimensional form of the harmonic approximation is known to lead to a set of coupled oscillators that can be decoupled by transforming to normal GCs associated with the point group symmetry of α_0.[14] There is also no problem in incorporating anharmonicities by GC-PT. As for the rotational part of the energy, the formalism of Section II.C strictly applies; that is,

$$f_R^{(0)}(\Omega) = \sum_K C_K D_{MK}^{J*}(\Omega) \qquad (5.25)$$

where the unknown coefficients C_K satisfy the matrix equation

$$\sum_L \left[H_{KL}^J - E_R^{(0)} \Delta_{KL}^J \right] C_L = 0 \qquad (5.26)$$

$$H_{KL}^J = \int D_{KL}^{J*}(\Omega) \langle \chi(\alpha_0) | \left[H - E(\alpha_0) \right] \mathcal{R}(\Omega) | \chi(\alpha_0) \rangle d\Omega \qquad (5.27)$$

$$\Delta_{KL}^J = \int D_{KL}^{J*}(\Omega) \langle \chi(\alpha_0) | \mathcal{R}(\Omega) | \chi(\alpha_0) \rangle d\Omega \qquad (5.28)$$

Thus at this level the rotational motion of the molecule is described by considering rotated equilibrium intrinsic states $\mathcal{R}(\Omega)\chi(\alpha_0)$ as basis functions. It has been shown how (5.26) can be solved approximately such that a rigid-rotor spectrum results.[76] However, a more involved procedure than a simple expansion of both kernels and Wigner functions was necessary. The spectrum resulting from (5.24) and (5.26) is

$$E^{(0)} = E(\alpha_0) + E_V^{(0)} + E_R^{(0)} \qquad (5.29)$$

where the particular forms of $E_V^{(0)}$ and $E_R^{(0)}$ may depend on the extra approximations used in solving the unperturbed problems.

Including rotation-vibration coupling presents no formal difficulties. Indeed, no formula in Section IV.D has to be modified, except that the single radial variable α is replaced by $3N - 6$ internal GCs μ and instead of the polar angle θ all three Euler angles are involved. It thus seems feasible to attempt the construction of GC term formula for the analysis of polyatomic spectra. From the results of Une et al.,[76,77] one can conclude that this is likely to lead to the use of differential geometry in the field of molecular spectroscopy.

D. Feasible Molecular Symmetries and Generator Coordinates

The symmetry group of the molecular Hamiltonian (neglecting overall translations in space) is the direct product of the rotation-inversion group with the permutation groups of all sets of identical particles. However, the observed spectral lines can in most cases be labeled by considering the irreducible representations of smaller groups. This has led Longuet-Higgins[78] to introduce the concept of so-called feasible symmetry transformations. These are associated with nuclear rearrangements that can be realized within the time scale of the experiment involved. The group formed by the feasible symmetry operators is called the *molecular symmetry group*. For the systems considered so far, (i.e., semirigid molecules in which the internal motion consists of small displacements from a single equilibrium configuration), the molecular symmetry group is the point group associated with the equilibrium geometry or the direct product of this group with the rotation-inversion group.[79] However, in a number of systems, the so-called flexible or nonrigid molecules, a larger group is needed to uniquely label the observed quantum states. Ammonia was the first example in which the classical concept of a definite structure failed.[80] Because equivalent potential minima for the nitrogen atom existed, an inversion splitting, associated with tunneling through the barrier separating the equivalent structures, was detected. For a review of other examples we refer to Berry.[79] Here we consider the following questions: Can the generator coordinate approximation be formulated consistent with all symmetries of H? What are the implications of considering feasible transformations only? The following paragraphs deal with partial answers to these problems.

The simplest type of nonrigidity occurs for homonuclear diatomic molecules; that is, in principle the identical nuclei in these molecules may change positions. This phenomenon was accounted for in Section IV.C by using parity-projected intrinsic states (4.32). However, in the diatomic case no inversion splittings are observed because for a given electronic state and total angular momentum, the nuclear parity is fixed by the sign factor in (4.37); that is, one of the levels of an inversion doublet is always nonexisting. However, the importance of nuclear permutation symmetry manifests itself in the appearance of the so-called ortho- and paramodifications that if separated, behave as essentially different forms of matter (e.g., *o*- and *p*-hydrogen[55]).

For polyatomic systems we have already established closure under the rotation-inversion group. Therefore, only the implications of particle statistics remain to be taken care of. For the permutations of electrons, antisymmetry of the total wave function is guaranteed by the choice of an antisymmetric electronic state. If the molecule contains identical boson and/or fermion nuclei, the nuclear intrinsic states may be symmetrized and/or antisymmetrized over these particles. This is particularly easy if we assume $\Phi(|R-\alpha|)$ to be built from one-particle functions $\Phi_k(|\mathbf{R}_i^{(k)} - \boldsymbol{\alpha}_i^{(k)}|)$ for nuclei of type k. Permutations of particle positions can then be translated in permutations of

GC vectors. Furthermore, the electronic Hamiltonian is invariant under interchange of like nuclei. For a nondegenerate electronic state this implies invariance of both the potential energy surface and the electronic eigenfunctions under these permutations.[81] In case of degeneracy, certain transformation properties of the electronic eigenstates may result. At any rate it seems feasible to construct the full intrinsic states such that conjugate to a permutation $\mathscr{P}$ of two identical nuclei a permutation P in the space of GCs exists for which

$$\mathscr{P}\chi(r, R|\alpha) = \chi(r, R|P(\alpha)) \tag{5.30}$$

That is, the intrinsic states can be chosen as closed under the molecular permutation group. The relations (5.6), (5.15), and (5.30) prove that the GC approximation can be made consistent with the full symmetry group of the molecular Hamiltonian.

Adapting the intrinsic state to the molecular permutation group generates extra terms in the GC integral kernels of the type already encountered in the discussion on inversion symmetry. To understand the physical meaning of these inversion- and exchange-type contributions to the Wheeler kernels, let us start from the semirigid picture—assuming, that is, that we have found a local minimum of the energy (5.20) in which the intrinsic state is not adapted to the full molecular symmetry group. If no combination of the inversion, a rotation, and permutations of identical nuclei brings the equilibrium GC structure α_0 into coincidence with itself, the inversion- and exchange- type contributions are vanishingly small around α_0 and can be neglected. However, if an element S of the full symmetry group can be found such that $S\chi(\alpha_0) = \pm\chi(\alpha_0)$, the corresponding term in the symmetry-adapted Wheeler kernels may be of similar magnitude as the direct part in the neighborhood of α_0, since in that case

$$|\langle\chi(\alpha)|H|\chi(\alpha_0)\rangle| = |\langle\chi(\alpha)|HS|\chi(\alpha_0)\rangle| \tag{5.31}$$

By definition the particular combination of symmetry operators is a feasible transformation if the associated inversion-exchange kernel has an effect on the energy spectrum, that is, if a nonrigidity of the geometric type exists (e.g., in methane[82]). If reconstructing the α_0 configuration can be performed only by also incorporating a large-amplitude vibrational motion, one speaks of a dynamical nonrigidity (e.g., cyclopentane[83]). Thus we have made a connection between the concept of feasible symmetry operators and the overall form of the symmetry-adapted Wheeler kernels.

Acknowledgments

L. Lathouwers acknowledges financial support from the I.I.K.W. (Interuniversitair Instituut voor Kernwetenschappen, Belgium) and the NATO Scientific Affairs Division under grant 1776.

References

1. D. L. Hill and J. A. Wheeler, *Phys. Rev.*, **89**, 1106 (1953).
2. L. Lathouwers et al., *Chem. Phys. Lett.*, **52**, 439 (1977).
3. M. M. Hessel and C. R. Vidal, *J. Chem. Phys.*, **70**, 4439 (1979).
4. L. Lathouwers and P. Van Leuven, *Chem. Phys. Lett.*, **70**, 410 (1980).
5. R. G. Woolley and B. T. Sutcliffe, *Chem. Phys. Lett.*, **45**, 393 (1977).
6. C. W. Wong, *Phys. Rep.*, **15**, 283 (1975).
7. See, e.g., S. T. Epstein, *The Variation Method in Quantum Chemistry*, Academic, New York, 1974.
8. P. O. Löwdin, in *Advances in Quantum Chemistry*, Vol. 5, Academic, New York, 1970.
9. J. J. Griffin and J. A. Wheeler, *Phys. Rev.*, **108**, 311 (1957).
10. I. Shavitt, *Methods in Computational Physics*, Vol. 2, Academic, New York, 1962.
11. E. Deumens and J. Broeckhove, *Z. Phys. A*, **292**, 243 (1979).
12. L. Lathouwers, *Ann. Phys.*, **102**, 347 (1976).
13. B. Laskowski and P. O. Löwdin, *Chem. Phys. Lett.*, **16**, 1 (1972).
14. D. M. Brink and A. Weiguny, *Nuclear Phys. A*, **120**, 59 (1968).
15. M. Bouten, *Physica*, **42**, 572 (1969).
16. A. Messiah, *Quantum Mechanics*, Vol. 2, North Holland, Amsterdam, 1960.
17. A. Kelemen and R. M. Dreizler, *Z. Phys. A*, **278**, 269 (1976).
18. See, e.g., F. G. Tricomi, *Integral Equations*, Interscience, New York, 1957.
19. L. Lathouwers and R. L. Lozes, *J. Phys. A*, **10**, 1465 (1977).
20. N. Wiener, *The Fourier Integral*, Cambridge University Press, Cambridge, 1933.
21. J. J. Griffin, *Phys. Rev.*, **108**, 328 (1957).
22. M. Born and J. R. Oppenheimer, *Ann. Phys.*, **84**, 457 (1927).
23. See, e.g., R. G. Woolley, *Adv. Phys.*, **25**, 27 (1976).
24. K. Scharzschild, *Ber. Preuss. Akad. Wiss.*, **1**, 548 (1916).
25. J. Franck, *Trans. Faraday Soc.*, **21**, 536 (1925).
26. J. M. Combes, *Proceedings of the Conference for the 50th Anniversary of the Schrödinger Equation*, Vienna, 1976, *Acta Phys. Austraiaca*, 1977.
27. M. Born and K. Huang, *Dynamical Theory of Crystal Lattices*, Clarendon Press, Oxford, 1954.
28. J. C. Browne, in *Proceedings of the Boulder Conference on Theoretical Chemistry*, Wiley, New York, 1972.
29. J. H. Van Vleck, *Rev. Mod. Phys.*, **23**, 213 (1951).
30. S. Epstein, *J. Chem. Phys.*, **44**, 836, 4062 (1966).
31. W. Kolos, in *Advances in Quantum Chemistry*, Vol. 5, Academic, New York, 1970.
32. E. B. Wilson, *Int. J. Quant. Chem. Symp.*, **13**, 5 (1979).
33. E. B. Wilson et al., *Molecular Vibrations*, McGraw-Hill, New York, 1955.
34. M. Moshinsky and C. Kittel, *Proc. Natl. Acad. Sci. (U.S.)*, **60**, 1110 (1968).
35. D. M. Bishop and L. M. Cheung, *Phys. Rev. A*, **16**, 640 (1977).
36. J. L. Dunham, *Phys. Rev.*, **41**, 721 (1932).
37. J. B. Howard and E. B. Wilson, *J. Chem. Phys.*, **4**, 260 (1936).
38. E. Hill and J. H. Van Vleck, *Phys. Rev.*, **32**, 250 (1928).
39. H. C. Allen and P. C. Cross, *Molecular Vib-Rotors*, Wiley, New York, 1963.
40. W. Demtröder, *Quantum Dynamics of Molecules*, Plenum, New York, 1980.
41. M. Born, *Göttinger Nachr. Math. Phys. Kl.*, **1** (1951).
42. K. Dressler et al., *J. Mol. Spectrosco.*, **75**, 205 (1979).
43. A. M. Halpern, *Phys. Rev.*, **186**, 14 (1969).
44. G. Hunter and H. O. Pritchard, *J. Chem. Phys.*, **46**, 215 (1967).
45. L. I. Ponomarev and S. I. Vinitsky, *J. Phys. B.*, **12**, 567 (1979).

46. W. H. Miller, Ed. *Dynamics of Molecular Collisions*, Part A, Plenum, New York, 1976.
47. W. H. Miller, Ed. *Dynamics of Molecular Collisions*, Part B, Plenum, New York, 1976.
48. L. Lathouwers and P. Van Leuven, *Quantum Dynamics of Molecules*, Plenum, New York, 1980.
49. W. H. Wing et al., *Phys. Rev. Lett.*, **36**, 1488 (1976).
50. D. M. Bishop and L. M. Cheung, *Phys. Rev. A*, **18**, 1846 (1978).
51. I. L. Thomas, *Phys. Rev. A*, **3**, 565 (1971), and references therein.
52. R. T. Pack and J. O. Hirschfelder, *J. Chem. Phys.*, **52**, 521 (1970).
53. I. Kovacs, *Rotational Structure in the Spectra of Diatomic Molecules*, American Elsevier, New York, 1969.
54. J. M. Combes and R. Seiler, *Quantum Dynamics of Molecules*, Plenum, New York, 1980.
55. G. Herzberg, *Spectra of Diatomic Molecules*, Van Nostrand, New York, 1950.
56. J. C. Slater, *Quantum Theory of Molecules and Solids*, McGraw-Hill, New York, 1963.
57. B. R. Judd, *Angular Momentum Theory for Diatomic Molecules*, Academic, New York, 1975.
58. L. D. Landau and E. M. Lifshitz, *Quantum Mechanics*, Pergamon, Oxford, 1964.
59. F. Reiche, *Z. Phys.*, **39**, 444 (1926).
60. L. Lathouwers, *J. Phys. A*, **13**, 2287 (1980).
61. H. W. Kroto, *Molecular Rotation Spectra*, Wiley, New York, 1975.
62. L. Lathouwers, *Phys. Rev. A*, **18**, 2150 (1978).
63. P. R. Bunker, *J. Mol. Spectrosc.*, **35**, 306 (1970).
64. S. Z. Moody and C. L. Beckel, Int. J. Quant. Chem. Symp., **3**, 469 (1970).
65. B. J. Verhaar, *Nuclear Phys.*, **54**, 641 (1964).
66. C. J. H. Schutte, *The Theory of Molecular Spectroscopy*, North Holland, Amsterdam, 1976.
67. J. H. Van Vleck, *Phys. Rev.*, **33**, 467 (1929).
68. J. Goodisman, *Diatomic Interaction Potential Theory*, Academic, New York, 1973.
69. L. Lathouwers, in preparation.
70. C. L. Beckel, *J. Chem. Phys.*, **50**, 2372 (1969).
71. C. Eckart, *Phys. Rev.*, **47**, 552 (1935).
72. E. P. Wigner, *Group Theory*, Academic, New York, 1959.
73. B. T. Sutcliffe, *Quantum Dynamics of Molecules*, Plenum, New York, 1980.
74. T. Oka and Y. Morino, *J. Mol. Spectrosc.*, **6**, 472 (1961).
75. T. Oka, *J. Chem. Phys.*, **47**, 5410 (1967).
76. N. Onishi and T. Une, *Prog. Theor. Phys.*, **53**, 504 (1975).
77. T. Une et al., *Prog. Theor. Phys.*, **55**, 498 (1976).
78. H. C. Longuet-Higgins, *Mol. Phys.*, **6**, 445 (1963).
79. R. S. Berry, *Quantum Dynamics of Molecules*, Plenum, New York, 1980.
80. D. M. Dennison and G. E. Uhlenbeck, *Phys. Rev.*, **41**, 313 (1931).
81. A. D. Liehr, *Prog. Inorg. Chem.*, **5**, 385 (1963).
82. J. T. Hougen, *J. Chem. Phys.*, **55**, 1122 (1971).
83. J. E. Kilpatrick et al., *J. Am. Chem. Soc.*, **69**, 2483 (1947).

A REVIEW OF QUANTUM-MECHANICAL APPROXIMATE TREATMENTS OF THREE-BODY REACTIVE SYSTEMS

MICHAEL BAER

Soreq Nuclear Research Center,
Yavne, Israel
and Department of Chemical Physics,
The Weizmann Institute of Science
Rehovot, Israel

CONTENTS

I. INTRODUCTION

The feature that makes the theoretical treatment of chemical reactions so attractive is the existence of more than one arrangement channel. This feature has stimulated the imagination of theoretical chemists for the past 50 years and has resulted in a harvest of beautiful physical models and mathematical techniques. Although some important steps were taken during the first three decades, still the rapid developments in experimental instrumentation on the one hand and the fast expansion of high-speed computers on the other enhanced interest in the theory of the chemical process and turned it, during the 1960s and 1970s, into not only one of the most sophisticated subjects in chemistry but also one of the major ones.

The review is represented in chronological order, to give some idea of the way the field has developed since the early days of quantum mechanics. This imposed some difficulties, which were partly removed by following a certain hierarchy. We start by discussing some early aspects of the potential energy surface, not only because it is closely related to the development of the dynamical aspects of the field but also because it enables us to introduce various concepts that are important for the understanding of the different approximations discussed further on. Next we present collinear models and different types of approximation, and we continue by discussing the three-dimensional approximations. Part of this review is devoted to the treatment of reactions that are accompanied by electronic transitions and consequently take place on more than one potential energy surface, according to the Born-Oppenheimer picture.

The present composition is in many aspects an additional link in a chain of books and reviews published in the past decade or so. Levine in 1969 made the first attempt, in his excellent textbook *Quantum Mechanics of Rate Processes* (Clarendon Press, Oxford, 1969) to present a quantum-mechanical approach to scattering of atomic and molecular species. Another excellent text on the same subject, but from a different point of view, was written by Child: *Molecular Collision Theory* (Academic Press, New York, 1974). This book, which appeared 5 years after Levine's, contained, among other topics, the more recent developments in the theory of reactive collisions.

Several reviews were published during this period; Levine in *MTP, International Review of Science*, (Vol. I, W. B. Brown, Ed., University Park Press, Baltimore, 1972, p. 229), Kouri in *Energy Structure and Reactivity*, (D. W. Smith and W. B. McRae, Eds., Wiley, New York, 1973, p. 26), George and Ross (*Annu. Rev. Phys. Chem.*, **24**, 269, 1973), and Micha (*Adv. Chem. Phys.*, **30**, 7, 1975) presented their points of view on the field in an efficient and delicate way. In 1977 Truhlar and Wyatt (*Adv. Chem. Phys.*, **36**, 141, 1977) wrote a review of the $H + H_2$ reactive system in the context of which they also referred to various developments in the field of reactive scattering.

Four reviews emphasizing different aspects in the theory and applications of reactive scattering were published in the rather thorough book edited by Bernstein, *Atom-Molecule Collision Theory: A Guide for the Experimentalist* (Plenum Press, New York, 1979); two by Light (one dealing with general quantal theories and the other with statistical theories), and two by Wyatt (one on the distorted wave Born approximation (DWBA) and the Franck-Condon models that originated from it, and the other dealing with exact treatments). Another, by Connor (*Comput. Phys. Commun.*, **17**, 117, 1979), reviews reactive collisions in general and also briefly refers to quantum-mechanical approximations.

To avoid being repetitious on those subjects that were treated in rather great detail in the more recent reviews just named, exact treatments are not

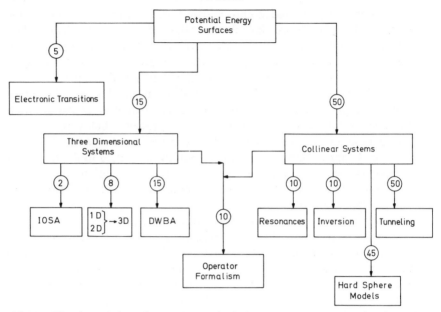

Fig. 1.   Flowchart of the main quantum-mechanical approximate treatments for chemical reactions. Numbers in circles represent years (to 1980) of work on the respective subjects.

described unless this is required for the sake of completeness; for the same
reason the DWBA is mentioned only briefly.

Figure 1 is a flowchart listing the main subjects treated since the early days
of quantum mechanics. The chapter closely follows this flowchart.

It is noted that 50 years have passed since the field in its present form was
founded. That is, about 50 years ago two papers were published: the first by
Eyring and Polanyi[1] describing the $H + H_2$ surface, and the second by Pelzer
and Wigner[2] analyzing the dynamics of it.

This chapter is dedicated to the memory of the late Prof. Michael Polanyi,
and to Profs. Henry Eyring and Eugene P. Wigner for the exciting heritage
they are leaving us.

References

1.   H. Eyring and M. Polanyi, Z. Phys. Chem. B, **12**, 279 (1931).
2.   H. Pelzer and E. P. Wigner, Z. Phys. Chem. B, **15**, 445 (1932).

II.   POTENTIAL ENERGY SURFACES

Two main subjects are discussed in this section:

1.   The Born-Oppenheimer approximation, which is the basis and justification for all later treatments of the reactive process.

2. The London-Eyring-Polanyi-Sato (LEPS) approximate method for calculating potential energy surfaces, which is still one of the most reliable and commonly used methods.

Other approximations, such as the diatomic in molecules (DIM), are not considered because they are close to the LEPS method in spirit. Also no mention is made of exact treatments, since they are beyond the scope of the present chapter.

A.   The Born-Oppenheimer Approximation[1]

The Hamiltonian describing the motion of the coupled system of the nuclei and the electrons is given in the form:

$$H = T_N + T_e + U(e, N) \tag{2.1}$$

where T_N and T_e are the kinetic energy operators of nuclei and electrons, respectively, $U(e, N)$ is the Coulomb potential between the various charged particles, and e and N stand for the electronic and nuclear coordinates. If $\psi(e, N)$ is the total wave function of the system

$$H\psi = E\psi \tag{2.2}$$

then it can be written in terms of a local adiabatic basis set $\zeta_j(e, N)$; $j = 1, 2, \ldots$, such that:

$$\psi(e, N) = \sum_{j=1}^{N} \zeta_j(e, N) \chi_j(N) \tag{2.3}$$

where $\zeta_j(e, N)$ fulfills the equation:

$$H_{el}\zeta_j(e, N) = V_j(N)\zeta_j(e, N) \tag{2.4}$$

Here,

$$H_{el} = T_e + U(e, N) \tag{2.5}$$

and $V_j(N)$ are the electronic eigenvalues that depend parametrically on the nuclear configuration and therefore are considered to be the potential energy surfaces (or hypersurfaces) of the system. Now, if T_N is written as a Laplacian in terms of some (nuclear) coordinates, namely:

$$T_N = -\frac{\hbar^2}{2\mu} \nabla^2 \tag{2.6}$$

where μ is a characteristic mass of the nuclei, then substituting (2.3) in (2.2), making use of (2.4) to (2.6), and projecting out the jth component, one obtains:

$$\left(-\frac{\hbar^2}{2\mu} \nabla^2 + V_j - E \right) \chi_j = \frac{\hbar^2}{2\mu} \left(\sum_l T_{jl}^{(2)} \chi_l + 2 \sum_l T_{jl}^{(1)} \nabla \chi_l \right) \qquad (2.7)$$

where

$$T_{jl}^{(1)} = \langle \zeta_j | \nabla \zeta_l \rangle ; \qquad T_{jl}^{(2)} = \langle \zeta_j | \nabla^2 \zeta_l \rangle \qquad (2.8)$$

The matrix elements $T_{jl}^{(k)}$; $k=1,2$, are the nonadiabatic coupling elements that couple the adiabatic surfaces $V_j(N)$ and $V_l(N)$. Now, assuming that $T_{jl}^{(k)}$; $k=1,2$, for $j=0$ are small compared to the difference $(V_0 - V_l)$, the right-hand part of (2.7) can be ignored and one is left with the Schrödinger equation, which describes the motion of the nuclei on the electronic ground state, that is,

$$\left(-\frac{\hbar^2}{2\mu} \nabla^2 + V - E \right) \chi = 0 \qquad (2.9)$$

where the index $j=0$ is omitted for convenience. Most of the rest of the chapter is concerned with this equation.

B.   The London-Eyring-Polanyi-Sato (LEPS) Potential Energy Surfaces

In 1929 F. London[2] presented his equation for the interaction of three and four monovalent atoms. The equation for the three-atom case (which is obtained as a special case of the four-atom case) takes the form:

$$V = Q - \left(\alpha_{AB}^2 + \alpha_{BC}^2 + \alpha_{AC}^2 - \alpha_{AB}\alpha_{BC} - \alpha_{AB}\alpha_{AC} - \alpha_{BC}\alpha_{AC} \right)^{1/2} \qquad (2.10)$$

where Q is the Coulombic binding energy between the three atoms A, B, and C, which would be obtained if classical electrodynamics were used, assuming the electrons to be diffuse clouds of quantum mechanics, and α_{AB}, α_{AC}, and α_{BC} to be the corresponding exchange integrals. The Coulombic energy is the sum of three terms:

$$Q = Q_{AB} + Q_{AC} + Q_{AC} \qquad (2.11)$$

where Q_{AB} is the Coulombic binding energy of the two atoms A and B, assuming that C is at infinity, and so on.

In 1931 Eyring and Polanyi[3] used this equation to calculate the first potential energy surfaces for three reactions, namely:

$$H + H_2^{para} \rightarrow H_2^{orth} + H$$

$$H + HBr \rightarrow H_2 + Br$$

$$H + Br_2 \rightarrow HBr + Br$$

To calculate these surfaces, the following steps were taken:

1. Reconsidering (2.10), it can be seen that if one of the three atoms is removed to infinity, the potential becomes:

$$V = Q_{XY} - \alpha_{XY} \qquad (2.12)$$

where X and Y are either A and B or A and C or B and C, when C or B or A, respectively, are removed to infinity.

2. Equation 2.12 describes the potential energy curve (Franck curve) of the two atoms X and Y. Morse, in 1929,[4] obtained a single function that yielded the binding energy of two atoms:

$$E(r) = D_r \left(e^{-2\beta(r - r_e)} - 2e^{-\beta(r - r_e)} \right) \qquad (2.13)$$

where D_e is the depth of the potential well, β is proportional to the frequency of the vibrational ground state, and r_e is the equilibrium distance.

Eyring and Polanyi[3] assumed that V given in (2.12) was identical to $E(r)$ given in (2.13), and consequently:

$$E(r) = Q_{XY} - \alpha_{XY} \qquad (2.14)$$

or, assuming Q_{XY} was known, then α_{XY} could be determined from the equation:

$$\alpha_{XY} = -E(r) + Q_{XY} \qquad (2.14')$$

3. Eyring and Polanyi[3] went one step further and assumed Q_{XY} to be a certain fraction of the total binding energy, namely:

$$Q_{XY} = \rho E(r) \qquad (2.15)$$

where ρ could be considered to be a free parameter.

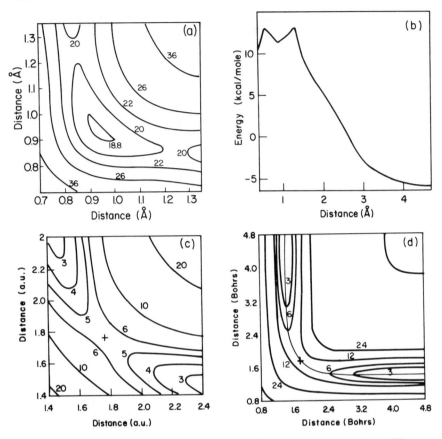

Fig. 2.   Collinear potential energy surfaces for the $H + H_2$ system. (a) The first LEP potential surface, due to Eyring and Polanyi.[3] (b) The minimum energy path of the first LEP potential surface.[5] (c) The LEPS surface due to Sato.[6] (d) The most recent "exact" surface, due to Siegbahn, Liu, Truhlar, and Horowitz (SLTH).[10, 11]

Equation 2.10, together with (2.13) to (2.15), form the scheme that Eyring and Polanyi[3] applied to calculate the three above-mentioned potential surfaces. In Fig. 2a we show the first potential energy surface for $H + H_2$, which was calculated assuming $\rho = 0.10$.[5]

One of the immediate results of the London-Eyring-Polanyi (LEP) procedure was that for many simple atom-diatom systems the minimum energy path lay in the linear arrangement. This curve for the $H + H_2$ system is shown in Fig. 2b.[5] Two main features should be mentioned:

1.   The existence of a barrier of about 18 kcal/mole in the close interaction region.
2.   The existence of a shallow basin on top of the barrier.

Although the basin was later found to be an artifact and the barrier height to be lower, the main features of this surface remain valid.

In 1955[6] Sato presented his version of potential energy surface calculations. He introduced two modifications:

1. He replaced (2.14) by a more accurate form devised by Heitler and London in 1927,[7, 8] namely,

$$E(r) = \frac{Q_{XY} - \alpha_{XY}}{1 + S_{XY}^2} \tag{2.16}$$

where S_{XY} is the corresponding overlap integral.

2. He made use[6] of the fact that the Heitler-London procedure yielded two solutions, which took the form:

$$E^{\pm}(r) = \frac{Q_{XY} \pm \alpha_{XY}}{1 \mp S_{XY}^2} \tag{2.17}$$

where $E^-(r)$ stands for the bonding state and $E^+(r)$ for the antibonding state. As before, the bonding state is presented by the Morse formula. For the antibonding state, Sato proposed the equation:

$$E^+(r) = \frac{D_e}{2} \left(e^{-2\beta(r-r_e)} + 2e^{-\beta(r-r_e)} \right) \tag{2.18}$$

which was found to be in good correspondence with the calculated data for the triplet state of the hydrogen molecule.[9] Here, D_e, β, and r_e are the same parameters that appear in the corresponding Morse function.

Equation (2.17), together with (2.13) and (2.18)—in fact, (2.18) replaces (2.15), which was justifiable for only a limited number of cases in a limited region of configuration space—form the two required independent equations for solving Q_{XY} and α_{XY} once S_{XY} is known. Usually, S_{XY} is not known, and consequently S_{AB}, S_{BC}, and S_{AC} serve as parameters that are varied in to permit the investigator to obtain certain properties of the potential energy surfaces. Figure 2c shows the corresponding surface for the $H + H_2$. For comparison, Fig. 2d shows the most recent Siegbahn-Liu-Truhlar-Horowitz (SLTH) ab initio surface,[10, 11] and the differences are seen to be quantitative; qualitatively, the three surfaces are similar.

References

1. M. Born and R. Oppenheimer, Ann. Phys., 84, 457 (1927).
2. F. London, Z. Electrochem., 35, 552 (1929).
3. H. Eyring and M. Polanyi, Z. Phys. Chem. B, 12, 279 (1931).
4. P. M. Morse, Phys. Rev., 34, 57 (1929).

5. H. Eyring, *J. Am. Chem. Soc.*, **53**, 2537 (1931).
6. S. Sato, *J. Chem. Phys.*, **23**, 592 (1955).
7. W. Heitler and F. London, *Z. Phys.*, **44**, 455 (1927).
8. Y. Sugiura, *Z. Phys.*, **45**, 484 (1927).
9. J. O. Hirschfelder and J. W. Linnett, *J. Chem. Phys.*, **18**, 130 (1950).
10. S. Liu and P. Siegbahn, *J. Chem. Phys.*, **68**, 2457 (1978).
11. D. G. Truhlar and C. J. Horowitz, *J. Chem. Phys.*, **68**, 2466 (1978).

III. THE COLLINEAR SYSTEM

In 1931 Eyring and Polanyi[1] analyzed the $H + H_2$ potential energy surface they had calculated and found that the minimum energy path lay in the linear configuration. Later, Pelzer and Wigner,[2] while using this surface to calculate reaction rates, confirmed the uniqueness and importance of the collinear arrangement, at least for the $H + H_2$ system. Most of the exact and approximate treatments in the 50 years that followed were related to the collinear arrangement. Section III.A analyzes the collinear system in some detail, discussing along the way the various coordinates usually employed in treating reactive systems and how one "moves" from one arrangement channel to the other. Section III.B is devoted to the study of one-dimensional (mathematically speaking) models, with the main emphasis on tunneling. Section III.C covers hard-sphere-type models and their impact on the field. The subject of Section III.D is the (vibrational) inversion in exothermic reactions, and the relevance of the curvature of the reaction coordinate, as well as the Franck-Condon type models, are considered. Section III.E deals with the possibility of the formation of Feschbach-type resonances due to shallow dips along the vibronic curves.

A.   The Schrödinger Equation

1.   Cartesian Coordinates

The Hamiltonian for three particles A, B, and C, in a collinear configuration, is written in the form:

$$H = \frac{1}{2m_A}\mathbf{P}_A^2 + \frac{1}{2m_B}\mathbf{P}_B^2 + \frac{1}{2m_C}\mathbf{P}_C^2 + V(\mathbf{r}_A, \mathbf{r}_B, \mathbf{r}_C) \qquad (3.1)$$

where $\mathbf{r}_A$, $\mathbf{r}_B$, and $\mathbf{r}_C$ are the distances from some reference point, and $\mathbf{P}_A$, $\mathbf{P}_B$, and $\mathbf{P}_C$ are the momenta of A, B, and C, respectively (see Fig. 3a). Eliminating the center of mass coordinate and expressing the kinetic energy in terms of interatomic distances $\mathbf{r}_{CB}$ $(=\mathbf{r}_C - \mathbf{r}_B)$ and $\mathbf{r}_{BA}$ $(=\mathbf{r}_B - \mathbf{r}_A)$ leads to:

$$H = \frac{1}{2\mu_{AB}}\mathbf{P}_{AB}^2 + \frac{1}{2\mu_{BC}}\mathbf{P}_{BC}^2 - \frac{1}{m_B}\mathbf{P}_{AB}\cdot\mathbf{P}_{BC} + V(\mathbf{r}_{BA}, \mathbf{r}_{BC}) \qquad (3.2)$$

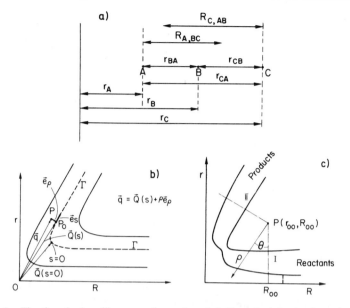

Fig. 3. The three-body collinear reactive system. (a) Absolute, interatomic, and center of mass distances. (b) The Marcus system of coordinates: s, the reaction coordinate; ρ, the vibrational one. (c) Polar coordinates with respect to an origin in the plateau region.

where μ_{AB} and μ_{BC} are the reduced masses:

$$\mu_{AB} = \frac{m_A m_B}{m_A + m_B}; \quad \text{and} \quad \mu_{BC} = \frac{m_B m_C}{m_B + m_C} \tag{3.3}$$

Now writing H in an operator form yields the corresponding Schrödinger equation[3]:

$$H = -\frac{\hbar^2}{2\mu_{AB}} \frac{\partial^2}{\partial r_{BA}^2} - \frac{\hbar^2}{2\mu_{BC}} \frac{\partial^2}{\partial r_{CB}^2} + \frac{\hbar^2}{2m_B} \frac{\partial}{\partial r_{BA}} \frac{\partial}{\partial r_{CB}} + V(\mathbf{r}_{BA}, \mathbf{r}_{BC}) \tag{3.4}$$

In a more common set of coordinates, as before, one coordinate is an interatomic distance, but the other is translational.[4] In the reagents channel (A, BC) we have the coordinates r_{CB} and R_A, the latter being defined as:

$$R_A = r_{BA} + \frac{m_C}{m_C + m_B} r_{CB} \tag{3.5}$$

and in the products channel (AB, C) we have r_{BA} and R_C defined as:

$$R_C = r_{CB} + \frac{m_A}{m_A + m_B} r_{BA} \tag{3.6}$$

The corresponding Hamiltonian in the (A, BC) channel is

$$H = -\frac{\hbar^2}{2\mu_{A,BC}} \frac{\partial^2}{\partial R_A^2} - \frac{\hbar^2}{2\mu_{BC}} \frac{\partial^2}{\partial r_{CB}^2} + V(R_A, r_{CB}) \tag{3.7}$$

where $\mu_{A,BC}$ is given as

$$\mu_{A,BC} = \frac{m_A(m_B + m_C)}{m_A + m_B + m_C} \tag{3.8}$$

A similar equation exists for the (AB, C) channel.

The disadvantage in this formulation is that in each channel, there are not only different coordinates but different masses as well. This yields a nonorthogonal transformation matrix from one arrangement channel to the other, and consequently kinematic effects are not well presented. In 1959 Smith[5] and Delves[6] each independently suggested scaling the coordinates in both channels in the following way:

$$r_I = \lambda_I r_{CB}, \qquad R_I = \lambda_I^{-1} R_A, \qquad \lambda_I = \left(\frac{\mu_{BC}}{\mu_{A,BC}} \right)^{1/4}$$

$$r_{II} = \lambda_{II} r_{BA}, \qquad R_{II} = \lambda_{II}^{-1} R_C, \qquad \lambda_{II} = \left(\frac{\mu_{AB}}{\mu_{C,AB}} \right)^{1/4} \tag{3.9}$$

where subscripts I and II stand for the reagent and the product channels, respectively. With these sets of coordinates the Hamiltonian in the two channels looks the same, that is,

$$H = -\frac{\hbar^2}{2\mu} \left(\frac{\partial^2}{\partial R^2} + \frac{\partial^2}{\partial r^2} \right) + V(R, r) \tag{3.10}$$

where $R = R_I, R_{II}$ and $r = r_I, r_{II}$ and μ, which from now on will be considered to be the characteristic mass of the system, takes the form:

$$\mu = \left(\frac{m_A m_B m_C}{m_A + m_B + m_C} \right)^{1/2} \tag{3.11}$$

With this scaling, the transformation matrix from the reagent coordinate to that of the product becomes orthogonal:

$$\begin{pmatrix} R_{II} \\ r_{II} \end{pmatrix} = \begin{pmatrix} \cos\beta & -\sin\beta \\ \sin\beta & \cos\beta \end{pmatrix} \begin{pmatrix} R_{I} \\ r_{I} \end{pmatrix} \qquad (3.12)$$

where:

$$\cos\beta = -\left(\frac{m_A m_C}{(m_A + m_B)(m_C + m_B)} \right)^{1/2},$$

$$\sin\beta = \left(\frac{m_A(m_A + m_B + m_C)}{(m_A + m_B)(m_C + m_B)} \right)^{1/2} \qquad (3.12')$$

It can be seen that whenever m_B (the central atoms) is much larger than m_A and m_C, β becomes close to $\pi/2$, and when m_B is much smaller than m_A and m_C, β approaches zero. As will be seen later on, this kinematic effect has a very important influence on the reaction process.

Since there are three arrangement channels, we have three different reactive processes, each of which is characterized by an angle β as described above. It can be shown that the sum of the three β angles is π.

2. The Reaction Coordinate

The idea of a coordinate that continuously leads from reagents to products is at least as old as the concept of the potential energy hypersurface and may be even older: in 1929 Langer[7] used it in his theory of chemical reactions without explicitly saying so. Most of the theories and processes that Wigner, Eyring, Polanyi, Hirschfelder, and others discussed[1, 2, 8-12] during the 1930s were based on this concept. However, no serious attempts were made to use this coordinate as *purely* a coordinate until the 1960s, when this possibility was first discussed and analyzed by Hofacker.[13] In 1966 Marcus presented his version of the Schrödinger equation for a collinear reactive system in which the reaction coordinate appears explicitly.[14, 15] The basic idea was using coordinates defined with respect to an origin which moved along a continuous curve, Γ.[16] As can be seen from Fig. 3b, the radius vector $\mathbf{q}$ describing the position of an arbitrary point P can be written as:

$$\mathbf{q} = \mathbf{Q}(s) + \rho\mathbf{e}_\rho \qquad (3.13)$$

where $\mathbf{Q}(s)$ is the radius vector of P with respect to a given origin; s is the distance between O' (the point for which s is defined as zero) and P_0, measured along the defined curve Γ: $\mathbf{e}_\rho$ is a unit vector perpendicular to the curve

Γ at P_0, and ρ is a scalar.* The Frenet formulas[16] for any smooth curve are given in the form:

$$\frac{d\mathbf{e}_\rho}{ds} = k\mathbf{e}_s, \qquad \frac{d\mathbf{e}_s}{ds} = -k\mathbf{e}_\rho \qquad (3.14)$$

where k is the curvature of Γ. Defining the conjugate momenta P_s and P_ρ, it can now be shown that the classical Hamiltonian takes the form[14]:

$$H = \frac{1}{2\mu}\left(\frac{P_s^2}{\eta^2} + P_\rho^2\right) + V(s,\rho) \qquad (3.15)$$

and the corresponding quantum-mechanical presentation is[15, 17]:

$$H = -\frac{\hbar^2}{2\mu}\left(\frac{1}{\eta}\frac{\partial}{\partial s}\frac{1}{\eta}\frac{\partial}{\partial s} + \frac{1}{\eta}\frac{\partial}{\partial \rho}\eta\frac{\partial}{\partial \rho}\right) + V(s,\rho) \qquad (3.16)$$

where η is defined as:

$$\eta = 1 + k\rho \qquad (3.17)$$

The convenient part of this presentation is that there is no need to consider each arrangement channel independently because it yields a smooth transition from the reagents to the products channel. In this sense the equation describing the reactive process becomes very similar to that describing the inelastic process. The similarity becomes even stronger when the curvature k, along the reaction path, is small. If that is the case, η can be assumed to be 1 and consequently the Hamiltonian in (3.16) takes the form

$$H = -\frac{\hbar^2}{2\mu}\left(\frac{\partial^2}{\partial s^2} + \frac{\partial^2}{\partial \rho^2}\right) + V(s,\rho) \qquad (3.16')$$

In general, (3.16') can be applied for most cases where the curvature is not too large. In case the curvature is large (e.g., in the heavy-light-heavy systems), the application of the Marcus equation becomes more complicated, because such difficulties as branch cuts and the multivalue character of the wave function[18] are encountered. Nevertheless, (3.16) has been applied successfully in many numerical and theoretical treatments, some of which are mentioned later.

*In Fig. 3b vectors are indicated by arrows over the respective symbols.

An important simplification was introduced by Johnson,[19] who suggested dividing all configuration space into sectors with constant curvatures. In this way, (3.16) takes a simpler form:

$$H = -\frac{\hbar^2}{2\mu} \left(\frac{\partial^2}{\partial \rho^2} + \frac{k}{\eta} \frac{\partial}{\partial \rho} + \frac{1}{\eta^2} \frac{\partial^2}{\partial s^2} \right) + V(s, \rho) \qquad (3.16'')$$

Johnson's modification yields, as a first division, three main regions: the entrance and exit channels, which are characterized by zero curvature, and the interaction region, where k is different from zero. The interaction region is then further divided into a number of subregions, each characterized by an arc of a circle of given radius (and, therefore, of given curvature). In another method, close in spirit to this approach, the entrance and exit channels are described by Cartesian coordinates [see (3.10)] and the interaction region by polar coordinates defined with respect to some point $P(r_{00}, R_{00})$ in the plateau region[20, 21] (see Fig. 3c). The relation between the polar and Cartesian coordinates is:

$$\rho \sin \theta = R_{00} - R$$
$$\rho \cos \theta = r_{00} - r \qquad (3.18)$$

and the corresponding Hamiltonian takes the form:

$$H = -\frac{\hbar^2}{2\mu} \left(\frac{1}{\rho^2} \frac{\partial^2}{\partial \theta^2} + \frac{\partial^2}{\partial \rho^2} + \frac{1}{\rho} \frac{\partial}{\partial \rho} \right) + V(\rho, \theta) \qquad (3.19)$$

Other types of reaction coordinate were used by Middleton and Wyatt[22] — the broken path model (see also earlier work by Blum[23]), by Connor and Marcus[24] — who mapped the skewed system of coordinates onto an infinite strip, with the help of the Schwartz-Christoffel formula (see also Rosenthal and Gordon[25]), and by Witriol et al.,[26] who constructed an orthogonal curvilinear coordinate system, based on a cannonical point transformation (see also McNutt and Wyatt[27]). As far as the last type of coordinate is concerned, it is interesting to mention that a similar type had been discussed by Hulburt and Hirschfelder as early as 1943.[28]

3. Exact Numerical Treatments

Most of the above-mentioned systems of coordinates were also applied in numerical calculations. Although we do not intend to go into any detail regarding the numerical techniques, still, for the sake of completeness, we add the following comments.

Most of the extant exact numerical treatments are based on solving the Schrödinger equation by propagating the wave functions either from one asymptotic region to the other[18, 19, 29, 30] or from one asymptotic region to the interaction region and back.[21] The corresponding T-matrix elements are derived following an asymptotic analysis (in case of propagation to the middle of the interaction region, the matching in the interaction region and the asymptotic analysis are performed simultaneously). In some of the techniques, the propagation is avoided by replacing the derivative in the Schrödinger equation by finite differences and obtaining the values of the wave functions at the grid points by solving the corresponding system of algebraic equations. A method such as that was devised by Diestler and McKoy.[31] A different approach (finite elements) was taken by Askar et al.,[32] who formulated an iterative procedure to obtain these values, thus circumventing the necessity for solving algebraic equations.

Other methods were devised by Manz[33] and by Light and Walker[34] to solve directly for S- and R-matrices, respectively. Again, these methods were based on propagation techniques, but instead of solving for the wave functions, the corresponding S- or R-matrix was obtained at each step. The main advantage in these methods is that the closed channels are handled more efficiently.

B. One-Dimensional Models

1. Derivation of the One-Dimensional Model

To reduce the (collinear) two-dimensional problem to a one (mathematical)-dimensional problem, we again consider the Marcus equation given in (3.16).

Three points will be made:

1. The potential is written in the form:

$$V(s, \rho) = V_1(s) + V_2(s, \rho) \qquad (3.20)$$

where $V_2(s, \rho)$ is a "Morse-type potential" that fulfills the condition:

$$V_2(s, \rho = 0) = 0 \qquad (3.21)$$

for any s. Consequently, one obtains

$$V(s, \rho = 0) = V_1(s) \qquad (3.22)$$

namely, $V_1(s)$ is the potential along a path that leads continuously from reagents to products. If the route is chosen such that

$$V_2(s, \rho=0)=\min(V(s,\rho)) \qquad (3.23)$$

for any s, then the route $\rho=\rho(s)=0$ is the minimum energy path.

2. If $\psi(s, \rho)$ is the total wave function, then making the substitution[35]:

$$\psi(s,\rho)=\eta^P\chi(s,\rho) \qquad (3.24)$$

leads to

$$\eta^P\left\{-\frac{\hbar^2}{2\mu}\left[\frac{1}{\eta^2}\frac{\partial^2\chi}{\partial s^2}+\frac{\partial^2\chi}{\partial\rho^2}+\frac{1}{\eta^2}M(s,\rho)\chi\right]+(V(s,\rho)-E)\chi\right\}=0$$

$$(3.25)$$

where $M(s, \rho)$ is:

$$M(s,\rho)=(2p-1)\frac{\rho}{\eta}\frac{dk}{ds}\frac{\partial}{\partial s}+p(p-2)\left(\frac{\rho}{\eta}\right)^2\frac{\partial^2}{\partial s^2}+p\frac{\rho}{\eta}\frac{d^2k}{ds^2}$$

$$+(2p+1)\eta k\frac{\partial}{\partial\rho}+p^2k^2 \qquad (3.26)$$

Equations 3.25 and 3.26 are valid for every p as long as the path $\rho=\rho(s)=0$ is chosen in such a way that $\eta\neq0$ in the region of interest. If, however, η becomes zero for certain s values (namely, when k is large), the choice of p is limited. In most treatments, p was taken to be either $-\frac{1}{2}$,[14, 15, 36, 37] 4,[29, 38, 39] or $+\frac{1}{2}$.[40-42] Different reasonings were suggested to justify the various choices. We shall refer to this later on.

3. Our next step will be to expand $\chi(s, \rho)$ in terms of a local basis set (Hirschfelder and Wigner[12] were the first to suggest using a local basis set), namely,

$$\chi(s,\rho)=\sum\xi_n(s)\phi_n(s,\rho)=\xi^*\phi \qquad (3.27)$$

where $\phi_n(s, \rho)$; $n=0,1,2,\ldots$, are the eigenfunctions of the equation:

$$\left(-\frac{\hbar^2}{2\mu}\frac{d^2}{d\rho^2}+V_2(s,\rho)-\varepsilon_n(s)\right)\phi_n(s,\rho)=0 \qquad (3.28)$$

and $\xi_n(s)$; $n=0,1,2,\ldots$, are the corresponding translational functions.

Substituting (3.27) in (3.25) and applying (3.28) leads to

$$\left(-\frac{\hbar^2}{2\mu}\frac{\partial^2}{\partial s^2}-2\tau^{(1)}\frac{\partial}{\partial s}-\tau^{(2)}+\mathbf{M}\right)\xi+\eta^{(2)}(\mathbf{U}-E\mathbf{I})\xi=0 \quad (3.29)$$

where

$$M_{nm}=\langle\phi_n|M|\phi_m\rangle$$
$$\eta^{(2)}_{nm}=\langle\phi_n|\eta^2|\phi_m\rangle$$
$$U_{nm}=(\varepsilon_n(s)+V_1(s))\delta_{nm} \quad (3.30)$$
$$\tau^{(1)}_{nm}=\left\langle\phi_n\left|\frac{\partial}{\partial s}\phi_m\right\rangle\right.$$
$$\tau^{(2)}_{nm}=\left\langle\phi_n\left|\frac{\partial^2}{\partial s^2}\phi_m\right\rangle\right.$$

Now, introducing the adiabatic-diabatic transformation[43-46] (see also Section VI):

$$\xi(s)=\mathbf{A}(s)\zeta(s) \quad (3.31)$$

and choosing $\mathbf{A}(s)$ to fulfill the equation:

$$\frac{d\mathbf{A}}{ds}+\tau^{(1)}\mathbf{A}=0 \quad (3.32)$$

it can be shown that $\mathbf{A}$ is the orthogonal and that the corresponding equation for $\xi(s)$ is:

$$-\frac{\hbar^2}{2\mu}\left(\frac{d^2}{ds^2}+\mathbf{M}_A\right)\zeta+\eta^{(2)}_A(\mathbf{W}-E)\zeta=0 \quad (3.33)$$

where

$$\mathbf{M}_A=\mathbf{A}^*\mathbf{M}\mathbf{A}, \qquad \eta^{(2)}_A=A^*\eta^{(2)}\mathbf{A}, \qquad \mathbf{W}=\mathbf{A}^*\mathbf{U}\mathbf{A} \quad (3.34)$$

The matrix $\mathbf{W}$ is called the adiabatic-diabatic potential matrix. When the curvature is small enough, one may ignore $\mathbf{M}_A$ and replace $\eta^{(2)}_A$ by a unity matrix so that the Schrödinger equation becomes:

$$\left(-\frac{\hbar^2}{2\mu}\frac{d^2}{ds^2}+(\mathbf{W}-E)\right)\zeta=0 \quad (3.35)$$

Below we discuss models and treatments that ignore the off-diagonal elements of W.

2. *Single-Curve Models*

The models relating to a single curve that starts at $s = -\infty$ and goes to $s = \infty$ are based on the equation:

$$\left(-\frac{\hbar^2}{2\mu} \frac{d^2}{ds^2} + W_{00}(s) - E \right) \zeta_0 = 0 \qquad (3.36)$$

In general, W_{00} is dependent on the nonadiabatic coupling terms and is a complicated function of these. However, for simplicity, we assume these terms to be weak and consequently:

$$W_{00} = \mathcal{E}_0(s) + V_1(s) \qquad (3.37)$$

In the following, we drop the indices from ζ and W.

Equation 3.36 was used in different studies on transmission probabilities and tunneling. One of the earliest studies is due to Langer,[7] who considered a potential curve as given in Fig. 4a. Langer assumed that every chemical reaction proceeds only because of what was called, somewhat later, "tunneling," and that a necessary and sufficient condition for a chemical reaction to occur is that the initial energy level of the "somewhat" distorted original AB molecule coincide with the "somewhat" distorted final energy level of the product molecule BC. Although the Langer theory is incorrect in general, still two basic concepts were introduced here: (1) The progress of a chemical reaction due to tunneling and (2) the nearly (vibrational) adiabatic behavior of a system when moving from reagents to products. Langer's two main errors were caused because, as was established 2 years later, the barrier in the interaction region was much lower than the dissociation energy, and therefore many reactions could occur classically without having to tunnel; moreover, the motion from reagents to products was described by two coordinates, at least one of which was unbound. Since the two modes of motion interacted with each other, the system could always arrange itself to be in any given open eigenstate. It is interesting to mention that in 1954 Bauer and Wu[47] considered a very similar potential form but assumed a different mechanism. Contrary to Langer, they ignored tunneling (but the existence of a barrier was essential for their model) and assumed the reaction process to be accompanied by a vibrational nonadiabatic process. The importance of the Bauer-Wu treatment lies not in the mechanism that, with the use of an exact treatment, was later found to be inappropriate, but in the fact that these authors were the first to apply the distorted wave Born approximation (DWBA) to reactive scattering.

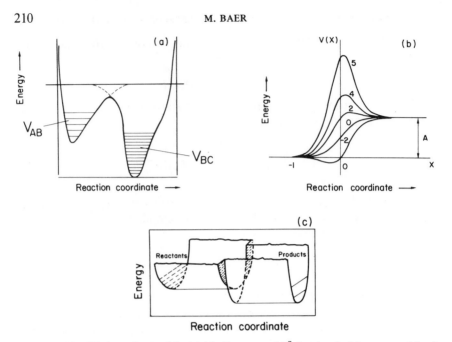

Fig. 4. Simplified reaction models. (a) The Langer model,[7] showing the Morse potentials of the reagent and the product molecules. (b) The Eckart model.[48] The numbers on the curves are the values of B/A. (c) The Hirschfelder-Wigner model.[12]

The first treatment of (3.36) was given by Eckart.[48] In his study, Eckart did not refer to a chemical reaction (he studied the penetration of a potential barrier by electrons). However, because of the form of the potential $U(s)$ that he used, Bell[49] and Wigner[50] recognized in 1933 that Eckart's treatment could apply to chemical reactions as well.

Eckart assumed a "reasonable" potential for which he was able to obtain an analytic expression for the reflection (and the transmission) coefficient. The function he considered has the form:

$$W(s) = -Ax/(1-x) - Bx/(1-x)^2 \qquad (3.38)$$

where

$$x = x(s) = -e^{2\pi s/l} \qquad (3.39)$$

and A, B, and l are constants. Here, A (>0) is the asymptotic value of the potential at $s = \infty$ (the potential becomes zero for $s = -\infty$). l is half the width of the transition region, and B is responsible for the actual form of the potential in that region, as can be seen from Fig. 4b. For instance, when $B>0$, $W(x)$ has a maximum given by:

$$W_m = \frac{(A+B)^2}{4B} \tag{3.40}$$

Substituting (3.38) for $W(s)$ in (3.36) and replacing s by x as given in (3.39) leads to a hypergeometric-type equation:

$$x^2\frac{d^2\zeta}{dx^2} + x\frac{d\zeta}{dx} + \frac{2\mu l^2}{\hbar^2}\left(Ax/(1-x) + Bx/(1-x)^2 + E\right)\zeta = 0 \tag{3.41}$$

Considering the asymptotic form of $\zeta(x)$ for $x \to -\infty$ (i.e., $s \to \infty$), it can be shown to be of the form:

$$\zeta(s) = a_1 e^{2\pi ix/\lambda} + a_2 e^{-2\pi ix/\lambda} \tag{3.42}$$

where

$$\lambda = \frac{h}{(2\mu E)^{1/2}} \tag{3.43}$$

Consequently the reflection coefficient ρ defined as

$$\rho = \left|\frac{a_2}{a_1}\right|^2 \tag{3.44}$$

becomes:

$$\rho = \left|\frac{\Gamma\left(\frac{1}{2}+i(\delta-\beta-\alpha)\right)\Gamma\left(\frac{1}{2}+i(-\delta-\beta-\alpha)\right)}{\Gamma\left(\frac{1}{2}+i(\delta-\beta+\alpha)\right)\Gamma\left(\frac{1}{2}+i(-\delta-\beta+\alpha)\right)}\right|^2 \tag{3.45}$$

Here

$$\alpha = \frac{l}{\lambda} = \frac{1}{2}\left(\frac{E}{C}\right)^{1/2}, \qquad \beta = \frac{l}{\lambda'} = \frac{1}{2}\left(\frac{E-A}{C}\right)^{1/2} : \delta = \frac{1}{2}\left(\frac{B-C}{C}\right)^{1/2} \tag{3.46}$$

where

$$\lambda' = h/(2\mu(E-A))^{1/2} : C = h^2/(8l^2\mu) \tag{3.47}$$

In analyzing this expression one should distinguish between two cases, namely, δ real and δ imaginary. When δ is real, it can be shown that ρ becomes:

$$\rho = \frac{\cosh[2\pi(\alpha-\beta)] + \cos[2\pi\delta]}{\cosh[2\pi(\alpha+\beta) + \cosh[2\pi\delta]}$$

(3.48)

and for an imaginary δ one has:

$$\rho = \frac{\cosh[2\pi(\alpha-\beta)] + \cos[2\pi|\delta|]}{\cosh[2\pi(\alpha+\beta)] + \cos[2\pi|\delta|]}$$

(3.49)

The values of ρ as a function E for $A=8B$ and various l values are shown in Fig. 5. Eckart also showed that the expression for the reflection coefficient based on Jeffrey's[51] approximate solution for the Schrödinger equation

$$\rho = 1 - \exp\left\{ -\frac{4\pi}{h} \int_{x_1}^{x_2} (2\mu(W-E))^{1/2}\, dx \right\}$$

(3.50)

fits reasonably well with his exact results as long as $E < W_m$ (here x_i, $i = 1, 2$ are the points for which $W(x) = E$).

As mentioned, Eckart's approach and results were adopted by Wigner[50] and Bell[49, 52, 53] to determine the importance of quantum effects on (heavy-particle) exchange reactions. However, whereas Wigner was looking only for first-order quantum corrections to the classical results, Bell extended Eckart's approach and applied it for the study of the temperature dependence of rate constants. Applying Eckart's potential and integrating analytically, Bell found large deviations between the classical and the quantum-mechanical

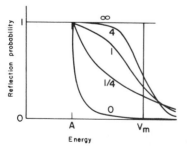

Fig. 5. The reflection coefficients (nonreactive transition probabilities) as a function of energy for the Eckart potential. The numbers are the values of A/C; V_m stands for the maximum of the potential.

rate constants for hydrogenic systems. For a parabolic barrier

$$W(x) = W_m\left(1 - \frac{x^2}{a^2}\right) \tag{3.51}$$

and using Jeffrey's formula,[51] Bell was able to show that the transmission coefficient τ is given by:

$$\tau = \exp\left[-\frac{2\pi^2 a}{h}\left(\frac{2\mu}{W_m}\right)^{1/2}(W_m - E)\right] \tag{3.52}$$

where $2a$ is the width of the parabola. Applying this expression, Bell derived an analytic form for the rate q of particles passing the barrier (i.e., reacting). To do that he extended the validity of the expression above for the whole E range of interest:

$$\tau(E) = \begin{cases} \exp\left[-\dfrac{2\pi^2 a}{h}\left(\dfrac{2\mu}{W_m}\right)(W_m - E)\right] & \text{for} \quad W_m \geq E \\ 1 & \text{for} \quad W_m \leq E \end{cases} \tag{3.53}$$

and found, by applying the definition for q (or q_Q — the index Q stands for "quantum-mechanical"):

$$q_Q = \frac{1}{kT}\int_0^\infty \tau(E)e^{-E/kT}dE \tag{3.54}$$

that

$$q_Q = (\beta - \alpha)^{-1}(\beta e^{-\alpha} - \alpha e^{-\beta}) \tag{3.55}$$

where:

$$\alpha = \frac{W_m}{kT} \quad \text{and} \quad \beta = \frac{2\pi^2 a(2\mu W_m)^{1/2}}{h} \tag{3.56}$$

The corresponding classical expression q_c is given as:

$$q_c = e^{-\alpha} \tag{3.57}$$

Figure 6 compares the classical and the quantum-mechanical results for different potential parameters. It can be seen that the deviations are largest for

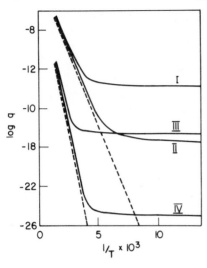

Fig. 6. The reaction flux (rate) according to Bell,[49] as a function of T^{-1} for different barrier heights and barrier widths. Solid curves, quantum-mechanical treatment; dashed curves, classical treatment.

low temperature T, but they become smaller as T increases. It can also be seen that the higher or the thinner the potential, the larger are the deviations.

Having both q_c and q_Q, we may define the tunneling correction factor Q as:

$$Q = \frac{q_Q}{q_c} \tag{3.58}$$

From (3.55), (3.57), and (3.58), we obtain:

$$Q = \frac{\beta}{\beta - \alpha} \left(1 - \frac{\alpha}{\beta} e^{-(\beta - \alpha)} \right) \tag{3.59}$$

As mentioned before, Wigner was able to derive a formula for Q in the high-energy (-temperature) limit:

$$Q = 1 + \frac{1}{2\pi} \left(\frac{\hbar \nu}{kT} \right)^2 \tag{3.60}$$

where

$$\nu = \left(-\frac{1}{\mu} \frac{d^2 W(s)}{ds^2} \right)^{1/2} \tag{3.61}$$

which in this case becomes:

$$v=\left(\frac{2W_m}{\mu a^2}\right)^{1/2} \tag{3.62}$$

On the other hand in the high-temperature region Bell's result takes the form:

$$Q=1+\frac{1}{24}\frac{h\nu}{kT}e^{-\beta} \tag{3.63}$$

and it can easily be seen that the derivations of Wigner and Bell differ. The reason for this is that Bell's approximate transition probability function is not accurate enough in this energy region.

About 20 years later, Shavitt[54] suggested reconsidering the Eckart potential. He assumed the potential to be symmetric and of the form:

$$W(s)=W_m\,\text{sech}\left(\frac{\pi s}{l}\right) \tag{3.64}$$

for which the transition coefficient is:

$$\tau=\left[\cosh(4\pi\alpha)-1\right]/\left[\cosh(4\pi\alpha)+\cos(2\pi\delta)\right] \tag{3.65}$$

where

$$\alpha=\frac{l}{\hbar}(2\mu E)^{1/2}:\delta=\frac{l}{\hbar}\left[8\mu\left(W_m-\frac{\hbar^2}{4l^2}\right)\right]^{1/2} \tag{3.66}$$

Applying the approximate form for τ, Shavitt[54] was able to show that his result coincides with Wigner's formula for a potential with a nonzero fourth derivative

$$Q=1+\frac{1}{2\pi}\left(\frac{h\nu}{kT}\right)^2\left(1-\frac{kT}{4}\frac{W_4}{W_2}\right) \tag{3.67}$$

where

$$W_l=\frac{d^l W}{ds^l}, \qquad l=2,4 \tag{3.68}$$

Shavitt also suggested fitting the potential of the $H+H_2$ system along the minimum energy path to an Eckart potential and obtaining reactive rate

216 M. BAER

constants in this way. The first attempt was made with respect to a H_3 potential obtained by Boys and Shavitt.[55] Treatments of a similar kind were proposed by Weston,[56] who fitted a parabolic potential to an LEPS potential, and again by Shavitt,[57] who fitted Eckart's curve to the Shavitt-Stevens-Min-Karplus (SSMK) potential.[58] More work along these lines was done by Truhlar and Kuppermann,[59] who solved numerically the one-dimensional Schrödinger equation for the potential along the minimum energy path, using the SSMK potential. There also were a few attempts to extend the ordinary one-dimensional treatments to cases that allowed the collinear system as a whole to rotate; see, for instance, Johnson and Rapp,[60] Child,[61] and Wyatt.[62]

C. Hard-Sphere-Type Models

Although the main concepts regarding the reactive process were available from the beginning of the 1930s, it was only in 1943 that Hulburt and Hirschfelder[28] carried out the first numerical treatment with respect to a reactive hypersurface. Earlier exact numerical treatments had consisted mainly of treating single-coordinate systems (see Section III.B) or two-coordinate type models that however were linear in the sense that a Cartesian coordinate changes from $-\infty$ to $+\infty$ (with zero curvature)[12] (see Fig. 4c). The model potential considered by Hulburt and Hirschfelder (HH) was simple enough to enable numerical results to be obtained and, moreover, close enough to a realistic potential hypersurface governing the motion of three particles constrained to move along a line (the collinear case). The importance of the HH study is not only historical; many of the present-day methods still use some of the techniques applied already in this very first treatment. HH considered a case that in present-day terminology would be called the "light-heavy-light" (LHL) case, where the "heavy" mass is infinitely heavy. As can be seen from Fig. 7b, the potential channel is divided into three regions: a reagent region (I), a product region (II), and the interaction region (III). The wave function in region I is presented in terms of incoming and outgoing waves, that is,

$$\psi_I(x,y)=e^{-ik_{n_0}^I x}\phi_{n_0}^I(y)+\sum_{n=0}^{\infty}\left(\frac{k_{n_0}^I}{k_n^I}\right)^{1/2}R_{nn_0}e^{ik_n^I x}\phi_n^I(y) \quad (3.69)$$

The wave function in region II is presented only in terms of outgoing waves, that is,

$$\psi_{II}(x,y)=\sum_{m=0}^{\infty}\left(\frac{k_{n_0}^I}{k_m^{II}}\right)^{1/2}T_{mn_0}e^{ik_m^{II}y}\phi_n^{II}(x) \quad (3.70)$$

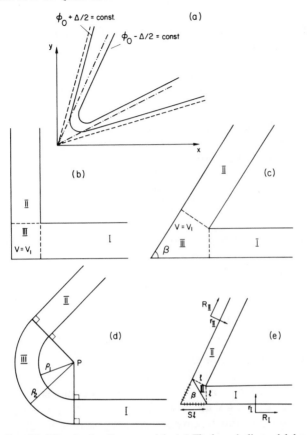

Fig. 7. Simplified "hard-sphere"-type models. (a) The hyperbolic model due to Hulburt and Hirschfelder.[28] (b) The rectangular Hulburt-Hirschfelder model (infinitely heavy central mass).[28] (c) The skewed (finite central mass) potential.[63] (d) The circular hard-sphere potential. (e) A modified skewed potential.[67]

and the wave function in region III is presented in terms of standing waves:

$$\psi_{III}(x, y) = \sum_{n'=0}^{\infty} \left[A_{n'n_0} \phi_{n'}^I(x)\sin(k_{n'}^I y) + B_{n'n_0} \phi_{n'}^{II}(y)\sin(k_{n'}^{II}x) \right] \quad (3.71)$$

Here

$$k_{n'}^l = \left(\frac{2\mu}{\hbar^2} \left(E - \mathcal{E}_{n'}^l \right) \right)^{1/2}, \quad l = I, II \quad (3.72)$$

where μ is the mass of the system, E is the total energy, $\mathcal{E}_{n'}^l$ and $\phi_{n'}^l(x)$; $l = I, II$, are respectively the nth eigenstate and the nth eigenfunction, both in the lth channel. HH chose the vibrational potential to be an infinite deep square-well-type potential, and consequently:

$$\phi_n^l(z) = \sin\left(\frac{n\pi}{L_l} z\right); \qquad z = x, y, l = \text{I}, \text{II} \qquad (3.73)$$

The R_{nn_0} is the nth nonreactive T-matrix element starting at an initial state n_0, and T_{mn_0} is the mth reactive T-matrix element. The conservation of particles is guaranteed by the relation:

$$\sum_n |R_{nn_0}|^2 + \sum_m |T_{mn_0}|^2 = 1 \qquad (3.74)$$

The coefficients T_{mn_0} and R_{nn_0}, as well as A_{nn_0} and B_{nn_0}, are solved by continuity requirement ("matching") at the boundaries of each region.

Thus, at $x = L_\text{I}$ we require:

$$\psi_\text{I}(x, y) = \psi_\text{III}(x, y), \qquad \left.\frac{\partial \psi_\text{I}}{\partial x}\right|_{x=L_\text{I}} = \left.\frac{\partial \psi_\text{III}}{\partial x}\right|_{x=L_\text{I}} \qquad (3.75)$$

and similar conditions must be met at $y = L_\text{II}$.

Now if the summations in (3.69) to (3.71) are truncated at $n = N$, the continuity conditions can be shown to yield $4N$ nonhomogeneous algebraic equations for each of the $4N$ unknowns $A_{n'n_0}$, $B_{n'n_0}$, R_{nn_0}, and T_{mn_0}.

Two modifications of this treatment were suggested about 25 years later. Tang, Kleinman, and Karplus (TKK)[63] and (independently) Wilson[64] showed how one can treat the same model but for any mass combination, and Baer and Kouri[65] removed the assumption concerning the infinite depth of the square-well potentials. TKK[63] followed the prescription offered by HH but used a different system of coordinates in region III. Instead of applying Cartesian coordinates, they used polar coordinates defined with respect to the origin:

$$x = R\cos\varphi \qquad \text{and} \qquad y = R\sin\varphi \qquad (3.76)$$

If β is the transformation angle between the two-arrangement channel coordinates (R_I, r_I) and $(R_\text{II}, r_\text{II})$ given by (3.12), then $\psi_\text{III}(R, \varphi)$ is defined as:

$$\psi_\text{III}(R, \varphi) = \sum_n (A_n F_n + B_n G_n) \qquad (3.77)$$

where

$$F_n(R,\varphi)=J_{nq}(\alpha R)\sin(nq\varphi)$$

$$G_n(R,\varphi)=J_{(n-1/2)q}(\alpha R)\sin\left(\left(n-\frac{1}{2}\right)q\varphi\right)$$

$$+\frac{J_{(n-1/2)q}(\alpha R_l)}{J_{(n+1/2)q}(\alpha R_l)}J_{(n+1/2)q}(\alpha R)\sin\left(\left(n+\frac{1}{2}\right)q\varphi\right) \quad (3.78)$$

Here $q=2\pi/\beta$, $\alpha=2\mu/\hbar^2(E-V_1)$ (V_1 is a step-function-type potential defined in region III only; see Fig. 7c), $R_l=l\sqrt{1+S^2}$ (see Fig. 7e), and J_k is the Bessel function of order k. The functions F_n and G_n are the two independent solutions of the Schrödinger equation in region III. In addition, they satisfy the following boundary conditions:

$$F_n(R,0)=G_n(R,0)=0$$

$$F_n(0,\varphi)=G_n(0,\varphi)=0$$

$$F_n(R,\beta)=G_n(R,\beta)=0$$

$$F_n\left(R_l,\frac{\beta}{2}\right)=G_n\left(R_l,\frac{\beta}{2}\right)=0 \quad (3.79)$$

TKK studied this model for two mass combinations, $m_A=m_C$, $m_\beta=\infty$ ($\beta=\pi/2$) and $m_A=m_B=m_C$ ($\pi=2\pi/3$) and for three different values of V_1, namely, $V_1=0$ (a flat interacting region) and $V_1=2.5(\pi/L)^2$ (an interaction region with a barrier) and $V_1=-1.5(\pi/L)^2$ (an interaction, with a well). From now on, the first mass combination will be termed light-heavy-light (LHL) and the second light-light-light (LLL).

The main findings are:

1. The LLL and LHL mass combination cases yield oscillatory transition probability functions, but their energy dependencies are rather different. In the LHL case, the oscillatory behavior is caused by Feschbach-type resonances,[66] whereas in the LLL case, because of the broadening of the interaction region, additional interference effects enhance the oscillatory behavior.

2. If there is a barrier, we notice the existence of the quantum-mechanical tunneling process, which becomes larger as the angle β decreases.

3. As far as threshold behavior is concerned, the case of a barrier and the case with a potential well exhibit very similar features.[66] In both cases, the initial slope is small and it increases gradually as the energy approaches the opening of the second vibrational state.

A treatment of the finite-mass case, which differs from the TKK treatment and is much closer in spirit to the HH treatment, was given by Wilson.[64] He suggested using the ordinary (vibrational and translational) coordinates in regions I and II, but in region III he recommended using the interatomic distances. Consequently, he was able to represent the wave function in region III in terms of trigonometric functions, thus avoiding the relatively complicated Bessel functions.

The finite (deep)-channel case (as opposed to the infinite-channel case) was handled by Baer and Kouri[65], who applied their coupled τ-method (see Section V). Since this method circumvents the need for matching of wave functions, one can treat a finite deep potential case as easily as the infinite deep case. The results show that the effect of the channel depths is secondary as long as the energy is far from a resonance. The treatment should be carried out with more care once the energy reaches the resonance region.

In 1943 Hulburt and Hirschfelder[28] recognized that to increase the flexibility and therefore the reliability of the hard-sphere model, a circular-type interaction region is more appropriate than a rectangular one. They suggested using a channel between two confocal hyperbolas where the potential energy is assumed to be zero inside and infinite outside (see Fig. 7a). Introducing the two elliptic coordinates θ and φ, which are related to the Cartesian coordinates x and y as:

$$x = R \cosh \theta \sin \varphi$$
$$y = R \sinh \theta \cos \varphi \qquad (3.80)$$

the Schrödinger equation takes the form:

$$-\frac{\hbar^2}{2\mu} \left(\frac{\partial^2 \psi}{\partial \theta^2} + \frac{\partial^2 \psi}{\partial \phi^2} \right) + R^2 (\cosh 2\theta + \cos 2\phi) \frac{E}{2} \psi = 0 \qquad (3.81)$$

In this model, φ serves as a vibrational type coordinate and θ as the translational one. Having these coordinates, HH found a way to construct a reaction coordinate s:

$$s = R \int_0^\theta \left(\sinh^2 \theta + \cos^2 \varphi \right)^{1/2} d\theta \qquad (3.82)$$

Now if $x(s)$ is defined as the translational part of the wave function, the corresponding Schrödinger equation becomes:

$$-\frac{\hbar^2}{2\mu} \frac{d^2 x}{ds^2} + (v(s) - E)\chi = 0 \qquad (3.83)$$

where $V(s)$ is a given function of s.

Another model of "circular" shape, but using Cartesian coordinates instead of circular ones, was studied by Baer.[67, 68] Regions I and II were as before, but region III was now defined by the lines $R_I = Sl$; $R_{II} = Sl$; $r_I = -S(R_I - Sl)$ (see Fig. 7e). The Cartesian coordinates used in region III were ζ and η, defined as:

$$\zeta = \frac{R_I - Sr_I}{(1+S^2)^{1/2}}, \quad \text{and} \quad \eta = \frac{SR_I - r_I}{(1+S^2)^{1/2}} \qquad (3.84)$$

and the corresponding wave function is

$$\psi_{III} = \left(\frac{2}{L}\right)^{1/2} \sum_{n'} \left(A_{n'} e^{ik_{n'}\zeta} + B_{n'} e^{-ik_{n'}\zeta} \right) \sin\frac{n'\pi}{L}(\eta - t) \qquad (3.85)$$

where t is

$$t = \frac{S^2}{(1+S^2)^{1/2}} L \qquad (3.86)$$

Very similar in spirit is the model in which region III is bounded by two circular arcs, as shown in Fig. 7d.[69] In this case, one employs polar coordinates [see (3.18)] and the Hamiltonian takes the same form as given in (3.19). The corresponding wave function is:

$$\psi_{III} = \frac{1}{\sqrt{\rho}} \sum \left(A_n e^{i\alpha_n\theta} + B_n e^{-i\alpha_n\theta} \right) \sin\frac{n\pi}{L}(\rho - \rho_1) \qquad (3.87)$$

where

$$L = \rho_2 - \rho_1 \qquad (3.88)$$

and α_n are determined by solving an algebraic eigenvalue problem. Here ρ_1 may or may not differ from zero.

In all previously described models, the vibrational potential was chosen to be a square-well potential. Almost no significant difficulties are expected when the vibrational potential is chosen to be harmonic. Gross and Korsch[70] did this when they considered a model in which only two regions are distinguished (instead of three); these are the reagent channel and the product channel, in each of which the potential $V(r, R)$ is assumed to be separable, namely,

$$V(r, R) = V^R(R) + V^r(r) \qquad R = R_I, R_{II} \qquad r = r_I, r_{II} \qquad (3.89)$$

The boundary between the two regions is defined by the line of intersection of the potentials:

$$V_{\rm I}(r_{\rm I}, R_{\rm I}) = V_{\rm II}(r_{\rm II}, R_{\rm II}) \tag{3.90}$$

In general, one could assume any form for $V^R(R)$ and $V^r(r)$. Gross and Korsch, in the numerical part of their work, considered only the case of V^R ($R=0$), and $V^r(r)$ was assumed to be a harmonic oscillator.

D. The Inversion Process in Chemical Reactions

Beutler and Polanyi[71] in 1928 had speculated on the possibility that a simple chemical reaction might yield highly vibrational, excited products. The interest in such reactions was immensely enhanced in the 1960s and 1970s when it was recognized that they could lead to the forming of vibrational inverted populations appropriate for use with (chemical) lasers. It was soon recognized not only that the exothermic reactions

$$F + H_2 \rightarrow HF + H$$
$$H + Cl_2 \rightarrow HCl + Cl$$
$$Cl + HBr \rightarrow HCl + Br$$
$$H + F_2 \rightarrow HF + F \tag{3.91}$$

could yield the required inverted populations; in fact, under appropriate conditions, some of them could yield the most powerful and efficient lasers. The intensive experimental research that was carried out was followed by the theory. Using the classical trajectories technique, the Polanyi group[72, 73] was the first to show that LEPS-type potential energy surfaces can yield outcomes that are close to the experimental results. As for quantum-mechanical treatments, it was only in 1973 that Wu et al.[74] obtained the inversion for $F + H_2$, then Baer[75] did the same for $H(D) + Cl_2$ (see results for $D + Cl_2$ in Fig. 8) and for $Cl + HBr$,[76] and Connor et al.[77] for $H + F_2$. A different treatment for $F + H_2$ by Schatz et al.[78] supported the preliminary findings of Wu et al.

Much effort was invested in devising simplified models to understand the inversion process. In general, one can distinguish between dynamic models that relate the inversion to the curvature of the reaction path, and static models that are based on Franck-Condon (FC) overlap integrals between two shifted harmonic (or other) potentials. The sections that follow discuss these two approaches to some extent.

1. Dynamic Approach

The first dynamic model was given by Hofacker and Levine in 1971.[80] They considered the Schrödinger equation in some general curvilinear coor-

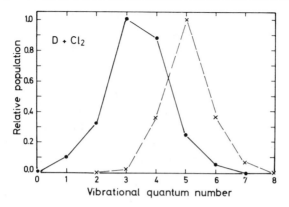

Fig. 8. Vibrational distribution for the reaction $D + Cl_2(v_i = 0) \rightarrow DCl(v_f) + Cl$. Solid curve, experimental results[79]; dashed curve, exact collinear quantum-mechanical calculation.[75]

dinates[13] and followed the Van Vleck transformation[81] to treat the remaining off-diagonal term by a first-order perturbation theory. The main outcome of this treatment is a coupling parameter, g, defined as:

$$g = \frac{E_k}{\hbar \omega} k \tilde{\rho} \qquad (3.92)$$

where E_K, ω, and $\tilde{\rho}$ are the kinetic energy, the frequency, and the oscillator width in the interaction region, respectively, and k is the curvature of the minimum energy path in the same region. It was argued that the larger g, the more pronounced the degree of inversion. This finding was confirmed to a certain degree by additional studies.[77, 82–86]

A similar approach was applied by Basilevsky et al.[36, 87, 88] who also reached the same conclusion. However, since Basilevsky was considering the Marcus equation, he (and later some others) recognized the limitations of the theory, which was found to be applicable only for small curvatures k and had no possibility of being extended in a straightforward way to large curvatures. Reconsidering the definition of g [see (3.91)] it can be seen that since k cannot grow large, the only way to obtain a large value for g is to make the energy E_k large enough. This outcome was confirmed to some extent by numerical studies.[75, 88]

As mentioned, the difficulty involved with large curvatures is best seen from the Marcus equation discussed earlier. It is not just a matter of terms that are too large to be handled by low-order perturbation theories; essential singularities appear in the equation itself. The singularities become apparent when $\eta(s)$ becomes zero or when ρ, the vibrational coordinate, becomes equal to $k(s)^{-1}$ [see (3.25)]. As long as $k(s)$ is small, $k(s)^{-1}$ is large and consequently $\eta(s)$ is different from zero in the region of physical significance. However, once $k(s)$ becomes large enough, $\eta(s)$ becomes zero and the equation is singular in the region of physical importance. Consequently, both the approximate and the exact treatments of this equation become rather complicated.

The study of dynamic models is difficult because, among other reasons, the concept of the curvature is not uniquely defined. The curvature function is linked to the form of the reaction coordinate. Therefore, as long as a reaction coordinate cannot be defined uniquely, the curvature function is almost entirely arbitrary. Part of this difficulty can be overcome by defining an average curvature:

$$\bar{k}(s)=\int \psi^*(s,\rho)k(s,\rho)\psi(s,\rho)\,d\rho \tag{3.93}$$

where $\psi(s,\rho)$ is the total wave function. To obtain an analytic expression for $k(s,\rho)$, Baer and Beswick[35] approximated the reaction coordinate in the vicinity of its maximal value by a hyperbola. Doing so, they showed that if $k(s,\rho)$ is

$$k(s,\rho)=K_0(\rho)\left(1-\tfrac{1}{2}K_2(\rho)s^2\right) \tag{3.94}$$

then $K_0(\rho)$ is

$$K_0(\rho)=\frac{1}{a+\rho}\cotan^2\left(\frac{\beta}{2}\right) \tag{3.95}$$

and $K_2(\rho)$—which is equal to $(1/K_0)(d^2k/ds^2)|_{s=0}$—is:

$$K_2(\rho)=\frac{3}{(a+\rho)^3}\sin^{-4}\left(\frac{\beta}{2}\right) \tag{3.96}$$

Here β is the rotation angle [see (3.12)] and a is the (closest) distance between the origin and the reference curve along which ρ is zero (see Fig. 9). To extend the expression for $k(s,\rho)$ to larger values of s, one may replace (3.94) by a Gaussian function

$$k(s,\rho)=K_0(\rho)\exp\left[-\tfrac{1}{2}K_2(\rho)s^2\right] \tag{3.97}$$

that guarantees that $k(s,\rho)$ becomes zero once $|s|$ is large enough (still, this form must be modified to account for the correct normalization[38, 88]). Baer and Beswick[35] also suggested linking the inversion to the second derivative

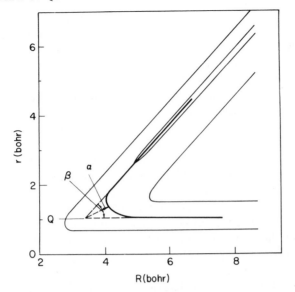

Fig. 9.   Parametric representation of the reaction coordinate in the vicinity of maximal curvature: α and β are parameters related to the hyperbola [see (3.35)].

of k with respect to s and consequently defining a different coupling term of the form:

$$\tilde{g} = \frac{\mathscr{E}_k}{\hbar\omega} k\tilde{\rho} \tag{3.98}$$

where $\mathscr{E}_k$ is equal to $(\hbar^2/2\mu)K_2$. It is seen that $\tilde{g}$ looks similar to g [see (3.92)] except that $\mathscr{E}_k$ is no longer equal to the kinetic energy but is related to the second derivative of the curvature function.

2.   Static Models

The static models are related to the distorted (or undistorted) wave Born approximation (DWBA), which for more than 30 years has been extensively applied in nuclear physics. The first attempt to apply this method in the study of chemical reactions was probably by Bauer and Wu in 1954.[47] This was followed 10 years later by a study by Micha.[89-91] However, it is only since the early 1970s that this method has become well established and thus widely used. Tang and Karplus[92] followed rigorously the formal expression for the required T-matrix element and made all the necessary theoretical and numerical efforts to obtain a meaningful result for the three-dimensional reactive cross-section for the $H+H_2$ reaction. The method was later modified

226 M. BAER

by Choi and Tang[93, 94] and extensively applied in their studies of $H+H_2$, $D+H_2$, and $F+H_2$.[95] The DWBA was also applied by other groups (e.g., in the studies on $He+H_2^+$,[96] $H+H_2$,[97] and $H+F_2^{98}$). Although most of the efforts were directed toward three-dimensional systems, some applications of this method were also tried with respect to collinear arrangements, mainly with the aim of testing the reliability of the method versus exact calculations.[99, 100] The approximate studies of McGuire and Mueller[101] and those of McGuire and Micha,[102] which extended and modified the former, may also be considered to be inspired by the DWBA.

This section is concerned mainly with a certain type of model that although closely related to the DWBA, can still be considered to be an independent group, namely, the Franck-Condon (FC) type of model. Whereas the DWBA treatment, as well as many other approximate treatments of reactive systems, preceded the exact treatment and was actually devised as an alternative for the exact (but complicated) treatment, the FC models are a direct outcome of the exact treatment and were devised mainly to provide a qualitative explanation of the exact results. While analyzing the exact results for the reaction $H(D)+Cl_2 \rightarrow H(D)Cl+Cl$, Baer[75] noted that the final vibrational distribution of the H(D)Cl molecule was strongly dependent on the initial vibrational state of the Cl_2 molecule and that a Gaussian-type vibrational distribution obtained for an initial ground state became strongly oscillatory, once started in an excited state. This observation was later confirmed by Connor et al.,[103] applying a somewhat different surface. Since no similar effects were observed in the classical results,[104-106] one can consider this phenomenon to be a pure quantum-mechanical effect. It was then suggested[75] that a Franck-Condon model could account for these observations, and it was shown qualitatively that such a model describes correctly the findings of the exact calculations.

A more quantitative approach was taken by Halavee and Shapiro,[107] who managed to derive, from the DWBA integral, an expression for the T-matrix elements that contained explicitly the corresponding FC overlap integral. Starting with the expression:

$$T_{n_\beta n_\alpha} = \int\int dR_\alpha dr_\alpha \psi_{n_\beta}^{(-)}(R_\beta, r_\beta)\psi_{n_\alpha}(R_\alpha, r_\alpha)$$ (3.99)

where α and β stand for the initial and final arrangement channels, they ended, following various approximations, with an expression that for an adiabatic behavior in each arrangement channel, takes the form:

$$T_{n_\beta n_\alpha} = A_{n_\beta n_\alpha}\int dr_\alpha \phi_{n_\beta}(R_\alpha - \gamma_\alpha r_\alpha)\phi_{n_\alpha}(r_\alpha)$$ (3.100)

where $r_\beta = R_\alpha - \gamma_\alpha r_\alpha$ is the vibrational coordinate in the β arrangement [γ_α stands for the mass ratio $m_C/(m_B + m_C)$] and $A_{n_\beta n_\alpha}$ is a coefficient that is connected with the translational part of the wave function. As can be seen, the overlap is between two shifted eigenfunctions, one related to the reagents and the other to the products. The shift d is by an amount

$$d = \frac{R_\alpha}{\gamma_\alpha} + r_{\alpha_0} - \frac{r_{\beta_0}}{\gamma_\alpha} \qquad (3.101)$$

where r_{α_0} and r_{β_0} are the equilibrium distances of the corresponding diatomics. In addition, the force constant k_β of the product molecule is changed and becomes $k_\beta \gamma_\alpha^2$. Figure 10 compares the exact and the model results for the reactions $H + Cl_2$ ($v_i = 0, 1$) $\rightarrow HCl + Cl$.

Other groups (e.g., Schatz and Ross[108] and Fischer and Venzl[109] derived similar expressions starting from the same Born integral. However, making other approximations, they obtained different values for $A_{n_\beta n_\alpha}$ and for d.

Schatz and Ross[110] also extended their treatment to three dimensions and obtained the corresponding expressions for the vibrational distributions. Good agreement was obtained for the $F + H_2(D_2)$ systems, but the agreement with respect to $H(D) + Cl_2$ was less satisfactory. Recently, Wong and Brumer[111] developed a Franck-Condon-type model, applying the Faddeev decomposition of the scattering wave functions. Judging from the agreement with exact calculations, their approach is most promising.

In this context one should mention several semiclassical theories that assuming the validity of the Franck-Condon model, were also able to account for the observations of the exact calculations.[112, 113]

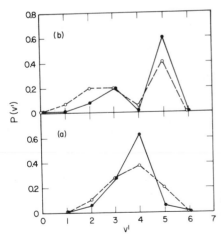

Fig. 10. Collinear vibrational distribution for the reaction $H + Cl_2(v_i) \rightarrow HCl(v_f) + Cl$. (a) $v_i = 0$. (b) $v_i = 1$. Solid curves, exact quantum-mechanical results[75]; dashed curves, Franck-Condon model results.[107]

M. BAER

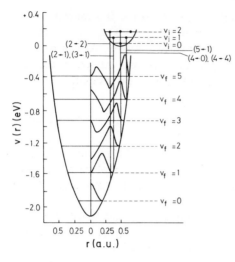

Fig. 11. The two shifted harmonic
potentials related to Cl_2 (the upper one)
and HCl. The straight lines indicate the
main possible Franck-Condon-type transitions.[75]

The different approach[114] presented now is similar in spirit to a model published by Hirschfelder and Wigner in 1939[12] (see Fig. 4c). The
Hirschfelder-Wigner model is applied, with one modification to account for
vibrational nonadiabaticity: the two channels (the reagent and the product
channels) that are to be matched are shifted by an amount b (see Fig. 11).

Once the model is solved,[114] and assuming the reflectivity to be negligibly
small, one finds for the reactive T-matrix elements, the following analytic
expression:

$$T_{mn} = \left(\frac{k_m}{k_n}\right)^{1/2} \langle \phi_m | \phi_n \rangle \qquad (3.102)$$

To check the relevance of this model, we applied it to the $H(D)+Cl_2$ systems in the following way. Using the Katriel formula[115] for the overlap integral σ_{mn} of two different shifted harmonic oscillator wave functions (see
also Refs. 116 and 117), a factorization for σ_{mn} exists which, for $n=0,1$; takes
the form:

$$\sigma_{m1} = \tau \frac{(2\chi)^{1/2}}{1+\chi} \sigma_{m0} - 2 \frac{(m\chi)^{1/2}}{1+\chi} \sigma_{m-10} \qquad (3.103)$$

where χ is the ratio between the two fundamental frequencies $\chi = \omega/\Omega$ and τ
is the shift b scaled with respect to the original harmonic oscillator half-width.
Combining (3.102) and (3.103) and assuming that:

$$\sigma_{mn} = \pm \left(\frac{k_n}{k_m} P_{mn}\right)^{1/2} \qquad (3.104)$$

the following linear relation is obtained:

$$\Gamma_{m1} = A + B\Gamma_{m0} \tag{3.105}$$

where A and B are constants:

$$A = \tau \frac{(2\chi)^{1/2}}{1+\chi} \left(\frac{k_0}{k_1} \right)^{1/2} ; \qquad B = -2 \frac{\chi^{1/2}}{1+\chi} \left(\frac{k_0}{k_1} \right)^{1/2} \tag{3.106}$$

and Γ_{m1} and Γ_{m0} are variables:

$$\Gamma_{m1} = \pm \left(\frac{P_{m1}}{P_{m0}} \right)^{1/2}, \qquad \Gamma_{m0} = \pm \left(\frac{k_m}{k_{m-1}} m \frac{P_{m-10}}{P_{m0}} \right)^{1/2} \tag{3.107}$$

Here, P_{m1} and P_{m0} are probabilities obtained by exact calculations. Figure 12 shows the results for the systems $H(D) + Cl_2$. Straight lines are obtained

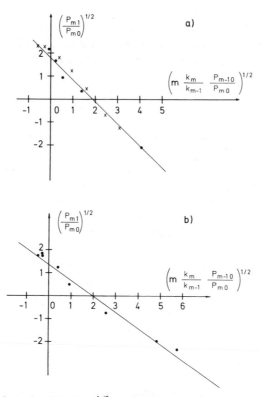

Fig. 12.   The values of $\pm(P_{m1}/P_{m0})^{1/2}$ as a function of $\pm[m(k_m P_{m-10})/(k_{m-1} P_{m0})]^{1/2}$ for $m=1,2,\dots$. The reactive transition probabilities P_{0m} and P_{1m} were obtained in exact calculation.[114] (a) Results for $H + Cl_2$: circles, values for $E = 0.268$ eV; crosses, values for $E = 0.50$ eV. (b) Results for $D + Cl_2$ (values are for $E = 0.27$ eV).

in both cases, thus supporting the relevance of the proposed model. It should be mentioned that factorization exists, not only with respect to T-matrix elements, but also for the corresponding probabilities (see Ref. 114).

E. The Resonances and the Time Delay

The theory of resonances was fully developed within the framework of nuclear physics and was transferred to the field of theoretical chemistry without any extension or modification. Although Child[118] (in 1967) was the first to consider the possibility of resonance effects in chemical reactions, it was only at the beginning of the 1970s that following the first exact quantum-mechanical calculations of the collinear reactive transition probabilities for $H + H_2$, resonances were studied in an extensive way. Levine et al.[74, 119] gave the first satisfactory explanation for the resonances in the $H + H_2$ reactive transition probability; their line of thinking was then followed and somewhat extended by Schatz and Kuppermann.[120] The concept of the resonance in relation to scattering was introduced around the mid-1930s. The energy dependence of the cross-section in the neighborhood of a resonance is given by the remarkably accurate Breit-Wigner formula,[121] which when written for the corresponding T-matrix element, takes the form:

$$T(E) = \frac{\Gamma/2}{E_0 - E + i\Gamma/2} e^{i\delta_0} \qquad (3.108)$$

Here, E_0 and Γ and δ_0 are assumed to be energy independent constants, E_0 being the resonance energy, Γ the width of the resonance curve at half its peak value, and δ_0 a phase shift associated with the nonresonant scattering. Although the theories developed by Teichman and Wigner[122] and others[123] account for this behavior, it was only in 1958 that Feschbach[124] presented his approach; his theory was free of parameters and led most straightforwardly from the compound nucleus assumption to the celebrated Breit-Wigner formula. The Feschbach theory is now the accepted explanation for the resonance structure and the corresponding resonances are accordingly termed Feschbach resonances. In the following, rather than discussing the details of the theory, which Feschbach himself[124] has given in a most comprehensive way, we deal only with its consequences and application to collinear reactive systems.

Since the findings of Levine and Wu,[119] resonances have been reported for other systems: $Cl + H_2$,[125, 126] $He + H_2^+$,[127, 128] $I + H_2$,[129, 130] $F + H_2$,[30, 42, 131-133] and $(Ar + H_2)^+$.[134] The search for resonances in the earlier studies[74, 119, 120] was performed in two ways: Reconsidering the Breit-Wigner formula for the T-matrix element, one may write it either in the form:

$$T(E) = \sin \delta_R(E) e^{i(\delta_R(E) + \delta_0)} \qquad (3.109)$$

where

$$\delta_R(E)=\tan^{-1}\left(\frac{\Gamma/2}{E-E_0}\right) \qquad (3.110)$$

or in the form:

$$T(E)=\mathrm{Re}(T(E))+i\,\mathrm{Im}(T(E)) \qquad (3.111)$$

where:

$$\mathrm{Re}(T(E))=\frac{\Gamma/2}{(E-E_0)^2+(\Gamma/2)^2}\left[(E_0-E)\cos\delta_0+\frac{\Gamma}{2}\sin\delta_0\right] \qquad (3.112)$$

and

$$\mathrm{Im}(T(E))=\frac{\Gamma/2}{(E-E_0)^2+(\Gamma/2)^2}\left[(E_0-E)\sin\delta_0-\frac{\Gamma}{2}\cos\delta_0\right] \qquad (3.113)$$

Applying the first presentation and plotting δ_R (the resonance phase shift) as a function of the energy, one may expect δ_R to change from zero to π, the main changes taking place in the vicinity of the resonance energy $\delta_R(E=E_0)$ $=\pi/2$. Considering the second presentation and making use of the Argand plot of the $\mathrm{Im}(T)$ against the $\mathrm{Re}(T)$ with the energy as a parameter, one may expect to obtain a counterclockwise-resonance quasicircle that becomes a perfect circle for energies very close to the resonance energy ($|E-E_0|\ll$ $\Gamma/2$).[135, 136] Such plots were made for the $H+H_2$ system,[74, 120] as well as for model systems[66] (see Fig. 13).

Although the early studies had recognized that the reactive resonances were the ordinary Feschbach resonances, it was only several years later that attempts were made to show that this was really the case. Following the unusual oscillatory reactive probability function obtained in the reactions given in (3.114)[126] and (3.115)[127, 128]

$$Cl+H_2\,(v_i=1)\rightarrow HCl+H \qquad (3.114)$$

$$He+H_2^+\,(v_i)\rightarrow HeH^++H \qquad (3.115)$$

Chapman and Hays[137, 138] made a major effort to relate this behavior to the formation of compound states. Their first studies were concerned with nonreactive systems, but in their more recent work they considered a reactive system that for reasons of convenience was taken to be the (endothermic) system $I+H_2$.[129] While considering the reactive transition probability $P(v_i$

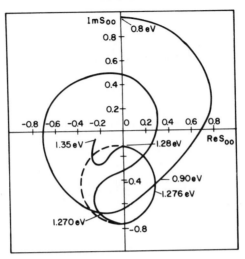

Fig. 13.   Argand plot of Im S_{00}^R versus Re S_{00}^R, with the energy as a parameter for the $H+H_2$ system.[120] The dashed part of the circle indicates the existence of a resonance circle in the energy region of $E=1.276$ eV.

$=1 \rightarrow v_f =0$), they found a markedly oscillatory behavior in the energy range of 1.890 to 1.975 eV. They identified four peaks, all of which could be directly attributed to four compound states of the H_2I system. It turned out that when plotted as a function of the reaction coordinate, the vibrational state $v_f =1$ in the product (HI) exit channel could support four bound states. Since the asymptotic $v_f =1$ state was still closed for the energy range above, based on the Feschbach theory and assuming a negligibly small interaction between the $v_f =1$ and the $v_f =0$ states, one would expect a strong oscillatory behavior for energies in the vicinity of these eigenstates. Applying a perturbative procedure, the corresponding eigenstates were calculated and compared with the scattering energies for which the probability function peaked. The deviations were less than 0.5%.

The study by Latham et al.[42] for the $F+H_2$ system can be considered to be somewhat similar in spirit to the foregoing. Although the importance of wells along the reaction coordinate for different states had been clearly demonstrated, they made no efforts to attribute the various resonances in this system to any specific compound eigenstate. Rather, they showed that the strong coupling between the different states in the interaction region, caused the final resonance to be a result of many interacting states.

A concept closely related to the concept of the resonance is the time delay. Although time delays were treated in only a few cases, they are interesting enough to be discussed here. The concept of time delay originated in the work of Eisenbud and Wigner.[139-142] Considering an incident beam that is a superposition of two monoenergetic beams of energy $h(\nu+\nu')$ and $h(\nu-\nu')$, where $\nu' \ll \nu$, the incident wave function is given by

$$\psi_{\text{inc}} = r^{-1}\left\{ \exp\left[-i\left((k+k')r - (\nu+\nu')t \right) \right] + \exp\left[-i\left((k-k')r - (\nu-\nu')t \right) \right] \right\} \tag{3.116}$$

and the outgoing wave is:

$$\begin{aligned} \psi_{\text{out}} = r^{-1}\Big\{ &\exp\left[i\left((k+k')r - (\nu+\nu')t + (\delta_R + \delta_R') \right) \right] \\ &+ \exp\left[i\left((k-k')r - (\nu-\nu')t + (\delta_R - \delta_R') \right) \right] \Big\} \end{aligned} \tag{3.117}$$

Here, $k+k'$ and $k-k'$ are the corresponding wave numbers ($k' \ll k$) and $(\delta_R + \delta_R')$ and $(\delta_R - \delta_R')$ are the phase shifts that correspond to the energies $h(\nu+\nu')$ and $h(\nu-\nu')$. The two components of the two waves considered are in phase when

$$2k'r + 2\nu't = 0 \tag{3.118}$$

for the incident wave, and when

$$2k'r - 2\nu't + 2\delta_R' = 0 \tag{3.119}$$

for the outgoing wave. Since $(\nu'/k') = d\nu/dk$ is the velocity of the center of the wave, it can be seen that the incident beam moves with the correct velocity [see (3.118)], but the outgoing beam is retarded by a stretch $d\delta/dk$ and arrives at point $r - d\delta/dk$ at the time when it should have arrived at r but was prevented from doing so by the action of the scattering center. To relate $(d\delta_R/dk)$ to time delay τ_R, we divide it by v (the velocity of the incident beam) so that:

$$\tau_R = \frac{1}{v}\frac{d\delta_R}{dk} = \hbar\frac{d\delta_R}{dE} \tag{3.120}$$

For the case of a Feschbach resonance the phase shift is given in the form:

$$\delta_R = \tan^{-1}\left(\frac{\Gamma/2}{E_0 - E} \right) \tag{3.121}$$

and consequently the time delay is

$$\tau = -\hbar\frac{\Gamma/2}{(E_0 - E)^2 + (\Gamma/2)^2} \tag{3.122}$$

or, recalling (3.106) we get:

$$\tau_R = -P(E)\frac{2\hbar}{\Gamma} \tag{3.123}$$

where $P(E)$ is the corresponding transition probability.

An extension of the concept of time delay was given in 1960 by Smith.[143] Applying the definition of the time residence in a given region and defining as the time delay (or lifetime) the difference between the two residence times with and without interaction, Smith was able to construct a general lifetime matrix Q. In the general case Q and S are related to each other as:

$$Q = i\hbar S \frac{dS^\dagger}{dE} \tag{3.124}$$

which, in view of the conjugate relation between time and energy, takes the form:

$$Q = StS^\dagger \tag{3.125}$$

Smith not only showed that his general approach reduces to the Wigner-Eisenbud formulation[139] for the elastic case, but also proved that the diagonal element of Q, (i.e., Q_{ii}) is the average time delay experienced by a particle starting at an initial state i:

$$Q_{ii} = (\Delta t_i)_{\text{avc}} = \sum S_{ij}^* S_{ij} \Delta t_{ij} \tag{3.126}$$

where Δt_{ij} is the time delay in the Wigner-Eisenbud sense.

As mentioned, only sporadic attempts were made to apply the concept of time delay in the studies of reaction dynamics.[66, 120] As far as I am aware, the only attempt to apply the Smith approach was by Eastes and Marcus[144] in a model study of inelastic (nonreactive) collisions.

References

1. H. Eyring and M. Polanyi, Z. Phys. Chem. B, 12, 279 (1931).
2. H. Pelzer and E. P. Wigner, Z. Phys. Chem. B, 15, 445 (1932).
3. N. Rosen, J. Chem. Phys., 1, 318 (1935).
4. E. Whittaker, A Treatise on the Analytical Dynamics of Particles and Rigid Bodies, 4th ed., Cambridge University Press, Cambridge, 1937.
5. F. T. Smith, J. Chem. Phys., 31, 1352 (1959).
6. L. M. Delves, Nuclear Phys., 9, 391 (1959).

7. R. M. Langer, *Phys. Rev.*, **34**, 92 (1929).
8. H. Eyring, *J. Am. Chem. Soc.*, **53**, 2537 (1931).
9. M. G. Evans and M. Polanyi, *Trans. Faraday Soc.*, **31**, 11 (1935).
10. H. Eyring, *Trans. Faraday Soc.*, **31**, 3 (1935).
11. E. P. Wigner, *Trans. Faraday Soc.*, **31**, 29 (1935).
12. J. O. Hirschfelder and E. P. Wigner, *J. Chem. Phys.*, **7**, 616 (1939).
13. G. L. Hofacker, *Z. Naturforschung*, **189**, 607 (1963).
14. R. A. Marcus, *J. Chem. Phys.*, **45**, 4493 (1966).
15. R. A. Marcus, *J. Chem. Phys.*, **45**, 4500 (1966).
16. H. C. Gorben and P. Stehle, *Classical Mechanics*, 2nd ed., Wiley, New York, 1960, pp. 319 ff.
17. B. Podolsky, *Phys. Rev.*, **32**, 812 (1928).
18. C. C. Rankin and J. C. Light, *J. Chem. Phys.*, **51**, 1701 (1969).
19. B. R. Johnson, *Chem. Phys. Lett.*, **13**, 172 (1972).
20. A. Kuppermann, G. C. Schatz, and M. Baer, *J. Chem. Phys.*, **65**, 4596 (1976).
21. Z. H. Top and M. Baer, *J. Chem. Phys.*, **66**, 1363 (1977).
22. P. B. Middleton and R. E. Wyatt, *J. Chem. Phys.*, **56**, 2702 (1972).
23. L. Blum, *J. Phys. Chem.*, **70**, 2758 (1966).
24. J. N. L. Connor and R. A. Marcus, *J. Chem. Phys.*, **53**, 3188 (1970).
25. A. Rosenthal and R. G. Gordon, *J. Chem. Phys.*, **64**, 1630 (1976).
26. N. M. Witriol, J. D. Stettler, M. A. Ratner, J. R. Sabin, and S. B. Trickey, *J. Chem. Phys.*, **66**, 1141 (1977).
27. J. F. McNutt and R. E. Wyatt, *J. Chem. Phys.*, **70**, 5307 (1979).
28. H. H. Hulburt and J. O. Hirschfelder, *J. Chem. Phys.*, **11**, 276 (1943).
29. S. F. Wu and R. D. Levine, *Mol. Phys.*, **22**, 881 (1971).
30. G. C. Schatz, J. M. Bowman, and A. Kuppermann, *J. Chem. Phys.*, **63**, 674 (1975).
31. D. J. Diestler and V. McKoy, *J. Chem. Phys.*, **48**, 2951 (1968).
32. A. Askar, A. S. Cakanak, and H. A. Rabitz, *Chem. Phys.*, **33**, 267 (1978).
33. J. Manz, *Mol. Phys.*, **28**, 399 (1974).
34. J. C. Light and R. B. Walker, *J. Chem. Phys.*, **65**, 4272 (1976).
35. M. Baer and J. A. Beswick, *Chem. Phys.*, **21**, 443 (1977).
36. M. V. Basilevsky, *Chem. Phys.*, **12**, 315 (1976).
37. R. E. Wyatt, *J. Chem. Phys.*, **51**, 3489 (1969).
38. M. S. Child, *Molecular Collision Theory*, Academic Press, London, 1974.
39. P. A. Madden and J. N. Murrel, *Mol. Phys.*, **31**, 1643 (1976).
40. J. T. Adams, R. L. Smith, and E. F. Hays, *J. Chem. Phys.*, **61**, 2193 (1974).
41. J. C. Light and R. B. Walker, *J. Chem. Phys.*, **65**, 1599 (1976).
42. S. L. Latham, J. F. McNutt, R. E. Wyatt, and M. J. Redmon, *J. Chem. Phys.*, **69**, 3746 (1978).
43. F. T. Smith, *Phys. Rev.*, **179**, 111 (1969).
44. M. Baer, *Chem. Phys. Lett.*, **35**, 112 (1975).
45. M. Baer, *Chem. Phys.*, **15**, 49 (1976).
46. M. Baer, G. Drolshagen, and J. P. Toennies, *J. Chem. Phys.*, **73**, 1690 (1980).
47. E. Bauer and T. Y. Wu, *J. Chem. Phys.*, **21**, 726 (1954).
48. C. Eckart, *Phys. Rev.*, **35**, 1303 (1930).
49. R. P. Bell, *Proc. R. Soc. London, Ser. A*, **139**, 466 (1933).
50. E. P. Wigner, *Z. Phys. Chem. B*, **19**, 203 (1933).
51. H. Jeffreys, *Proc. Math. Soc.*, **23**, 428 (1924).
52. R. P. Bell, *Proc. R. Soc. London, Ser. A*, **148**, 241 (1935).
53. R. P. Bell, *Proc. R. Soc. London, Ser. A*, **158**, 128 (1959).

54. I. Shavitt, *J. Chem. Phys.*, **31**, 1359 (1959).
55. S. F. Boys and I. Shavitt, University of Wisconsin Naval Research Laboratory Technical Report WIS-AF-13, 1959.
56. W. E. Weston, Jr., *J. Chem. Phys.*, **31**, 892 (1959).
57. I. Shavitt, *J. Chem. Phys.*, **49**, 4048 (1968).
58. I. Shavitt, R. M. Stevens, F. L. Minn, and M. Karplus, *J. Chem. Phys.*, **47**, 3961 (1967).
59. D. G. Truhlar and A. Kuppermann, *J. Am. Chem. Soc.*, **93**, 1840 (1971).
60. H. S. Johnson and D. Rapp, *J. Am. Chem. Soc.*, **83**, 1 (1961).
61. M. S. Child, *Mol. Phys.*, **12**, 401 (1967).
62. R. E. Wyatt, *J. Chem. Phys.*, **51**, 3489 (1969).
63. K. T. Tang, B. Kleinman, and M. Karplus, *J. Chem. Phys.*, **50**, 1119 (1969).
64. D. J. Wilson, *J. Chem. Phys.*, **51**, 5008 (1969).
65. M. Baer and D. J. Kouri, *J. Chem. Phys.*, **56**, 4840 (1972).
66. J. O. Hirschfelder and K. T. Tang, *J. Chem. Phys.*, **64**, 760 (1977).
67. M. Baer, *J. Chem. Phys.*, **54**, 3670 (1971).
68. B. M. Mahan, *J. Chem. Educ.*, **51**, 377 (1974).
69. M. Baer, unpublished results.
70. M. Gross and H. J. Korsch, *Chem. Phys. Lett.*, **28**, 573 (1974).
71. H. Beutler and M. Polanyi, *Z. Phys. Chem. B*, **1**, 3 (1928).
72. P. J. Kuntz, E. M. Nemeth, J. C. Polanyi, S. D. Rosner, and C. E. Young, *J. Chem. Phys.*, **44**, 1168 (1966).
73. K. G. Anlauf, P. J. Kuntz, D. H. Maylotte, P. D. Pacey, and J. C. Polanyi, *Faraday Discuss. Chem. Soc.*, **44**, 83 (1967).
74. S. F. Wu, B. R. Johnson, and R. D. Levine, *Mol. Phys.*, **25**, 839 (1973).
75. M. Baer, *J. Chem. Phys.*, **60**, 1057 (1974).
76. M. Baer, *J. Chem. Phys.*, **62**, 305 (1975).
77. J. N. L. Connor, W. Jakubetz, and J. Manz, *Chem. Phys. Lett.*, **44**, 516 (1976).
78. G. C. Schatz, J. M. Bowman, and A. Kuppermann, *J. Chem. Phys.*, **56**, 1024 (1973).
79. K. G. Anlauf, D. S. Home, R. G. MacDonald, J. C. Polanyi, and K. B. Woodall, *J. Chem. Phys.*, **57**, 1561 (1972).
80. G. L. Hofacker and R. D. Levine, *Chem. Phys. Lett.*, **9**, 617 (1971).
81. J. H. Van Vleck, *Phys. Rev.*, **33**, 467 (1929).
82. G. L. Hofacker and N. Rösch, *Ber. Bunsenges. Phys. Chem.*, **77**, 661 (1973).
83. G. L. Hofacker and K. W. Michel, *Ber. Bunsenges. Phys. Chem.*, **78**, 174 (1974).
84. G. L. Hofacker and R. D. Levine, *Chem. Phys. Lett.*, **33**, 404 (1975).
85. K.-D. Hänsel, *Ber. Bunsenges. Phys. Chem.*, **79**, 285 (1975).
86. J. W. Duff and D. G. Truhlar, *J. Chem. Phys.*, **62**, 2477 (1975).
87. M. V. Basilevsky, *Mol. Phys.*, **28**, 617 (1974).
88. M. V. Basilevsky and V. M. Rayaboy, *Chem. Phys.*, **41**, 461, 477, 489 (1979).
89. D. Micha, *Arkh. Fys.*, **30**, 411 (1965).
90. D. A. Micha, *Arkh. Fys.*, **30**, 425 (1965).
91. D. A. Micha, *Arkh. Fys.*, **30**, 437 (1965).
92. K. T. Tang and M. Karplus, *Phys. Rev. A*, **4**, 1844 (1971).
93. B. H. Choi and K. T. Tang, *J. Chem. Phys.*, **62**, 3642 (1975).
94. B. H. Choi and K. T. Tang, *J. Chem. Phys.*, **62**, 3642 (1975).
95. Y. Shan, B. H. Choi, R. T. Poe, and K. T. Tang, *Chem. Phys. Lett.*, **57**, 379 (1978).
96. C. Zuhrt, F. Schnider, and L. Zülicke, *Chem. Phys. Lett.*, **43**, 571 (1976).
97. D. C. Clary and J. N. L. Connor, *Chem. Phys.*, **48**, 175 (1980).
98. D. C. Clary and J. N. L. Connor, *Chem. Phys. Lett.*, **66**, 493 (1979).

99. R. B. Walker and R. E. Wyatt, *Chem. Phys. Lett.*, **16**, 52 (1972).
100. R. B. Gilbert and T. F. George, *Chem. Phys. Lett.*, **20**, 189 (1973).
101. P. McGuire and C. R. Mueller, *Phys. Rev. A*, **3**, 1338 (1973).
102. P. McGuire and D. A. Micha, *Mol. Phys.*, **25**, 1335 (1973).
103. J. N. L. Connor, A. Lagana, J. C. Whitehead, W. Jakubetz, and J. Manz, *Chem. Phys. Lett.*, **62**, 479 (1979).
104. H. Essen, G. D. Billing, and M. Baer, *Chem. Phys.*, **17**, 443 (1976).
105. D. G. Truhlar, J. A. Merrick, and J. W. Duff, *J. Am. Chem. Soc.*, **98**, 677 (1971).
106. J. C. Gray, D. G. Truhlar, and M. Baer, *J. Phys. Chem.*, **83**, 1045 (1979).
107. U. Halavee and M. Shapiro, *J. Chem. Phys.*, **64**, 2826 (1976).
108. G. C. Schatz and J. Ross, *J. Chem. Phys.*, **66**, 1021 (1977).
109. S. Fischer and G. Venzl, *J. Chem. Phys.*, **67**, 1335 (1977).
110. G. C. Schatz and J. Ross, *J. Chem. Phys.*, **66**, 1037 (1977).
111. J. K. C. Wong and P. Brumer, *Chem. Phys. Lett.*, **68**, 517 (1979).
112. M. S. Child and K. B. Whaley, *Faraday Discuss. Chem. Soc.*, **67**, 57 (1979).
113. G. D. Billing, B. C. Eu, N. Garisto-Zaritsky, and C. Nyeland, *J. Chem. Phys.*, **73**, 1627 (1980).
114. M. Baer, to be published.
115. J. Katriel, *J. Phys. B*, **3**, 1315 (1970).
116. E. Hutchisson, *Phys. Rev.*, **36**, 410 (1930).
117. K. Husimi, *Prog. Theor. Phys.*, **9**, 381 (1953).
118. M. S. Child, *Mol. Phys.*, **12**, 401 (1967).
119. R. D. Levine and S.-F. Wu, *Chem. Phys. Lett.*, **11**, 557 (1971).
120. G. C. Schatz and A. Kuppermann, *J. Chem. Phys.*, **59**, 964 (1973).
121. G. Breit and E. P. Wigner, *Phys. Rev.*, **49**, 519 (1936).
122. T. Teichman and E. P. Wigner, *Phys. Rev.*, **87**, 123 (1952).
123. P. I. Kapur and P. Peierls, *Proc. R. Soc. London, Ser. A*, **166**, 277 (1937).
124. H. Feschbach, *Ann. Phys.*, **5**, 357 (1958).
125. M. Baer, *Mol. Phys.*, **27**, 1429 (1974).
126. M. Baer, U. Halavee, and A. Perski, *J. Chem. Phys.*, **61**, 5122 (1974).
127. D. J. Kouri and M. Baer, *Chem. Phys. Lett.*, **24**, 37 (1974).
128. J. T. Adams, *Chem. Phys. Lett.*, **33**, 275 (1975).
129. F. M. Chapman and E. F. Hays, *J. Chem. Phys.*, **66**, 2554 (1977).
130. R. C. Liedtke, D. L. Knisk, and E. F. Hays, *Int. J. Quant. Chem. Symp.*, **11**, 337 (1977).
131. G. C. Schatz, J. M. Bowman, and A. Kuppermann, *J. Chem. Phys.*, **63**, 685 (1975).
132. J. N. L. Connor, W. Jakubetz, and J. Manz, *Mol. Phys.*, **29**, 347 (1975); **35**, 1301 (1978); **39**, 799 (1980).
133. J. S. Hutchison and R. E. Wyatt, *J. Chem. Phys.*, **70**, 3509 (1979).
134. M. Baer and J. A. Beswick, *Phys. Rev. A*, **19**, 1559 (1979).
135. K. Adair, *Phys. Rev.*, **113**, 338 (1959).
136. R. K. Dalitz, *Annu. Rev. Nuclear Sci.*, **13**, 346 (1964).
137. F. M. Chapman, Jr., and E. F. Hays, *J. Chem. Phys.*, **62**, 4400 (1975).
138. F. M. Chapman, Jr., and E. F. Hays, *J. Chem. Phys.*, **65**, 1032 (1976).
139. E. P. Wigner and L. Eisenbud, *Phys. Rev.*, **72**, 29 (1947).
140. L. Eisenbud, dissertation, Princeton University, June 1948.
141. E. P. Wigner, *Phys. Rev.*, **98**, 145 (1955).
142. D. Bohm, *Quantum Theory*, Prentice-Hall, Englewood Cliffs, N.J., 1951, pp. 260–261.
143. F. T. Smith, *Phys. Rev. A*, **118**, 349 (1960).
144. W. Eastes and R. A. Marcus, *J. Chem. Phys.*, **45**, 57 (1973).

IV. THE THREE-DIMENSIONAL SYSTEM

A. Introduction

Most of the theoretical efforts during the past decade have been directed toward the development of a reliable and computationally feasible method to obtain what is called an "exact cross-section for a chemical reaction in three dimensions." Most of the approaches were based on treatment of the Schrödinger equation, and mainly for historical reasons, the subject for calculation was the system $H + H_2$.[1-11] Other approaches based on treatment with the Lippman-Schwinger equation were also tried,[12, 13] and these are discussed in Section V. The details of the (differential) methods that finally yielded the desired cross-sections for the "realistic" $H + H_2$ system were recently reviewed by Wyatt,[14] and there is no justification for going over them again in this chapter.

Although more than 5 years have elapsed since the first exact cross-sections for the $H + H_2$ system were published, no "exact" treatment for any other system has been reported, except for one coplanar calculation.[15] This contrasts with the case of the classical trajectory method which, once it was fully developed, became an everyday tool for both theoreticians and experimentalists and enabled the study of many systems.

There are several difficulties associated with the three-dimensional quantum-mechanical computations. (*1*) Even today, when chemists have access to the most efficient computers, the computations are time-consuming, necessitating tens or even hundreds of computer hours. (*2*) From the numerical point of view, it is still unclear how one should bifurcate in case of a nonsymmetric three-channel system. (*3*) Since the programs themselves are large, complicated, and cumbersome, users should be experts and know most of the details.

Each of the difficulties just mentioned is enough to discourage further work in this direction. If, to top all this, one must acknowledge the (little known) fact that these various methods (which, incidentally, are all very similar in many technical details) are applicable only for collinear-dominant systems, one can hardly expect any activity at all in this area.

This situation obviously calls for a change, and two possibilities are at hand:

1. Developing a new method, based on a different, much simpler, more efficient approach, applicable for a broader range of cases.
2. Developing reliable approximations, as close as possible to the exact treatments and containing most of their features, but circumventing the above-listed difficulties.

This section discusses the second possibility, presenting two types of approximation: the infinite-order sudden approximation (IOSA) and the one-dimensional to three-dimensional ($1D \rightarrow 3D$) transformation technique.

B. The Infinite-Order Sudden Approximation

The basic aim of the IOSA, as used in the context of molecular collisions, is avoiding having to consider all the quantum-mechanical states that must be included in an "exact" treatment. In any simple inelastic collision taking place between nonhydrogenic species, it turns out that the number of rotational states increases dramatically with the energy and runs into the hundreds even before vibrational excitation is reached. In reactive scattering, the situation may seem not to be so bad, because reactive cross-sections are usually small and consequently the orbital quantum numbers that must be taken into account are quite small. Although this belief has some grounds and may apply to thermoneutral collinear-dominant reactions, it certainly cannot apply to exothermic (endothermic) reactions for which at least one channel has enough energy to yield a large number of open rotational states.

To date, the IOSA has only been applied with the aim of reducing the rotational effort (the only exception to that is a study[16] that applied this approach to the vibrational coordinate). In this sense, it probably is the final link in a series of successful approximations devised for this purpose.[17-22] A complete list of references can be found in the review by D. J. Kouri.[23] The IOSA was suggested by Takanayaki,[24] extended by Curtiss,[25] studied by Pack and co-workers,[26, 27] and given a beautiful form by Secrest.[28] Again, a detailed description of this method was offered by Kouri.[23] Since Pack's first study,[29] the IOSA was repeatedly tested against exact close-coupling and coupling-state results, and the comparisons led in most cases to very encouraging conclusions.[30, 31] As a final point we mention the analytical extension made by Goldflam et al.[32] and Khare,[33] who showed that all cross-sections can be expressed as the sum of products of (system-independent) spectroscopic factors and (system-dependent) dynamical factors, the latter applying only to transitions out of the rotational ground state.

It was only very recently that the IOSA was introduced into reactive scattering.[34-41] Three theoretical papers have been published so far, and they all follow the "exact" procedure due to Kuppermann, Schatz, and Baer (KSB), but differ in various details. Whereas two theories due to Khare, Kouri, and Baer (KKB) and Bowman and Lee (BL) employ l-labeling, the third theory, due to Barg and Drolshagen (BD), makes use of J-labeling. The two labelings yield different expressions for the differential cross-section, whereas for the total integral cross-sections the expressions are the same with both. There are other differences, mainly with respect to the transformation from one arrangement channel to the other. Whereas according to the KKB theory the transformation is done without introducing an approximation and in this sense each γ-fixed problem is solved "exactly," in the other two theories additional (and unnecessary) approximations, which might affect the results as the energy becomes larger and the reaction less collinear, are employed. Sections IV.B.1 and IV.B.2 present the main points of the IOSA procedure, as given by KKB.

1. The IOS Reactive Cross-Sections

In general one distinguishes between three representations for the S-matrix elements: (1) the S-matrix within the body-fixed (BF) formation—the formulation most appropriate for the matching of the solutions related to the different channels, that is, $S^{J}(\lambda v_{\lambda} j_{\lambda} \Omega_{\lambda} | \nu v_{\nu} j_{\nu} \Omega_{\nu})$, where λ and ν denote arrangement channels, v_{α} and j_{α}, $\alpha = \lambda, \nu$, are the corresponding vibrational and rotational quantum numbers, and Ω_{α} is the projection of the total angular momentum along the αz-axis (Ω_{α} is also equal to the projection of rotor angular momentum j_{α} along the αz-axis); (2) the S-matrix within the (coupled) space-fixed (SF) representation, namely, $S^{J}(\lambda v_{\lambda} j_{\lambda} l_{\lambda} | \nu v_{\nu} j_{\nu} l_{\nu})$, where l_{α}, $\alpha = \lambda, \nu$, are the orbital angular momenta; and (3) the S-matrix within the (uncoupled) SF representation, namely, $S(\lambda v_{\lambda} l_{\lambda} m_{\lambda} j_{\lambda} s_{\lambda} | \nu v_{\nu} l_{\nu} m_{\nu} j_{\nu} s_{\nu})$, where m_{α} and s_{α}, $\alpha = \lambda, \nu$, are the projections of l_{α} and j_{α}, $\alpha = \lambda, \nu$, along the single SF z-axis, respectively. The three types of matrix element are related to one another by the transformations (to shorten the notations, we discard part of the quantum numbers whenever it is clear that no confusion can be caused):

$$S^{J}(j_{\lambda}\Omega_{\lambda} | j_{\nu}\Omega_{\nu}) = \sum_{l_{\nu}l_{\lambda}} \frac{\sqrt{[l_{\lambda}][l_{\nu}]}}{[J]} \langle l_{\lambda} 0 j_{\lambda}\Omega_{\lambda} | J\Omega_{\lambda}\rangle$$
$$\times \langle l_{\nu} 0 j_{\nu}\Omega_{\nu} | J\Omega_{\nu}\rangle S^{J}(j_{\lambda}l_{\lambda} | j_{\nu}l_{\nu}) \quad (4.1)$$

and

$$S^{J}(j_{\lambda}l_{\lambda} | j_{\nu}l_{\nu}) = \sum_{\substack{m_{\lambda}m_{\nu} \\ s_{\lambda}s_{\nu}}} \langle l_{\lambda}m_{\lambda} j_{\lambda}s_{\lambda} | JM\rangle$$
$$\times \langle l_{\nu}m_{\nu} j_{\nu}s_{\nu} | JM\rangle S(l_{\lambda}m_{\lambda} j_{\lambda}s_{\lambda} | l_{\nu}m_{\nu} j_{\nu}s_{\nu}) \quad (4.2)$$

where [X] stands for $(2X+1)$. To relate the IOSA angle-dependent S-matrix elements to the ordinary ones, we must consider the general coordinate representations of these elements. In the atom-diatom case, we distinguish between four angles. These can be taken either as four SF coordinates

$(\theta_\alpha, \varphi_\alpha, \theta_{\alpha r}, \varphi_{\alpha r})$, $\alpha = \lambda, \nu$, or as the three Euler coordinates $(\varphi_\alpha, \theta_\alpha, \chi_\alpha)$ and the relative angle γ_α. Here, θ_α and φ_α are the angles of $\mathbf{R}_\alpha$ and $\theta_{\alpha r}$ and $\varphi_{\alpha r}$ are the angles of $\mathbf{r}_\alpha$ ($\mathbf{R}_\alpha$ and $\mathbf{r}_\alpha$ are the corresponding translational and vibrational coordinates in the α arrangement, with respect to a fixed system of coordinates. Thus, the coordinate representation that corresponds to $S(l_\lambda m_\lambda j_\lambda s_\lambda | l_\nu m_\nu j_\nu s_\nu)$ is:

$$
\begin{aligned}
S(\theta_\lambda \varphi_\lambda \theta_{\lambda r} \varphi_{\lambda r} | \theta_\nu \varphi_\nu \theta_{\nu r} \varphi_{\nu r}) = \sum_{\substack{l_\lambda m_\lambda j_\lambda s_\lambda \\ l_\nu m_\nu j_\nu s_\nu}} & Y_{l_\nu m_\nu}(\theta_\nu, \varphi_\nu) Y_{j_\nu s_\nu}(\theta_{\nu r}, \varphi_{\nu r}) \\
& \times S(l_\lambda m_\lambda j_\lambda s_\lambda | l_\nu m_\nu j_\nu s_\nu) Y^*_{l_\lambda m_\lambda}(\theta_\lambda, \varphi_\lambda) Y^*_{j_\lambda s_\lambda}(\theta_{\lambda r}, \varphi_{\lambda r})
\end{aligned}
$$

$$(4.3)$$

where the $Y_{pq}(\theta, \varphi)$ are the spherical harmonics that stand for the unitary transformation from the (θ, φ) representation to the p, q representation. A similar expression can be obtained for the S-matrix elements in terms of the Euler angles $\Lambda_\alpha (= \varphi_\alpha, \theta_\alpha, \chi_\alpha)$ and the internal angles γ_α, $\alpha = \lambda, \nu$:

$$
\begin{aligned}
S(\Lambda_\lambda \gamma_\lambda | \Lambda_\nu \gamma_\nu) = \sum_{\substack{M\Omega_\lambda \Omega_\nu \\ J j_\lambda j_\nu}} & D^{J*}_{M\Omega_\nu}(\Lambda_\nu) Y^*_{j_\nu \Omega_\nu}(\gamma_\nu, 0) \\
& \times S^J(j_\lambda \Omega_\lambda | j_\nu \Omega_\nu) D^J_{M\Omega_\lambda}(\Lambda_\lambda) Y_{j_\lambda r_\lambda}(\gamma_\lambda, 0)
\end{aligned}
\qquad (4.4)
$$

Here we use Rose's[43] convention for the rotation matrix, namely:

$$
D^p_{q'q}(\alpha, \beta, \gamma) = e^{-iq'\alpha} d^p_{q'q}(\beta) e^{-iq\gamma}
\qquad (4.5)
$$

In (4.4), the summation over M can be performed immediately, because the body-fixed S^J-matrix elements are independent of M. Therefore:

$$
\begin{aligned}
S(\Lambda_\lambda \gamma_\lambda | \Lambda_\nu \gamma_\nu) = \sum_{\substack{\Omega_\lambda \Omega_\nu \\ j_\lambda j_\nu J}} & D^J_{\Omega_\nu \Omega_\lambda}(\Lambda_\nu^{-1} \Lambda_\lambda) \\
& \times Y^*_{j_\nu \Omega_\nu}(\gamma_\nu, 0) S^J(j_\lambda \Omega_\lambda | j_\nu \Omega_\nu) Y_{j_\lambda \Omega_\lambda}(\gamma_\lambda, 0)
\end{aligned}
\qquad (4.6)
$$

where $\Lambda_\nu^{-1} \Lambda_\lambda$ is the relative Euler angle, which is equal to $\Delta_{\nu\lambda}$ (see Fig. 14a). If we now introduce the mixed representation of the S-matrix, namely, $S^J_{\Omega_\nu \Omega_\lambda}(\gamma_\lambda | \gamma_\nu)$, then:

$$
S^J_{\Omega_\nu \Omega_\lambda}(\gamma_\lambda | \gamma_\nu) = \sum_{j_\nu j_\lambda} Y^*_{j_\nu \Omega_\nu}(\gamma_\nu, 0) S^J(j_\lambda \Omega_\lambda | j_\nu \Omega_\nu) Y_{j_\lambda \Omega_\lambda}(\gamma_\lambda, 0)
\qquad (4.7)
$$

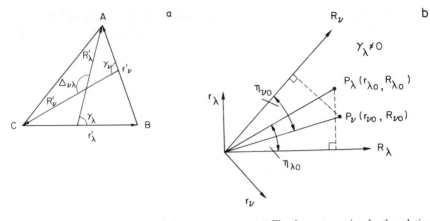

Fig. 14. The three-dimensional three-atom system. (*a*) The three-atom triangle: the relation between the λ and the ν coordinates. (*b*) The fixed $\gamma(\neq 0)$ planes. The reagent channel for fixed γ_λ is bound in the interaction region by the line $\eta_\lambda = \eta_{\lambda 0}$ and the product channel for the corresponding fixed $\gamma_\nu (= \gamma(B_{\gamma\lambda}, \gamma_\lambda))$ is bound by the line $\eta_\nu = \eta_{\nu 0}$.

then the ordinary BF S-matrix element is given in the form:

$$S^J(j_\lambda \Omega_\lambda | j_\nu \Omega_\nu) = 4\pi^2 \int \int d(\cos \gamma_\lambda) \, d(\cos \gamma_\nu) \, Y_{j_\nu \Omega_\nu}(\gamma_\nu, 0)$$
$$\times S_{\Omega_\nu \Omega_\lambda}(\gamma_\lambda | \gamma_\nu) Y^*_{j_\lambda \Omega_\lambda}(\gamma_\lambda, 0) \qquad (4.8)$$

The aim now is finding an approximate expression for $S^J_{\Omega_\nu \Omega_\lambda}(\gamma_\lambda | \gamma_\nu)$. To do that we should consider the system of equations that must be solved, and introduce the required simplification. The general form of the equation to be solved is:

$$\left[-\frac{\hbar^2}{2\mu} \left(\nabla^2_{R_\alpha} + \nabla^2_{r_\alpha} \right) + V(r_\alpha, R_\alpha, \gamma_\lambda) - E \right] \psi_\alpha(\mathbf{r}_\alpha, \mathbf{R}_\alpha) = 0, \qquad \alpha = \lambda, \nu$$

$$(4.9)$$

where it is understood that the vectors $\mathbf{R}_\alpha$ and $\mathbf{r}_\alpha$ are already scaled with respect to the appropriate masses. In what follows, we also drop the arrangement index, whenever this can be done without causing confusion. Since the total angular momentum $\mathbf{J}$ and the z-component of the total angular momentum $\mathbf{J}_z$ are constants of the motion, we may expand $\psi(\mathbf{r}, \mathbf{R})$ in terms of eigenfunctions of J^2 and J_z:

$$\psi(\mathbf{r}, \mathbf{R}) = \sum_{J=0}^{\infty} \sum_{M=-J}^{J} C_{JM} \psi_{JM}(\mathbf{r}, \mathbf{R}) \qquad (4.10)$$

The next two steps are to further expand $\psi_{JM}(\mathbf{r}, \mathbf{R})$ in terms of various basis sets. Since the numerical calculations are best carried out in the BF coordinates, where each z-axis is oriented along the corresponding R-vector, the first transformation is connected with the elimination of the two angles φ and θ that define $\mathbf{R}$. Thus:

$$\psi_{JM}(\mathbf{r}, \mathbf{R}) = \sum_{\Omega} \sqrt{\frac{[J]}{4\pi}} \; D_{M\Omega}^{J}(\varphi, \theta, 0) \psi_{J\Omega}(R, r, \gamma, \chi) \qquad (4.11)$$

where Ω is the projection of J on the BF z-axis. The second transformation is connected with the elimination of the two other angles, γ and χ:

$$\psi_{J\Omega}(R, r, \gamma, \chi) = \frac{1}{rR} \sum_{|j|>\Omega} Y_{j\Omega}(\gamma, \chi) F_{Jj\Omega}(R, r) \qquad (4.12)$$

so that one is left with the problem of solving for $F_{Jj\Omega}(r, R)$. However, the integration can also be performed in an SF system, and since introducing the IOSA is best in such a system, we could in fact ignore the BF formulation altogether. But since the matching of the wave functions related to the different arrangement channels is performed in the BF system, we must consider the transformation matrix elements between the two representations. Thus, if $F_{Jjl}(R, r)$ is the radial part of the wave function in the SF system, then:

$$F_{Jjl}(R, r) = \sum_{\Omega} \sqrt{\frac{[l]}{[J]}} \; \langle l0j\Omega | J\Omega \rangle F_{Jj\Omega}(R, r) \qquad (4.13)$$

The corresponding equation for $F_{Jjl}(R, r)$ is:

$$\left[\left(\frac{\partial^2}{\partial R^2} + \frac{\partial^2}{\partial r^2} \right) + \frac{2\mu E}{\hbar^2} - \frac{l(l+1)}{R^2} - \frac{j(j+1)}{r^2} \right] F_{Jjl}(R, r)$$

$$= \frac{2\mu}{\hbar^2} \sum_{l'j'} \langle lj|V^J|l'j' \rangle F_{Jj'l'}(R, r)$$

$$(4.14)$$

where the potential matrix elements are given in the form:

$$\langle lj|V^J|l'j' \rangle = 2\pi \sum_{|\bar{\Omega}|\leq J} \frac{\sqrt{[l][l']}}{[J]} \langle l0j\bar{\Omega}|J\bar{\Omega} \rangle \langle l'0j'\bar{\Omega}|J\bar{\Omega} \rangle$$

$$\times \int_0^{\pi} d(\cos\gamma) Y_{j\bar{\Omega}}(\gamma, 0) V(r, R, \gamma) Y_{j\bar{\Omega}}(\gamma, 0) \qquad (4.15)$$

Equation (4.14) is in the appropriate form for IOSA introduction. In general, one distinguishes between two types of IOS, namely, the J-labeled and the l-labeled IOS. However, as recently shown[42] for nonreactive problems, the J-labeled IOS leads to nonphysical processes such as m transitions for spherical symmetric potentials; we shall, therefore, limit ourselves to the l-labeled IOS.

Considering (4.15), one notes the relation between the potential $V(r, R, \gamma)$ and the corresponding matrix element $\langle lj|V^J|l'j'\rangle$. Similarly one may define a relation between the matrix elements $F_{Jjl}(R, r)$ and the function $F_{Jj\bar{l}\bar{\Omega}}$ (R, r, γ), where, for the time being, $\bar{j}$ and $\bar{l}$ are undefined constants:

$$F_{Jjl}(R,r)=2\pi\sum_{\bar{\Omega}}\sqrt{\frac{[l]}{[J]}}\,\langle L0j\bar{\Omega}|J\bar{\Omega}\rangle\int d(\cos\gamma)\,Y_{j\bar{\Omega}}(\gamma,0)F_{Jj\bar{l}\bar{\Omega}}(R,r,\gamma)$$

$$(4.16)$$

Then it can easily be shown that $F_{Jj\bar{l}\bar{\Omega}}(R, r, \gamma)$ is the solution of the equation:

$$\left[\left(\frac{\partial^2}{\partial R^2}+\frac{\partial^2}{\partial r^2}\right)-\frac{\bar{l}(\bar{l}+1)}{R^2}-\frac{\bar{j}(\bar{j}+1)}{r^2}+\frac{2\mu}{\hbar^2}(E-V(R,r,\gamma))\right]$$
$$\times F_{Jj\bar{l}\bar{\Omega}}(R,r,\gamma)=0$$

$$(4.17)$$

Although it may seem that the two quantum numbers J and $\bar{\Omega}$ are redundant, in fact the dependence on them enters through the boundary conditions. Equation (4.17) is the basic differential equation to be solved within the IOS framework. It is solved independently in each arrangement channel, and as a final step the solutions are matched. The details of the matching procedure are discussed in the next section. Equation (4.17) has three parameters γ, $\bar{j}$, and $\bar{l}$; γ is the angle, and the equation must be solved for each value of γ before the final S-matrix elements can be obtained; the other two parameters are still arbitrary and can therefore be chosen to minimize the deviations from the exact solutions introduced by the approximation. The common choice is $\bar{j}=0$ and $\bar{l}=l$-initial (l-in), but other choices are possible.

To find the relation between the γ-dependent S-matrix elements as obtained by solving (4.17) and the true BF S-matrix elements in the quantum number representation [i.e., $S^J(j_\lambda\Omega_\lambda|j_\nu\Omega_\nu)$], one should consider the asymptotic form of the wave functions. Requiring that the IOS wave function be identical to the exact physical wave function in the asymptotic region leads to a unique, well-defined transformation. This has been analyzed in detail by KKB and we quote here only the final results. We mainly consider the choice $\bar{l}=l$-in, but we shall also discuss the extension to a somewhat more realistic, but also more complicated, labeling, namely $\bar{l}=(l_\nu + l_\lambda)/2$, which is called the l-average (l-av) labeling.

According to KKB, the "mixed" BF S-matrix element $S^J_{\Omega_\nu\Omega_\lambda}(\gamma_\lambda|\gamma_\nu)$ is given by

$$S^J_{\Omega_\nu\Omega_\lambda}(\gamma_\lambda|\gamma_\nu)=\sum_{jl\Omega}\frac{[l]}{[J]}\langle l0j\Omega|J\Omega\rangle\langle l0j\Omega_\lambda|J\Omega_\lambda\rangle$$
$$\times Y_{j\Omega}(\gamma_\lambda(\gamma_\nu),0)d_{\Omega_\nu\Omega}(\Delta_{\nu\lambda}(\gamma_\lambda))S_l(\gamma_\nu)Y_{j\Omega_\lambda}(\gamma_\lambda,0) \quad (4.18)$$

Here we distinguish between two kinds of γ_λ's, the first independent and the other dependent on γ_ν and on the matching surface through $B_{\nu\lambda}$ (for details concerning $B_{\nu\lambda}$ and how it is related to the matching, see Section IV.B.2). A different way to look at this matter is to assume that γ_λ, like γ_ν, is always an independent variable, but that every choice of γ_λ and γ_ν leads to a unique value for $B_{\nu\lambda}$ determined by the equation:

$$\gamma_\lambda(B_{\nu\lambda},\gamma_\nu)=\gamma_\lambda \quad (4.19)$$

The two different approaches determine how one should continue. If γ_λ and γ_ν are both chosen to be independent, (4.18) is employed and, after the double integral is numerically integrated, the required BF S-matrix elements are derived. If on the other hand, $B_{\nu\lambda}$ is fixed, then one γ_λ is dependent on γ_ν and $B_{\nu\lambda}$, whereas the other is still independent and, since nothing else depends on it, the integration in (4.18) on γ_λ can be performed trivially and we are left with an integral over γ_ν only. The final result for the BF S-matrix element in this case is:

$$S^J(j_\lambda\Omega_\lambda|j_\nu\Omega_\nu)=\sum_{l\Omega}2\pi\frac{[l]}{[J]}\langle l0j_\lambda\Omega|J\Omega\rangle\langle l0j_\lambda\Omega_\lambda|J\Omega_\lambda\rangle\int d(\cos\gamma_\lambda)$$
$$\times Y^*_{j_\lambda\Omega}[\gamma_\lambda(\gamma_\nu,B_{\nu\lambda}),0]d^J_{\Omega_\nu\Omega}(\Delta_{\nu\lambda}(\gamma_\nu,B_{\nu\lambda}))S_l(\gamma_\nu)Y_{j_\nu\Omega}(\gamma_\nu,0)$$
$$(4.20)$$

For the simple case when $B_{\nu\lambda}=1$, it can be shown (see next section) that $\gamma_\lambda=\pi-\gamma_\nu$ and (4.20) takes the form:

$$S^J(j_\lambda\Omega_\lambda|j_\nu\Omega_\nu)=\sum_{l\Omega}2\pi(-1)^{j_\lambda-\Omega}\frac{[l]}{[J]}\langle l0j_\lambda\Omega|J\Omega\rangle\langle l0j_\lambda\Omega_\lambda|J\Omega_\lambda\rangle$$
$$\times\int d(\cos\gamma)Y_{j_\nu\Omega}(\gamma,0)d^J_{\Omega_\nu\Omega}(\Delta_{\nu\lambda}(\gamma))S_l(\gamma)Y_{j_\lambda\Omega}(\gamma,0) \quad (4.21)$$

To write the expression for a more general choice of $\bar{l}$ one must employ the SF S-matrix rather than the BF S-matrix. Thus, for a general l-labeling, one can show, by analogy with the nonreactive case, that:

$$S^J(j_\lambda l_\lambda | j_\nu l_\nu) = i^{l_\lambda + l_\nu - 2\bar{l}} 2\pi \sum_{\Omega_\lambda \Omega_\nu} \frac{\sqrt{[l_\lambda][l_\nu]}}{[J]} (-1)^{j_\nu - \Omega_\nu} \langle l_\lambda 0 j_\lambda \Omega_\lambda | J\Omega_\lambda \rangle$$

$$\times \langle l_\nu 0 j_\nu \Omega_\nu | J\Omega_\nu \rangle \int d(\cos\gamma) Y_{j_\nu \Omega_\nu}$$

$$\times (\gamma, 0 | d^J_{\Omega_\lambda \Omega_\nu}(\Delta(\gamma)) S^-_{\bar{l}}(\gamma) Y_{j_\lambda \Omega_\lambda}(\gamma, 0) \qquad (4.22)$$

To find $S^J(j_\lambda \Omega_\lambda | j_\nu \Omega_\nu)$, one substitutes (4.22) in (4.1), where the summation over l_λ and l_ν must be done with care because $\bar{l}$ may also depend on them.

Having obtained the BF S-matrix elements, one is in a position to calculate the various measurable quantities ranging from state-to-state differential cross-sections to the integral total cross-section. The exact differential scattering amplitude for reaction (assuming distinguishable atoms) is given by:

$$f(j_\nu \Omega_\nu | j_\lambda \Omega_\lambda | \theta, \varphi) = \frac{i^{j_\lambda - j_\nu + 1}}{2k_{j\lambda}} \sum_J [J] d^J_{\Omega_\lambda \Omega_\nu}(\theta) S^J(j_\nu \Omega_\nu | j_\lambda \Omega_\lambda) \quad (4.23)$$

Recalling that $d\sigma/d\Omega = |f|^2$, one may substitute (4.22) in (4.1), which subsequently is substituted in (4.23), from which the cross-sections can then be calculated. Almost no simplification can be achieved for a general l-labeling or even for the case of $\bar{l} = l$-av labeling. However, the expressions can be simplified once the choice of $\bar{l} = l$-in is made, as was found previously in the nonreactive case. Thus, for the differential cross-section, one has[39]:

$$\frac{d\sigma}{d\theta}(j_\lambda : \lambda \to \nu | \theta) = \frac{\pi}{4k^2_{j\lambda}} \sum_L d^L_{00}(\theta) \sum_{ll'} [l][l'] |\langle l0l'0 | L0 \rangle|^2$$

$$\times \int_0^{\pi/2} d(\cos\gamma) d^L_{00}(\Delta_{\nu\lambda}(\gamma)) S_l(\gamma) S^*_{l'}(\gamma) \quad (4.24)$$

and for the corresponding integral cross-section:

$$\sigma(j_\lambda : \lambda \to \nu) = \frac{\pi}{2k^2_j} \sum_l [l] \int_0^{\pi/2} d(\cos\gamma) |S_l(\gamma)|^2 \qquad (4.25)$$

The final expressions are surprisingly simple. The independence of both the integral and the differential cross-sections on the initial rotor state (except for the $k_{j_\lambda}^2$ term in front of the summation sign) is of particular interest. In addition, the disappearance of the term containing the angle $\Delta_{\nu\lambda}(\gamma_\nu)$ (which is the transformation angle from the λ to the ν arrangement) from the expression for the total cross-section is somewhat unexpected.

2. The Matching Procedure Within the IOSA Framework

One of the more complicated problems encountered in the theory of two- and three-dimensional reactive scattering has to do with the transition from one arrangement channel to the other. The accepted procedure for overcoming this difficulty is to treat each channel independently (which means obtaining a general solution of the Schrödinger equation in each arrangement channel) and to smoothly match the wave functions of the various channels on *surfaces* that separate the respective channels.[6] In principle, the procedure is similar to the one discussed in Section III except that here the surfaces are more complicated forms.

In this review, we follow mainly the procedure and notations of KSB[6] and KKB.[34, 39] Starting with a mass-scaled, center of mass system of coordinates $(\mathbf{r}_\lambda, \mathbf{R}_\lambda)$ and $(\mathbf{r}_\nu, \mathbf{R}_\nu)$, one can show that the two sets of vectors are related by a transformation matrix $\mathbf{M}$:

$$\begin{pmatrix} \mathbf{R}_\nu \\ \mathbf{r}_\nu \end{pmatrix} = \mathbf{M} \begin{pmatrix} \mathbf{R}_\lambda \\ \mathbf{r}_\lambda \end{pmatrix} \tag{4.26}$$

where $\mathbf{M}$ is a 2×2 orthogonal matrix [see (3.12)]:

$$\mathbf{M} = \begin{pmatrix} \cos\beta & -\sin\beta \\ \sin\beta & \cos\beta \end{pmatrix} \tag{4.27}$$

Introducing the internal angle γ_λ defined through:

$$\cos\gamma_\lambda = (\hat{R}_\lambda \cdot \hat{r}_\lambda) \tag{4.28}$$

and applying (4.26), one obtains the trigonometric relation:

$$r_\nu^2 = R_\lambda^2 \sin^2\beta + r_\lambda^2 \cos^2\beta + r_\lambda R_\lambda \cos\gamma_\lambda \sin 2\beta \tag{4.29}$$

The equation of a surface in three dimensions in terms of the three coordinates R_λ, r_λ, and γ_λ can be written in general as:

$$F(R_\lambda, r_\lambda, \gamma_\lambda) = 0 \tag{4.30}$$

Recalling (4.29) and imposing an additional relation between r_ν and the λ

coordinates, that is,

$$r_\nu = g(R_\lambda, r_\lambda, \gamma_\lambda) \qquad (4.31)$$

the two equations (4.29) and (4.31) form an equation for a surface:

$$F(R_\lambda, r_\lambda, \gamma_\lambda) = g^2(R_\lambda, r_\lambda, \gamma) - \left[R_\lambda^2 \sin^2\beta + r_\lambda^2 \cos^2\beta + r_\lambda R_\lambda \cos\gamma_\lambda \sin 2\beta\right] \qquad (4.32)$$

A convenient and simple choice for $g(R_\lambda, r_\lambda, \gamma_\lambda)$ is:

$$g(R_\lambda, r_\lambda, \gamma_\lambda) = B_{\nu\lambda} r_\lambda \qquad (4.33)$$

where $B_{\nu\lambda}$ is a constant. Substitution of (4.33) in (4.32) leads to the equation:

$$R_\lambda^2 \sin^2\beta + r^2\left(\cos^2\beta - B_{\nu\lambda}^2\right) + r_\lambda R_\lambda \cos\gamma_\lambda \sin 2\beta = 0 \qquad (4.34)$$

If the two "Cartesian" coordinates R_λ and r_λ are replaced by two polar coordinates ζ_λ and η_λ (see Refs. 6, 34):

$$\zeta_\lambda = \sqrt{R_\lambda^2 + r_\lambda^2}$$
$$\tan\eta_\lambda = \frac{r_\lambda}{R_\lambda} \qquad (4.35)$$

then (4.34) becomes:

$$\tan^2\eta_\lambda\left(\cos^2\beta - B_{\nu\lambda}^2\right) + \tan\eta_\lambda \cos\gamma_\lambda \sin 2\beta + \sin^2\beta = 0 \qquad (4.36)$$
$$\zeta_\lambda = \text{free}$$

The two equations (4.36) form a surface that contains straight lines passing through the origin and can be identified with a vibrational coordinate. Thus, if the wave function $\psi(R_\lambda, r_\lambda, \gamma_\lambda)$ is expressed in terms of ζ_λ, η_λ, and γ_λ (not necessarily on the surface), η_λ is the "propagation" (translational) coordinate and ζ_λ is the vibrational coordinate. The surface defined in (4.36) is conelike; its vertex is at the origin and its axis lies along the R_λ axis (it becomes a perfect cone for the case when the central mass m_B [see (3.12′)] becomes infinite).

Within the IOSA framework, one treats a decoupled–fixed-angle Schrödinger equation like the one given in (4.17). According to (4.36), each value of γ_λ determines a line on the surface separating the λ channel from the ν channel. Thus, for every given value of γ_λ we not only have a Schrödinger equation but we are also able to define the boundary line for the λ channel. This boundary line is a straight line that goes through the origin and forms an angle $\eta_{\lambda 0}$ with the R-axis given by the equation (see Fig. 14b, and also Fig. 3c):

$$\tan \eta_{\lambda 0} = \frac{\sin \beta}{B_{\nu\lambda}^2 - \cos^2 \beta} \left[\cos \beta \cos \gamma_\lambda + \left(B_{\nu\lambda}^2 - \sin^2 \gamma_\lambda \cos^2 \beta \right)^{1/2} \right] \quad (4.37)$$

The angle γ_λ and $B_{\nu\lambda}$ also determine the boundary line of the ν channel [in fact, these two boundary lines are the same lines in space, but become different lines once projected on the (R_λ, r_λ) or the (R_ν, r_ν) plane]. This line is again a straight line going through the origin and forming an angle $\eta_{\nu 0}$ with the R_ν axis given by (see Fig. 14b):

$$\tan \eta_{\nu 0} = \frac{B_{\nu\lambda}}{\left(\left(1 - B_{\nu\lambda}^2 \right) \tan^2 \eta_{\lambda 0} + 1 \right)^{1/2}} \tan \eta_{\lambda 0} \quad (4.38)$$

The main advantage of this procedure is that once γ_λ and $B_{\nu\lambda}$ are assigned numerical values, everything else is uniquely determined. Thus, among other things, a fixed γ_λ value yields a unique γ_ν value to be used in the ν channel:

$$\cos \gamma_\nu = \frac{-\cos \gamma_\lambda + \left(1 - B_{\nu\lambda}^2 \right) \tan \eta_{\lambda 0} \cotan \beta}{B_{\nu\lambda} \left(1 + \left(1 - B_{\nu\lambda}^2 \right) \tan^2 \eta_{\lambda 0} \right)^{1/2}} \quad (4.39)$$

One additional relation is still required, and that has to do with the angle $\Delta_{\nu\lambda}$, which is the transformation angle between the λ and the ν arrangements (see Fig. 14a):

$$\cos \left(\Delta_{\nu\lambda} \right) = \left(\hat{R}_\nu \cdot \hat{R}_\lambda \right) \quad (4.40)$$

or:

$$\cos \left(\Delta_{\nu\lambda} \right) = \frac{\cos \beta - \cos \gamma_\lambda \sin \beta \tan \eta_{\lambda 0}}{\left(1 + \left(1 - B_{\nu\lambda}^2 \right) \tan^2 \eta_{\lambda 0} \right)^{1/2}} \quad (4.41)$$

The relations above are valid for any $B_{\nu\lambda}$. Simplified expressions are obtained for $B_{\nu\lambda} = 1$. Thus,

$$\gamma_\nu = \pi - \gamma_\lambda$$

$$\eta_{\nu 0} = \eta_{\lambda 0} \qquad (4.42)$$

$$\tan \eta_{\lambda 0} = \cotan \beta \cos \gamma_\lambda + \left(\cosec^2 \beta \cos^2 \gamma_\lambda + \sin^2 \gamma_\lambda \right)^{1/2}$$

$$\cos \left(\Delta_{\nu\lambda} \right) = \cos \beta - \cos \gamma_\lambda \sin \beta \tan \eta_{\lambda 0}$$

All the relations given so far are sufficient to solve a fixed-γ collinear-type equation like the one given in (4.17) and to derive the corresponding $S_l^-(\gamma)$-matrix element, which is then used for the derivation of the BF S-matrix elements.

3. Numerical Studies Within the IOSA Framework

The IOSA has so far been applied to two reactive systems:

$$H + H_2(v_i = 0, 1) \rightarrow H_2(v_f) + H$$

$$F + H_2(v_i = 0) \rightarrow HF(v_f) + H \qquad (4.43)$$

The main emphasis in these studies was on the relevance of the numerical results, as compared to those due to other treatments. For the $H + H_2$ system, the comparison can be done with exact quantum-mechanical results, although for only a limited range of energies.[9] There also are quantum-mechanical results[44] for $F + H_2$ that could be termed "exact," but since they were derived with only a reduced basis set (because of numerical limitations), they are in fact only approximate results.

A much more extensive source of numerical results is supplied by the classical trajectory studies.[45-50] The two systems have been exposed several times to this kind of treatment and this enables making meaningful quantitative comparisons. Other kinds of treatment are also available. Doll et al.[51] performed a numerical treatment of the $H + H_2$ system within the classical S-matrix approach, but they did this only in the low-energy region, where tunneling is the dominant process. The distorted wave Born approximation was applied several times to the three-dimensional $H + H_2$ and to the $F + H_2$ systems (see Refs. 52–54 and 55, respectively; it was also recently applied to the $H + F_2$ system[56]), but since only sporadic results are available, the comparison has limited value.

a. The $H_2 + H \rightarrow H_2 + H$ Reaction. This system was treated independently by two groups, Bowman and Lee[36, 40] (BL) and Baer, Khare, and Kouri[35, 38, 39] (BKK). Whereas BL performed the calculation only within the l-in IOSA, BKK performed it within both the l-in and the l-av IOSA. As far

TABLE I
Total Integral Degeneracy Averaged Cross-Sections (bohr^2)^a for the Reaction
$$H_2(v_i=0)+H\rightarrow H+H_2$$

E_{total} (eV)	l-in	l-av	Exact quantum-mechanical results[9]	Exact classical results[45]
0.40	1.0 (−4)	8.5 (−5)	2.5 (−6)	—
0.45	5.1 (−3)	4.7 (−3)	1.8 (−4)	—
0.50	6.6 (−2)	6.3 (−2)	5.0 (−3)	—
0.55	0.26	0.30	6.0 (−2)	—
0.60	0.54	0.70	0.35	0.2
0.70	1.2	1.8	1.5	1.4
0.80	1.9	2.9	—	2.4
0.90	2.4	3.8	—	3.2
1.00	2.9	5.0	—	3.7
1.20	3.7	6.7	—	4.4

^{a}Numbers in parentheses are powers of 10.

as the l-in treatments are concerned, the results of the two groups are the same for all practical purposes, although the two formal procedures differ in certain details. Therefore whenever l-in results are discussed, they can be attributed to each of the two groups, except for those shown in Table I, which are due to BKK.

Table I gives the results for the reaction:

$$H_2(v_i=0)+H\rightarrow H_2+H \qquad (4.44a)$$

as a function of total energy. These results are compared with the classical results[45] and with quantum-mechanical results.[9]

Table II gives the results for the reaction:

$$H_2(v_i=1)+H\rightarrow H_2(v_f=0,1)+H \qquad (4.44b)$$

as compared with the classical results due to Mayne.[46] The comparison is also shown in Figs. 15 and 16. (In Fig. 15 the DWBA results are added.) From Fig. 15 it can be seen that the fit among the different results is satisfactory in the low-energy region, but as the energy increases the DWBA result becomes much too high, with a trend to becoming even higher for larger energy values. The l-av are the next highest (sometimes as much as 50% higher than the corresponding classical results), but they seem to follow the same trend as the quantum-mechanical results, and the two lowest sets are the classical and l-in results, which differ by 20%. A very encouraging fit is shown in Fig. 16. Here, the l-av and the classical results are, for all practical purpo-

TABLE II

Total Integral Degeneracy Averaged Cross-Sections (bohr^2) for the Reactions: $H_2(v_i=1)+H \rightarrow H+H_2(v_f)$[a]

E_{total} (eV)	$v_i=1 \rightarrow v_f=1$			$v_i=1 \rightarrow v_f=0$			$v_i=1 \rightarrow$ total		
	l-in	l-av	Classical[46]	$l=$in	l-av	Classical[46]	$l=$in	l-av	Classical[46]
0.85	8.6 (−5)	3.6 (−4)	—	1.1 (−3)	6.9 (−4)	—	1.2 (−3)	1.1 (−3)	
0.90	0.10	0.054	—	0.11	0.079	—	0.21	0.13	
0.95	0.58	0.49	0.60	0.23	0.26	0.37	0.81	0.75	0.97
1.00	1.2	1.5	1.7	0.29	0.59	0.77	1.5	2.1	2.5
1.10	2.5	3.2	2.9	0.44	0.80	1.4	2.9	4.0	4.3
1.20	3.3	4.7	4.0	0.57	1.1	1.9	3.9	5.8	5.9

[a]Numbers in parentheses are powers of 10.

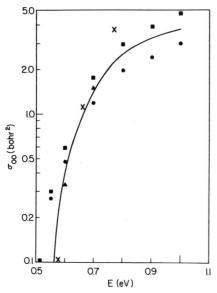

Fig. 15. Total integral (reactive) cross-sections for the reaction $H_2(v_i = 0) + H_I \rightarrow HH_I + H$. Curve indicates classical results. Triangles, exact quantum-mechanical results[9]; circles, l-in IOSA results[34]; squares, l-av IOSA results[39]; crosses, DWBA results (Ref. 53).

ses, identical, whereas the l-in are much below them. The figure also shows the branching ratio $\lambda(1,0)$, defined as:

$$\lambda(1,0) = \frac{\sigma(v_i = 1 \rightarrow v_f = 1)}{\sigma(v_i = 1 \rightarrow v_f = 0)} \tag{4.45}$$

and the two versions of the IOSA are seen to yield much larger branching ratios than the classical treatment. In this respect, it should be emphasized that in a recent quantum-mechanical calculation performed by Walker et al.,[57] the branching ratio for $J = 0$ is about 4 in this energy range; this is in agreement with the l-av results.

Comparisons of more detailed quantities, such as differential cross-sections, are also encouraging. Figure 17 shows differential cross-sections for the reactions:

$$H_2(v_i = 0; j_i = 0) + H \rightarrow H + H_2(v_f = 0; j_f) \tag{4.46}$$

where $j_f = 0, 1, 2$, compared with the corresponding exact quantum-mechanical magnitudes. The comparison is also extended to rate constants defined as[45]:

$$K = N_A \left(\frac{8}{\mu \pi} \right)^{1/2} \frac{1}{kT} \int E_t \sigma(E_t) e^{-E_t/kT} dE_t \tag{4.47}$$

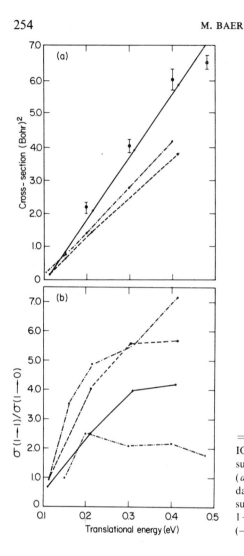

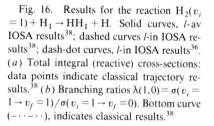

Fig. 16.   Results for the reaction $H_2(v_i = 1) + H_1 \rightarrow HH_1 + H$. Solid curves, l-av IOSA results[38]; dashed curves l-in IOSA results[38]; dash-dot curves, l-in IOSA results[36]. (*a*) Total integral (reactive) cross-sections: data points indicate classical trajectory results.[38] (*b*) Branching ratios $\lambda(1,0) = \sigma(v_i = 1 \rightarrow v_f = 1)/\sigma(v_i = 1 \rightarrow v_f = 0)$. Bottom curve ($-\cdots-\cdots$), indicates classical results.[38]

where E_t is the translational energy, T is the temperature, and N_A is Avogadro's number.

From the corresponding Arrhenius plots, we obtained activation energies for the reactions given in (4.44a). Thus, the classical trajectory result[45] is 7.4 kcal/mole, the exact quantum-mechanical result[9] is 6.4 kcal/mole, and the IOSA result is 5.2 kcal/mole, reflecting the larger cross-sections in the low-energy region (see Table I).

b.   The $F + H_2 \rightarrow HF + H$ Reaction.   As for the $F + H_2$, there are only preliminary results derived by Jellinek et al.[41] Figure 18 compares the re-

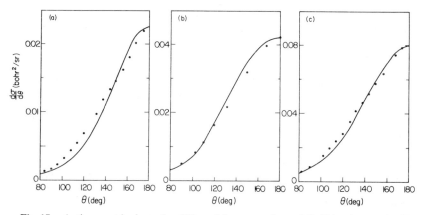

Fig. 17.   Antisymmetrized reactive differential cross-section at 0.7 eV for the reaction $H_2(v_i = 0, j_i = 0) + H_I \rightarrow HH_1(v_f = 0, j_f)$. Dots indicate exact quantum-mechanical results[9] scaled to agree with the IOSA results at $\theta = \pi$. Solid curve indicates l-av IOSA.[39] $(a) j_f = 0$, $(b) j_f = 1$, (c) $j_f = 2$.

sults of two IOSA versions and results due to other treatments. As can be seen, the l-in results seem to follow the quantum-mechanical results of Redmon and Wyatt[44] (although the latter were done on a somewhat different surface), whereas the l-av results are much larger and therefore closer to the classical results[49] as performed on the same surface (the other classical results are done on similar surfaces). Table III gives the branching ratios

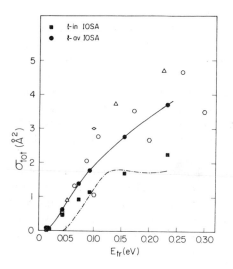

Fig. 18.   Total integral (reactive) cross-sections for the reaction $F + H_2(v_i = 0) \rightarrow HF + H$. Broken curve indicates quantum-mechanical results done with a limited basis set[44]; curve with solid dots, l-av IOSA results.[41] Solid squares, l-in IOSA results[41]; triangles, classical results[49]; diamonds, classical results[50]; open circles, classical results[47]; hexagons, classical results.[48]

TABLE III

Branching Ratios of State-to-State Cross-Sections $\sigma(0 \to 3)/\sigma(0 \to 2)$ for Different Translational Energies E_{tran} for the Reaction $F + H_2(v_i = 0) \to HF(v_f) + H$

E_{tran} (eV)	Komornicki, Morokuma, and George[48]	Redmon and Wyatt[44]	Connor, Jakubetz, Manz, and Whitehead[49]	IOSA	
				l-in	l-av
0.013				-0	-0
0.018				-0	-0
0.043				0.0003	0.00024
0.052			0.68		
0.073		0.06		0.004	0.004
0.093		0.06		0.015	0.014
0.100	0.60				
0.139			0.63		
0.156		0.29		0.14	0.14
0.200	0.54				
0.226			0.63		
0.233		0.71		0.293	0.308
0.300	0.42				

$\lambda(3,2)$ defined as:

$$\lambda(3,2) = \frac{\sigma(v_i = 0 \to v_f = 3)}{\sigma(v_i = 0) \to v_f = 2)} \tag{4.48}$$

which are the more significant ones.

Here, in contrast to the absolute values of the cross-sections, the l-av and l-in results are almost identical and are qualitatively similar to the quantum-mechanical results of Redmon and Wyatt.[44] However, they differ considerably from the various classical results. The difference in branching ratio between classical and quantum-mechanical results has already been mentioned in connection with the results of the $H + H_2$ system. Thus the classical approach seems to be quite relevant for total cross-sections, but it seems to be inadequate where more detailed magnitudes, such as vibrational distributions, are concerned.

Before the end of this section, it should be mentioned that the DWBA results[55] are different from any of the results presented here, although they were done on the same surface. For $E_{\text{tran}} = 0.1$ eV, the total cross-sections were found to be 10 Å^2, which is five times as large as any other cross-section, and the branching ratio $\lambda(3,2)$ is about 4, whereas the largest value found in any of the other treatments is much less than 1.

C. One- to Three-Dimensional Transformation

In the transformation from one to three dimensions one applies pure collinear results (i.e., $\gamma = 0$ results only) to obtain three-dimensional measurable magnitudes. Although it is commonly accepted that collinear treatments cannot account for three-dimensional processes, the fact that in many cases the vibrational distributions obtained are in qualitative agreement with experiment (see Fig. 8) encouraged the search for the right "push" to bring the collinear results into quatitative agreement with experiment as well.

In general, one distinguishes between two approaches. The first, due to Bernstein and Levine,[58] is based on the information theoretic approach, which was later modified and extensively used.[59–61] The other, due to Baer,[62, 63] is based on the fact that potential *independent* selection rules govern the rotational distribution, which in turn affects the final "collinear" potential-*dependent* vibrational distribution. Approaches based on other arguments are those due to Walker and Wyatt,[64] who presented a transformation from two to three dimensions to Polanyi and Schreiber,[65] who discussed a one- to three-dimensional transformation in a more qualitative way, and to Venzl and Fischer, who presented a statistical approach.[66]

1. The Information Theoretic Approach

The information theoretic approach is based on the assumption that to be able to estimate three-dimensional reaction cross-sections from one-dimensional reactive transition probabilities, one must account for the difference in the available volume in phase space between one- and three-dimensional systems. According to Bernstein and Levine, the effect of the "dimensionality bias" is reduced by considering the function:

$$I_{nD} = P_{nD}(f_I) | P_{nD}^0(f_I) \qquad (4.49)$$

where $P_{nD}(f_I)$, $n = 1,3$, is the nD reactive transition probability, $P_{nD}^0(f_I)$ is the nD, $n = 1,3$, reactive statistical probability, and f_I stands for the fraction of total internal energy, namely, E_I/E. Now, making the assumption that:

$$I_{3D}(f_I) = \gamma I_{1D}(f_I) \qquad (4.50)$$

the contact between the two "worlds" is made. Here γ is an f_I-independent but energy-dependent constant to account for the normalization. The second assumption has to do with the rotational distribution: it is assumed that for each vibrational state v, the actual rotational distribution $P(j|v)$ is equal to the statistical distribution $P^0(j|v)$, where j stands for the rotational state.

This procedure was applied by Connor et al.[59] but was found to be inconsistent because it yields negative vibrational probabilities. Consequently, the second assumption was replaced by the requirement that for a given (i.e., experimental) average rotation energy $\langle E_R \rangle$, the vibrational-rotational distribution is forced to fulfill the relation:

$$\sum_{v,j} P(j,v) f(j,v) = \langle f_R \rangle \qquad (4.51)$$

where

$$f(j,v) = \frac{E_{jv}}{E}, \qquad P(j,v) = P(v)P(j|v) \qquad (4.52)$$

The probabilities $P(j,v)$ are then synthesized by minimizing the entropy deficiency[67]:

$$\Delta S = R \sum_{j,v} P(j,v) \ln\left[P(j,v)|P^0(j,v) \right] \qquad (4.53)$$

subject to the dynamic constraints (4.50) and (4.51).

This method was applied to various reactions: $F+H_2$, $H(D)+Cl_2$, and $H(D,T,Mu)+F_2$ (Refs. 59, 59 and 60, and 60, respectively). However its importance is not clearly demonstrated, since the modified 3-D distributions hardly differ from the original 1-D ones.

The first version of the Bernstein-Levine theory was also applied by White.[68] There, 1-D, 2-D, and 3-D reactive probabilities for the reactions $H+Br_2$ and $Br+H_2$ were calculated using the classical trajectory method. The exact 3-D results were compared with those obtained from the 1-D and 2-D results following the Levine-Bernstein transformation. It was concluded that whereas 3-D results obtained from 2-D results can be considered to be reliable, those obtained from the collinear arrangement cannot be trusted.

2. The Rotation–Decoupled-Vibration Model (RDVM)

The rotation–decoupled-vibration model (RDVM)[62, 63] is based on the assumption that in general one may distinguish between two types of process, which are assumed to be independent in the zeroth approximation: vibrational transitions that are (strongly) potential dependent, and rotational transitions that are energy and potential independent but are governed by rotational selection rules.

In general, it is well accepted that vibrational transitions are potential dependent, but the importance of selection rules for rotational distributions is less familiar. Therefore, let us consider a few examples:

1. Assigning L for "light" and H for "heavy," let us examine a reaction:

$$L + HL' \rightarrow LH + L' \qquad (4.54)$$

Assuming HL' to be in its ground state, l_i to be the initial orbital angular momentum, and j_f to be the final internal angular momentum of the product molecule LH, then because of conservation of total angular momentum—and as a result, the poor efficiency in the transfer of angular momentum between "heavy" and "light" particles—it was shown[69] that the relation

$$j_f \sim l_i \qquad (4.55)$$

holds, at least approximately.

A similar selection rule was found and confirmed by experiment[70, 71] for the reaction:

$$L + HH' \rightarrow LH + H' \qquad (4.56)$$

2. Another example is the reaction

$$H + LH' \rightarrow HL + H' \qquad (4.57)$$

In this case, again since "heavy" particles hardly exchange angular momentum with "light" particles, one would expect:

$$l_f \sim l_i \qquad (4.58)$$

These and similar ideas led Schulten and Gordon[72] to look for a general approach that would yield the relation:

$$j_f = j_f(J, l_i, j_i) \qquad (4.59)$$

Such a formula was found only after several additional assumptions were made; these could be justified in certain cases.

A classical approach along the same lines was recently given by Venzl and Fischer.[73]

Now returning to the two processes (i.e., vibrational and rotational transitions) mentioned at the beginning of this section and assuming the two to be decoupled, one could treat them independently and then write, *a posteriori*, the overall vibrational-rotational distribution. As a relevant model for the vibrational distribution, one could choose the collinear system, for instance.

The rotation distribution could be obtained in different ways, such as using a statistical approach[74] or using Franck-Condon-type models.[72, 75]

Next, the RDVM assumes that the rotational selection rules are dominant; consequently the rotational distribution is determined "first" and the vibrational distribution is formed "on top of it." Consequently, any vibrational-translational distribution obtained by ignoring the existence of rotation must be modified to account for the fact that the final rotational distribution is *already fixed*. The modification according to the RDVM is through conservation of total energy, namely, the only combination of allowable states is that in which these states do not violate conservation of energy.

To fit these ideas into a theory,[63] two *normalized* distribution functions should be defined: one is the rotational probability function $P(E_R, E)$ (for total energy E) and the other is the distribution function $P_X^{(m)}(E_x, G)$, where E_x stands for E_V or E_T, m stands for "model," and $G(\geq E_x)$, the total energy available in the model, is assumed to be dependent on E and E_R. Having these two functions and assuming that $G = E - E_R$ the "three-dimensional" (normalized) X-ational probability function is obtained from:

$$P_X(E_X|E) = \int_0^E dE_R\, P_X^{(m)}(E_x|E - E_R) P(E_R) \qquad (4.60)$$

Equation 4.60 is similar to (4.25), which represents the l-in IOSA reactive cross-sections. Moreover, there as here, one notices an integration and a summation over weighted "collinear" reactive transition probabilities. To use (4.60), one must form collinear-type reactive probabilities for different energies $(E - E_R)$ and then perform the integral. It should be mentioned that recently Venzl and Fischer[73] derived a very similar equation by applying somewhat different arguments.

Equation 4.60 is the basic equation in the theory and constitutes the starting point for any theoretical analysis or numerical treatment. Thus, for instance, if E_I is defined as the internal energy of the three-dimensional product molecule, the $P_I(E_I|E)$ is found to be[63]:

$$P_I(E_I|E) = \int_0^E dE_R\, P_v^{(m)}(E_I - E_R|E - E_R) P(E_R|E) \qquad (4.61)$$

Equation 4.60 was found to be particularly reliable for highly exothermic reactions, and tests were performed to establish its reliability. Figure 19 presents vibrational distributions for two systems:

$$H + Br_2 \rightarrow HBr + Br$$
$$F + H_2 \rightarrow HF + H \qquad (4.62)$$

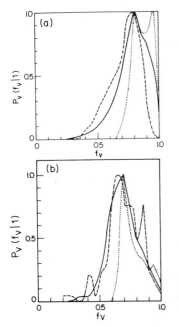

Fig. 19. Vibrational distribution obtained in the transformation from one dimension to three dimensions. (a) $F + H_2 \rightarrow HF + H$. (b) $H + Br_2 \rightarrow HBr + Br$.[63] Dotted curves, one-dimensional vibrational distribution obtained in a trajectory calculation; dashed curves, three-dimensional vibrational distribution obtained in a 3-D trajectory calculation; solid curves, three-dimensional vibrational distribution as obtained from the one-dimensional classical vibrational distribution applying (4.60).

The collinear transition probabilities and the three-dimensional rotational and vibrational distributions were obtained from classical trajectory studies (by White[68, 76] for $H + Br_2$ and by Whitehead[77] for $F + H_2$). As can be seen, the fit between the exact three-dimensional vibrational distribution and the distribution derived applying a simplified version[63] of (4.60) is satisfactory.

References

1. W. H. Miller, *J. Chem. Phys.*, **50**, 407 (1969).
2. G. Wolken, Jr., and M. Karplus, *J. Chem. Phys.*, **60**, 351 (1974).
3. R. P. Saxon and J. C. Light, *J. Chem. Phys.*, **56**, 3874 (1972).
4. R. P. Saxon and J. C. Light, *J. Chem. Phys.*, **56**, 3885 (1972).
5. A. Altenberger-Siczek and J. C. Light, *J. Chem. Phys.*, **61**, 4373 (1974).
6. A. Kuppermann, G. C. Schatz, and M. Baer, *J. Chem. Phys.*, **65**, 4596 (1976).
7. G. C. Schatz and A. Kuppermann, *J. Chem. Phys.*, **65**, 4624 (1976).
8. G. C. Schatz and A. Kuppermann, *J. Chem. Phys.*, **65**, 4642 (1976).
9. G. C. Schatz and A. Kuppermann, *J. Chem. Phys.*, **65**, 4668 (1976).
10. A. B. Elkovitz and R. E. Wyatt, *J. Chem. Phys.*, **63**, 702 (1975).
11. R. B. Walker, J. C. Light, and A. Altenberger-Siczek, *J. Chem. Phys.*, **64**, 1166 (1976).
12. M. Baer and D. J. Kouri, *J. Chem. Phys.*, **56**, 1758 (1972).
13. M. Baer and D. J. Kouri, *J. Chem. Phys.*, **57**, 3991 (1972).
14. R. E. Wyatt, in *Atom-Molecule Collision Theory: A Guide for the Experimentalist*, R. B. Bernstein, Ed., Plenum Press, New York, 1979.
15. M. Baer, *J. Chem. Phys.*, **65**, 493 (1976).

16. J. M. Bowman, G. Drolshagen, and J. P. Toennies, *J. Chem. Phys.*, **71**, 2270 (1979).
17. H. Rabitz, *J. Chem. Phys.*, **57**, 1718 (1972).
18. G. Zarur and H. Rabitz, *J. Chem. Phys.*, **59**, 943 (1973).
19. P. McGuire and D. J. Kouri, *J. Chem. Phys.*, **60**, 2488 (1974).
20. R. T. Pack, *J. Chem. Phys.*, **60**, 633 (1974).
21. M. Tamir and M. Shapiro, *Chem. Phys. Lett.*, **31**, 166 (1975).
22. A. E. de Pristo and M. H. Alexander, *J. Chem. Phys.*, **64**, 3009 (1976).
23. D. J. Kouri, in *Atom-Molecule Collision Theory: A Guide for the Experimentalist*, R. B. Bernstein, Ed., Plenum Press, New York, 1979.
24. K. Takayanaki, *Prog. Theor. Phys. (Kyoto)*, Suppl. **25**, 1 (1963).
25. C. F. Curtiss, *J. Chem. Phys.*, **49**, 1952 (1968).
26. T. P. Tsien and R. T. Pack, *Chem. Phys. Lett.*, **6**, 54 (1970).
27. T. P. Tsien, G. A. Parker, and R. T. Pack, *J. Chem. Phys.*, **59**, 5373 (1973).
28. D. Secrest, *J. Chem. Phys.*, **62**, 710 (1970).
29. R. T. Pack, *J. Chem. Phys.*, **60**, 633 (1974).
30. D. E. Fitz and P. McGuire, *Chem. Phys. Lett.*, **44**, 503 (1976).
31. G. Pfeffer and D. Secrest, *J. Chem. Phys.*, **67**, 1394 (1977).
32. R. Goldflam, S. Green, and D. J. Kouri, *J. Chem. Phys.*, **67**, 4149 (1977).
33. V. Khare, *J. Chem. Phys.*, **68**, 4631 (1978).
34. V. Khare, D. J. Kouri, and M. Baer, *J. Chem. Phys.*, **71**, 1188 (1979).
35. M. Baer, V. Khare, and D. J. Kouri, *Chem. Phys. Lett.*, **68**, 378 (1979).
36. J. M. Bowman and K. T. Lee, *Chem. Phys. Lett.*, **64**, 29 (1979).
37. G. D. Barg and G. D. Drolshagen, *Chem. Phys.*, **47**, 209 (1980).
38. M. Baer, H. R. Mayne, V. Khare, and D. J. Kouri, *Chem. Phys. Lett.*, **72**, 269 (1980).
39. D. J. Kouri, V. H. Khare, and M. Baer, *J. Chem. Phys.*, **75**, 1179 (1981).
40. J. M. Bowman and K. T. Lee, *J. Chem. Phys.*, **72**, 5071 (1980).
41. J. Jellinek, M. Baer, V. Khare, and D. J. Kouri, *Chem. Phys. Lett.*, **75**, 460 (1980).
42. V. Khare and D. J. Kouri, *J. Chem. Phys.*, **69**, 4916 (1978).
43. M. E. Rose, *Elementary Theory of Angular Momentum*, Wiley, New York, 1957.
44. M. J. Redmon and R. E. Wyatt, *Chem. Phys. Lett.*, **63**, 209 (1979).
45. M. Karplus, R. N. Porter, and R. D. Sharma, *J. Chem. Phys.*, **43**, 3259 (1965).
46. H. R. Mayne, *Chem. Phys. Lett.*, **66**, 487 (1979).
47. J. T. Muckerman, *J. Chem. Phys.*, **56**, 2997 (1972).
48. A. Komornicki, K. Morokuma, and T. F. George, *J. Chem. Phys.*, **67**, 5012 (1977).
49. J. N. L. Connor, W. Jakubetz, J. Manz, and J. C. Whitehead, *Chem. Phys. Lett.*, **62**, 479 (1979).
50. D. F. Feng, E. R. Grant, and R. W. Root, *J. Chem. Phys.*, **64**, 3450 (1976).
51. J. D. Doll, T. F. George, and W. H. Miller, *J. Chem. Phys.*, **58**, 1343 (1973).
52. K. T. Tang and M. Karplus, *Phys. Rev. A*, **4**, 1844 (1971).
53. B. H. Choi and K. T. Tang, *J. Chem. Phys.*, **65**, 5161 (1976).
54. D. C. Clary and J. N. L. Connor, *Chem. Phys.*, **48**, 175 (1980).
55. Y. Shan, B. H. Choi, R. T. Poe, and K. T. Tang, *Chem. Phys. Lett.*, **57**, 379 (1974).
56. D. C. Clary and J. N. L. Connor, *Chem. Phys. Lett.*, **66**, 493 (1979).
57. R. B. Walker, E. B. Stechel, and J. C. Light, unpublished (see Ref. 36).
58. R. B. Bernstein and R. D. Levine, *Chem. Phys. Lett.*, **29**, 314 (1974).
59. J. N. L. Connor, W. Jakubetz, and J. Manz, *Chem. Phys.*, **17**, 451 (1976).
60. J. N. L. Connor, W. Jakubetz, and J. Manz, *Ber. Bunsenges. Phys. Chem.*, **81**, 165 (1977).
61. J. N. L. Connor, W. Jakubetz, and J. Manz, *Chem. Phys.*, **28**, 219 (1978).
62. M. Baer, *J. Chem. Phys.*, **62**, 4545 (1975).
63. M. Baer, *Chem. Phys. Lett.*, **57**, 316 (1978).

64. R. B. Walker and R. E. Wyatt, *Mol. Phys.*, **28**, 101 (1974).
65. J. C. Polanyi and J. L. Schreiber, *Chem. Phys. Lett.*, **29**, 314 (1974).
66. G. Venzl and S. F. Fischer, *Chem. Phys.*, **33**, 305 (1978).
67. R. D. Levine and R. B. Bernstein, in *Modern Theoretical Chemistry, Vol. II, Dynamics of Molecular Collisions*, W. H. Miller, Ed., Plenum Press, New York, 1976, p. 323.
68. J. M. White, *J. Chem. Phys.*, **65**, 3674 (1976).
69. M. Baer, *Mol. Phys.*, **26**, 369 (1973).
70. C. Maltz, N. D. Weinstein, and D. R. Herschbach, *Mol. Phys.*, **24**, 133 (1972).
71. J. E. Mosch, S. A. Safron, and J. P. Toennies, *Chem. Phys. Lett.*, **29**, 7 (1974).
72. K. Schulten and R. Gordon, *J. Chem. Phys.*, **64**, 2918 (1976).
73. G. Venzl and S. F. Fischer, *J. Chem. Phys.*, **71**, 4175 (1979).
74. A. Ben-Shaul, R. D. Levine, and R. B. Bernstein, *J. Chem. Phys.*, **57**, 5427 (1972).
75. G. C. Schatz and J. Ross, *J. Chem. Phys.*, **66**, 2943 (1977).
76. J. M. White, unpublished.
77. J. C. Whitehead, unpublished.

V. INTEGRAL EQUATION APPROACHES TO REACTIVE SCATTERING

A. Introduction

Most of the analytical approaches, as well as the numerical results in the field of reactive scattering, were based on the treatment of the Schrödinger wave equation. In the literature of the past 50 years, one may find only sporadic attempts to apply the operator formalism that leads to integral equations.[1-5] The only exception is possibly Miller's variational treatment.[6] Although this approach does not result from an operator formalism, still its equations in their final form bear some similarity to those obtained by this approach. Recently, Garrett and Miller[7] applied this method to the $H + H_2$ system and were able to obtain the correct results.

The failure to apply the available method for obtaining "exact" reactive cross-sections to systems other than the $H + H_2$ during the past 5 years seems to be a clear indication that one of the ultimate goals in theoretical chemistry, namely, developing a practical method to calculate exact reactive cross-sections, has not yet been reached. Under these circumstances there seems to be no alternative but to invest more effort in deriving workable methods based on the operator formalism. The main bottleneck in any attempt to extend the existing methods to other systems is connected with the problem of the *matching*. The main advantage of the methods based on the operator formalism is that the matching problem is circumvented because of the inherent structure of these methods.

In the past decade, two approaches close in spirit to the above were formulated: the first, the coupled-channel operator approach, due to Baer and Kouri[1, 3, 8] was later modified and somewhat extended by Kouri and

Levin[9, 10] and by Tobocman[11, 12] (this approach is hereafter referred to as the BKLT approach); the other is the many-body approach, due to Micha.[13, 14]

The BKLT approach yields an effective general procedure to couple Lippmann-Schwinger integral equations for all (nondissociative) open-arrangement channels. In other words, the final outcome of the theory is a well defined set of integral equations, obtained by simultaneously imposing the boundary conditions. The many-body approach is based on the Faddeev-Watson equations,[15-17] but it can result in a tractable numerical method only when the interaction potential is given as the sum of pair potentials. In molecular systems, however, the interaction potential is dominated by three-body terms if simply expressed in terms of the coordinates of the three atoms, which suffices to prevent any further attempt to apply these equations. Micha took advantage of the fact that any molecular potential can be constructed up to a certain accuracy by a sum of two-body *spin-dependent* potentials (as, e.g., the LEPS surface; see Section II) and can, in this way, eventually circumvent the inherent limitations of this approach. The method has been applied in some cases,[4, 5] but all were in the high-energy region, using simplified expressions. More details on the method can be found in Ref. 18.

The rest of this section is devoted to the BKLT approach, which at this stage seems to be the most advantageous because it has already been applied successfully in some realistic cases[3, 19, 20] and does not seem to possess any inherent difficulties.

B. Baer-Kouri-Levin-Tobocman (BKLT) Theory

Since their first publication, the BKLT equations have changed form several times, although their basic structure and the main idea behind them are maintained.

Given a Hamiltonian H, one may partition it in different ways:

$$H = H_\gamma + V_\gamma, \qquad \gamma = \lambda, \nu, k \tag{5.1}$$

where $\gamma = \lambda, \nu, k$ stand for the various arrangement channels, H_γ is the unperturbed Hamiltonian, and V_γ is the interaction potential, such that:

$$\lim_{R_\gamma \to \infty} V_\gamma(R_\gamma \cdots) = 0 \tag{5.2}$$

Having introduced these definitions, we are now in a position to consider the channel operator $T_{\gamma'\gamma}$, which obeys the equations[21]:

$$\bar{T}_{\gamma'\gamma} = V_\gamma + V_{\gamma'} G^+ V_\gamma \tag{5.3a}$$

$$T_{\gamma'\gamma} = V_{\gamma'} + V_{\gamma'} G^+ V_\gamma \tag{5.3b}$$

where

$$G^+ = (E - H + i\varepsilon)^{-1} \tag{5.4}$$

is the full outgoing Green function operator for the system. It is important to emphasize that the T-matrix elements formed from $\bar{T}_{\gamma'\gamma}$ and $T_{\gamma'\gamma}$ are exactly identical.[21] Next, with each arrangement channel γ'' one may associate an outgoing channel Green function operator:

$$G_{\gamma''} = (E - H_{\gamma''} + i\varepsilon)^{-1} \tag{5.5}$$

so that G and $G_{\gamma''}$ are related by the equations:

$$G = G_{\gamma''} + G_{\gamma''} V_{\gamma''} G \tag{5.6}$$

First, let us consider (5.3a). Substitution of (5.6) in (5.3a) leads to:

$$\bar{T}_{\gamma'\gamma} = V_\gamma + V_{\gamma'} \left[G_{\gamma''} + G_{\gamma''} V_{\gamma''} G \right] V_\gamma$$

or

$$\bar{T}_{\gamma'\gamma} = V_\gamma + V_{\gamma'} G_{\gamma''} \left[V_\gamma + V_{\gamma''} G V_\gamma \right] \tag{5.7}$$

and recalling (5.3a), one immediately finds that:

$$\bar{T}_{\gamma'\gamma} = V_\gamma + V_{\gamma'} G_{\gamma''} \bar{T}_{\gamma''\gamma} \tag{5.8}$$

Equation (5.8) is valid for any γ'' and therefore also for any linear combination of these,[3] namely:

$$\bar{T}_{\gamma'\gamma} = V_\gamma + V_{\gamma'} \sum_{\gamma''} W_{\gamma'\gamma''} G_{\gamma''} \bar{T}_{\gamma''\gamma'} \tag{5.9}$$

where $W_{\gamma'\gamma''}$ are numbers that fulfill the relation[3]:

$$\sum_{\gamma''} W_{\gamma'\gamma''} = 1 \qquad \text{for any } \gamma' \tag{5.10}$$

Next, we consider (5.3b); again, substitution of (5.6) in (5.3b) leads to:

$$T_{\gamma'\gamma} = V_{\gamma'} + V_{\gamma'} \left[G_{\gamma''} + G_{\gamma''} V_{\gamma''} G \right] V_\gamma$$

or:

$$T_{\gamma'\gamma} = V_{\gamma'} + V_{\gamma'} G_{\gamma''} \left[V_{\gamma''} + V_{\gamma''} G V_\gamma \right] + V_{\gamma'} G_{\gamma''} (V_\gamma - V_{\gamma''}) \tag{5.11}$$

and recalling (5.3b), one obtains:

$$T_{\gamma'\gamma} = V_{\gamma'}\left(1 + G_{\gamma''}\left(V_{\gamma} - V_{\gamma''}\right)\right) + V_{\gamma'}G_{\gamma''}T_{\gamma''\gamma'} \tag{5.12}$$

It can be shown that for those cases where three-body bound states do not occur, the following relation[22] holds:

$$1 + G_{\gamma''}\left(V_{\gamma} - V_{\gamma''}\right) = \delta_{\gamma\gamma''} \tag{5.13}$$

and consequently (5.12) takes the form:

$$T_{\gamma'\gamma} = V_{\gamma'}\delta_{\gamma\gamma'} + V_{\gamma'}G_{\gamma''}T_{\gamma''\gamma'} \tag{5.14}$$

Again multiplying by $W_{\gamma'\gamma''}$ and summing over γ'' leads to:

$$T_{\gamma'\gamma} = V_{\gamma'}W_{\gamma'\gamma} + V_{\gamma'}\sum_{\gamma''} W_{\gamma'\gamma''}G_{\gamma''}T_{\gamma''\gamma} \tag{5.15}$$

Equations 5.9 and 5.15 are formally different but are expected to yield identical numerical results. Also none of them depends on any particular choice of W, as long as (5.10) holds. However, it was argued by Tobocman[11] that to make the kernel of the integral equations $K_{\gamma'\gamma''}$, defined as:

$$K_{\gamma'\gamma''} = V_{\gamma'}W_{\gamma\gamma''}G_{\gamma''} \tag{5.16}$$

connected, so that the resulting equations appear to be a suitable basis for many-body calculations, one has to impose:

$$W_{\gamma\gamma'} = \begin{cases} 1, & \gamma' = \gamma + 1 \\ 1, & \gamma = N, \gamma' = 1 \\ 0, & \text{otherwise} \end{cases} \tag{5.17}$$

Thus, for an N-channel problem, there are $(N-1)!$ different W matrices.[23] In case of $N=2$ we have one W matrix:[3]

$$W = \begin{pmatrix} 0 & 1 \\ 1 & 0 \end{pmatrix} \tag{5.18}$$

and for $N=3$ we have two:

$$W = \begin{pmatrix} 0 & 1 & 0 \\ 0 & 0 & 1 \\ 1 & 0 & 0 \end{pmatrix} \quad \text{and} \quad W = \begin{pmatrix} 0 & 0 & 1 \\ 1 & 0 & 0 \\ 0 & 1 & 0 \end{pmatrix} \tag{5.19}$$

In further studies conducted by Kouri et al.[20] it was shown that formally both sets of equations obey the optical theorem; however, the two behave differently when a truncated basis set is considered. It was found that (5.15) still obey the optical theorem, whereas (5.9) do not. As a final point in this approach, we present the matrix forms of (5.9), that is,

$$T = UV + VWGT \qquad (5.20)$$

and the matrix form of (5.15), that is,

$$T = VW + VWGT \qquad (5.21)$$

Here,

$$
\begin{aligned}
(T)_{\gamma\gamma'} &= T_{\gamma\gamma'} \\
(V)_{\gamma\gamma'} &= V_\gamma \delta_{\gamma\gamma'} \\
(G)_{\gamma\gamma'} &= G_\gamma \delta_{\gamma\gamma'} \qquad\qquad (5.22) \\
U_{\gamma\gamma'} &= 1 \qquad \text{for any } \gamma \text{ and } \gamma' \\
(W)_{\gamma\gamma'} &= W_{\gamma\gamma'}
\end{aligned}
$$

It should be mentioned that some attempts to present coupled equations for rearrangement collisions had already been made prior to the present ones,[24, 25] but only in a limited context.

For the purpose of completeness, let us also discuss the differential approach. (The numerical advantages of such an approach in general were discussed by Diestler and Kruger.[26])

If the total wave function ψ is written in the form:

$$\psi = \sum_\gamma \psi_\gamma \qquad (5.23)$$

where ψ_γ is defined such that for $R \to \infty$ it has the correct physical form, then reemploying the W-matrix elements, one may form the following sytem of coupled equations:

$$(E - H_\gamma)\psi_\gamma = \sum_{\gamma'} V_{\gamma'}\psi_{\gamma'}W_{\gamma'\gamma} \qquad (5.24)$$

where H_γ and V_γ have the same meaning as in (5.1). To show that ψ, defined in (5.23) along with the ψ_γ's determined from (5.24), is really the solution of the equation:

$$(E-H)\psi=0 \tag{5.25}$$

one may sum over γ in (5.24):

$$\sum_{\gamma}(E-H_{\gamma})\psi_{\gamma}=\sum_{\gamma\gamma'}V_{\gamma'}\psi_{\gamma'}W_{\gamma'\gamma} \tag{5.26}$$

Now, since

$$\sum_{\gamma}W_{\gamma'\gamma}=1 \tag{5.27}$$

one obtains

$$\sum_{\gamma}\left(E-\left(H_{\gamma}+V_{\gamma}\right)\right)\psi_{\gamma}=0 \tag{5.28}$$

and recalling (5.1) and (5.23), we reach (5.25). Imposing Tobocman's[11] choice for the W-matrix elements, one obtains the following coupled system of equations:

$$\left(E-H_{\gamma}\right)\psi_{\gamma}=V_{\gamma+1}\psi_{\gamma+1}, \qquad \gamma=1\cdots N-1$$
$$\left(E-H_{N}\right)\psi_{N}=V_{1}\psi_{1} \tag{5.29}$$

As a final subject in this section, we present an extended form for the BKLT equations that should be numerically tractable for both exact[27] and approximate[28] treatments.

To do that, (5.1) must be replaced by

$$H=H_{\gamma}+U_{\gamma}+V_{\gamma}, \qquad \gamma=\lambda,\nu,k \tag{5.30}$$

where V_{γ}'s are the arrangement channel potentials and are nonreactive distortion potentials chosen so that the partial wave functions χ_{γ} decrease to zero at an appropriate rate when the distance between the relevant interacting particles become small enough. With such a choice of the potential, one faces the well-known problem of scattering by two potentials.[29] Thus one may write

$$T_{\gamma\gamma'}=T_{\gamma}^{nr}\delta_{\gamma\gamma'}+T_{\gamma\gamma'}^{r}, \qquad \gamma,\gamma'=\lambda,\nu,k \tag{5.31}$$

Here, the $T_{\gamma\gamma'}^{r}$ (r stands for "reactive") fulfills the ordinary BKLT equations, namely, (5.9) or (5.15), except that the corresponding Green functions are no longer expressed in terms of the free Hamiltonian wave functions (i.e., Bessel and Neumann functions) but in terms of the distorted functions $|\chi_\gamma\rangle$ obtained from the corresponding Lippmann-Schwinger equations:

$$|\chi_\gamma\rangle = |\phi_\gamma\rangle + G_{0\gamma} U_\gamma |\chi_\gamma\rangle$$
$$G_{0\gamma} = (E - H_\gamma)^{-1} \tag{5.32}$$

where $|\phi_\gamma\rangle$ is the free Hamiltonian wave function. As for the T_γ^{nr}'s ("nr" stands for "nonreactive"), they obey the ordinary Lippmann-Schwinger equation:

$$T_\gamma^{\mathrm{nr}} = U_\gamma + U_\gamma G_{0\gamma} T_\gamma^{\mathrm{nr}} \tag{5.33}$$

This completes our presentation.

In the spirit of this recent development, one may consider a new way of forming a DWBA.[28] Employing the BKLT equations, it was rigorously shown by Shapiro and Zeiri[28] that the reactive DWBA T-matrix element can be written as:

$$T = \langle \chi_\lambda^a | U_\lambda^a | \chi_\gamma^a \rangle, \qquad \gamma \neq \lambda \tag{5.34}$$

where λ is the initial channel, U_λ^a stands for the "exchange" potential

$$U_\lambda^a = V_\nu + V_k \tag{5.35}$$

and $|\chi_\lambda^a\rangle$, $\gamma = \lambda, \nu, k$, are the distorted wave functions obtained from:

$$|\chi_\gamma\rangle = |\phi_\gamma\rangle + G_{0\gamma} U |\chi_\gamma\rangle \tag{5.36}$$

where

$$U = V - (V_\lambda + V_\nu + V_k) \tag{5.37}$$

Here V is the entire potential of the system.

C. Numerical Applications

The BKLT equations were subjected to only a limited number of "exact" numerical treatments,[1-3, 19, 30-36] mainly with the aim of studying the relevance of their solutions. Two kinds of system were considered: the Hulburt-Hirschfelder type models[37] and the e-H system (with exchange). In what follows, we mention the main results.

1. The Hulburt-Hirschfelder (HH) model

The HH potential is shown in Fig. 7b. The first treatment of this system applying the BK equations (6.9) was carried out by Baer and Kouri[2] for the simplest form of this model. The results were then compared with those obtained by Tang et al.[38] (TKK) and a satisfactory agreement was found. This study was then extended by Evers and Kouri[33] to more complicated versions of this model (the flat reactive channel was replaced by potentials that in the interaction zone had either potential barriers or potential wells). Special emphasis was placed on flux conservation and microreversibility. It was established that except for a few "ill-conditioned" cases, these equations conserved flux (total probability added up to 1) and fulfilled the requirement with respect to microreversibility once the results were converged. Here, too, the final converged results were compared with those of TKK and again the agreement was found to be satisfactory.

A similar study by Lewanski and Tobocman[36] led to different conclusions. Not only did the results they obtained from the BK equations differ from those obtained from the KL equations (5.15), but it was claimed that the BK results did not fulfill flux conservation, which, as mentioned above, is contrary to what was found by Evers and Kouri for very similar models.

A modified HH model was extended by Baer and Kouri[1, 30–32] to three dimensions, and the numerical results constitute the first three-dimensional, quantum-mechanical reactive cross-sections. The modifications were done with respect to the interaction region. There, the flat potential was replaced by the following one:

$$V_1(R_I, R_{II}, \gamma) = v_0 + v_1 \cos \gamma, \qquad 0 \leq R_I \leq R_0, \qquad 0 \leq R_{II} \leq R_0 \quad (5.38)$$

where R_I and R_{II} are the two interatomic distances (the model is devised for the light-(infinitely) heavy-light mass combination), and R_0 is a magnitude related to the dimensions of the interaction region. Figure 20a presents results for $v_1 = 0$ and $v_0 = 0, \pm 2.3$ (the energy unit is the ground state $n = 1$), and Fig. 20b gives results for $v_0 = 2.3$ and $v_1 = 0, 1.33$.

2. The e-H System

The e-H system (including exchange) is a "reactive" system and therefore constitutes a physically realistic case for which the present approach can be tested. This system was considered many times in the past and served as a test case for many numerical methods. The results due to Schwartz[39] are considered to be exact with respect to all published digits. To do the calculations, only S-waves were considered, and therefore the results are expected to be valid only for low enough energies. Table IV compares the singlet and triplet phase shifts, as obtained by Baer and Kouri,[3] with those obtained by other methods.[40] As can be seen, the comparison between BK results and

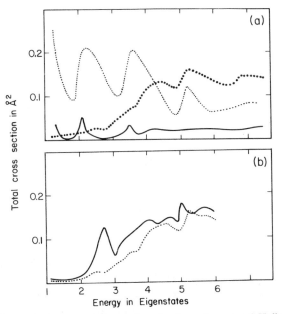

Fig. 20. Total integral reactive cross-sections for three-dimensional Hulburt-Hirschfelder-type potentials.[1] (a) Angular independent potential. Small dots, $v_0 = 0$ (flat potential); big dots, $v_0 = 2.3$ (potential with a barrier); solid curve, $v_0 = -2.3$ (potential with a well). (b) Comparison of results for an angular-dependent and an angular-independent potential. Dots, angular-independent results ($V_0 = 2.3$; $V_1 = 0.0$); solid curve, angular-dependent results ($V_0 = 2.3$; $V_1 = 1.33$).

TABLE IV

Singlet and Triplet S-Wave Phase Shifts for Elastic Scattering of Electrons by Hydrogen Atoms

Energy (ryd)	Hartree-Fock[a]	Massey-Moiseiwitsch[a]	Schwartz[a]	Baer and Kouri[3]
Singlet				
0.01	2.396	2.484	2.553	2.541
0.04	1.871	2.003	2.067	2.039
0.09	1.508	1.649	1.696	1.646
0.25	1.031	1.250	1.202	1.080
0.36	0.869	—	1.041	0.871
Triplet				
0.01	2.908	2.909	2.939	2.935
0.04	2.679	2.680	2.717	2.737
0.09	2.461	2.447	2.500	2.564
0.25	2.070	2.029	2.105	2.264
0.36	1.901	1.909	1.933	2.170

[a]See Ref. 40.

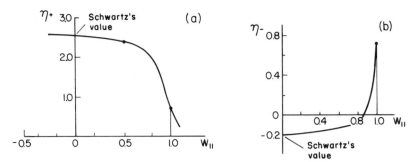

Fig. 21. The S-wave phase shifts for elastic scattering of electrons by hydrogen atoms[35] as a function of w_{11} for $E=0.01$ ryd. The Schwartz results are obtained for $w_{11}=0$. (a) Singlet, (b) Triplet.

those by Schwartz[39] is very encouraging. In a more recent study, Kouri et al.[19] performed a similar study applying the Kouri-Levin equations, and as far as the phase shifts were concerned, the results were identical to those obtained previously. A very instructive and important study was then performed by Kouri et al.,[20] who reconsidered the e-H case, calculating the singlet and triplet phase shifts for different choices of W (the choice in the previous treatments had been $W_{11}=W_{22}=0$; $W_{12}=W_{21}=1$). Figures 21a and 21b present the phase shifts as a function of W_{11} for the energy $E=0.01$ ryd., and it is seen that Schwartz's results best fit those when $W_{11}=0$ [see (5.17) and (5.18)].

References

1. M. Baer and D. J. Kouri, *J. Chem. Phys.*, **56**, 1758 (1972).
2. M. Baer and D. J. Kouri, *J. Chem. Phys.*, **56**, 4840 (1972).
3. M. Baer and D. J. Kouri, *J. Math. Phys.*, **14**, 1637 (1973).
4. D. A. Micha and P. McGuire, *Chem. Phys. Lett.*, **17**, 207 (1972).
5. D. A. Micha and J. M. Yuan, *J. Chem. Phys.*, **63**, 5642 (1975).
6. W. H. Miller, *J. Chem. Phys.*, **50**, 407 (1969).
7. B. C. Garrett and W. H. Miller, *J. Chem. Phys.*, **68**, 4051 (1978).
8. M. Baer and D. J. Kouri, *Phys. Rev. A*, **4**, 1924 (1971).
9. D. J. Kouri and F. S. Levin, *Phys. Lett.* **B**, **501**, 421 (1974).
10. D. J. Kouri and F. S. Levin, *Nuclear Phys. A*, **250**, 127 (1975).
11. W. Tobocman, *Phys. Rev. C*, **9**, 2466 (1974).
12. W. Tobocman, *Phys. Rev. C*, **12**, 1146 (1975).
13. D. A. Micha, *J. Chem. Phys.*, **57**, 2184 (1972).
14. D. A. Micha, *Int. J. Quant. Chem. Symp.*, **10**, 259 (1976).
15. L. D. Faddeev, *Sov. Phys. (J. Exp. Theor. Phys.)*, **12**, 1014 (1961).
16. L. D. Faddeev, *Sov. Phys. Dokl.*, **7**, 600 (1963).
17. K. M. Watson and J. Nuttall, *Topics in Several-Particle Dynamics*, Holden-Day, San Francisco, 1967.
18. D. A. Micha, *Adv. Chem. Phys.*, **61**, 17 (1974).

19. D. J. Kouri, M. Craigie, and D. Secrest, *J. Chem. Phys.*, **60**, 1851 (1974).
20. D. J. Kouri, F. S. Levin, M. Craigie, and D. Secrest, *J. Chem. Phys.*, **61**, 17 (1974).
21. R. G. Newton, *Scattering Theory of Waves and Particles*, McGraw-Hill, New York, 1966, p. 485.
22. B. A. Lippmann, *Phys. Rev.*, **102**, 204 (1956).
23. S. Rabitz and H. Rabitz, *J. Chem. Phys.*, **67**, 2964 (1977).
24. L. Eyges, *Phys. Rev.*, **115**, 1643 (1959).
25. Y. Hahn, *Phys. Rev.*, **169**, 794 (1968).
26. D. J. Diestler and H. Kruger, *Chem. Phys.*, **4**, 151 (1974).
27. M. Baer and D. J. Kouri, in a proposal to the U.S.–Israel Binational Science Foundation, 1979 (unpublished).
28. M. Shapiro and Y. Zeiri, submitted for publication.
29. R. G. Newton, *Scattering Theory of Waves and Particles*, McGraw-Hill, New York, 1966, p. 193.
30. M. Baer and D. J. Kouri, *Chem. Phys. Lett.*, **11**, 238 (1971).
31. M. Baer and D. J. Kouri, *J. Chem. Phys.*, **57**, 3441 (1972).
32. M. Baer and D. J. Kouri, *J. Mol. Phys.*, **22**, 289 (1971).
33. N. S. Evers and D. J. Kouri, *J. Chem. Phys.*, **58**, 1955 (1973).
34. W. G. Cooper, N. S. Evers, and D. J. Kouri, *Mol. Phys.*, **27**, 707 (1974).
35. D. J. Kouri, F. S. Levin, M. Craigie, and D. Secrest, *J. Chem. Phys.*, **65**, 2756 (1976).
36. A. J. Lewanski and W. Tobocman, *Phys. Rev. C*, **17**, 423 (1978).
37. H. Hulburt and J. O. Hirschfelder, *J. Chem. Phys.*, **11**, 273 (1943).
38. K. T. Tang, B. Kleinman, and M. Karplus, *J. Chem. Phys.*, **50**, 1119 (1969).
39. C. Schwartz, *Phys. Rev.*, **124**, 1468 (1961).
40. N. F. Mott and H. S. W. Massey, *The Theory of Atomic Collisions*, Oxford University Press, Oxford, 1965, p. 530.

VI. ELECTRONIC TRANSITIONS IN CHEMICAL REACTIONS

A. Introduction

The treatment of electronic transitions taking place during atomic collisions has been considered to be important since the early days of quantum mechanics. Born and Oppenheimer[1] were the first to study this problem, and their theory still serves as the starting point for any other treatment or approximation. We refer to this later, though not extensively.

Strictly speaking, electronic nonadiabatic transitions occur because of the breakdown of the Born-Oppenheimer (BO) approximation. In general one may distinguish between two forms of electronic nonadiabatic transition: (*1*) those originating from the radial motion, which in the atom-atom case arises from the translational motion, and in more general cases from vibrational and angular motions as well, and (*2*) those originating from the rotation of the body axis of a group of atoms with respect to an axis fixed in space.

Radial coupling was first treated by Zener,[2] Landau,[3] and Stückelberg.[4] They found that the BO approximation breaks down when two adiabatic molecular states of the same symmetry very closely approach each other. Two cases are to be distinguished, according to the strength of the nonadiabatic coupling term. The first (treated by Zener,[2] Landau,[3] Stückelberg,[4] and others[5-9]) is characterized by the intersection of the two diabatic curves or surfaces. The second, treated mainly by Demkov,[10] is characterized by a weak nonadiabatic coupling term (e.g., spin-orbit coupling) which yields two non-intersecting diabatic curves (or surfaces).

Rotational coupling was first discussed by Kronig[11] who found that to properly treat electrons and nuclei on the whole, one should deal with two different systems of coordinates, the one fixed in space (SF) and the other fixed in the molecule (BF). He found that the transformation from SF to BF yields additional nonadiabatic couplings between various electronic states, because of the conservation of total angular momentum. The nature of this coupling was studied extensively by Kronig[11] and others.[12-18]

The main difference between the two types of coupling resides in the following properties: radial coupling may cause transitions only between states of the same symmetry, whereas rotational coupling can mix states of the same and different symmetries.

The literature mentioned in the preceding paragraphs was concerned with either diatomic molecules or atom (ion)-atom collisions. Electronic nonadiabatic transitions induced in three-body collisions (atom-diatom) were, until very recently, mainly treated by applying approximate models. Such were the studies of ion-molecule reactions[19-22] and inelastic (nonreactive) atom-molecule scattering.[23-26] However, the introduction (in the late 1960s) of the trajectory surface-hopping model (TSHM) on the one hand, and the rapid development of efficient quantum-mechanical methods to treat inelastic and reactive collisions on the other, evoked extensive activity in the direction of more rigorous treatments. In the TSHM, Bjerre and Nikitin[27] coupled the Landau-Zener formula with the ordinary classical trajectory method and studied the quenching of an excited sodium ion by nitrogen. Four years later this method was extended to reactive systems and applied to reactive charge transfer processes between D_2 and H^+.[28] The calculated cross-reactions agreed with experiment.[29, 30] A different approach was used by Miller and George,[31] who developed a semiclassical method coupling the Stückelberg method for single coordinates with the classical trajectory method to be performed on a complex plane. This method was later applied to various systems.[32, 33]

Treatment of electronic nonadiabatic transitions in atom-molecule collisions started not too long ago, and the field has hardly been reviewed. Still, some aspects were mentioned by Micha in 1975.[34] The TSHM and related topics have been discussed by Tully,[35] and a unified picture of the field was recently given by Child.[36]

B. The General Approach

Following the first semiclassical treatments, attempts were made to apply quantum-mechanical methods to reactive systems undergoing electronic transitions. To be able to treat this case rigorously, one should repeat the BO treatment and ignore the approximation Born and Oppenheim introduced. To do that, the Hamiltonian H written in the following form should be considered:

$$H = T_n + H_e \qquad (6.1)$$

where T_n is the nuclear kinetic energy and H_e the electronic part containing the electronic kinetic energy and the electronic potential energy, depending parametrically on the nuclear coordinates. If $\psi(e, n)$ is the total wave function of the electrons and the nuclei (e stands for electronic coordinates and n for nuclear coordinates), it can be presented in terms of an electronic basis set $\zeta_i(e; n_0)$, $i = 1, 2, \ldots$, that is,

$$\psi(e, n) = \sum_i \zeta_i(e; n_0) \chi_i(n) \qquad (6.2)$$

In this expression, the $\chi_i(n)$, $i = 1, 2, \ldots$, are the nuclear wave functions and $\zeta_i(e; n_0)$ are solutions of the eigenvalue problem:

$$H_e \zeta_i(e; n_0) = V_i(n_0) \zeta_i(e; n_0), \qquad i = 1, 2, \ldots \qquad (6.3)$$

where $V_i(n_0)$, $i = 1, 2, \ldots$, are the corresponding electronic eigenvalues and n_0 stands for a set of nuclear coordinates that may or may not be equal to n. Next, we distinguish between two representations, namely the diabatic[5, 37] and the adiabatic.[8]

1. The Diabatic Representation

In the diabatic representation, n_0 is assumed to be constant; thus $\psi(e, n)$ is presented in terms of an electronic basis set that is unaffected by changes in the nuclear coordinates.

Recalling that ψ is the solution of the equation

$$H\psi = R\psi \qquad (6.4)$$

by substituting (6.2) in (6.4) and applying (6.3) we obtain that:

$$T_n \chi + (\mathsf{U} - E)\chi = 0 \qquad (6.5)$$

where χ stands for a vector column of the functions $(\chi_1, \chi_2, \ldots)$ and U is a potential matrix with the elements:

$$U_{ij} = \langle \zeta_i | v(e; n) - v(e; n_0) + V_j(n_0) | \zeta_j \rangle \qquad (6.6)$$

Here, $v(e; n)$ is the total electronic potential made up of Coulomb potential terms that depend parametrically on the nuclear coordinates n. It is important to note that the "bra" and "ket" notations are applied with respect to integration over electronic coordinates both here and hereafter.

For simplicity, we consider next a collinear system. The Schrödinger equation that governs the motion of three particles in a line is given in the following form (see Section III):

$$\left[-\frac{\hbar^2}{2\mu_{BC}} \frac{\partial^2}{\partial r^2} - \frac{\hbar^2}{2\mu_{A,BC}} \frac{\partial^2}{\partial R^2} + U(r, R) - E \right] \chi(r, R) = 0 \qquad (6.7)$$

where r and R are the vibrational and translational coordinates, μ_{BC} and $\mu_{A,BC}$ are the reduced masses of the diatomic molecule BC and the atom A with respect to BC, respectively. Accordingly, the corresponding equation, in case electronic transitions are included, is:

$$\left[-\frac{\hbar^2}{2\mu_{BC}} \frac{\partial^2}{\partial r^2} - \frac{\hbar^2}{2\mu_{A,BC}} \frac{\partial^2}{\partial R^2} + (U_{ii}(r, R) - E) \right] \chi_i(r, R)$$
$$+ \sum_{i \neq j} U_{ij}(r, R) \chi_j(r, R) = 0 \qquad (6.8)$$

In principle, to solve this set of equations one could apply the ordinary close-coupling methods developed for the single-surface case.[38, 39] However, it should be emphasized that this can be done only when the off-diagonal elements U_{ij} are small enough, that is,

$$|U_{ij}| \ll |U_i - U_j| \qquad (6.9)$$

When these conditions are not fulfilled, numerical instabilities necessitate further work.

2. The Adiabatic Representation

In the adiabatic representation, one applies an electronic basis set that depends on the nuclear configuration. This amounts to replacing n_0 by n in (6.2) and (6.3). The advantage of the adiabatic representation is twofold:

1. The electronic basis now depends on the nuclear coordinates; thus it contains an important part of the nuclear information as well. There-

fore, the need for an extended basis set, such as is usually required in diabatic representations, is reduced. Of course, when infinite sets are used, the information is identical in diabatic and adiabatic representations. It is very likely that the need for an adiabatic basis set is much stronger in case of reactive systems where at least two asymptotic regions are encountered.

2. Based on (6.6), one sees that, by replacing n_0 by n, a diagonal potential matrix can be obtained and thus the numerical instabilities mentioned above can be avoided.

In addition, *ab initio* treatments such as Configuration Interaction calculations yield adiabatic surfaces; for such cases, therefore, the adiabatic approach is obligatory. The disadvantage in applying the adiabatic basis set lies in the capacity of T_n, which here is a differential nuclear operator, to act also on the electronic wave function. In this way, the nonadiabatic coupling terms that cannot easily be calculated are formed. The nonadiabatic terms are responsible for the transitions between adiabatic surfaces; in this sense they are similar to the off-diagonal terms of the potential matrix in the diabatic representation.

For the sake of clarity, let us reconsider the collinear case.[40] First, we introduce the following notation:

1. ∇ will stand for the vectorial operator

$$\nabla = \left(\frac{\partial}{\partial r}, \frac{\partial}{\partial R} \right) \tag{6.10}$$

2. The scalar product $\mathbf{A} \cdot \mathbf{B}$ will be defined as

$$\mathbf{A} \cdot \mathbf{B} = A_r B_r + A_R B_R \tag{6.11}$$

where A_r (B_r) and A_R (B_R) are the vibrational and translational components of $\mathbf{A}$ ($\mathbf{B}$). Also, the representation in terms of mass-scaled coordinates for T_n will be preferred. Thus, replacing r and R in (6.7) with $\lambda^{-1}r$ and λR [$\lambda = (\mu_{A,BC}/\mu_{BC})^{1/4}$) leads to:

$$T_n = -\frac{\hbar^2}{2\mu} \left(\frac{\partial^2}{\partial R^2} + \frac{\partial^2}{\partial r^2} \right) \tag{6.12}$$

where

$$\mu = (\mu_{A,BC}\mu_{BC})^{1/2} = \left(\frac{m_A m_B m_C}{m_A + m_B + m_C} \right)^{1/2} \tag{6.13}$$

Writing $\psi(e, n)$ as:

$$\psi(e, n) = \sum_i \zeta_i(e, n) \chi_i(n) \tag{6.14}$$

where

$$(H_e - V_i(n))\zeta_i(e, n) = 0, \qquad i = 1, 2, \ldots \tag{6.16}$$

Since

$$T_n = -\frac{\hbar^2}{2\mu} \nabla^2 \tag{6.17}$$

it can be shown that:

$$\left\langle \zeta_i \middle| T_n \sum_i \chi_i \zeta_i \right\rangle = -\frac{\hbar^2}{2\mu} \left[\nabla^2 \chi_j + 2 \sum_j \tau_{ji}^{(1)} \cdot \nabla \chi_j + \sum_j \tau_{ji}^{(2)} \chi_i \right] \tag{6.18}$$

where

$$\tau_{ji}^{(1)} = \langle \zeta_j | \nabla \zeta_i \rangle \qquad \tau_{ji}^{(2)} = \langle \zeta_j | \nabla^2 \zeta_i \rangle \tag{6.19}$$

Combining (6.16) with (6.19) leads to the final set of equations, expressed in a matrix form:

$$\nabla^2 \chi + 2 \tau^{(1)} \cdot \nabla \chi + \tau^{(2)} \chi = \frac{2\mu}{\hbar^2} (V - E) \chi \tag{6.20}$$

where V is a diagonal matrix that contains the adiabatic potential energy surfaces.

Now, comparing the two sets of equations—that for the diabatic case (6.8) and that for the adiabatic case (6.20)—the former undoubtedly seems to be more attractive.

3.   The Adiabatic-Diabatic Transformation[8, 40–42]

With (6.20) available, one should be able just to apply the close-coupling method, that is, form a whole matrix from *each element* of V, $\tau^{(1)}$, and $\tau^{(2)}$ and solve the resulting set of equations by some method. At this stage, however, severe difficulties are likely to arise. First, the equations contain first-order derivatives (the usual equations for the single surface case contain only second-order derivatives), and no efficient method for integration of these equations is available now. Second, some of the $\tau^{(1)}$ matrix elements are known to behave rather sharply, and this is likely to cause numerical instabilities in any further treatment. Cases such as these were encountered in the $(H+H_2)^+$ system[43] and in the $(Ar+H_2)^+$ system[44] (see Figs. 22 and 23). Moreover, for this kind of system, Top and Baer[43] showed that the off-diagonal element $\tau_{12}^{(1)}$ (the vibrational nonadiabatic coupling element that couples the ground and the first excited electronic states) behaves asymptotically as

$$\lim_{R \to \infty} \tau_{12}(r, R) = \frac{\pi}{2} \delta(r - r_s) \qquad (6.21)$$

Here, $r = r_s$ is the value of r where for $R \to \infty$ the two adiabatic surfaces come infinitely close to each other and where the two diabatic surfaces intersect (which amounts to the same thing). To exploit the efficiency of the diabatic representation and yet avoid the above-mentioned deficiencies, Smith[8] devised a procedure that transforms the adiabatic representation of the Schrödinger equation (and its corresponding wave function) into a diabatic representation. Since the two representations contain exactly the same information, the solutions and the transition probabilities are identical. This procedure, originally suggested for the atom-atom case (a single internal co-

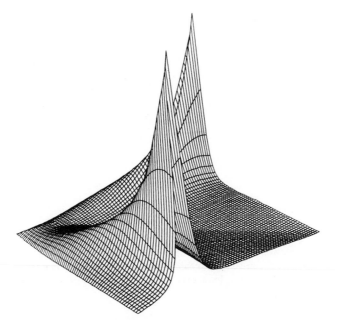

Fig. 22. A three-dimensional figure for the *vibrational* nonadiabatic coupling term as a function of the interatomic distances in the $(H_2 + H)^+$ reactive system.[43]

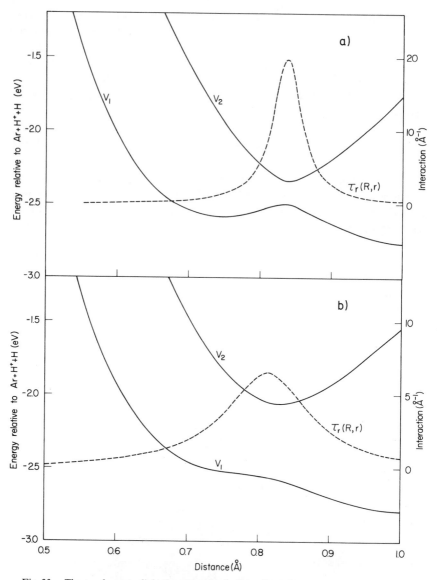

Fig. 23. The two lowest adiabatic energy levels $V_1(r, R)$ and $V_2(r, R)$ (solid curves) and the corresponding vibrational non-adiabatic coupling term $\tau_r(R, r)$ (dashed curves), as a function of the interatomic distance $r = r_{HH}$ for a given fixed value of R. (a) $R = 10$ a.u. (b) $R = 8$ a.u. The curves are for the $(Ar + H_2)^+$ system.[44]

ordinate case), was extended by Baer[40, 41] for the atom-molecule case (first for the collinear case and then for the three dimensions), and by Top and Baer[42] for reactive systems. (A similar method was successfully applied recently for vibrational inelastic collisions.[45]).

To demonstrate the transformation for a case of more than one (internal) coordinate, we return in more detail to the collinear atom-molecule case involving two coordinates, R and r. Reconsidering (6.20), making the transformation

$$\chi = A\eta \qquad (6.22)$$

and performing the required algebra, one obtains:

$$A\nabla^2\eta + 2(\nabla A + \tau^{(1)}A)\cdot\nabla\eta$$

$$+ \left[(\tau^{(2)} + \nabla^2 + 2\tau^{(1)}\cdot\nabla)A - \frac{2\mu}{\hbar^2}(V-E)A\right]\eta = 0 \qquad (6.23)$$

The careful selection of the matrix A will ensure that the coefficient of $\nabla\eta$ vanishes. This implies that A has to be a solution of the equation:

$$\nabla A + \tau^{(1)}A = 0 \qquad (6.24)$$

This (vectorial) equation was shown to have a unique (and orthogonal) solution, provided the two components of $\tau^{(1)}$ (i.e., $\tau_r^{(1)}$ and $\tau_R^{(1)}$) fulfilled the condition:

$$\frac{\partial\tau_r^{(1)}}{\partial R} - \frac{\partial\tau_R^{(1)}}{\partial r} = \left[\tau_r^{(1)}, \tau_R^{(1)}\right] \qquad (6.25)$$

Equation 6.24 can be replaced by an integral equation—a presentation that is convenient for computational purposes. One way of doing this (there are several other ways) is:

$$A(r_1, R_1) = A(r_0, R_0) - \int_{R_0}^{R_1}\tau_R^{(1)}(r_0, R)A(r_0, R)\,dR$$

$$- \int_{r_0}^{r_1}\tau_r^{(1)}(r, R_1)A(r, R_1)\,dr \qquad (6.26)$$

where $A(r_0, R_0)$ can be chosen to be unity.

Returning to (6.23) and applying (6.24), it can be shown that the equation for η simplifies dramatically:

$$\nabla^2 \eta - \frac{2\mu}{\hbar^2}(W-E)\eta = 0 \qquad (6.27)$$

where W is now a usual diabatic potential matrix given in the form:

$$W = A^*VA \qquad (6.28)$$

Thus, one may say that the adiabatic-diabatic transformation shifts the nonadiabatic coupling term into the potential matrix. It should be emphasized that this diabatic potential matrix is different in general from the ordinary diabatic matrix that is based, as described above, on some (asymptomatic) basis set; however, when the adiabatic basis set is large enough, the two coincide (this was demonstrated in a model study by Baer et al.[45]).

The two-state case[40] deserves a more detailed treatment, because the equations simplify considerably. As mentioned, A is orthogonal and can therefore be written as:

$$A(r, R) = \begin{pmatrix} \cos\gamma & \sin\gamma \\ -\sin\gamma & \cos\gamma \end{pmatrix} \qquad (6.29)$$

where $\gamma = \gamma(r, R)$. Since $\tau^{(1)}$ are antisymmetric matrices, they take the form:

$$\tau_R^{(1)} = \begin{pmatrix} 0 & \tau_R \\ -\tau_R & 0 \end{pmatrix}, \qquad \tau_r^{(1)} = \begin{pmatrix} 0 & \tau_r \\ -\tau_r & 0 \end{pmatrix} \qquad (6.30)$$

Substituting (6.29) and (6.30) in (6.24) leads to:

$$\frac{\partial\gamma}{\partial R} = -\tau_R, \qquad \frac{\partial\gamma}{\partial r} = -\tau_r \qquad (6.31)$$

and therefore $\gamma(r, R)$ becomes [see (6.26)]:

$$\gamma(r, R) = \gamma(r_0, R_0) - \int_{R_0}^{R} \tau_R(r_0, R)\,dR - \int_{r_0}^{r} \tau_r(r, R)\,dr \qquad (6.32)$$

where $\gamma(r_0, R_0)$ can be assigned an arbitrary value.

Having constructed A, we can proceed by presenting the element of W:

$$W_{11} = V_1 \cos^2\gamma + V_2 \sin^2\gamma$$
$$W_{22} = V_1 \sin^2\gamma + V_2 \cos^2\gamma \qquad (6.33)$$
$$W_{12} = W_{21} = (V_2 - V_1)\tfrac{1}{2}\sin 2\gamma$$

As presented so far, the method seems to be relevant only when the electronic information is obtained from *ab initio* treatments. This is not the case, and in fact its application cannot be avoided in many other cases where the potential is obtained by semiempirical methods (or by any other methods). As an example, let us consider the diatom-in-molecule (DIM) method.[46-54] Here one constructs a potential matrix with dimensions equal to the number of electronic states that are assumed to be relevant. The most direct procedure is applying the ordinary close-coupling method and solving the corresponding set of differential equations. Such an approach may or may not be feasible, depending on the dimensionality of the potential matrix, but in many cases it is not efficient. Alternatively, one could diagonalize the potential matrix (the eigenvalues are the adiabatic potential energy surfaces) and form the corresponding nonadiabatic coupling terms. Thus, the scattering problem would be treated within the framework of the adiabatic representation for which the dimensionality of the problem could be significantly reduced. To apply the adiabatic-diabatic transformation, which will bring us back to a diabatic representation but with reduced dimensionality, the nonadiabatic coupling terms are needed. These can be obtained using the Hellmann-Feynman theorem in the form[48]:

$$\tau^{(1)}_{x_{ij}} = \frac{C_i^*(\partial U/\partial x)C_j}{V_i - V_j}, \qquad x = r, R \qquad (6.34)$$

where C_i and C_j are the ith and jth eigenvectors of the DIM potential matrix and V_i and V_j are the corresponding eigenvalues. Once the adiabatic potential surfaces and the corresponding nonadiabatic coupling terms are available, the lowest adiabatic states (or any other states) can be selected and the relevant scattering problem solved.

4.   The Adiabatic-Diabatic Transformation in Three Dimensions

So far, we have presented the adiabatic-diabatic transformation for the general case and applied it to the collinear (nonreactive) system. It is of interest and of importance to see how this transformation evolves in a more realistic situation, and whether it still simplifies the equations.

The Hamiltonian for the nuclei and electrons is written as usual, that is,

$$H = T_n + T_e + v(e; n) \tag{6.35}$$

where T_e and $v(e; n)$ are given as before. However T_n should be considered in more detail: T_n is the kinetic energy operator that takes different forms according to the system of coordinates applied. Usually, two such systems—namely, the space-fixed and the body-fixed systems—can be distinguished. Each has its advantages, and we shall not elaborate. However, when electronic transitions are included in the scattering process, the body-fixed system is preferable by far because the electronic wave functions are expressed in terms of these coordinates. The starting point is the space-fixed system and its transformation to the body-fixed system. This transformation should, however, include the kinetic effect of the electron cloud that moves along with the atom-molecule system. The transformation has been done by Baer,[41] and the corresponding form for T_n is:

$$T_n = -\frac{\hbar^2}{2\mu} \left[\frac{1}{r^2} \frac{\partial}{\partial r} r^2 \frac{\partial}{\partial r} + \frac{1}{R^2} \frac{\partial}{\partial R} R^2 \frac{\partial}{\partial R} \right]$$
$$+ \frac{1}{2\mu} \left(\frac{1}{r^2} + \frac{1}{R^2} \right) \mathbf{j}'^2 + \frac{1}{2\mu R^2} \left[\mathbf{K}^2 - 2\mathbf{K}_z^2 + \mathbf{K}_+ \mathbf{j}'_- + \mathbf{K}_- \mathbf{j}'_+ \right] \tag{6.36}$$

Here $\mathbf{K}$ is the total angular momentum of the system, namely:

$$\mathbf{K} = \mathbf{j} + \mathbf{l} + \mathbf{L} \tag{6.37}$$

where $\mathbf{j}$ is the internal angular momentum, $\mathbf{l}$ is the orbital angular momentum, and $\mathbf{L}$ is the total electronic angular momentum. One should be somewhat careful with respect to $\mathbf{j}$, which differs from $\mathbf{j}'$ by $\mathbf{L}_z$, thus:

$$\mathbf{j}' = \mathbf{j} + \mathbf{L}_z \tag{6.38}$$

This makes $\mathbf{j}'_z$ equal to $\mathbf{K}_z$, as in the case of absence of electrons. The other operators, $\mathbf{K}_+$, $\mathbf{j}'_+$, $\mathbf{K}_-$, and $\mathbf{j}'_-$, are the usual raising and lowering operators, respectively.

Our next step is to consider the Schrödinger equation:

$$(T_n + H_e)\psi_M^K = E\psi_M^K \tag{6.39}$$

where ψ_M^K are eigenfunctions of $\mathbf{K}^2$ and $\mathbf{K}_z$. Expressing ψ_M^K in terms of $D_{MM'}^K$, the coefficients of the irreducible representation of the rotation group introduces ψ_μ^K which is the wave function in terms of body-fixed coordinates (M

and M' are the z-components of K along the space-fixed and the body-fixed y-axes, respectively):

$$\psi_M^K(e,n) = \sum_{\mu'} D_{MM'}^K(\Lambda)\psi_{M'}^K(e,n) \qquad (6.40)$$

where Λ stands for the three Euler angles. Substituting (6.40) in (6.35), applying the relations

$$\mathbf{K}^2 D_{MM'}^K = \hbar^2 K(K+1)D_{MM'}^K, \qquad \mathbf{K}_z D_{MM'}^K = \hbar M' D_{MM'}^K$$

$$K_\pm D_{MM'}^K = \hbar \lambda_\pm^{KM'} D_{MM'}^K \qquad (6.41)$$

where

$$\lambda_\pm^{KM'} = \sqrt{K(K+1) - M'(M' \pm 1)} \qquad (6.42)$$

using the fact that the $D_{MM'}^K$ are linearly independent and replacing $\psi_{M'}^K(e,n)$ by

$$\psi_{M'}^K(e,n) = \frac{1}{Rr}\phi_{M'}^K(e,n) \qquad (6.43)$$

we finally obtain[55, 56]:

$$T_{MM}\phi_M^K + T_{MM+1}\phi_{M+1}^K + T_{MM-1}\phi_{M-1}^K + \frac{2\mu}{\hbar^2}(H_e - E)\phi_M^K = 0 \quad (6.44)$$

where the prime sign is deleted from M. Here,

$$T_{MM} = -\left(\frac{\partial^2}{\partial R^2} + \frac{\partial^2}{\partial r^2}\right) + \frac{1}{R^2}\left(K(K+1) - 2M^2\right)$$

$$- \left(\frac{1}{r^2} + \frac{1}{R^2}\right)\left(\frac{\partial^2}{\partial\theta^2} + \cot\theta\frac{\partial}{\partial\theta} - \frac{M^2}{\sin^2\theta}\right)$$

$$T_{MM\pm1} = -\frac{1}{R^2}\lambda_\pm^{KM}\left((M\pm1)\cot\theta \pm \frac{\partial}{\partial\theta}\right) \qquad (6.45)$$

and θ is defined as the angle between $\mathbf{R}$ and $\mathbf{r}$.

To attain our goal without unnecessary complications, we assume below that the total electronic angular momentum is zero and remains zero during the collision. In this way, all angular momenta of the system are due to the nuclei, therefore $\mathbf{K}$ becomes equal to $\mathbf{J}$ ($=\mathbf{l}+\mathbf{j}$). Applying this assumption, we may expand $\phi_M^K(e; r, R, \theta)$ in terms of the electronic basis set

$$\phi_M^K(e; r, R, \theta) = \sum \zeta_j(e; r, R, \theta) \chi_j^{KM}(r, R, \theta) \qquad (6.46)$$

where the ζ_j's are the eigensolutions of the equation:

$$(H_e - V_j(r, R, \theta)) \zeta_j(e; r, R, \theta) = 0 \qquad (6.47)$$

Substituting (6.46) in (6.44) and (6.45), applying (6.47), multiplying from the left by the electronic wave function, and integrating over electronic coordinates leads to the equation:

$$T_{MM}\chi^{KM} + T_{MM+1}\chi^{KM+1} + T_{MM-1}\chi^{KM-1} + \frac{2\mu}{\hbar^2}(V - E)\chi^{KM}$$
$$+ 2\tau^{(1)} \cdot \nabla \chi^{KM} + \tau^{(2)}\chi^{KM} + \underset{\sim}{\Gamma} Q\chi^{KM} = 0$$

$$(6.48)$$

where ∇ is a vectorial operator

$$\nabla = \left(\frac{\partial}{\partial R}, \frac{\partial}{\partial r}, \frac{\sqrt{R^2 + r^2}}{rR} \frac{\partial}{\partial \theta} \right) \qquad (6.49)$$

and Q is an operator defined as:

$$Q\chi^{KM} = \frac{1}{R^2} \left(\lambda_+^{KM}\chi^{KM+1} - \lambda_-^{KM}\chi^{KM-1} + \frac{R^2 + r^2}{r^2} \cot \theta \chi^{KM} \right) \quad (6.50)$$

and the elements for $\tau^{(1)}$, $\tau^{(2)}$, and Γ are:

$$\tau_{ij}^{(1)} = -\langle \zeta_i | \nabla \zeta_j \rangle$$
$$\tau_{ij}^{(2)} = -\langle \zeta_i | \nabla^2 \zeta_j \rangle \qquad (6.51)$$
$$\Gamma_{ij} = -\left\langle \zeta_i \left| \frac{\partial}{\partial \theta} \zeta_j \right. \right\rangle$$

Here, as previously, the scalar product $A \cdot B$ between two vectors is defined:

$$\mathbf{A} \cdot \mathbf{B} = A_R B_R + A_r B_r + A_\theta B_\theta \qquad (6.52)$$

We now continue the procedure just as in the collinear case; thus an adiabatic-diabatic transformation $\mathbf{A}$ matrix is introduced, together with a new set of nuclear wave functions η^{KM}:

$$\chi^{KM} = A\eta^{KM} \tag{6.53}$$

Assuming that A is a solution of the vector equation (and therefore is also orthogonal)

$$\nabla A - \tau^{(1)}A = 0 \tag{6.54}$$

it can be shown[41] that the corresponding equation for η^{KM} is:

$$T_{MM}\eta^{KM} + T_{MM+1}\eta^{KM+1} + T_{MM-1}\eta^{KM-1}$$
$$+ (2\mu/\hbar^2)(W-E)\eta^{KM} = 0 \tag{6.55}$$

where W is now the diabatic potential matrix obtained by:

$$W = A^*VA \tag{6.56}$$

Thus, as in the collinear case, the nuclear kinetic operator used in single-surface problems remains unchanged when the number of electronic states is greater than one and the diabatic potential matrix W is obtained, by means of a similar orthogonal transformation, from a diagonal adiabatic potential matrix. This outcome was obtained by ignoring altogether total electronic angular momentum.

Further work is necessary if a "non-zero electronic angular momentum case" is to be considered. In such a situation, the adiabatic electronic basis set and eigenvalues are insufficient to uniquely define the system. In addition, those functions would have to be expanded in terms of eigenfunctions of the electronic angular momentum. These eigenfunctions, in turn, are not always known, signifying that difficulties should be anticipated. For more details about this possibility, see Refs. 57 and 58.

C.   Studies on Specific Systems

1.   Introduction

Several of the reactive treatments were carried out within the diabatic representation, justifying neither the origin of the diabatic surfaces, nor the origin of the diabatic coupling term. Therefore, in this sense all these studies can be considered at most to be model studies. The quantum-mechanical treatment of the electronic problem coupled with the existence of a reactive channel was first attempted by Haas et al.,[59] who formulated for this purpose a two-surface model similar to the one-surface model of Hofacker and

Levine.[60] They attributed electronic excitations to Franck-Condon factors that resulted from the different dynamic displacements of the reaction paths on the two electronic surfaces. The second study was by Nakamura,[61] who performed a theoretical investigation of the conditions for electronic transitions, using the distorted wave approximation formulation.

The first exact collinear reactive treatment on two surfaces was carried out by Top and Baer.[62, 63] In these model studies, two H_3-type surfaces located one on top of the other, were coupled by a vibrational nonadiabatic coupling term that was either a constant or a Gaussian-type function, peaking at the point where the two surfaces came the nearest to each other. As long as the adiabatic coupling was not too strong, the nuclear processes taking place on each surface could be decoupled from the electronic process. Consequently, the application of the distorted wave Born approximation was found most adequate. Transition probabilities obtained using the DWBA were found to be in excellent agreement with the exact ones. It was also verified that as was assumed in early models,[6] the vibronic states play a dominant role in the electronic transition from one surface to the other. For instance, the Massey parameter[6] w defined as:

$$w = \frac{\Delta\mathcal{E}}{v\tau} \tag{6.57}$$

where $\Delta\mathcal{E}$ is the energy gap between two adiabatic curves in the atom-atom case, v is the radial velocity, and τ, the adiabatic coupling term, is defined, in the atom-molecule case, between two *vibronic states* (one on each surface), not between two *electronic surfaces*. Since the smaller w is, the larger is the expected transition probability, the Massey parameter tells us that if two vibronic states are nearing each other, enhanced transition probabilities may be expected, as in a regular resonance situation. The exact calculations for reactive systems[62, 63] confirm this assumption.

Two studies still within the diabatic framework were performed with respect to more specific systems. Bowman et al.[64] performed a calculation for a collinear model for the system:

$$Ba + N_2O \rightarrow \begin{cases} BaO(x'\Sigma) + N_2 \\ BaO(a^3\Sigma) + N_2 \end{cases} \tag{6.58}$$

treating N_2 as a mass point. The calculations were performed in two different ways; an exact quantum-mechanical calculation and a quasiclassical trajectory surface-hopping calculation.[27, 28] The two sets of results were compared, and it was concluded that the classical model is inadequate for the description of processes in this system. Zimmerman et al.[65] performed a calculation for the reactive $F + H_2$ system in which they considered the two spin states of the F atom (see also Ref. 68):

$$\left. \begin{array}{c} F\left(^2P_{3/2}\right) \\ F\left(^{1/2}P_{1/2}\right) \end{array} \right\} + H_2 \rightarrow HF + H \qquad (6.59)$$

Here, the lower (ground) surface was taken to be the Muckerman V surface,[66] whereas the upper is a modified version of the valence bond potential worked out by Blais and Truhlar.[67] The interaction originating from the spin-orbit coupling is given in the form:

$$H_{so} = \frac{\lambda}{3} \begin{pmatrix} 1 & -\sqrt{2} \\ \sqrt{2} & 2 \end{pmatrix} \qquad (6.60)$$

where λ is the spin-orbit splitting of fluorine (i.e., $\lambda = 0.051$ eV). The adiabatic vibronic states (see Section VI.D.2) related to the two surfaces are given as a function of the reaction coordinates, in Fig. 24c. For more details on the system, see also Ref. 68. The main findings are:

1. The effect of the upper surface on the processes taking place on the lower surface is minor.
2. Electronic nonadiabatic processes are of limited importance; this includes also the reactive process with an excited fluorine atom, which, according to this treatment, practically does not exist.

Another study on reactive $F + H_2$ was performed by Wyatt and Walker.[69] In this three-dimensional treatment the effect of the (missing) product channel was simulated by applying outgoing boundary conditions in the vicinity of the origin. The authors found, as in the collinear study mentioned previously, that nonadiabatic electronic transitions are of secondary importance in this system.

Below, we give a somewhat more detailed discussion of two additional reactive systems, the $(H_2 + H)^+$ and the $(Ar + H_2)^+$, which were treated within the adiabatic framework and therefore also exposed the adiabatic-diabatic transformation to numerical application.

2. The $(H_2 + H)^+$ System

To study the $(H_2 + H)^+$ system, Top and Baer[43] used a DIM potential.[48] Figure 22 presents the *vibrational* nonadiabatic coupling term τ_r (the translational nonadiabatic coupling term is small and we ignore it for the rest of this discussion). This term is small for small R values, but it increases sharply

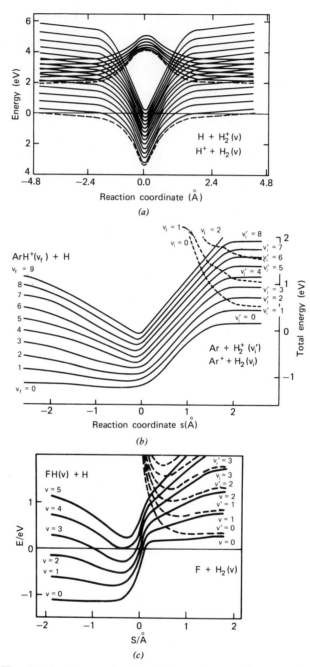

Fig. 24. The vibronic states as a function of the reaction coordinate. (a) The $(H+H_2)^+$ system. (b) The $(Ar+H_2)^+ \to ArH^+ + H$ system. (c) The $F(^2P_{3/2}, {}^2P_{1/2}) + H_2 \to HF + H$ system.

with R and, as mentioned before, turns into a δ-function located at the vibrational coordinate value $r=r_s$, where the two diabatic surfaces intersect. Further details on the system are given in Fig. 24a, where (adiabatic) vibronic states of the two surfaces along the reaction coordinate are presented. Among other things, it is noted that the threshold for charge transfer is around 2 eV.

To obtain some further insight with respect to $\gamma(r, R)$, the rotation angle [see (6.29)], we present in Fig. 25, $\gamma(r, R)$ as a function of r for different R values. Here the γ-function becomes more rectangular as R increases. This kind of behavior follows from the shape of $\tau_r(r, R)$ as R increases. Considering (6.32), ignoring the first integral because it is assumed that $\tau_R \approx 0$, and making $\gamma(r_0, R_0)$ equal to zero, leads to:

$$\gamma(r, R) = \int_{r_0}^{r} \tau_r(r, R)\, dr \qquad (6.61)$$

Substituting τ_r given in (6.21) in (6.61) leads to a step function:

$$\lim_{R \to \infty} \gamma(r, R) = \frac{\pi}{2}\int_{r_0}^{r} \delta(r - r_s)\, dr = \begin{cases} 0, & r < r_s \\ \dfrac{\pi}{2}, & r > r_s \end{cases} \qquad (6.62)$$

which can clearly be seen in Fig. 25.

Starting a reaction with a hydrogen molecule and a proton, four alternative reaction outcomes could result, provided the energy is above 2 eV, the minimal energy necessary to reach the upper surface (see Fig. 24a), yet re-

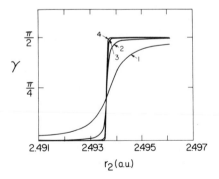

Fig. 25. The adiabatic-diabatic transformation angle $\gamma(r, R)$ for the $(H_2 + H)^+$ system, as a function of the vibrational coordinate r for various values of R. Curves 1 to 3 refer to three finite values of R (not large); curve 4 presents the asymptotic case.[43]

main below dissociation:

$$H_2(v_i)+H_I^+ \rightarrow \begin{cases} H_2(v_f)+H_I^+ & \text{vibrational inelastic} \\ H_2^+(v_f)+H_I & \text{electronic nonadiabatic} \\ HH_I(v_f)+H^+ & \text{reactive} \\ HH_I^+(v_f)+H & \text{reactive+electronic nonadiabatic} \end{cases}$$

(6.63)

Top and Baer[43] presented results concerning all four channels. The results given here for the purpose of illustration involve electronic nonadiabatic processes; therefore no distinction is made between reactive and nonreactive processes. Thus $P_a(v_i, v_f)$ is the probability for transition from an initial state v_i to a final state v_f on the same surface ("a" stands for "adiabatic") and $P_{na}(v_i, v_f')$ is the probability for a transition to a final vibrational state v_f' on the upper surface ("na" stands for "nonadiabatic"). Having obtained these probabilities, we introduce the total adiabatic and nonadiabatic probability functions:

$$P_a(v_i) = \sum_{v_f} P_a(v_i, v_f)$$

$$P_{na}(v_i) = \sum_{v_f'} P_{na}(v_i, v_f')$$

(6.64)

and define the nonadiabatic branching ratio $\Gamma(v_i)$ as:

$$\Gamma(v_i) = \frac{P_{na}(v_i)}{P_a(v_i)}$$

(6.65)

Figure 26 shows $\Gamma(v_i)$ as a function of v_i for different total energies. It is clear from the figure that $\Gamma(v_i)$ is much smaller than 1 for $v_i \leq 3$, but once $v_i \geq 4$, $\Gamma(v_i)$ is in the vicinity of 1. This result indicates the existence of vibrational-to-electronic resonance transitions for the higher vibrational states. By considering the vibrational states along the reaction coordinate (see Fig. 24a), $v_i = 4, 5, 6$ can be seen to be above threshold for charge transfer and $v_i \leq 3$ below it.

3. The $(Ar+H_2)^+$ System

To study the $(Ar+H_2)^+$ system, Baer and Beswick[44, 70] used a DIM potential.[71] Four electronic states were included in the DIM matrix, whereas in the scattering calculation the two lowest adiabatic surfaces, were em-

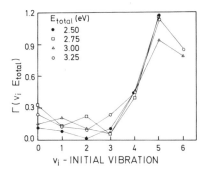

Fig. 26. Electronic branching ratio for the (H + H$_2$)$^+$ system, $\Gamma(v_i) = [P_{na}(v_i)]/[P_a(v_i)]$, as a function of the initial state v_i for different values of (total) energy.[43]

ployed, together with the corresponding nonadiabatic coupling terms. Some information on coupling terms and surfaces is given in Figs. 23 and 24b. Figure 23 shows the vibrational coupling term and the corresponding adiabatic potentials as a function of r, for two different R values. It is seen that as R increases, the function $\tau_r(R, r)$ peaks more sharply and the two adiabatic surfaces near each other at the seam. Figure 24b shows the vibronic curves of the two surfaces, along the reaction coordinate. From this drawing, the ArH$^+$ ion is seen to correlate with H$_2^+$, and so the reaction:

$$Ar + H_2^+ \rightarrow ArH^+ + H \tag{6.66}$$

is direct and does not involve an electronic nonadiabatic transition. On the other hand, the surface governing the reagents Ar$^+$ and H$_2$ is nonreactive and consequently, for the reaction

$$Ar^+ + H_2 \rightarrow ArH^+ + H \tag{6.67}$$

to occur, an electronic nonadiabatic process should take place.

The two reactions are exothermic, the first by 1.30 eV and the second by 1.65 eV. The study by Baer and Beswick mainly emphasizes the latter reaction. Threshold behavior and vibrational distribution were also treated. The main findings were:

1. The transition probability function demonstrates a threshold behavior at $E_{total} = 0.06$ eV (see Fig. 27).
2. A sharp spike in the probability function is encountered at $E_{total} = 0.11$ eV, indicating a possible complex with a lifetime of 10^{-12} sec (see Fig. 27).
3. As for the vibrational distribution, the reaction is found to lead to a highly inverted population, peaking at $v_f = 5$ (see Fig. 28a).

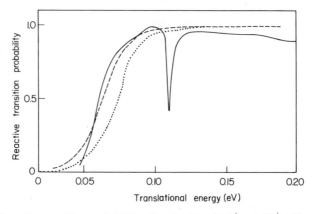

Fig. 27. Reactive transition probabilities for the $(Ar + H_2)^+ \rightarrow ArH^+ + H$ system, as a function of translational energy.[44] Solid curve, exact close-coupling results; dotted curve, results for a two-state model; dashed curve, results for a three-state model.

Findings 1 and 3 are a result of the intersection of the $v_i = 0$ vibrational state of the $Ar^+ + H_2$ system with the $v' = 2$ of the $Ar^+ + H_2$ system. The threshold energy of 0.06 eV is due to a barrier of 0.06 eV formed as a result of the diabatic coupling between the two interacting states. This intersection is discussed in further detail in Section VI.D. The vibrational distribution is seen to be very similar to a vibrational distribution obtained starting with

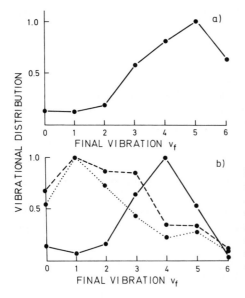

Fig. 28. Normalized vibrational distribution of the ArH^+ system.[70] (a) Results for the reaction $Ar^+ + H_2(v_i = 0) \rightarrow ArH^+(v_f) + H$, which evolves through an electronic nonadiabatic transition (a two-surface calculation). (b) Single-surface results for the adiabatic reaction $Ar + H_2^+ (v_i') \rightarrow ArH^+ (v_f) + H$. Dotted curve, results for $v_i' = 0$; dashed curve, results for $v_i = 1$; solid curve, results for $v_i' = 2$.

$v_i' = 2$ from the lower surface. The vibrational distributions for $v_i' = 0, 1, 2$ of the lower surface are also added for comparison, and they support this outcome (see Fig. 28b).

Only the result concerning the long-lived complex seems to be unrelated to the electronic processes taking place in the entrance channel. It could be attributed, however, to the shallow potential wells formed by the higher vibrational states in the interaction region, which probably support bound states (see Fig. 24b). This possibility was recently supported by experimental evidence presented by Bilotta et al.[72, 73]

D. Electronic Nonadiabatic Processes Among Potential Curves

1. Introduction

A network of potential curves usually stands for vibrational-rotational states related to one or more potential energy surfaces. The use of potential curves as a model for atom-molecule collision goes back to the early days of quantum mechanics. It started when Pelzer and Wigner[74] established the uniqueness and the importance of the minimum energy path in the $H + H_2$ potential surface generated by Eyring and Polanyi.[75] One year earlier, Eckart[76] had constructed a potential curve to study the penetration of electrons through a potential barrier and had derived the corresponding transmission probability. The Eckart potential was flexible enough to fit also the minimum energy path of the $H + H_2$ surface and was taken over by Wigner,[90] Bell,[91] and others for use in the study of chemical reactions (see Section III.B).

The idea of replacing atom-diatom surfaces by curves was considerably extended by Bauer, Fischer, and Gilmore (BFG),[23] who used it in studying the quenching reaction of $Na(3^2P)$ by N_2. To explain the large experimental cross-sections for this and other, similar processes, BFG devised a network of potential curves related to three different potential energy surfaces: the initial, correlated with the excited Na, the final, correlated with the Na ground state, $Na(3^2S)$, and an intermediate ionic potential energy surface that was assumed to couple the two and correlated with Na^+ and N_2^-. The various curves form a grid of crossing points, to each of which is attributed a diabatic and an adiabatic transition probability, calculated using the Landau-Zener formula[2, 3]:

$$P_d = \exp(-q)$$
$$P_a = 1 - \exp(-q) \qquad (6.68)$$

where

$$q = \frac{2\pi u^2}{\Delta Fhv} \qquad (6.69)$$

Here u is the diabatic coupling term, v is the velocity, and ΔF stands for the difference in the gradients of the two intersecting curves:

$$\Delta F = (F_1 - F_2) \qquad F_i = \frac{\partial V_{ii}}{\partial R}, \qquad i = 1,2 \qquad (6.70)$$

all calculated at the crossing point. This model is elaborated on in Section VI.D.3.

2. Reducing the Surface-Crossing Problem to a Curve-Crossing Problem

Usually, the curve-crossing problem is treated within the diabatic framework, and consequently we assume that the Schrödinger equation for the surface-crossing problem is within the diabatic representation (for simplicity, the collinear case is considered):

$$-\frac{\hbar^2}{2\mu}\left(\frac{\partial^2}{\partial R^2} + \frac{\partial^2}{\partial r^2}\right)\chi_n + (W_{nn} - E)\chi_n + \sum_{n \neq n'} W_{nn'}\chi_{n'} = 0, \qquad n = 1,2,\dots$$

$$(6.71)$$

The next step is expanding $\chi_n(R, r)$, $n = 1,2,\dots$, in terms of a vibrational basis set. In dealing with a reactive system the vibrational basis has to be adiabatic (an adiabatic basis set is also preferred in many nonreactive problems), and therefore we assume the vibrational basis set to be R-dependent, that is,

$$\chi_n(r, R) = \sum_{i=1}^{M} \eta_{ni}(R)\phi_{ni}(r, R) \qquad (6.72)$$

where $\phi_{ni}(r, R)$ are solutions of the eigenvalue problem:

$$\left(-\frac{\hbar^2}{2\mu}\frac{d^2}{dr^2} + W_{nn}(r, R) - \mathcal{E}_{ni}(R)\right)\phi_{ni}(r, R) = 0 \qquad (6.73)$$

$$n = 1,2,\dots, N$$
$$i = 1,2,\dots, M$$

(for simplicity M is assumed to be independent of n).

Substituting (6.71) in (6.70) and applying (6.73) leads to:

$$-\frac{\hbar^2}{2\mu}\frac{d^2}{dR^2}\eta_n + (V_n - E)\eta_n + \sum_{n' \neq n} V_{nn'}\eta_{n'} - \left(2\tau_n^{(1)}\frac{d}{dR} + \tau_n^{(2)}\right)\eta_n = 0$$

$$(6.74)$$

where $\boldsymbol{\eta}_n$ is a column vector of the form:

$$\boldsymbol{\eta}_n = \begin{pmatrix} \eta_{n1} \\ \eta_{n2} \\ \vdots \\ \eta_{nM} \end{pmatrix} \tag{6.75}$$

and V_n is a diagonal matrix having the eigenvalues $\mathcal{E}_{ni}$ in its diagonal, $V_{nn'}$ is a matrix with the elements of the form

$$V_{nin'i'} = \langle \phi_{ni} | W_{nn'} | \phi_{n'i'} \rangle \tag{6.76}$$

and $\tau_n^{(1)}$ and $\tau_n^{(2)}$ are the usual nonadiabatic terms:

$$\tau_{nii'}^{(1)} = \frac{\hbar^2}{2\mu} \left\langle \phi_{ni} \left| \frac{d}{dR} \phi_{ni'} \right. \right\rangle$$

$$\tau_{nii'}^{(2)} = \frac{\hbar^2}{2\mu} \left\langle \phi_{ni} \left| \frac{d^2}{dR^2} \phi_{ni'} \right. \right\rangle \tag{6.77}$$

Continuing now as in Section VI.B.3, namely, replacing $\boldsymbol{\eta}_n$ by $\boldsymbol{\zeta}_n$:

$$\boldsymbol{\eta}_n = \mathsf{A}_n \boldsymbol{\zeta}_n \tag{6.78}$$

where A_n is assumed to be a solution of

$$\frac{\hbar^2}{2\mu} \frac{d}{dR} \mathsf{A}_n + \tau_n^{(1)} \mathsf{A}_n = 0 \tag{6.79}$$

one obtains:

$$-\frac{\hbar^2}{2\mu} \frac{d^2}{dR^2} \boldsymbol{\zeta}_n(R) + (\overline{\mathsf{W}}_n - E)\boldsymbol{\zeta}(R) + \sum_{n' \neq n} \overline{\mathsf{W}}_{nm'}\boldsymbol{\zeta}_{n'} = 0 \tag{6.80}$$

where

$$\overline{\mathsf{W}}_n = \mathsf{A}_n^* V_n \mathsf{A}_n \quad \text{and} \quad \overline{\mathsf{W}}_{nn'} = \mathsf{A}_n^* V_{nn'} \mathsf{A}_{n'} \tag{6.81}$$

For more details regarding this approach, see Ref. 45.

Equation (6.80) is solved subject to the following boundary conditions:

$$\lim_{R \to \infty} \boldsymbol{\zeta}_{ni}(R) = \delta_{nn_0}\delta_{ii_0}e^{-ik_{ni}R} + \left(\frac{k_{n_0i_0}}{k_{ni}} \right)^{1/2} T_{nin_0i_0}^{(R)}e^{ik_{ni}R} \tag{6.82}$$

In dealing with a nonreactive problem, we have on the "other side":

$$\lim_{R\to 0} \zeta_{ni}(R)=0 \tag{6.83}$$

but in a reactive problem we then have:

$$\lim_{R\to -\infty} \zeta_{ni}(R)= \left(\frac{k_{n_0 i_0}}{k'_{ni}} \right)^{1/2} T^{(T)}_{n i n_0 i_0} e^{-ik'_{ni}R} \tag{6.84}$$

Here:

$$k_{ni}= \left[\frac{2\mu}{\hbar^2}(E-\mathscr{E}_{ni}(R=\infty)) \right]^{1/2}$$

$$k'_{ni}= \left[\frac{2\mu}{\hbar^2}(E-\mathscr{E}_{ni}(R=-\infty)) \right]^{1/2} \tag{6.85}$$

the indices (R) and (T) on the T-matrix elements stand for "reflected" and "transmitted," respectively. Equation 6.80 can also be written in the form:

$$\frac{d^2}{dR^2}\zeta + (\overline{W}-E)\zeta = 0 \tag{6.86}$$

where $\overline{W}$ stands for a "super" potential matrix and ζ for a "super" wave function vector.

As an example, we shall apply this approach to the $(Ar+H_2)^+$ system.[44] Since the electronic interaction takes place in the entrance channel only before the strong vibrational interaction region is reached (see Fig. 24b), we avoid using the above-mentioned adiabatic transformation and use only the asymptotic basis set (i.e., a diabatic basis set). The various diabatic potential matrix elements are:

$$\begin{aligned}
\overline{W}_{11} &=0.5+0.19\exp(-4.5(R-4.4)) \\
\overline{W}_{22} &=0.68\tanh(2.11(R-3.5)) \\
\overline{W}_{33} &=\overline{W}_{22}+0.24 \\
\overline{W}_{12} &=\overline{W}_{13}=14.52\exp(-1.15R) \\
\overline{W}_{23} &=0
\end{aligned} \tag{6.87}$$

The energies are given in electron volts and the distances in angstrom units.

Here $\overline{W}_{11}$ stands for the vibrational state $v_i = 0$ of the upper surface and $\overline{W}_{22}$ and $\overline{W}_{33}$ for $v_i' = 2$ and $v_i' = 3$ of the lower surface, respectively (Fig. 29). The $\overline{W}_{23}$, which is the coupling between two states of the same surface ($v_i' = 2$ and $v_i' = 3$), is negligibly small in the asymptotic region and is therefore set equal to zero.

The set of differential equations is solved twice; once including only two curves (one for each surface), and once including three curves (one for the upper surface, $v_i' = 0$ and two for the lower, $v_i' = 2, 3$). The reactive probability functions for the process:

$$\mathrm{Ar}^+ + \mathrm{H}_2(v_i = 0) \rightarrow \mathrm{ArH}^+ + \mathrm{H} \qquad (6.88)$$

were given in Fig. 27, which showed three curves: one obtained by the extensive close-coupling calculations (i.e., the "exact"), and the other two obtained by applying the two- and three-curve models, respectively. For all practical purposes, the three-curve model results seem to overlap with the exact results.

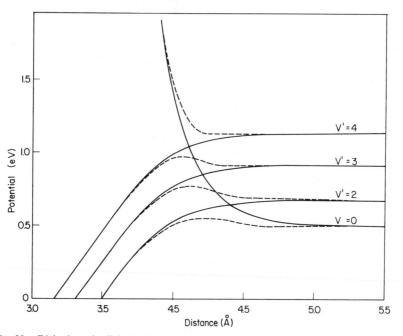

Fig. 29.   Diabatic and adiabatic vibronic states for the model studies of the $(\mathrm{Ar} + \mathrm{H}_2)^+ \rightarrow$ $\mathrm{ArH}^+ + \mathrm{H}$ system, as a function of translational energy.[44]

In this context we mention that several groups have studied numerically and analytically the reactive two-state curve-crossing system (Laing et al.,[79] Baer and Child,[80] and Korsch and Kruger[81]).

3. A Curve-Crossing Model for Reactions

Now we consider the treatment of a network of reactive curves that stand for the vibronic states of two surfaces crossing each other along a seam.[82] This study serves to test the reliability and to determine the range of validity of two simplified models: the Bauer-Fischer-Gilmore (BFG) model[23] modified for reactive systems, and a Franck-Condon-type model.

a. The Model. The model, which contains two surfaces, is defined in terms of two displaced harmonic oscillators with equal force constants. As for the R-dependences, one is assumed to be constant and the other to increase exponentially with R.

Thus,

$$V_1(r, R) = \tfrac{1}{2}k(r-r_1)^2, \qquad -\infty \le R \le \infty$$

$$V_2(r, R) = \tfrac{1}{2}k(r-r_2)^2 + A\left(1 - \exp\left(-\frac{R}{R_0}\right)\right) \qquad (6.89)$$

The coupling is taken to be constant over the effective crossing region. Scaling the various energy and length variables and following the procedure described in the preceding subsection, one ends up with the following set of equations:

$$\left(\frac{1}{\rho_0^2}\frac{d^2}{dR^2} + E\right)\phi_{\nu i} = \sum_{\gamma' i'} V_{\nu i \nu' i'}\phi_{\nu' i'} \qquad (6.90)$$

where E is the total energy (in units of $\hbar\omega$). ρ_0 is given by:

$$\rho_0 = \frac{R_0}{r_0} \qquad (6.91)$$

r_0 being half the width of the oscillator, ν and i being indices for the surfaces and the vibronic states, respectively, and $V_{\nu i \nu' i'}$ given as:

$$V_{\nu i \nu' i'} = \begin{cases} i\delta_{ii'}, & \nu=\nu'=1 \\ (\alpha(1-e^{-\rho})+i)\delta_{ii'}, & \nu=\nu'=2 \\ V_0 S_{ii'}(\xi_0), & \nu\neq\nu' \end{cases} \qquad (6.92)$$

Here, U_0 is the scaled coupling term V_0, $S_{ii'}(\xi_0)$ is the overlap integral between two harmonic wave functions shifted by a scaled distance ξ_0, and α is equal to $A/\hbar\omega$.

Equations (6.90) were solved exactly by a propagator method, subject to the usual (reactive) boundary conditions.

b. The Simplified Models

The Bauer-Fischer-Gilmore (BFG) Model. The BFG model is a simple and straightforward extension of the Landau-Zener model to a network of curves. Thus for each grid point (i, j) (the crossing of curve i belonging to surface 1, with curve j belonging to surface 2), one attributes a diabatic probability $P_d(i, j)$ and an adiabatic probability $P_a(i, j)$, both as given by the Landau-Zener formula [see (6.68)]. It is assumed that the system evolves from right ($R = \infty$) to left ($R = -\infty$) without reflection. If $\sigma_1(i, j)$ is the probability for the system to pass the grid point (i, j) and to be on surface 1, and $\sigma_2(i, j)$ is defined similarly but with respect to surface 2, the equations for $\sigma_\nu(i, j)$, $\nu = 1, 2$ (taking into account the previous corresponding probabilities) are:

$$\sigma_1(i, j) = \sigma_1(i, j-1)P_d(i, j) + \sigma_2(i+1, j)P_a(i, j)$$
$$\sigma_2(i, j) = \sigma_1(i, j-1)P_a(i, j) + \sigma_2(i+1, j)P_d(i, j) \qquad (6.93)$$

The chain of equations just presented is solved subject to a given set of initial (boundary) conditions.

The Franck-Condon (FC) Model. In the context of the Franck-Condon model, one assumes that the transition from one surface to the other is sudden, with no intermediate transitions as in the BFG model, and that the system moves from a given vibrational state in one surface to some vibrational state of the other. The transition probability, therefore, involves a Landau-Zener branching ratio between the *surfaces*, multiplied by an appropriate Franck-Condon overlap factor:

$$P_{I,j}^{a} = P_a |S_{I,j}|^2 \qquad (6.94)$$

where I stands for the initial vibronic state on surface 1. As mentioned, P_a is a Landau-Zener probability function, calculated at the point where the two curves I and j intersect.

In general the two models yield different results (their results become identical when either the coupling term or the overlap integral becomes small enough), and the question is whether a single parameter can be found that will enable one to determine the range of validity of each model. Obviously such a parameter is the energy. The FC model becomes more relevant the higher the energy is, and because it is based on a step-by-step transition process, the BFG model is expected to be more relevant for low energies. However, this parameter is not sharp enough, because the relevance of the models is also related to the way the energy is distributed among the various degrees of freedom as well as on the relative position of the seam at the crossing point. Child and Baer[82] were able to construct a parameter γ in which all this information was lumped together.

Considering the equation of the seam

$$G(r, R) = V_1(r, R) - V_2(r, R) = 0 \qquad (6.95)$$

it was shown that γ takes the form:

$$\gamma = \frac{v_R}{v_r} \frac{\Delta G_R}{\Delta G_r} \qquad (6.96)$$

where

$$\Delta G_S = \frac{\partial G}{\partial S}, \qquad S = r, R \qquad (6.97)$$

and v_R and v_r are the translational and vibrational velocities, respectively. Next, it was established that when $\gamma \gg 1$, the FC model becomes exact, but when $\gamma \ll 1$ the BFG model should be preferred.

Figure 30 compares the exact results with those obtained with the models, and the parameter γ can be seen to enable distinguishing between the relevance of the two models (all the results were obtained with the same energy).

E.   Electronic Nonadiabatic Processes in Strong Laser Fields

The study of electronic nonadiabatic transitions caused by strong radiation fields (laser radiation) operating during a molecular collision was recently introduced by Zimmerman et al.[83] Following the stimulating research on the effect of laser fields on atomic collisions,[84-86] Zimmerman et al. performed their study on atom-molecule collisions, with particular emphasis on instances of vibrational-rotational level spacing within the collision complex passing through a resonance with the field quantum, as the encounter proceeds. Although no experimental evidence has been published regarding the effect of a laser field on an atom-diatom collision process, still it is of interest to briefly discuss here the main concepts and possible outcomes. As in the Born-Oppenheimer treatment,[1] we start with the Hamiltonian that con-

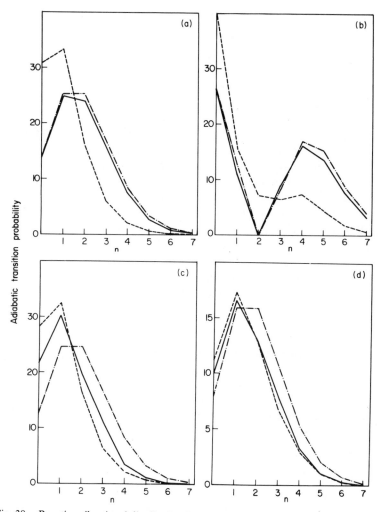

Fig. 30. Reactive vibrational distribution for the multicurve crossing model. All the results were calculated for the same total energy.[82] Solid curves, exact results; dashed curves, Bauer-Fischer-Gilmore model results; dot-dash curves, Franck-Condon model results. (a) The probability for the (reactive) transition $(\nu=1, i=0) \rightarrow (\nu'=2, i'=n)$. The value of γ is 50; the total nonadiabatic transition probability is 0.9. (b) The probability for the (reactive) transition $(\nu=1, i=1) \rightarrow (\nu'=2, i'=n)$. The value of γ is 50; the total nonadiabatic transition probability is 0.9. (c) The probability for the (reactive) transition $(\nu=1, i=0) \rightarrow (\nu'=2, i'=n)$. The value of γ is 2; the total nonadiabatic probability is 0.9. (d) The probability for the reactive transition $(\nu=1, i=0) \rightarrow (\nu'=2, i'=n)$. The value of γ is 2; the total nonadiabatic probability is 0.5.

303

tains the interactions of the electrons, the nuclei and the radiation field:

$$H' = \frac{1}{2m} \sum_l \left(\frac{\hbar}{i} \nabla_l^{(e)} - eA(q_l) \right)^2 + \sum_j \frac{1}{2M_j} \left(\frac{\hbar}{i} \nabla_j^{(n)} - Z_j eA(Q_j) \right)^2$$

$$+ v(q, Q) + \hbar \omega a^+ a \tag{6.98}$$

Here $(\hbar/i)\nabla_l^{(e)}$ and $(\hbar/i)\nabla_j^{(n)}$ are the linear momenta of electrons and nuclei, respectively, A is the vector potential through which the radiation field is described, q and Q are electronic and nuclear coordinates, e and m are the electronic charge and mass, $Z_j e$ and M_j are the nuclear charge and mass, $v(q, Q)$ is the Coulomb interaction between the various charges, and the last term stands for the energy of the free radiation field, where a and a^+ are the annihilation and creation operators, and ω is the frequency associated with the field. In the following, a single-mode field is assumed.

The Hamiltonian above is inconvenient for practical application, and therefore one performs a contact transformation[87-89]:

If ψ stands for the solution of the equation

$$H'\psi = E\psi \tag{6.99}$$

then, applying a new wave function χ such that

$$\psi = S\chi \tag{6.100}$$

it is readily seen that χ is a solution of the equation:

$$H\chi = E\chi \tag{6.101}$$

where

$$H = S^{-1}H'S \tag{6.102}$$

In the following, it is assumed that the wave number k is small compared to the range of the interaction, so that A can be written as (the dipole approximation):

$$A = iA_0(a - a^+) \tag{6.103}$$

Now, if S is written as product:

$$S = \prod_t S_t \tag{6.104}$$

where t runs over all the charged particles (both electrons and nuclei) in the system, a convenient choice for S_t is:

$$S_t = \exp\left[\frac{Z_t e}{\hbar}(\mathbf{A}_0 \cdot \boldsymbol{\rho}_t)(a - a^+)\right] \tag{6.105}$$

where the product in parentheses is a scalar product and ρ_t stands for the coordinate of the tth particle.

Applying this transformation and retaining only first-order terms, H takes the more familiar form:

$$H = -\frac{\hbar^2}{2m}\sum_l \nabla_l^{(e)2} - \frac{\hbar^2}{2}\sum_j \frac{1}{M_j}\nabla_j^{(n)2} + v(q,Q) + \hbar\omega a^+ a$$

$$+ e\left[\mathbf{Z}_l(\mathbf{E}\cdot\mathbf{q}_l) - \sum_j Z_j(\mathbf{E}\cdot\mathbf{Q}_j)\right] \tag{6.106}$$

where $\mathbf{E}(\rho)$, the electric component of the field, is given as:

$$\mathbf{E} = \frac{\partial \mathbf{A}}{\partial t} = -\frac{i}{\hbar}[H_{\text{rad}}, \mathbf{A}] \tag{6.107}$$

or [see (6.103)]:

$$\mathbf{E} = \omega\mathbf{A}(O)(a + a^+) \tag{6.108}$$

The next step is to employ the usual close-coupling technique. Accordingly, the total wave function is written in the form:

$$\psi(q,Q) = \sum_n |n\rangle \zeta^* \chi_n \tag{6.109}$$

where $|n\rangle$ is the nth eigenstate of the radiation field associated with the eigenvalue $n\hbar\omega$, ζ^* is the electronic wave function vector, which may (adiabatic) or may not (diabatic) depend on the nuclear coordinates, and χ_{ni} is the nuclear wave function associated with both the ith electronic state and the nth state of the radiation field. (Since the only effect of the field is to shift each state by $n\hbar\omega$, the electronic eigenfunctions are not affected and consequently do not depend on n.)

Substitution of (6.109) in (6.106) and integration over the electronic and radiation field coordinates leads to the coupled equations for the nuclear part:

$$T_N \chi_n + (n\hbar\omega - E)\chi_n + \mathsf{U}\chi_n + (\mathbf{E}\cdot\boldsymbol{\mu})\left[\sqrt{n+1}\,\chi_{n+1} + \sqrt{n}\,\chi_{n-1}\right] = 0 \tag{6.110}$$

where for simplicity the diabatic representation is employed. Here, μ is the dipole moment matrix, defined as:

$$\mu_{ij} = e \sum_l \langle \zeta_i | \mathbf{q}_l | \zeta_j \rangle \qquad (6.111)$$

T_N is the nuclear kinetic energy operator and U is the diabatic potential matrix. In all the applications so far, two "photonic" states were considered.[90] These studies indicated once again that resonance transitions are favorable. Thus, if a certain vibrational level in a given electronic state is shifted (by an amount $\hbar\omega$) such that it comes energetically close to a vibrational state in the other electronic state, this particular $(v, e, n) \rightarrow (v', e', n-1)$ transition is enhanced considerably.[83]

Most of the treatments were carried out for nonreactive systems, and only one reactive model study has been reported.[91] An interesting idea was discussed recently by Orel and Miller,[92] who considered the possibility of enhancing the $H + H_2$ reaction by lowering the activation energy by means of an interaction between the radiation field and the molecular system. It was argued that such an interaction could occur when the three H atoms were in close proximity and behaved for a short time as a (collinear) three-atom molecule, which could become infrared active as a result of the asymmetric stretch. This process is plausible, even though neither the reagents nor the products are infrared active. The possibility of the potential barrier of a reactive system being lowered because of the existence of a (temporary or permanent) upper electronic state was discussed and treated in detail in Refs. 62 and 63 and would apply to this and other similar studies.[91]

References

1. M. Born and J. R. Oppenheimer, *Ann. Phys.* (*Leipzig*), **84**, 457 (1927).
2. C. Zener, *Proc. R. Soc. London, Ser. A*, **137**, 696 (1932).
3. L. D. Landau, *Phys. Z. Sov. Union*, **2**, 46 (1932).
4. E. C. G. Stückelberg, *Helv. Phys. Acta*, **5**, 369 (1932).
5. W. Lichten, *Phys. Rev.*, **131**, 229 (1963).
6. E. E. Nikitin, in *Chemische Elementar Prozesse*, H. Hartman, Ed., Springer-Verlag, Berlin, 1968.
7. N. Rosen and C. Zener, *Phys. Rev.*, **40**, 502 (1932).
8. F. T. Smith, *Phys. Rev.*, **179**, 111 (1969).
9. M. S. Child, *Mol. Phys.*, **20**, 171 (1970).
10. Yu. N. Demkov, *Sov. Phys.* (*J. Exp. Theor. Phys.*), **18**, 138 (1964).
11. R. de L. Kronig, *Band Spectra and Molecular Structure*, Cambridge University Press, New York, 1930, pp. 6–16.
12. D. R. Bates, *Proc. R. Soc. London, Ser. A*, **240**, 437 (1957); **243**, 15 (1958).
13. D. R. Bates, *Proc. R. Soc. London, Ser. A*, **257**, 22 (1960).
14. W. R. Thorson, *J. Chem. Phys.*, **34**, 1744 (1961); **42**, 3878 (1965).
15. D. J. Kouri and C. F. Curtiss, *J. Chem. Phys.*, **44**, 2120 (1966).

16. R. T. Pack and J. O. Hirschfelder, *J. Chem. Phys.*, **49**, 4009 (1968); **52**, 521 (1970).
17. C. Gaussorgues, C. Le Sech, F. Mosnow-Seeuws, R. McCarroll, and A. Riera, *J. Phys. B, At. Mol. Phys.*, **8**, 239 (1975).
18. C. Gaussorgues, C. Le Sech, F. Mosnow-Seeuws, R. McCarroll, and A. Riera, *J. Phys. B, At. Mol. Phys.*, **8**, 253 (1975).
19. G. Gioumousis and D. P. Stevenson, *J. Chem. Phys.*, **29**, 294 (1958).
20. F. S. Klein and L. Friedman, *J. Chem. Phys.*, **41**, 1789 (1964).
21. A. Henglein, K. Lackmann, and B. Knoll, *J. Chem. Phys.*, **43**, 1048 (1965).
22. E. A. Gislason, *J. Chem. Phys.*, **57**, 3396 (1972).
23. E. Bauer, E. R. Fischer, and F. R. Gilmore, *J. Chem. Phys.*, **51**, 4173 (1969).
24. E. E. Nikitin and S. Ya. Umanski, *Disc. Faraday Soc., Chem.*, **53**, 7 (1972).
25. M. S. Child, *Disc. Faraday Soc., Chem.*, **55**, 30 (1973).
26. M. S. Child, *Molecular Collision Theory*, Academic Press, London, 1974, pp. 161–179.
27. A. Bjerre and E. E. Nikitin, *Chem. Phys. Lett.*, **1**, 179 (1967).
28. J. C. Tully and R. K. Preston, *J. Chem. Phys.*, **55**, 562 (1971).
29. J. R. Krenos, R. K. Preston, R. Wolfgang, and J. C. Tully, *J. Chem. Phys.*, **60**, 1634 (1974).
30. G. Ochs and E. Teloy, *J. Chem. Phys.*, **61**, 4930 (1974).
31. W. H. Miller and T. F. George, *J. Chem. Phys.*, **56**, 5637 (1972).
32. Y.-W. Lin, T. F. George, and M. Morokuma, *J. Chem. Phys.*, **60**, 4311 (1974).
33. J. R. Laing, T. F. George, I. H. Zimmerman, and Y.-W. Lin, *J. Chem. Phys.*, **63**, 842 (1975).
34. D. A. Micha, *Adv. Chem. Phys.*, **30**, 7 (1975).
35. J. C. Tully, in *Dynamics of Molecular Collisions*, Part B, W. H. Miller, Ed., Plenum Press, New York, 1976, Chapter 5.
36. M. S. Child, in *Atom-Molecule Collision Theory*, R. B. Bernstein, Ed., Plenum Press, New York, 1979, Chapter 13.
37. T. F. O'Malley, *Adv. At. Mol. Phys.*, **7**, 223 (1971).
38. R. G. Gordon, *J. Chem. Phys.*, **51**, 14 (1969).
39. W. N. Sams and D. J. Kouri, *J. Chem. Phys.*, **51**, 4815 (1969).
40. M. Baer, *Chem. Phys. Lett.*, **35**, 112 (1975).
41. M. Baer, *Chem. Phys.*, **15**, 49 (1976).
42. Z. H. Top and M. Baer, *J. Chem. Phys.*, **66**, 1363 (1977).
43. Z. H. Top and M. Baer, *Chem. Phys.*, **25**, 1 (1977).
44. M. Baer and J. A. Beswick, *Phys. Rev. A*, **19**, 1559 (1979).
45. M. Baer, G. Drolshagen, and J. P. Toennies, *J. Chem. Phys.*, **73**, 1690 (1980).
46. F. O. Ellison, *J. Am. Chem. Soc.*, **85**, 3540, 3544 (1963).
47. F. O. Ellison and J. C. Patel, *J. Am. Chem. Soc.*, **86**, 2155 (1964).
48. R. K. Preston and J. C. Tully, *J. Chem. Phys.*, **54**, 4297 (1971).
49. P. J. Kuntz and A. C. Roach, *J. Chem. Soc., Faraday Trans. II*, **68**, 259 (1972).
50. J. C. Tully, *J. Chem. Phys.*, **58**, 1396 (1973).
51. Y. Zeiri and M. Shapiro, *J. Chem. Phys.*, **70**, 5264 (1979).
52. I. Last and M. Baer, *J. Chem. Phys.*, **75**, 288 (1981).
53. I. Last and M. Baer, *Chem. Phys. Lett.*, **73**, 514 (1980).
54. I. Last, *Chem. Phys.*, in press.
55. D. J. Vezzetti and S. I. Rubinow, *Ann. Phys.*, **35**, 373 (1955).
56. W. H. Miller, *J. Chem. Phys.*, **43**, 2373 (1968).
57. P. L. DeVries and T. F. George, *J. Chem. Phys.*, **76**, 1293 (1968).
58. D. L. Miller and R. E. Wyatt, *J. Chem. Phys.*, **67**, 1302 (1977).
59. Y. Haas, R. D. Levine, and G. Stein, *Chem. Phys. Lett.*, **15**, 7 (1972).
60. G. L. Hofacker and R. D. Levine, *Chem. Phys. Lett.*, **9**, 617 (1971).
61. H. Nakamura, *Mol. Phys.*, **26**, 673 (1973).

62. Z. H. Top and M. Baer, *Chem. Phys.*, **10**, 95 (1975).
63. Z. H. Top and M. Baer, *Chem. Phys.*, **16**, 447 (1976).
64. J. M. Bowman, S. C. Leasure, and A. Kuppermann, *Chem. Phys. Lett.*, **43**, 374 (1976).
65. I. H. Zimmerman, M. Baer, and T. F. George, *J. Chem. Phys.*, **71**, 4132 (1979).
66. D. F. Feng, E. R. Grantland, and J. W. Root, *J. Chem. Phys.*, **64**, 3450 (1976).
67. N. C. Blais and D. G. Truhlar, *J. Chem. Phys.*, **69**, 846 (1978).
68. I. H. Zimmerman and T. F. George, *Chem. Phys.*, **7**, 323 (1975).
69. R. E. Wyatt and R. B. Walker, *J. Chem. Phys.*, **70**, 1501 (1979).
70. M. Baer and J. A. Beswick, *Chem. Phys. Lett.*, **51**, 360 (1977).
71. P. J. Kuntz and A. C. Roach, *J. Chem. Soc. Faraday Trans., II*, **68**, 258 (1972).
72. R. M. Billota, R. N. Preuminger, and J. M. Farrar, *Chem. Phys. Lett.*, **74**, 95 (1980).
73. R. M. Billota, F. N. Preuminger, and J. M. Farrar, *J. Chem. Phys.*, **73**, 1637 (1980).
74. H. Pelzer and E. P. Wigner, *Z. Phys. Chem. B*, **15**, 445 (1932).
75. H. Eyring and M. Polanyi, *Z. Phys. Chem. B*, **12**, 279 (1931).
76. C. Eckart, *Phys. Rev.*, **35**, 1303 (1930).
77. E. P. Wigner, *Z. Phys. Chem. B*, **19**, 203 (1933).
78. R. P. Bell, *Proc. R. Soc. London, Ser. A*, **139**, 466 (1933).
79. J. R. Laing, J. M. Yuan, I. H. Zimmerman, P. L. DeVries, and T. F. George, *J. Chem. Phys.*, **66**, 2801 (1977).
80. M. Baer and M. S. Child, *Mol. Phys.*, **36**, 1449 (1978).
81. H. J. Korsch and H. Kruger, *Mol. Phys.*, **39**, 51 (1980).
82. M. S. Child and M. Baer, *J. Chem. Phys.*, **74**, 2832 (1981).
83. I. H. Zimmerman, J. M. Yuan, and T. F. George, *J. Chem. Phys.*, **66**, 2638 (1977).
84. N. M. Kroll and K. M. Watson, *Phys. Rev. A*, **8**, 804; **13**, 1018 (1976).
85. A. M. F. Lau, *Phys. Rev. A*, **13**, 139 (1976); **14**, 279 (1976).
86. J. I. Gersten and M. H. Mittleman, *Phys. Rev. A*, **12**, 1840 (1975).
87. M. Goppert Mayer, *Ann. Phys. (Leipzig)*, **9**, 273 (1931).
88. P. I. Richards, *Phys. Rev.*, **73**, 254 (1949).
89. E. A. Power and S. Zienau, *Phil. Trans. R. Soc.*, **251A**, 427 (1959).
90. T. F. George, I. H. Zimmerman, P. L. DeVries, J. M. Yuan, Kai-shae Lam, J. C. Bellum, H. W. Lee, M. S. Slatsky, and J. T. Lin, in *Chemical and Biochemical Applications of Lasers*, Vol. IV, C. B. Moore, Ed., Academic Press, New York, 1979.
91. J. C. Light and A. Altenberger-Siczek, *J. Chem. Phys.*, **70**, 4108 (1979).
92. A. E. Orel and W. H. Miller, *Chem. Phys. Lett.*, **57**, 362 (1978).

VII. CONCLUDING REMARKS

Before ending this review, I shall point out a few possible directions of research, representing my personal viewpoint. Nevertheless, these suggestions are based largely on very recently published work or on ongoing studies of which I am aware.

The main efforts in the study of the chemical reaction in its full dimensionality are expected to shift toward developing reliable and efficient approximations. The infinite-order sudden approximation (IOSA) is certainly one reasonable way to treat chemical reactions. Although the IOSA method has been rather successfully applied to two reactions, the physical contents of this approximation are still far from being understood and the range of its validity is yet unknown. And although it developed within the framework of quantum mechanics, there is no difficulty in extending the use of the IOSA to classical mechanics. Moreover, this extension is most desirable because once such calculations have been done, the results can be compared with those obtained from exact, three-dimensional (classical) treatments.

In addition, one should expect much activity in the treatment of electronic transitions taking place during a chemical reaction. However, since studies of this type are more involved than studies of ordinary chemical reactions, it is anticipated that cruder approximations will be incorporated. Therefore, it is important to gain a good understanding of the various processes (e.g., by studying simplified models). As possible approximations, I would consider the distorted wave Born approximation, which yielded reasonable results in the few cases to which it was applied, and the IOSA, which could be of great help if applied correctly.

In this context, the adiabatic-diabatic transformation should be mentioned. Here, two kinds of study are possible. One has to do with the understanding of this transformation when it is applied in three dimensions in the most general case, and the other with its exposure to various approximations.

A relatively new subject, closely connected to the one just noted, is the study of chemical reactions taking place in strong laser fields. Although most of the insight and techniques gained from the studies of electronic transitions can be directly applied here, still further theoretical research will contribute to a better understanding of the coupling between strong electromagnetic fields and colliding atoms and molecules.

Finally, much work is being devoted to developing efficient numerical techniques to treat dissociative collisions. Several groups are tackling that aspect, and the first reports have already appeared in print. It is expected that during the next few years, the study of these processes will become an important subject in the framework of reactive collisions. In this context, I believe, many theoretical studies dealing with the threshold for dissociation and how a system prepares itself for this "catastrophic" event will soon be in progress.

Acknowledgments

I thank Profs. H. Kruger and P. McGuire for inviting me to stay at the Department of Physics of the University of Kaiserslautern, Kaiserslautern, Germany, from July to December, 1980, during which time this composition was completed. I also thank the Deutsche Forschungsgemeinschaft, which within the framework of the Sonderforschungsbereich 91 *Energietransfer bei Atomaren und Molekularen Stossprozessen*, partially supported this work. I am most grateful to Profs. R. D. Levine and D. J. Kouri for many helpful discussions, and to Mrs. S. Sapir from the Soreq Nuclear Research Center for her devoted work in doing most of the drawings. Finally, I thank my wife Malvine for editing this review and making the project come true.

NONEQUILIBRIUM PHASE TRANSITIONS AND CHEMICAL INSTABILITIES

D. WALGRAEF,* G. DEWEL,* AND P. BORCKMANS*

Service de Chimie-Physique II
Université Libre de Bruxelles
Brussels, Belgium

CONTENTS

I. INTRODUCTION

The physical and biological worlds provide us with an overabundance of systems presenting instabilities leading to highly ordered states. Therefore the understanding of why order may appear spontaneously and which structures are selected among a large manifold of possibilities has become a major branch of research both experimentally and theoretically.[1-3]

Among the most analyzed systems are the laser transitions, electrical current, hydrodynamic, and nonlinear chemical instabilities. Various biophysical situations have also been considered.

*Chercheurs Qualifiés au Fonds National de la Recherche Scientifique de Belgique.

In the threshold region, because of the length and time-scale separation in the dynamics of the modes undergoing instabilities as compared to the other macroscopic variables, analogies with phase transitions have often been stressed.[1,2]

One main line of theoretical approach is based on this similarity, sometimes even isomorphism (see, however, Refs. 4 and 5), with equilibrium phase transition phenomena, the study of which has met in the recent years with much success.

For instance, the theory of the renormalization group has furnished techniques to assess the influence of the inhomogeneous fluctuations near critical points.[6] Besides computational techniques for the critical indices, it leads to a deeper understanding of the concept of universality and shows that the degree to which order is modified by the fluctuations depends strongly on the spatial dimensions of the system.[7]

The study of degenerate ordered systems exhibiting the breaking of a continuous symmetry has also been clarified, and the concept of spontaneous symmetry breaking has allowed the derivation of general relations for the correlation functions independently of any approximation.[8,9]

Because of the separation in the time scales, the dynamics near instability is purely that of the unstable mode σ, which plays the role of *order parameter*. The fluctuations of this mode are shown in many cases to satisfy an equation of motion similar to the time-dependent Ginzburg-Landau equation, which is used to describe critical dynamics near equilibrium.[6]

$$\partial_t \sigma_{\mathbf{k}} = -\frac{\delta \mathscr{F}}{\delta \sigma_{-\mathbf{k}}} + \eta_{\mathbf{k}} \qquad (|\mathbf{k}| < \Lambda) \tag{1.1}$$

Here the noise term $\eta_{\mathbf{k}}$ represents the contributions of the modes of length scale smaller than the cutoff length Λ.

For example, the general potential functional $\mathscr{F}$ is in the simplest case (d = space dimensionality):

$$\mathscr{F} = \int dr^d \left[\frac{1}{2} \left(\frac{A_c - A}{A_c} \right) \sigma^2(\mathbf{r}) + \frac{1}{4} u \sigma^4(\mathbf{r}) + \frac{1}{2} c (\nabla \sigma(\mathbf{r}))^2 \right] \tag{1.2}$$

Retaining only the quadratic terms in σ in (1.2) defines the so-called *Gaussian model*, in which case (1.1) is a linear equation.

In the *mean field approximation* the value of the order parameter is determined by the minimum of $\mathscr{F}$. Its behavior near instability is then

$$|\sigma| = 0, \qquad A < A_c$$

$$|\sigma| \simeq \left(\frac{A - A_c}{A_c} \right)^{1/2} \qquad A > A_C \tag{1.3}$$

where A is the externally controlled parameter that characterizes the instability and thus plays the role of temperature in equilibrium systems. One also gets the "critical slowing down" of the fluctuations of the order parameter, that is, the inverse lifetime τ behaves as $|A-A_c|^{-1}$. The enhancement of these fluctuations ($\langle |\tau|^2 \rangle \simeq |A-A_c|^{-1}$) is exhibited by "critical opalescence," and the appearance of long-range order as the correlation length goes as $\xi \simeq \xi_0 (A_c/|A-A_c|)^{1/2}$. It is mainly at this level that the formal analogy with equilibrium phase transitions has been developed.

However one expects deviations from the mean field behavior due to the inhomogeneous fluctuations when A approaches A_c. Indeed there is experimental evidence that in laserlike systems the transition is rounded off in the threshold region,[10] whereas in the case of the Bénard instability the effects seem to depend strongly on the aspect ratio.[11] Controllable systems are scarce in chemistry and biology,[1,12] and quantitative results are not yet at hand.

But not knowing whether the effects of inhomogeneous fluctuations will ultimately be observed should not deter us from studying their influence, to learn whether they modify the picture qualitatively. That is, do they shift the location of the instability point, and do they introduce further selection mechanisms among the various possible structures?

It may be shown for model (1.2) that the contributions of the fluctuations to quantities showing critical behavior go as follows:

$$\frac{u}{c^{d/2}} \left| \frac{A-A_c}{A_c} \right|^{(d-4)/2} \tag{1.4}$$

This is the so-called *Ginzburg criterion*,[6] which tells us that for $d>4$ the fluctuations are negligible, even in the close vicinity of the critical point ($A \to A_c$), while for $d<4$, when one approaches the critical point, the mean field theory breaks down when

$$\left| \frac{A_c-A}{A_c} \right| \ll \left(\frac{u}{c^{d/2}} \right)^{2/(4-d)} = r_L$$

The fluctuations then become important, leading to the now well-known nonclassical behavior in equilibrium critical phenomena.

The finiteness of the size of the system may also influence this behavior. Indeed the correlation length is bounded by the spatial extension of the system

$$\xi^2 \simeq \frac{cA_c}{|A-A_c|} < L^2 \tag{1.5}$$

It follows that when $|(A-A_c)/A_c| < c/L^2$, the system behaves coherently

314 D. WALGRAEF, D. DEWEL, AND P. BORCKMANS

over the entire container; this is the so-called *zero-dimensional regime* defined by $|(A-A_c)/A_c| \ll r_{0d} = c/L^2$, which exhibits rounding-off effects near the transition, whereas it is essentially analogous to mean field elsewhere. For most equilibrium situations r_{0d}, which is the squared ratio of a microscopic (the characteristic length of the interactions) to a macroscopic length, is much smaller than r_L, which is related to the ratio of microscopic quantities. Moreover, if r_E is the extreme experimental accessible value of $|(A-A_c)/A_c|$, one has $r_{0d} < r_E < r_L$. The deviations from the mean field theory are thus essential in this case because they are of experimental relevance. Let us however remark that for the superconducting transition one has $r_{0d} < r_L \sim r_E$, and for this transition the mean field theory is well adapted to all experimental situations.

For laserlike systems, which may also be described by Ginzburg-Landau potentials. ξ_0 is related to the coherence length of the electromagnetic field, which is macroscopic, and one has $r_E < r_{0d}$ leading to the zero-dimensional behavior of such systems, whereas the hydrodynamic instabilities require further clarification.

In the following we try to assess the influence of inhomogeneous fluctuations on the behavior of nonlinear chemical systems around various classes of instabilities. Because the concept of universality seems to be extendable, our results may be of some relevance for instabilities of other types, as well.

A. The Reaction-Diffusion Equations

We thus consider open systems in which a set of chemical reactions is occurring. For the sake of simplicity we suppose isothermal and mechanical equilibrium. The phenomonological rate equations for the local concentrations of the intermediate species $\{X_i\}$ then take the form

$$\partial_t X_i = f_i(\{X_j\}, \lambda) + D_i \nabla^2 X_i \qquad (1.6)$$

where $\{D_i\}$ are the diffusion coefficients, $\{f_i\}$ the rates of change of $\{X_i\}$ due to chemical reactions, and λ stands for a set of parameters describing the external constraints. We assume natural boundary conditions (infinite systems or periodic conditions). When the external constraints are increased beyond the linear regime around thermal equilibrium, which corresponds for the homogeneous steady state $\{X_i^0\}$ to the law of mass action, various instabilities may arise.

These are located by analyzing the linear stability of the reference state $\{X_i^0\}$. Linearizing (1.6), one obtains

$$\partial_t x_{\mathbf{q}}^i(t) = L_{ij}(q^2, \lambda) x_{\mathbf{q}}^j(t) \qquad (1.7)$$

where

$$x_{\mathbf{q}}^i(t) = X_{\mathbf{q}}^i(t) - X_i^0 = \frac{1}{L^{d/2}} \int dr^d e^{i\mathbf{q}\cdot\mathbf{r}} x^i(\mathbf{r})$$

with

$$L_{ij}(q^2, \lambda) = \left(\frac{\partial f_i}{\partial x_j} \right)^0 - q^2 D_i$$

The stationary state is asymptotically stable if all the eigenvalues of the matrix L_{ij} have negative real parts. Among the various instabilities that may develop when λ is varied, this chapter analyzes the following three cases:

1. L_{ij} has one eigenvalue $\omega_{\mathbf{q}}(\lambda) = 0$ for $q = 0$ and $\lambda = \lambda_c$ corresponding to transitions between homogeneous steady states.
2. L_{ij} has one eigenvalue $\omega_{\mathbf{q}}(\lambda) = 0$ for $|\mathbf{q}| = q_c \neq 0$ and $\lambda = \lambda_c$, this soft-mode instability induces the formation of stationary periodic structures (Turing instability).
3. L_{ij} has two complex conjugate eigenvalues with zero real part when $\lambda = \lambda_c$, $q = 0$. This hard-mode instability may lead to time-periodic solutions of the limit-cycle type.

In the region where the instability develops many authors have stressed the need to include the local fluctuations in the description of systems in which a coherent behavior takes place on the macroscopic level.[1,2] In chemical systems, the fluctuations have been mainly analyzed using the master equation approach. The main idea is to appeal explicitly to the chemical and diffusion mechanism underlying (1.6) and to construct a Markov process in an appropriate phase space. The usual rules for constructing this process are to model diffusion as a random walk and to view chemical reactions as birth and death processes. One then obtains a multivariate master equation that is presently the subject of intensive studies.[13] We here adopt another point of view and the method that has been successfully used to describe equilibrium phase transitions.[6] Following the Landau-Lifshitz theory of fluctuating hydrodynamics,[14] we add appropriate Langevin noise terms $\eta_i(r, t)$ to the balance equations (1.6). Because they take care of the rapid small-scale effects, we assume that

$$\langle \eta_i(\mathbf{r}, t)\eta_j(\mathbf{r}', t') \rangle = 2\left[\Gamma_{ij}^c(\mathbf{r}) + \nabla_i^2 \Gamma_{ij}^d(\mathbf{r}) \right] \delta(\mathbf{r} - \mathbf{r}')\delta(t - t') \qquad (1.8)$$

and also that the system remains in a state of local equilibrium throughout the entire transition region. The coefficients Γ^c and Γ^d may be determined by a local fluctuation dissipation theorem.[15] Indeed the nonequilibrium phase transitions involve macroscopic disturbances that do not modify the fluctuations on a microscopic scale where the short-scale equilibrating processes remain efficient in maintaining local equilibrium. This line of reasoning then transforms (1.6) into an equation of the structure of (1.1), and the methods outlined in the introductory paragraphs are at hand to pursue the analysis.

This review is written in the same spirit as the recent one by P. H. Richter et al.[16]; however the analogy with equilibrium phase transitions is explicitly exploited.

II.   CRITICAL PHENOMENA IN MULTIPLE STEADY-STATE SYSTEMS

Second-order-type transitions between homogeneous steady states may arise in various nonlinear chemical schemes. As a prototype of hard transition we consider the Schlögl model,[1] defined as

$$A + 2X \underset{k_2}{\overset{k_1}{\rightleftharpoons}} 3X$$

$$B + X \underset{k_4}{\overset{k_3}{\rightleftharpoons}} C \tag{2.1}$$

The concentration x of species X is the only variable, since the other concentrations are kept constant by appropriate feeding. The reactions are assumed to be isothermic.

Using standard notations,[18] the deterministic rate equation for the Fourier transform $x_q(t)$ of the local concentration may be written in the case of an ideal mixture:

$$\partial_{\bar{\tau}} x_q(\bar{\tau}) = -\left[(3+\delta)a^2 + q^2 \overline{D}\right] x_q + L^{-d/2} 3a \sum_k x_{q-k} x_k$$

$$- L^{-d} \sum_k \sum_{k'} x_k x_{k'} x_{q-k-k'} + L^{d/2}(1+\delta')a^3 \delta_{q,0} \tag{2.2}$$

where

$$\bar{\tau} = k_2 t; \qquad \frac{k_1}{k_2} = 3, \qquad \frac{k_3 b}{k_2} = (3+\delta)a^2, \qquad \frac{k_4 C}{k_2} = (1+\delta')a^3$$

$\overline{D} = D_x/k_2$ is the diffusion coefficient of the intermediate species, and a, b are the concentrations of the constraint species A, B.

The nonequilibrium critical point, where the three stationary solutions of (2.2) coalesce, is characterized by $\delta = \delta' = 0$, $x_c = a$. The fluctuations $\sigma_q =$

$(x_q/a)-\delta_{q,0}$ around this point obey the following equation:

$$\partial_\tau \sigma_q = -\left[\delta+q^2 D\right]\sigma_q + L^{d/2}(\delta'-\delta)\delta_{q,0}$$

$$+L^{-d}\sum_k \sum_{k'}\sigma_k \sigma_{k'}\sigma_{q-k-k'}+\eta_q(\tau) \qquad \left(a^2\bar{\tau}=\tau,\, D=\frac{\overline{D}}{a^2}\right) \quad (2.3)$$

The noise $\eta_q(\tau)$ was introduced as discussed in the preceding section.

Because the contribution of the diffusion processes to the noise in (1.8) is negligible in the long-wavelength limit ($q\to 0$), it may be assumed that

$$\langle \eta_q(\tau)\eta_{q'}(\tau')\rangle = 2\Gamma \delta_{q,-q'}\delta(\tau-\tau') \qquad (2.4)$$

with $\Gamma \sim a^{-1}=$constant. Then, together with (2.3), this leads to the following Fokker-Planck equation for the distribution of the fluctuations:

$$\partial_\tau P(\{\sigma_k\},\tau)=\sum_k \frac{\partial}{\partial \sigma_k}(\delta+Dk^2)\sigma_k P+L^{-d}\sum_{kk'k''}$$

$$\times \frac{\partial}{\partial \sigma_k}\sigma_{k-k'-k''}\sigma_{k'}\sigma_{k''}P - \frac{\partial}{\partial \sigma_0}L^{-d/2}(\delta'-\delta)P + \sum_k \frac{\partial^2}{\partial \sigma_k \partial \sigma_{-k}}\Gamma P$$

$$(2.5)$$

Its stationary solution takes the form

$$P_{st}(\{\sigma_k\})=\mathfrak{N}\exp\left\{-\mathfrak{F}(\{\sigma_k\})\right\} \qquad (2.6)$$

where $\mathfrak{N}$ is a normalization constant and

$$\mathfrak{F}(\{\sigma_k\})=\frac{1}{\Gamma}\left\{\sum_k \left[\frac{\delta}{2}+\frac{Dk^2}{2}\right]|\sigma_k|^2 \right.$$

$$\left. +\frac{1}{4}L^{-d}\sum_k \sum_{k'}\sum_{k''}\sigma_k \sigma_{k'}\sigma_{k''}\sigma_{-k-k'-k''}+L^{d/2}(\delta-\delta')\sigma_0\right\} \quad (2.7)$$

Recently, Nicolis and Malek-Masour[10] analyzed the multivariate master equation for the same model, using a singular perturbation approach. They have shown that near the critical point, the system may be described by a generalized potential identical to (2.7). The kinetic potential obtained by integrating the differential form $d_x P$ occurring in the Glansdorff-Prigogine[19] criterion also reduces to the functional (2.7), sufficiently close to the instability ($\delta, \delta' \to 0$).

A. Critical Behavior[20]

The analogy of the potential (2.7) with the Ginzburg-Landau Hamiltonian is striking. As a result of this isomorphism, the results of the renormalization group methods can straightforwardly be applied to evaluate the nonclassical exponents characterizing the critical behavior for d less than the critical dimensionality $d_c = 4$.[6,7] For instance, at the order two we get in $\varepsilon = 4 - d$:

$$\langle \sigma_q \rangle = (-\delta)^\beta \delta_{q,0} \quad \text{if} \quad \delta < 0, \quad \delta = \delta'$$

and

$$\beta = \frac{1}{2} - \frac{\varepsilon}{6} + \frac{\varepsilon^2}{18} + O(\varepsilon^3)$$

$$\langle \sigma_q \rangle = (\delta')^{1/\bar{\delta}} \delta_{q,0} \quad \text{for} \quad \delta = 0$$

with

$$\bar{\delta} = 3 + \varepsilon + \frac{25}{54}\varepsilon^2 + O(\varepsilon^3) \tag{2.8}$$

Also as $\delta \to 0$ ($\delta = \delta'$) the correlation function $g(q) = \langle \sigma_q \sigma_{-q} \rangle$ satisfies the following scaling relation:

$$g(q) = \frac{1}{\delta^\gamma} D\left(\frac{q^2}{\delta^{2\gamma}} \right)$$

with

$$\gamma = 1 + \frac{\varepsilon}{6} + \frac{25}{36}\varepsilon^2 + O(\varepsilon^3)$$

and

$$\nu = \frac{1}{2} + \frac{\varepsilon}{12} + \frac{7}{142}\varepsilon^2 + O(\varepsilon^3) \tag{2.9}$$

A similar analysis can be applied to other multiple steady-state systems such as photothermal instabilities,[21] or the Edelstein model,[22] where an adiabatic elimination[2] of the stable modes must be performed to derive the generalized potential. In principle one can transpose the concept of universality classes to nonequilibrium critical phenomena: each static class of universality is determined by the space dimensionality and the number n of unstable modes at the critical point. However all the multiple steady-state chemical models studied up to now belong to the class $n = 1$, isomorphic to the Ising model. The domain of validity of the mean field theory can be estimated by the Ginzburg criterion (1.4). In the notation of (2.7) it takes the form:

$$\frac{\Gamma}{l_0^3 \delta^{1/2}} < 1 \qquad (2.10)$$

where $l_0 = D^{1/2} = (D_x \cdot \tau_{chem})^{1/2}$ and $\tau_{chem} = (a^2 k_2)^{-1}$ is a characteristic chemical relaxation time.

As a consequence, the width of the nonclassical critical region, $\delta_L \sim \Gamma^2 l_0^{-6}$, is reduced with respect to the case of equilibrium phase transitions. Indeed as we argued in the introduction, Γ is of the same order of magnitude as in equilibrium systems. On the contrary, the correlation length, far from the instability, is much greater because of the macroscopic character of the transition.

Only for very fast chemical reactions and slow diffusion might nonclassical behavior be experimentally attainable. Nevertheless, because of the large variability of the chemical rate constants, chemical instabilities remain the best candidate for the observation of nonequilibrium, nonclassical indices.[23]

B. Finite Size Effects

When diffusion becomes more efficient than the chemical reactions, l_0 is very large and finite size effects rapidly become important. Indeed when $\delta = \delta_{0d} = (l_0/L)^2$ (L being a characteristic length of the reactor), the main contribution in (2.7) comes from the spatially uniform fluctuations ($\sigma_{q=0}$). The order parameter then behaves coherently over the whole system. The static correlation function may be calculated exactly with the zero-dimensional potential

$$\mathcal{F}_0 = \frac{L^d}{\Gamma} \left[\frac{\delta}{2} \sigma^2 + \frac{1}{4} \sigma^4 + (\delta - \delta')\sigma \right] \qquad (2.11)$$

where $\sigma = L^{-d/2} \sigma_{q=0}$.

From this, when $\delta = \delta'$, we get

$$\langle \sigma^2 \rangle = \frac{1}{2} \left(\frac{2\Gamma}{L^d} \right)^{1/2} \frac{D_{-3/2}\left(\delta \left(L^d/2\Gamma \right)^{1/2} \right)}{D_{-1/2}\left(\delta \left(L^d/2\Gamma \right)^{1/2} \right)} \qquad (2.12)$$

where $D_{-n/2}(x)$ is a parabolic cylinder function.[24]

Using the asymptotic expansions of these functions, one obtains

$$\text{for}\quad \delta\left(\frac{L^d}{2\Gamma}\right)^{1/2}\gg 1,\quad \begin{aligned}&\delta>0:\langle\sigma^2\rangle\simeq\frac{\Gamma}{\delta L^d}\\&\delta<0:\langle\sigma^2\rangle\simeq|\delta|\end{aligned}$$

$$\text{for}\quad \delta\left(\frac{L^d}{2\Gamma}\right)^{1/2}\ll 1,\quad \langle\sigma^2\rangle\simeq 2\left(\frac{\Gamma}{L^d}\right)^{1/2}\frac{\Gamma(\frac{3}{4})}{\Gamma(\frac{1}{4})}\tag{2.13}$$

This $L^{-d/2}$ dependence has previously been derived for the model using the master equation approach.[18] Away from the critical point, this zero-dimensional transition looks like mean field, however it becomes rounded when $\delta\sim(\Gamma/L^d)^{1/2}$. Similar results have been obtained in the zero dimension for complex-order parameters in the cases of the single-mode laser threshold, small superconducting particles, and the Bénard instability.[25,30]

In the presence of an "external field" ($\delta-\delta'\neq 0$), the mean value of the order parameter is different from zero and may be evaluated in various limits. Here also $(\Gamma/L^d)^{1/2}$ appears as the borderline of the mean field behavior. However it may be seen that the mean value of the order parameter obeys the following finite-length scaling relation

$$\langle\sigma\rangle=(\delta'-\delta)^{1/3}f\left(\frac{L^d\delta^2}{\Gamma},\frac{(\delta'-\delta)^{2/3}}{\delta}\right)\tag{2.14}$$

For most equilibrium phase transitions, one has $\delta_{0_d}\ll\delta_L$ and the size effects may be safely neglected, whereas for many nonequilibrium instabilities one expects $\delta_L\ll\delta_{0_d}$, which may prevent the observation of the effects due to inhomogeneous fluctuations.

III.   PHASE TRANSITIONS TO NONUNIFORM STEADY STATES[26]

We now consider the soft-mode instability of frozen spatial periodic structures. Because it is a symmetry-breaking transition, the fluctuations are supposed to play an important role. Indeed at equilibrium long-range fluctuations develop in the ordered phase of systems which have undergone a phase transition with the breakdown of a continuous symmetry group of the Hamiltonian.[9] These fluctuations deeply influence the properties of the system. For instance, they suppress the long-range order in one and two dimensions.[27] To study the possibility of occurrence of similar effects in nonequilibrium systems, we consider a simple two-variable model: the trimolecular model,[1] which corresponds to the "Brusselator" reaction scheme:

$$A \xrightarrow{k_1} X$$

$$B + X \xrightarrow{k_2} Y + D$$

$$2X + Y \xrightarrow{k_3} 3X$$

$$X \xrightarrow{k_4} E$$

The system is driven by keeping A in excess, whereas D and E are instantly removed. The concentration of the injected B then plays the role of λ, the control parameter. If we use the standard scaled variables, the rate equations for the local concentrations $(\bar{x}_q, \bar{y}_q)$ are:

$$\partial_t \bar{x}_q(t) = a\delta_{q,0} - (b + 1 + q^2 D_x)\bar{x}_q(t) + L^{-d}\sum_{k,k'} \bar{x}_k(t)\bar{x}_{k'}(t)\bar{y}_{-k-k'}(t)$$

$$\partial_t \bar{y}_q(t) = b\bar{x}_q(t) - q^2 D_y \bar{y}_q(t) - L^{-d}\sum_{k,k'} \bar{x}_k(t)\bar{x}_{k'}(t)\bar{y}_{-k-k'}(t) \qquad (3.1)$$

The physical mechanism responsible for the instability is the enhancement by the autocatalytic step of local fluctuations in the concentration of X; on the other hand, the diffusion term tends to reestablish the homogeneous state. The dynamic equilibrium between these two competing processes leads to spatial periodic structures. This simple model constitutes a prototype for the onset of symmetry-breaking instabilities. Turing's instability has been extensively studied by Prigogine et al.[1] It has been proposed as an explanation for the onset of order in biological systems. Indeed these autocatalytic mechanisms are found frequently in biochemical systems.

The inhomogeneous fluctuations around the uniform steady state $x_0 = a$, $y_0 = b/a$ satisfy the following kinetic equations:

$$\partial_t \begin{pmatrix} x_q \\ y_q \end{pmatrix} = K_q \begin{pmatrix} x_q \\ y_q \end{pmatrix} + \begin{pmatrix} +1 \\ -1 \end{pmatrix} N_q \qquad (3.2)$$

where

$$x_q = \bar{x}_q - a\delta_{q,0} \quad \text{and} \quad y_q = \bar{y}_q - \left(\frac{b}{a}\right)\delta_{q,0}$$

$$K_q = \begin{pmatrix} b - 1 - q^2 D_x & a^2 \\ -b & -a^2 - q^2 D_y \end{pmatrix}$$

$$N_q = L^{-d/2}\sum_k x_k \left(\frac{b}{a}x_{q-k} + 2a y_{q-k}\right) + L^{-d}\sum_{k,k'} x_k x_{k'} y_{q-k-k'} \qquad (3.3)$$

322   D. WALGRAEF, D. DEWEL, AND P. BORCKMANS

when

$$\left(\frac{D_x}{D_y}\right)^{1/2}=\eta<\frac{1}{a}\left[(1+a^2)^{1/2}-1\right]$$

linear stability analysis predicts that a soft transition will appear as the first instability.

Indeed for $b=b_c=(1+a\eta)^2$, $q^2=q_c^2=a/(D_xD_y)^{1/2}$, one of the roots of the dispersion equation

$$\omega^2+\omega\left[\frac{a}{\eta}(1+a\eta)(1-\eta^2)+(q^2-q_c^2)(D_x+D_y)+b_c-b\right]$$
$$+\left[(q^2-q_c^2)^2D_xD_y+(b_c-b)q^2D_y\right]=0 \quad (3.4)$$

vanishes, while the other one remains negative. If one calls these frequencies ω_S and ω_R, respectively, the corresponding eigenmodes are:

$$S_{\mathbf{q}}=(\omega_R(\mathbf{q})-\omega_S(\mathbf{q}))^{-1}\left[(b-1-q^2D_x-\omega_S(\mathbf{q}))y_{\mathbf{q}}+bx_{\mathbf{q}}\right]$$
$$R_{\mathbf{q}}=(\omega_S(\mathbf{q})-\omega_R(\mathbf{q}))^{-1}\left[(b-1-q^2D_x-\omega_R(\mathbf{q}))y_{\mathbf{q}}+bx_{\mathbf{q}}\right] \quad (3.5)$$

In the vicinity of the instability point ω_S and ω_R behave as

$$\omega_S(\mathbf{q})\simeq-\frac{1+a\eta}{1-\eta^2}\left[\frac{b_c-b}{b_c}+\frac{D_x}{q_c^2b_c}(q^2-q_c^2)^2\right]$$
$$\omega_R(\mathbf{q})\simeq\frac{a}{\eta}(1+a\eta)(1-\eta^2) \quad (3.6)$$

and the slow mode

$$S_{\mathbf{q}}=\left[\frac{1+a\eta}{a\eta}x_{\mathbf{q}}+y_{\mathbf{q}}\right]\frac{\eta^2}{\eta^2-1} \quad (3.7)$$

becomes unstable and shows critical slowing down.

In this Gaussian description, the transition behavior is characteristically second order.[6] As in similar problems, we expect that the dynamics of the system will entirely be governed, for sufficiently long times, by the slow-mode dynamics. The critical modes are all the $S_{\mathbf{q}}$ with $|\mathbf{q}|$ near q_c and are denoted $\bar{S}_{\mathbf{q}}$. Indeed the phase space associated with these modes is described in the reciprocal space by a thin spherical shell defined by $|q^2 - q_c^2| \ll q_c^2$. After diagonalization of the linear evolution matrix, the kinetic equations (4.2) become

$$\partial_t S_{\mathbf{q}} = \omega_S(\mathbf{q}) S_{\mathbf{q}} + \frac{\omega_S(\mathbf{q}) + 1 + q^2 D_x}{\omega_R(\mathbf{q}) - \omega_S(\mathbf{q})} NL_{\mathbf{q}}(\{S_{\mathbf{k}}, R_{\mathbf{k}}\})$$

$$\partial_t R_{\mathbf{q}} = \omega_R(\mathbf{q}) R_{\mathbf{q}} + \frac{\omega_R(\mathbf{q}) + 1 + q^2 D_x}{\omega_S(\mathbf{q}) - \omega_R(\mathbf{q})} NL_{\mathbf{q}}(\{S_{\mathbf{k}}, R_{\mathbf{k}}\}) \qquad (3.8)$$

The nonlinear term NL_q couples slowly and rapidly decaying modes $(|\omega_S(q \sim q_c, b \sim b_c)| \ll |\omega_R(q \sim q_c, b \sim b_c)|)$. The fast modes may be adiabatically eliminated on the characteristic time scale of the evolution of the slow modes. The dynamics of the system will then be given by the kinetic equation for $\bar{S}_{\mathbf{q}}$:

$$\partial_t \bar{S}_{\mathbf{q}} = -\frac{1 + a\eta}{1 - \eta^2} \left[\frac{b_c - b}{b_c} + \frac{D_x}{q_c^2 b_c} (q^2 - q_c^2)^2 \right] \bar{S}_{\mathbf{q}}$$

$$- \frac{\eta}{a(1 - \eta^2)} NL_{q \sim q_c}(\{\bar{S}_{\mathbf{k}}\}) \qquad (3.9)$$

To extract the slow time behavior of NL_q, let us first remark that there are two kinds of fast mode:

1. When q belongs to the critical shell, the only fast modes are the $\bar{R}_{\mathbf{q}} = R_{\mathbf{q}}(|\mathbf{q}| \sim q_c)$.
2. When q does not belong to the critical shell, either the $S_{\mathbf{q}}$ or the $R_{\mathbf{q}}$ are fast modes because

$$|\omega_S(\mathbf{q})|, |\omega_R(\mathbf{q})| \gg (\omega_S(|\mathbf{q}| \sim q_c))$$

From (3.8) one deduces that

$$\partial_t \bar{R}_{\mathbf{q}} = \omega_R(q_c) \bar{R}_{\mathbf{q}} - \frac{\omega_R + 1 + q^2 D_x}{\omega_S + 1 + q^2 D_x} (\omega_S \bar{S}_{\mathbf{q}} - \partial_t \bar{S}_{\mathbf{q}})$$

or

$$\bar{R}_{\mathbf{q}}(t) = \left[\bar{R}_{\mathbf{q}}(0) - \frac{\omega_R + 1 + q^2 D_x}{\omega_S + 1 + q^2 D_x} \bar{S}_{\mathbf{q}}(0) \right] \exp \omega_R t$$

$$+ \frac{\omega_R + 1 + q^2 D_x}{\omega_S + 1 + q^2 D_x} \left[\bar{S}_{\mathbf{q}}(t) - (\omega_S - \omega_R) \int_0^t d\tau \bar{S}_{\mathbf{q}}(t - \tau) \exp \omega_R \tau \right]$$

$$(3.10)$$

On the time scale $t \sim |\omega_S|^{-1} \gg |\omega_R^{-1}|$ the Markovian approximation leads to

$$\bar{R}_q(t) = -\frac{\omega_S(\omega_R + 1 + q^2 D_x)}{\omega_R(\omega_S + 1 + q^2 D_x)} \bar{S}_q(t) + O(\exp \omega_R t) \qquad (3.11)$$

Using this result, the nonlinear terms NL_q may be split into two parts: the first, $\overline{NL_q}$, depends only on the slow modes $\bar{S}_k$; the other, NL'_q, depends only on the modes R_k and S_k for which k does not belong to the critical shell.

Therefore after the elimination of the $\bar{R}_q$, the following equations are to be considered:

$$\partial_t S_q = \omega_S(q) S_q + \frac{\omega_S(q) + 1 + q^2 D_x}{\omega_R(q) - \omega_S(q)} \left[\overline{NL_q} + NL'_q\right] \qquad (3.12)$$

$$\partial_t R_q = \omega_R(q) R_q + \frac{\omega_R(q) + 1 + q^2 D_x}{\omega_S(q) - \omega_R(q)} \left[\overline{NL_q} + NL'_q\right] \qquad (3.13)$$

$$\partial_t \bar{S}_q = \omega_S(q \simeq q_c) \bar{S}_q + \frac{\omega_S(q \simeq q_c) + 1 + q_c^2 D_x}{\omega_R(q \simeq q_c) - \omega_S(q \simeq q_c)}$$
$$\times \left[\overline{NL}_{q \simeq q_c}(\{\bar{S}_k\}) + NL'_{q \simeq q_c}(\{S_k, R_k\})\right] \qquad (3.14)$$

Because NL'_q contains only the rapidly decaying modes, on the critical time scale $[\exp(\omega_R t) \ll 1]$, (3.12) and (3.13) become

$$S_q = -\frac{\omega_R(\omega_S + 1 + q^2 D_x)}{\omega_S(\omega_R + 1 + q^2 D_x)} R_q, \quad R_q = \frac{\omega_S + 1 + q^2 D_x}{\omega_S(\omega_R - \omega_S)} \overline{NL_q}(\{\bar{S}_k\}) \quad (3.15)$$

which leads directly to:

$$S_q = \frac{\omega_S + 1 + q^2 D_x}{\omega_S(\omega_R - \omega_S)} \left[\frac{a}{\eta^2} \left(\frac{1 - a\eta}{1 + a\eta} - r_0 \right) L^{-d/2} \sum_k \bar{S}_{q-k} \bar{S}_k \right.$$
$$\left. + \frac{a^2}{\eta^2} \frac{1}{1 + a\eta} L^{-d} \sum_k \sum_{k'} \bar{S}_{q-k-k'} \bar{S}_k \bar{S}_{k'} + \cdots \right] \qquad (3.16)$$

These slowly evolving parts of S_q and R_q are then used to evaluate $NL'_{q \simeq q_c}$. Symbolically (3.14) then becomes

$$\partial_t \bar{S}_q = \omega_S(q \simeq q_c) \bar{S}_q + \frac{\omega_S(q_c) + 1 + q_c^2 D_x}{\omega_R(q_c) - \omega_S(q_c)}$$
$$\times \left[NL_{q \simeq q_c}(\{\bar{S}_k\}) + NL'_{q \simeq q_c}(\{\overline{NL}_k\}) + \cdots \right] \qquad (3.17)$$

where the various nonlinear terms can be systematically calculated using (3.2), (3.3), and (3.5).

So on the time scale associated with $\omega_S[|r_0|,(|q^2-q_c^2|)/q_c^2 \ll 1]$, the dynamics of the system is given in the Langevin form by the following time-dependent Ginzburg-Landau equations for the critical modes:

$$\partial_\tau \sigma_q = -\Gamma \frac{\delta \mathcal{F}}{\delta \sigma_{-q}} + \eta_q(\tau) \tag{3.18}$$

where the noise takes care of short-time-scale processes and local events.

The Brazovskii generalized potential $\mathcal{F}$ is

$$\mathcal{F} = \frac{1}{\Gamma} \left\{ \frac{1}{2} \sum_k \left[r_0 + D(k^2 - q_c^2)^2 \right] |\sigma_k|^2 + L^{-d/2} \frac{v}{3!} \sum_k \sum_{k'} \sigma_k \sigma_{k'} \sigma_{-k-k'} \right.$$

$$\left. + L^{-d} \sum_k \sum_{k'} \sum_{k''} \frac{u(k+k')}{4!} \sigma_k \sigma_{k'} \sigma_{k''} \sigma_{-k-k'-k''} \right\}$$

with

$$\sigma_q = \gamma \frac{(\omega_R - \omega_S)(1 + q^2 D_x)}{\omega_R(\omega_S + 1 + q^2 D_x)} \bar{S}_q = \gamma y_q$$

$$\tau = \gamma^{-2} t, \qquad \gamma^{-1} = \left(\frac{q_c^2 b_c D_y}{|\omega_R|} \right)^{1/2} = \left(\frac{1 + a\eta}{1 - \eta^2} \right)^{1/2}$$

$$D = \frac{D_x}{q_c^2 b_c}, \qquad r_0 = \frac{b_c - b}{b_c}$$

$$\langle \eta_q(\tau) \eta_{q'}(\tau') \rangle = 2\Gamma \delta(\tau - \tau') \delta_{q, -q'}$$

$$v = \frac{2\gamma}{|\omega_R(q_c)|} \left[\frac{a}{\eta^2}(1 + a\eta) \right] \left[\frac{1 - a\eta}{1 + a\eta} - r_0 \right] + O\left(\frac{q^2 - q_c^2}{q_c^2}, r_0 \right)$$

$$u(k+k') = \frac{6a^2}{|\omega_R(q_c)|\eta^2(1 + a\eta)} \left[1 + 2\left(\frac{1 - a\eta}{a\eta} \right)(1 - \delta_{|k+k'|, q_c}) \right.$$

$$\left. \cdot \frac{a\eta + 2(1 + \phi_{kk'})((a\eta)^2 - a\eta - 1)}{a\eta(2(1 + \phi_{kk'}) - 1)^2} \right] + O\left(\frac{q^2 - q_c^2}{q_c^2}, r_0 \right)$$

$$\phi_{k,k'} = \frac{k \cdot k'}{q_c^2} \quad (3.19) \tag{3.19}$$

The essential feature of the potential that has occurred in a variety of situations[28-30] lies in its degeneracy. There exists an infinite number of equivalent-order parameters, each associated with the choice of a set of wave vectors of length q_c that depends only on intrinsic parameters of the system $(q_c^2 = a/(D_x D_y)^{1/2})$, not on geometrical variables determining the external boundary conditions. Each set is characterized by the number of vectors and their relative orientations. It may be shown that the nonlinear terms provide a pattern-selection mechanism, already in the mean field picture. One must then test the influence of fluctuations on such structures.

We consider here only the stationary problem, the question of the nucleation of these structures remaining an open problem.

A. The Landau Theory

In the mean field approximation the most probable configurations (which minimize $\mathcal{F}$) are such that $q^2 = q_c^2$. They are therefore characterized by an order parameter built on m pairs of wave vectors $\mathbf{q}_i, -\mathbf{q}_i$

$$\bar{\sigma}_\mathbf{q} = \sum_{i=1}^m \left(\sigma_i \delta_{\mathbf{q}, \mathbf{q}_i} + \sigma_{-i} \delta_{\mathbf{q}, -\mathbf{q}_i} \right)$$

or

$$\bar{\sigma}(\mathbf{r}) = \sum_{i=1}^m 2 \operatorname{Re} \sigma_i \exp i q_c (\mathbf{1}_{q_i} \cdot \mathbf{r})$$

$$= 2 \sum_{i=1}^m |\sigma_i| \cos\left[q_c (\mathbf{1}_{q_i} \cdot \mathbf{r}) + \varphi \right] \qquad (3.20)$$

where the phase φ has to be determined by lateral boundary conditions. Because the system has infinite space extension, the phase is chosen to be $\varphi = 0$ in the following discussion. Then the functional (3.19) becomes

$$V = \frac{r_0}{2} L^{-d/2} \sum_{\mathbf{k}\mathbf{k}'} \bar{\sigma}_\mathbf{k} \bar{\sigma}_{\mathbf{k}'} \delta_{\mathbf{k}+\mathbf{k}',0} + \frac{v}{3!} L^{-d} \sum_{\mathbf{k}\mathbf{k}'\mathbf{k}''} \bar{\sigma}_\mathbf{k} \bar{\sigma}_{\mathbf{k}'} \bar{\sigma}_{\mathbf{k}''}$$

$$\cdot \delta_{\mathbf{k}+\mathbf{k}'+\mathbf{k}'',0} + L^{-3d/2} \sum_{\mathbf{k}\mathbf{k}'\mathbf{k}''\mathbf{k}'''} u(\mathbf{k}+\mathbf{k}') \bar{\sigma}_\mathbf{k} \bar{\sigma}_{\mathbf{k}'} \bar{\sigma}_{\mathbf{k}''} \bar{\sigma}_{\mathbf{k}'''} \delta_{\mathbf{k}+\mathbf{k}'+\mathbf{k}''+\mathbf{k}''',0}$$

$$(3.21)$$

Let us note that the wave vector-dependent terms in V are proportional to $(1 - a\eta)/a\eta$, which must be small to preserve the validity of the adiabatic elimination procedure, and in a first approximation these terms can be neglected and be taken constant. Later it is shown that in this regime these terms do not affect the following results.

1. Second-Order Transitions

We first consider situations where the m pairs are independent. This means that

$$\delta_{k_\alpha + k_\beta + k_\gamma, 0} = 0 \quad \text{and} \quad \delta_{k_\alpha + k_\beta + k_\gamma + k_\delta, 0} = \delta_{k_\alpha + k_\beta, 0} \cdot \delta_{k_\gamma + k_\delta, 0}$$

or permutations thereof. Then putting (3.20) in (3.21) and collecting all terms, one gets

$$V_m = r_0 \sum_{i=1}^{m} |\sigma_i|^2 + \frac{u}{4} \sum_{i=1}^{m} |\sigma_i|^4 + u \sum_{i \neq j} |\sigma_i|^2 |\sigma_j|^2 \tag{3.22}$$

which straightforwardly leads to the equation of state

$$\frac{\partial V_m}{\partial \sigma_{-i}} = h_i = r_0 \sigma_i + \frac{u}{2} |\sigma_i|^2 \sigma_i + u \sum_{j \neq i} \sigma_i |\sigma_j|^2 \tag{3.23}$$

(where the h_i are fictitious symmetry-breaking fields) and the elements of the inverse susceptibility matrix

$$\frac{\partial V_m}{\partial \sigma_i \partial \sigma_{-i}} = r_{i,i} = r_D = r_0 + u \sum_{j=1}^{m} \sigma_j^2$$

$$\frac{\partial V_m}{\partial \sigma_{-i} \partial \sigma_{-i}} = r_{-i,i} = r_{ND} = \frac{u}{2} \sigma_i^2$$

$$\frac{\partial V_m}{\partial \sigma_i \partial \sigma_j} = r_{i,j} = u \sigma_{-i} \sigma_{-j} \tag{3.24}$$

for all $h_i = 0$ the amplitudes are equal to

$$\sigma_i = \sigma = \begin{cases} 0, & b < b_c \\ \left[\dfrac{-2r_0}{(2m-1)u} \right]^{1/2}, & b > b_c \end{cases} \tag{3.25}$$

which is again characteristic of a second-order transition, as indeed the susceptibility diverges at $b = b_c$.

$$r_{i,i} = r_{i,-i} = \frac{1}{2} r_{i,j} = -\frac{r_0}{2m-1} \tag{3.26}$$

From (3.20) and (3.23), the stationary excess concentrations $x(\mathbf{r})$ and $y(\mathbf{r})$ for these structures can be reconstructed, namely,

$$y(\mathbf{r})=2\sum_{i=1}^{m}\left[-\frac{2r_0}{(2m-1)u}\right]^{1/2}\cos\mathbf{q}_i\cdot\mathbf{r}+2\sum_{i,j=1}^{m}\frac{1+(\mathbf{q}_i+\mathbf{q}_j)^2D_x}{D_x\left[(\mathbf{q}_i+\mathbf{q}_j)^2-q_c^2\right]}$$

$$\times\left[\frac{a}{\eta^2}\frac{1-a\eta}{1+a\eta}\right]\left[-\frac{2r_0}{(2m-1)u}\right]\cos(\mathbf{q}_i+\mathbf{q}_j)\cdot\mathbf{r}+\cdots \qquad (3.27)$$

This result shows that the contribution of the higher harmonics to the patterns is negligible in the vicinity of the instability point because they behave as $|r_0|,|r_0|^{3/2},\ldots$.

This behavior predicted by Eckhaus[31] has been confirmed experimentally[32] in the Bénard problem, which may be described by the same kind of generalized potential.[29]

The relative stability of the various phases may be calculated by comparison of the corresponding potentials (3.22). (In this case the uniform phase with $\sigma=0$ has $V_0=0$).

$$V_m=-\frac{m}{2m-1}\frac{r_0^2}{u}<V_0 \qquad (3.28)$$

and

$$V_m>V_{m-1}$$

Therefore the stablest phase corresponds to $m=1$.

However let us remark that for $h_i=0$ the inverse susceptibility matrix elements are such that

$$r_{i,i}=r_D=\frac{u}{2}\sigma^2=r_{ND}=r_{i,-i}$$

$$r_{i,j}=r=u\sigma^2. \qquad (3.29)$$

As a consequence, the determinant of the second derivatives of the potential vanishes in all the ordered phases owing to the breakdown of the translational symmetry. One must then consider the question of the stability of these structures. In the mean field picture the linear stability analysis of the patterns against homogeneous perturbations goes through the study of the eigenvalues of the matrix associated with the linear part of the kinetic equations, that is,

$$
\begin{vmatrix}
\begin{array}{cc|cc|c}
\omega + r_D & r_{ND} & r & r & \cdots \\
r_{ND} & \omega + r_D & r & r & \cdots \\
\hline
r & r & \omega + r_D & r_{ND} & \cdots \\
r & r & r_{ND} & \omega + r_D & \\
\hline
\vdots & \vdots & \vdots & &
\end{array}
\end{vmatrix}
$$

$$
= \omega^m \left[\omega + 2(r_D - r) \right]^{m-1} \left[\omega + 2r_D + 2(m-1)r \right] = 0 \tag{3.30}
$$

For $m > 1$, positive roots ($\omega = -2(r_D - r) = u\sigma^2$) appear, implying the instability of these structures.

The only stable pattern corresponds to the case where the concentration varies periodically in the direction conjugate to q_1, and this corresponds to the roll pattern in the case of the Bénard instability. This structure is also stable against small inhomogeneous perturbations of wave vectors $\mathbf{k}$, since in this case the dispersion equation is:

$$
\omega^2 + \omega \left[2r_D + 8(\mathbf{q}_i \cdot \mathbf{k})^2 + 2k^4 \right] + r_D \left[2k^4 + 8(\mathbf{q}_i \cdot \mathbf{k})^2 \right] + \left[k^4 - (2\mathbf{q}_i \cdot \mathbf{k})^2 \right]^2 = 0 \tag{3.31}
$$

The two roots are negative; the greater behaves to dominant order in k^2 as $\omega(\mathbf{k}) = -[4(q_i k)^2 \cos^2 \theta + k^4]$, θ being the angle between $\mathbf{q}_i$ and $\mathbf{k}$. This behavior is characteristic of systems that have undergone a symmetry-breaking transition, and in this case, owing to the degeneracy of the order parameter, the fluctuations of wave vectors perpendicular to q_i are the slowest to decay.

What is the effect of the k-dependent terms on the picture outlined above? They modify the state equation as follows:

$$
h_i = r_0 \sigma_i + \frac{u_i}{2} |\sigma_i|^2 \sigma_i + \sum_j u_{ij} |\sigma_j|^2 \sigma_i \tag{3.32}
$$

where

$$
u_i = \tfrac{1}{3} \left[2u(0) + u(2\mathbf{q}_i) \right]
$$

$$
u_{ij} = \tfrac{1}{3} \left[u(0) + u(\mathbf{q}_i + \mathbf{q}_j) + u(\mathbf{q}_i - \mathbf{q}_j) \right]
$$

To guarantee stability u must be positive. If it is not, one must go further in the adiabatic elimination to get at least sixth-order terms in the potential, leading to the possibility of multicritical behavior. However we do not consider this case because the positiveness of u is preserved by the condition $(1-a\eta)/a\eta \ll 1$. The smallest values of u are obtained for symmetric arrangements of the q_i vectors in space. The corresponding patterns are thus the stablest, and in this case one obtains for the inverse susceptibility matrix elements

$$r_i = r_D = r_{ND} = \frac{u_i}{2}\sigma^2$$

$$r_{i,j} = u_{ij}\sigma^2 \tag{3.33}$$

with

$$a_i = \sigma = \left[-\frac{2r_0}{2\sum\limits_{j\neq i} u_{ij} + u_i} \right]^{1/2}, \qquad b > b_c$$

The dispersion equation associated to the homogeneous stability analysis is now

$$\omega^m \left(\omega + 2r_D + 2 \sum\limits_{j\neq i} r_{i,j} \right) f\left(\omega, r_D, \{r_{i,j}\} \right) = 0 \tag{3.34}$$

Since the sum of the roots that correspond to the trace of the matrix is

$$\sum\limits_{i=1}^{2m} \omega_i = -2mr_D < 0$$

and

$$\omega_1 = -2r_D - 2\sum\limits_{j\neq i}^{m} r_{i,j} < 0$$

the sum of the remaining roots appearing in f is $2[\sum_{j\neq i} r_{i,j} - (m-1)r_D]$: this is positive because r_{ij} is always greater than r_D. So except for $m=1$ there is at least one positive root; again these structures are unstable, and the previous conclusions are recovered.

When $r_0 < 0$, structures with wave vectors $\mathbf{q}$ of length different from q_c may also appear, provided $r_0 + D(q^2 - q_c^2)^2 < 0$. Their stability properties may be studied as when $|\mathbf{q}_i| = q_c$ through the state equation, which becomes

$$h_i = \left[r_0 + D\left(q^2 - q_c^2\right)^2 \right] \sigma_i + \frac{u}{2} |\sigma_i|^2 \sigma_i + u \sum_{j \neq i} |\sigma_j|^2 \sigma_i \qquad (3.35)$$

and the inverse susceptibility matrix elements are

$$r_{i,i} = r_0 + D\left(q^2 - q_i^2\right)^2 + u \sum_j \sigma_j^2$$

$$r_{i,-i} = \frac{u}{2} \sigma_i^2$$

$$r_{i,j} = u\sigma_{-i}\sigma_{-j}. \qquad (3.36)$$

Moreover when $h_i = 0$, the amplitude of the patterns is given by

$$\sigma_i = \left[\frac{-2\left(r_0 + D\left(q^2 - q_i^2\right)^2\right)}{(2m-1)u} \right]^{1/2} \qquad (3.37)$$

while the potential is

$$V_m = -\frac{m}{2m-1} \frac{\left(r_0 + D\left(q^2 - q_c^2\right)^2\right)^2}{u} \qquad (3.38)$$

which is minimum for $m = 1$ and $q = q_c$.

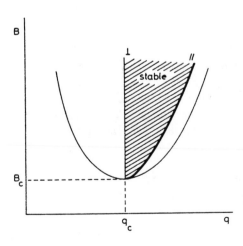

Fig. 1. Sketch of the stability diagram for the structures periodic in one dimension. The stability zone is limited by transverse ($\perp$) and longitudinal ($\parallel$) fluctuations.

Following the same lines as above it may be shown directly that the only stable structure against inhomogeneous fluctuations is the $m=1$ "roll" pattern. The stability of this structure against inhomogeneous fluctuations may be studied through the following dispersion equation:

$$\omega^2 + \omega\left[-2\left(r_0 + D\left(q^2 - q_c^2\right)^2\right) + 2D\left(2k^2\left(q^2 - q_c^2\right)\right.\right.$$
$$\left.+ 4(qk)^2\cos^2\theta + k^4\right)\Big] - k^2\Big[\left(r_0 + D\left(q^2 - q_c^2\right)^2\right)4D$$
$$\left(q^2 - q_c^2 + 2q^2\cos^2\theta\right) + 4D^2\left(q^2 - q_c^2\right)^2\left(2q\cos\theta\right)^2\Big] + O(k^4) = 0 \quad (3.39)$$

It follows that this structure is stable in the region schematized in Fig. 1, which is limited by the line $q^2 = q_c^2$ (stability against transverse fluctuations) and $r_0 = -D(q^2 - q_c^2)^2 (7q^2 - q_c^2)/(3q^2 - q_c^2)$ (stability against longitudinal fluctuations). In conclusion, this structure is the stablest one in the mean field analysis.

2. First-Order Transitions

We consider now the case of patterns in which sets of vectors arrange to form equilateral triangles that lead to nonvanishing contributions of cubic terms of the potential. The simplest one occurs for $m=3$, the triangle itself. The corresponding equation of state in zero field is:

$$h_i = h = \sigma\left(r_0 + v\sigma + \tfrac{5}{2}u\sigma^2\right) = 0 \quad (3.40)$$

(u is in this case $a^2/[|\omega_R|\eta^2(1+a\eta)]$ as $|\mathbf{k}+\mathbf{k}'| = q_c$ for this structure) and leads to first-order subcritical behavior, giving rise to rodlike structures with two-dimensional hexagonal periodicity (cf. Fig. 2):

$$\bar{\sigma}_{tr}(\mathbf{r}) = 2\sigma\left[\cos q_c x + \cos\frac{q_c}{2}\left(x + \sqrt{3}\,y\right) + \cos\frac{q_c}{2}\left(x - \sqrt{3}\,y\right)\right] \quad (3.41)$$

It is expected on the basis of Landau's symmetry arguments that there will be no critical point. The new phase appears for $r_0 = r_{tr} = v^2/10\ u$. However the potential is now $V_{tr} = V_3 + v\sigma^3$, and the stability exchange with the uniform phase takes place at $r_0 = r'_{tr} = 4v^2/45u$. The stability of this structure may also be determined along the same lines as above. Indeed from the following relations, we can write:

$$r_D = r_0 + 3u\sigma^2 = \tfrac{1}{2}u\sigma^2 - v\sigma$$
$$r_{ND} = \tfrac{1}{2}u\sigma^2$$
$$r = u\sigma^2$$
$$\bar{r} = u\sigma^2 + v\sigma \quad (3.42)$$

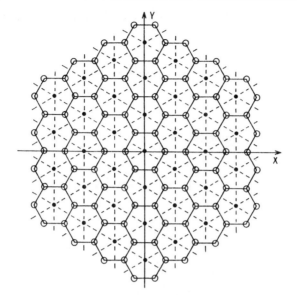

Fig. 2.   Maxima of concentration [cf. (3.41)] in real space ($d=2$) of the structure defined by an equilateral triangle in q-space ($m=3$). Circles and dots correspond to $v>0$ and $v<0$, respectively.

($\bar{r}$ characterizes wave vectors linked in a triangular relation) and $(r_D - r_{ND}) = (r - \bar{r})$. The dispersion relation is

$$\left[\omega + (r_D + r_{ND}) + 2(r + \bar{r})\right]\left[\omega + (r_D + r_{ND}) - (r + \bar{r})\right]^2$$
$$\times \omega^2 \left[\omega + (r_D - r_{ND}) + 2(r - \bar{r})\right] = 0 \quad (3.43)$$

and its roots are $\omega_1 = 0$, $\omega_2 = 3v\sigma$, $\omega_3 = -\frac{2}{5}(r_0 - 4v\sigma)$, $\omega_4 = 2r_0 + v\sigma$. This implies that the solution $\sigma_- = [-v - \sqrt{v^2 - 10r_0 u}]/5u$ is the only stable one against homogeneous fluctuations from $b_1 = b_c - v^2/10u$ to $b_2 = b_c + 8v^2/u$.

This pattern is reminiscent of the Bénard hexagonal cell structure. This analogy may even be carried further. Depending on the sign of v, the maxima of the concentration define, respectively, a triangular ($v<0$) or a honeycomb ($v>0$) lattice corresponding in the convection problem to l- or g-hexagons, respectively, as defined by Busse.[33] Furthermore, for small r_0, the amplitude behaves as $\sigma_- = -2v/5u + r_0/2v$. This linear dependence in r_0 has been experimentally observed by Bergé et al.[34]

For $m=6$, the triangles can be chosen to form an octahedron. The equation of state is then

$$h = \sigma\left(r_0 + 2v\sigma + \frac{11}{2}u\sigma^2\right) = 0 \quad (3.44)$$

It leads also subcritically to structures with three-dimensional cubical symmetry:

$$\bar{\sigma}_{\text{oct}}(\mathbf{r}) = 2\sigma\left[\cos\frac{q_c}{\sqrt{2}}x\cos\frac{q_c}{\sqrt{2}}y + \cos\frac{q_c}{\sqrt{2}}x\cos\frac{q_c}{\sqrt{2}}z + \cos\frac{q_c}{\sqrt{2}}y\cos\frac{q_c}{\sqrt{2}}z\right]$$

$$(3.45)$$

The corresponding parameters are:

$$r_{\text{oct}} = \frac{2}{11}\frac{v^2}{u} \qquad r'_{\text{oct}} = \frac{16}{99}\frac{v^2}{u}$$

while

$$r_D = r_0 + 6u\sigma^2 = \tfrac{1}{2}u\sigma^2 - 2v\sigma$$
$$r_{ND} = \tfrac{1}{2}u\sigma^2$$
$$r = u\sigma^2 \qquad\qquad (3.46)$$
$$\bar{r} = u\sigma^2 + v\sigma$$
$$r_D - r_{ND} = 2(r - \bar{r})$$

The dispersion equation for the homogeneous stability problem is:

$$[\omega + (r_D + r_{ND} + 6r + 4\bar{r})][\omega + (r_D + r_{ND} - 2\bar{r})]^2$$
$$\times[\omega + (r_D - r_{ND} - 2r)]^3[\omega + r_D - r_{ND}]^2$$
$$\times[\omega + r_D - r_{ND} + 2(r - \bar{r})]^2\omega^2 = 0 \qquad (3.47)$$

The study of the roots shows that the solution

$$\sigma_- = \frac{1}{11u}\left[-2v - \sqrt{4v^2 - 22r_0 u}\right]$$

is the only stable one, and this from $b_1 = b_c - 4v^2/22u$ to $b_2 = b_c + 18v^2/u$.

When $v < 0$, the pattern of the concentration maxima forms a body-centered cubic lattice, whereas $v > 0$ leads to a filamental structure with cubic symmetry. These structures are the first to appear, a situation that presents analogies with the theory of the freezing transition, where from experimental observation almost all metals on the left-hand side of the periodic table are known to be body-centered cubic near the melting line at low pressures.[35] The mean field theory predicts the occurrence of a great variety of patterns.

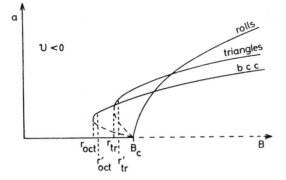

Fig. 3.   Schematic bifurcation diagram in the mean field approximation.

The stability analysis shows that a selection between them occurs. It shows that when $1 - a\eta \neq 0$, various structures may appear through successive first-order transitions from patterns having cubical symmetry to the formation of rolls, while when $a\eta = 1$ the second-order mean field transition occurs from the homogeneous state to roll patterns (Fig. 3). Many analogous analyses for various systems have been carried out; among these we mention only the Bénard problem[33, 36] and morphogenesis,[2] because of their close relation to this work. Moreover we would like to show that the structures emerging from our description are the same as those obtained by a bifurcation analysis of the problem.

B.   The Bifurcation Approach

Bifurcation theory provides another possible approach to the study of small-amplitude periodic patterns[37] and has been widely applied to various far-from-equilibrium situations, including the Brusselator in small reactors.[38]

It is based on the following expansion in powers of the amplitude of the branching solutions for b in the immediate neighborhood of b_c:

$$\begin{pmatrix} x \\ y \end{pmatrix} = \begin{pmatrix} x_0 \\ y_0 \end{pmatrix} \varepsilon + \begin{pmatrix} x_1 \\ y_1 \end{pmatrix} \varepsilon^2 + \cdots$$

$$b = b_c + \varepsilon b^{(1)} + \varepsilon^2 b^{(2)} + \cdots \qquad (3.48)$$

When substituted in (3.1), these expansions generate a sequence of linear inhomogeneous equations

$$L_c \begin{pmatrix} x_n \\ y_n \end{pmatrix} = \begin{pmatrix} -I_n \\ +I_n \end{pmatrix} \qquad (3.49)$$

where

$$L_c = \begin{pmatrix} b_c - 1 - D_x \nabla^2 & a^2 \\ -b_c & -a^2 + D_y \nabla^2 \end{pmatrix} \tag{3.50}$$

and the first few coefficients I_n are

$$I_0 = 0$$

$$I_1 = b^{(1)} x_0 + \frac{b_c}{a} x_0^2 + 2a x_0 y_0$$

$$I_2 = b^{(2)} x_0 + \left(b^{(1)} + \frac{2b_c}{a} x_0 + 2a y_0 \right) x_1$$

$$+ 2a x_0 y_1 + b^{(1)} \frac{x_0^2}{a} + x_0^2 y_0 \tag{3.51}$$

At the first step

$$L_c \begin{pmatrix} x_0 \\ y_0 \end{pmatrix} = 0 \tag{3.52}$$

and therefore $\begin{pmatrix} x_0 \\ y_0 \end{pmatrix}$ is proportional to the right eigenvectors of L_c with zero eigenvalue, that is, to the critical mode (at $b = b_c, |\mathbf{k}_i| = q_c$)

$$e^{i\mathbf{k}_i \cdot \mathbf{r}} \begin{pmatrix} -\dfrac{a}{\eta(1+a\eta)} \\ 1 \end{pmatrix} \tag{3.53}$$

However, because of the degeneracy already mentioned, $\begin{pmatrix} x_0 \\ y_0 \end{pmatrix}$ is written as a linear combination of such eigenvectors[33] as

$$\begin{pmatrix} x_0 \\ y_0 \end{pmatrix} = \sum_i S_{\mathbf{k}_i} e^{i\mathbf{k}_i \cdot \mathbf{r}} \begin{pmatrix} -\dfrac{a}{\eta(1+a\eta)} \\ 1 \end{pmatrix} \tag{3.54}$$

with $S^0_{-\mathbf{k}_i} = S^{*0}_{\mathbf{k}_i}$ and $\Sigma_i |S_{\mathbf{k}_i}|^2 = 1$, for $\begin{pmatrix} x_0 \\ y_0 \end{pmatrix}$ to be real.

At the next order

$$L_c \begin{pmatrix} x_1 \\ y_1 \end{pmatrix} = \begin{pmatrix} -I_1 \\ +I_1 \end{pmatrix} \tag{3.55}$$

and the solvability condition requires that the inhomogeneous term be orthogonal to the left eigenvectors of L_c with zero eigenvalue, thus

$$\int d\mathbf{r} \frac{\eta^2}{\eta^2-1} e^{i\mathbf{k}_i \cdot \mathbf{r}} \left(\frac{1+a\eta}{a\eta}, 1 \right) \begin{pmatrix} -I_1 \\ +I_1 \end{pmatrix} = 0 \tag{3.56}$$

leading to

$$b^{(1)} S_\mathbf{p} - \left(\frac{1-a\eta}{a\eta} \right) \sum_{\mathbf{q}\mathbf{q}'} S_\mathbf{q} S_{\mathbf{q}'} \delta_{\mathbf{q}+\mathbf{q}',\mathbf{p}} = 0 \tag{3.57}$$

which determines $b^{(1)}$. Therefore nonvanishing values of $b^{(1)}$ are possible only when triangular conditions between wave vectors are satisfied.

Next one solves (3.55) using the solvability conditions (3.56), obtaining

$$L_c \begin{pmatrix} x_1 \\ y_1 \end{pmatrix} = \begin{pmatrix} -1 \\ +1 \end{pmatrix} \frac{a}{\eta^2} \frac{1-a\eta}{1+a\eta} \sum_{ij} S_{\mathbf{k}_i} S_{\mathbf{k}_j} e^{i(\mathbf{k}_i+\mathbf{k}_j)\cdot\mathbf{r}}$$

$$\cdot \left[1 - \delta_{|\mathbf{k}_i+\mathbf{k}_j|, q_c} \right] \tag{3.58}$$

Because of the presence of the Laplacian operator in L_c and because of the structure of the inhomogeneous term, $\begin{pmatrix} x_1 \\ y_1 \end{pmatrix}$ is written as

$$\begin{pmatrix} x_1 \\ y_1 \end{pmatrix} = \sum_i \sum_j \begin{pmatrix} F_x(\phi_{ij}) \\ F_y(\phi_{ij}) \end{pmatrix} e^{i(\mathbf{k}_i+\mathbf{k}_j)\cdot\mathbf{r}} S_{\mathbf{k}_i} S_{\mathbf{k}_j} \tag{3.59}$$

where $\phi_{ij} = (\mathbf{k}_i \cdot \mathbf{k}_j)/q_c^2$.

This immediately leads to

$$F_x(\phi_{ij}) = \frac{2a^2}{\eta^3} \frac{1-a\eta}{1+a\eta} \frac{1+\phi_{ij}}{\Delta_{ij}} \left[1 - \delta_{|\mathbf{k}_i+\mathbf{k}_j|, q_c} \right]$$

$$F_y(\phi_{ij}) = -\frac{a}{\eta^2} \frac{1-a\eta}{1+a\eta} \frac{1+2a\eta(1+\phi_{ij})}{\Delta_{ij}} \left[1 - \delta_{|\mathbf{k}_i+\mathbf{k}_j|, q_c} \right]$$

where

$$\Delta_{ij} = a^2 \left[1 - 2(1+\phi_{ij}) \right]^2 \tag{3.60}$$

One then proceeds along similar lines to calculate the second solvability condition, which determines $b^{(2)}$. Summing these two conditions, writing $\bar{a}_i = \varepsilon S_{\mathbf{k}_i}$, and dividing by $(1+a\eta)^2$, one obtains

$$
\frac{b-b_c}{b_c}\bar{a}_p + \sum_{n,n'}\frac{1}{(1+a\eta)\eta}\left[\frac{1-a\eta}{1+a\eta}-\frac{b_c-b}{b_c}\right]\bar{a}_n\bar{a}_{n'}\delta_{\mathbf{k}_n+\mathbf{k}_{n'},\mathbf{k}_p}
$$

$$
+\frac{a}{\eta(1+a\eta)^3}\sum_{n,n',n''}\left[1+2\frac{1-a\eta}{\eta^2}\frac{a\eta+2(1+\phi_{nn'})[(a\eta)^2-a\eta-1]}{\Delta_{nn'}}\right]
$$

$$
\cdot\bar{a}_n\bar{a}_{n'}\bar{a}_{n''}\delta_{\mathbf{k}_n+\mathbf{k}_{n'}+\mathbf{k}_{n''},\mathbf{k}_p}=0 \qquad (3.61)
$$

This equation may also be written using the notation of (3.19)

$$
r_0\bar{a}_p + \sum_{nn'}\frac{v}{2}\gamma\bar{a}_n\bar{a}_{n'}\delta_{\mathbf{k}_n+\mathbf{k}_{n'},\mathbf{k}_p}+\sum_{nn'n''}\frac{u}{3!}\gamma^2\bar{a}_n\bar{a}_{n'}\bar{a}_{n''}\delta_{\mathbf{k}_n+\mathbf{k}_{n'}+\mathbf{k}_{n''},\mathbf{k}_p}=0
$$

$$(3.62)$$

Since $\sigma_i = \gamma\bar{a}_i$, one obtains the mean field approximation of the equation of state with zero external field, and (3.27) for the amplitude may be easily recovered. A similar bifurcation analysis of inhomogeneous solutions to nonlinear reaction-diffusion equations in the infinite plane has recently been carried out by L. Pismen.[39]

C. Effects of Fluctuations

As stated above, it is well known that fluctuations may alter the mean field picture significantly when the system undergoes a transition with a broken symmetry. As a result of the breakdown of the translational symmetry, exact relations may be derived. On the one hand, identities (Ward-Takahashi[40]) between the elements of the correlation matrix of the order parameter express the response of the system to an infinitesimal homogeneous translation. On the other hand, inequalities (Bogoliubov[41]) for the correlations of the long-range fluctuations of the order parameter characterize the response of the system to an infinitesimal inhomogeneous translation. These relations may be explicitly derived from the description of the system by the Landau-Ginzburg-Brazovskii potential (3.19).

The generating functional of the correlation functions $g_{\mathbf{q},\mathbf{q}'}=\langle\sigma_{\mathbf{q}}\sigma_{\mathbf{q}'}^*\rangle$ is

$$
\exp-\mathcal{G}=\int\mathcal{D}\sigma\exp[-\Phi], \qquad \left(\Phi=\mathcal{F}-\sum_{\mathbf{q}}h_{\mathbf{q}}\sigma_{-\mathbf{q}}\right) \qquad (3.63)
$$

where the h_q are fictitious symmetry-breaking fields. We obviously have:

$$\langle \sigma_q \rangle = \frac{\delta \mathcal{G}}{\delta h_{-q}}, \qquad \langle \sigma_q \sigma_{q'}^* \rangle = \frac{\delta^2 \mathcal{G}}{\delta h_{-q} \delta h_{q'}} \tag{3.64}$$

Since Φ is translationally invariant in r-space, for an infinitesimal translation defined by the vector $\boldsymbol{\xi}$ one has

$$\sigma_q' = \sigma_q + i\mathbf{q} \cdot \boldsymbol{\xi} \sigma_q \tag{3.65}$$

then

$$\mathcal{F}\left(\{\sigma_q'\}\right) = \mathcal{F}\left(\{\sigma_q\}\right)$$

and

$$\mathcal{G}\left(\{\sigma_q\}, \{h_q\}\right) = \mathcal{G}\left(\{\sigma_q'\}, \{h_q - i\mathbf{q} \cdot \boldsymbol{\xi} h_q\}\right) \tag{3.66}$$

Let us consider the correlation functions $g_{\mathbf{q}_i, \mathbf{q}_j}$ associated to an arbitrary structure $(|\mathbf{q}_i| = |\mathbf{q}_j| = q_c; \mathbf{q}_i, \mathbf{q}_j \in \{\mathbf{q}_1, -\mathbf{q}_1; \mathbf{q}_2, -\mathbf{q}_2, \cdots, \mathbf{q}_m, -\mathbf{q}_m\})$ with m pairs of wave vectors. Differentiating functionally (3.63) with respect to the symmetry-breaking fields associated with the structure and putting all other fields equal to zero leads to

$$\sum_{i:1}^{2m} \frac{\delta \mathcal{G}}{\delta h_{\mathbf{q}_i}} \mathbf{q}_i \cdot \boldsymbol{\xi} h_{\mathbf{q}_i} = 0 \tag{3.67}$$

and furthermore for any j:

$$\sum_{i:1}^{2m} \left[\frac{\delta^2 \mathcal{G}}{\delta h_{\mathbf{q}_i} \delta h_{\mathbf{q}_j}} \mathbf{q}_i \cdot \boldsymbol{\xi} h_{\mathbf{q}_i} - \frac{\delta \mathcal{G}}{\delta h_{\mathbf{q}_j}} \mathbf{q}_j \cdot \boldsymbol{\xi} \right] = 0 \tag{3.68}$$

In the case of m independent pairs of vectors, this relation has to be satisfied for any $\boldsymbol{\xi}$ and any relative orientation of the $\mathbf{q}_i$; it follows that ($h_{\mathbf{q}_i} = h, \langle \sigma_{\mathbf{q}_i} \rangle = \sigma$)

$$g_{\mathbf{q}_i, \mathbf{q}_i} - g_{\mathbf{q}_i, -\mathbf{q}_i} = \frac{\sigma}{h} = g_D - g_{ND} \tag{3.69}$$

and

$$g_{\mathbf{q}_i, \mathbf{q}_j} = g_{\mathbf{q}_i, -\mathbf{q}_j} = \bar{g}, \qquad \mathbf{q}_i \neq \mathbf{q}_j$$

Moreover the following relation holds ($r_{i,j} = (g^{-1})_{i,j}$):

$$r_D - r_{ND} = (g_D - g_{ND})^{-1} = \frac{h}{\sigma} \qquad (3.70)$$

Since for small $h, \langle \sigma_{q_i} \rangle$ behaves as $\sigma + \alpha h^\varepsilon (\varepsilon > 0)$, the eventual divergence of $g_D + g_{ND}$ or $\bar{g}$ is weaker than the divergence of $g_D - g_{ND}$. It follows that in the limit $h \to 0$, g_D, and g_{ND} behave as h^{-1} while $\bar{g} \ll g_D$.

When the structure is constructed on m pairs of linked vectors (triangles, octahedrons, etc) the situation is somewhat different. For example, in the case $m = 3$ (hexagonal structures) one has

$$(g_D - g_{ND}) + \frac{1}{2}(g_{1,2} - g_{1,-2}) + \frac{1}{2}(g_{1,3} - g_{1,-3}) = \frac{\sigma}{h}$$

$$(g_{1,2} - g_{1,-2}) - (g_{1,3} - g_{1,-3}) = 0 \qquad (3.71)$$

$$g_{1,2} = g_{1,3} = g_{2,3} = g \neq \bar{g} = g_{1,-2} = g_{1,-3} = g_{2,-3}$$

$\mathbf{q}_1, \mathbf{q}_2, \mathbf{q}_3$ being related by the triangle relation

$$\mathbf{q}_1 + \mathbf{q}_2 + \mathbf{q}_3 = 0$$

The Ward identities read in this case:

$$g_D - g_{ND} + (g - \bar{g}) = \frac{\sigma}{h} = [(r_D - r_{ND}) - (r - \bar{r})]^{-1} \qquad (3.72)$$

For $m = 6$ one obtains:

$$g_D - g_{ND} + 2(g - \bar{g}) = \frac{\sigma}{h} = [(r_D - r_{ND}) + 2(r - \bar{r})]^{-1} \qquad (3.73)$$

In these cases g_D, g_{ND}, g, and $\bar{g}$ behave as h^{-1} for small h. Among the structures considered, only those corresponding to the rolls ($m = 1$), the triangles ($m = 3$), and the octahedrons ($m = 6$) are stable as discussed in the preceding paragraph. Let us now study the response of these systems to a long-wavelength deformation of the structure[8]

$$\sigma'(\mathbf{r}) = \sigma(\mathbf{r} - \xi \cos \mathbf{k} \cdot \mathbf{r})$$

where ξ is infinitesimal and $|\mathbf{k}| \ll q_c$, or

$$\sigma'_\mathbf{q} = \sigma_\mathbf{q} + \frac{i}{2}(\mathbf{q} \cdot \xi)(\sigma_{\mathbf{q}+\mathbf{k}} + \sigma_{\mathbf{q}-\mathbf{k}})$$

The corresponding variation of the potential $\delta\Phi_m$ (with u constant) is due to the Laplacian term and to the symmetry-breaking term, but the quantities $\int d\mathbf{r}\,\sigma''(\mathbf{r})$ are invariant for this transformation,

$$
\begin{aligned}
\delta\Phi_m &\equiv U_k(\xi) \\
&= \sum_{j:1}^{m} \frac{i\mathbf{q}_j\cdot\xi}{2}\left\{ h_{\mathbf{q}_j}(\sigma_{-\mathbf{q}_j+\mathbf{k}} + \sigma_{-\mathbf{q}_j-\mathbf{k}}) - h_{-\mathbf{q}_j}(\sigma_{\mathbf{q}_j+\mathbf{k}} + \sigma_{\mathbf{q}_j-\mathbf{k}}) \right\} \\
&\quad + \frac{1}{2}\sum_{\mathbf{q}} D\big(|\mathbf{q}|^2 - q_c^2\big)^2 \frac{i\mathbf{q}\cdot\xi}{2}\left\{ \sigma_{\mathbf{q}}(\sigma_{-\mathbf{q}+\mathbf{k}} + \sigma_{-\mathbf{q}-\mathbf{k}}) \right. \\
&\quad \left. -\sigma_{-\mathbf{q}}(\sigma_{\mathbf{q}-\mathbf{k}} + \sigma_{\mathbf{q}+\mathbf{k}}) \right\}
\end{aligned}
\tag{3.74}
$$

Considering the averaged product of two variables as a scalar product

$$
\langle A,B\rangle = \mathfrak{N}^{-1}\int \mathfrak{D}\sigma A(\sigma)B(\sigma)e^{-\Phi(\sigma)}
\tag{3.75}
$$

where $\mathfrak{N} = \int \mathfrak{D}\sigma e^{-\Phi(\sigma)}$, one derives the following Schwartz inequality:

$$
\langle \sigma_{\mathbf{q}_i+\mathbf{k}}\sigma_{-\mathbf{q}_i-\mathbf{k}}\rangle\langle U_k(\xi)U_k^*(\xi)\rangle \geq |\langle \sigma_{-\mathbf{q}_i-\mathbf{k}}U_k(\xi)\rangle|^2
\tag{3.76}
$$

Using the following relation ($|\xi|\ll 1$)

$$
\begin{aligned}
\langle A(\sigma)U_k\rangle &= \mathfrak{N}^{-1}\int \mathfrak{D}\sigma A(\sigma)e^{-\Phi(\sigma)} - \mathfrak{N}^{-1}\int \mathfrak{D}\sigma A(\sigma)e^{-[\Phi(\sigma)+\delta\Phi(\sigma)]} \\
&= \mathfrak{N}^{-1}\int \mathfrak{D}\sigma A(\sigma)e^{-\Phi(\sigma)} - \mathfrak{N}^{-1}\int \mathfrak{D}\sigma A(\sigma)e^{-\Phi(\sigma')} \\
&= \mathfrak{N}^{-1}\int \mathfrak{D}\sigma\big[A(\sigma) - A((\sigma')^{-1})\big]e^{-\Phi(\sigma)}
\end{aligned}
$$

where $(\sigma')^{-1}$ is the inverse transformation of σ and the fact that $\langle\sigma_{\mathbf{q}_i}\rangle$ is nonzero only for $\mathbf{q}_i \in \{\mathbf{q}_1, -\mathbf{q}_1; \cdots; \mathbf{q}_m, -\mathbf{q}_m\}$ leads to

$$
|\langle \sigma_{-\mathbf{q}_i-\mathbf{k}}U_k\rangle|^2 = (\mathbf{q}_i\cdot\xi)^2\sigma^2
\tag{3.77}
$$

Introducing (3.77) into (3.76) leads to the following inequality for $h\to 0$

$$
\begin{aligned}
g_{\mathbf{q}_i+\mathbf{k},\mathbf{q}_i+\mathbf{k}} \geq (\mathbf{q}_i\cdot\xi)^2\sigma^2 \Bigg[&\sum_{j:1}^{m}\big(\kappa_j^{+2} + \kappa_j^{-2}\big)(\mathbf{q}_j\cdot\xi)^2\sigma^2 D \\
&+ \sum_{\mathbf{q}}\Big(\big(|\mathbf{q}-\mathbf{k}|^2 - q_c^2\big)^2 - \big(q^2 - q_c^2\big)^2\Big)D(\mathbf{q}\cdot\xi)^2 g_{\mathbf{q},\mathbf{q}} \Bigg]^{-1}
\end{aligned}
$$

where $(|\mathbf{q}_i|=q_c)$

$$\kappa_i^{+2}+\kappa_i^{-2}=\left[\left(|\mathbf{q}_i+\mathbf{k}|^2-q_c^2\right)^2+\left(|\mathbf{q}_i-\mathbf{k}|^2-q_c^2\right)^2\right]$$

$$=\left[2k^4+4(\mathbf{q}_i\cdot\mathbf{k})^2\right] \tag{3.78}$$

From (3.78) it results that $g_{\mathbf{q}_i+\mathbf{k},\mathbf{q}_i+\mathbf{k}}$ behaves at least as k^{-2} for small k. Thus the integration over k of the right-hand side of (3.69) diverges at $k=0$ for $d\leq2$. As in the corresponding case of the crystalline order,[41] these diverging correlations are incompatible with the existence of a positional long-range order. This contradiction can be removed only if the amplitude σ equals zero. In two dimensions the divergence is weak enough to allow finite systems to display order. On the other hand, because of this weakness, other types of ordering might also exist.[42] For $d=3$, the problem needs further discussion. Using the rotational symmetry in q-space the Ward identities give

$$g_{\mathbf{q}'_i,\mathbf{q}'_i}=g_{\mathbf{q}_i,\mathbf{q}_i}=\frac{\sigma}{h} \quad \text{for} \quad h\to0 \quad \text{and} \quad |\mathbf{q}'_i|=|\mathbf{q}_i|=q_c.$$

For an infinitesimal rotation $\mathbf{q}'_i=\mathbf{q}_i+\mathbf{k}_\perp([\mathbf{q}_i\cdot\mathbf{k}_\perp]=0)$

$$g_{\mathbf{q}_i+\mathbf{k},\mathbf{q}_i+\mathbf{k}}=g_{\mathbf{q}_i,\mathbf{q}_i}=\frac{\sigma}{h}$$

This leads to the following-stronger inequalities:

1. For the $m=1$ pattern, since the problem is invariant under the rotation of $\mathbf{q}_i$, one obtains $(|\mathbf{k}|\ll q_c)$

$$g_{\mathbf{q}_i+\mathbf{k},\mathbf{q}_i+\mathbf{k}}\geq\frac{1}{8(\mathbf{q}_i\cdot\mathbf{k})^2+2k^4} \tag{3.79}$$

and since

$$\int d^3k g_{\mathbf{g}_i+\mathbf{k},\mathbf{q}_i+\mathbf{k}}\geq\frac{1}{2}\int_{|\mathbf{k}|\ll q_c}d^3k\frac{1}{k^4+4q_c^2k^2\cos^2\theta}$$

$$=\pi\int_{|\mathbf{k}|\ll q_c}dk\int_{-1}^{+1}d\cos\theta\frac{1}{4q_c^2\cos^2\theta+k^2}\to\infty$$

The structure is also destroyed by the long-range fluctuations.

2. For the $m=3$ or $m=6$ pattern, the correlation matrix is invariant only for the rotation of the q-structure as a whole, and the Bogoliubov inequality becomes

$$g_{\mathbf{q}_i+\mathbf{k},\mathbf{q}_i+\mathbf{k}} \geq (\mathbf{q}_i \cdot \boldsymbol{\xi})\sigma^2 \left[\sum_{j:1}^m \left(\kappa_j^{+2} + \kappa_j^{-2}\right)(\mathbf{q}_j \cdot \boldsymbol{\xi})^2 \sigma^2\right]^{-1}$$

$$\propto \left[mk^4 + 4q_c^2 k^2 \sum_{j:1}^m \cos^2\theta_j\right]^{-1} \qquad (3.80)$$

Since the $\cos\theta_j$ do not vanish simultaneously, this correlation matrix behaves as k^{-2} and $\int d^d k g_{\mathbf{q}_i+\mathbf{k},\mathbf{q}_i+\mathbf{k}}$ diverges no more for $d \geq 3$. Let us note that these results are not affected by the vector dependence in u because this leads to an extra k^{-2}-dependence in the correlation functions.

As a consequence of the Bogoliubov inequalities, the $m=1$ structure is destroyed by the fluctuations in infinite three-dimensional systems. This result is thus analogous to the impossibility of the existence of one-dimensional crystals in three-dimensional space as argued by Landau[43] and Peierls.[44] Therefore only the structures characterized by wave vectors that satisfy definite angular relations, body-centered cubic, hexagonal prisms, remain. In these cases the first-order transition induced by the cubic terms ($v \neq 0$) is only slightly affected by fluctuations as long as $v^2/u \ll r$.

D. Finite Size Effects

The $m=1$ structures may be stabilized in three dimensions if the system is of finite dimensions because the long-range fluctuations are then inhibited.

If we take into account the inhomogeneous fluctuations, the equation of state may be written at the Hartree approximation

$$h_{\mathbf{q}_0} = r_0 \langle \sigma_{\mathbf{q}_0} \rangle + \frac{u}{2} \langle \sigma_{\mathbf{q}_0} \rangle^2 \langle \sigma_{-\mathbf{q}_0} \rangle$$

$$+ \frac{u}{2} \langle \sigma_{\mathbf{q}_0} \rangle \int d^d k \langle \sigma_{\mathbf{k}} \sigma_{-\mathbf{k}} \rangle$$

$$+ \frac{u}{2} \langle \sigma_{-\mathbf{q}_0} \rangle \int d^d k \langle \sigma_{\mathbf{q}_0+\mathbf{k}} \sigma_{\mathbf{q}_0-\mathbf{k}} \rangle \qquad (3.81)$$

where the integrals run over the phase space associated with the critical mode characterized by the wave vector $\mathbf{q}_0(|\mathbf{q}_0| = \mathbf{q}_c)$. The first correlated term of

(3.81) may be split as

$$\int d^d k \langle \sigma_{\mathbf{k}} \sigma_{-\mathbf{k}} \rangle = \int' d^d k \langle \sigma_{\mathbf{k}} \sigma_{-\mathbf{k}} \rangle$$

$$+ \int'' d^d k' \langle \sigma_{\mathbf{q}_0 + \mathbf{k}'} \sigma_{-\mathbf{q}_0 - \mathbf{k}'} \rangle \qquad (3.82)$$

to emphasize the contributions arising from vectors $\mathbf{k}$ of the critical shell near $\mathbf{q}_0$, that is, $|\mathbf{k}'/\mathbf{q}_0| \ll 1$. Recalling (3.29) and (3.69), and because $\langle \sigma_{\mathbf{q}_0} \rangle = \langle \sigma_{-\mathbf{q}_0} \rangle$, (3.81) becomes

$$h_{\mathbf{q}_0} = r_0 \langle \sigma_{\mathbf{q}_0} \rangle + \frac{u}{2} \langle \sigma_{\mathbf{q}_0} \rangle^3$$

$$+ \frac{u}{2} \langle \sigma_{\mathbf{q}_0} \rangle \left[\int' d^d k \langle \sigma_{\mathbf{k}} \sigma_{-\mathbf{k}} \rangle + \int'' d^d k \, g_D(\mathbf{k}) + \int'' d^d k \, g_{ND}(\mathbf{k}) \right]$$

$$(3.83)$$

These functions may be expressed at the same order of approximation, in terms of the inverse susceptibility matrix, which now has the elements

$$r_D = r_0 + u \langle \sigma_{\mathbf{q}_0} \rangle^2 + \frac{u}{2} \int' d^d k \langle \sigma_{\mathbf{k}} \sigma_{-\mathbf{k}} \rangle$$

$$+ \frac{u}{2} \int'' d^d k \, g_D(\mathbf{k})$$

$$r_{ND} = \frac{u}{2} \langle \sigma_{\mathbf{q}_0} \rangle^2 + \frac{u}{2} \int d^d k \, g_{ND}(\mathbf{k}) \qquad (3.84)$$

Two kinds of contribution are therefore present in (3.83). First we mention those pertaining to g_D and g_{ND}, the integrals of which are now finite (the weak logarithmic divergence having been removed by considering a finite system) and small because the integration runs over a very small part of the critical shell. They may thus be neglected in first approximation.

The other correlations $\langle \sigma_{\mathbf{k}} \sigma_{-\mathbf{k}} \rangle$ concern fluctuations the wave numbers of which belong to the critical shell; however they are not close to $\mathbf{q}_0$, that is, $\mathbf{k} = \mathbf{q}_0 + \boldsymbol{\kappa}$ where $|\boldsymbol{\kappa}/\mathbf{q}_0| \nleqslant 1$. There exists no corresponding nondiagonal contribution because $\mathbf{q}_0 - \boldsymbol{\kappa}$ necessarily lies outside the critical shell. These functions thus behave as

$$\langle \sigma_{\mathbf{k}} \sigma_{-\mathbf{k}} \rangle \simeq \frac{1}{r + D(k^2 - q_c^2)^2} \qquad (3.85)$$

where ($\alpha = \pi q_c^{d-1} S_q / (2\pi)^d$, S_q = surface area of the unit sphere)

$$r = r_0 + u \langle \sigma_{q_0} \rangle^2 + \alpha u r^{-1/2} D^{-1/2} \tag{3.86}$$

This then leads, when $h = 0$, to the equation of state

$$r + r_0 + \alpha u D^{-1/2} r^{-1/2} = 0 \tag{3.87}$$

with

$$r \langle \sigma_{q_0} \rangle - \frac{u}{2} \langle \sigma_{q_0} \rangle^3 = 0 \tag{3.88}$$

Therefore, the corresponding nonuniform state arises with a finite amplitude

$$\langle \sigma_{q_0} \rangle = \left(\frac{2 r_1}{u} \right)^{1/2} \tag{3.89}$$

When

$$-r_0 \geq r_1 = \left(\frac{\alpha u}{2 D^{1/2}} \right)^{2/3} \tag{3.90}$$

the structure becomes stabler than the uniform state for $r_1 < r_1' < \sqrt{2} r_1$, where r_1' defines the *first*-order transition temperature. This again provides an example of fluctuations modifying the character of the transition qualitatively.

However, because of the finite size, when

$$r_0 \simeq \frac{D}{L^2} \simeq \frac{1}{L^2 q_c^2} = r_{0d} \tag{3.91}$$

the problem becomes zero-dimensional with all the consequences discussed previously.

It is also worth mentioning that the contributions of g_D and g_{ND} neglected above, although very small, may play an important role insofar as they depend on the shape of the system.

The relations (3.87) and (3.88) completed with the g_D and g_{ND} contributions become

$$r + r_0 + \alpha u D^{-1/2} r^{-1/2} + \frac{u}{2} \int d^d k \left[g_D(\mathbf{k}) + g_{ND}(\mathbf{k}) \right] = 0 \tag{3.92}$$

$$\langle \sigma_{q_0} \rangle^2 = \frac{2}{u} \left[r - \int d^d k \left[g_D(\mathbf{k}) + g_{ND}(\mathbf{k}) \right] \right] \tag{3.93}$$

D. WALGRAEF, D. DEWEL, AND P. BORCKMANS

where

$$g_D(\mathbf{k}) + g_{ND}(\mathbf{k}) = \frac{1}{2\left[r + 4(\mathbf{q}_0 \cdot \mathbf{k})^2 + O(k^4)\right]} \qquad (3.94)$$

Considering a two-dimensional rectangular box of lateral dimensions L_X and $L_Y (L_X > L_Y)$, (3.92) reads

$$r + \alpha u D^{-1/2} r^{-1/2} + r_0' = 0 \qquad (3.95)$$

where

$$r_0' = r_0 + \frac{1}{2} u \int_{L_X^{-1}}^{1} dk_x \int_{L_Y^{-1}}^{1} dk_y \, \frac{1}{r + 4(\mathbf{q}_0 \mathbf{k})^2} \qquad (3.96)$$

and r_0' depends on the orientation of q_0 in the k-space associated with the container (L_X, L_Y), and (3.90) becomes $-r_0' \geq r_1$.

Furthermore, since $\langle \sigma_{\mathbf{q}_0} \rangle^2$ grows with r, it can be shown that the most probable structure is that where q_0 is oriented in such a way that $r_0' - r_0$ is minimum. This occurs when $(\mathbf{q}_0)_y = 0$ and the structure is thus aligned along the small side of the container. This discussion could straightforwardly be applied to the Bénard problem in the Swift-Hohenberg description.[29]

To conclude, we stress that whereas the system shows many solutions, the pattern selection mechanism is twofold. First the nonlinear terms already remove the pattern degeneracy in infinite systems. Then the remaining orientational degeneracy for each pattern is resolved in finite systems by the geometry of the boundaries.

We also note the appearance of two classes of dissipative structures. In the first, because they must be stabilized by the finite dimensions of these systems, the general theorems resulting from the breakdown of the translation and rotation symmetry are not applicable. On the contrary, in the second class the instability is analogous to equilibrium first-order transitions such as crystallization.

IV. HARD-MODE TRANSITION TO CHEMICAL OSCILLATIONS

When $1 + a^2 < (1 + a\eta)^2$, a hard-mode transition to chemical oscillations occurs in the Brusselator at $b = b_c' = 1 + a^2$. Near this instability point the linear stability analysis of the homogeneous steady state gives as characteristic eigenfrequencies:

$$\omega_q = \tfrac{1}{2}\left\{ b - b_c' - q^2(D_x + D_y) \pm 2ia\left[1 + q^2(D_x - D_y) \right.\right.$$
$$\left.\left. - \frac{1}{4a^2}\left((b - b_c') - q^2(D_x - D_y) \right)^2 \right]^{1/2} \right\} \qquad (4.1)$$

The corresponding eigenvectors (at the lowest order in $b - b_c'$ and q^2) are:

$$T_q = \frac{i - a}{2i}\left[\frac{a + i}{a} x_q + y_q \right], \qquad S_q = T_q^* \qquad (4.2)$$

The associate nonlinear Langevin equations may easily be written in the rotating wave approximation,[45] which is valid near the instability point where $(b - b_c')/a \ll 1$ and for times much greater than a^{-1}, as

$$\partial_\tau \sigma_k = -[r + ck^2]\sigma_k - uL^{-d}\sum_{k'}\sum_{k''} \sigma_{k'}\sigma_{k''}^* \sigma_{k-k'-k''} + \eta_k \qquad (4.3)$$

with $\sigma_k = T_k e^{ia\tau}$, the usual hypothesis concerning the noise

$$\langle \eta_k(\tau)\eta_{k'}^*(\tau') \rangle = 2\Gamma\delta(\tau - \tau')\delta_{k+k',0} \qquad (4.4)$$

and

$$r = r_1 + ir_2$$
$$u = u_1 + iu_2$$
$$c = c_1 + ic_2$$

At the lowest order in the nonlinearity

$$r_1 = \frac{b_c' - b}{b_c'} \qquad r_2 = \frac{1}{2a}\frac{(b - b_c')^2}{b_c'}, \qquad c_1 = \frac{D_x + D_y}{b_c'}$$

$$c_2 = 2a\frac{D_x - D_y}{b_c'} \qquad u_1 = \frac{a^2 + 2}{2a^2 b_c'}, \qquad u_2 = \frac{4 - 7a^2 + 4a^4}{3a^3 b_c'} \qquad (4.5)$$

$$\tau = 2b_c't$$

A. Mean Field Behavior

Equation 4.3 exhibits no evident gradient structure, and the critical behavior can be obtained through direct use of the dynamic renormalization group.[46] However let us begin with the study of the homogeneous problem that corresponds to the zero-dimensional regime for the system.

Then (4.3) may be written as ($\sigma_0 = L^{-d/2}\sigma_{q=0}$)

$$\partial_\tau \sigma_0 = -\Gamma \frac{\delta \mathcal{F}_0}{\delta \sigma_0^*} - i\big[r_2\sigma_0 + u_2|\sigma_0|^2\sigma_0\big] + \eta_0 \qquad (4.6)$$

where

$$\mathcal{F}_0 = \frac{1}{\Gamma}\Big[r_1|\sigma_0|^2 + \frac{u_1}{2}|\sigma_0|^4\Big] \qquad (4.7)$$

is the standard homogeneous potential for a system with a two-component order parameter such as the $n=2$ Time Dependent Ginzburg Landau (TDGL) model (XY model)[47] or the single-mode laser.

Its shape, a Gaussian well in the disordered phase, becomes craterlike in the ordered region (Fig. 4). An analogous conclusion was obtained in the master equation approach.[48] The steady-state probability distribution thus has a maximum (for $\sigma_0 = \bar{\sigma}_0$) describing the orbit of the limit cycle corresponding to (4.3).

Therefore all these systems have static behavior belonging to the same universality class, whereas the dynamics of the Brusselator will differ because of the presence of a divergence free current $V_0 = i[r_2\sigma_0 + u_2\sigma_0|\sigma_0|^2]$ in the equation of motion.

Indeed at this level of description, the mean field theory gives the following values for the order parameter $\bar{\sigma}_0$

$$\bar{\sigma}_0(\tau) = R(\tau)e^{i\phi(\tau)}$$

with

$$R^2(\tau) = \begin{cases} -\dfrac{r_1 R^2(0)}{u_1 R^2(0) - \big(r_1 + u_1 R^2(0)\big)\exp 2r_1\tau}, & r_1 \neq 0 \\[4mm] \dfrac{R^2(0)}{1 + 2u_1 R^2(0)\tau}, & r_1 = 0 \end{cases} \qquad (4.8)$$

leading for $\tau \to \infty$ to the stationary values

$$R^2(\tau \to \infty) = \overline{R^2} = \begin{cases} 0, & r_1 \geq 0 \\[2mm] \dfrac{|r_1|}{u}, & r_1 < 0 \end{cases} \qquad (4.9)$$

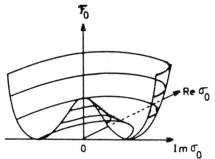
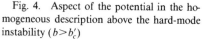

Fig. 4. Aspect of the potential in the homogeneous description above the hard-mode instability $(b > b'_c)$

characteristic of a mean field second-order transition, $\phi(\tau)$ being in this case

$$\phi(\tau) = \begin{cases} \phi(0) - \dfrac{u_2}{2u_1}\ln\left(1 + \dfrac{u_1 R^2(0)}{r_1}\right) - r_2\tau, & r_1 > 0 \\[4mm] \phi(0) - \dfrac{u_2}{u_1}\ln\dfrac{R(0)}{\overline{R}} - \left(r_2 + u_2\overline{R^2}\right)\tau, & r_1 < 0 \end{cases} \tag{4.10}$$

Therefore any perturbation, even of the radius alone, leads to a phase shift in the rotation of the order parameter of amplitude $\phi(0) - (u_2/u_1)$ $\ln(R(0)/\overline{R})$. The corresponding isochrons may be visualized from Fig. 5. Let us recall that the concept of isochron was introduced by A. Winfree[43] to denote the locus of all the points in the $(\mathrm{Re}\,\sigma, \mathrm{Im}\,\sigma)$ plane that come into phase

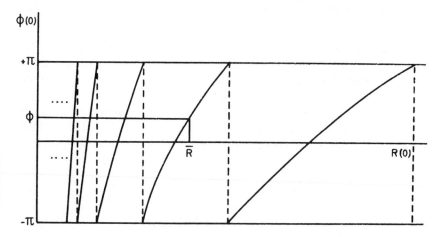

Fig. 5. The isochrons near the transition point. The curves represent all $(\phi(0), R(0))$ points that come into phase synchrony with one another at the $(\phi, \overline{R})$ point on the limit cycle.

synchrony with one another on the limit cycle for $t \to \infty$. One may easily deduce from (4.10) that there is a 2π phase discontinuity for $R_0^{(k)} = R\exp(u_1/u_2)[\phi_0 + 2k\pi]$. In this mean field approximation the concentrations of X and Y can be deduced from (4.3) and (4.8) to (4.10) and read

$$x(t) = \left[\frac{a^2(b-b_c')}{a^2+2} \right]^{1/2} \cos\Omega t$$

$$+ \frac{2(b-b_c')(6a^2-1-a^4)^{1/2}}{3(a^2+2)} \cos(2\Omega t + \psi_1) + \cdots$$

$$\Omega = a - \frac{4-7a^2+4a^4}{6a(a^2+2)}(b-b_c')$$

$$\psi_1 = \arctan\left(\frac{a^2-1}{2a} \right)$$

$$y(t) = \left(\frac{b-b_c'}{2+a^2} \right)^{1/2} (1+a^2)^{1/2}\cos(\Omega t + \theta) + \frac{2(a^2-1)}{a(a^2+2)}(b-b_c')$$

$$+ \frac{(b-b_c')(1+6a^2+9a^4+4a^6)^{1/2}}{3(2+a^2)} \cos(2\Omega t + \varphi_1)$$

$$\theta = \arctan\left(-\frac{1}{a} \right)$$

$$\varphi_1 = \arctan\frac{2a(2-a^2)}{1-5a^2} \tag{4.11}$$

The static correlation functions with respect to the rotating frame can be calculated from the probability distribution, giving identical results to those for the single-mode laser. This second-order-like phase transition takes place with a symmetry breaking of the phase.

However the diffusion term of the associated Fokker-Planck equation leads to a symmetry-restoring mechanism, the so-called phase diffusion. Indeed the Fokker-Planck equation may be written in polar coordinates ($\sigma_0 = \rho e^{i\theta}$)

$$\partial_t P(\rho,\theta,t) = \frac{1}{2}\left[\frac{1}{\rho}\partial_\rho(r_1\rho^2+u_1\rho^4)P(\rho,\theta,t) \right.$$

$$+ \partial_\theta(r_2+u_2\rho^2)P(\rho,\theta,t)\bigg]$$

$$+ \frac{\Gamma}{4}\left(\frac{1}{\rho}\partial_\rho\rho\partial_\rho + \frac{1}{\rho^2}\partial_\theta^2 \right)P(\rho,\theta,t)$$

and we have

$$\partial_t \langle e^{i\theta} \rangle_t = - \left\langle \left[i(r_2 + u_2 \rho^2) + \frac{\Gamma}{4\rho^2} \right] e^{i\theta} \right\rangle_t \tag{4.12}$$

In the ordered phase and in the mean field regime we obtain

$$\langle e^{i\theta} \rangle_t = \langle e^{i\theta'} \rangle \exp - \left[i(r_2 + u_2 \bar{R}^2) + \frac{\Gamma}{4\bar{R}^2} \right] t$$

$$\left(\theta' = \theta - \frac{u_2}{u_1} \ln \frac{R(0)}{\bar{R}} \right) \tag{4.13}$$

and this quantity goes to zero in the long time limit where the phase symmetry is restored. Nevertheless in finite systems when $\Gamma \simeq \Omega^{-1}$ is sufficiently small, the phase may be considered to be locked in experimental time scales, and one may safely assume that

$$\langle e^{i\theta} \rangle_t \simeq e^{i\theta_0} \exp - i(r_2 + u_2 \bar{R}^2) t \tag{4.14}$$

For greater values of Γ (small systems), quantitative modifications to this behavior may occur in the zero-dimensional regime as discussed, for instance, in Ref. 50.

B. Effects of Inhomogeneous Fluctuations

Because of the structure of (4.3), one expects inhomogeneous fluctuations to affect the behavior of the system for space dimensionality less than four.[46] Following an analysis analogous to that used in Section IV.A, (4.3) may be written as

$$\partial_t \sigma_k = -\Gamma \frac{\delta \mathcal{F}}{\delta \sigma_{-k}^*} + V_k + \eta_k \tag{4.15}$$

Again the generalized potential and the divergence free current are such that

$$\sum_k \left(\partial_{\sigma_k} V_k e^{-\mathcal{F}} + \partial_{\sigma_k^*} V_k^* e^{-\mathcal{F}} \right) = 0 \tag{4.16}$$

and

$$\mathcal{F} = \frac{1}{\Gamma} \left[\sum_k (r_1 + c_1 k^2) |\sigma_k|^2 + \frac{1}{2L^d} \sum_{kk'k''} u(k, k', k'') \sigma_k \sigma_{k'}^* \sigma_{k''} \sigma_{-k-k'-k''}^* \right]$$

where

$$u(\mathbf{k},\mathbf{k}',\mathbf{k}'') = u_1 + \frac{i(c_1 u_2 - c_2 u_1)\beta(\mathbf{k},\mathbf{k}',\mathbf{k}'')}{\alpha(\mathbf{k},\mathbf{k}',\mathbf{k}'') + ic_2\beta(\mathbf{k},\mathbf{k}',\mathbf{k}'')}$$

$$\beta(\mathbf{k},\mathbf{k}',\mathbf{k}'') = k^2 - k'^2 + k''^2 - (\mathbf{k}+\mathbf{k}'+\mathbf{k}'')^2$$

$$\alpha(\mathbf{k},\mathbf{k}',\mathbf{k}'') = 4r_1 + c_1\left(k^2 + k'^2 + k''^2 + (\mathbf{k}+\mathbf{k}'+\mathbf{k}'')^2\right) \quad (4.17)$$

while

$$V_{\mathbf{k}} = -i\left(r_2 + c_2 k^2\right)\sigma_{\mathbf{k}}$$

$$+ L^{-d}\sum_{\mathbf{k}'\mathbf{k}''}\left[\frac{i(c_1 u_2 - u_1 c_2)\beta(\mathbf{k},\mathbf{k}',\mathbf{k}'')}{\alpha(\mathbf{k},\mathbf{k}',\mathbf{k}'') + ic_2\beta(\mathbf{k},\mathbf{k}',\mathbf{k}'')} - iu_2\right]\sigma_{\mathbf{k}'}\sigma_{\mathbf{k}''}^{*}\sigma_{\mathbf{k}-\mathbf{k}'-\mathbf{k}''}$$

$$(4.18)$$

It may easily be checked that the condition (4.16) is verified up to terms of the sixth order in σ, which is consistent with this approximation scheme in the vicinity of the instability point. The stationary distribution of the fluctuations is given by the potential $\mathcal{F}$ which for $c_2 = u_2 = 0$ is exactly the Ginzburg-Landau potential of a system with a two-component order parameter—for example, the XY model of ferromagnetism. Moreover, the critical behavior associated with the potential (4.17) ($c_2, u_2 \neq 0$) will be identical because the k-dependent terms in u have a critical dimensionality of two and are therefore irrelevant to the critical behavior for $d > 2$ systems.

This is in agreement with the result of Hentschel,[46] who directly applied the dynamic renormalization group to (4.3). For example, the order parameter behaves as $|\sigma_0| \sim (b - b_c^*)^{\beta}$ with $\beta = \frac{1}{2} - 3\epsilon/20$ for $\epsilon = 4 - d$ systems, while the static correlation function behaves as

$$\langle\sigma_{-\mathbf{k}}^{*}\sigma_{\mathbf{k}}\rangle \simeq \frac{\Gamma}{c_1 k^2 + \xi_1^{-2}}, \, \xi_1 \simeq |b - b_c^*|^{-\nu_1}, \, \nu_1 = \frac{1}{2} + \frac{\epsilon}{10} \quad (4.19)$$

b_c^* being the true instability point, since the mean field instability point b_c' is shifted by the fluctuations.

In the ordered phase, an interesting conceptual aspect of the analogy between the asymptotic behavior of the limit cycle (in the vicinity of the instability) and systems with a two-dimensional order parameter is the behavior of its correlation functions.

Because of the existence of a broken continuous symmetry (here the phase symmetry), the methods used in the preceding section also lead to Bogoliubov inequalities. They induce the following dependence for the correlation function of the fluctuations of wave vectors $\mathbf{k}$, transverse to the radius of the limit cycle

$$g_\perp(\mathbf{k}) \geq |\mathbf{k}|^{-2} \qquad (4.20)$$

This leads again to the destruction of long-range order for $d \leq 2$ infinite systems. This mechanism should not be confused with the so-called phase diffusion, which restores the symmetry at any dimensions for finite systems.

The possible existence of topological order[42] is not discussed here. One also notes that this leads to the following behavior in all the ordered phases for the correlation functions of fluctuations of wave vectors $\mathbf{k}$, along the radius of the limit cycle:

$$g_\parallel(\mathbf{k}) \simeq |\mathbf{k}|^{d-4}$$

because of its coupling to the transverse fluctuations.[9]

In the disordered phase the dynamic correlation functions behave as[46]

$$\langle \sigma^*_{-\mathbf{k}}(t)\sigma_{\mathbf{k}}(0) \rangle \simeq 2\langle \sigma^*_{-\mathbf{k}}\sigma_{\mathbf{k}} \rangle \exp - \Gamma t/\tau(\mathbf{k}) \qquad (4.21)$$

with

$$\tau^{-1}(\mathbf{k}) = c_1 k^2 + \xi_1^{-2} + i\left(c_2 k^2 + c_2\xi_1^{-2} + \xi_2^{-2}\right)$$

where

$$\xi_2 \simeq |b - b_c^*|^{-\nu_2}, \qquad \nu_2 = \tfrac{1}{2}$$

However for $b > b_c^*$, the Goldstone modes alter the dynamics. In the case of n components TDGL models describing equilibrium systems ($u_2 = c_2 = 0$), the transverse modes exhibit a diffusive behavior and have an infinite lifetime for $k \to 0$ that is associated with their symmetry-restoring character. The longitudinal modes show a nonexponential decay as extensively discussed by Mazenko.[51]

In our case it may easily be seen that u_2 and c_2 couple the longitudinal and transverse fluctuations whose kinetic equations are:

$$\partial_t U_{\mathbf{q}}'' = -\left[2u_1 R^2 + c_1 q^2\right]U_{\mathbf{q}}'' + c_2 q^2 U_{\mathbf{q}}^\perp + u_1 R\left[3U''^2 + U^{\perp 2}\right]_{\mathbf{q}}$$
$$+ 2u_2 R(U''U^\perp)_{\mathbf{q}} - u_1\left[(U''^2 + U^{\perp 2})U''\right]_{\mathbf{q}}$$
$$+ u_2\left[(U''^2 + U^{\perp 2})U^\perp\right]_{\mathbf{q}} + \eta_{\mathbf{q}}''$$
$$\partial_t U_{\mathbf{q}}^\perp = -c_1 q^2 U_{\mathbf{q}}^\perp - \left[2u_2 R^2 + c_2 q^2\right]U_{\mathbf{q}}'' - 2u_1 R(U''U^\perp)_{\mathbf{q}} - u_2 R$$
$$\times\left[3U''^2 + U^{\perp 2}\right]_{\mathbf{q}} - u_1\left[(U''^2 + U^{\perp 2})U^\perp\right]_{\mathbf{q}}$$
$$- u_2\left[(U''^2 + U^{\perp 2})U''\right]_{\mathbf{q}} + \eta_{\mathbf{q}}^\perp$$
$$\sigma_{\mathbf{q}} = \left(R\delta_{\mathbf{q},0} + U_{\mathbf{q}}'' + iU_{\mathbf{q}}^\perp\right)e^{i(\omega t + \phi)} \qquad (4.22)$$

354 D. WALGRAEF, D. DEWEL, AND P. BORCKMANS

The stability analysis shows that the behavior of the system depends on the sign of $u_1 c_1 + u_2 c_2$. When $u_1 c_1 + u_2 c_2$ is positive, the homogeneous limit cycle is stable and either the transverse or the longitudinal fluctuations are symmetry restoring. They have a diffusive behavior with infinite lifetime for $k \rightarrow 0$. When $u_1 c_1 + u_2 c_2$ is negative, the homogeneous limit cycle is unstable against inhomogeneous fluctuations, and this may lead to chemical phase turbulence.[52]

We have emphasized that the stationary distribution of the fluctuations around the limit cycle of the Brusselator close to the critical point resembles closely a two-component Ginzburg-Landau system (e.g., the XY model of ferromagnetism). Owing to the spontaneous phase symmetry breaking down, the inhomogeneous fluctuations destroy the long-range order for $d \leq 2$ infinite systems. The specific aspects of the Brusselator appear in the dynamic behavior of the system, even in the zero-dimensional regime where, for example, the phase behavior is by no means analogous to otherwise comparable systems such as single-mode lasers or superfluids. Once again, the mean field description is shown to be affected by the inhomogeneous fluctuations.

Acknowledgments

We thank Profs. I. Prigogine and G. Nicolis for their constant interest and encouragement. We are also indebted to Dr. J. Wallenborn for critical comments on the manuscript.

References

1. G. Nicolis and I. Prigogine, *Self-Organization in Nonequilibrium Systems*, Wiley, New York, 1977.
2. H. Haken, *Synergetics*, 2nd ed., Springer-Verlag, Berlin, 1978.
3. For a recent review, see *Proceedings of the Seventeenth Solvay Conference on Physics: Order and Fluctuations in Equilibrium and Nonequilibrium Statistical Mechanics*, G. Nicolis, G. Dewel, and J. W. Turner, Eds., Wiley, New York, (1981).
4. D. L. Stein, *J. Chem. Phys.*, **72**, 2869 (1980).
5. D. Walgraef, G. Dewel, and P. Borckmans, *J. Chem. Phys.*, **74**, 755 (1981).
6. S. K. Ma, *Modern Theory of Critical Phenomena*, Benjamin, New York, 1976.
7. M. E. Fisher, *Rev. Mod. Phys.*, **46**, 597 (1974).
8. N. N. Bogoliubov, *Lectures in Quantum Statistics*, Vol. 2, MacDonald, London, 1971.
9. V. L. Pokrovskii, *Adv. Phys.*, **28**, 597 (1979).
10. F. T. Arecchi, *Proceedings of the Enrico Fermi International School of Physics—Quantum Optics*, Academic, New York, 1969.
11. P. Bergé, in *Fluctuations, Instabilities and Phase Transitions*, T. Riste, Ed., NATO Advanced Studies Series, Plenum, New York, 1975.
12. A. Boiteux and B. Hess, "Kinetics of Physicochemical Oscillations," paper presented at the meeting of Deutsche Bunsegesellschaft für Physikalische Chemie, Aachen, 1979; K. Showalter, *J. Chem. Phys.*, to appear.
13. G. Nicolis and M. Malek-Mansour, *J. Stat. Phys.*, **22**, 495 (1980).
14. L. Landau and E. M. Lifshitz, *Fluid Mechanics*, Addison-Wesley, Reading, Mass., 1959.

15. S. Grossman, *J. Chem. Phys.*, **65**, 2007 (1976).
16. P. Richter, I. Proccacia, and J. Ross, *Adv. Chem. Phys.*, **43**, 217 (1980).
17. F. Schlögl, *Z. Phys.*, **253**, 147 (1972).
18. G. Nicolis and J. W. Turner, *Physica*, **89A**, 326 (1977).
19. P. Glansdorff and I. Prigogine, *Thermodynamics of Structure, Stability and Fluctuations*, Wiley, New York, 1974.
20. G. Dewel, D. Walgraef, and P. Borckmans, *Z. Phys. B*, **28**, 235 (1977).
21. C. L. Creel and J. Ross, *J. Chem. Phys.*, **65**, 3779 (1976).
22. B. B. Edelstein, *J. Theor. Biol.*, **29**, 57 (1970).
23. A. Nitzan, *Phys. Rev. A*, **17**, 1513 (1978).
24. M. Abramowitz and I. Stegun, *Handbook of Mathematical Functions*, Dover, New York, 1966.
25. S. Grossman and P. M. Richter, *Z. Phys.* **242**, 458 (1971), B. Mühlschlegel, D. J. Scalapino, and R. Denton, *Phys. Rev. B*, **6**, 1767 (1972).
26. D. Walgraef, G. Dewel, and P. Brockmans, *Phys. Rev. A*, **21**, 397 (1980).
27. N. D. Mermin, *Phys. Rev.*, **176**, 250 (1968).
28. S. A. Brazovskii, *Sov. Phys. (J. Exp. Theor. Phys.)*, **41**, 85 (1975).
29. J. Swift and P. C. Hohenberg, *Phys. Rev. A*, **15**, 319 (1977).
30. R. Graham, *Phys. Rev. A*, **10**, 1762 (1974).
31. W. Eckhaus, *Studies in Non-Linear Stability Theory*, Springer-Verlag, New York, 1965.
32. P. Bergé and M. Dubois, "*Les Instabilitiés Hydrodynamiques en Convection Libre, Forcée et Mixte*," Vol. 72, *Lecture Notes in Physics*. Springer-Verlag, Berlin, 1978.
33. F. H. Busse, *Rep. Prog. Phys.*, **41**, 1930 (1978).
34. M. Dubois, P. Bergé, and J. Wesfreid, *J. Phys.*, **39**, 1253 (1978).
35. S. Alexander and J. McTague, *Phys. Rev. Lett.*, **41**, 702 (1978).
36. C. Normand, Y. Pomeau, and M. Velarde, *Rev. Mod. Phys.*, **49**, 581 (1977).
37. D. Sattinger, *Topics in Stability and Bifurcation*, Vol. 309, *Lecture Notes in Mathematics*, Springer-Verlag, Berlin, 1973.
38. G. Nicolis, T. Erneux, and M. Herschkowitz-Kaufman, *Adv. Chem. Phys.*, **72**, 1900 (1980).
39. L. M. Pismen, *J. Chem. Phys.*, **72**, 1900 (1980).
40. E. S. Abers and B. W. Lee, *Phys. Rep.*, **9**, 1 (1973).
41. D. Mermin, *J. Phys. Soc. Japan*, **26**, S203 (1969).
42. J. M. Kosterlitz and D. J. Thouless, *J. Phys.*, **C6**, 1181 (1973).
43. L. D. Landau and E. M. Lifshitz, *Statistical Physics*, Addison-Wesley, Reading, Mass., 1969.
44. R. Peierls, *Ann. Inst.* Henri Poincaré, **5**, 177 (1935).
45. Y. Kuramato and T. Tsuzuki, *Progr. Theor. Phys.*, **54**, 687 (1975).
46. H. Hentschel, *Z. Physik.*, **B31**, 401, (1978).
47. P. C. Hohenberg and B. I. Halperin, *Rev. Mod. Phys.*, **49**, 435, (1977).
48. J. W. Turner in *Dynamics of Synergetic Systems*, H. Haken, Ed., N. Y., Springer-Verlag, 1980.
49. A. Winfree, *J. Math. Biol.*, **1**, 73 (1974).
50. P. Schranner, S. Grossmann, and P. H. Richter, *Z. Physik.*, **B35**, 363 (1979).
51. G. F. Mazenko, *Phys. Rev.*, **B14**, 3933 (1976).
52. Y. Kuramoto, *Progr. Theor. Phys.*, **56**, 679 (1976).

STRESS AND STRUCTURE IN FLUID INTERFACES*

H. T. DAVIS

*Department of Chemical Engineering and Materials Science and
Department of Chemistry
University of Minnesota
Minneapolis, Minnesota*

L. E. SCRIVEN

*Department of Chemical Engineering and Materials Science
University of Minnesota
Minneapolis, Minnesota*

CONTENTS

*This work was supported by the National Science Foundation, the U.S., Department of Energy, and the University of Minnesota Computer Center.

I. INTRODUCTION

The modern theory of interfaces and capillarity has its roots in the works of Rayleigh[1] and of van der Waals[2] and his followers.[3] Rayleigh argued that the density variation across an interface is continuous, contrary to the discontinuous structure assumed by Young and by Laplace. Both Rayleigh and van der Waals derived versions of the so-called gradient theory of interfaces, which has been the most revealing and practical tool used during the past two decades for the theoretical analysis of inhomogeneous equilibrium fluid. Van der Waals's work, much more complete than any others of the time, was based on statistical thermodynamics, the starting point being an approximate expression for the free energy of the inhomogeneous fluid in the form of a functional of the fluid density distribution.

The gradient theory of inhomogeneous fluids, rediscovered by Cahn and Hilliard,[4] has led to an understanding of the mechanism of spinodal decomposition.[5] In another early application the behavior of fluids near the critical point was studied using gradient theory.[2, 6] For example, Fisk and Widom[7-9] have obtained from gradient theory the scaling laws characteristic of the current theory of critical phenomena.

The next major advance in the fundamental theory of interfaces came some 50 years after the van der Waals era. This was what this chapter designates as the mechanical theory of the inhomogeneous fluid. Since interfacial tension is determined by the fluid stress, it is a mechanical quantity whose definition does not depend on the existence of thermodynamic laws. Kirkwood and Buff[10-12] and Kirkwood and co-workers[13, 14] derived molecular theoretical expressions for the stress tensor of an inhomogeneous fluid and related the interfacial tension rigorously, for the first time, to intermolecular forces and the distribution functions of the inhomogeneous fluid. The distribution functions can in principle be determined from the solution to the heirarchy of equations of hydrostatic equilibrium[11] (i.e., the Yvon-Born-Green equations[15]).

In the early 1960s, the techniques of functional differentiation were exploited in several seminal papers[16-22] that provide the statistical thermodynamic basis for relating the equilibrium properties of inhomogeneous fluids to density distributions and direct correlation functions of the type introduced originally by Ornstein and Zernike.[23] From this theory, a rigorous formula for interfacial tension emerges[24, 25] that is similar in form to the original Rayleigh–van der Waals gradient formula. The exact elements of the gradient theory of density distributions and correlation functions have been related,[26, 27] with the aid of the general theory, to certain homogeneous fluid quantities. An important aspect of the density functional theory is that no particular assumptions are made concerning the nature of the intermolecular interactions. In fact, in formulas for the thermodynamic functions of the inhomogeneous fluid, molecular forces do not appear explicitly[27]—all information is carried by the direct correlation functions of inhomogeneous fluid.

As is unfortunately almost always the case with rigorous results for fluids, neither the exact mechanical nor the exact density functional thermodynamic theory yields tractable algorithms for actually computing fluid properties. Nevertheless, a rigorous theory does provide the right starting point for introducing simplifying approximations and, in fact, can be suggestive of new approximations. General expressions for inhomogeneous fluids such as the Yvon-Born-Green equations and the density functional free-energy equation have been rendered tractable by such intuitive assumptions as discontinuous interfaces and locally homogeneous correlation functions and by such systematic approximations as truncated virial expansions and gradient theory.

This chapter attempts to put into perspective (1) the present state of the rigorous theory of the existence and stability of inhomogeneous fluid structures, (2) the derivation and application of tractable models, and (3) the physical interpretation of interfacial structure and thermodynamics. Over the past two decades, activity in the study of the theory of fluid interfaces has exploded. Although we have tried to cover the major ideas and results, we cannot hope to provide a balanced discussion of the works of everyone involved. For a broader review of the literature than that provided here, we recommend the older article by Ono and Kondo[28] and the recent articles by Widom,[9] Croxton,[29] and Jhon and Dahler,[30] Rowlinson,[31] and Evans.[32]

II. STATE OF STRESS AND MECHANICAL EQUILIBRIUM

A. Pressure Tensor

Pressure (or its negative, stress) is an intensive property of matter that measures the force exerted by matter on one side of a unit area on matter on the other side of the same unit area. We denote by $t_{\hat{\varepsilon}} dA$ the force exerted through the infinitesimal area dA (see Fig. 2.A.1), with unit normal $\hat{\varepsilon}$, by material above (positive $\hat{\varepsilon}$-direction) the element. The quantity $t_{\hat{\varepsilon}}$, called the traction vector, has units of force per unit area and of course obeys the law of equal action and reaction (i.e., $t_{\hat{\varepsilon}} dA = -t_{-\hat{\varepsilon}} dA$). The traction vector depends on the location r and orientation $\hat{\varepsilon}$ of dA in the material. However, it is closely related to the pressure tensor P, which is a function of the location but not the orientation of dA; therefore P is a material property.

To establish the connection between the pressure tensor and traction vector, consider the tetrahedron in Fig. 2.A.1. The center of the tetrahedron is located at r and $\hat{i}, \hat{j}, \hat{k}$ are an orthogonal set of unit vectors, so that the areas of the faces of the tetrahedron with normals $-\hat{i}$, $-\hat{j}$, and $-\hat{k}$ are $dA_1 =$

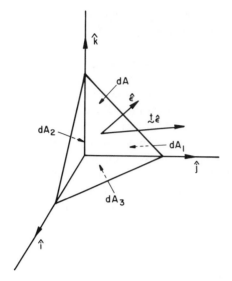

Fig. 2.A.1. Cauchy tetrahedron for re-
solving traction force $\mathbf{t}_{\hat{\varepsilon}}$ on dA.

$\hat{\varepsilon} \cdot \hat{i}\,dA$, $dA_2 = \hat{\varepsilon} \cdot \hat{j}\,dA$, and $dA_3 = \hat{\varepsilon} \cdot \hat{k}\,dA$. Newton's second law of motion applied to the tetrahedron is

$$\frac{d^2}{dt^2}(\rho\mathbf{U}\,dV) = \mathbf{t}_{\hat{\varepsilon}}\,dS - \mathbf{t}_{i}\hat{\varepsilon}\cdot\,dS - \mathbf{t}_{j}\hat{\varepsilon}\cdot\hat{j}\,dS - \mathbf{t}_{k}\hat{\varepsilon}\cdot k\,dS + (\rho\,dV)\hat{\mathbf{F}}_e \quad (2.A.1)$$

where ρ and $\mathbf{U}$ are the mass average density and velocity of the contents of the tetrahedron and $\hat{\mathbf{F}}_e$ is the external (e.g., gravity, centrifugal) force per unit mass exerted on the contents of the tetrahedron. Dividing (2.A.1) by dS and then shrinking dS to zero, noting that dV/dS goes to zero in such a limit, we deduce

$$\mathbf{t}_{\hat{\varepsilon}} = -\hat{\varepsilon}\cdot\mathbf{P} \quad (2.A.2)$$

where the pressure tensor $\mathbf{P}(\equiv -\hat{i}\mathbf{t}_i - \hat{j}\mathbf{t}_j - \hat{k}\mathbf{t}_k)$ emerges as a possible function of position $\mathbf{r}$ but not of $\hat{\varepsilon}$. According to (2.A.2), the traction force on an arbitrarily oriented element dA is generated by projection of the negative pressure tensor along the normal of dA.

In an isotropic fluid, the force on dA is $-\hat{\varepsilon}P\,dA$, so that the pressure tensor of isotropic fluid is $\mathbf{P} = P\mathbf{I}$, $\mathbf{I}(= \hat{i}\hat{i} + \hat{j}\hat{j} + \hat{k}\hat{k})$ being the unit tensor. Because an inhomogeneous fluid is not isotropic, $\mathbf{P}$ in general can have three distinct principal values, and as many as nine components in coordinate representations. In a Cartesian coordinate basis

$$\mathbf{P} = \sum_{\nu,\mu=1}^{3} P_{\nu\mu} \hat{e}_{\nu} \hat{e}_{\mu} \qquad (2.A.3)$$

where $\hat{e}_{\nu}$, $i=1,2,3$, are an orthogonal set of unit vectors and $P_{\nu\mu}$, ν, $\mu=1,\ldots,3$; the components of $\mathbf{P}$. In the special case of an isotropic fluid $P_{\nu\nu} = P$ and $P_{\nu\mu} = 0$, $\nu \neq \mu$. The next section demonstrates that the experimental fact of the existence of interfacial tension establishes that $\mathbf{P}$ is not isotropic in the interfacial region between isotropic bulk phases.

If one considers in a fluid an arbitrary fixed volume V with boundary A, then a momentum balance yields

$$\frac{d}{dt} \int_{V} \rho \mathbf{U} \, dV = \int_{A} [-\rho \hat{\varepsilon} \cdot \mathbf{U} - \hat{\varepsilon} \cdot \mathbf{P}] \, dA + \int_{V} \rho \hat{\mathbf{F}}_{e} \, dV \qquad (2.A.4)$$

that is, the change in momentum in V equals that due to fluid flow across the boundary plus that arising from surface and body forces. Using the property $d/dt \int_{V} \rho \mathbf{U} \, dV = \int_{V} (\partial \rho \mathbf{U} / \partial t) \, dV$, the divergence theorem $\int_{A} \hat{\varepsilon} \cdot \mathbf{a} \, dA = \int_{V} \nabla \cdot \mathbf{a} \, dV$, and the mass continuity equation $\partial \rho / \partial t = -\nabla \cdot (\rho \mathbf{U})$, one can rearrange (2.A.4) to the form $\int_{V} I \, dV = 0$. Since V is arbitrary, the quantity $I = 0$, which from the definition of I yields the equation of motion

$$\rho \frac{\partial \mathbf{U}}{\partial t} + \rho \mathbf{U} \cdot \nabla \mathbf{U} = -\nabla \cdot \mathbf{P} + \rho \hat{\mathbf{F}}_{e} \qquad (2.A.5)$$

At equilibrium $\mathbf{U} = 0$, and (2.A.5) reduces to the equation of hydrostatics

$$0 = -\nabla \cdot \mathbf{P} + \rho \hat{\mathbf{F}}_{e} \qquad (2.A.6)$$

For a one-component fluid, if $\mathbf{P}$ is given as a function or a functional of density, the density distribution of an equilibrium fluid can be determined with appropriate boundary conditions by solution of the equation of hydrostatics. For a multicomponent fluid, the overall equation of hydrostatics, which is what (2.A.6) is, is insufficient. The equations of hydrostatics of individual species must be used. The contributions to the species hydrostatic equations are given precise meaning in Section III, where the molecular formula for stress is considered. The equation of hydrostatics of species α is[14,]

$$0 = -\nabla \cdot \mathbf{P}_{\alpha} + B_{\alpha} + \rho_{\alpha} \hat{\mathbf{F}}_{e}^{\alpha} \qquad (2.A.7)$$

where $\mathbf{P}_{\alpha}$ and ρ_{α} are the pressure tensor and mass density of species α, $\mathbf{B}_{\alpha}$ the density of asymmetric force on α arising from the other species, and $\hat{\mathbf{F}}_{e}^{\alpha}$ the external force per unit mass on species α. The total pressure tensor is $\Sigma_\alpha \mathbf{P}_\alpha$ and the overall asymmetric force vanishes (i.e., $\Sigma_\alpha \mathbf{B}_\alpha = 0$), so that the sum of the species hydrostatic equations yields the overall hydrostatic equation. Equations 2.A.7 are known in molecular theory as the Yvon-Born-Green equations.

B. Interfacial Tension and the Pressure Tensor

The most commonly considered fluid interface is a horizontal interface between two isotropic fluid phases, the lighter of which is on top because of the earth's gravity. Aside from ordering the phases with the less dense on top, the force of gravity has a negligible effect on the fluid properties over distances of several multiples of the width of the interfacial zone (sufficiently near the critical point, there may be exceptions to this). Thus the equation of hydrostatics becomes $\nabla \cdot \mathbf{P} = 0$ in the neighborhood of planar interface in the absence of external fields other than gravity; and because of the planar geometry, the fluid density, composition, and pressure tensor depend only on the distance x perpendicular to the flat interface.

With planar conditions, the equation of hydrostatics expressed in components becomes

$$\frac{dP_{x\nu}}{dx} = 0, \qquad \nu = x, y, z \tag{2.B.1}$$

This implies that P_{xx}, P_{xy}, P_{xz} are constant in a planar system. Far above or below the interface, $\mathbf{P}$ is isotropic and equal to $P_B\mathbf{I}$, where P_B is the pressure of the equilibrium bulk phases. Hydrostatic equilibrium for a planar interface thus leads to the conclusions $P_{xx} = P_B$ and $P_{xy} = P_{xz} = 0$.

With the further condition of symmetry of the pressure tensor at equilibrium (i.e., $P_{\nu\mu} = P_{\mu\nu}$, $\nu \neq \mu$), it follows that $P_{yx} = P_{zx} = 0$. And finally, the transverse or cylindrical isotropy of a flat interface implies $P_{yy} = P_{zz}$, $P_{yz} = P_{zy} = 0$.

A planar interface then has only two distinct principal stresses, a *normal* one $P_N \equiv P_{xx}$ and a *transverse* one $P_T \equiv P_{zz}$, where P_N is constant and equal to the bulk pressure, and P_T can be a function only of x. In terms of a Cartesian basis with $\hat{i}$ normal to the interface, the pressure tensor is

$$\mathbf{P} = P_N \hat{i}\hat{i} + P_T(\hat{j}\hat{j} + \hat{k}\hat{k}) \tag{2.B.2}$$

Since $\mathbf{P} = P_B\mathbf{I}$ in the bulk phase above or below the interfacial zone, it follows that $P_T = P_N$ when $x \gg l_w$, where l_w is the width of the interfacial zone.

From the conditions stated so far—hydrostatic equilibrium and planarity, symmetry, and transverse isotropy of the pressure tensor—one can conclude only that the transverse pressure can depend on position in the interfacial zone. The experimental facts that surface tension exists and that interfacial zones are microscopically thin, however, allow stronger conclusions to be made.

Imagine a transducer of height L and width w (into the page) and positioned as illustrated in Fig. 2.B.1a either in the bulk phase or across the interfacial zone. Suppose $x=0$ lies in the interfacial zone and that the transducer in the interfacial zone is centered on $x=0$. Then the force detected by the transducer face is $F_I = -\int_{-L/2}^{L/2} P_T \, w \, dx$. If the transducer is in the bulk phase, the force detected is $F_B = -P_N w L \equiv -\int_{-L/2}^{L/2} P_N \, w \, dx$. The difference between F_I and F_B is the increase in the force exerted on transducer caused by the "tension" of the interface. The force of tension is equal to the product of the tension γ and the length w of interface sampled by the transducer. Thus the tension is given by $(F_I - F_B)/w$, or

$$\gamma \equiv \int_{-L/2}^{L/2} (P_N - P_T) \, dx \qquad (2.B.3)$$

One normally measures pressures and tensions with macroscopic devices for which L is much greater than the width l_w of the interfacial zone. Since $P_N - P_T = 0$ outside the interfacial zone, the limits of the integral in (2.B.3) can be extended to $\pm\infty$ without loss of generality, with the result that

$$\gamma = \int_{-\infty}^{\infty} (P_N - P_T) \, dx \qquad (2.B.4)$$

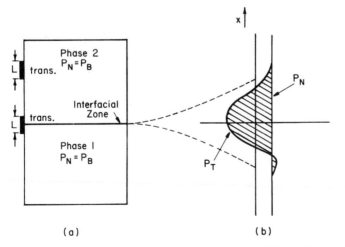

(a)   (b)

Fig. 2.B.1.   (a) Two-phase system separated by a planar interface. (b) Normal, P_N, and transverse, P_T, pressure profiles across the interfacial zone.

The experimental fact that the tension of a stable interface is positive implies that P_T must be a function of x and that through at least part of the interfacial zone P_T is less than P_N. From the experimental fact that the widths of low-vapor-pressure, high-tension interfaces are of the order of tens or hundreds of angstroms, it follows that *the transverse stress is negative over part of the interfacial zone.* Possible profiles of the pressure components across an interface are sketched in Fig. 2.B.1*b*. The tension of the interface is equal to the difference between the magnitudes of the cross-hatched areas to the left and to the right of the P_N-line. As we shall see in a later section, a P_T-profile such as that shown is indeed predicted from molecular models. An interesting feature is that there can be regions of tension, $P_T < P_N$, and of compression, $P_T > P_N$, in the interfacial zone. Stability of the interface requires only that the integral of $P_N - P_T$ across the zone be positive.

For curved interfaces neither the normal nor transverse components of pressure are constant across the interface. For example, for a spherical drop or bubble surrounded by isotropic fluid, the pressure tensor has the form

$$\mathbf{P} = P_N(r)\hat{r}\hat{r} + P_T(r)(\mathbf{I} - \hat{r}\hat{r}) \tag{2.B.5}$$

where r is the distance from the center of the spherical structure and $\hat{r}$ is a unit vector directed from the center. For this case the equation of hydrostatics yields

$$\frac{dP_N}{dr} = -2\frac{P_N - P_T}{r} \tag{2.B.6}$$

Suppose that the spherical structure is surrounded by an interfacial zone in the vicinity of $r = R$. In the center of the drop $P_N = P_T$, since pressure cannot physically have an infinite derivative. Far from the drop (i.e., $r \gg R$), $P_N = P_T$, since the pressure tensor has the isotropic form $P_B\mathbf{I}$ sufficiently far from the interfacial zone. At intermediate distances P_N and P_T are coupled through (2.B.6). The difference between the normal pressure at the center of the structure and far outside the interfacial zone is given by

$$\Delta P \equiv P_N(0) - P_N(\infty) = \int_{-R}^{\infty} \frac{P_N - P_T}{x + R} dx \tag{2.B.7}$$

This result is obtained by integrating (2.B.6) over dr from zero to infinity and making use of the transformation $r = x + R$.

In the special case that the drop is large compared to the interfacial width, that is, if R is large compared to the range of x over which $P_N - P_T$ is nonzero, (2.B.7) is well approximated by

$$\Delta P = \frac{2}{R}\gamma^{(0)} \tag{2.B.8}$$

where $\gamma^{(0)} \equiv \int_{-R}^{\infty}(P_N - P_T)\,dx$. If the drop or bubble is sufficiently large (i.e., if $R \gg l_w$), the interfacial zone is independent of the curvature of the interfacial zone and $\gamma^{(0)}$ is equal to the tension of a planar interface. In such circumstances (2.B.8) is a special case of the *Young-Laplace* equation of capillarity, which states that

$$\Delta P = 2H\gamma \tag{2.B.9}$$

in which ΔP is the pressure drop across an interface of tension γ and mean curvature $H \equiv \frac{1}{2}(R_1^{-1} + R_2^{-1})$, R_i being the principal radii of curvature at any point on the surface defining an interface.

As is made clear by (2.B.7), the Young-Laplace equation is a special limit in which the magnitudes of the radii of curvature are large compared to the width of the interfacial zone separating the phases on either side of the interface.

C. Molecular Theory of the Pressure Tensor and Mechanical Equilibrium

One of the aims of this chapter is to investigate the detailed behavior of the pressure tensor in interfacial zones with the aid of molecular models. The formal molecular theory of the pressure tensor was developed some time ago by Irving and Kirkwood[13] and Bearman and Kirkwood[14] for classical molecules interacting via pairwise additive central forces. We shall briefly review their results and add to them the results for central three-body forces. The molecular origin of the anisotropy of the pressure tensor in an interface is brought out clearly by the theory.

Consider first a one-component system. Denote by $n(\mathbf{r})$ and $N_2(\mathbf{r},\mathbf{r}')$ respectively, the density and the doublet density of particles. Then $n(\mathbf{r})\,d^3r$ is the probable number of particles in the volume element d^3r and $n_2(\mathbf{r},\mathbf{r}')$ $d^3r\,d^3r'$ is the probable number of pairs with one member of the pair in d^3r and the other in d^3r'; d^3r is centered on $\mathbf{r}$ and d^3r' on $\mathbf{r}'$.

Imagine a surface element dA (Fig. 2.C.1) of orientation $\hat{\varepsilon}$ and fixed at a position $\mathbf{r}$. If there were no forces between the molecules, the pressure exerted on dA would be purely kinetic—that is, it would be $n(\mathbf{r})kT$ in magnitude and normal to the surface element—so that the pressure tensor would be

$$\mathbf{P}^K \equiv nkT\,\mathbf{I} \tag{2.C.1}$$

for a system at thermal equilibrium.

The kinetic contribution to the pressure tensor must be added to the contribution $\mathbf{P}^u$ arising from intermolecular forces that molecules exert on one another across the surface element dA. Particle 1 at position $\mathbf{r}'$ and particle 2

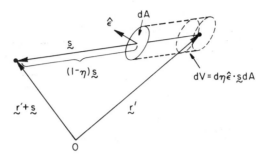

Fig. 2.C.1.   Particles at $\mathbf{r}'$ and $\mathbf{r}'+\mathbf{s}$ exerting a force on one another across the area element dA with normal $\hat{\varepsilon}$ and located at $\mathbf{r}$.

at $\mathbf{r}'+\mathbf{s}$ in Fig. 2.C.1 act across dA only if their connecting vector $\mathbf{s}$ intersects the surface at some distance $|\eta s|$ $(0\leq\eta\leq1)$ along $\mathbf{s}$. The resulting force on the positive $\hat{\varepsilon}$ side of dA is $\mathbf{s}/s[du(s)/ds]$, where $u(s)$ denotes the potential energy between the pair of particles. For fixed $\mathbf{s}$, the volume in which particle 1 must lie is $dV=dA\hat{\varepsilon}\cdot\mathbf{s}\,d\eta$ and particle 2 must lie in a volume d^3s fixed on $\mathbf{s}$. The probable number of such pairs is $\mathbf{n}_2(\mathbf{r}',\mathbf{r}'+\mathbf{s})\,(dA\hat{\varepsilon}\cdot\mathbf{s}\,d\eta)\,d^3s$ and the probable force is $\mathbf{s}/s[du(s)/ds]$ times the probable number of pairs. Noting that $\mathbf{r}'=\mathbf{r}-\eta\mathbf{s}$, we determine the total force exerted across dA by integrating over $d\eta$ and d^3s with the constraints $0\leq\eta\leq1$ and $\mathbf{s}\cdot\hat{\varepsilon}>0$, ensuring that the interacting pairs are separated by dA and that the force is that exerted by particles on the positive $\hat{\varepsilon}$ side on those of the other side. The result is

$$. dA\hat{\varepsilon}\cdot\left\{\int_{\mathbf{s}\cdot\hat{\varepsilon}>0}\left[\int_0^1\frac{\mathbf{s}\mathbf{s}}{s}\frac{du}{ds}n_2(\mathbf{r}-\eta\mathbf{s},\mathbf{r}-\eta\mathbf{s}+\mathbf{s})\,d\eta\right]d^3s\right\}\equiv-\hat{\varepsilon}\cdot\mathbf{P}^u\,dA$$

$$(2.C.2)$$

With the symmetry condition $\eta_2(\mathbf{r},\mathbf{r}')=\eta_2(\mathbf{r}',\mathbf{r})$ and the coordinate transformations $\mathbf{s}'=-\mathbf{s}$, $\eta'=1-\eta$, the tensor in the curly brackets of (2.C.2) transforms into the same formal expression except the constraint on the vector $\mathbf{s}$ becomes $\mathbf{s}\cdot\hat{\varepsilon}<0$. Taking half the sum of (2.C.2) and its transformed version enables one to identify a formula for $\mathbf{P}^u$ in which the volume integral is unconstrained. Addition of this expression to $\mathbf{P}^K$ yields the Irving-Kirkwood formula for the pressure tensor:

$$\mathbf{P}=nkT\mathbf{I}-\frac{1}{2}\iint_0^1\frac{\mathbf{s}\mathbf{s}}{s}\frac{du}{ds}n_2(\mathbf{r}-\eta\mathbf{s},\mathbf{r}-\eta\mathbf{s}+\mathbf{s})\,d\eta\,d^3s\qquad(2.C.3)$$

It is straightforward to include centrally symmetric three-body forces in the pressure tensor. Denote by $u^{(3)}$ (r_{12},r_{13},r_{23}) the three-body intermolecular potential of particles 1, 2, and 3; r_{ij} is the magnitude of the vector $\mathbf{r}_{ij}\equiv\mathbf{r}_i$

$-\mathbf{r}_j$, and $\mathbf{r}_i$ is the position of the ith particle. The three-body force exerted on particle 1 by particles 2 and 3 is

$$-\nabla_{r_1}u^{(3)} = \frac{\mathbf{r}_{21}}{r_{12}}f_{12}(r_{12},r_{13},r_{23}) + \frac{\mathbf{r}_{31}}{r_{31}}f_{13}(r_{12},r_{13},r_{23}) \qquad (2.C.4)$$

where

$$f_{12} \equiv \left(\frac{\partial u^{(3)}}{\partial r_{12}}\right)_{r_{13},r_{23}}, \qquad f_{13} \equiv \left(\frac{\partial u^{(3)}}{\partial r_{13}}\right)_{r_{12},r_{23}} \qquad (2.C.5)$$

Adding to the right-hand side of (2.C.2) the contributions from the three-body force whose lines of action pass through dA at $\mathbf{r}$, using the fact that the particles are identical, and again symmetrizing to get rid of $\hat{\varepsilon}$ in the volume integrals, we obtain

$$\mathbf{P} = \text{rhs } (2.C.3) - \iiint_0^1 \frac{\mathbf{ss}}{s}f_{12}(s,s',|s'-s|)$$

$$\times n_3(\mathbf{r}-\eta\mathbf{s},\mathbf{r}-\eta\mathbf{s}+\mathbf{s},\mathbf{r}-\eta\mathbf{s}+\mathbf{s}')\,d\eta\,d^3s\,d^3s' \qquad (2.C.6)$$

where n_3 is the triplet density function.

The multicomponent versions of (2.C.3) and (2.C.6) are

$$\mathbf{P} = \sum_\alpha n_\alpha kT\mathbf{I} - \sum_{\alpha,\beta}\frac{1}{2}\iint_0^1 \frac{\mathbf{ss}}{s}\frac{du_{\alpha\beta}}{ds}n_2^{\alpha\beta}(\mathbf{r}-\eta\mathbf{s},\mathbf{r}-\eta\mathbf{s}+\mathbf{s})\,d\eta\,d^3s \qquad (2.C.7)$$

and

$$\mathbf{P} = \text{rhs } (2.C.7) - \sum_{\alpha,\beta,\gamma}\iiint_0^1 \frac{\mathbf{ss}}{s}f_{12}^{\alpha\beta\gamma}(s,s',|s'-s|)$$

$$\times n_3^{\alpha\beta\gamma}(\mathbf{r}-\eta\mathbf{s},\mathbf{r}-\eta\mathbf{s}+\mathbf{s},\mathbf{r}-\eta\mathbf{s}+\mathbf{s}')\,d\eta\,d^3s\,d^3s' \qquad (2.C.8)$$

We note from the results above that spherical symmetry of interparticle interactions requires symmetry of the pressure tensor.

From its molecular derivation, the origin of pressure tensor anisotropy in inhomogeneous fluid becomes clear. First of all, the anisotropy must arise from intermolecular forces, since the kinetic part of the pressure tensor is isotropic. Consider a one-component fluid whose density varies in the x-direction and imagine two identical circular cones of very small solid angle $\delta\Omega$ originating from a molecule placed at point O, the axis of one cone being along the x-axis and that of the other along the z-axis. The situation is illustrated in Fig. 2.C.2. Suppose that the length of each cone is equal to the

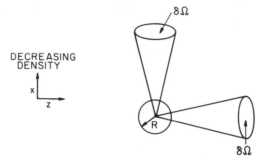

Fig. 2.C.2. Normal and tranverse cones of particles interacting with a reference particle in a planar fluid.

range of intermolecular force between the fluid particles. The force between the reference particle and the contents of a cone is proportional to the number of particles in the cone. Since the density varies with x, the number of particles in the normal cone (along the x-axis) is different from the number in the transverse cone. Thus the magnitude of the force exerted on the reference particle depends on the orientation of the cone. If a small sphere of radius R is drawn to include just the reference molecule, then the force exerted by the contents of a cone on the reference molecule divided by the area $R^2\delta\Omega$ is equal to the potential energy contribution to the pressure component along the cone axis in the limit that $\delta\Omega$ shrinks to zero. Since the force depends on orientation of the cone axis, so also do the pressure components, and at least one of the principal stresses must be distinct.

It follows then that the stress anisotropy in the interfacial zone of a one-component fluid is the consequence of the density variation across the zone. The nature of the density variation across the interfacial zone is governed by the condition that the kinetic part of the normal pressure, nkT, balance with the potential part, P_N^u, so that the sum remains constant, which is the condition of hydrostatic equilibrium.

In a multicomponent fluid, the anisotropy of the pressure tensor arises not only from density variations but also from the composition variations required to maintain hydrostatic equilibrium for each species.

The equations of hydrostatic equilibrium for each species were presented formally in Section II.A. Let us now turn to the derivation of these equations and the molecular formulas for $\mathbf{P}_\alpha$ and $\mathbf{B}_\alpha$, the input needed for the equations.

Since the hydrostatic equations are equilibrium stand-ins for the momentum balance equations, the logical way to derive them is through the momentum balance. It is convenient to introduce $f_\alpha(\mathbf{r}, \mathbf{v}, t)$ and $f_2^{\alpha\beta}(\mathbf{r}, \mathbf{r}', \mathbf{v}, \mathbf{v}', t)$, the singlet and doublet velocity distribution functions. They have the meaning that $f_\alpha(\mathbf{r},\mathbf{v},t)\,d^3v$ is the number density at $\mathbf{r}$ of particles of species α having velocity between $\mathbf{v}$ and $\mathbf{v}+d\mathbf{v}$ and that $f_2^{\alpha\beta}(\mathbf{r},\mathbf{r}',\mathbf{v},\mathbf{v}',t)\,d^3v\,d^3v'$ is the doublet density of a pair of particles of species α and β at $\mathbf{r}$ and $\mathbf{r}'$ and having respective velocities in the ranges $\mathbf{v}$ to $\mathbf{v}+d\mathbf{v}$ and $\mathbf{v}'$ to $\mathbf{v}'+d\mathbf{v}'$. The density $h_\alpha(\mathbf{r},t)$ and doublet density $n_2^{\alpha\beta}(\mathbf{r},\mathbf{r}',t)$ are generated by integrating over velocities:

$$n_\alpha(\mathbf{r},t)=\int\!\int f_\alpha d^3v; \qquad n_2^{\alpha\beta}(\mathbf{r},\mathbf{r}',t)=\int\!\int f_2^{\alpha\beta}d^3v\,d^3v' \qquad (2.\text{C}.9)$$

The local average velocity $\mathbf{U}_\alpha$ of species α is given by

$$n_\alpha(\mathbf{r},t)\mathbf{U}_\alpha(\mathbf{r},t)=\int \mathbf{v}f_\alpha(\mathbf{r},\mathbf{v},t)\,d^3v \qquad (2.\text{C}.10)$$

and the local mass average velocity by

$$\mathbf{U}=\sum_\alpha \frac{m_\alpha n_\alpha \mathbf{U}_\alpha}{\sum_\beta m_\beta n_\beta} \qquad (2.\text{C}.11)$$

m_α being the molecular mass of species α.

For a system of ν different species of classical particles interacting by way of pairwise additive centrally symmetric forces, the laws of mechanics require that f_α obey the kinetic equation[33]

$$\frac{\partial f_\alpha}{\partial t}+\mathbf{v}\cdot\nabla f_\alpha+\frac{\mathbf{F}_e^\alpha}{m_\alpha}\cdot\frac{\partial f_\alpha}{\partial \mathbf{v}}=\sum_{\beta=1}^\nu \frac{1}{m_\alpha}\int \nabla u_{\alpha\beta}(|\mathbf{r}-\mathbf{r}|)\cdot\frac{\partial f_2^{\alpha\beta}}{\partial \mathbf{v}}d^3r'\,d^3v'$$

$$(2.\text{C}.12)$$

where $\mathbf{F}_e^\alpha$ is the external force acting on a particle of species α and $\partial/\partial\mathbf{v}$ denotes the gradient with respect to velocity.

Multiplying (2.C.12) by $\mathbf{v}\,d^3v$, integrating, and rearranging (using the property $\int(\partial h/\partial\mathbf{v})\,d^3v=0$ for $h=f_\alpha$ or $f_2^{\alpha\beta}$), we obtain the following momentum balance equation for species α:

$$m_\alpha\frac{\partial(n_\alpha \mathbf{U}_\alpha)}{\partial t}+m_\alpha\nabla\cdot[n_\alpha(\mathbf{U}\mathbf{U}_\alpha+\mathbf{U}_\alpha\mathbf{U}-\mathbf{U}\mathbf{U})]=\mathbf{S}_\alpha+n_\alpha \mathbf{F}^\alpha \qquad (2.\text{C}.13)$$

where $\mathbf{S}_\alpha$ is the local force per unit volume arising from interaction of particles. By definition

$$\mathbf{S}_\alpha\equiv-\nabla\cdot\mathsf{P}_\alpha^k-\sum_\beta\int\nabla u_{\alpha\beta}(|\mathbf{r}'-\mathbf{r}|)n_2^{\alpha\beta}(\mathbf{r},\mathbf{r}',t)\,d^3r' \qquad (2.\text{C}.14)$$

where P_α^k is the local kinetic pressure tensor exerted by species α:

$$P_\alpha^k = m_\alpha \int (\mathbf{v}-\mathbf{U})(\mathbf{v}-\mathbf{U}) f_\alpha \, d^3v \qquad (2.C.15)$$

For a fluid at equilibrium $\mathbf{U}=0$, f_α is a Maxwell-Boltzmann distribution, and $P_\alpha^k = n_\alpha kT \mathbf{I}$. For brevity, the variable time t will no longer be displayed as an explicit argument of the quantities dealt with here.

Because (2.C.13) is a species balance equation, S_α cannot be expressed entirely as a divergence of a tensor. However, it can be cast into the form of a divergence of what is called the species pressure tensor P_α *plus* a residue termed the asymmetric force B_α, which arises from fluid inhomogeneity.[15] To accomplish this division, note that the second term of (2.C.14) can be rearranged to

$$\sum_\beta \int \nabla u_\alpha(|\mathbf{r}'-\mathbf{r}|) n_2^{\alpha\beta}(\mathbf{r},\mathbf{r}') \, d^3r'$$

$$= \sum_\beta \frac{1}{2} \int \frac{\mathbf{s}}{s} \frac{du_{\alpha\beta}(s)}{ds} \left[n_2^{\alpha\beta}(\mathbf{r},\mathbf{r}+\mathbf{s}) - n_2^{\alpha\beta}(\mathbf{r},\mathbf{r}-\mathbf{s}) \right] d^3s \qquad (2.C.16)$$

Equation 2.C.16 is obtained by introduction of the variable transformation $\{\mathbf{r},\mathbf{r}'\} \to \{\mathbf{r},\mathbf{s}=\mathbf{r}'-\mathbf{r}\}$ and addition of the transformed integral to itself under the further transformation $\mathbf{s} \to -\mathbf{s}$. The result divided by 2 yields (2.C.16). Now with the aid of the Taylor series

$$n_2^{\alpha\beta}(\mathbf{r},\mathbf{r}-\mathbf{s}) = \sum_{k=0}^{\infty} \frac{(-\mathbf{s}\cdot\nabla)^k}{k!} n_2^{\alpha\beta}(\mathbf{r}+\mathbf{s},\mathbf{r})$$

and the particle-and-position-interchange condition $n_2^{\alpha\beta}(\mathbf{r}+\mathbf{s},\mathbf{r})=n_2^{\beta\alpha}(\mathbf{r},\mathbf{r}+\mathbf{s})$, we obtain, with some manipulation,

$$S_\alpha = -\nabla \cdot P_\alpha + B_\alpha \qquad (2.C.17)$$

where the pressure tensor for species α is

$$P_\alpha \equiv P_\alpha^k - \sum_{\beta=1}^{\nu} \frac{1}{2} \int \frac{\mathbf{ss}}{s} \frac{du_{\alpha\beta}}{ds} \left[\sum_{k=0}^{\infty} \frac{(-\mathbf{s}\cdot\nabla)^k}{(k+1)!} \right] n_2^{\alpha\beta}(\mathbf{r},\mathbf{r}+\mathbf{s}) \, d^3s \qquad (2.C.18)$$

and the asymmetric force on species α is[15]

$$B_\alpha = \sum_\beta \frac{1}{2} \int \frac{\mathbf{s}}{s} \frac{du_{\alpha\beta}}{ds} \left[n_2^{\alpha\beta}(\mathbf{r}-\mathbf{s},\mathbf{r}) - n_2^{\beta\alpha}(\mathbf{r}-\mathbf{s},\mathbf{r}) \right] d^3s \qquad (2.C.19)$$

Plainly this force vanishes in homogeneous fluid, for then $n_2^{\alpha\beta}(\mathbf{r},\mathbf{r}')=n_2^{\beta\alpha}(\mathbf{r},\mathbf{r}')$. In the linear gradient approximation, Bearman and Kirkwood[14] introduced in the context of transport processes the partial stress of species α (their $\underset{\sim}{\sigma}_\alpha$) and the asymmetric force (their $\overline{\mathbf{F}}_\alpha^{(1)}*$). In the present chapter we are more concerned with inhomogeneous fluids at equilibrium. Equations 2.C.13 and 2.C.17 would, however, be the appropriate starting points if transport in strongly inhomogeneous systems (e.g., interfacial transport) were of interest.

It remains to show that the pressure tensor given by (2.C.18) is the same as that derived directly from force analysis. From the Taylor series

$$n_2^{\alpha\beta}(\mathbf{r}-\eta\mathbf{s},\mathbf{r}-\eta\mathbf{s}+\mathbf{s})=\sum_{k=0}^{\infty}\eta^k\frac{(-\mathbf{s}\cdot\nabla)^k}{k!}n_2^{\alpha\beta}(\mathbf{r},\mathbf{r}+\mathbf{s}) \quad (2.C.20)$$

and the relation $\int_0^1\eta^k\,d\eta=1/(k+1)$, it follows that the formula

$$\mathbf{P}_\alpha=\mathbf{P}_\alpha^k-\sum_{\beta=1}^{\nu}\frac{1}{2}\iint\frac{\mathbf{s}\mathbf{s}}{s}\int_0^1\frac{du_{\alpha\beta}}{ds}n_2^{\alpha\beta}(\mathbf{r}-\eta\mathbf{s},\mathbf{r}-\eta\mathbf{s}+\mathbf{s})\,d\eta\,d^3s \quad (2.C.21)$$

is the same as (2.C.18), in agreement with the result (cf. 2.C.3) from force analysis. Although (2.C.18) is useful for deriving a formula for interfacial tension or when trying to approximate $\mathbf{P}_\alpha$ by a few terms of a gradient expansion, (2.C.21) is the preferred form for direct computation of pressure since it does not require convergence (not to mention term-by-term evaluation) of a Taylor series.

For a system in mechanical equilibrium, $\mathbf{U}_\alpha=\mathbf{U}=0$ and (2.C.13) reduces to the condition of hydrostatic equilibrium for species α, namely,

$$\mathbf{S}_\alpha+n_\alpha\mathbf{F}_e^\alpha=-\nabla\cdot\mathbf{P}_\alpha+\mathbf{B}_\alpha+n_\alpha\mathbf{F}_e^\alpha=0 \quad (2.C.22)$$

the form indicated earlier at (2.A.7). (Since $\mathbf{F}_e^\alpha$ is a force per molecule, $n_\alpha\mathbf{F}_e^\alpha=m_\alpha n_\alpha(\mathbf{F}_e^\alpha/m_\alpha)=\rho_\alpha\hat{\mathbf{F}}_e^\alpha$.) At equilibrium $\mathbf{P}_\alpha^k=n_\alpha kT\mathbf{I}$ so that (2.C.22), with the form of $\mathbf{S}_\alpha$ given by (2.C.14), becomes

$$-kT\nabla n_\alpha-n_\alpha\nabla u_e^\alpha=\sum_\beta\int\nabla u_{\alpha\beta}(|\mathbf{r}'-\mathbf{r}|)n_2^{\alpha\beta}(\mathbf{r},\mathbf{r}')\,d^3r' \quad (2.C.23)$$

where $\mathbf{F}_e^\alpha\equiv-\nabla u_e^\alpha$. Equation 2.C.23 is the familiar Yvon-Born-Green (YBG) equation. Its significance as a means of representing the hydrostatic condition for fluid interfaces was long ago recognized,[11] although its equivalence to the more physical momentum balance, (2.C.22), is sometimes overlooked in recent work on interfacial theory.

If three-body, spherically symmetric forces were included, the only change would be three-body additions to the species pressure tensor and the asymmetric force. Omitting the details, we give here the resulting equilibrium YBG equations and the total equilibrium pressure tensor for this case:

$$-kT \nabla n_\alpha - n_\alpha \nabla u_e^\alpha = \sum_\beta \int \nabla u_{\alpha\beta}(|\mathbf{r}'-\mathbf{r}|) n_2^{\alpha\beta}(\mathbf{r},\mathbf{r}') \, d^3 r'$$

$$+ \sum_{\beta,\gamma} \iint \nabla u_{\alpha\beta\gamma}^{(3)} n_3^{\alpha\beta\gamma}(\mathbf{r},\mathbf{r}',\mathbf{r}'') \, d^3 r' d^3 r'' \quad (2.C.24)$$

and

$$P = \sum_\alpha n_\alpha kT \mathbf{I} - \sum_{\alpha,\beta} \frac{1}{2} \int \frac{\mathbf{ss}}{s} \frac{du_{\alpha\beta}}{ds} \left[\sum_{k=0}^\infty \frac{(-\mathbf{s}\cdot\nabla)^k}{(k+1)!} \right] n_2^{\alpha\beta}(\mathbf{r},\mathbf{r}+\mathbf{s}) \, d^3 s$$

$$- \sum_{\alpha,\beta,\gamma} \int \frac{\mathbf{ss}}{s} f_{12}^{\alpha\beta\gamma}(s,s'|\mathbf{s}'-\mathbf{s}|) \left[\sum_{k=0}^\infty \frac{(-\mathbf{s}\cdot\nabla)^k}{(k+1)!} \right] n_3^{\alpha\beta\gamma}(\mathbf{r},\mathbf{r}+\mathbf{s},\mathbf{r}+\mathbf{s}') \, d^3 s \, d^3 s'$$

$$(2.C.25)$$

This formula for P is equivalent to that obtained at (2.C.8) from force analysis.

We have also studied the case for which the interactions between the particles depend on particle orientation as well as position, but the densities n_α and external force depend only on particle position. For this case, considering only two-body forces, we obtain the YBG equations,

$$-kT \nabla n_\alpha - n_\alpha \nabla u_e^\alpha = \sum_\beta \int \nabla u_{\alpha\beta}(|\mathbf{r}'-\mathbf{r}|,\mathbf{e},\mathbf{e}') n_2^{\alpha\beta}(\mathbf{r},\mathbf{r}',\mathbf{e},\mathbf{e}') \frac{d^3 r' d^3 e \, d^3 e'}{(8\pi^2)^2}$$

$$(2.C.26)$$

and the pressure tensor,

$$P = \sum_\alpha n_\alpha kT \mathbf{I} - \sum_{\alpha,\beta} \frac{1}{2} \int \mathbf{s} \nabla_s u_{\alpha\beta}(\mathbf{s},\mathbf{e},\mathbf{e}') \left[\sum_{k=0}^\infty \frac{(-\mathbf{s}\cdot\nabla)^k}{(k+1)!} \right]$$

$$\times n_2^{\alpha\beta}(\mathbf{r},\mathbf{r}+\mathbf{s},\mathbf{e},\mathbf{e}') \frac{d^3 s \, d^3 e \, d^3 e'}{(8\pi^2)^2} \quad (2.C.27)$$

where $\mathbf{e}$ represents the particle orientation and $d^3 e = \sin\theta \, d\theta \, d\phi \, d\psi$, where θ, ϕ, and ψ are Eulerian angles $(0<\theta<\pi, 0<\phi<2\pi, 0<\psi<2\pi)$.

The interfacial tension can be computed from the formula given at (2.B.4). Taking the difference between the normal (P_{xx}) and transverse (P_{zz}) components of (2.C.25) and integrating the difference over dx from $-\infty$ to $+\infty$ with the condition that densities and pressure components approach constant bulk values at the limits, we obtain

$$\gamma = \sum_{\alpha,\beta} \frac{1}{2} \iint_{-\infty}^{\infty} \frac{s_z^2 - s_x^2}{s} \frac{du_{\alpha\beta}}{ds} n_2^{\alpha\beta}(\mathbf{r}, \mathbf{r}+\mathbf{s}) \, dx \, d^3s$$

$$+ \sum_{\alpha,\beta,\gamma} \iiint_{-\infty}^{\infty} \frac{s_z^2 - s_x^2}{s} f_{12}^{\alpha\beta\gamma}(s, s', |\mathbf{s}'-\mathbf{s}|) n_3^{\alpha\beta\gamma}(\mathbf{r}, \mathbf{r}+\mathbf{s}, \mathbf{r}+\mathbf{s}') \, dx \, d^3s \, d^3s'$$

$$(2.C.28)$$

The two-body version of this formula was first obtained by Kirkwood and Buff,[10] and the one-component version has recently been published by Present et al.[34]

For the case represented by (2.C.27), it follows that[35, 36]

$$\gamma = \sum_{\alpha,\beta} \frac{1}{2} \iint_{-\infty}^{\infty} \left[s_z \frac{\partial u_{\alpha\beta}}{\partial s_z} - s_x \frac{\partial u_{\alpha\beta}}{\partial s_x} \right] n_2^{\alpha\beta}(\mathbf{r}, \mathbf{r}+\mathbf{s}, \mathbf{e}, \mathbf{e}') \, dx \, d^3s \frac{d^3e \, d^3e'}{(8\pi^2)^2}$$

$$(2.C.29)$$

To use the theoretical formulas presented in this section, one must have correlation functions of inhomogeneous fluid. In principle, these can be obtained either by solving a YBG hierarchy,[33, 37] which relates the rth-order distribution function to the $(r+1)$st-order function, or by evaluating a density functional expansion calling for direct correlation functions of homogeneous fluid to every order. Both these possibilities are impractical and have generally been circumvented by approximating the correlation functions of inhomogeneous fluid by the correlation functions of homogeneous fluid evaluated at local density or densities.[15, 29, 38-40] Even with such an approximation, the YBG equation is an integrodifferential equation that is difficult to solve.[38, 39]

A very profitable simplification of the theory of inhomogeneous fluid is the so-called gradient theory, an approximation going back to Rayleigh[1] and van der Waals.[2] The gradient theory of the YBG equations and the stress tensor is outlined in the next section. Succeeding sections present evidence that gradient theory fairly accurately predicts density profiles and tensions of planar interfaces.

D. Gradient Theory

Most of the arguments in this section are put forth for a one-component fluid, but multicomponent results are stated.

In homogeneous fluid the pressure depends on the particle forces, the fluid density (or densities in multicomponent fluids), and the correlation functions, which in turn also depend on particle forces and fluid density. Since density is the same throughout the fluid, there is no problem of deciding what density to use in computing pressure of homogeneous fluid. In inhomogeneous fluid, on the other hand, density and the pressure tensor vary in space. Because intermolecular forces act at distance, the local correlation functions and the pressure tensor $P(r)$ depend not only on the local density $n(r)$, but also on the density distribution in the vicinity of r. Thus the correlation functions and the pressure tensor are functionals of density.

The neighborhood about r in which the density distribution contributes to the local pressure tensor is quite small, being determined by the range of intermolecular forces. This leads to the expectation that the local pressure tensor can be characterized by the local density and a finite number of local derivatives of density, that is, that P is a function of the form

$$P = P(n, \nabla n, \nabla \nabla n, \ldots, \nabla \cdots \nabla n) \qquad (2.D.1)$$

Rayleigh and van der Waals applied such a *gradient* representation to the free energy of inhomogeneous fluid. As far as we have ascertained, Korteweg[41] was the first to apply gradient theory of the pressure tensor of fluid interfaces. As stated in the introduction, this approximation has been applied extensively during the last two decades.[42]

Sufficiently near the critical point, the density is a weak function of density, and only the first few derivatives of n are necessary for estimating the local pressure tensor. Numerical results, some to be discussed briefly in the next section, and others presented elsewhere,[26, 43] indicate that even for the sharp interfaces far from the critical point, properties of an interface can be estimated accurately (to, say, within perhaps 10%) by keeping only the first two derivatives of n. Thus, this will be the level of the gradient theory described here.

A Taylor expansion of P about $\nabla n = \nabla \nabla n = \cdots = 0$ yields

$$P = P_0 + A_1 \cdot \nabla n + A_2 : \nabla n \nabla n + B_1 : \nabla \nabla n + \ldots \qquad (2.D.2)$$

where P_0 is isotropic because it is the pressure tensor of fluid in the absence of density gradients. Thus, $P_0 = P_0(n)I$, where $P_0(n)$ is the pressure of homogeneous fluid at density n. The quantities A_1, A_2, $B_1, \ldots$ are also isotropic functions of n, since they are determined by the properties of fluid in the absence of density gradients. Since A_1 is a third-rank isotropic tensor, it is identically zero. And the fourth-rank isotropic tensors A_2 and B_1 have the property that $T : ab = T_1 ab + T_2(a \cdot b)I$, T_1 and T_2 being the only components of the fourth-rank isotropic tensor T. Through second order in ∇, then, the pressure tensor has the form

$$P(r) = P_0 I + l_{11} \nabla^2 n I + l_{12} \nabla \nabla n + l_{21}(\nabla n)^2 I + l_{22} \nabla n \nabla n \quad (2.D.3)$$

where P_0 and l_{ij} are scalar functions of the local density $n(r)$. Equation 2.D.3 is valid through third order in ∇ because the coefficients of the odd powers of ∇ vanish owing to the isotropy of a gradient-free fluid.

We shall refer to formulas such as (2.D.3) as those of gradient theory. The relationship between the gradient theory of stress, (2.D.3), and the thermodynamic gradient theory of free energy has recently been discussed by Serrin[44] from the continuum-mechanical point of view. According to gradient theory, the state of stress can be determined for a given state of inhomogeneity once the functions $P_0(n)$ and $l_{ij}(n)$ are known. These functions must be determined either experimentally or from a molecular theoretical model. Numerous empirical and theoretical models exist for the equation of state $P_0(n)$ of homogeneous fluid. This is not the case for the l_{ij}, for inhomogeneous fluid has not received the attention that has been accorded homogeneous fluid.

An equation of state, much used because it is simple and yet captures most of the qualitative features of a real fluid, is the van der Waals (VDW) equation

$$P_0(n) = \frac{nkT}{1 - nb} - n^2 a \quad (2.D.4)$$

where b and a are molecular size and energy parameters of the model. Isotherms of P_0 versus n^{-1} predicted by this equation are plotted in Fig. 2.D.1.

The density states under the spinodal envelope (dashed dome) indicated in Fig. 2.D.1 cannot be realized in homogeneous bulk phase, because such states of bulk phase are unstable. For this reason, the density dependence of $P_0(n)$ can be investigated in homogeneous fluid only in the density domain lying outside the spinodal dome. However, in principle, the density dependence of $P_0(n)$ and $l_{ij}(n)$ for the densities under the spinodal can be determined experimentally in inhomogeneous fluid (e.g., in interfacial zones), by applying (2.D.3) as the constitutive equation for the pressure tensor. It might be difficult to measure P within an interface, but this is not relevant. The point is that density states unstable in homogeneous fluid can be stabilized by formation of density gradients, whose effects on the pressure tensor are measured by the l_{ij} in (2.D.3).

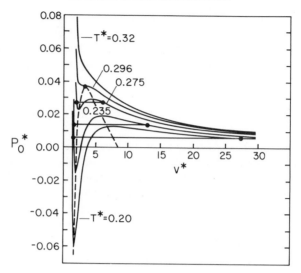

Fig. 2.D.1.   Pressure-volume isotherms of a van der Waals fluid. The solid dome locates the liquid-vapor coexistence envelope and the dashed dome is the spinodal envelope enclosing unstable homogeneous states: $P_0^* = P_0 b^2/a$, $v^* = 1/n^*$, $n^* = nb$, $T^* - bkT/a$.

Whether the values of $P_0(n)$ in the spinodal density domain are correctly predicted by analytical continuation of an equation of state derived theoretically or determined empirically for stable or metastable homogeneous fluid is an important unresolved theoretical question. It has been assumed traditionally that the analytical continuation is valid. For example, it is generally accepted that Maxwell tie-line constructions of coexisting phases are correct and that thermodynamic functions can be computed by integrating the equation of state over any desired density domain. In what follows we accept without further question the validity of this analytic continuation of homogeneous fluid thermodynamic functions. Our point of view is that there is no reason to expect these functions to change character in the spinodal density or composition domain; rather, we hold that stability requirements are met merely by formation of stabilizing density or composition gradients with a corresponding change of P from an isotropic to an anisotropic state. The consequences of not assuming analytical continuation of the thermodynamic functions of homogeneous fluid have been discussed recently by Serrin.[44]

For a planar interface, $n = n(x)$, the normal and transverse principal pressures are

$$P_N = P_0(n) + (l_{11} + l_{12}) \frac{d^2 n}{dx^2} + (l_{21} + l_{22}) \left(\frac{dn}{dx} \right)^2 \qquad (2.D.5)$$

and

$$P_T = P_0(n) + l_{11}\frac{d^2n}{dx^2} + l_{21}\left(\frac{dn}{dx}\right)^2 \qquad (2.\text{D}.6)$$

The interfacial tension, computed from the integral over dx of the difference $P_N - P_T$, can be rearranged to the form

$$\gamma = \int_{-\infty}^{\infty} c\left(\frac{dn}{dx}\right)^2 dx \qquad (2.\text{D}.7)$$

where the *influence parameter*[26] c is defined by

$$c = l_{22} - \frac{\partial l_{12}}{\partial n} \qquad (2.\text{D}.8)$$

As will become apparent below, the influence parameter plays a key role in the statistical mechanical gradient theory of inhomogeneous fluid. Here we see that it measures the contribution of density gradient to interfacial tension.

Equation 2.D.7 was obtained by Rayleigh and given full interpretation by van der Waals. According to the expression, the smaller the density gradient between phases, the lower the tension. In subsequent analysis we shall see that small gradients and broad interfaces occur near critical points and as a consequence tension is small near critical points.

Equations 2.D.5 to 2.D.7 with the boundary conditions $n \to n_l$ as $x \to -\infty$ and $n \to n_g$ as $x \to +\infty$ completely determine the density and pressure profiles and the tension of a planar interface. Since P_N is constant and equal to $P_B = P_0(n_l) = P_0(n_g)$, (2.D.5) is a second-order differential equation from which $n(x)$ can be determined. Then P_T versus x can be computed from (2.D.6) and the tension from (2.D.7). As input to the theory we need $P_0(n)$ as well as the l_{ij}. We turn now to the molecular theory of these quantities.

No special assumptions regarding molecular properties have been made at this point of development of gradient theory. Let us now, however, restrict our attention to a fluid of classical particles interacting via pairwise additive centrally symmetric forces. It is convenient to express the doublet density $n_2(\mathbf{r}, \mathbf{r} + \mathbf{s})$ in terms of the pair correlation function $g(\mathbf{r}, \mathbf{r} + \mathbf{s})$ through the definition

$$n_2(\mathbf{r}, \mathbf{r} + \mathbf{s}) \equiv n(\mathbf{r})n(\mathbf{r} + \mathbf{s})g(\mathbf{r}, \mathbf{r} + \mathbf{s}) \qquad (2.\text{D}.9)$$

Consistent with (2.D.1), we assume the functional dependence of g on the density distribution can be characterized by a local density, say $n(\mathbf{r})$, and a few of its derivatives. Since g is a scalar function of $\mathbf{s}$, $n(\mathbf{r})$, and gradients of $n(\mathbf{r})$, the general form of the gradient expansion of g is

$$g(\mathbf{r},\mathbf{r}+\mathbf{s})=g_0\big(s;n(\mathbf{r})\big)+q_N\hat{s}\cdot\nabla n+\big[t_N\hat{s}\hat{s}+t_T(\mathbf{I}-\hat{s}\hat{s})\big]\;:\;\nabla n\nabla n$$
$$+\big[v_N\hat{s}\hat{s}+v_T(\mathbf{I}-\hat{s}\hat{s})\big]\;:\;\nabla\nabla n \qquad (2.D.10)$$

where $q_N, t_N, \ldots, v_T$ are scalar functions of s and $n(\mathbf{r})$. These functions depend in a rather complicated way on two-, three-, and four-body correlation functions of homogeneous fluids. Their structures are discussed briefly in Section III.D. The subscript N or T indicates gradient-induced correlations in particles whose line of centers is parallel or perpendicular to the gradient vector.

Expansion of $n(\mathbf{r}+\mathbf{s})$ in a Taylor series about $\mathbf{s}=0$ and combination of (2.D.10), (2.D.9), and (2.C.18) leads, to second order in gradients, to a pressure tensor of the form

$$\mathbf{P}=P_0(n)\mathbf{I}+l_{11}\nabla^2n\mathbf{I}+2(l_{11}-l_T)\nabla\nabla n+l_{21}(\nabla n)^2\mathbf{I}+2(l_{21}-m_T)\nabla n\nabla n, \qquad (2.D.11)$$

where

$$l_{11}\equiv-\frac{1}{30}\int s\frac{du(s)}{ds}\left[\frac{s^2}{6}\frac{\partial(n^2g_0)}{\partial n}-\frac{s^2}{2}n^2q_N+n^2v_N+4n^2v_T\right]d^3s \qquad (2.D.12)$$

$$l_{21}\equiv-\frac{1}{30}\int s\frac{du(s)}{ds}\left[\frac{s^2}{6}\frac{\partial^2(n^2g_0)}{\partial n^2}-\frac{sn^2}{2}\frac{\partial q_N}{\partial n}+n^2t_N+4n^2v_T\right]d^3s \qquad (2.D.13)$$

$$l_T\equiv-\frac{1}{6}\int s\frac{du(s)}{ds}v_Td^3s \qquad (2.D.14)$$

$$m_T\equiv-\frac{1}{6}\int s\frac{du(s)}{ds}t_Td^3s \qquad (2.D.15)$$

The multicomponent versions of (2.D.10) and (2.D.11) are

$$g^{\alpha\beta}=g_0^{\alpha\beta}+\sum_\gamma q_N^{\alpha\beta\gamma}\hat{s}\cdot\nabla n_\gamma+\sum_{\gamma,\delta}\big[t_N^{\alpha\beta\gamma\delta}\hat{s}\hat{s}+t_T^{\alpha\beta\gamma\delta}(\mathbf{I}-\hat{s}\hat{s})\big]\;:\;\nabla n_\gamma\nabla n_\delta$$
$$+\sum_\gamma\big[v_N^{\alpha\beta\gamma}\hat{s}\hat{s}+v_T^{\alpha\beta\gamma}(\mathbf{I}-\hat{s}\hat{s})\big]\;:\;\nabla\nabla n \qquad (2.D.16)$$

$$\mathbf{P}=P_0\mathbf{I}+\sum_{\alpha,\beta}\big[l_{11}^{\alpha\beta}\nabla^2n_\beta+l_{21}^{\alpha\beta}\nabla n_\alpha\cdot\nabla n_\beta\big]\mathbf{I}$$
$$+2\sum_{\alpha,\beta}\big[(l_{11}^{\alpha\beta}-l_T^{\alpha\beta})\nabla\nabla n_\beta+(l_{21}^{\alpha\beta}-m_T^{\alpha\beta})\nabla n_\alpha\nabla n_\beta\big] \qquad (2.D.17)$$

The coefficients of density gradients in $\mathbf{P}$ can be related to those of $g^{\alpha\beta}$, but since the results are much more complicated than those for the one-component fluid, they are not given here. The interfacial tension corresponding to (2.D.17) is

$$\gamma = \sum_{\alpha,\beta} \int c_{\alpha\beta} \frac{dn_\alpha}{dx} \frac{dn_\beta}{dx} dx \qquad (2.D.18)$$

where the influence parameters are

$$c_{\alpha\beta} = l_{11}^{\alpha\beta} + l_{11}^{\beta\alpha} - l_T^{\alpha\beta} - l_T^{\beta\alpha} + \frac{\partial}{\partial n_\gamma} \sum_\gamma \left[\frac{\partial}{\partial n_\alpha} \left(l_{21}^{\gamma\beta} - m_T^{\gamma\beta} \right) + \frac{\partial}{\partial n_\beta} \left(l_{21}^{\gamma\alpha} - m_T^{\gamma\alpha} \right) \right]$$

$$(2.D.19)$$

Unfortunately, the density distributions n_α for a planar interface cannot be determined by the condition that P_N be constant. The species hydrostatic equations (2.C.22, the YBG equations) must be invoked; these become a set of coupled third-order differential equations for the density distributions. Summation of the YBG equations yields the condition of hydrostatic equilibrium (2.A.6), from which constancy of P_N follows for a planar, field-free interface. For a discussion of the gradient theory of the multicomponent YBG equations, the reader is referred to the work of Carey et al.[15]

From statistical thermodynamics,[45] the coefficients of the gradient theoretical formula for g can be computed from the doublet, triplet, and quadruplet direct correlation functions of homogeneous fluid. Thus all the inputs, P_0 and l_{ij}, of gradient theory can be computed from homogeneous fluid properties. The special properties of inhomogeneous fluid manifest themselves as derivatives of density distributions.

Because the gradient theory of the pair correlation function demands triplet and quadruplet direct correlation functions, the theoretical inputs of hydrostatic (YBG) theory of inhomogeneous fluid are still not easy to estimate. Approximate pair correlation functions of homogeneous fluid again provide the only simple description of interfaces. However, in the free-energy approach of statistical thermodynamics, presented in Section III, only the equation of state and the doublet direct correlation function of homogeneous fluid are needed in gradient theory. The statistical thermodynamical result is valid independently of the nature of the intermolecular forces; that is, they need not be two-body or even central, as long as density depends only on position in space. Moreover, gradient theoretical equations of multicomponents of fluids are second-order differential equations rather than third order as are the YBG gradient equations. Thus although the mechanical theory of inhomogeneous fluid, based on force balances, is perhaps more appealing physically, the statistical thermodynamical approach is mathematically simpler and less demanding of theoretical inputs. However, the true pressure tensor can be identified only by the mechanical theory (see Section III.D).

In many numerical studies of fluid interfaces, the exact pair correlation function has been approximated by the pair correlation function of homogeneous fluid at some local density or densities. It can be argued that since the pair correlation function enters the hydrostatic equations as a product with intermolecular force, only the configurations for which the particles are close together contribute appreciably to the theory; therefore pair correlations can be estimated as those of homogeneous fluid at a density or densities characteristic of the neighborhood of the correlated particles. Reasonable candidates for such an approximation are:

Case 1.   Density at mean location:

$$g(\mathbf{r}, \mathbf{r}') = g_0\left(|\mathbf{r} - \mathbf{r}'|; \, n\left(\frac{\mathbf{r} + \mathbf{r}'}{2}\right)\right) \tag{2.D.20}$$

Case 2.   Mean density:

$$g(\mathbf{r}, \mathbf{r}') = g_0\left(|\mathbf{r}' - \mathbf{r}'|; \, \frac{n(\mathbf{r}) + n(\mathbf{r}')}{2}\right) \tag{2.D.21}$$

Case 3.   Mean function:

$$g(\mathbf{r}, \mathbf{r}') = \tfrac{1}{2}\left[g_0(|\mathbf{r} - \mathbf{r}'|; \, n(\mathbf{r})) + g_0(|\mathbf{r} - \mathbf{r}'|; \, n(\mathbf{r}'))\right] \tag{2.D.22}$$

Carey et al.[15] have shown for a simple model fluid that gradient theory is not very sensitive to which of these approximation is used, and McCoy et al.[46] have shown for a 6–12 Lennard–Jones model fluid that the density profiles and tensions predicted with these three approximations agree well with those predicted by exact gradient theory. In these works stable phase pair correlation functions are analytically continued into the unstable density region.

The feature of Cases 1 to 3 and related approximations to be emphasized here is that the terms t_T and v_T are zero, and correspondingly l_T and m_T are zero. When these latter quantities are zero, the gradient theoretical transverse pressure in a planar interface obeys the revealing relation

$$P_T = \tfrac{1}{3} P_N + \tfrac{2}{3} P_0(\mathbf{n}) \tag{2.D.23}$$

and the tension is

$$\gamma = \frac{2}{3} \int_{-\infty}^{\infty} \left[P_N - P_0(\mathbf{n})\right] dx \tag{2.D.24}$$

These results bring out the intimate relationship between the stress in and tension of an interface and the isotherm of pressure of homogeneous fluid along the interfacial composition path, which of necessity traverses the spinodal domain. For a one-component fluid, $P_N - P_T(n)$ equals two-thirds of the vertical distance from $P_0(n)$ to the level line $P_N = P_0(n_l) = P_0(n_g)$ of an isotherm of P_0 versus density. These formulas are predicted by any gradient correlation model that neglects transverse, gradient-induced correlations, $t_T^{\alpha\beta\gamma\delta}$ and $v_T^{\alpha\beta\gamma}$. This includes, for example, the model that Lekner and Henderson[47] have found so fruitful recently, namely,

$$g(\mathbf{r},\mathbf{r}') = \frac{\left[n\left(\dfrac{\mathbf{r}+\mathbf{r}'}{2}\right)-n_g\right]g_0(|\mathbf{r}-\mathbf{r}'|;n_l)+\left[n_l-n\left(\dfrac{\mathbf{r}+\mathbf{r}'}{2}\right)\right]g_0(|\mathbf{r}-\mathbf{r}'|;n_g)}{n_l-n_g}$$

(2.D.25)

and other similar models used earlier.[29, 40] One trouble with a model such as (2.D.25) is that, fluid at a solid surface or spherical drop is not treatable by such a model correlation function. According to rigorous theory,[45] the transverse, gradient-induced correlations are not zero. However, the success of approximations such as Cases 1 to 3 and (2.D.25) implies that the neglected correlations do not contribute strongly to fluid tension. It is interesting to compare the normal component P_N of the pressure tensor predicted by gradient theory for Cases 1 to 3 and (2.D.25). Respectively, these are

$$P_N = P_0(n) - \left[\frac{n^2}{16}\frac{\partial u_2(n)}{\partial n} + \frac{n}{2}u_2(n)\right]\frac{d^2n}{dx^2}$$
$$- \left[\frac{n^2}{16}\frac{\partial^2 u_2(n)}{\partial n^2} + \frac{n}{4}\frac{\partial u_2(n)}{\partial n} - \frac{1}{4}u_2(n)\right]\left(\frac{dn}{dx}\right)^2 \qquad (2.D.26)$$

$$P_N = \text{rhs } (2D.26) - \frac{3}{16}n^2\frac{\partial u_2(n)}{\partial n}\frac{d^2n}{dx^2} \qquad (2.D.27)$$

$$P_N = \text{rhs } (2.D.26) - \frac{3}{16}n^2\frac{\partial u_2(n)}{\partial n}\frac{d^2n}{dx^2} - \frac{3}{16}n^2\frac{\partial^2 u_2(n)}{\partial n^2}\left(\frac{dn}{dx}\right)^2 \qquad (2.D.28)$$

and

$$P_N = P_0(n) - \left\{\frac{9n^2}{16}\left[\frac{u_2(n_l)-u_2(n_g)}{n_l-n_g}\right] + \frac{n}{2}\left[\frac{n_l u_2(n_g)-n_g u_2(n_l)}{n_l-n_g}\right]\right\}\frac{d^2n}{dx^2}$$
$$- \frac{1}{4}\left[\frac{n_g u_2(n_l)-n_l u_2(n_g)}{n_l-n_g}\right]\left(\frac{dn}{dx}\right)^2 \qquad (2.D.29)$$

The quantity $u_2(n)$, defined by

$$u_2(n)=\frac{1}{15}\int s^3 u'(s)g_0(s;n)\,d^3s \tag{2.D.30}$$

plays a crucial role in the gradient stress tensor in all four cases. An advantage of the last one is that it calls for the pair correlation function only at the coexistence densities n_g and n_l rather than across the entire density region between n_g and n_l. In the limit that u_2 is independent of density, all four cases yield the same formula,

$$P_N=P_0(n)-\frac{nu_2}{2}\frac{d^2n}{dx^2}+\frac{u_2}{4}\left(\frac{dn}{dx}\right)^2 \tag{2.D.31}$$

That u_2 is predicted to be relatively insensitive to density in simple fluids[15, 46] indicates that the stress and density profiles might not be strong tests of a model pair correlation function.

We have not incorporated thermodynamic principles in the formal development of the gradient theory of stress. An important theoretical question is whether (or when) enforcement of hydrostatic equilibrium automatically yields thermodynamic equilibrium (i.e., chemical as well as mechanical equilibrium). The boundary conditions on the density profile $n(x)$ for a planar interface are

$$n(x=-\infty)=n_l$$
$$n(x=+\infty)=n_g \tag{2.D.32}$$

where n_l and n_g are the densities of the bulk liquid and vapor phases. The condition of mechanical equilibrium, namely, that P_N be constant, has the following requirement.

$$P_N=P_0(n_l)=P_0(n_g) \tag{2.D.33}$$

Gradient theory with the boundary conditions (2.D.32) yields (2.D.33), of course, but appears to imply more when $P_0(n)$ is identified as the thermodynamic pressure function of homogeneous fluid. For example, consider the case of constant u_2, differentiate the gradient equation (2.D.31) with respect to x, and use the thermodynamic identity $dP_0(n)=nd\mu_0(N)$ to rearrange the resulting equation to the form

$$\frac{d}{dx}\left[\frac{u_2}{2}\frac{d^2n}{dx^2}-\mu_0(n)\right]=0 \tag{2.D.34}$$

This equation can be integrated to obtain

$$\frac{u_2}{2}\frac{d^2n}{dx^2} - \mu_0(n) = \text{constant} \qquad (2.D.35)$$

This is the equation obtained originally by Rayleigh[1] and van der Waals[2] from the thermodynamic gradient theory of planar interfaces. With this result, the boundary conditions (2.D.32) lead to

$$\mu_0(n_g) = \mu_0(n_l) \qquad (2.D.36)$$

the condition of chemical equilibrium between the liquid and vapor phases.

Thus, at least for gradient theory with constant u_2, *the condition of hydrostatic equilibrium, $dP_N/dx = 0$ (and therefore the YBG equation) between bulk phases obeying thermodynamic laws implies both mechanical and chemical equilibrium.* Without restriction on u_2, the same conclusion follows for the gradient theory with the pair correlation function model used by Lekner and Henderson. To show this, we differentiate (2.D.29), use again $dP_0 = nd\mu_0$, rearrange terms, and integrate the resulting expression to obtain

$$0 = \int_{-\infty}^{\infty} \frac{d}{dx}\left\{\mu_0(n) - \frac{1}{2}\left[\frac{n_l u_2(n_g) - n_g u_2(n_l)}{n_l - n_g}\right]\frac{d^2n}{dx^2}\right\}dx$$

$$-\frac{9}{16}\left[\frac{u_2(n_l) - u_2(n_g)}{n_l - n_g}\right]\int_{-\infty}^{\infty}\frac{1}{n}\frac{d}{dx}\left\{n^2\frac{d^2n}{dx^2}\right\}dx \qquad (2.D.37)$$

With the boundary conditions (2.D.32), this equation reduces to

$$0 = \mu_0(n_g) - \mu_0(n_l) \qquad (2.D.38)$$

the condition of chemical equilibrium.

Gradient theory with the pair correlation function approximated by Cases 1 to 3 does not appear to enforce the chemical equilibrium as a by-product of hydrostatic equilibrium except when u_2 is constant. If one assumes, as we believe one should, that thermodynamic equilibrium should be automatically imposed by the hydrostatic equations (i.e., the YBG equations), once P_0 is identified as the thermodynamic pressure, then the failure of the pair correlation functions of Cases 1 to 3 to do this can be taken as a measure of the inadequacy of these model correlation functions. However, the fact that chemical equilibrium is not automatically imposed causes no difficulty in applying gradient theory to planar interfaces since the infinity of density pairs (n_l, n_g) satisfying the mechanical equilibrium condition (2.D.33) contains the unique pair that coincides with chemical equilibrium (2.D.36). Thus, one simply chooses the unique pair for solution of the gradient profile equations.

E. Numerical Study of Structure and Stress of a Planar Interface

The purpose of this section is twofold. First, we demonstrate with a model calculation that even for a sharp interface, the density profile predicted by gradient theory agrees well with that predicted by the integrodifferential YBG equation. Second, we illustrate the qualitative aspects of the density and stress profiles and the tension of planar interfaces with the aid of gradient theory. An approximate pair correlation function, Case 1, is chosen for the computations.

As an interaction potential we use the square-well model:

$$
\begin{aligned}
u(s) &= \infty, & s &< d \\
&= -\varepsilon, & d &< s < Rd \\
&= 0, & s &> Rd
\end{aligned}
\tag{2.E.1}
$$

With the Case 1 pair correlation function, the normal principal pressure for a square-well fluid can be rearranged to

$$
P_N = nkT - 2\pi \int_0^\infty \int_0^1 n(x - \eta s_x) n(x - \eta s_x + s_x) K\left(s_x; n\left(x + \tfrac{1}{2}s_x\right)\right) s_x^2 \, d\eta \, ds_x
\tag{2.E.2}
$$

where

$$
K(s_x, n) \equiv -g_0(d_+; n) H(d - s_x) + (e^{\varepsilon/kT} - 1) g_0(Rd_+; n) H(Rd - s_x)
\tag{2.E.3}
$$

Here $H(x)$ is the Heaviside step function; $g_0(d_+; n)$ and $g_0(Rd_+; n)$ are values of the pair correlation function at the point of contact of the rigid cores of the molecules and at the lip of the attractive well.

The gradient theoretical normal pressure is [see (2.D.11)]

$$
P_N = P_0(n) + l_1(n) \frac{d^2 n}{dx^2} + l_2(n) \left(\frac{dn}{dx}\right)^2
\tag{2.E.4}
$$

where

$$
l_1 \equiv 3 l_{11}, \quad l_2 \equiv 3 l_{21} \quad \text{and} \quad l_\perp = m_\perp = 0
\tag{2.E.5}
$$

for pair correlation functions depending only on particle separation $|\mathbf{r} - \mathbf{r}'|$ and mean position $\frac{1}{2}(\mathbf{r} + \mathbf{r}')$. For the Case 1 correlation function it follows from (2.D.26) that

$$l_1 = \frac{n^2}{16}\frac{\partial u_2}{\partial n} - \frac{n}{2}u_2 \qquad (2.E.6)$$

$$l_2 = \frac{n^2}{16}\frac{\partial^2 u_2}{\partial n^2} - \frac{n}{4}\frac{\partial u_2}{\partial n} + \frac{1}{4}u_2 \qquad (2.E.7)$$

where u_2 is defined by (2.D.30). The terms l_1 and l_2 can be deduced from (2.D.27) to (2.D.29) for Cases 2 and 3 and for Lekner and Henderson's model.

For a square-well fluid

$$u_2 = \frac{4\pi}{15}d^5kT\left[-g_0(d_+) + R^5)e^{\varepsilon/kT} - 1)g_0(Rd_+)\right] \qquad (2.E.8)$$

and

$$P_0(n) = nkT\left\{1 + \frac{2\pi}{3}nd^3\left[g(d_+) - R^3(e^{\varepsilon/kT} - 1)g_0(Rd_+)\right]\right\} \qquad (2.E.9)$$

With the condition of mechanical (hydrostatic) equilibrium, $P_N = P_0(n_g) = P_0(n_l)$, Equation 2.E.2 becomes an integral equation determining the density profile between bulk phases at densities n_g and n_l. Similarly, (2.E.4) is a differential equation for the density profile. The condition of mechanical equilibrium of an isothermal system can be satisfied by a range of density pairs, n_g and n_l — any pair for which a constant pressure line cuts a $P_0(n)$ versus n^{-1} isotherm more than once satisfies mechanical equilibrium. At thermodynamic equilibrium, a unique vapor-liquid pair (n_h, n_l) is obtained by adding to the condition of mechanical equilibrium the condition of chemical equilibrium, namely, $\mu_0(n_g) = \mu_0(n_l)$, where $\mu_0(n)$ is the chemical potential of homogeneous fluid. Given the equation of state for $P_0(n)$, $\mu_0(n)$ can be obtained through the thermodynamic identity $(d\mu_0)_T = n^{-1}(dP_0)_T$.

Toxvaerd solved the integral equation (2.E.2) for $n(x)$ by successive substitutions.[38] He used formulas for $g_0(d_+)$ and $g_0(Rd_+)$ that give an equation of state in agreement with computer simulations. Using Toxvaerd's correlation function, we solved, by a method to be outlined shortly, the differential equation (2.E.4) for $n(x)$. The resulting density profiles predicted by the integral and gradient theories agree very well (Fig. 2.E.1), even though the interface is fairly sharp and the temperature $(kT/\varepsilon = 1)$ is far from critical $(kT_c/\varepsilon = 1.4$ for Toxvaerd's model). The result is typical of the various

SQUARE-WELL FLUID

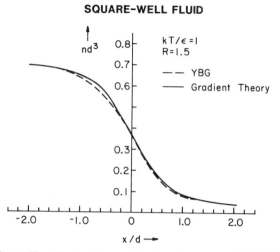

Fig. 2.E.1.   Density profile of the liquid-vapor interface of a square-well fluid. Comparison of gradient theory and the Yvon-Born-Green (YBG) equation. (From Ref. 38.)

comparisons of integral models and gradient approximations that we have made.[15, 26, 43] Since solutions to integral versions of the YBG equation can be quite expensive computationally,[40] it is of practical significance that gradient theory agrees well with integral models.

Let us now consider only the gradient theory, which is mathematically much simpler than the more rigorous YBG theory and, from the example above, a good approximation to it. Instead of Toxvaerd's model for $g_0(d_+)$ and $g_0(Rd_+)$, we use the simpler model introduced by Carey et al.,[15] namely,

$$g_0(d_+)=\frac{4-2\eta+\eta^2}{4(1-\eta)^3}, \qquad g_0(Rd_+)=1 \qquad (2.E.10)$$

where $\eta=(\pi/6)nd^3$. The chemical potential determined by $P_0(n)$ for this model is

$$\mu_0(n)=\mu^+(T)+kT\left[8\eta(1-e^{\epsilon/kT})R^3-\ln\left(\frac{1}{\eta}-1\right)\right.$$

$$\left.+\frac{\eta}{1-\eta}+\frac{3}{2(1-\eta)^3}+\frac{3\eta}{(1-\eta)^2}\right] \qquad (2.E.11)$$

where $\mu^+(T)$ is a constant of integration.

The gradient theoretical version of the transverse pressure is, as shown in the preceding section, of the form

$$P_T = \tfrac{1}{3}P_N + \tfrac{2}{3}P_0(\mathbf{n}) \qquad (2.D.23)$$

for the Case 1 pair correlation function. The isotherms of $P_0(\mathbf{n})$ plotted versus density in Fig. 2.E.2 for the present square-well model reveal that as anticipated earlier from the temperature dependence of surface tension, the transverse pressure can have large negative values in the interfacial zone and can have values greater than P_N in part of the zone. And according to (2.D.23), the behavior of $P_0(\mathbf{n})$ in the unstable bulk phase region (spinodal) plays a major role in the physics at an interface.

With the transformations

$$\frac{dn}{dx} = m(n) \qquad \text{and} \qquad \frac{d^2n}{dx^2} = m\frac{dm}{dn} = \frac{1}{2}\frac{dm^2}{dn} \qquad (2.E.12)$$

the normal pressure differential equation can be rearranged and solved, with

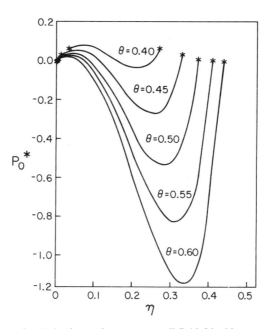

Fig. 2.E.2. Pressure-density isotherms for a square-well fluid. Liquid-vapor coexistence states are located by asterisks: $P_0^* = P_0 d^3/\varepsilon$, $\eta = (\pi/6)nd^3$, $\theta = \varepsilon/kT$, $R = 1.85$.[15]

the boundary condition $m(n_l)=0$, to get

$$m^2(n)=2\int_{n_l}^{n}\exp\left[-\int_{\tau}^{n}\frac{2l_2(\xi)}{l_1(\xi)}d\xi\right]\frac{P_0(\tau)-P_N}{l_1(\tau)}d\tau \qquad (2.E.13)$$

Once m versus n is computed, the density profile can be computed from

$$x-x_a=\int_{n(x_a)}^{n}\frac{dn}{m(n)} \qquad (2.E.14)$$

the surface tension from

$$\gamma=\int_{-\infty}^{\infty}c\left(\frac{dn}{dx}\right)^2 dx=\int_{n_l}^{n_g}c(n)m(n)\,dn \qquad (2.E.15)$$

and the transverse pressure from (2.D.23). For Case 1, the influence parameter is

$$c=\frac{1}{2}u_2+\frac{1}{4}n\frac{\partial u_2}{\partial n} \qquad (2.E.16)$$

Density and transverse pressure profiles are presented in Figs. 2.E.3 and 2.E.4 for several temperatures. The corresponding surface tensions are given in Fig. 2.E.4.

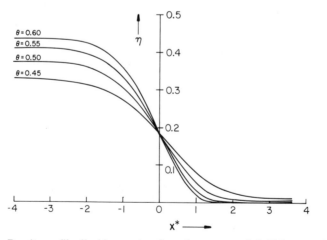

Fig. 2.E.3. Density profiles liquid-vapor interface of a square-well fluid for several temperatures: $\eta=(\pi/6)nd^3$, $\theta=\varepsilon/kT$, $R=1.85$.[15]

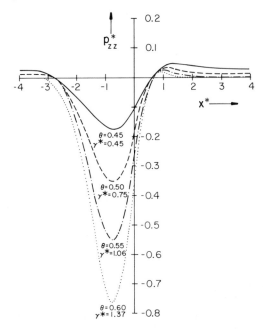

Fig. 2.E.4. Transverse pressure profiles of liquid-vapor interface of a square-well fluid for several temperatures. Tensions γ^* ($=\gamma d^2/\varepsilon$) are displayed on the graph; $\theta=\varepsilon/kT$, $R=1.85$.[15]

As temperature increases, the width l_w of the interfacial zone—defined, for example, as the inverse of $\int_{-\infty}^{\infty}(dn/dx)^2\,dx/(n_l-n_g)$—increases, the transverse pressure tends to flatten out, and the interfacial tension decreases.

The gradient theory with the present equation of state predicts that $l_w \propto (T_c-T)^{-0.5}$ and $\gamma \propto (T_c-T)^{1.5}$ for temperatures near the critical point. The prediction for l_w agrees with experiment, whereas experimentally it has been found that $\gamma \propto (T_c-T)^{1.3}$. The disagreement is not the fault of gradient theory, but rather arises from the well-known[48] failure near the critical point of equations of state of the type used here.

The monotonicity predicted for the density profile has been borne out by computer simulations[49, 50] and by the numerical predictions of numerous free-energy models.[26, 38, 39, 43, 51-53] The lack of monotonicity reported earlier[54] turned out to be a computer artifact.[49]

F. Discontinuous Interface

The opposite of the limit in which gradient theory is favored is the limit of an interface so sharp that it is well approximated as a step function in density and composition. Such an approximation goes back to Young and

Laplace and was exploited in modern terms by Fowler[55] and by Kirkwood and Buff.[10] In the Fowler-Kirkwood-Buff version of the approximation, a low-vapor-pressure interface of a one-component fluid is assumed to have a density profile obeying $n(x)=n_l$, $x<0$, and $n(x)=0$, $x>0$. The pair correlation function is similarly assumed to be of the form $g(\mathbf{r},\mathbf{r}')=g_0(|\mathbf{r}-\mathbf{r}'|;n_l)$, x, $x'<0$. With these assumptions, the one-component, two-body interaction limit of (2.C.28) yields the formula

$$\gamma=\frac{n_l^2}{32}\int s^2 u(s)g_0(s;n_l)\,d^3r \qquad (2.F.1)$$

For argon at low temperatures where the interface is sharp, this expression, on the basis of a 6-12 Lennard-Jones potential, agrees fairly well with experiment (at $T=84°K$, $\gamma_{\exp}=13.3$ dynes/cm,[56] whereas (2.F.1) predicts[57] 13.3 dynes/cm). As the temperature increases, Equation 2.F.1 becomes worse, as expected, experiment and theory giving[57] 5.1 and 10.1 dynes/cm at 115°K, for example.

Present et al.[34] have estimated for a discontinuous interface the three-body contribution to the surface tension of low-pressure argon. The Fowler-Kirkwood-Buff reduction of the three-body term in (2.C.28) is

$$\gamma^{(3)}=n_l^3\int\int\int_0^\infty\frac{s_z^2-s_x^2}{s}f_{12}(s,s'|s'-s|)H(s_x+x)$$

$$\times H(s_x'+x)g_3^0(s,s',|s'-s|;n_l)\,dx\,d^3s\,d^3s' \qquad (2.F.2)$$

where $g_3^0(\equiv n_3^0/n^3)$ is the three-body correlation function of homogeneous fluid. Using the superposition approximation for $g_3^0(g_3^0(r_{12},r_{13},r_{23})=g_0(r_{12})g_0(r_{13})g_0(r_{23}))$ and the Axilrod-Teller triple-dipole interaction for the three-body potential, Present et al. estimate $\gamma^{(3)}=-4.5$ dyne/cm for argon at 85°K. Such a correction greatly worsens the agreement just noted between experiment and (2.F.1) for low-vapor-pressure argon. There could be a number of reasons for this, of which we note two. First, the superposition approximation might introduce large errors—three-body effects might cancel out with the correct correlation function. Second, and more important in our opinion, three-body effects are already included in some sense in the 6-12 Lennard-Jones potential. Salter and Davis[57] chose the parameters of the potential to get good fits between the theoretical and experimental vapor pressure of argon (their choice was $\varepsilon/k=116.41°K$ and $\sigma=3.37$ Å). Computer simulations, with and without three-body potentials, would be useful for resolving this problem.

A multicomponent generalization of the discontinuous interface model for a fluid of molecules with pair-additive, orientation-dependent forces reduces (2.C.29) to[36]

$$\gamma = \sum_{\alpha,\beta} \left[n_\alpha^{(1)} n_\beta^{(1)} B_{\alpha\beta}^{(1,1)} + n_\alpha^{(2)} n_\beta^{(2)} B_{\alpha\beta}^{(2,2)} - n_\alpha^{(1)} n_\beta^{(2)} B_{\alpha\beta}^{(1,2)} - n_\alpha^{(2)} n_\beta^{(1)} B_{\alpha\beta}^{(2,1)} \right] \quad (2.F.3)$$

where

$$B_{\alpha\beta}^{(i,j)} \equiv \frac{1}{2} \int_{s_x>0} s_x \left[s_x \frac{\partial u_{\alpha\beta}}{\partial s_x} - s_z \frac{\partial u_{\alpha\beta}}{\partial s_z} \right] g_{i,j}^{\alpha\beta} d^3 s \frac{d^3 e \, d^3 e'}{(8\pi^2)^2} \quad (2.F.4)$$

When $i=j$, indicating that both particles lie in the ith phase, $g_{i,i}^{\alpha\beta} = g_0^{\alpha\beta}(\mathbf{s}, \mathbf{e}, \mathbf{e}'; \mathbf{n}^{(i)})$ $\mathbf{n}^{(i)}$ being the composition of the ith bulk phase. The pair correlation functions for particles of species α and β lying in phase 1 and 2 and 2 and 1, respectively, are denoted by $g_{1,2}^{\alpha\beta}$ and $g_{2,1}^{\alpha\beta}$.

For two interesting special cases (2.F.3) has been used to generate estimates of the tension of multicomponent systems. One is the limit of a two-component, liquid-liquid system in which component 1 is almost immiscible in phase 1 and component 2 is almost immiscible in phase 2. Then $n_1^{(2)} \simeq n_2^{(1)} \simeq 0$, and

$$\gamma \simeq \left(n_1^{(1)} \right)^2 B_{11}^{(1)} + \left(n_2^{(2)} \right)^2 B_{22}^{(2)} - 2 n_1^{(1)} n_2^{(2)} B_{12}^{(1,2)} \quad (2.F.5)$$

Identifying $(n_1^{(i)})^2 B_{ii}^{(i)}$ with the surface tension $\gamma_i^{(0)}$ of pure liquid i and approximating $B_{12}^{(1,2)}$ by $(B_{11}^{(1)} B_{22}^{(2)})^{1/2} \Phi$, $\Phi = 4 d_1 d_2 / (d_1 + d_2)^2$, $d_i^{-3} = n_i^{(i)}$, Girifalco and Good[58] and Davis[59] have estimated the interfacial tensions between immiscible liquids in terms of pure fluid properties. The estimate works well for quite a few fluids.

The other special case is a low-vapor-pressure, liquid-vapor system for which $n_\alpha^{(g)} \ll n_\alpha^{(l)}$, so that

$$\gamma \simeq \sum_{\alpha,\beta} n_\alpha^{(l)} n_\beta^{(l)} B_{\alpha\beta}^{(l)} \quad (2.F.6)$$

Winterfeld et al.[60] investigated the mixing rule $B_{\alpha\beta}^{(l)} = (B_{\alpha\alpha}^{(l)} B_{\beta\beta}^{(l)})^{1/2}$, under the assumption that the $B_{\alpha\beta}$'s are independent of density and composition. In this case (2.F.6) becomes

$$\gamma = \sum_{\alpha,\beta} \frac{n_\alpha^{(l)} n_\beta^{(l)}}{n_\alpha^0 n_\beta^0} \gamma_\alpha^0 \gamma_\beta^0 \quad (2.F.7)$$

where n_α^0 and γ_α^0 are the liquid density and surface tension of pure α and $n_\alpha^{(l)}$

is the density of α in the mixture. Estimating the mixture density from Amagat's law, $n^{-1} = \Sigma_\alpha x_\alpha (n_\alpha^0)^{-1}$, x_α mole fraction, Winterfeld et al. applied (2.F.6) to predict the composition dependence of the surface tension of 32 binary mixtures involving mixtures of polar and nonpolar hydrocarbons, alcohols, and water. Without the water-alcohol mixtures, the average error of the predictions was about 1% (never more than 3%). For the methanol-water and ethanol-water solutions the average errors were 18 and 32%.

On the basis of a lattice model and the discontinuous density profile, the authors[61] derived the approximate formula

$$\gamma = \frac{1}{8n_l^{1/3}} \left(\frac{\partial U}{\partial V} \right)_{T,N} \tag{2.F.8}$$

for the surface tension of a one-component liquid at low vapor pressure. Here U is the thermodynamic energy of the liquid. When applied to benzene, n-heptane, diethyl ether, argon, nitrogen, methane, and carbon tetrachloride at various temperatures, the formula had an average absolute error of 17%. Predicted values were low in most cases, and errors were largest for methane and oxygen.

As indicated by the computations reviewed here, the discontinuous interface approximation is often useful to estimate the tension of very sharp interfaces. However, it has the drawbacks that it is limited to virtually totally immiscible fluids or low-vapor-pressure, liquid-vapor systems. Furthermore, as is easily verified, the normal pressure component P is *not* constant across the interface whenever the density profile is discontinuous, and so the physics of an interface cannot be correctly represented by the approximation. Thus, although the discontinuous profile theory and the gradient theory are approximations that seem to bracket the extremes of interfacial behavior, the former fails qualitatively but the latter does not.

G. Sum Rules of the Yvon-Born-Green Equation

Lekner and Henderson[47] have shown that the YBG equation for a one-component fluid can be integrated many different ways, generating moment equations or what they call sum rules. On the basis of these sum rules and the pair correlation function model given at (2.D.25), they obtained simple relations between the coexistence densities and the vapor pressure of a one-component fluid. Because such relations can be used to test model correlation functions their work is discussed briefly in this section.

For a planar interface the YBG equation is

$$kT \frac{dn(x)}{dx} = \int u'(|\mathbf{r}'-r|) \frac{(x'-x)}{|\mathbf{r}'-\mathbf{r}|} n_2(\mathbf{r},\mathbf{r}') \, d^3r' \tag{2.G.1}$$

As pointed out in Section II.C, (2.G.1) is equivalent to the hydrostatic equation $\nabla \cdot \mathbf{P} = 0$, which for a planar fluid is simply $dP_N/dx = 0$. Thus, if one integrates (2.G.1) from one end $(x = -L)$ of the system to the other end $(x = L)$, the result must be equivalent to the condition that the pressure P_N at $x = -L$ is the same as that at $x = L$. Consider a planar system of cross-section A normal to the interface. Integration of (2.G.1) over the volume of the system yields

$$AkT[n(L) - n(-L)] = \int\int u'(|r-r'|) \frac{(x'-x)}{|\mathbf{r}'-\mathbf{r}|} n_2(\mathbf{r},\mathbf{r}') \, d^3r \, d^3r' \quad (2.G.2)$$

With the symmetry condition $n_2(\mathbf{r},\mathbf{r}') = n_2(\mathbf{r}',\mathbf{r})$, the integrand on the right-hand side of this equation is antisymmetric and, therefore, vanishes, so that $n(L) = n(-L)$. This result would seem to imply that the YBG equation does not allow a liquid-vapor interface, since $n(-L) = n_l$ and $n(L) = n_g$ is not allowed. This is, however, not so. Since the wall forces at the ends of the system are accounted for only by terminating the integration over x, the pressure equality $P_N(-L) = P_N(L)$ can lead only to identical densities at the wall. The situation can be remedied by putting wall forces in the YBG equation, but this is obviously undesirable when we are interested only in fluid properties.

Lekner and Henderson circumvented the difficulty by integrating the YBG equation from a position x_l deep in the bulk liquid, but not to a wall, to a position x_g deep in the bulk vapor but not to a wall. Thus they obtained

$$n_g - n_l = \frac{2\pi}{kT} \int_{x_l}^{x_g} dx \, n(x) \int_{-L}^{L} dx' \, n(x')(x-x') \int_{|x-x'|}^{\infty} ds \, g(s; x, x') u'(s)$$

$$= \frac{2\pi}{kT} \int_{x_l}^{x_g} dx \, n(x) \left\{ \int_{-L}^{x_l} + \int_{x_l}^{x_g} + \int_{x_g}^{L} \right\} dx' \, n(x'x)(x-x')$$

$$\times \int_{|x-x'|}^{\infty} ds \, g(s; x, x') u'(s) \quad (2.G.3)$$

The contribution of the middle term on the right-hand side of (2.G.3) vanishes because of antisymmetry of the integrand. In the range in which the integrand of the first term is nonzero, the densities and pair correlation function are at their bulk liquid values n_l and $g_0(s; n_l)$. Similarly, in the third term the densities and pair correlation function are at n_g and $g_0(s; n_g)$ in the nonzero range of the integrand. Thus with the relationship

$$n_l^2 \int_0^\infty dx \int_{-\infty}^0 dx'(x-x') \int_{|x-x'|}^\infty ds \, g_0(s; n_l) u'(s)$$

$$= \frac{2\pi}{3} n_l^2 \int_0^\infty ds \, s^3 g_0(s; n_l) u'(s) \qquad (2.G.4)$$

and a corresponding one for the other term, (2.G.3) can be rearranged to

$$P_0(n_g) = P_0(n_l) \qquad (2.G.5)$$

appropriately expressing the condition of mechanical equilibrium between the liquid and vapor phases. Equation 2.G.5 follows for any pair n_g, n_l of densities for which $P_0(n)$ is the same.

Another relationship between n_l and n_g can be obtained by multiplying (2.G.1) by $n(x)^{-1}$ and integrating from x_l to x_g. Again dividing the interval of dx' into the three regions, Lekner and Henderson obtained, after rearrangement of terms,

$$kT \ln \frac{n_g}{n_l} = P_0(n_l) \left(\frac{1}{n_g} - \frac{1}{n_l} \right) + \pi \int_{x_l}^{x_g} dx \int_{x_l}^{x_g} dx' \left[n(x) - n(x') \right] (x-x')$$

$$\times \int_{|x'-x|}^\infty ds \, g(s; x, x') u'(s) \qquad (2.G.6)$$

Since multiplication by $n(x)^{-1}$ destroys the antisymmetry of the integrand in the YBG equation, the term involving $\int_x^{x_g} dx'(\ )$ does not vanish in (2.G.6). Because of the short range of the integrand of (2.G.6), x_l and x_g can be extended to $-\infty$ and $+\infty$ without loss of generality. Multiplication of the YBG equation by $[n(x)]^{-m}$, $m>1$, and integration yields the general "sum rule":

$$kT(1-m^{-1})(n_l^m - n_g^m) = P_0(n_l)(n_l^{m-1} - n_g^{m-1}) + \pi \int_{x_l}^{x_g} dx \int_{x_l}^{x_g} dx' \, n(x) n(x')$$

$$\times \left[n^{m-1}(x) - n^{m-1}(x') \right] (x-x')$$

$$\times \int_{|x-x'|}^\infty ds \, g(s; x, x') u'(s) \qquad (2.G.7)$$

The sum rules could serve to test ad hoc theoretical models if enough were known about the interfacial structure. Alternatively, a given model can be used in connection with the sum rules to obtain predictions from the model. This approach was used by Lekner and Henderson. They assumed that $g(s; x, x') = g(s; (x + x')/2)$, expanded $n(x)$ and $n(x')$ in (2.G.6) in a Taylor series about $(x + x')/2$, and collected terms to obtain

$$kT\ln\frac{n_l}{n_g} = P_0(n_l)\left(\frac{1}{n_g} - \frac{1}{n_l}\right) - \pi(n_l - n_g)\sum_{k=0}^{\infty}\frac{2^{-2k}}{(2k+3)(2k+1)!}$$

$$\times \int_0^{\infty} ds\, s^{2k+3} u'(s)\int_{-\infty}^{\infty} d\bar{x}\left[\frac{d^{2k+1}}{d\bar{x}^{2k+1}}n(\bar{x})\right]g(s,\bar{x}) \quad (2.G.8)$$

With the pair correlation function given at (2.D.25), all the terms in the sum over k vanish except the one corresponding to $k=0$. The surviving term yields, after rearrangement, the coexistence curve equation.

$$P_0(n_l) = \frac{n_l n_g kT}{(n_l - n_g)^2}\left\{n_l + n_g - \frac{2n_l n_g}{n_l - n_g}\ln\frac{n_l}{n_g}\right\} \quad (2.G.9)$$

Equation 2.G.9 predicts a critical compressibility factor $P_c/n_c kT_c = \frac{1}{3}$, compared to the observed value 0.29 for argon, krypton, and xenon. It also fits the vapor pressure curve of xenon surprisingly well,[47] deviating of course as the critical point is approached.

Different sum rules yield with (2.D.25) different coexistence curve equations, but the same critical compressibility factor. Presumably, with the exact pair correlation function the different sum rules would yield the same coexistence curve equations, so a comparison between the curves generated by different sum rules would provide a test of the pair correlation function model. Lekner and Henderson found for the $m=2$ sum rule the formula,

$$P_0(n_l) = \frac{kTn_l n_g(n_l + n_g)}{n_l^2 + 4n_l n_g + n_g^2} \quad (2.G.10)$$

a result that looks quite different from (2.G.9) but agrees with it fairly well for xenon.

The sum rules, connecting as they do the bulk fluid properties and the correlation function in the interfacial zone, offer promising tests for model interfacial correlation functions. The discriminating value of such tests is yet to be demonstrated and ought to be the aim of future research on the sum rules.

III. STATISTICAL THERMODYNAMICS OF INHOMOGENEOUS FLUID

A. Thermodynamics

In Section II the theory of inhomogeneous fluid was built around the concept of the pressure tensor, a mechanical quantity that is well defined quite apart from the existence of thermodynamical laws. It was not required that

entropy or free-energy functions exist and obey the second law of thermodynamics. The straightforward nature of a purely mechanical theory is attractive, and its independence of thermodynamics is of interest for its generality. However, the gain in generality must be balanced against the loss of the
considerable power provided by the thermodynamical laws. To harness this
power we assume in what follows that the entropy function exists for inhomogeneous, equilibrium systems and obeys the second law, namely, that the
internal states (i.e., density and composition distributions) of an isolated
system change in any spontaneous process in such a way that entropy increases to a maximum value at equilibrium. In other words, we assume that
the entropy statement of the second law of thermodynamics is the same
whether the system is homogeneous or inhomogeneous. This postulate is
tacitly accepted in the usual thermodynamical treatment of interfaces and in
statistical mechanical ensemble theory of matter in the presence of external
fields and containing walls. In fact, nothing in the axioms of statistical thermodynamics, which relate entropy to the density of quantum states, draws a
distinction between homogeneous and inhomogeneous systems.

If the entropy function S of a system exists, so also does the Helmholtz
free energy F, defined by $F = U - TS$. An alternative statement of the second law is that the internal states of a closed, isothermal system take on values that minimize the Helmholtz free energy at equilibrium. This minimum
principle provides a convenient basis for investigation of the existence and
stability of the microstructure of inhomogeneous fluid.

The thermodynamic state of a homogeneous system can be fully specified
by temperature, volume V, and overall composition $\mathbf{N} \equiv \{N_1, N_2, \ldots, N_\nu\}$, N_i
being the number of particles of species i in V. The thermodynamic quanti-
ties can be expressed as functions of these variables. In inhomogeneous fluid
the density distributions $\mathbf{n}(\mathbf{r}) \equiv \{n_1(\mathbf{r}), n_2(\mathbf{r}), \ldots, n_\nu(\mathbf{r})\}$ must be specified as
well. In the theory developed here we presume that the thermodynamic
quantities are functionals of the density distributions, functional depen-
dence being indicated by the notation $F(\{\mathbf{n}\})$. Since $N_\alpha = \int_V n_\alpha(\mathbf{r}) \, d^3r$, the
overall composition of inhomogeneous fluid is an especially simple func-
tional of the density distributions.

Of particular significance in the theory of inhomogeneous fluid are the
chemical potentials $\mu_1, \ldots, \mu_\nu$ of the components. In homogeneous fluid μ_α
can be computed as the Helmholtz free-energy change upon addition of a
molecule of species α, while holding constant the temperature, volume, and
amounts of the other species. There is no question of where the molecule is
added. In inhomogeneous fluid, on the other hand, since densities vary in
space, the change in a thermodynamic function will depend on where the
molecule of species α is added. Addition of δN_α molecules to a small volume
τ fixed on the point $\mathbf{r}_0$ is illustrated in Fig. 3.A.1. The addition is accomplished by a small increase δn_α of the density of α in τ so that $\delta N_\alpha =$

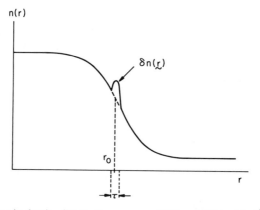

Fig. 3.A.1.   Change in density distribution n upon addition of δN $(=\int_\tau \delta n d^3 r)$ molecules to a small volume element τ fixed on the position r_0.

$\int_\tau \delta n_\alpha d^3 r$. If this is done while holding temperature, volume, and the density distributions of the other species fixed, the chemical potential at $\mathbf{r}_0$ can be computed from

$$\mu_\alpha(\mathbf{r}_0) = \lim_{\delta N_\alpha \to 0} \frac{F(\{\mathbf{n}+\delta n_\alpha\}) - F(\{\mathbf{n}\})}{\delta N_\alpha}$$

$$= \lim_{\tau, \delta n_\alpha \to 0} \frac{F(\{\mathbf{n}+\delta n_\alpha\}) - F(\{\mathbf{n}\})}{\int_\tau \delta n_\alpha d^3 r} \qquad (3.A.1)$$

The second form of the right-hand side of (3.A.1) defines a functional derivative of F, which we shall denote by $\delta F/\delta n_\alpha(\mathbf{r}_0)$. Thus the chemical potential measured at some position $\mathbf{r}_0$ is given by

$$\mu_\alpha(\mathbf{r}_0) = \frac{\delta F}{\delta n_\alpha(\mathbf{r}_0)} \qquad (3.A.2)$$

where the functional differentiation is performed with respect to $n_\alpha(\mathbf{r})$ while holding fixed the other distributions $n_\beta(\mathbf{r})$, $\beta \neq \alpha$, the temperature and the volume.

The condition of chemical equilibrium between homogeneous bulk phases is that the chemical potential of each species is the same in both phases. One might hope that the chemical potential as defined by (3.A.2) is a constant everywhere in an inhomogeneous fluid at equilibrium. Then (3.A.2), rather than being an expression from which the chemical potential can be computed with given density distributions, becomes instead an equation (integral, differential, or integrodifferential, depending on the functional form of F) from which the density distributions allowed at equilibrium can be computed. The chemical potentials of inhomogeneous fluid do indeed turn out to be everywhere constant at equilibrium, as we shall now prove.

Consider a closed, isothermal system. At equilibrium, the density distributions $\mathbf{n}(\mathbf{r})$ are such that F is a minimum. Thus,

$$F(\{\mathbf{n}+\boldsymbol{\varepsilon}\cdot\mathbf{v}\})<F(\{\mathbf{n}\}) \tag{3.A.3}$$

for any small fluctuation $\boldsymbol{\varepsilon}\cdot\mathbf{v}$, $\boldsymbol{\varepsilon}\cdot\mathbf{v}\equiv\{\varepsilon_1 v_1,\ldots,\varepsilon_\nu v_\nu\}$, consistent with the closed-system constraints of constant N_α,

$$N_\alpha=\int_V \left(n_\alpha+\varepsilon_\alpha v_\alpha\right)d^3r, \qquad \alpha=1,\ldots,\nu \tag{3.A.4}$$

The conditions that $F(\mathbf{n}+\boldsymbol{\varepsilon}\cdot\mathbf{v})$ be an extremal at $\varepsilon_\alpha=0$, $\alpha=1,\ldots,\nu$, subject to the constraints (3.A.4) of particle conservation are

$$\left\{\frac{\partial}{\partial\varepsilon_\alpha}\left[F(\{\mathbf{n}+\boldsymbol{\varepsilon}\cdot\mathbf{v}\})-\sum_\beta\lambda_\beta\int_V\left(n_\beta+\varepsilon_\beta v_\beta\right)d^3r\right]\right\}_{\varepsilon=0}=0 \tag{3.A.5}$$

for $\alpha=1,\ldots,\nu$ and for arbitrary $\mathbf{v}$. The λ_β's are constants known as Lagrange multipliers.

To evaluate (3.A.5), it is convenient to expand $F(\{\mathbf{n}+\boldsymbol{\varepsilon}\cdot\mathbf{v}\})$ in a functional Taylor series about the distributions $\mathbf{n}$, that is,

$$F(\{\mathbf{n}+\boldsymbol{\varepsilon}\cdot\mathbf{v}\})=F(\{\mathbf{n}\})+\sum_\alpha\int\frac{\delta F}{\delta n_\alpha(\mathbf{r})}\varepsilon_\alpha v_\alpha(\mathbf{r})\,d^3r$$

$$+\sum_{\alpha,\beta}\frac{1}{2!}\int\int\frac{\delta^2 F(\{\mathbf{n}\})}{\delta n_\alpha(\mathbf{r})\,\delta n_\beta(\mathbf{r}')}\varepsilon_\alpha v_\alpha(\mathbf{r})\varepsilon_\beta v_\beta(\mathbf{r}')\,d^3r\,d^3r'+\cdots$$

$$\tag{3.A.6}$$

where the δ's denote functional derivatives. One way of determining a functional derivative is a limiting process such as that used in (3.A.1). Another way is implied by (3.A.6), from which it follows that

$$\left[\frac{\partial F(\{\mathbf{n}+\boldsymbol{\varepsilon}\cdot\mathbf{v}\})}{\partial\varepsilon_\alpha}\right]_{\varepsilon=0}=\int\frac{\delta F}{\delta n_\alpha(\mathbf{r})}v_\alpha(\mathbf{r})\,d^3r \tag{3.A.7}$$

$$\left[\frac{\partial^2 F(\{\mathbf{n}+\boldsymbol{\varepsilon}\cdot\mathbf{n}\})}{\partial\varepsilon_\alpha\partial\varepsilon_\beta}\right]_{\varepsilon=0}=\int\int\frac{\delta^2 F}{\delta n_\alpha(\mathbf{r})\,\delta n_\beta(\mathbf{r}')}v_\alpha(\mathbf{r})v_\beta(\mathbf{r}')\,d^3r\,d^3r' \tag{3.A.8}$$

and so on. Thus, to obtain the functional derivative of a given functional $F(\{\mathbf{n}\})$ with respect to $n_\alpha(\mathbf{r})$, $n_\beta(\mathbf{r}')$,..., take the corresponding ordinary derivative of $F(\{\mathbf{n}+\boldsymbol{\varepsilon}\cdot\mathbf{v}\})$ with respect to $\varepsilon_\alpha, \varepsilon_\beta$,..., set $\boldsymbol{\varepsilon}=\mathbf{0}$, rearrange the results so that the product $v_\alpha(\mathbf{r})v_\beta(\mathbf{r}')\cdots$ appears as an integrand with integration over $d^3r\,d^3r'\ldots$, and finally identify the coefficient of $v_\alpha(\mathbf{r})v_\beta(\mathbf{r}')\cdots d^3r\,d^3r'\ldots$ in the integrand as the desired functional derivative of $F(\{\mathbf{n}\})$. The functional derivatives of functions are defined by admitting generalized functions in the definitions of functional derivatives. References in the bibliography of Ref. 22 contain more detailed discussions of the theory of functional derivatives.

Comparing (3.A.5) and (3.A.7), we find $\lambda_\alpha=\delta F/\delta n_\alpha(\mathbf{r})$. Thus, the chemical potential $\mu_\alpha(\mathbf{r})$ is equal to the Lagrange multiplier, *proof that the chemical potentials of the components of a fluid are everywhere constant in an inhomogeneous fluid at equilibrium if F obeys the minimum principle, the second law.*

Equation 3.A.5 is of course only the condition that $\mathbf{n}$ represent an extremal of F. The sufficient condition that $\mathbf{n}$ represent a minimum of F is that δ^2F be positive for arbitrary $\boldsymbol{\varepsilon}\cdot\mathbf{v}$, consistent with the constraints (3.A.4), where

$$\delta^2F=\sum_{\alpha,\beta}\frac{1}{2}\left[\frac{\partial^2F(\{\mathbf{n}+\boldsymbol{\varepsilon}\cdot\mathbf{v}\})}{\partial\varepsilon_\alpha\partial\varepsilon_\beta}\right]_{\boldsymbol{\varepsilon}=0}\varepsilon_\alpha\varepsilon_\beta$$

$$=\sum_{\alpha,\beta}\frac{1}{2}\int\int\frac{\delta^2F(\{\mathbf{n}\})}{\delta n_\alpha(\mathbf{r})\,\delta n_\beta(\mathbf{r}')}\varepsilon_\alpha v_\alpha(\mathbf{r})\varepsilon_\beta v_\beta(\mathbf{r}')\,d^3r\,d^3r' \quad (3.A.9)$$

The closed-system constraints are of course $\int_V v_\alpha(\mathbf{r})\,d^3r=0$, $\alpha=1,\ldots,\nu$.

Without loss of generality, since ε_α and v_α occur in δ^2F only as the product $\varepsilon_\alpha v_\alpha$, we can set $\varepsilon_\alpha\equiv 1$, $\alpha=1,\ldots,\nu$, and express the sufficiency condition, more commonly called the thermodynamic stability condition, as follows: the equilibrium state $\mathbf{n}$ represents a locally stable equilibrium if

$$\delta^2F=\sum_{\alpha,\beta}\frac{1}{2}\int\int\frac{\delta^2F(\{\mathbf{n}\})}{\delta n_\alpha(\mathbf{r})\,\delta n_\beta(\mathbf{r}')}v_\alpha(\mathbf{r})v_\beta(\mathbf{r}')\,d^3r\,d^3r'>0 \quad (3.A.10)$$

for arbitrary v_α, not all zero, obeying the closed-system constraints. If $\delta^2F>0$ for all allowed v_α, then $\mathbf{n}$ is locally stable; if $\delta^2F<0$ for any v_α, then $\mathbf{n}$ is unstable; but if $\delta^2F=0$ for some v_α and is greater than zero otherwise, higher order variations of F must be investigated to decide on stability.

The stability criterion based on δ^2F can be replaced by an eigenvalue problem. Let us define a matrix function $\mathbf{K}(\mathbf{r},\mathbf{r}')$ with elements $\delta^2F(\{\mathbf{n}\})/\partial n_\alpha(\mathbf{r})\,\partial n_\beta(\mathbf{r}')$ and a column vector function $\mathbf{v}(\mathbf{r})$ with elements $v_1(\mathbf{r}),\ldots,v_\nu(\mathbf{r})$.

Then (3.A.10) becomes

$$\delta^2 F = \frac{1}{2} \int \int \mathbf{v}^T(\mathbf{r}) \mathsf{K}(\mathbf{r},\mathbf{r}') \mathbf{v}(\mathbf{r}') \, d^3r \, d^3r' > 0 \qquad (3.A.11)$$

for arbitrary allowed vectors $\mathbf{v}(\mathbf{r})$, not all of whose components are zero. Here $\mathbf{v}^T$ is the transpose of $\mathbf{v}$; that is, it is a row vector. Let $\mathsf{K}(\mathbf{r},\mathbf{r}')$ be the kernel of an integral operator, the stability operator for $\mathbf{n}$, whose eigenvectors $\mathbf{u}$ and eigenvalues λ are defined by

$$\int \mathsf{K}(\mathbf{r},\mathbf{r}') \mathbf{u}(\mathbf{r}') \, d^3r' = \lambda \mathbf{u}(\mathbf{r}) \qquad (3.A.12)$$

Under the right boundary conditions in a suitably chosen function space, whose specification will depend on the functional form of F, the operator becomes a self-adjoint one whose eigenvectors form a complete orthonormal set. Thus any vector $\mathbf{v}(\mathbf{r})$ can be expanded in the series $\mathbf{v} = \Sigma_\lambda a_\lambda \mathbf{u}_\lambda$, with $a_\lambda = \int \mathbf{v}^T(\mathbf{r}) u_\lambda(\mathbf{r}) \, d^3r$, where the $\mathbf{u}_\lambda$ have the property $\int \mathbf{u}_\lambda^T(\mathbf{r}) \mathbf{u}_\lambda(\mathbf{r}) \, d^3r = \delta_{\lambda\lambda}$. With the orthonormal properties of $\mathbf{u}_\lambda$, (3.A.11) can be rewritten as

$$\delta^2 F = \tfrac{1}{2} \sum_\lambda \lambda a_\lambda^2 > 0 \qquad (3.A.11')$$

The distribution $\mathbf{n}$ is stable if (3.A.11') is obeyed for any fluctuation $\mathbf{v}$ satisfying particle conservation, $\int \mathbf{v} \, d^3r = 0$, the constraints of a closed system. Stability or instability of $\mathbf{n}$ can be conveniently classified according to the signs of λ_0 and λ_1, the smallest and next-smallest eigenvalues. The possibilities are set forth in Table 3.A.1. The simplest situations are those in which λ_0 is positive, in which case $\mathbf{n}$ is stable, and in which λ_0 and λ_1 are negative, in which case $\mathbf{n}$ is unstable.

If $\lambda_0 = 0$ is a simple eigenvalue and $\int \mathbf{u}_{\lambda_0} d^3r \neq 0$, then $\mathbf{n}$ is stable because any $\mathbf{v}$ satisfying particle conservation will have a contribution from at least one eigenvector $\mathbf{u}_\lambda$, $\lambda > 0$. When $\lambda_0 = 0$ and λ_0 is not simple or $\int \mathbf{u}_{\lambda_0} d^3r = 0$, higher variations of F must be considered for stability analysis.

The situation becomes more complicated when $\lambda_0 < 0$ and $\lambda_1 \geq 0$. When $\lambda_0 < 0$, $\mathbf{n}$ is unstable if $\int \mathbf{u}_{\lambda_0} d^3r = 0$. If $\lambda_0 < 0$ and $\lambda_1 = 0$, $\mathbf{n}$ is unstable if $\int \mathbf{u}_{\lambda_1} d^3r \neq 0$ since a particle-conserving linear combination of $\mathbf{u}_{\lambda_0}$ and $\mathbf{u}_{\lambda_1}$ can then be constructed. To obtain further stability criteria for the case $\lambda_0 < 0$ and $\lambda_1 \geq 0$, we have to find the set of a_λ's that minimize $\tfrac{1}{2}\Sigma_\lambda \lambda a_\lambda^2$ subject to the particle-conserving constraint $\int \mathbf{v} \, d^3r = \Sigma_\lambda a_\lambda \int \mathbf{u}_\lambda \, d^3r = 0$. The set is given by

$$\lambda a_\lambda - \mathbf{l}^T \int \mathbf{u}_\lambda \, d^3r = 0 \qquad (3.A.13)$$

TABLE 3.A.1
Eigenvalue Criteria for the Stability of **n**

Situation[a]	Implication for **n**
$\lambda_0 > 0$	Stable
$\lambda_0 = 0$	Stable if λ_0 is simple and $\int \mathbf{u}_{\lambda_0} d^3 r \neq 0$; otherwise higher variations of F must be considered
$\lambda_0 < 0, \lambda_1 > 0$	Stable if $\int \mathbf{u}_{\lambda_0} d^3 r \neq 0$ and $\Sigma_\lambda (1/\lambda)(\mathbf{l}^T \int \mathbf{u}_\lambda d^3 r)^2 > 0$ for all $\mathbf{l}^T$; unstable otherwise
$\lambda_0 < 0, \lambda_1 = 0$	Stable if $\int \mathbf{u}_{\lambda_0} d^3 r \neq 0$, $\int \mathbf{u}_{\lambda_1} d^3 r = 0$, and $\Sigma_{\lambda \neq \lambda_1}(1/\lambda)(\mathbf{l}^T \int \mathbf{u}_\lambda d^3 r)^2 > 0$ for all $\mathbf{l}^T$; unstable otherwise
$\lambda_0 < 0, \lambda_1 < 0$	Unstable

[a] The smallest and next-smallest eigenvalues of the stability operator are denoted by λ_0 and λ_1, respectively.

where $\mathbf{l}^T$ is a row vector whose elements $l_1, \ldots, l_\nu$ are the Lagrange multipliers. If $\int \mathbf{u}_{\lambda_0} d^3 r \neq 0$, $\int \mathbf{u}_{\lambda_1} d^3 r = 0$, and $\lambda_0 < 0$, $\lambda_1 = 0$, then **n** is stable if $\Sigma_{\lambda \neq \lambda_1}(1/\lambda)(\mathbf{l}^T \int \mathbf{u}_\lambda d^3 r)^2 > 0$ for arbitrary $\mathbf{l}^T$ and is unstable if $\Sigma_{\lambda \neq \lambda_1}(1/\lambda)(\mathbf{l}^T \int \mathbf{u}_\lambda d^3 r)^2 < 0$ for any $\mathbf{l}^T$. If $\int \mathbf{u}_{\lambda_0} d^3 r \neq 0$ and $\lambda_0 < 0$, $\lambda_1 > 0$, then **n** is stable if $\Sigma_\lambda (1/\lambda)(\mathbf{l}^T \int \mathbf{u}_\lambda d^3 r)^2 > 0$ for arbitrary $\mathbf{l}^T$ and is unstable $\Sigma_\lambda (1/\lambda)(\mathbf{l}^T \int \mathbf{u}_\lambda d^3 r)^2 < 0$ for any $\mathbf{l}^T$. See Table 3.A.1.

For a planar interface defined on an infinite domain ($\mathbf{n} = \mathbf{n}(x)$, $-\infty < x < \infty$) $\lambda = 0$ is an eigenvalue. This is the consequence of the arbitrariness of the position of the interface on an infinite domain; that is, if $\mathbf{n}(x)$ satisfies the equilibrium condition, (3.A.5), so also does $\mathbf{n}(x + a)$ for arbitrary a. From this translational invariance, it follows that $\delta F(\{\mathbf{n}(x + a)\})/\delta n_\alpha(\mathbf{r}) = \delta F(\{\mathbf{n}(x)\})/\delta n_\alpha(\mathbf{r})$. Expanding $\delta F(\{\mathbf{n}(\mathbf{x} + \mathbf{a})\})/\delta n_\alpha(\mathbf{r})$ in a functional Taylor series about $\mathbf{n}(x)$, we obtain

$$\frac{\delta F}{\delta n_\alpha(\mathbf{r})}\left(\{\mathbf{n}(x + a)\}\right) = \frac{\delta F}{\delta n_\alpha(\mathbf{r})}\left(\{\mathbf{n}(x)\}\right) + \sum_\beta \int \frac{\delta^2 F(\{\mathbf{n}(x)\})}{\delta n_\alpha(\mathbf{r})\,\delta n_\beta(\mathbf{r}')}$$

$$\times \left[n_\beta(x' + a) - n_\beta(x')\right] d^3 r' + \mathcal{O}(n(x + a) - n(x))^2$$

$$(3.A.14)$$

The left-hand side and the first term on the right-hand side of this expression cancel. Dividing the resulting expansion by a passing to the limit $a = 0$, we obtain

$$\sum_\beta \int \frac{\delta^2 F(\{\mathbf{n}\})}{\delta n_\alpha(\mathbf{r})\,\delta n_\beta(\mathbf{r}')} \frac{dn_\beta(x')}{dx'} d^3 r' = 0 \qquad (3.A.15)$$

Thus, if in (3.A.10) v_α is set equal to dn_α/dx, where $n_\alpha(x)$, $\alpha = 1, \ldots, \nu$, are the density profiles of a planar interface, it follows from (3.A.15) that $\delta^2 F = 0$. In other words, a zero-eigenvalue solution to the eigenvalue problem, (3.A.12), is a vector $\mathbf{u}$ whose components are dn_α/dx. Since

$$\int_{-\infty}^{\infty} \frac{dn_\alpha}{dx} dx = n_\alpha^{(2)} - n_\alpha^{(1)} \neq 0,$$

this eigenvector does not constitute a particle-conserving fluctuation, so that other eigenvalues must be examined to determine stability of the planar interface. This can only be done with explicit models.

Let us now turn to the problem of generating from the free-energy formulas for computing the pressure tensor and the tension of a planar interface. For this purpose it is convenient to introduce the Helmholtz free-energy density $f(\mathbf{r})$:

$$F = \int_V f(\mathbf{r}) \, d^3r \tag{3.A.16}$$

where $f(\mathbf{r})$ is a functional of $\mathbf{n}$ if F is.

Consider the planar one-component system illustrated in Fig. 3.A.2. Density n varies in the x-direction, and the cross-section normal to the x-direction is denoted by A. A very narrow piston, with its normal in the z-direction, is positioned at a point x_0 and is allowed to move isothermally an increment to change the cross-sectional area from A to $A + \Delta A(x_0)$. For a sufficiently small change, the reversible work or the free-energy change of the system is given by

$$\Delta F = -P_T \Delta V + \mathbb{O}(\Delta V)^2 \tag{3.A.17}$$

where $\Delta V = \Delta A(x_0) \Delta x_0$; Δx_0 is the width of the piston in the x-direction.

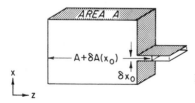

Fig. 3.A.2.   Work experiment defining a transverse pressure of a planar interface.

The transverse pressure at x_0 can be computed from

$$P_T = -\lim_{\Delta V \to 0} \frac{\Delta F}{\Delta V}$$

$$= -\lim_{\Delta V \to 0} \frac{\int_V \left[f(\{n + \Delta n\}) - f(\{n\}) \right] d^3 r}{\Delta V}$$

$$- \lim_{\Delta V \to 0} \frac{\int_{\Delta V} f(\{n\}) \, d^3 r}{\Delta V} \tag{3.A.18}$$

The second member on the right-hand side of (3.A.18) reduces to $-f(x_0)$ in the limit $\Delta V = 0$. To evaluate the first member, we note that particle conservation yields $\int_{V + \Delta V} \Delta n \, d^3 r = -\int_{\Delta V} n \, d^3 r = -n(x_0) \Delta V + \mathcal{O}(\Delta V)^2$ so that ΔV can be replaced by $-\int_{V + \Delta V} \Delta n \, d^3 r / n(x_0)$. The resulting expression has the form of the functional derivative (3.A.1) defining the chemical potential. Thus (3.A.18) yields for arbitrary x_0

$$P_T(x_0) = n(x_0)\mu - f(x_0) \tag{3.A.19}$$

a result published previously by Hill.[51] The same formula has been established for spherical drops.[28] Because P_T is determined from a functional, its form is not unique. However it yields a unique interfacial tension. This point is discussed further in Section III.D.

The tension of a planar interface computed from (2.B.4) with the aid of (3.A.19) is

$$\gamma = \int_{-\infty}^{\infty} \left[P_N + f(x) - n(x)\mu \right] dx \tag{3.A.20}$$

where $P_N = P_T(x) = n(x)\mu - f(x)$ for x in either bulk phase far from the interfacial zone.

The multicomponent versions of (3.A.19) and (3.A.20) are formed by replacing $n\mu$ by $\sum_\alpha n_\alpha \mu_\alpha$.

The tension of a planar interface can be computed directly from the Helmholtz free energy. If the small piston in Fig. 3.A.2 is replaced by a piston wide compared to the width of the interfacial zone, the free-energy change upon infinitesimal, isothermal displacement ΔA of the piston is

$$\Delta F = -\left[\int_{-\Delta x_0/2}^{\Delta x_0/2} P_T(x) \, dx \right] \Delta A = -P_N \Delta V + \left[\int_{-\infty}^{\infty} (P_N - P_T) \, dx \right] \Delta A$$

$$= -P_N \Delta V + \gamma \Delta A \tag{3.A.21}$$

The limits of x have been extended to $\pm\infty$ because $P_N - P_T = 0$ outside the narrow interfacial zone. Equation 3.A.21 yields the well-known thermodynamic expression

$$\gamma = \left(\frac{\partial F}{\partial A}\right)_{T,V,N} \tag{3.A.22}$$

for the tension of a planar interface.

If a fluid is subject to a conservative external force whose potential for each species is $u_e^\alpha(\mathbf{r})$, the Helmholtz free energy is

$$F = \int \left[f(\mathbf{r}') + \sum_\alpha n_\alpha u_e^\alpha(\mathbf{r}') \right] d^3r' \tag{3.A.23}$$

and the chemical potential obeys the equation

$$\mu_\alpha = \int \frac{\delta f(\mathbf{r}')}{\delta n_\alpha(\mathbf{r})} d^3r' + u_e^\alpha(\mathbf{r}) \tag{3.A.24}$$

where the functional derivative is taken with temperature, volume, and external field fixed. Since μ_α is constant, its gradient is zero and the gradient of (3.A.24) generates

$$-\nabla \int \frac{\delta f(\mathbf{r}')}{\delta n_\alpha(\mathbf{r})} d^3r' = \nabla u_e^\alpha(\mathbf{r}) \tag{3.A.25}$$

This result is the thermodynamic equivalent of the Yvon-Born-Green equation (2.C.23) expressing hydrostatic equilibrium of species α.

This completes the formal thermodynamics of inhomogeneous fluid. Next we turn to statistical thermodynamical theory.

B. Statistical Thermodynamics

Of the functional derivative theoretical works referenced above[16-22] those of Lebowitz and Percus[20] and of Percus[21] are especially convenient for laying out the statistical thermodynamic theory of classical inhomogeneous fluids. Distribution functions and thermodynamic functions are cast with the theory into appropriate functionals of the density distributions. Yang, Fleming, and Gibbs[27] have used the approach of Lebowitz and Percus to develop the theory of the Helmholtz free energy.

The starting point of the theory is the grand canonical ensemble partition function $\underline{H}$ of a classical inhomogeneous fluid in the presence of conservative external forces. For a one-component fluid

$$\overline{\underline{H}} = \sum_{N=0}^{\infty} \frac{q_I^N}{N!} \int_V \cdots \int \exp\left[-\beta u_N - \beta \sum_{i=1}^{N} \phi(\mathbf{r}_i)\right] d^3 r_1 \ldots d^3 r_N \quad (3.B.1)$$

where $\beta \equiv 1/kT$ and

$$\phi(\mathbf{r}) \equiv u_e(\mathbf{r}) - \mu \quad (3.B.2)$$

$u_e(\mathbf{r})$ being the external potential on a particle at $\mathbf{r}$ and μ the chemical potential. The contribution to the canonical partition function of the kinetic and internal motions of a single molecule is $q_I(T)$. Since q_I drops out in computations of phase and interfacial properties, its form need not be considered further. In what follows, no particular assumptions are made about the nature of the intermolecular forces except that u_N depends only on particle positions. The results are also valid for molecules with classical rotational and vibrational degrees of freedom, although if the density of particles at position $\mathbf{r}$ *and* with given internal coordinates is sought, these coordinates must be included in the external potential and the theory extended to cover these degrees of freedom.

The s-body density distribution function for the grand ensemble is

$$n_s(\mathbf{r}_1,\ldots,\mathbf{r}_s) = \sum_{N=s}^{\infty} \frac{q_I^N}{\overline{\underline{H}}(N-s)!}$$

$$\times \int \cdots \int \exp\left[-\beta u_N - \beta \sum_{i=1}^{N} \phi(r_i)\right] d^3 r_{s+1} \cdots d^3 r_N \quad (3.B.3)$$

A special property of functional derivatives enables one to compute the density distribution functions from functional derivatives of the partition function with respect to the potential $\phi(r)$. This property is

$$\frac{\delta\phi(\mathbf{r}_i)}{\delta\phi(\mathbf{r})} = \delta(\mathbf{r}_i - \mathbf{r}) \quad (3.B.4)$$

where $\delta(\mathbf{r}_i - \mathbf{r})$ is the Dirac delta function. With this property it is not difficult to show that

$$\frac{\delta\Omega}{\delta\phi(r)} = -n(\mathbf{r}) \quad (3.B.5)$$

$$\frac{\delta^2\Omega}{\delta\phi(\mathbf{r})\delta\phi(\mathbf{r}')} = -\frac{\delta n(r)}{\delta\phi(r')} = \beta\left[n_2(\mathbf{r},\mathbf{r}') - n(\mathbf{r})n(\mathbf{r}') + n(\mathbf{r})\delta(\mathbf{r}' - \mathbf{r})\right]$$

$$(3.B.6)$$

where Ω is the thermodynamic potential

$$\Omega \equiv kT \ln \underline{\underline{H}} \qquad (3.B.7)$$

In this manner n_s is determined by the first s functional derivatives of Ω or the first $s-1$ functional derivatives of $n(\mathbf{r})$. Note that Ω is not equal to PV for inhomogeneous fluid, since pressure is a tensor in inhomogeneous fluid and PV is not well defined; Ω is equal to PV for homogeneous fluid.

The thermodynamic potential, the density, and the density distribution functions are functionals of the potential $\phi(\mathbf{r})$. By inversion of functional relationships (i.e., by transformation of independent functional variables), the thermodynamic potential, the potential $\phi(\mathbf{r})$, and the density distribution functions become functionals of the density $n(\mathbf{r})$. This is the most important step in the density functional theory of inhomogeneous fluids.

Transformation of functional variables is accomplished with the aid of the chain rule,

$$\frac{\delta h}{\delta n(\mathbf{r})} = \int \frac{\delta h}{\delta \phi(\mathbf{r}')} \frac{\delta \phi(\mathbf{r}')}{\delta n(\mathbf{r})} d^3 r' \qquad (3.B.8)$$

for an arbitrary functional h. Choosing $h = n(\mathbf{r}'')$, we obtain

$$\delta(\mathbf{r}'' - \mathbf{r}) = \int \frac{\delta n(\mathbf{r}'')}{\delta \phi(\mathbf{r}')} \frac{\delta \phi(\mathbf{r}')}{\delta n(\mathbf{r})} d^3 r' \qquad (3.B.9)$$

an expression showing that $\delta \phi(\mathbf{r}')/\delta n(\mathbf{r})$ is the operator inverse of $\delta \phi(\mathbf{r}')/\delta n(\mathbf{r})$; that is, if $\delta n(\mathbf{r}'')/\delta \phi(\mathbf{r}')$ is the kernel of the integral operator K, then $\delta \phi(\mathbf{r}')/\delta n(\mathbf{r})$ is the kernel of the inverse of K.

In the limit of no correlations, $\delta n(\mathbf{r})/\delta \phi(\mathbf{r}) = -\beta n(\mathbf{r})\delta(\mathbf{r}' - \mathbf{r})$ and by inspection of (3.B.8) it follows that $\delta \phi(\mathbf{r}')/\delta n(\mathbf{r}'') = -[\beta n(\mathbf{r}')]^{-1}\delta(\mathbf{r}'' - \mathbf{r}')$. Defining $\beta^{-1}C(\mathbf{r}'',\mathbf{r}')$ as the part due to correlations, that is,

$$\frac{\delta \phi(\mathbf{r}')}{\delta n(\mathbf{r}'')} = -[\beta n(\mathbf{r}')]^{-1}\delta(\mathbf{r}'' - \mathbf{r}') + \beta^{-1}C(\mathbf{r}'',\mathbf{r}') \qquad (3.B.10)$$

combining (3.B.6), (3.B.9), and (3.B.8), and using the definition $g(\mathbf{r}',\mathbf{r}) \equiv n_2(\mathbf{r}',\mathbf{r})/n(\mathbf{r})n(\mathbf{r}')$, we obtain the well-known equation

$$g(\mathbf{r},\mathbf{r}'') - 1 = C(\mathbf{r},\mathbf{r}'') + \int C(\mathbf{r},\mathbf{r}')n(\mathbf{r}')[g(\mathbf{r}',\mathbf{r}'') - 1] d^3 r' \qquad (3.B.11)$$

Thus, computation of the inverse of $\delta n(\mathbf{r})/\delta\phi(\mathbf{r}'')$ is the problem of determining the *direct correlation function* $C(\mathbf{r},\mathbf{r}')$ from the *pair correlation function* $g(\mathbf{r},\mathbf{r}')$.

Let us define the singlet direct correlation function by

$$C_1(\mathbf{r}')\equiv\ln\left\{n(\mathbf{r}')e^{\beta[\phi(\mathbf{r}')+\mu^+(T)]}\right\}\qquad(3.B.12)$$

where $\mu^+(T)$, the contribution of q_I to the chemical potential, has been added so that C_1 vanishes in the limit of zero density. From (3.B.10) it follows that

$$C(\mathbf{r},\mathbf{r}')=\frac{\delta C_1(\mathbf{r}')}{\delta n(\mathbf{r})}\qquad(3.B.13)$$

a formula suggestive of the s-body direct correlation function, namely,[21]

$$C_s(\mathbf{r}_1,\dots,\mathbf{r}_s)=\frac{\delta C_{s-1}}{\delta n(\mathbf{r}_s)}(\mathbf{r}_1,\dots,\mathbf{r}_{s-1})=\frac{\delta^{s-1}C_1(\mathbf{r}_1)}{\delta n(\mathbf{r}_2)\dots\delta n(\mathbf{r}_s)}\qquad(3.B.14)$$

By functional integration Yang, Fleming, and Gibbs related the chemical potential and the Helmholtz free energy to the two-body direct correlation function $C(\mathbf{r},\mathbf{r}')$ of inhomogeneous fluid, a significant result because of the connection between C and g and, therefore, between C and the X-ray diffraction structure function of the fluid. The identity

$$\frac{dh(\{\varepsilon n\})}{d\varepsilon}=\int\frac{\delta h(\{\varepsilon n\})}{\delta(\varepsilon n(\mathbf{r}))}n(\mathbf{r})\,d^3r\qquad(3.B.15)$$

which can be verified from (3.A.7) by relabeling variables, can be integrated to obtain

$$h(\{n\})-h(\{0\})=\int d^3r\int_0^1 d\varepsilon\frac{\delta h(\{\varepsilon n\})}{\delta(\varepsilon n(r))}n(\mathbf{r})\qquad(3.B.16)$$

This is the basic formula of functional integration.

With $h=C_1(\mathbf{r})$, $\delta h/\delta n(\mathbf{r}')=C(\mathbf{r},\mathbf{r}')$, and (3.B.15) yields

$$C_1(\mathbf{r})=\int d^3\mathbf{r}'\int_0^1 d\varepsilon\, C(\mathbf{r},\mathbf{r}';\{\varepsilon n\})n(\mathbf{r}')\qquad(3.B.17)$$

But from (3.B.12) it follows that

$$\mu = \mu^+(T) + kT\ln n - kTC_1(\mathbf{r}) + u_e(\mathbf{r}) \qquad (3.B.18)$$

or, with (3.B.17),

$$\mu = \mu^+(T) + kT\ln n - kT\int d^3r' \int_0^1 d\varepsilon\, C(\mathbf{r},\mathbf{r}'; \{\varepsilon n\})n(\mathbf{r}') + u_e(\mathbf{r})$$

$$(3.B.19)$$

one of the desired results.

Taking next $h = F$, noting that $\delta F/\delta n(\mathbf{r}')$, and substituting μ from (3.B.19) into (3.B.16), we obtain

$$F = \int \left[f(\mathbf{r}) + n(\mathbf{r})u_e(\mathbf{r}) \right] d^3r \qquad (3.B.20)$$

where

$$f(\mathbf{r}) = n(\mathbf{r})\left\{ \mu^+(T) + kT[\ln n(\mathbf{r}) - 1] \right\}$$
$$- n(\mathbf{r})kT\int_0^1 \int (1-\varepsilon)C(\mathbf{r},\mathbf{r}'; \{\varepsilon n\})n(\mathbf{r}')\, d\varepsilon\, d^3r' \qquad (3.B.21)$$

the second of the desired results. We have used the relationship

$$\int_0^1 d\varepsilon\, \varepsilon \int_0^1 d\varepsilon'\, C(\mathbf{r},\mathbf{r}'; \{\varepsilon\varepsilon'n\}) = \int_0^1 d\varepsilon(1-\varepsilon)\, C(\mathbf{r},\mathbf{r}'; \{\varepsilon n\}) \qquad (3.B.22)$$

obtained by change of variables and order of integration.

The importance of (3.B.19) and (3.B.21) is that the chemical potential and the free energy of inhomogeneous fluid are determined by the two-body direct correlation function. Of course, this does not solve the problem, $C(\mathbf{r},\mathbf{r}')$ not being known for inhomogeneous fluid, but it does focus on what theoretical quantity we should seek, and it is suggestive of an approximate free-energy model that is discussed in a later section. Moreover, under gradient theory the chemical potential becomes a function only of density, its gradients, and the two-body direct correlation function of *homogeneous* fluid, about which a lot is known.

A representation of the transverse pressure of a planar interface can be computed from free energy with the formula $P_T = n\mu - f$, and P_N can be obtained from the limit $P_N = P_T(x = \pm\infty)$. Equations 3.B.19 and 3.B.21 yield

$$P_T = n(x)kT\left\{ 1 - \iint_0^1 \varepsilon C(\mathbf{r},\mathbf{r}'; \{\varepsilon n\})n(x')\, d\varepsilon\, d^3r' \right\} \qquad (3.B.23)$$

The interfacial tension can be computed from (2.B.4) in principle. An alternative and simpler formula has been derived by Triezenberg and Zwanzig[24] and by Lovett et al.[25] The derivation of the latter investigators follows easily from the approach presented here. Imagine a spherical interface in the presence of an external force that depends only on radial distance. The normal component of the equation of hydrostatics is $dP_N/dr = (P_T - P_N)/r + n\,d\phi/dr$. Suppose the radius R of the interfacial zone is very large compared to the thickness of the interfacial zone, and integrate the equation of hydrostatics along the x-axis ($y = z = 0$) to obtain

$$\Delta P_N = \frac{2\gamma}{R} + \int_{-\infty}^{\infty} n \frac{\partial \phi}{\partial x} dx \qquad (3.B.24)$$

the Young-Laplace equation for an interface in an external field. It has been assumed here that the potential ϕ goes to zero outside the interfacial zone. The trick of Lovett et al. is to treat $\phi(\mathbf{r})$ as a functional of $n(\mathbf{r})$, which is zero for a planar interface and which curves the planar interface so that $\Delta P_N = 0$, or

$$\gamma = -\frac{R}{2} \int_{-\infty}^{\infty} n \frac{\partial \phi}{\partial x} dx \qquad (3.B.25)$$

If $n^s(r)$ denotes the density distribution of the spherical interface and $n(x)$ that of a planar interface, then for y and z not too far from zero, the difference $n^s(r) - n(x)$ is small ($x = 0$ centered on R in both cases) and the density functional Taylor series of $\phi(\mathbf{r})$ about $n(x)$ can be truncated after the first term, that is,

$$\phi(\mathbf{r}) \simeq \int \left[\frac{\delta\phi(r)}{\delta n(r')} \right]_{n=n(x)} \left[n^s(\mathbf{r}') - n(x') \right] d^3r'$$

$$= \int \left[-\frac{kT}{n(x')} \delta(\mathbf{r}' - \mathbf{r}) + kTC(\mathbf{r}, \mathbf{r}'; \{n(x')\}) \right] \left[n^s(\mathbf{r}) - n(x') \right] d^3r'$$

$$(3.B.26)$$

For the tension of the curved interface to equal that of the planar fluid, the density of the curved interface must be of the form $n^s(r) = n([z^2 + y^2 + (x + R)^2]^{1/2} - R)$. Thus, expansion of n^s about $y = z = 0$ yields $n^s(r') - n(x') = [(y'^2 + z'^2)/2R][dn(x')/dx'] + \mathcal{O}(R^{-2})$. Putting this into (3.B.26), inserting the result into (3.B.25), and rearranging terms, we get the desired result:

$$\gamma = \frac{kT}{4} \iint_{-\infty}^{\infty} (s_y^2 + s_z^2) C(\mathbf{r}, \mathbf{r} + \mathbf{s}; \{n\}) \frac{dn(x)}{dx} \frac{dn(x + s_x)}{dx} dx\, d^3s$$

$$(3.B.27)$$

This rigorous expression for the interfacial tension is bilinear in density gradients across the interface. The weighting factor is essentially the mean-square range of direct correlations. The similarity of this result to gradient theory (2.D.18) probably accounts for the success of the gradient theory in predicting tension.

We also recall that the interfacial tension can be computed as the area derivative of the Helmholtz free energy. If one uses a canonical ensemble, $F = -kT \ln Q$, then the tension formulas generated[11, 28] are the Kirkwood-Buff expressions obtained directly from the pressure tensor in Section II. Thus, aside from verifying the consistency of the mechanical and thermodynamic approaches, this definition of tension generates nothing new at this point.

The kernel of the stability operator of inhomogeneous fluid is determined by the two-body direct correlation function, namely,

$$\frac{\delta^2 F}{\delta n(\mathbf{r}) \delta n(\mathbf{r}')} = [\beta n(\mathbf{r})]^{-1} \delta(\mathbf{r}' - \mathbf{r}) - \beta^{-1} C(\mathbf{r}, \mathbf{r}') \qquad (3.\text{B}.28)$$

so that the eigenvalue problem (3.A.12) for stability of n becomes

$$\frac{kT}{n(\mathbf{r})} u(\mathbf{r}) - \int C(\mathbf{r}, \mathbf{r}'; \{n\}) u(\mathbf{r}') d^3 r' = \lambda u(\mathbf{r}) \qquad (3.\text{B}.29)$$

The direct correlation functions of inhomogeneous fluid can be related to the direct correlation functions of homogeneous fluid by expanding the former in a functional Taylor series about a fixed density n^0. For the two-body case the series is

$$C(\mathbf{r}_1, \mathbf{r}_2) = C(\mathbf{r}_1, \mathbf{r}_2)|_{n=n^0}$$

$$+ \sum_{k=1}^{\infty} \frac{1}{k!} \int \cdots \int \frac{\delta^k C(\mathbf{r}_1, \mathbf{r}_2)}{\delta n(\mathbf{r}_3) \ldots \delta_n(\mathbf{r}_{k+2})} \bigg|_{n=n^0} \prod_{i=3}^{k+2} [n(\mathbf{r}_i) - n^0] d^3 r_i$$

$$= C^0(\mathbf{r}_1, \mathbf{r}_2; n^0) + \sum_{k=1}^{\infty} \frac{1}{k!} \int \cdots \int C_{k+2}^0(\mathbf{r}_1, \ldots, \mathbf{r}_{k+2}; n^0)$$

$$\times \prod_{i=3}^{k+2} [n(\mathbf{r}_i) - n^0] d^3 \mathbf{r}_i \qquad (3.\text{B}.30)$$

where $C_s^0(\mathbf{r}_1, \ldots, \mathbf{r}_s; n^0)$ denotes the s-body direct correlation function of homogeneous fluid at density n^0. For the special limit $n^0 = 0$, (3.B.30) becomes the virial expansion of the direct correlation function of inhomogeneous fluid. For the purposes of gradient theory and truncated expansions,

n^0 can be set to $n(\mathbf{r}_1)$, $n[(\mathbf{r}_1+\mathbf{r}_2/2]$, $\frac{1}{2}[n(\mathbf{r}_1)+n(\mathbf{r}_2)]$, and so on, as demanded by the particular approximation. Equation 3.B.30 is then an expansion of the direct correlation function about a reference state of locally homogeneous fluid. Iterative solution of (3.B.11) for g in terms of C yields the functional Taylor's expansion of the pair correlation function when (3.B.30) is substituted for C.

Derivation of the multicomponent versions of the results above is essentially a matter of keeping a record of species indices. The pertinent generalizations are:

$$C_1^\alpha(\mathbf{r})=\ln\left\{n_\alpha(\mathbf{r})e^{\beta[\phi_\alpha(r)+\mu^+(T)]}\right\} \tag{3.B.31}$$

$$C_s^{\alpha\beta\cdots\omega}(\mathbf{r}_1,\ldots,\mathbf{r}_s)=\frac{\delta^{s-1}C_1^\alpha(\mathbf{r}_1)}{\delta n^\beta(\mathbf{r}_2)\ldots\delta n^\omega(\mathbf{r}_s)} \tag{3.B.32}$$

$$g^{\alpha\beta}(\mathbf{r},\mathbf{r}'')-1=C^{\alpha\beta}(\mathbf{r},\mathbf{r}'')+\sum_\gamma\int C^{\alpha\gamma}(\mathbf{r},\mathbf{r}')n_\gamma(\mathbf{r}')\left[g^{\gamma\beta}(\mathbf{r}',\mathbf{r}'')-1\right]d^3r' \tag{3.B.33}$$

$$\mu_\alpha=\mu_\alpha^+(T)+kT\ln n_\alpha-\sum_\beta kT\int d^3r'\int_0^1 d\varepsilon C^{\alpha\beta}(\mathbf{r},\mathbf{r}';\{\varepsilon n\})n_\beta(\mathbf{r}') \tag{3.B.34}$$

$$f(\mathbf{r})=\sum_\alpha n_\alpha(\mathbf{r})\left\{\mu^+(T)+\left[\ln n_\alpha(\mathbf{r})-1\right]\right\}$$

$$-\sum_{\alpha,\beta}\frac{1}{2}n_\alpha(\mathbf{r})\int_0^1\int(1-\varepsilon)C(\mathbf{r},\mathbf{r}';\{\varepsilon n\})n_\beta(\mathbf{r}')\,d^3r' \tag{3.B.35}$$

$$\gamma=\frac{kT}{4}\sum_{\alpha,\beta}\iint_{-\infty}^\infty\left(s_y^2+s_z^2\right)C^{\alpha\beta}(\mathbf{r},\mathbf{r}+\mathbf{s};\{n\})\frac{dn_\alpha(x)}{dx}\frac{dn_\beta(x+s_x)}{dx}\,dx\,d^3s \tag{3.B.36}$$

$$\frac{\delta^2 F}{\delta n_\alpha(\mathbf{r})\delta n_\beta(\mathbf{r}')}=\frac{kT}{n_\alpha(\mathbf{r})}\delta_{\alpha\beta}\delta(\mathbf{r}'-\mathbf{r})-kTC^{\alpha\beta}(\mathbf{r},\mathbf{r}') \tag{3.B.37}$$

The multicomponent version of the expansion of $C^{\alpha\beta}(\mathbf{r}_1,\mathbf{r}_2)$ is obtained from (3.B.30) by adding to every factor of $n(\mathbf{r}_i)-n^0$ is a species index putting the corresponding indices on C_s^0 and summing over all species labels except α and β. $\delta_{\alpha\beta}$ is the Kronecker delta function.

The functional Taylor series for $C_1^\alpha(\mathbf{r})$ and $C^{\alpha\beta}(\mathbf{r},r')$ can be combined to obtain the statistical thermodynamic equivalent of the YBG equation

(2.C.23). As outlined by Wertheim[62] and by Sullivan and Stell,[63] the combination of (3.B.30) for $C^{\alpha\beta}$ and the gradient of the corresponding Taylor series for C_1^{α} leads to the formula (cf. 2.C.23)

$$-kT \nabla n_\alpha - n_\alpha \nabla u_e^\alpha = -\sum_\beta n_\alpha kT \int C^{\alpha\beta}(\mathbf{r},\mathbf{r}') \nabla n_\beta(\mathbf{r}') d^3 r' \quad (3.B.38)$$

This result, also published by Lovett et al.,[64] plays an important role in the connection established by Sullivan and Stell between a wall-fluid density profile and the fluid interfacial density profile. This opens a promising approach to the problem of fluid structure. Because of the unfortunate strictures of time and space, we do not review this work here.

It is important to note[27] that the component singlet direct correlation function $C_1^\alpha(\mathbf{r})$ can be generated as the density functional derivative of a thermodynamic function Φ, defined by

$$\Phi \equiv \sum_\gamma \int n_\gamma(\mathbf{r}) \left[\ln n_\gamma(\mathbf{r}) - 1 + \mu_\gamma^+(T) + \beta\phi_\gamma(\mathbf{r}) \right] d^3 r + \beta\Omega \quad (3.B.39)$$

By functional differentiation, with the aid of the chain rule for $h(\{\mathbf{n}\})$,

$$\frac{\delta h(\{\mathbf{n}\})}{\delta n_\alpha(\mathbf{r})} = \sum_\gamma \int \frac{\delta h(\{\mathbf{n}\})}{\delta \phi_\gamma(\mathbf{r}')} \frac{\delta \phi_\gamma(\mathbf{r}')}{\delta n_\alpha(\mathbf{r})} d^3 r' \quad (3.B.40)$$

and the property $\delta\Omega/\delta n_\alpha(\mathbf{r}) = -n_\alpha(\mathbf{r})$, it is easy to demonstrate that $\delta\phi/\delta n_\alpha(\mathbf{r}) = C_1^\alpha(\mathbf{r})$. This result implies that

$$C_s^{\alpha\beta\gamma\cdots\omega}(\mathbf{r}_1,\mathbf{r}_2,\mathbf{r}_3,\ldots,\mathbf{r}_s) = \frac{\delta^s \Phi}{\delta n^\alpha(\mathbf{r}_1)\,\delta n^\beta(\mathbf{r}_2)\,\delta n^\gamma(\mathbf{r}_3)\cdots\delta n^\omega(\mathbf{r}_s)}$$

$$(3.B.41)$$

a relationship whose significance is that $C_s^{\alpha\beta\cdots}$ has the symmetries of an s-order functional derivative, that is, $C_s^{\alpha\beta\gamma\cdots\omega}(\mathbf{r}_1,\mathbf{r}_2,\mathbf{r}_3,\ldots,\mathbf{r}_s) = C^{\beta\alpha\gamma\cdots\omega}(\mathbf{r}_2,\mathbf{r}_1,\mathbf{r}_3,\ldots,\mathbf{r}_s) = C^{\gamma\alpha\beta\cdots\omega}(\mathbf{r}_3,\mathbf{r}_1,\mathbf{r}_2,\ldots,\mathbf{r}_s)$, and so on.

The remarkable thing about the theory outlined here is that the phase behavior and the equilibrium density distributions and stability of classical, inhomogeneous fluids are completely determined by the two-body direct correlation function. Interactions among fluid particles do not appear explicitly in the free energy, chemical potential, or stability matrix. The results are not restricted to fluids with pairwise additive intermolecular forces; the only restriction is that intermolecular forces be determined by particle positions. Extensions of the theory to ordered fluids in which the forces and particle densities depend on particle position and orientation are under investigation.[65] Such systems are not treated in this chapter.

The difficulty with the theory is that there presently exists no tractable theory of the direct correlation function of inhomogeneous fluid. Even (3.B.30), with which $C(\mathbf{r}, \mathbf{r}')$ can be computed from the density distribution and direct correlation functions of homogeneous fluid, is not practicable because the theory demands homogeneous fluid direct correlation functions of all orders and the performance of numerous multiple integrals. Further development of the theory has consequently had to rely on introduction of correlation function approximations or gradient theory, both of which were discussed in Section II in connection with solutions of the YBG equations. The remainder of the chapter is devoted to approximate theories and their applications.

C. Model Free Energies

The oldest attempt to model the Helmholtz free energy was that of van der Waals.[1] In its modernized form his approach amounts to a mean field theory. The updated or modified van der Waals (MVDW) model introduced by Bongiorno and Davis[66] (see also Ref. 43) yields density profiles and tensions in reasonable agreement (on the order of 25% error) with computer simulations of 6-12 Lennard-Jones fluids. A similar model introduced earlier by Toxvaerd[52] also agrees with computer results.[54] The MVDW model can be set forth on the basis of two related assumptions. The first is that the entropy S of inhomogeneous fluid is given by $\int s_0(n(\mathbf{r}))\, d^3r$, where $s_0(n)$ is the entropy density of homogeneous fluid at density n. The second is that the pair correlation function of inhomogeneous fluid can be approximated by that of homogeneous fluid at a density or densities near that of the pair of correlated particles. We denote such an approximate correlation function by $g_0(|\mathbf{r}'-\mathbf{r}|; \bar{n})$ and recall Cases 1 to 3, (2.D.20) to (2.D.22) as typical choices for the approximate correlation function. Probably $\bar{n} = n[(\mathbf{r}+\mathbf{r}')/2]$ and $\bar{n} = \frac{1}{2}[n(\mathbf{r})+n(\mathbf{r}')]$ are the choices of local density most often used in interfacial profile and tension calculations. For these assumptions to make sense, the entropy and the pair correlation of fluids must be determined primarily by the short-range forces between molecules. Evidence that this is true, at least at higher densities, is found in the successes of the scaled-particle theory[67] and the theory of Zwanzig[68] and its successors such as the perturbation theory of Weeks, Chandler, and Andersen[69]: from these theories it follows that the entropy and the pair correlation function of dense fluid depend mostly on the short-range, strongly repulsive forces between molecules. Although the Weeks, Chandler, and Andersen theory breaks down at sufficiently low densities, in the limit of a dilute gas the pair correlation function is independent of density, so that the pair correlation function of the MVDW model becomes exact in this limit.

The free energy of the MVDW model for a one-component fluid of particles interacting with pair, central forces is then

$$F = \int \left\{ f_0(n(\mathbf{r})) + \frac{1}{2} \int \left[n(\mathbf{r})n(\mathbf{r}')g_0(|\mathbf{r}'-\mathbf{r}|;\bar{n}) \right. \right.$$
$$\left. \left. - n^2(\mathbf{r})g_0(|\mathbf{r}'-\mathbf{r}|;n(\mathbf{r})) \right] u(|\mathbf{r}'-\mathbf{r}|) d^3r' \right\} d^3r \qquad (3.C.1)$$

The usual additions of species indices and summations yield the multicomponent version of the model.

A free-energy model, which we call the approximate density functional (ADF) model, has been introduced recently by Ebner et al.[53] Note that in the low-density limit $C(\mathbf{r},\mathbf{r}') = C_0(|\mathbf{r}'-\mathbf{r}|)$, $C_0(|\mathbf{r}'-\mathbf{r}|)$ being the density-independent direct correlation function of dilute gas; thus the density functional free energy, (3.B.20), can be rearranged to obtain

$$F = \int \left\{ f_0(n(\mathbf{r})) + \frac{kT}{4} \int [n(\mathbf{r}') - n(\mathbf{r})]^2 C_0(|\mathbf{r}'-\mathbf{r}|) d^3r' \right\} d^3r, \qquad (3.C.2)$$

where $f_0(n)$ is the Helmholtz free-energy density of dilute homogeneous fluid. Ebner et al. extend (3.C.2) to a dense fluid by replacing $C_0(|\mathbf{r}'-\mathbf{r}|)$ by $C_0(|\mathbf{r}'-\mathbf{r}|;\bar{n})$, the direct correlation function of homogeneous fluid at some density characteristic of the correlated pair and by taking the dense fluid version of $f_0(n)$. The ADF free energy is then

$$F = \int \left\{ f_0(n) + \frac{kT}{4} \int [n(\mathbf{r}') - n(\mathbf{r})]^2 C_0(|\mathbf{r}'-\mathbf{r}|;\bar{n}) d^3r' \right\} d^3r \qquad (3.C.3)$$

Both the MVDW model and the ADF model must be completed by specifying the local correlation functions $g_0(\bar{n})$ and $C_0(\bar{n})$. Ebner et al. choose Case 2, or $\bar{n} = \frac{1}{2}[n(\mathbf{r}) + n(\mathbf{r}')]$. For this choice the chemical potential of the MVDW model is

$$\mu = \mu_0(n(\mathbf{r})) + \int \left[n(\mathbf{r}')g_0(|\mathbf{r}'-\mathbf{r}|;\bar{n}) - n(\mathbf{r})g_0(|\mathbf{r}'-\mathbf{r}|;n(\mathbf{r})) \right] u(|\mathbf{r}'-\mathbf{r}|) d^3r'$$
$$+ \frac{1}{2} \int \left[n(\mathbf{r})n(\mathbf{r}') \frac{\partial g_0}{\partial \bar{n}}(|\mathbf{r}'-\mathbf{r}|;\bar{n}) - n^2(\mathbf{r}) \frac{\partial g_0}{\partial n}(|\mathbf{r}'-\mathbf{r}|;n(\mathbf{r})) \right] u(|\mathbf{r}'-\mathbf{r}|) d^3r'$$
$$(3.C.4)$$

and of the ADF model is

$$\mu = \mu_0(n(\mathbf{r})) + \frac{kT}{2}\int[n(\mathbf{r}') - n(\mathbf{r})]C_0(|\mathbf{r}' - \mathbf{r}|; \bar{n})\,d^3r'$$

$$+ \frac{kT}{4}\int[n(\mathbf{r}') - n(\mathbf{r})]^2\frac{\partial C_0}{\partial \bar{n}}(|\mathbf{r}' - \mathbf{r}|; \bar{n})\,d^3r' \qquad (3.\mathrm{C}.5)$$

where $\mu_0(n)$ is the chemical potential of homogeneous fluid at density n. These are integral equations that determine the possible equilibrium fluid microstructures of inhomogeneous fluid: planar solutions, solutions with spherical symmetry, solutions with various spatial symmetries, and so on. Solutions can be sought and, if found, are candidates for stable equilibrium structures, stability being determined by the second functional derivative of F as outlined in Section III.A. This is a major area of our ongoing theoretical studies.

McCoy and Davis[43] compared the density profiles and tensions predicted for the vapor-liquid interface of a 6-12 Lennard-Jones fluid by the MVDW and the ADF models with $\bar{n} = \frac{1}{2}[n(\mathbf{r}) + n(\mathbf{r}')]$. The calculations were done at $kT/\varepsilon = 1$. The free-energy and correlation functions of homogeneous fluid were computed from the Weeks, Chandler, and Andersen theory. The density profile was computed by minimizing the free-energy function with the six-parameter trial function introduced by Ebner et al. McCoy and Davis verified that the trial function is accurate by using it with the gradient theoretical model for which the density profile problem can be solved analytically.

The density profiles and tensions obtained are shown in Fig. 3.C.1. The profiles of the two models agree fairly well, and the predicted interfacial width is quite similar to those observed in computer simulations of the 6-12 Lennard-Jones (LJ) fluid.

Although no computer data are available for the interface of a 6-12 LJ fluid at $kT/\varepsilon = 1$, there are data for interfaces at higher and at lower temperatures. From a smooth curve of tension versus temperature, drawn from what appears to be the most accurate simulations, those of Chapela et al., one interpolates a tension of $\gamma\varepsilon/\sigma^2 = 0.54$. In view of the approximate nature of f_0, g_0, and C_0, we deem this to be acceptable agreement with the predictions of the ADF and MVDW models ($\gamma\varepsilon/\sigma^2 = 0.601$ and 0.676). Some appreciation of the source of error from the homogeneous fluid theory comes from the fact that the predicted difference between the liquid-vapor densities is higher than that of the computer simulations. For example, at $kT/\varepsilon = 0.7$, $n_l\sigma^3$ and $n_g\sigma^3$ of the simulated fluid are 0.788 and 0.010, whereas the Weeks-Chandler-Andersen model used by McCoy and Davis yields 0.848 and 0.00196. Since tension is roughly proportional to $n_l - n_g$, one can anticipate that the theoretical predictions will be somewhat high.

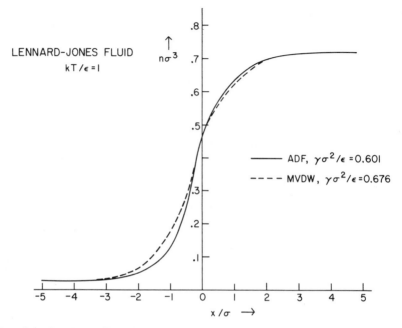

Fig. 3.C.1.   Density profile of the vapor-liquid interface of a 6-12 Lennard–Jones fluid. Comparison of ADF and MVDW models.[43]

Whether the MVDW model or the ADF has greater validity cannot be proved a priori. Both are ad hoc extensions of theoretical formulas valid in certain limits: the MVDW model is exact in the limit of attractive forces long ranged and repulsive forces short ranged compared to the scale of density variation— the ADF model is exact for fluids at sufficiently low densities that the second virial coefficient suffices for computing thermodynamic properties. The fact that the two models yield similar results also gives little reason to choose one over the other. What is more attractive about the ADF model is that it is based on a theory not restricted to two-body forces. And, as will be seen, the gradient version of the ADF model agrees with exact gradient theory.

D.   Gradient Theory: General Aspects

In the gradient theory of stress discussed in Section II.D, there remained the problem of determining the functional density dependence of the correlation functions entering the pressure tensor. This problem can be avoided with the aid of the free-energy theory presented in Section III.B. In the gradient free-energy theory the densities and tensions are governed by differential equations, similar to the YBG equations but simpler; in these equations the only required input consists of the Helmholtz free-energy density and the two-body direct correlation function of homogeneous fluid.

The sacrifice of rigor made in replacing the exact free energy by its gradient theoretical limit is handsomely rewarded. The physics of inhomogeneous fluid is made fairly simple in gradient theory. Differential equations are easier to deal with than integral ones. Moreover, whereas at the gradient level the theoretical inputs are $f_0(n)$ and $C_0(s; n)$ — which are well-defined, much studied properties of homogeneous fluid — at the exact level no tractable theory is available for the direct correlation function of inhomogeneous fluid. Several model studies, such as that presented in Section II.E, and comparisons between theory and experiment indicate that for unconstrained fluid microstructures, for example planar interfaces and spherical drops, gradient theory is a good approximation even far from a critical point. Comparisons are discussed in Section III.E at some length. There are indications, however, that the detailed structure of a fluid near a confining solid wall is poorly represented by gradient theory.[53]

Analogously to what was done in Section II.D with the pressure tensor, the free-energy gradient approximation is obtained by assuming that $f(\mathbf{r})$ is a function of $\mathbf{n}(\mathbf{r})$ and all its derivatives at $\mathbf{r}$ and expanding about the homogeneous state to obtain through third order in gradients the Cahn-Hilliard result[4]:

$$f(\mathbf{r}; \{\mathbf{n}\}) = f_0(\mathbf{n}) + \sum_\alpha \frac{1}{2} A_\alpha \nabla^2 n_\alpha + \sum_{\alpha,\beta} \frac{1}{2} B_{\alpha\beta} \nabla n_\alpha \cdot \nabla n_\beta \quad (3.D.1)$$

where A_α and $B_{\alpha\beta}$ are properties of homogeneous fluid. Third-order terms vanish because of isotropy of homogeneous fluid. The free energy, computed from $F = \int f d^3 r$, can be expressed in the form

$$F = \int \left[f_0(\mathbf{n}) + \sum_{\alpha,\beta} \frac{1}{2} c_{\alpha\beta} \nabla n_\alpha \cdot \nabla n_\beta \right] d^3 r \quad (3.D.2)$$

The influence parameters $c_{\alpha\beta}$ are related to A_α and $B_{\alpha\beta}$ by $c_{\alpha\beta} = B_{\alpha\beta} - \partial A_\alpha / \partial n_\beta$. Equation 3.D.2 is obtained integration by parts with the condition that all density gradients vanish on the boundary of the system. If the density gradients do not vanish on the boundary, the term $-\nabla \cdot [\sum_\alpha \frac{1}{2} A_\alpha \nabla n_\alpha]$ must be added to the integrand of (3.D.2). Also, if $f(\mathbf{r})$ is identified with the integrand of (3.D.2), the transverse pressure computed from $P_T = n\mu - f$ differs from that obtained with (3.D.1) for f. The difference however does not contribute to tension, since it involves boundary terms that vanish upon integration.

The chemical potential of species α, $\mu_\alpha = \delta F/\delta n_\alpha(\mathbf{r})$, computed from (3.D.2), obeys a second-order differential equation that can be rearranged to the form

$$\sum_\beta \nabla \cdot \left(c_{\alpha\beta} \nabla n_\beta \right) - \frac{1}{2} \sum_{\gamma,\beta} \frac{\partial c_{\gamma\beta}}{\partial n_\alpha} \nabla n_\gamma \cdot \nabla n_\beta = \frac{\partial \omega}{\partial n_\alpha} \qquad (3.D.3)$$

where ω is a thermodynamic potential defined by

$$\omega(\mathbf{n}) \equiv f_0(\mathbf{n}) - \sum_\alpha n_\alpha \mu_\alpha \qquad (3.D.4)$$

Equation (3.D.3) is also the Euler equation of extrema of F for a closed system.

With (3.D.3) the determination of a given fluid microstructure becomes a nonlinear boundary value problem. The existence of a fluid microstructure (e.g., planar interface, spherical drop, thin film, periodic structure) is governed by ω (and therefore the Helmholtz free-energy density of homogeneous fluid) and the influence parameters $c_{\alpha\beta}$, which affect the stability and characteristic length scales of microstructures.

For a planar system, $n_\alpha = n_\alpha(x)$, (3.D.3) can be multiplied by dn_α/dx, summed, and integrated to yield

$$\sum_{\alpha,\beta} \frac{1}{2} c_{\alpha\beta} \frac{dn_\alpha}{dx} \frac{dn_\beta}{dx} = \omega(\mathbf{n}) + K \qquad (3.D.5)$$

where K is a constant of integration.

For a planar interface, the boundary conditions $\mathbf{n}(x = -\infty) = \mathbf{n}^{(1)}$ and $\mathbf{n}(x = \infty) = \mathbf{n}^{(2)}$, $\mathbf{n}^{(i)}$ the bulk composition of phase i, imply from (3.D.3) and (3.D.5) the equilibrium conditions $(\partial\omega/\partial n_\alpha)(\mathbf{n}^{(i)}) = 0$, $\omega(\mathbf{n}^{(i)}) + K = 0$, for $i = 1$ and 2, or from the definition of $\omega(\mathbf{n})$, $\mu_\alpha = \mu_\alpha^0(\mathbf{n}^{(i)})$ and $P_N = P_0(\mathbf{n}^{(i)})$ for $i = 1$ and 2. These are the usual thermodynamic conditions of phase equilibria. With $K = -P_N$, (3.D.5) can be used to eliminate $f_0(\mathbf{n})$ from (3.D.2) to obtain

$$F = \sum_\alpha N_\alpha \mu_\alpha - P_N V + A \int_{-\infty}^{\infty} \sum_{\alpha,\beta} c_{\alpha\beta} \frac{dn_\alpha}{dx} \frac{dn_\beta}{dx} \qquad (3.D.6)$$

from which we find

$$\gamma = \sum_{\alpha,\beta} \int_{-\infty}^{\infty} c_{\alpha\beta} \frac{dn_\alpha}{dx} \frac{dn_\beta}{dx} dx \qquad (3.D.7)$$

the form of the gradient theoretical tension established at (2.D.18) in Section II.

Expanding the integrand of (3.B.36) about derivatives of $\mathbf{n}(x)$ and comparing the result with (3.D.7), we find the exact formula for the influence parameters,

$$c_{\alpha\beta} = \frac{kT}{6} \int s^2 C_0^{\alpha\beta}(s;\mathbf{n}) \, d^3s \qquad (3.\text{D}.8)$$

established by Bongiorno, Scriven, and Davis[26] and by Fleming, Yang, and Gibbs.[27] This significant result means that through gradient theory the equilibrium distributions of classical inhomogeneous fluid are determined by the homogeneous fluid quantities f_0 and $C_0^{\alpha\beta}$. We note here that as mentioned earlier, the gradient expansion of the ADF model also yields (3.D.8).

In the presence of an external field, the gradient free energy is modified by adding $\sum n_\alpha(\mathbf{r})u_e^\alpha(\mathbf{r})$ to the integrand of (3.D.2). Then the chemical potential obeys (3.D.3) with u_e^α added to the right-hand side. The resulting expression can be used to derive the equation of hydrostatics and identify a pressure tensor. To do this, we add u_e^α to the right-hand side of (3.D.3), multiply by ∇n_α, sum over α, and rearrange the results to obtain

$$\nabla \cdot \left[\sum_{\alpha,\beta} c_{\alpha\beta} \nabla n_\alpha \nabla n_\beta - \tfrac{1}{2} \sum_{\alpha,\beta} c_{\alpha\beta} (\nabla n_\alpha \cdot \nabla n_\beta)\mathbf{I} - \left(\omega + \sum_\alpha n_\alpha u_e^\alpha \right)\mathbf{I} \right]$$

$$= -\sum_\alpha n_\alpha \nabla u_e^\alpha \qquad (3.\text{D}.9)$$

Comparing (3.D.9) with the equation of hydrostatics (2.A.6), we identify a pressure tensor:

$$\mathbf{P}' = -\left(\omega + \sum_\alpha n_\alpha u_e^\alpha \right)\mathbf{I} + \sum_{\alpha,\beta} c_{\alpha\beta} \nabla n_\alpha \nabla n_\beta - \tfrac{1}{2} \sum_{\alpha,\beta} c_{\alpha\beta} \nabla n_\alpha \cdot \nabla n_\beta \mathbf{I}$$

$$(3.\text{D}.10)$$

This procedure, involving as it does the divergence of $\mathbf{P}'$, does not yield a unique formula for the pressure tensor, however. If to $\mathbf{P}'$ one adds a quantity $\underset{\sim}{\delta}$ whose divergence is zero, the result still satisfies (3.D.9). Nevertheless, for a planar interface, inasmuch as (3.D.10) equals $P_0(\mathbf{n})\mathbf{I}$ in a homogeneous fluid, the normal component of any $\underset{\sim}{\delta}$ would be zero and the integral $\int \delta_T dx$ of the transverse component of $\underset{\sim}{\delta}$ would vanish when integrated across the interface. Thus the pressure tensor of (3.D.10) is adequate for computing interfacial tension, although its transverse component does not generally agree with the transverse component of (2.D.17), which is the true pressure tensor, since it was derived directly from force considerations.

That there are terms missing from (3.D.10) can be illustrated for a one-component fluid in the special case that the pair and direct correlation functions in the gradient terms are the dilute gas ones, that is, in the l_{ij}'s, $g_0(s)$ $=e^{-\beta u(s)}$, and in c, $C_0(s)=e^{-\beta u(s)}-1$. In this special case $l_{11}=\frac{1}{2}l_{12}=\frac{1}{3}nc$, $l_{21}=\frac{1}{2}l_{22}=-\frac{1}{6}c$, $c=(kT/6)\int s^2(e^{-\beta u(s)}-1)d^3s$ and the difference between (2.D.11) and (3.D.10) is

$$\underset{\sim}{\delta}=\tfrac{2}{3}c\left[n\nabla^2 n\mathbf{I}-n\nabla\nabla n+(\nabla n)^2\mathbf{I}-\nabla n\nabla n\right]\qquad(3.\text{D}.11)$$

As anticipated, δ has a zero divergence and does not contribute to fluid tension, δ_N is zero, and the integral of δ_T across the interface is zero.

The stability operator of gradient theory is a differential instead of an integral operator. Computing $\delta^2 F$ from the first form of (3.A.9) ($\varepsilon_\alpha=1$), we find from (3.D.2), after integration by parts and setting the v_α equal to zero on the boundary of the system,

$$\delta^2 F=\frac{1}{2}\sum_{\alpha,\beta}\int v_\alpha\left\{-\nabla\cdot c_{\alpha\beta}\nabla v_\beta-\sum_\delta\left[\nabla\cdot\left(v_\beta\frac{\partial c_{\alpha\delta}}{\partial n_\beta}\nabla n_\delta\right)-v_\beta\frac{\partial c_{\beta\delta}}{\partial n_\alpha}\nabla n_\beta\cdot\nabla n_\delta\right]\right.$$

$$\left.+\left[\frac{1}{2}\sum_{\gamma,\delta}\frac{\partial^2 c_{\gamma\delta}}{\partial n_\alpha\partial n_\beta}\nabla n_\gamma\cdot\nabla n_\delta+\frac{\partial^2 f_0}{\partial n_\alpha\partial n}\right]v_\beta\right\}d^3r\qquad(3.\text{D}.12)$$

The eigenvalue problem associated with (3.D.12) is

$$\sum_\beta\left\{-\nabla\cdot\left(c_{\alpha\beta}\nabla u_\beta\right)+a_{\alpha\beta}u_\beta\right.$$

$$\left.-\sum_\delta\left[\nabla\cdot\left(u_\beta\frac{\partial c_{\alpha\delta}}{\partial n_\beta}\nabla n_\delta\right)-\frac{\partial c_{\beta\delta}}{\partial n_\alpha}\nabla u_\beta\cdot\nabla n_\delta\right]\right\}=\lambda u_\alpha\qquad(3.\text{D}.13)$$

where the notation $a_{\alpha\beta}\equiv(\partial^2 f_0/\partial n_\alpha\partial n_\beta)+\frac{1}{2}\sum_{\gamma,\delta}(\partial^2 c^{\gamma\delta}/\partial n_\alpha\partial n_\beta)\nabla n_\gamma\cdot\nabla n_\delta$ has been introduced. If the boundary conditions are chosen to make the operator in (3.D.13) self-adjoint, then the eigenvalue stability criteria given in Table 3.A.1 hold. With constant influence factors, (3.D.13) simplifies greatly to $-c\nabla^2 u+au=\lambda u$.

Frequently ions are present in microstructured fluids, and the associated double layers are important actors in the behavior of the systems. The gradient theoretical Helmholtz free energy when external fields and electrostatic fields are present can be expressed in the form

$$F = \int \left[f_0(\mathbf{n}) + \sum_{\alpha,\beta} \frac{1}{2} c_{\alpha\beta} \nabla n_\alpha \cdot \nabla n_\beta + \sum_\alpha n_\alpha u_e^\alpha(\mathbf{r}) + \frac{1}{8\pi} \varepsilon E^2 \right] d^3 r$$

$$(3.D.14)$$

where $\mathbf{E}(\mathbf{r})$ is the electric field at $\mathbf{r}$ and ε is the dielectric constant. The electric field is the negative gradient of the voltage ψ ($\mathbf{E} = -\nabla \psi$); ψ obeys the Poisson equation

$$\nabla \cdot (\varepsilon \nabla \psi) = -4\pi \sum_\alpha z_\alpha e n_\alpha \qquad (3.D.15)$$

where e is the unit electronic charge and z_α the valence of species α.

Consistent with the gradient approximation to free energy, $\varepsilon(\mathbf{r})$ is assumed to be the dielectric constant of homogeneous fluid at composition $\mathbf{n}(\mathbf{r})$. The equation of equilibrium predicted by (3.D.14) for species α is then

$$\sum_\beta \nabla \cdot (c_{\alpha\beta} \nabla n_\beta) - \frac{1}{2} \sum_{\gamma,\beta} \frac{\partial c_{\gamma\beta}}{\partial n_\alpha} \nabla n_\gamma \cdot \nabla n_\beta = \frac{\partial \omega}{\partial n_\alpha} + u_e^\alpha - \frac{1}{8\pi} \frac{\partial \varepsilon}{\partial n_\alpha} E^2 + z_\alpha e \psi$$

$$(3.D.16)$$

The quantity $-(1/8\pi)(\partial \varepsilon/\partial n_\alpha) E^2$ represents the contribution of electrostriction to the chemical potential. It is usually ignored in double-layer theories because it is small except at very high electric fields. Sufficiently near a critical point, this contribution could be appreciable compared to $\partial \omega/\partial n_\alpha$ and could therefore play an important role in near-critical microstructures.

As before, multiplication of (3.D.16) by ∇n_α, summation over α, and comparison with the equation of hydrostatics leads to identification of a pressure tensor:

$$\mathbf{P}' = P_0(\mathbf{n})\mathbf{I} - \frac{\varepsilon}{4\pi} \left[\mathbf{EE} - \frac{1}{2} \left(1 - \sum_\alpha n_\alpha \frac{\partial \ln \varepsilon}{\partial n_\alpha} \right) E^2 \mathbf{I} \right]$$

$$+ \sum_{\alpha,\beta} \left[c_{\alpha\beta} \left(\nabla n_\alpha \nabla n_\beta - \frac{1}{2} \nabla n_\alpha \cdot \nabla n_\beta \mathbf{I} \right) \right.$$

$$\left. - n_\alpha \nabla \cdot (c_{\alpha\beta} \nabla n_\beta)\mathbf{I} + \frac{1}{2} \sum_\gamma n_\alpha \frac{\partial c_{\gamma\beta}}{\partial n_\alpha} \nabla n_\gamma \cdot \nabla n_\beta \mathbf{I} \right] \qquad (3.D.17)$$

In the limit that $\mathbf{E} = 0$, (3.D.17) reduces to (3.D.10) when ω is replaced by

$f_0(\mathbf{n}) - \Sigma_\alpha n_\alpha \mu_\alpha$ and μ_α is eliminated with the aid of the equilibrium equation (3.D.3).

For a planar interface with $\mathbf{E}$ in the x-direction and $E = 0$ at $x = \pm\infty$, the interfacial tension is given by

$$\gamma = \gamma_D + \gamma_{El} \tag{3.D.18}$$

where γ_D is the tension of dispersion forces, (3.D.7), and γ_{El} is the double-layer contribution

$$\gamma_{El} = -\frac{1}{4\pi} \int_{-\infty}^{\infty} \varepsilon E^2 \, dx \tag{3.D.19}$$

a quantity that is always negative. Although γ_D and γ_{El} have usually been estimated as if they were independent, they are not. The density profiles and the electric field should be determined simultaneously through solution of (3.D.15) and (3.D.16).

The gradient theoretical direct correlation function can be obtained from (3.B.30) by setting $n^0 = n(\mathbf{r}_1)$ and expanding $n(\mathbf{r}_i) - n(\mathbf{r}_1)$ about $\mathbf{r}_i = \mathbf{r}_1$. The result is[45]

$$\begin{aligned}
C(\mathbf{r}_1, \mathbf{r}_2) = {} & C_0(r_{12}; n) + \frac{1}{2}\frac{\partial C_0}{\partial n}(r_{12}; n)\mathbf{r}_{21} \cdot \nabla n \\
& + \left[B_N \hat{r}_{21}\hat{r}_{21} + (\hat{r}_{21}\hat{r}_{21} - \mathbf{I})B_T \right] : \nabla \nabla n \\
& + \left[D_N \hat{r}_{21}\hat{r}_{21} + (\hat{r}_{21}\hat{r}_{21} - \mathbf{I})D_T \right] : \nabla n \nabla n
\end{aligned} \tag{3.D.20}$$

where

$$B_N \equiv \frac{1}{4}\int C_3^0 \left[r_{31}^2 - 3\frac{(\mathbf{r}_{20}\mathbf{r}_{31})^2}{r_{21}^2} \right] d^3r_3 \tag{3.D.21}$$

$$B_T \equiv \frac{1}{2}\int C_3^0 \frac{(\mathbf{r}_{21}\cdot\mathbf{r}_{31})^2}{r_{21}^2} d^3r_3 \tag{3.D.22}$$

$$D_N \equiv \frac{1}{4}\int C_4^0 \left[\frac{3(\mathbf{r}_{21}\cdot\mathbf{r}_{31})(\mathbf{r}_{21}\cdot\mathbf{r}_{41})}{r_{21}^2} - \mathbf{r}_{31}\cdot\mathbf{r}_{41} \right] d^3r_3 \, d^3r_4 \tag{3.D.23}$$

and

$$D_T \equiv \frac{1}{2}\int C_4^0 \frac{(\mathbf{r}_{21}\cdot\mathbf{r}_{31})(\mathbf{r}_{21}\cdot\mathbf{r}_{41})}{r_{21}^2} d^3r_3 \, d^3r_4 \tag{3.D.24}$$

The combination of (3.D.20) and the Ornstein-Zernicke equation yields, to second order in derivatives of density, an expansion of the form indicated by (2.D.10) for the pair correlation function. The coefficients of $g(\mathbf{r}_1, \mathbf{r}_2)$ in (2.D.10) can be determined as functions of $C_0, B_N, \ldots, D_T$. However, the formulas are quite complicated and do not appear to be tractable at this point. The Percus-Yevick approximation, $g(\mathbf{r}_1, \mathbf{r}_2) = \{1 - \exp\beta u(r_{12})\}^{-1} C(\mathbf{r}_1, \mathbf{r}_2)$, however, provides estimates of the coefficients of the gradient theoretical pair correlation function. In particular, the coefficients of the various density derivatives in (2.D.10) can be estimated as $\{1 - \exp\beta u(r_{12})\}^{-1}$ times the coefficient of the corresponding density derivative in (3.D.20). From the forms of the B's and D's, the normal (N) and transverse (T) coefficients appear to be of the same order of magnitude. Thus, the agreement between density profiles and tensions computed from approximations such as Cases 1 to 3 and those computed from exact gradient theory indicates cancellation of error or insensitivity of profiles and tension to transverse, gradient-induced correlations.

E. Gradient Theory: Applications to Planar Interfaces

Gradient theory is appealing because it is mathematically simple, is rigorous to the order of the terms retained, and presents the physics of interfaces plainly. The determining factors of tension separate nicely into the homogeneous fluid free energy and the influence parameters of inhomogeneous fluid. The origins of low tensions are clearly identified. The theoretical formulas are suggestive of semiempirical models for interfacial properties along the lines that have been successful for engineers in modeling phase behavior of complex fluids. Of course, these advantages are important only if it can be established that gradient theory is not restricted to near-critical interfaces.

One test of the accuracy of gradient theory is to compare the predictions of model free energies with their corresponding gradient approximations. This has been done for a Van der Waals free-energy model, namely, (3.C.1) with $g_0(s) = 0$, $s < d$, and $g_0(s) = 1$, $s > d$, with $u(s) = -4\varepsilon d^6/s^6$, $s > d$, and with the VDW free-energy formula. The liquid-vapor density profiles predicted by the VDW model agree well with those predicted by the corresponding gradient approximation even down to temperatures at which the liquid would solidify. Predicted surface tensions differ by less than 10% over the entire liquid-vapor coexistence range.[26] The MVDW and the ADF models, described in Section III.C, were compared to their gradient approximations,[43] with $\bar{n} \equiv \frac{1}{2}[n(\mathbf{r}') + n(\mathbf{r})]$, for a 6-12 Lennard-Jones fluid at kT/ε, a temperature at which the density profile is quite sharp. The comparison for the ADF model (whose influence parameter is the exact one) is presented in Fig. 3.E.1. The density profiles are quite similar and the tensions agree to within 16%. Similar results were obtained with the MVDW model.[43]

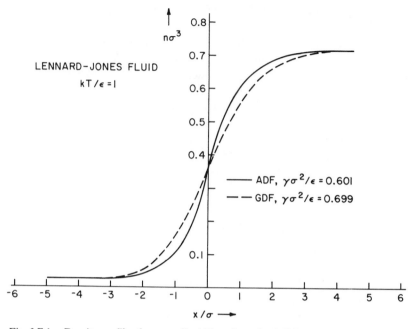

Fig. 3.E.1.   Density profile of a vapor-liquid interface of a 6-12 Lennard–Jones fluid. Comparison of approximate density functional theory (solid curve) and gradient theory (dashed curve).[43]

Another test of the gradient theory is to compare theory with computer simulations. The computer density profile and gradient theoretical prediction are plotted in Fig. 3.E.2.[43] The system is a 6-12 Lennard-Jones fluid at $kT/\varepsilon = 0.703$. The exact formula, (3.D.8), was used to compute the influence parameter; f_0 and g_0 were calculated from the Verlet-Weis improvement of the Weeks-Chandler-Andersen theory, and the direct correlation function was determined from $C_0 = (1 - e^{\beta u})g_0$, the Percus-Yevick approximation. The surface tension determined from the computer simulations is 16.5 ± 2.3 dynes/cm, to be compared with 19.6 dynes/cm predicted by gradient theory with $\varepsilon/k = 119°K$ and $\sigma = 3.4 \text{Å}$.[54]

On the basis of the evidence discussed above, plus successful applications[70-74] of the gradient theory to hydrocarbons and their mixtures, gradient theory appears to provide not only a qualitative but also a quantitative (within a tolerance of some 10 to 20%) description of interfacial phenomena. The remainder of this section is devoted to examination and applications of gradient theory.

The boundary conditions of a planar interface are $\mathbf{n}(x) \to \mathbf{n}^{(1)}$ as $x \to -\infty$ and $\mathbf{n}(x) \to \mathbf{n}^{(2)}$ as $x \to +\infty$, where $\mathbf{n}^{(i)}$ is the bulk composition of phase i. As mentioned previously, these boundary conditions, through (3.D.3) and

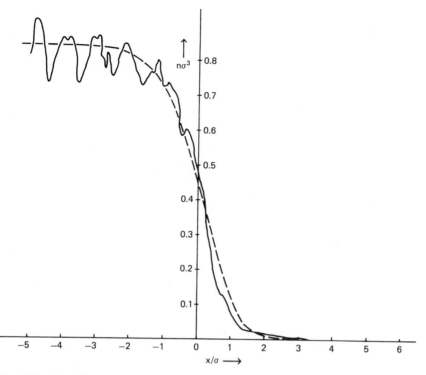

Fig. 3.E.2. Liquid-vapor density profile of 6-12 Lennard–Jones fluid. Comparison of computer simulation (solid curve) with gradient theory (dashed curve); $kT/\varepsilon = 0.703$.

(3.D.5), are equivalent to the usual equilibrium conditions: $\mu_\alpha = \mu_\alpha^0(\mathbf{n}^{(1)}) = \mu_\alpha^0(\mathbf{n}^{(2)})$ and $\omega_B = \omega(\mathbf{n}^{(1)}) = \omega(\mathbf{n}^{(2)}) = -P_N$, where μ_α and P_N are constant at thermodynamic equilibrium. For a one-component fluid, (3.D.5) can be solved directly to get

$$dx = \sqrt{\frac{c}{2}} \, \frac{dn}{\sqrt{\Delta\omega(n)}} \qquad (3.\text{E}.1)$$

where $\Delta\omega(n) \equiv \omega(n) - \omega_B$. With this result, the tension formula, (3.D.7), can be put into the profile-independent form

$$\gamma = \sqrt{2} \int_{n_g}^{n_l} \left[c \, \Delta\omega(n) \right]^{1/2} dn \qquad (3.\text{E}.2)$$

For a 6-12 Lennard-Jones fluid, with C_0 given by the Percus-Yevick approximation and g_0 by the Weeks-Chandler-Anderson–Verlet-Weis model, the influence factor is positive.[46]. In fact it is almost independent of temperature and density: between vapor and liquid densities $c/\varepsilon\sigma^5$ ranges from 6.40 to 6.07, 6.30 to 6.20, 6.56 to 6.34 at the temperatures $kT/\varepsilon = 0.8$, 1.0, and 1.3, respectively. From this it follows that acting through the quantity $\Delta\omega(n)$, the free energy of homogeneous fluid has the major role in the surface tension of a one-component fluid.

Geometrically $\Delta\omega(n)$ can be represented as the vertical distance between the curve of $f_0(n)$ versus n and a straight line touching f_0 at the vapor and liquid densities, n_g and n_l. The construction of $\Delta\omega(n)$ from $f_0(n)$ is illustrated in Fig. 3.E.3: $\Delta\omega(n)$ is the Helmholtz free-energy density of homogeneous fluid measured relative to the Helmholtz free energy of bulk liquid and vapor at the overall density n.

As the critical point T_c is approached, $\Delta\omega(n)$ flattens out rapidly between n_g and n_l (Fig. 3.E.3). Thus, the surface tension goes to zero as the critical point is approached not only because the vapor and liquid densities approach each other, but also because $[\Delta\omega(n)]^{1/2}$ becomes smaller as the free-energy curve flattens out between n_g and n_l. As it seems unlikely that c will vary appreciably with temperature, flattening of the free-energy curve is probably the only mechanism of low tensions in one-component fluids. Interfaces near the critical point become broad because $\Delta\omega(n) \to 0$ and $dx \propto \Delta\omega^{-1/2}\, dn$ according to (3.E.1).

For a multicomponent fluid, it is convenient to introduce a path function $\mathfrak{p}$, which increases monotonically, according to $(d\mathfrak{p})^2 \equiv \Sigma_{\alpha,\beta} G_{\alpha\beta}\, dn_\alpha\, dn_\beta$ as the composition $\mathbf{n}$ advances from the bulk value $\mathbf{n}^{(1)}$ of phase 1 to the bulk value $\mathbf{n}^{(2)}$ of phase 2 across the interfacial zone. The coefficients $G_{\alpha\beta}$ can be freely chosen to assure monotonicity of $\mathfrak{p}$. If one of the components, say component 1, increases monotonically across the interfacial zone, then $\mathfrak{p} = n_1$ is acceptable. If no density is monotonic or if it is not known whether any is monotonic, then the arc length, $(d\mathfrak{p})^2 = \Sigma_\alpha (dn_\alpha)^2$ is a good choice for $\mathfrak{p}$. In any case, a suitable path function can be constructed. In terms of it, (3.D.5)

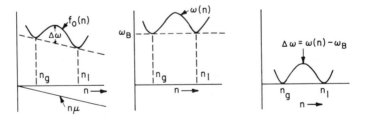

Fig. 3.E.3.   Helmholtz free-energy density f_0 of a one-component homogeneous fluid and thermodynamic potential difference $\Delta\omega(n)$ of inhomogeneous fluid.

can be rearranged to

$$dp = \left(\frac{2\Delta\omega(\mathbf{n})}{\displaystyle\sum_{\alpha,\beta} c_{\alpha\beta}(dn_\alpha/dp)(dn_\beta/dp)} \right)^{1/2} dx \equiv \sqrt{\mathcal{H}(p)}\, dx \quad (3.\text{E}.3)$$

and (3.D.7) to

$$\gamma = \sqrt{2} \int_{p^{(1)}}^{p^{(2)}} \left\{ \left[\sum_{\alpha,\beta} c_{\alpha\beta} \frac{dn_\alpha}{dp} \frac{dn_\beta}{dp} \right] \Delta\omega(\mathbf{n}) \right\}^{1/2} dp \quad (3.\text{E}.4)$$

To perform this integral, the composition $\mathbf{n}(p)$ along the path p must be determined. This is done by transforming the independent variable of the differential profile equations, (3.D.3), from x to p using the relations

$$\frac{dn_\alpha}{dx} = \sqrt{\mathcal{H}} \frac{dn_\alpha}{dp}, \quad \frac{d^2 n_\alpha}{dx^2} = \sqrt{\mathcal{H}} \frac{d}{dp} \left[\sqrt{\mathcal{H}} \frac{dn_\alpha}{dp} \right] \quad (3.\text{E}.5)$$

The differential profile equations then become

$$\sum_\beta c_{\alpha\beta} \sqrt{\mathcal{H}} \frac{d}{dp} \left[\sqrt{\mathcal{H}} \frac{dn_\beta}{dp} \right] + \frac{\mathcal{H}}{2} \sum_{\gamma,\beta} \frac{\partial c_{\gamma\beta}}{\partial n_\alpha} \frac{dn_\gamma}{dp} \frac{dn_\beta}{dp} = \frac{\partial\omega}{\partial n_\alpha} \quad (3.\text{E}.6)$$

$\alpha = 1, 2, \ldots, \nu$. Both the original [(3.D.3)] and transformed [(3.E.6)] profile equations are two-point boundary value problems. The former, however, involves boundaries at $x = \pm\infty$, whereas the latter are at finite values, $p^{(1)}$ and $p^{(2)}$. Depending on the availability of asymptotic solutions, the transformed equations may be more convenient for numerical solution. From several numerical studies of two- and three-component liquid-vapor and liquid-liquid interfaces, Carey[74] concluded that the transformed equations are indeed to be preferred.

When the choice $dp = dn_1$ can be made, the ν equations, (3.E.6), reduce to $\nu - 1$ equations ($\alpha = 2, \ldots, \nu$). For a single-component fluid, this choice leads immediately to the analytic solution, (3.E.1), of the profile equation. This yields the particularly simple formula, (3.E.2), for the surface tension. For the multicomponent case, similar simplification results if one chooses $G_{\alpha\beta} = c_{\alpha\beta}$, in which case $\sum_{\alpha,\beta} c_{\alpha\beta}(dn_\alpha/dp)(dn_\beta/dp) = 1$, $\mathcal{H} = 2\Delta\omega$, and

$$\gamma = \int_0^{p_B} \sqrt{2\Delta\omega(\mathbf{n}(p))}\, dp \quad (3.\text{E}.7)$$

where p_B is the influence parameter-scaled composition path length between phases 1 and 2:

$$\mathfrak{p}_B \equiv \int_{\substack{\text{path} \\ \text{integral}}} \left(\sum_{\alpha, \beta} c_{\alpha\beta} \frac{dn_\alpha}{d\mathfrak{p}} \frac{dn_\beta}{d\mathfrak{p}} \right)^{1/2} d\mathfrak{p} \qquad (3.E.8)$$

In practice an iterative method must be used to determine the path length $\mathfrak{p}_B$ and $\mathbf{n}$ as a function of the path parameter. For example, we could guess $\mathfrak{p}_B$. Then with the boundary conditions $\mathbf{n}(\mathfrak{p}=0)=\mathbf{n}^{(1)}$ and $\mathbf{n}(\mathfrak{p}_B)=\mathbf{n}^{(2)}$, (3.E.6) could be solved for $\mathbf{n}(\mathfrak{p})$. A new value of $\mathfrak{p}_B$ could then be computed from (3.E.8). The process could be iterated until successive values of $\mathfrak{p}_B$ agreed to within some prescribed limit.

In a multicomponent fluid $\Delta\omega(\mathbf{n})$ is the vertical distance from the tangent plane parallel to $\Sigma_\alpha n_\alpha \mu_\alpha$ in n_α space and touching the $f_0(\mathbf{n})$ surface at bulk phase points. Again $\Delta\omega(\mathbf{n})$ corresponds to the Helmholtz free-energy density of homogeneous fluid measured relative to the free energy of the two-phase system at the overall composition $\mathbf{n}$. The $\Delta\omega(\mathbf{n})$ follows the composition path across the interface in the integral defining γ. Figure 3.E.4 illustrates for a strictly regular solution the free-energy surface and $\Delta\omega(\mathbf{n})$ along an interfacial composition path $\mathfrak{p}$ (hypothetical) between two bulk phases. The free-energy surface flattens out rapidly as T_c is approached, so that $\Delta\omega(\mathbf{n})$ tends to zero between bulk phases and drives the tension to zero, according to (3.E.4).

If gradient theory is to be useful, it will have to be pushed beyond computer fluids of simple 6-12 Lennard-Jones particles. Most real fluids are composed of complicated polyatomic molecules. Application of the theory to them is presently hampered by the lack of theoretical equations of state and correlation functions for homogeneous polyatomic fluids. In practical calculations of phase behavior this limitation is frequently circumvented by introduction of semiempirical models, that is, equations of state based on theory but whose structure and parameters are adjusted to better accommodate real fluid behavior. Three-parameter descendants of the Van der Waals equation of state (e.g., the Soave[75] and Peng-Robinson[76] models) and the three-parameter lattice fluid equation derived by Sanchez and Lacombe[77] have been quite successful in predicting phase behavior of nonpolar fluids. Such equations, with similar semiempirical modeling of the influence parameters, provide a practical means of computing tensions of real fluids. We demonstrate this by reviewing briefly some applications by Carey et al.[70, 71, 74] and by Poser and Sanchez.[72, 73]

The Peng-Robinson (PR) equation of state of a one-component fluid is

$$P_0(n) = \frac{nkT}{1-nb} - \frac{n^2 a}{\left[1+2nb-(nb)^2\right]} \qquad (3.E.9)$$

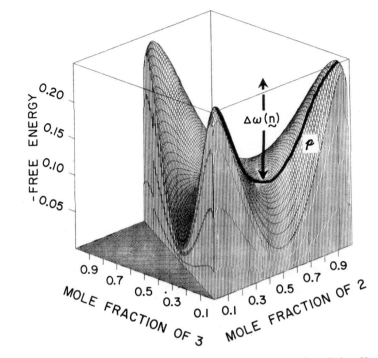

Fig. 3.E.4. Negative free-energy density surface for a symmetrical regular solution. Hypothetical composition path $\mathfrak{p}$ and corresponding $\Delta\omega(\mathbf{n})$ are illustrated.

where $a \cong a_c \xi$,

$$\xi = \left\{ 1 + \left[1 - \left(\frac{T}{T_c} \right)^{1/2} \right] \left(0.37464 + 1.54226\,\omega_a - 0.26992\,\omega_a^2 \right) \right\}^2$$

$$(3.E.10)$$

Here ω_a is the acentric factor and b and a are volume and interaction parameters determined from critical point data with the aid of the formulas $b = 0.07780 k T_c / P_c$ and $a_c = 0.45724 k^2 T_c^2 / P_c$. By integration of the thermodynamic relation $d(f_0 / n - f_0^{\mathrm{Id}} / n) = [(P_0 - nkT) / n^2]\, dn$ at constant temperature, the free-energy density of homogeneous fluid can be determined from (3.E.9). Then the chemical potential of homogeneous fluid is computed from the formula $\mu_0(n) = \partial f_0^{(n)} / \partial n$.

For the Van der Waals model, $b = \frac{2}{3}\pi d^3$, $a = -\frac{1}{2}\int_{s>d} u(s)\, d^3 s$, and $c = -\frac{1}{6}\int_{s>d} s^2 u(s)\, d^3 s$. With the power law potential $u(s) = -4\varepsilon (d/s)^\nu$, ν a positive integer, the VDW model yields $c / ab^{2/3} = [0.61(\nu - 3) / [3(\nu - 5)]$.

With the London dispersion force, $\nu=6$ and $c/ab^{2/3}=0.61$, Carey et al.[70] argued that although the ratio $c/ab^{2/3}$ for real fluids might not equal the value predicted by the VDW theory, the ratio might nevertheless be constant for a class of similar fluids. To test this idea, the ratio $c/ab^{2/3}$ was determined by fitting gradient theory to experiment with the Peng-Robinson equation of state. In Fig. 3.E.5, the values of c so determined are plotted versus $ab^{2/3}$ for normal alkanes ranging in carbon number from 5 to 15. The ratio $c/ab^{2/3}$ is indeed nearly constant and has a value of about 0.3. A least-squares straight-line fit of the points plotted in Fig. 3.E.5 yields $c=0.27$ $ab^{2/3}+2\times10^{-67}$ J·m^5. This correlation between c and $ab^{2/3}$ predicts the surface tension versus temperature as shown in Fig. 3.E.6 for a number of normal alkanes. The agreement is quite good.

For a density-independent influence factor, the quantity $\gamma^*=\gamma b^2/\sqrt{ac}$ is a universal function of $T^*=bkT/a$ with the PR equation of state; γ^* is plotted versus T^* in Fig. 3.E.7 The approximation $c/ab^{2/3}\simeq0.33$ leads to $\gamma=0.57\,ab^{-5/3}\gamma^*$. This formula has been tried for the normal alkanes ranging from a carbon number of 5 to 20 and for one-, two-, and three-methylated butane, pentane, and heptane. The predictions are within 1 or 2% of experimental tensions.

The importance of the work just described is that the surface tension of nonpolar fluids can be computed entirely from homogeneous fluid parameters.

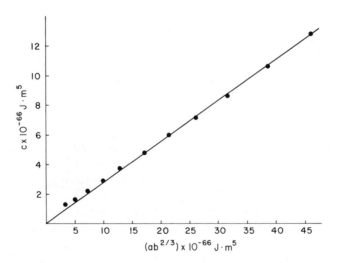

Fig. 3.E.5.   Correlation between the influence parameter c and the homogeneous fluid parameter $ab^{2/3}$ for normal alkanes.[70]

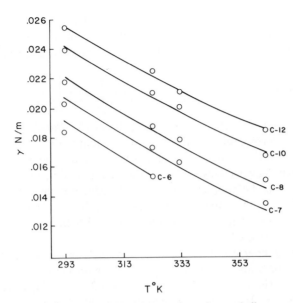

Fig. 3.E.6.   Comparison of experimental surface tensions of normal alkanes and predictions of gradient theory with the correlation shown in Fig. 3.E.5.[70]

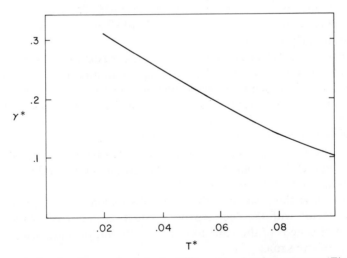

Fig. 3.E.7.   Reduced surface tension ($\gamma^* = b\gamma/\sqrt{ac}$) versus reduced temperature ($T^* = bkT/a$) predicted by gradient theory with the Peng-Robinson equation of state.[70]

The lattice fluid (LF) equation of state derived by Sanchez and Lacombe[77] has been used by Poser and Sanchez[72] to predict from gradient theory surface tensions of normal hydrocarbons and polymer melts. The LF equation is

$$P_0(n) = -\frac{kT}{v^*}\left[\ln(1-nrv^*) + \left(1-\frac{1}{r}\right)nrv^*\right] - n^2\varepsilon^*r^2v^* \quad (3.\text{E}.11)$$

where ε^*, v^*, and r are the parameters of the model. Sanchez and Lacombe recommend determination of them by least-squares fit of theory to vapor pressure data. In the model a molecule is postulated to behave as a polymer on a lattice having vacancies; ε and v^* are the interaction energy and volume of a mer, and r is the number of mers constituting a molecule. An attractive feature of this model is that the parameter r might be convenient for representing association in hydrogen-bonded fluids. Whether this is the case has yet to be demonstrated, however.

Poser and Sanchez estimate ε^* and c from the van der Waals formulas $\varepsilon^* = -\frac{1}{2}\int_{s>d} u(s)d^3s$ and $c = -\frac{1}{6}\int_{s>d} u(s)s^2 d^3s$, with $u(s) = -\varepsilon_0(d/s)^\nu$, $s>d$, and $d \equiv v^{*1/3}$. Their result is $c/\varepsilon^*v^{*2/3} = (\nu-3)/3(\nu-5)$, which for $\nu=6$ yields $c = \varepsilon^*v^{*2/3}$. In comparing their model calculations to experimental surface tensions of a large number of hydrocarbons (including normal and branched alkanes, benzene, the xylenes, diethyl ether, and carbon tetrachloride), they find that with $c/\varepsilon^*v^{*2/3} = 1.24$ the theory predicts the tension to within 5% error over a wide range of temperatures. The LF model also applies to polymer melts ($r=\infty$) and with $c/\varepsilon^*v^{*2/3} = 1.10$ gives good predictions for several polymers (Fig. 3.E.8).

For low-molecular-weight fluids there is little reason to choose between the PR equation[78] (or its relative, the Soave equation) and the LF equation for applications of gradient theory. However, the LF equation has the advantages that its derivation is less empirical and that it applies naturally to polymers, whereas the PR equation has not been demonstrated to be valid for polymers. On the other hand, the rules for extending the PR equation to mixtures are simple and have been demonstrated by a number of applications to be useful for nonpolar fluids, whereas the mixture rules for the LF equation are not yet as well established.[73]

Let us close this section with a brief discussion of the application of gradient theory with the Peng-Robinson equation to binary solutions. This equation of state is of the same form as (3.E.9), but the parameters a and b obey the mixture rules

$$a = \sum_{\alpha,\beta} x_\alpha x_\beta a_{\alpha\beta} = \sum_{\alpha,\beta} x_\alpha x_\beta(1-\delta_{\alpha\beta})\sqrt{a_{\alpha\alpha}a_{\beta\beta}}$$

$$b = \sum x_\alpha b_\alpha \quad\quad\quad\quad\quad\quad\quad\quad\quad (3.\text{E}.12)$$

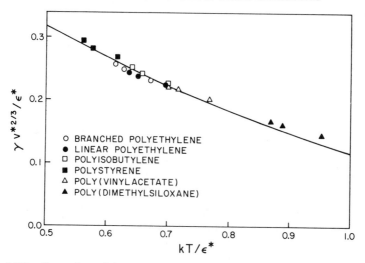

Fig. 3.E.8. Comparison of the lattice-fluid, gradient-theoretical and the experimental surface tensions of several polymers.[72]

where x_α is the mole fraction of species α and b_α and $a_{\alpha\alpha}$ are the PR parameters of pure fluid α. The $\delta_{\alpha\beta}$ are parameters representing the deviation of $a_{\alpha\beta}$ from the geometric mean $\sqrt{a_{\alpha\alpha}a_{\beta\beta}}$. The homogeneous fluid free energy of the PR model is

$$f_0(\mathbf{n}) = -\sum_\alpha n_\alpha \left\{ kT \left[\ln\left(\frac{1-nb}{n_\alpha} \right) + 1 \right] + \mu_\alpha^+(T) \right\}$$
$$- \frac{na}{2\sqrt{2}\,b} \ln\left[\frac{1+nb(1+2)}{1+nb(1-2)} \right] \qquad (3.E.13)$$

and $\mu_0 = \partial f_0 / \partial n$.

For mixtures of nonpolar hydrocarbons, the parameters $\delta_{\alpha\beta}$ are small enough that the mixing rule $a_{\alpha\beta} = \sqrt{a_{\alpha\alpha}b_{\beta\beta}}$ is often adequate. Similarly, Carey et al.[71, 74] found that the geometric mixing rule $c_{\alpha\beta} = \sqrt{c_{\alpha\alpha}c_{\beta\beta}}$, $c_{\alpha\alpha}$ being the influence parameter of pure α, yielded the best results for several binary nonpolar hydrocarbon mixtures. This mixing rule for the influence parameters is fortunate, because then the determinant of the influence matrix $\mathbf{c}$ is zero and the differential profile equations, (3.D.3), can be rearranged to a set of algebraic equations for n. In particular, the set of equations rearrange to

$$\sqrt{c_{11}} \left[\mu_0^1(\mathbf{n}) - \mu \right] = \sqrt{c_{\alpha\alpha}} \left[\mu_0^\alpha(\mathbf{n}) - \mu_\alpha \right], \alpha = 2, \dots, \nu \qquad (3.E.14)$$

Thus, the n_α can be computed algebraically as functions of n_1 and the tension can be computed from (3.E.4).

With the geometric mixing rule, Carey et al.[71] predicted the composition dependence of the surface tension of several binary hydrocarbon systems using the PR equation and fitting the pure fluid influence parameter to experimental data (the results would be about the same if the correlation between c and $ab^{2/3}$ were used). Their results are summarized in Table 3.E.1. Measures of the deviation between predicted and experimental tensions are $\bar{\Delta}$ and Δ_{max}, where $\Delta_i = [\gamma_{exp}(i) - \gamma_{pred}(i)]/\gamma_{exp}(i)$, $i = 1, \ldots, M$, $\bar{\Delta} = (1/M)\Sigma_i \Delta_i$, and $\Delta_{max} = \max_i(\Delta_i)$, where M is the number of mole fractions at which Δ_i is computed (M was usually 6 or more). Typical results are shown in Fig. 3.E.9. Predictions are also fairly accurate (less than 3% error) for the three-component system, hexane, octane, and hexadecane. The results are displayed in Fig. 3.E.10 in a triangular phase diagram (mole fraction coordinates) showing contours of constant tension and points of experimental data. Similar results were obtained for toluene, cyclohexane, and isooctane.

It is interesting that some of the components of a three-component system of normal hydrocarbons exhibit surface activity. This is illustrated in Fig. 3.E.11, where hexane and octane densities are seen to rise in the interface

TABLE 3.E.1

Average and Maximum Error of Surface Tension Predicted for Binary Solutions by Gradient Theory

	$T(°K)$	$\gamma_1^0(N/m)$	$\gamma_2^0(N/m)$	$\bar{\Delta}$	Δ_{max}
Cyclopentane–carbon tetrachloride	298	.02185	.02613	.00	.00
Cyclopentane-tetrachloroethylene	298	.02185	.03130	.01	.01
Cyclopentane-benzene	298	.02185	.02820	.02	.02
Cyclopentane-toluene	298	.02185	.02794	.01	.01
Cyclohexane-benzene	293	.02438	.02886	.01	.02
Cyclohexane-(cis)-decalin	298	.02438	.03224	.00	.01
Cyclohexane-(trans)-decalin	298	.02438	.02997	.01	.01
Cyclohexane-toluene	298	.02438	.02794	.02	.02
Cyclohexane-(n)-hexadecane	298	.02438	.01980	.00	.01
Isooctane-cyclohexane	303	.01789	.02377	.00	.00
Isooctane-dodecane	303	.01789	.02447	.01	.07
Isooctane-benzene	303	.01789	.02753	.01	.02
(n)-hexane-(n)-dodecane	298	.01794	.02469	.00	.00
(n)-hexane-(n)-dodecane	313	.01638	.02342	.01	.01
Methanol-(n)-butanol	298	.02210	.02418	.04	.05
Methanol-(t)-butanol	298	.02210	.02011	.07	.12

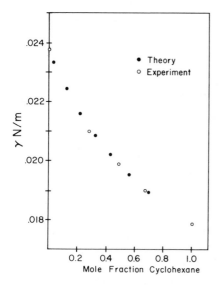

Fig. 3.E.9. Comparison of experimental surface tensions of a binary mixture with predictions of gradient theory from the Peng-Robinson equation of state.[71]

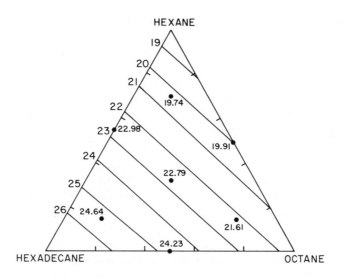

Fig. 3.E.10. Comparison of experimental surface tensions of a ternary mixture with predictions of gradient theory equation of state. Lines correspond to predicted mole fractions for fixed surface tension.[74]

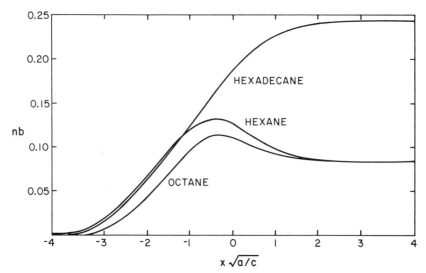

Fig. 3.E.11.   Gradient theoretical density profiles of a vapor-liquid interface of a ternary mixture, predicted from the Peng-Robinson equation of state.[74]

above their bulk values in the liquid phase. More dramatic is the surface activity of a solubilizing component in a three-component liquid-liquid system. Using the VDW equation for simplicity, applying the geometric mixing rules to $a_{\alpha\beta}$ and $c_{\alpha\beta}$, and adjusting the parameters so that the solubility behavior of the components mimics a system of hydrocarbon, water, and alcohol, Carey[74] predicted the density profiles of the liquid-liquid interface shown in Fig. 3.E.12. The excess alcohol in the interfacial region is of course accompanied by a decrease in the interfacial tension.

Attempts to predict quantitatively the tensions of alcohol mixtures and alcohol, water, and oil mixtures by means of PR equation and the geometric mixing rules were not successful. Perhaps this is because the PR equation is not sufficiently accurate for hydrogen-bonded fluids. Other mixing rules, especially admission of composition-dependent influence parameters, might lead to better results. Carey[74] explored this possibility, but without conclusive results. Thus, the challenge for applications of gradient theory at this point is to discover equations of state and mixing rules for the influence parameters that are quantitatively accurate for polar and hydrogen-bonded fluids and their mixtures. The equation of state proposed by Fuller[79-81] for water may be a possibility, and the LF equation, properly extended to mixtures of hydrogen-bonded and nonpolar fluids, may be another.

Stability analysis of the planar interface has been completed so far only for constant influence factors.[74] A one-component interface is stable for any

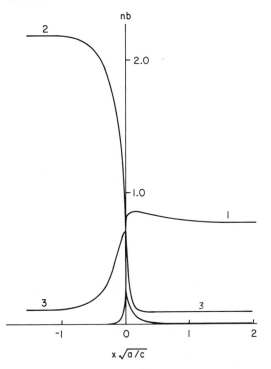

Fig. 3.E.12. Gradient theoretical liquid-liquid density profiles of a ternary mixture predicted from a van der Waals equation of state qualitatively representing oil, water, and alcohol.[74]

positive constant c, a result that follows from the fact that $n(x)$ is monotonic and dn/dx is the eigenvector of (3.D.13) corresponding to $\lambda=0$. Since dn/dx has no zeros, it is the eigenvector corresponding to the minimum eigenvalue of (3.D.13) according to the oscillation theorem for Sturm-Liouville equations. If for a multicomponent fluid the matrix of influence parameters $c_{\alpha\beta}$ has a negative eigenvalue, then a planar interface is unstable, quite possibly with respect to two or more simultaneously admissible interface structures, in some circumstances at least. Otherwise, stability depends on the details of the curvature matrix $\partial^2 f_0/\partial n_\alpha \partial n_\beta$ of $f_0(\mathbf{n})$ and the influence parameter matrix.

F. Gradient Theory: Origins of Low Tension

It is well known that the tension between phases is low near a critical point. For a one-component fluid, the behavior of $\Delta\omega(n)$ as the critical point is approached is illustrated in Fig. 3.F.1 for a 6-12 Lennard-Jones fluid ($kT_c/\varepsilon=$ 1.45 for this fluid). As the temperature approaches the critical point, two

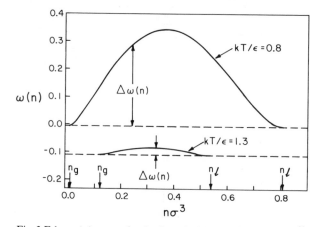

Fig. 3.F.1. $\omega(n)$ versus density for a 6-12 Lennard–Jones fluid.[46]

things happen: (1) the free-energy curve flattens out, and therefore $\Delta\omega$ approaches zero, and (2) the coexisting densities approach one another. As shown in Fig. 3.E.1, $\sqrt{\Delta\omega}$ goes to zero faster than $n_l - n_g$ as T_c is approached. The ratio $\sqrt{\Delta\omega_{max}}/(n_l - n_g)$ equals 0.7 at $kT/\varepsilon = 0.8$ and 0.4 at $kT/\varepsilon = 1.3$.

The significance of this to the tension is apparent in (3.E.2). Sufficiently near the critical point the tension is given approximately by

$$\gamma \simeq q\sqrt{2c(n_u)}\,\sqrt{\Delta\omega_{max}}\,(n_l - n_g) \qquad (3.F.1)$$

where q is a number of the order of $\frac{1}{2}$ and $\Delta\omega_{max}$, the maximum value of $\Delta\omega$, is located at n_u.

According to the modern theory of critical phenomena,[48, 82] the maximum value of $\Delta\omega(n)$ and the difference $n_l - n_g$ approach zero according to the "scaling laws,"

$$\Delta\omega_{max} \propto (T_c - T)^{\zeta} \qquad \text{and} \qquad n_l - n_g \propto (T_c - T)^{\beta} \qquad (3.F.2)$$

as the temperature approaches its critical value. The "critical exponents" ζ and β are universal positive numbers; that is, they are independent of the properties of a particular fluid. From known[83] properties of the direct correlation function, it follows that the influence parameter is finite at the critical point (for simple models it is almost independent of temperature and density). Thus, near the critical point the tension obeys the scaling law

$$\gamma = \gamma_0 \left(\frac{T_c - T}{T_c} \right)^{\nu} \qquad (3.F.3)$$

where

$$\nu = \tfrac{1}{2}\zeta + \beta \qquad (3.F.4)$$

and γ_0 is the tension scale factor; γ_0 is a function of the subject fluid properties, but ν is universal. Such a scaling law was first suggested for tension by Fisk and Widom.[7-9] The law is expected to hold for one-component fluids and for multicomponent fluids approaching T_c at the critical composition.

Gradient theory provides expressions for the scale factor γ_0 and relates the critical exponent ζ to other critical exponents. Expanding $\Delta\omega(n_u)$ about n_g, we obtain to second order in $n_u - n_g$

$$\Delta\omega_{max} \approx \frac{1}{2}(n_u - n_g)^2 \frac{\partial\mu_0}{\partial n}(n_g, T) \approx \frac{(n_l - n_g)^2}{8n_g^2 \kappa_T(n_g, T)} \qquad (3.F.5)$$

where $\kappa_T(n_g, T)$ is the isothermal compressibility of the coexistent gas phase at temperature T. The second form of (3.F.5) is obtained by noting that $n_u = 1/2(n_l - n_g)$ plus terms contributing higher order to (3.F.5) sufficiently close to T_c.[84] The isothermal compressibility also obeys a critical scaling law,

$$\kappa_T \propto (T_c - T)^{-\delta} \qquad (3.F.6)$$

where δ is a positive critical exponent. Combining (3.F.1) through (3.F.5), we get the following relationships among critical exponents:

$$\zeta = 2\beta + \delta \qquad \text{and} \qquad \nu = \frac{\delta}{2} + 2\beta \qquad (3.F.7)$$

For carbon dioxide, it has been found that $\delta = 1.20 \pm 0.02$ and $\beta = 0.3447 \pm 0.0007$, and so $\zeta = 1.89 \pm 0.02$. Thus the flattening of $\Delta\omega$ drives tension to zero as the 0.95 power of $T_c - T$ and the convergence of densities drives it to zero as the 0.3447 power, the former clearly being the more important effect. From the formula $\nu = \tfrac{1}{2}\delta + 2\beta$, we obtain $\nu = 1.29$. Experimentally it has been observed that surface tension obeys (3.F.3), a value of $\nu = 1.253$ for carbon dioxide[85] and of $\nu = 1.302$ for xenon.[86] Observed values[87] of δ and β for xenon yield the prediction $\nu = \delta/2 + 2\beta = 1.290$.

Gradient theory then, *with correctly scaled thermodynamic functions*, correctly predicts the asymptotic behavior of tension as the critical temperature is approached. Mean field theories of the van der Waals type predict,[6] on the other hand, that $\Delta n \propto (T_c - T)^{0.5}$ and $\kappa_T \propto (T_c - T)^{-1}$, yielding the result $\gamma \propto (T_c - T)^{1.5}$ as T approaches T_c. The fault lies in the mean field equation of state of homogeneous fluid, not in the gradient theory of tension. The derivation of equations of state predicting the right critical exponents involves renormalization theory[48] and lies outside the scope of this chapter.

Since $dx \propto (\Delta \omega)^{-1/2} dn$, it follows from gradient theory that an interface becomes increasingly broad as the critical point is approached. For convenience let us define the interfacial width l_w of a one-component fluid by

$$l_w = \frac{(\Delta n)^2}{\int_{-\infty}^{\infty} (dn/dx)^2 dx} = \frac{(\Delta n)^2}{\int_{n_g}^{n_l} \sqrt{(2/c)\Delta \omega}\, dn} \qquad (3.F.8)$$

With arguments similar to those used to obtain (3.F.1) and (3.F.5) we find that sufficiently near the critical point

$$l_w \simeq \frac{n_g \sqrt{c(n_u)\kappa_T}}{q\sqrt{2}} \qquad (3.F.9)$$

The critical exponent ρ for l_w, $l_w \propto (T_c - T)^{-\rho}$, is predicted to be $\frac{1}{2}$ by mean field theory. From the experimental value $\delta \simeq 1.2$, (3.F.9) predicts $\rho \simeq 0.6$. From reflectance measurements in the critical region, Huang and Webb[88] found a critical exponent of 0.67 for interfacial thickness in a binary mixture. [Actually, the quantity Huang and Webb measured is l_2 defined by (3.F.13).]

In homogeneous fluid slightly above the critical point, the intensity $I(k)$ of scattered light at small wave vector $\mathbf{k}$ is proportional to[27] $I(k=0)/[1 + \xi^2 k^2]$, where ξ is the length of long-range intermolecular correlations. Since $I(k)$ is the Fourier transform of $g_0(s; n) - 1$, the expansion of $I(k)$ in a Taylor series about $k = 0$ and the use of the homogeneous fluid version of (3.B.11) lead to the result $\xi = n\sqrt{c(n)\kappa_T(n)}$. From this it follows that the width of a near-critical interface is approximately equal to the characteristic length of long-range intermolecular correlations. To turn this around, the relationship implies that the light-scattering correlations near the critical point arise from the formation of local liquid-vapor interfaces.

If (3.F.1), (3.F.4), (3.F.5), and (3.F.9) are combined to eliminate $c(n_u)$, the result, $\gamma = 2q^2 l_w (\Delta n)^2 / n_g^2 \kappa_T$, implies the relationship

$$\nu = 2\beta + \delta - \rho \qquad (3.F.10)$$

between the critical exponents of the quantities γ, Δn, κ_T, and l_w, given some time ago by Fisk and Widom[7] and recently by Yang et al.[27] Of course, from (3.F.9) it follows that $\rho = \delta/2$, which when inserted into (3.F.10) yields one of the expressions at (3.F.7).

Lovett et al.[25] derived a near-critical formula for tension on the basis of the exact formula, (3.B.27). They assumed that the direct correlation function of inhomogeneous fluid can be approximated by its density-independent dilute gas limit, and approximated the distribution function $p(x)$

$$p(x) \equiv -\frac{1}{\Delta n}\frac{dn}{dx} \qquad (3.F.11)$$

by its cumulant expansion truncated at the fourth moment, that is,

$$p(x) = \frac{1}{\sqrt{2\pi l_2^2}} e^{-x^2/2l_2^2} \qquad (3.F.12)$$

where

$$l_2^2 \equiv \int_{-\infty}^{\infty} x^2 p(x)\, dx \qquad (3.F.13)$$

The origin of the coordinate system for x was chosen so that the first moment of $p(x)$ would be zero. With (3.F.12) they obtained the following limiting formulas:

$$\gamma \simeq \frac{kT}{8}(\Delta n)^2 \int s C_0(s)\, d^3 s, \qquad l_2 \text{ small}$$

$$\simeq \frac{(\Delta n)^2 c}{2\sqrt{\pi}\, l_2}, \qquad l_2 \text{ large} \qquad (3.F.14)$$

With critical exponents of 0.34 for Δn and of 0.6 for l_2 ($l_2 \propto |T_c - T|^{-0.6}$), (3.F.14) predicts for γ the critical exponent 1.28, in good agreement with experiment. The formula for small l_2 is not relevant to near-critical interfaces, but it provides a simple estimate for the tension of low temperature and sharp interfaces.

In a multicomponent fluid the interfacial tension is also proportional to $\sqrt{\Delta\omega(\mathbf{n})}$. Here again low tensions near a critical point result from the flattening out of the free-energy surface as T_c is approached. This is illustrated in Fig. 3.F.2 for a model system that separates into three coexisting equilibrium phases. As the temperature varies from T_1 to T_2 the three phases approach a tricritical point. For illustration we consider a solution for which density is everywhere constant at a given temperature: $\Delta\omega(\mathbf{n})$ is the vertical distance from the free-energy surface to the top planes in Figs. 3.F.2a and 3.F.2b. The value of $\Delta\omega(\mathbf{n})$ decreases rapidly as temperature approaches the critical point and the tension decreases and interfacial width increases accordingly. This near-critical, low-tension mechanism is the same as that of a one-component interface. It is the only means for a one-component system to achieve low tensions, whereas other possibilities exist for multicomponent systems.

Two phases at equilibrium can be made miscible by titration with a solubilizer. For example, water and a hydrocarbon split into two phases but

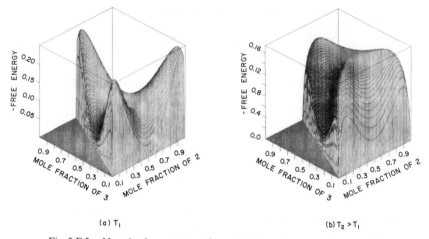

(a) T_1 (b) $T_2 > T_1$

Fig. 3.F.2.   Negative free-energy surfaces of a three-component regular solution.

can be titrated with alcohol until they become miscible. With the right titration path, the sequence of equilibrium phases passes through a *plait point*, a point at which the tie-line between the phases becomes of zero length and the phases become identical. Figure 3.F.3 shows a ternary phase diagram and its associated free-energy surface. The domes in the triangular phase diagram represent the binodal curve of two-phase states. Lines 1 and 2 represent tie lines, and 3, a plait point. For the titration path, t, shown on the free energy surface, the system passes through the plait point (3). The plait point

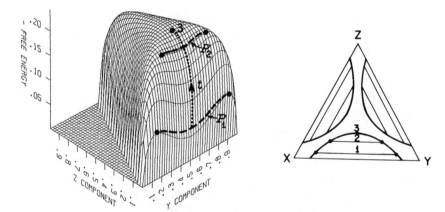

Fig. 3.F.3.   The negative-free energy, interfacial composition paths þ and ternary phase diagram of a three-component regular solution.

is a critical point. From the composition paths indicated on the free-energy surface, it follows that $\Delta\omega$ will become increasingly flatter as the plait point is approached. Also the compositions of the phases approach each other. Thus, since $\gamma \propto \sqrt{\Delta\omega_{max}} \, \mathfrak{p}_B$, according to (3.E.7), the tension approaches zero as the square root of $\Delta\omega_{max}$ and as the first power of the composition differences (measured by $\mathfrak{p}_B$).

According to scaling law theory,[89] if a critical point is approached by varying a field variable other than temperature, the asymptotic formulas for $\Delta\omega$ and $\mathfrak{p}_B$ are the same with respect to temperature. Field variables are those intensive variables having the same value for all phases in equilibrium[89, 90] (e.g., temperature, pressure, and chemical potential). Thus, if μ_R is the chemical potential of the chemical species added to titrate to the plait point (where $\mu_R = \mu_{RC}$), then $\Delta\omega_{max} \propto |\mu_R - \mu_{RC}|^\zeta$ and $\mathfrak{p}_B \propto |\mu_R - \mu_{RC}|^\beta$ near the plait point, and so[91]

$$\gamma = \gamma_0 \left| \frac{\mu_R - \mu_{RC}}{\mu_{RC}} \right|^\nu \qquad (3.F.15)$$

near the plait point.

From the values of ζ and β (1.89 and 0.34) established above, it again follows that the tension becomes small as the plait point is approached as a result of the two factors, $\sqrt{\Delta\omega_{max}} \propto |\mu_R - \mu_{RC}|^{0.95}$ and $\mathfrak{p}_B \propto |\mu_R - \mu_{RC}|^{0.34}$, the first factor arising from flattening of the free-energy surface and the second from convergence of phase density and compositions. Because of the larger critical exponent, the flattening of the free-energy surface is the dominant factor in near-critical, low-tension behavior. This explains how the low tensions observed between apparently quite different phases can nevertheless be the result of near-critical phenomena. The early failure to associate ultralow-tension microemulsion behavior[92] with near-critical phenomena probably derived from lack of appreciation of these factors. Fleming et al.,[91] however, demonstrated that tensions between phases occurring in salinity scans (thus μ_R = chemical potential of salt) of microemulsion systems of oil, aqueous brine, and surfactant mixtures obey well (3.F.15) with the critical exponent $\nu = 10/7$ predicted for the three-dimensional Ising model. This agreement and other evidence, summarized elsewhere,[92] on phase volumes, compositions, and light scattering provide a fairly convincing case for concluding that ultralow-tension phase pairs involving microemulsions are near-critical systems.

An interesting situation that can arise is illustrated by Fig. 3.F.4, in which the surface $f_0(\mathbf{n})$ is constructed over a triangular plot of volume (or mole) fractions of the species of a ternary system. In the absence of component 3, the free energy surface has a high mountain between the composition of the two phases in equilibrium. Since $\Delta\omega(\mathbf{n})$ is large along the composition path between the phases, the tension of the interface is large. If, however, enough of component 3 is added to produce a third phase, then along a path from phase α to δ and along another path from δ to β the free-energy surface is flat, $\Delta\omega(\mathbf{n})$ small, and the interfacial tensions $\gamma_{\alpha\delta}$ and $\gamma_{\delta\beta}$ are small. What is remarkable is that neither of the valleys in which the respective paths run appears to be near a plait point. It is likely that the situation illustrated by Fig. 3.F.4 occurs when to water and oil is added a surfactant under conditions leading to formation of a third phase having low tension against both water and oil. Such systems have received a lot of attention recently as candidates for enhanced oil recovery.[92, 93] Small amounts of the third phase can go undetected, with result that one concludes that the tension between phases α and β is very small. Such a misinterpretation was in fact common for certain oil-water-surfactant systems until recently, when the situation was clarified.[94]

For all the situations discussed above, the shape of the free-energy surface $f_0(\mathbf{n})$ holds the key to low-tension behavior. In principle, the influence parameters $c_{\alpha\beta}$ can also lead to, or at least greatly assist, the formation of

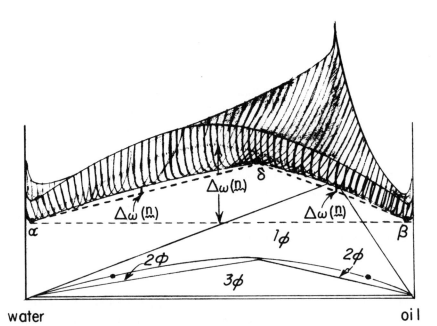

water   oil

Fig. 3.F.4.   Free-energy surface illustrating the small $\Delta\omega(n)$, low-tension paths between oil and water provided by a surfactant-rich third phase.

low-tension interfaces. If somehow their values can be adjusted (e.g., by design of molecular segments or choice of species of mixtures) so that everywhere along a composition path the quantity $\Sigma_{\alpha,\beta} c_{\alpha\beta}(dn_\alpha/d\wp)(dn_\beta/d\wp)$ is very small, the tension according to (3.E.4) will be small. In such a case, however, according to (3.E.3), the interface will be sharp. In the limit of a very sharp interface, the model of Lovett et al.[25] is applicable if no component has substantial surface activity. For the multicomponent case, their model yields

$$\gamma \simeq \frac{kT}{8} \sum_{\alpha,\beta} \Delta n_\alpha \Delta n_\beta \int s C_0^{\alpha\beta}(s)\, d^3 s \qquad (3.F.16)$$

The quantity $\int s C_0^{\alpha\beta}(s)\,d^3 s$ replaces the influence parameter $c_{\alpha\beta}$ in this approximation, and a low-tension interface could be obtained by adjusting conditions so that the sum $\Sigma_{\alpha,\beta}\Delta n_\alpha \Delta n_\beta \int s C_0^{\alpha\beta}(s)\,d^3 s$ is small. As an abstract example, we imagine a two-component system of fixed density so that $\Delta n_1 = -\Delta n_2$ and (3.F.16) becomes $\gamma = (kT/8)(\Delta n_1)^2 \int s[C_0^{11} + C_0^{22} - 2C_0^{12}]\,d^3 s$. If the correlations were such that $C_0^{12} = \frac{1}{2}[C_0^{11} + C_0^{22}]$, or the integral of $2sC_0^{12}$ equals that of $s[C_0^{11} + C_0^{22}]$, the tension of the interface would be zero.

Although it has not been experimentally confirmed that there are low-tension interfaces resulting from a judicious set of influence parameters, as opposed to a flat free-energy surface, the possibility is plainly indicated by (3.D.7) and (3.E.4). It is quite possible that the lowering of tension by a monolayer of surfactant at the interface is the result of adjustment of influence parameters. To consider this situation, imagine components 1 and 3, which are totally immisible, and a surfactant component 2 that concentrates at the interface of 1 and 3. Approximate the interfacial profiles as follows

$$\{n_1, n_2, n_3\} = \{n_1^{(\alpha)}, 0, 0\}, \qquad x < -l$$

$$= \left\{-n_1^{(\alpha)}\frac{x}{l}, n_2^{(\alpha)}\left(\frac{x+l}{l}\right), 0\right\}, \qquad -l < x < 0$$

$$= \left\{0, n_2^{(\alpha)}\frac{l-x}{l}, n_3^{(\beta)}\frac{x}{l}\right\}, \qquad 0 < x < l$$

$$= \{0, 0, n_3^{(\beta)}\}, \qquad x > l \qquad (3.F.17)$$

This profile mimics a surfactant monolayer between two immisible phases. The corresponding tension, computed from (3.D.7), is

$$\gamma = \frac{1}{l}\left[c_{11}\left(n_1^{(\alpha)}\right)^2 + c_{33}\left(n_3^{(\beta)}\right)^2 + 2c_{22}\left(n_2^{(\alpha)}\right)^2 - 2\left(c_{12}n_1^{(\alpha)} + c_{23}n_3^{(\beta)}\right)n_2^{(1)}\right]$$

$$(3.F.18)$$

The design of tension-lowering surfactants involves, according to (3.F.18), adjusting molecular groups so that the contributions of the cross-influence parameters, c_{12} and c_{23}, reduce the contributions of the direct parameters $c_{ii}\ldots$.

The search for models and fluid systems allowing independent variations of the free energy of homogeneous fluid and of the influence parameters is presently underway in our laboratory. More rational design of tension-lowering agents is the practical goal of this endeavor.

G.   Gradient Theory: Stability of Planar Interfaces

As Section III.A proved, the stability equation, (3.D.13), has a zero eigenvalue corresponding to the eigenvector $u_\alpha = dn_\alpha/dx$, $\alpha = 1, \ldots, \nu$. Thus, to determine stability, other eigenvalues must be considered. The stability analysis given here differs from that of Carey[74] in that his was restricted to constant influence parameters and ours is not. For a one-component fluid, (3.D.13) becomes

$$-c\nabla^2 u - \frac{dc}{dx}\frac{du}{dx} + q(x)u = \lambda u \qquad (3.G.1)$$

where

$$q(x) \equiv \frac{\partial^2 f_0(n(x))}{\partial n^2} + \frac{1}{2}\frac{\partial^2 c(n(x))}{\partial n^2}\left(\frac{dn}{dx}\right)^2 - \frac{d^2 c(n(x))}{dx^2} \qquad (3.G.2)$$

and $n(x)$ is the equilibrium density of a planar interface at position x. Since all the coefficients in (3.G.1) depend only on the position x, it is convenient to Fourier-transform u in the yz-plane. Equation 3.G.1 transforms to

$$\mathcal{L}\tilde{u} \equiv -c\frac{d^2\tilde{u}}{dx^2} - \frac{dc}{dx}\frac{d\tilde{u}}{dx} + q(x)\tilde{u} = (\lambda - k^2)\tilde{u} \qquad (3.G.3)$$

where

$$\tilde{u}(x) \equiv \frac{1}{2\pi}\int\int e^{i(k_y y + k_z z)}u(x, y, z)\,dy\,dz \qquad (3.G.4)$$

and $k^2 = k_y^2 + k_z^2$.

The linear operator $\mathcal{L}$ defined by (3.G.3) is a Sturm-Liouville operator in which c, dc/dx, and $q(x)$ are continuous, bounded functions of x. With appropriate boundary conditions (which can be that $\tilde{u} = 0$ on the boundary of the system), the following oscillation theorem follows: $\mathcal{L}$ possesses an infinite number of eigenvalues $\mu_0, \mu_1, \mu_2, \ldots$, forming a monotone sequence with $\mu_n \to \infty$ as $n \to \infty$. Moreover, the eigenfunction corresponding to λ_n has exactly n zeros.

As we established in Section III.A, dn/dx is an eigenfunction of $\mathcal{L}$ having zero eigenvalue. Since dn/dx has no zeros, it follows from the theorem above that zero is the minimum eigenvalue (i.e., $\mu_0 = 0$). This implies that a planar interface is stable. An arbitrary density fluctuation v can be expressed in the form

$$v = \sum_{n=0}^{\infty} \sum_{\mathbf{k}} a_{n,\mathbf{k}} \frac{\exp\left[i\left(k_y y + k_z z\right)\right]}{2\pi} \tilde{u}_n$$

yielding

$$\delta^2 F = \frac{1}{2} \sum_{n=0}^{\infty} \sum_{\mathbf{k}} \left(\mu_n + k^2\right)|a_{n,\mathbf{k}}|^2 \qquad (3.G.5)$$

Since $\tilde{u}_0 = dn/dx$ is not a particle-conserving fluctuation, $\delta^2 F > 0$ for any allowed density fluctuation. It also follows that fluctuations in the yz-plane ($\mathbf{k} \neq 0$) always increase $\delta^2 F$ over its value for a fluctuation in the x-direction ($\mathbf{k} = 0$). Thus, according to gradient theory, three-dimensional interfacial structures (capillary waves) are damped, hence can occur only as fluctuations.

Stability analysis is more complicated for multicomponent systems and must be considered case by case for each set of influence parameters $c_{\alpha\beta}$ and free-energy density function f_0. One general result that can be stated is that a planar interface will be unstable if the matrix $\mathbf{c}$ of influence parameters fails to be positive definite on any interval (a, b) of x in the interfacial zone. For if $\mathbf{c}$ is not positive definite in (a, b), then the matrix has at least one negative eigenvalue, ζ. In the interval (a, b), we define the functions $[x=0$ is chosen to be in the interval $(a, b)]$

$$v_\alpha = a_\alpha \sin 2\pi m x; \qquad 0 \le x \le \frac{1}{m}$$
$$= 0; \qquad \text{otherwise} \qquad (3.G.6)$$

where m is a large integer and a_α is an as-yet unspecified constant. Consider the integral

$$I = \frac{1}{2} \sum_{\alpha, \beta} c_{\alpha\beta}(\mathbf{n}(x)) \frac{dv_\alpha}{dx} \frac{dv_\beta}{dx} dx$$

$$= 2\pi^2 m^2 \sum_{\alpha, \beta} \int_0^{1/m} c_{\alpha\beta}(\mathbf{n}(x)) a_\alpha a_\beta \sin^2 2\pi m x \, dx$$

$$= \pi^2 m \sum_{\alpha, \beta} c_{\alpha\beta}(\mathbf{n}(0)) a_\alpha a_\beta \left[1 + \mathcal{O}\left(\frac{1}{m}\right)\right]. \qquad (3.G.7)$$

We now choose the set $a_1, \ldots, a_\nu$ to be the components of an eigenvector of $\mathbf{c}$ corresponding to the negative eigenvalue ζ. Then

$$\lim_{m \to \infty} I = -\infty \tag{3.G.8}$$

It is not difficult to show that for sufficiently large m, the integral I is the part of $\delta^2 F$, (3.D.12), having the largest magnitude. Thus it follows that a planar interface is unstable if $\mathbf{c}$ is not positive definite everywhere in the interfacial zone.

H. Gradient Theory: Capillary Waves

The gradient theoretical Helmholtz free energy of an interface in the presence of gravity is

$$F = \int \left[f_0(n(\mathbf{r})) + \tfrac{1}{2} c (\nabla n)^2 + m n(\mathbf{r}) g x \right] d^3 r \tag{3.H.1}$$

where g is the acceleration of gravity and x is the vertical direction. The planar interface, $n = n(x)$, represents a local minimum in F and therefore is the equilibrium configuration. However, thermal fluctuations can lead to instantaneous deviations of the local density profile from the planar solution. Especially important are small-amplitude fluctuations of large extent in the yz-plane. These resemble ripples and can be regarded as thermally induced capillary waves. Since they can be created with small expense in energy, they are likely to occur.

Some years ago Buff et al.[95, 25] and Lovett[96] introduced a version of capillary wave theory to describe a planar interface as a fluctuating discontinuous interface. They proposed capillary wave theory as an alternative to theories such as gradient theory for computing interfacial density profiles and tensions. Recently, capillary waves have been the subject of a computer simulation[97] and several theoretical papers [98-101] and have created some controversy.[32] For a full appreciation of the status of this theory, we refer the reader to the evolving literature on the subject. The brief account here is intended only to highlight the issues.

Capillary waves tend to make a planar interface more diffuse. For a flat interface, a dividing surface can be defined as the position $x = \zeta_p$ at which

$$\int_{-L_l}^{\zeta_p} x[n_l - n(x)] \, dx = \int_{\zeta_p}^{L_g} x[n(x) - n_g] \, dx$$

where $L_l + L_g$ is the height of the system. We assume for convenience that the system is a cube. By choice of the origin of the coordinate system, ζ_p can be set equal to zero. The instantaneous dividing surface of an interface perturbed by a thermal fluctuation can be represented by a dividing surface function $\zeta(y, z)$. If the magnitudes of the principal radii of curvature associated with the surface $\zeta(y, z)$ are large compared to the intrinsic interfacial width, then the density profile dependence on s, the distance normal to the interface, is the same as that of a planar interface. The dividing surface can then be located by the equation

$$\int_{-L_l}^{\zeta(y,z)} \left[n\left(\frac{x}{\left(1+\zeta_y^2+\zeta_z^2\right)^{1/2}} \right) - n_l \right] x\,dx$$

$$= \int_{\zeta(y,z)}^{L_g} \left\{ n_g - n\left[\frac{x}{\left(1+\zeta_y^2+\zeta_z^2\right)^{1/2}} \right] \right\} x\,dx \qquad (3.\mathrm{H}.2)$$

where ζ_y and ζ_z are the y and z partial derivatives of $\zeta(y, z)$ and $x/(1+\zeta_y^2 + \zeta_z^2)^{1/2}$ normal distance s from the interface corresponding to the point (x, y, z). With (3.H.2), the coordinate transformation, $d^3r=dx\,dy\,dz=(1+\zeta_y^2+\zeta_z^2)^{1/2}\,ds\,dy\,dz$, and the relation $(\nabla n)^2 =(dn/ds)^2$, the free energy of the interface over a plane of area A becomes

$$F=F_B + W; \qquad W=\gamma \int_A \left(1+\zeta_y^2+\zeta_z^2\right)^{1/2} dy\,dz + \tfrac{1}{2}\Delta\rho g \int_A \zeta^2\,dy\,dz$$

$$(3.\mathrm{H}.3)$$

where $F_B = V\omega(n_l)+N\mu+\tfrac{1}{2}(M_gL_g - M_lL_l)g$, $\gamma=\int_{-\infty}^{\infty}c(dn/ds)^2\,ds$, and $\Delta\rho =m(n_l-n_g)$. M_g and M_l are the masses of the vapor and liquid phases, respectively.

For a flat interface, $\zeta=0$, and $W=\gamma A$. Otherwise, $W>\gamma A$. Since the probability of a large-amplitude fluctuation is small, the quantity $(1+\zeta_y^2+\zeta_z^2)^{1/2}$ can be approximated by $1+\tfrac{1}{2}(\zeta_y^2+\zeta_z^2)$. The surface $\zeta(y, z)$ can be expressed as a superposition of capillary waves, that is,

$$\zeta(y, z)= \sum_{k>0} \mathcal{C}(\mathbf{k})e^{i\mathbf{k}\cdot\mathbf{t}} \qquad (3.\mathrm{H}.4)$$

where $\mathbf{t}= y\hat{i}+ z\hat{j}$ with $\mathbf{k}$ the wave vector of a capillary wave ($k_y, k_z = \pm 2\pi/L, \pm 4\pi/L,\ldots$) and $\mathcal{C}(\mathbf{k})$ its amplitude. The energy W then becomes

$$W=\gamma A + \tfrac{1}{2} \sum_{k>0} |\mathcal{C}(\mathbf{k})|^2\left[\Delta\rho g + \gamma k^2\right] A \qquad (3.\mathrm{H}.5)$$

The mean-square dispersion of the interfacial zone due to a given fluctuation is $\overline{\zeta^2} \equiv 1/A \int \zeta^2 dy\, dz = \Sigma_{k>0} |\mathcal{Q}(k)|^2$. The probability of a given fluctuation is proportional to $e^{-\beta W} d\mathcal{Q}(k_1)\, d\mathcal{Q}(k_2)\ldots$. Thus the thermally averaged mean-square dispersion of the interfacial zone is

$$\sigma_\zeta^2 = 2 \sum_{k>0} \left\{ \beta L^2 \left[\Delta \rho g + \gamma k^2 \right] \right\}^{-1} \qquad (3.H.6)$$

This summation actually diverges as k tends to infinity. However, there are two physical arguments for cutting off the sum after some value k_{max}. First, the assumption that the fluctuation does not alter the intrinsic properties (tension γ and density profile normal to the surface) breaks down for capillary waves of sufficiently small wavelength ($2\pi k^{-1}$). Second, as argued by Buff et al., representing the fluctuation energy by (3.H.5) amounts to the introduction of capillary waves as collective coordinates to describe particle fluctuations. Since a finite number of particles are involved, the number of independent capillary waves, which equals $\Sigma_{k>0} 1$, should also be finite. We then truncate the sum for $k > k_m$, which, when approximated by an integral, yields

$$\sigma_\zeta^2 = \frac{1}{4\pi\gamma\beta} \ln \left[\frac{\chi^2 + k_m^2}{\chi^2 + (2\pi/L)^2} \right] \qquad (3.H.7)$$

where $\chi^2 = \Delta \rho g / 2\gamma$.

Buff et al. suggest that a consistent approximation for k_m is to set it equal to $2\pi/\sigma_\zeta$ or some fraction of this. With this approximation and the typical parameters $\gamma = 30$ dynes/cm, $\Delta\rho = 1$ g/cm^3, $T = 300°$K, and $L = 1$ cm, (3.H.7) yields $\sigma_\zeta \simeq 7.3 \times 10^{-8}$ cm. The intrinsic width of such an interface predicted from the planar profile $n(x)$ may be typically a few molecular diameters, or some tens of angstroms. Thus the capillary wave dispersion moment σ_ζ of the interfacial zone is of the order of magnitude of the intrinsic interfacial width. If the cutoff wave vector k_m for the example just considered is taken to be $2\pi/(10^{-4}$ cm) instead of $2\pi/(7.3 \times 10^{-8}$ cm), the predicted value of σ_ζ is 6×10^{-8} cm, indicating considerable insensitivity to the value of k_m. Thus the major contribution to σ_ζ comes from the long-wavelength (small-k) capillary waves.

Even quite near the critical point, for $k_m \simeq 2\pi/\sigma_\zeta$, the term k_m^2 is much greater than χ^2. Thus, for an infinite system, $\sigma_\zeta^2 \simeq (1/4\pi\gamma\beta)\ln(8\pi^2\gamma/\Delta\rho\sigma_\zeta^2)$, a relationship implying that σ_ζ approaches infinity roughly as $\gamma^{-1/2}$ (or as $(T_c - T)^{-0.65}$ as $T \to T_c$. Thus the critical index of σ_ζ is quite similar to that of l_w of gradient theory and agrees with 0.67 observed by Huang and Webb for interfacial width.

Alternatively to the approximation $k_m = 2\pi/\sigma_\xi$, one can estimate k_m by equating the number of capillary waves, $\Sigma_k 1 \simeq L^2 k_m^2/4\pi$, to the number of particles, $\bar{n}l_w L^2$, in the interfacial zone, where $\bar{n}$ is the average number density across the interfacial zone. In this case, $k_m \simeq 2\sqrt{\pi \bar{n}l_w}$, and again even quite close to the critical point the quantity χ^2 is much smaller than k_m^2 and, consequently, σ_ξ goes as $|T_c - T|^{-0.65}$ near the critical point.

In the absence of gravity, (3.H.7) predicts that σ_ξ increases as $\ln L$ as the size of the system increases. This trend was observed in the computer simulations of Chapela et al.[50] As suggested by the work of Wertheim,[98] Kalos et al.[97] ascribe to capillary waves the long-range pair correlations observed transverse to the interface in computer simulations.

The argument of Evans[32] against introduction of capillary waves appears to derive in part from the uncertainty of the value, and indeed the meaning, of the cutoff parameter k_m in the theory and from the fact that the capillary dispersion σ_ξ depends on the interfacial area and the gravitational constant. The uncertainty of k_m and the breakdown of a capillary wave picture for short-wavelength fluctuations are valid reasons for questioning the meaning of (3.H.7). However, Lovett et al.[24] state without proof that the value of k_m can be established with capillary wave pair density theory. Evans also dismisses as fortuitous the agreement Lovett et al.[25] obtained between experimental interfacial width and σ_ξ when they set in (3.H.7) $k_m = 0.543\pi/\sigma_\xi$, replaced by γ_0 and estimated it from $\gamma_0 = \gamma_{exp} + 3k_m^2/16\pi\beta$, the expression Buff et al.[95] presented to relate the tension γ_0 of a discontinuous reference interface and the observed tension γ of an interface.

Our position at this writing is that (1) the tension and intrinsic width of an interface are those given by the profile theories described in preceding sections and (2) fluctuations about the lowest free-energy state tend to make a planar interface more diffuse, and (3.H.7) provides an estimate of this effect. Weeks[99] argues on the other hand that small capillary wave corrections should be added to the tension. That long-wavelength capillary waves exist as excitations is supported by the observation that light scattered from them can be used to deduce interfacial tensions in agreement with those obtained by other means.[102] Future computer simulations and theoretical work ought to settle the controversial aspects of capillary wave behavior.

References

1. Lord Rayleigh, *Phil. Mag.*, **33**, 208 (1892).
2. J. D. van der Waals and P. Kohnstamm, *Lehrbuch der Thermodynamik*, Vol. 1, Mass and van Suchtelen, Leipzig, 1908.
3. For a discussion of the early history of the theory of capillarity, see F. Drowan, *Proc. R. Soc. (London), Ser A*, **316**, 473 (1970).
4. J. W. Cahn and J. E. Hilliard, *J. Chem. Phys.*, **28**, 258 (1958).
5. J. W. Cahn, *J. Chem. Phys.*, **42**, 93 (1965).

6. N. G. van Kampen, *Phys. Rev. A*, **135**, 362 (1964).
7. S. Fisk and B. Widom, *J. Chem. Phys.*, **50**, 3219 (1969).
8. B. Widom, "Surface Tension of Fluids," in *Phase Transitions and Critical Phenomena*, Vol. 2, C. Domb and M. S. Green, Eds., Academic Press, New York, 1972), Chapter 3.
9. B. Widom, "Structure and Thermodynamics of Interfaces," in *Statistical Mechanics and Statistical Methods in Theory and Application*, U. Landman, Ed. Plenum Press, New York, 1976, pp. 33–71.
10. J. G. Kirkwood and F. Buff, *J. Chem. Phys.*, **17**, 338 (1949).
11. F. P. Buff, *Z. Elektrochem.*, **56**, 311 (1952).
12. F. P. Buff, *J. Chem. Phys.*, **19**, 1591 (1951); **23**, 419 (1955); **25**, 146 (1956); F. P. Buff and H. Saltsburg, *J. Chem. Phys.*, **26**, 23 (1957).
13. J. H. Irving and J. G. Kirkwood, *J. Chem. Phys.*, **18**, 817 (1950).
14. R. Bearman and J. G. Kirkwood, *J. Chem. Phys.*, **28**, 136 (1958).
15. B. S. Carey, L. E. Scriven, and H. T. Davis, *J. Chem. Phys.*, **69**, 5040 (1978).
16. K. Hiroike, *J. Phys. Soc. Japan*, **12**, 864 (1957); 771 (1960).
17. T. Morita and K. Hiroike, *Prog. Theor. Phys.*, **25**, 537 (1961).
18. C. De Dominicis, *J. Math. Phys.*, **3**, 983 (1962).
19. F. H. Stillinger and F. P. Buff, J. Chem. Phys., **37**, 1 (1962).
20. J. L. Lebowitz and J. K. Percus, *J. Math Phys.*, **4**, 116 (1963).
21. J. K. Percus, in *The Equilibrium Theory of Classical Fluids*, H. L. Frisch and J. L. Lebowitz, Eds., Benjamin, New York, pp. II33–II70.
22. For a short discussion of the history of the use of functional differentiation in statistical mechanics, see G. Stell, in *The Equilibrium Theory of Classical Fluids*, H. L. Frisch and J. L. Lebowitz, Eds., Benjamin, New York, 1966, pp. II–71–II266.
23. L. S. Ornstein and F. Zernike, *Proc. Acad. Sci. Amsterdam*, **17**, 793 (1914).
24. D. G. Triezenberg and R. Zwanzig, *Phys. Rev. Lett.*, **28**, 1183 (1972).
25. R. Lovett, P. W. DeHaven, J. J. Vieceli, Jr., and F. P. Buff, *J. Chem. Phys.*, **48**, 1880 (1973).
26. V. Bongiorno, L. E. Scriven, and H. T. Davis, *J. Colloid Interface Sci.*, **57**, 462 (1976).
27. A. J. M. Yang, P. D. Fleming III, and J. H. Gibbs, *J. Chem. Phys.*, **64**, 3732 (1976); P. D. Fleming III, A. J. M. Yang, and J. H. Gibbs, *J. Chem. Phys.*, **65**, 7 (1976).
28. S. Ono and S. Kondo, *Handbuch der Physik*, Vol. X, S. Flügge, Ed., Springer-Verlag, Berlin, 1960, p. 134.
29. C. A. Croxton, *Adv. Phys.*, **22**, 385 (1973); in *Progress in Liquid Physics*, C. A. Croxton, Ed., Wiley, New York, 1978, p. 42.
30. M. S. Jhon, J. S. Dahler, and R. C. Desai, *Adv. Chem. Phys.*, **46**, 279 (1981).
31. J. Rowlinson, *Chem. Soc. Rev.*, **7**, 329 (1978).
32. R. Evans, *Adv. Phys.*, **28**, 143 (1979).
33. J. G. Kirkwood, *J. Chem. Phys.*, **14**, 180 (1946); **15**, 72 (1947).
34. R. D. Present, Chia C. Shih, and Yea H. Uang, *Phys. Rev. A*, **14**, 863 (1976).
35. H. T. Davis, *J. Chem. Phys.*, **62**, 3412 (1975).
36. C. G. Gray and K. E. Gubbins, *Mol. Phys.*, **30**, 179 (1975).
37. T. L. Hill, *Statistical Mechanics*, McGraw-Hill, New York, 1956.
38. S. Toxvaerd, *J. Chem. Phys.*, **64**, 2863 (1976).
39. S. Toxvaerd, in *Statistical Mechanics*, Vol. 2, K. Singer, Ed., Chemical Society, London, 1975, p. 256.
40. K. U. Co, J. J. Kozak, and K. D. Luks, *J. Chem. Phys.*, **66**, 1002 (1977).
41. D. J. Korteweg, *Archives Neerl. Sci. Exacts Nat.*, **6**, 1 (1904).
42. See Ref. 15 and Sections III.E and III.F of this chapter for further discussion of such applications.
43. B. F. McCoy and H. T. Davis, *Phys. Rev. A*, **20**, 1201 (1979).

44. J. Serrin, in *Recent Methods in Nonlinear Analysis and Applications*, ed. by A. Canfora, S. Rionero, C. Sbordone, and G. Trombetti, published by Istituto di Matematica, Universitá di Napoli (1980).
45. H. T. Davis and L. E. Scriven, *J. Chem. Phys.*, **69**, 5215 (1978).
46. B. F. McCoy, L. E. Scriven, and H. T. Davis, *J. Chem. Phys.*, (to appear 1981).
47. J. Lekner and J. R. Henderson, *Mol. Phys.*, **39**, 1437 (1980).
48. S.-K. Ma, *Modern Theory of Critical Phenomena*, Benjamin, New York, 1976.
49. F. F. Abraham, D. E. Schreiber, and J. A. Barker, *J. Chem. Phys.*, **62**, 1958 (1975).
50. G. A. Chapela, G. Saville, S. M. Thompson, and J. S. Rowlinson, *J. Chem. Soc., Trans. Faraday II*, **73**, 1133 (1977).
51. T. L. Hill, *J. Chem. Phys.*, **20**, 141 (1952).
52. S. Toxvaerd, *J. Chem. Phys.*, **55**, 3116 (1971).
53. C. Ebner, W. F. Saam, and D. Stroud, *Phys. Rev. A*, **14**, 2264 (1976).
54. J. K. Lee, J. A. Barker, and G. M. Pound, J. Chem. Phys., **60**, 1976 (1974).
55. R. H. Fowler, *Proc. R. Soc. London, Ser. A*, **159**, 229 (1937).
56. J. J. Jasper, *J. Phys. Chem. Ref. Data*, **1**, 841 (1972).
57. S. J. Salter and H. T. Davis, *J. Chem. Phys.*, **63**, 3298 (1975).
58. L. A. Girifalco and R. J. Good, *J. Phys. Chem.*, **61**, 904 (1957).
59. B. W. Davis, *J. Colloid Interface Sci.*, **52**, 150 (1975).
60. P. H. Winterfeld, L. E. Scriven, and H. T. Davis, *Am. Inst. Chem. Eng. J.*, **24**, 1010 (1978).
61. H. T. Davis and L. E. Scriven, *J. Phys. Chem.*, **80**, 2805 (1976).
62. M. S. Wertheim, *J. Chem. Phys.*, **65**, 2377 (1976).
63. D. E. Sullivan and G. Stell, *J. Chem. Phys.*, **67**, 2567 (1977).
64. R. Lovett, C. Y. Mou, and F. P. Buff, *J. Chem. Phys.*, **65**, 570 (1976).
65. J. Stecki and A. Kloczkowski, *J. Phy. Colloq. C3*, Suppl. 4, **40**, C3-360 (1979).
66. V. Bongiorno and H. T. Davis, *Phys. Rev. A*, **12**, 2213 (1975).
67. H. Reiss, *Adv. Chem. Phys.*, **9**, 1 (1965).
68. R. Zwanzig, *J. Chem. Phys.*, **22**, 1420 (1954).
69. J. D. Weeks, D. Chandler, and H. C. Andersen, *J. Chem. Phys.*, **54**, 5237 (1971).
70. B. S. Carey, L. E. Scriven, and H. T. Davis, *Am. Inst. Chem. Eng. J.*, **24**, 1076 (1978).
71. B. S. Carey, L. E. Scriven, and H. T. Davis, *Am. Inst. Chem. Eng. J.*, **26**, 705 (1980).
72. C. I. Poser and I. C. Sanchez, *J. Colloid Interface Sci.*, **69**, 539 (1979).
73. C. I. Poser, Ph.D. thesis, University of Massachusetts, 1979.
74. B. S. Carey, Ph.D. thesis, University of Minnesota, 1979.
75. A. L. Horvath, *Chem. Eng. Sci.*, **29**, 1334 (1974).
76. D. Y. Peng and D. B. Robinson, *Ind. Eng. Chem. Fundam.*, **15**, 59 (1976).
77. I. C. Sanchez and R. H. Lacombe, *J. Phys. Chem.*, **80**, 2352 (1976); **80**, 2568 (1976).
78. K. K. Mohanty, M. Dombrowski, and H. T. Davis, *Chem. Eng. Commun.*, **5**, 85 (1980).
79. G. G. Fuller, *Ind. Eng. Chem. Fundam.*, **15**, 254 (1976).
80. M. Guerrero and H. T. Davis, *Ind. Eng. Chem. Fundam.*, **19**, 309 (1980).
81. M. Guerrero, *Ind. Eng. Chem. Fundam.*, to be published.
82. H. E. Stanley, *Introduction to Phase Transitions and Critical Phenomena*, Clarendon Press, Oxford University Press, Oxford 1971.
83. A. Münster, *Statistical Thermodynamics*, Vol. I, Academic Press, New York, 1969.
84. P. D. Fleming III and J. E. Vinatieri, *Am. Inst. Chem. Eng. J.*, **25**, 493 (1979).
85. M. A. Bouchiat and J. Meunier, *J. Phys. (Paris) Suppl.*, **C1**, 141 (1972).
86. J. Zollweg, G. Hawkins, and G. B. Benedek, *Phys. Rev. Lett.*, **27**, 1182 (1971).
87. For a summary of critical exponent data, see Refs. 48 and 82.
88. J. S. Huang and W. W. Webb, *J. Chem. Phys.*, **50**, 3677 (1969).
89. R. B. Griffiths and J. C. Wheeler, *Phys. Rev. A*, **2**, 1047 (1970).

90. L. Tisza, *Generalized Thermodynamics*, MIT Press, Cambridge, Mass., 1966.
91. P. D. Fleming III, J. E. Vinatieri, and G. R. Glinsmann, *J. Phys. Chem.*, **84**, 1526 (1980).
92. For example, see article by R. L. Reed and R. N. Healy in *Improved Oil Recovery by Surfactant and Polymer Flooding*, ed. by D. O. Shah and R. S. Schechter, Academic Press, N.Y. (1977).
93. (a) W. R. Foster, *J. Pet. Technol.*, **205**; *Trans. AIME*, **255** (1973).
93. (b) G. P. Ahearn and W. W. Gale, U.S. Patent No. 3,302,713 (1967).
94. E. I. Franses, J. E. Puig, Y. Talmon, W. G. Miller, L. E. Scriven, and H. T. Davis, *J. Phys. Chem.*, **84** 1547 (1980).
95. F. P. Buff, R. A. Lovett, and F. H. Stillinger, *Phys. Rev. Lett.*, **15**, 621 (1965).
96. R. A. Lovett, Ph.D. thesis, University of Rochester, 1965.
97. M. H. Kalos, J. K. Percus, and M. Rao, *J. Stat. Phys.*, **17**, 111 (1977).
98. M. S. Wertheim, *J. Chem. Phys.*, **65**, 2377 (1976).
99. J. D. Weeks, *J. Chem. Phys.*, **67**, 3106 (1977).
100. H. T. Davis, *J. Chem. Phys.*, **67**, 3636 (1977); erratum in *J. Chem. Phys.* **70**, 600 (1979).
101. M. W. Cole, *Phys. Rev. A*, **1**, 1838 (1970).
102. See, for example, S. Hard, Y. Hamnerius, and O. Nilsson, *J. Appl. Phys.*, **47**, 2433 (1976).

THE NATURE AND STRUCTURAL PROPERTIES OF GRAPHITE INTERCALATION COMPOUNDS

S. A. SOLIN

Department of Physics
Michigan State University
East Lansing, Michigan
U.S.A.

CONTENTS

I. INTRODUCTION

The subject of graphite intercalation compounds (GICs) is more than 3000 years old[1] and spans the fields of physics, chemistry, materials science, and others. In recent times, the period between 1930 and 1960 was one of extreme productivity in GIC research and was dominated by chemists and physical chemists who synthesized and characterized many new and unusual compounds. Excellent reviews of the 1930–1960 period by Rudorff,[2] Ubbelohde and Lewis[3], Hennig,[4] and others[5, 6] have been prepared. A central feature of the early twentieth-century work on GICs was the use of graphite powder as the starting material for the synthesis of compounds. Such starting material has several obvious drawbacks, not the least of which is the difficulty of reproducing samples in different laboratories.

Following a 15-year lull between 1960 and 1975 there was an explosive rebirth of interest in GIC research not only on the part of chemists and chemical physicists, but also among solid-state physicists. This renewed interest can in part be traced to the availability after 1960 of a synthetic carbon commonly referred to as highly oriented pyrolytic graphite (HOPG).[7] This material, which is discussed in more detail later, made possible new and interesting physical measurements of GICs and did not suffer as severely from the reproducibility problems of powder graphite alluded to above.

Two additional reasons, one practical and one fundamental, can easily be identified as sources of post-1975 interest in GICs. Of practical consideration was the possibility of synthesizing compounds from plentiful materials (e.g., carbon), which were lightweight and had very high conductivities, of the order of or greater than that of copper.[8] Such compounds might then be fabricated into wires and thus replace more costly copper or aluminum conductors. Of fundamental consideration was the quasi-two-dimensional character of graphite and its intercalation compounds. It was recognized by several researchers that GICs might constitute an interesting area for an examination of current ideas on two-dimensional physics.

Since 1975, the electronic properties,[9, 10] chemical properties,[11, 12] vibrational excitations,[13, 14] and some of the structural properties[15] of GICs have been the subjects of review papers. However, there has been no extensive review of the rich variety of structural phase transitions that can be induced in GICs by temperature and/or pressure. Moreover, many new and exciting results in this area have been reported only recently, within the past 2 to 3 years. During this period several myths concerning the structures and stoichiometries of GICs have also been dispelled. Therefore, a review of these most recent developments is clearly warranted, especially if the new results are contrasted and evaluated with respect to previously reviewed work. To present such a review, addressed to the general reader who is not yet familiar with GICs, I have included several elements of background and/or introductory material.

II. DEFINING AND CHARACTERIZING GRAPHITE INTERCALATION COMPOUNDS

Graphite is one of the prototypical lamellar compounds.[16] It occurs in nature in two forms. The structure of the hexagonal form (Fig. 1) is characterized by the layer stacking sequence ABAB · · · . Notice that the nearest-neighbor carbon—carbon distance within a plane is 1.42 Å, whereas the nearest-neighbor interplanar carbon—carbon distance is much larger, namely, 3.35 Å. It is this disparity in nearest-neighbor distances that is responsible for the lamellar properties and weak interplanar bonding. The other form of graphite that occurs in nature is rhombohedral graphite, and it is characterized by an ABCABC · · · layer stacking sequence. This form is quite rare compared to the hexagonal form and it has not been employed as a host material for GICs. Therefore it is not discussed further in this chapter. Hereinafter, the term "graphite" is used to describe the hexagonal form only.

Natural graphite single crystals are readily available in sizes ranging from ≈ 1 mm to 1 μm, the latter being constituents of graphite powder. However, the larger crystals are invariably imperfect and contain many defects, including stacking faults and twins. Such crystals cannot conveniently be employed as host material in many bulk physical measurements of interest in GIC research.

As noted above, a relatively new form of hexagonal graphite has been synthetically prepared. Highly oriented pyrolitic graphite (HOPG) can be

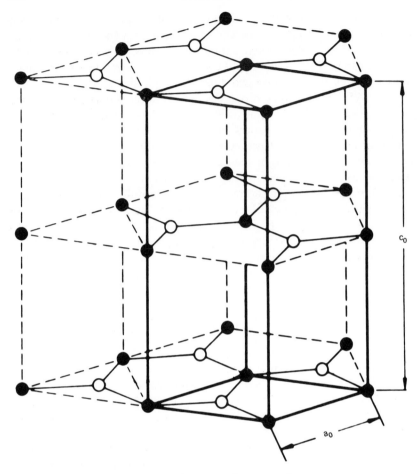

Fig. 1.   Crystal structure of graphite, $a_0 = 2.45$ Å, $c_0 = 6.70$ Å. Both the solid and the open circles correspond to carbon atoms.

produced with dimensions of the order of several centimeters. The HOPG form consists of compacted microcrystallites of typical dimension 1 μm arranged so that their c-axes are all well aligned to within $\leq 1°$, whereas their a-axes are randomly oriented in the basal plane. Thus, in an X-ray diffraction experiment with the scattering vector $\mathbf{q}$ parallel to the c-axis, a measurement of the $(00l)$ reflections of HOPG will yield a single crystal pattern. In contrast, a measurement of the $(hk0)$ reflections with $\mathbf{q}$ in the basal plane will yield the pattern of a two-dimensional powder.

Highly oriented pyrolytic graphite has been used extensively since 1960 as a host material for the preparation and study of GICs. It, like other graphitic forms, can ingest whole layers of atoms, ions, or molecules into the interlayer spaces between the carbon hexagonal planes. This process is referred to as intercalation. The ingested species is called the intercalant; the graphite is called the host. It has generally been assumed that the host carbon layers remain essentially undistorted after intercalation,[17] that is, that the planar hexagonal structure is preserved over distances of hundreds of angstroms, albeit with possible small ($\approx 1\%$) changes in intraplanar carbon—carbon bond distances.[18] I shall show later that the carbon layers of a GIC contain several distortions including puckering, edge dislocations, and intraplanar kinks. For the present discussion, however, it is conceptually convenient to consider a GIC to consist of undistorted flat carbon layers of infinite two-dimensional extent with ideal layers of intercalant strategically occupying the carbon interlayer spaces.

A. Classification Schemes

Several criteria can be used to characterize a GIC, and some of the most important are discussed below.

1. Donors and Acceptors

Naturally, the properties of a given GIC depend on the specific chemical species that occupies the carbon interlayer space. Intercalant species are generally classified as donors or acceptors according, respectively, to whether they give up electrons to the graphite layers or extract electrons from the graphite layers.

The known donor species are relatively limited[19] and consists of the alkali metals including lithium, potassium, rubidium, and cesium, and several divalent metals such as barium, calcium, and strontium, as well as transition metals such as europium and ytterbium. All known donor species enter the graphite matrix as partially ionized atoms to form two-dimensional monionic structures.

The number of acceptor species is vast[20] and thus much larger than the number of available donors. All known acceptor species are molecular (i.e., multiatomic). Moreover, most acceptors retain this multiatomic character after intercalation into the graphite host. Among the most widely studied acceptor intercalants are ferric chloride ($FeCl_3$), bromine (Br_2), nitric acid (HNO_3), and arsenic pentafluoride (AsF_5). Many other metal chlorides and fluorides as well as acid salts intercalate graphite. In general the acceptor intercalants occupy a given carbon layer interspace either as isolated essentially uncoupled molecules (AsF_5 at $300°K$)[21] or as two-dimensional crystals with a well-defined formula unit ($FeCl_3$ at $T \leq 300°K$).[22]

Recent reviews of intercalation chemistry have been written by Ebert[11] and also by Hérold.[20] The reader is referred to these reviews for up-to-date lists of the known intercalants. A thorough but somewhat dated review of alkali metal intercalation compounds has also been prepared by Novikov and Vol'pin.[23]

In the discussion above I have tacitly assumed that GICs are binary compounds composed of a carbon host matrix and a single intercalant donor or acceptor species. In fact, many ternary GICs have also been prepared and studied for several years.[24, 25] Nevertheless, there is still a paucity of information on their physical properties. Therefore, this chapter focuses exclusively on binary compounds.

2. Staging

One of the most intriguing properties of GICs is the property of staging, which is exhibited schematically in Fig. 2. As can be seen from the figure, a stage n compound is one in which each pair of intercalant layers is sep-

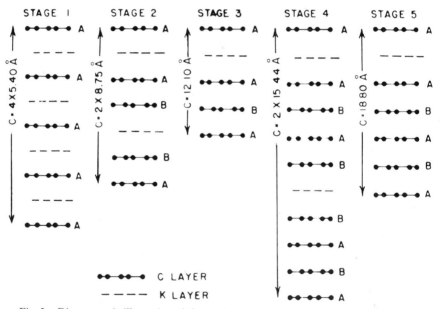

Fig. 2. Diagrammatic illustration of the c-axis stacking sequences in graphite and stages 1 to 5 potassium-graphite.[31] Dashed and solid lines represent the potassium and the carbon layers, respectively. In stage 1 KC_8 the former are ordered at room temperature. In stage n, KC_{12n} ($n \geq 2$) the potassium layers are disordered at room temperature. The letters A and B indicate the stacking arrangement of individual carbon layers, the projections of which are indicated by solid circles.

arated by exactly n carbon layers. When this stacking configuration is regularly repeated along the c-axis so that there is one-dimensional long-range order in that direction, the compound is labeled a "pure" stage. Note particularly that the stage of a given GIC is independent of the intraplanar structure of the intercalant species. Thus, for example, a compound in which the intercalant species form disordered liquidlike planar layers at room temperature can undergo a disorder-order transition at reduced temperatures without a change of stage.

The property of staging is apparently unique to the graphite intercalates, and several recent theories have been put forth to account for it.[27, 28] These are discussed in Sections II and IV. It is important to point out here that in principle, the intercalant-intercalant interlayer interaction can be controlled and altered stepwise continuously by changing the stage of a GIC prepared from a particular chemical species of intercalant. Thus GICs offer a unique possibility for the study of dimensionality effects and crossover phenomena. For high-stage compounds the intercalant layers are separated and screened by many carbon layers. Thus the intercalant-intercalant interlayer interaction will be negligible; however, the intercalant-graphite interaction may or may not be negligible compared to intralayer interactions.

3. Layer and Bulk Stoichiometries

Graphite intercalation compounds are also characterized by their layer and bulk stoichiometries. The layer stoichiometry is determined by the ratio of the number of intercalant atoms or molecules in a layer of intercalant to the number of carbon atoms in an adjacent carbon layer. In contrast the bulk stoichiometry has its usual definition. For stage 1 compounds, the bulk and layer stoichiometries are identical. For example, the stage 1 heavy alkali metal GICs have an ideal stoichiometry of MC_8 ($M = K$, Rb, or Cs). However, the higher stage compounds may have bulk and layer stoichiometries that differ significantly. Thus for $n \geq 2$, the ideal stoichiometries of heavy alkali GICs are MC_{12n}, $n = 2, 3, 4, \ldots$ ($M = K$, Rb, or Cs), whereas the layer stoichiometries for ideal compositions are constant and given by MC_{12}.

In a series of classic experiments Hérold showed[29] that for each species of intercalant there were preferred or ideal compositions, the realization of which set both the stage number and layer stoichiometries. An example of his results is shown as the isobar (solid curve) in Fig. 3 for potassium-graphite. In the famous "two-bulb" experiments[29] from which the data of Fig. 3 were obtained, graphite and the alkali metal to be intercalated are sealed in an evacuated glass tube and heated to different temperatures, t_2 and t_1, respectively. As can be seen from Fig. 3, the isobar exhibits plateaus of composition that correspond to specific stages. In the regions between the

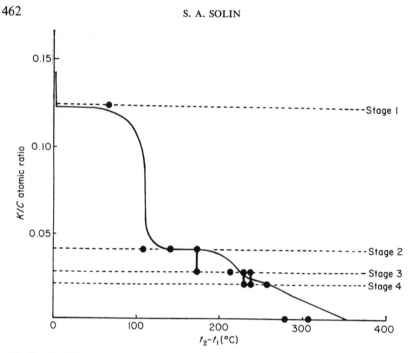

Fig. 3.   Potassium-graphite reaction products identified in Ref. 30; solid dots compared with the absorption isobar measured by Hérold; solid curve, from Ref. 29. Mixtures of two stages are represented by vertical lines linking the stages involved; $t_2 - t_1$ is the temperature difference between the sample (t_2) and the potassium reservoir (t_1) for each product. The t_1 was constant at 250°C for each preparation.

plateaus, mixed stages were observed in subsequent experiments carried out by Nixon and Parry.[30] However, these workers concluded that mixed stages existed only in samples that were microscopically inhomogeneous.

B.   Departures from Ideality

The foregoing descriptions of staging and stoichiometry cover the ideal or model situations. Unfortunately, many workers in GIC research take such descriptions too literally when in fact the samples they study in their experiments are considerably less than ideal. Moreover, the attendant nonideality can critically alter and render useless many attempts to interpret results using idealized theoretical models. For these reasons, it is important to separate reality from fiction, and I now show how this can be accomplished when one carefully examines the staging properties of GICs.

1.   Pure and Impure Stages

Since staging is so basic to GICs, three questions concerning the ramifications and limitations of this concept immediately come to mind: How is the stage determined? How pure is the stage of a given compound? How can we quantify the purity of the stage?

Historically, a phenomenological or "classical" model of staging[31-37] has evolved[31-37] and it has the following property. The characteristic c-axis repeat distance, d_c, of a stage n GIC is given by the formula

$$d_c = N[(n-1)d_1 + d_2], \qquad n = 1, 2, \dots \qquad (1)$$

Here N is an integer, d_1 is the distance between adjacent carbon planes, and d_2 is a package thickness equal to the spacing along the c-axis between the two carbon layers adjacent to any given intercalant layer. The key features of the classical model are that d_2 is assumed to be stage independent for a given species and that d_1 is assumed to be constant for all species and equal to 3.35 Å, the carbon—carbon interplanar distance in pristine graphite.[16] The features of the classical model are illustrated in Fig. 4. Notice that when the intercalant layers form a disordered two-dimensional structure, $N = 1$, since the repeat distance cannot be enhanced by stacking arrangements of the intercalant layers. In contrast, when the intercalant layers are two-dimensionally ordered, there will necessarily be three-dimensional, long-range order in the intercalant positions and the planar positions of the intercalant atoms will be correlated along the c-axis. In this case, various stacking sequences of intercalant layers are possible and they would correspond to values of $N \geq 1$.

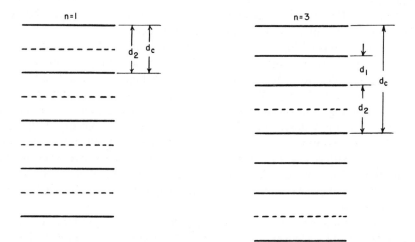

Fig. 4. Schematic illustration of the parameters of the classical model of staging employed in (1); d_1 is the carbon layer interlayer distance, d_2 is the package thickness, and d_c is the c-axis repeat distance. Examples given are for stages 1 and 3.

To the extent that (1) is accurate, one can use that equation in conjunction with a measurement of the $(00l)$ X-ray diffraction pattern of a GIC to establish its stage number. Note, however, that for GICs with ordered intercalants, the integer N and therefore the c-axis repeat distance cannot be deduced from the $(00l)$ reflections alone and can be determined only in measurements involving the h and/or k indicies. A typical series of $(00l)$ X-ray measurements for several stages of the graphite-FeCl$_3$ system is shown in Fig. 5,[38] which evidences the clear dependence of the diffraction pattern on stage. The patterns of Fig. 5 have been indexed[38] and the resulting distances used in (1) to determine the stage number n, assuming that $N = 1$ and that the FeCl$_3$ layers are uncorrelated.

If the GIC is an inhomogeneous mixture of macroscopic regions (≥ 500 Å) of several "pure" stages, the X-ray $(00l)$ reflections will be sharp (≈ 1–$2°$ for HOPG intercalation compounds with an MoKα X-ray source) but will reflect the macroscopic inhomogeneity. In that case two or more distinguishable sets of sharp X-ray reflections, each set of which can be separately indexed to a given c-axis repeat distance, will occur. If, however, the sample is an inhomogeneous microscopic mixture of stages in which stacking disorder manifests itself as a statistical spatial variation of stage number, the X-ray pattern will exhibit reflections corresponding to the average stage, but those reflections will be significantly broadened because of the stacking disorder. This latter point has been elegantly explored in a series of papers by Metz and co-workers,[39–41] who studied stacking disorder in a series of graphite-FeCl$_3$ intercalation compounds prepared from single crystal hosts. The effect of stacking disorder on the $(00l)$ X-ray reflections is well illustrated in Fig. 6, which compares $(00l)$ reflections from "stage" 2 FeCl$_3$ GICs exhibiting different degrees of stacking order.

The reflections of Fig. 6a are sharp and their half-widths are instrument limited. Therefore, Fig. 6a corresponds to a sample with a very high degree of stacking order. In contrast, the reflections in Fig. 6b are relatively broad and asymmetric, although they occur at approximately the same 2θ values as those of Fig. 6a. The pattern of Fig. 6b thus corresponds to a stage 2 sample with a low degree of stacking order. Note also from Fig. 5 that the width of the X-ray reflections increases with increasing stage number, even though all patterns of Fig. 5 were recorded under the same experimental conditions. This variation is to be expected, since the higher the stage, the larger the number of possible stacking arrangements and the wider the statistical spread.

The quantitative determination of the degree of stacking order in GICs has been described by Metz and co-workers,[39–41] who used well-established X-ray techniques and a standard theoretical approach. They defined a parameter

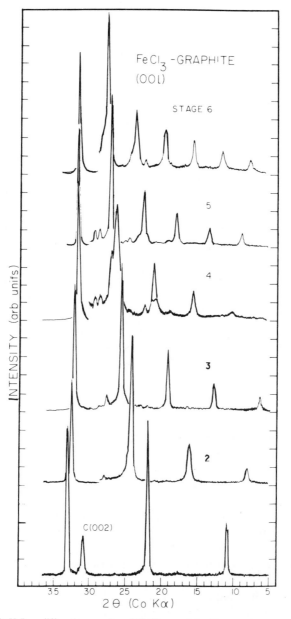

Fig. 5.   (00*l*) X-Ray diffraction spectra of FeCl₃-graphite. The spectra were recorded using CoKα radiation at room temperature from samples prepared with single-crystal graphite. From Refs. 38 and 13.

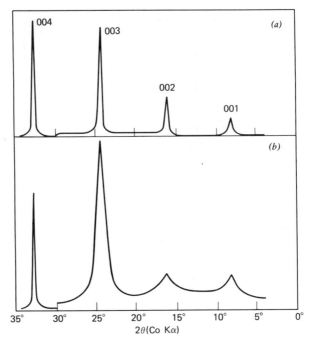

Fig. 6.   Room temperature (00l) X-ray diffraction spectra of (a) pure and (b) disordered stage
2 FeCl$_3$-graphite recorded with a CoKα source. From Ref. 39.

Q_n, which is a measure of the degree of stacking order. For a pure stage n compound $Q_n = 1$, while for a compound whose average composition is close to that of stage n but for which there is a purely random distribution of the number of carbon layers separating pairs of intercalant layers, $Q_n = 0$. By measuring the half-width of the X-ray (00l) profiles or and/or by fitting the observed (00l) diffraction pattern with a calculated pattern in which Q_n is an adjustable parameter, the degree of stacking order can be determined.

Thus to properly characterize an intercalation compound using (00l) X-ray diffraction, one must consider not only the number of reflections and their positions but also their profiles. However such a characterization, while necessary, may be far from sufficient because many experimental techniques that have been usefully applied to GICs involve probes that interact with the surface region of the sample only. For example, the depth of the interaction region in X-ray photoelectron spectroscopy (XPS)[42] or ultraviolet photoelectron spectroscopy (UPS)[43] measurements is of the order of 5 Å while for infrared reflection measurements[44] or light scattering[45] a depth of the order of 1000 Å is sampled.

It is known that the sample surface may differ drastically from the bulk in composition as well as other properties.[46] Therefore an X-ray diffraction study carried out with a "hard" source such as $MoK\alpha$ that penetrates deep into the sample (≈ 0.5 mm) may properly serve to characterize the bulk yet yield little information about the surface. But "hard" X-ray sources are most often employed to study GICs because such sources readily penetrate common encapsulating materials such as fused silica. Therefore, unless the optical probe is self-calibrating vis-à-vis sample characterization, it may be necessary to characterize samples with soft X-ray sources and/or other surface-sensitive probes such as the electron microprobe.[46]

2. Distorted Carbon Layers

Some of the properties of GICs will be significantly affected if the carbon layers are distorted rather than perfectly planar as indicated in the idealized models of Figs. 2 and 4. In fact, definitive evidence against planar carbon layers with macroscopic (≈ 1 mm) dimensions was presented many years ago by Daumus and Hérold.[47]

Consider the dynamics of the intercalation reaction. It is known that for many intercalants, particularly the heavy alkali metals, intercalation proceeds progressively by the successive formation of lower and lower stages.[30, 48] The equilibrium stage is obtained when a stoichiometry corresponding to the temperature difference ($t_1 - t_2$) (see Fig. 3) between the metal vapor and GIC is reached in a two-bulb experiment.[29] The desorption process also occurs by way of the successive formation of higher and higher stages.[30, 48] If the carbon layers were indeed flat, then to transform from an odd-stage number to an even-stage number, whole interlayer spaces containing intercalant must be entirely emptied and other interlayer spaces originally devoid of intercalant must be filled.

In an extremely clever experiment, Daumas and Hérold[49] heated stage 2 KC_{24} in a tube containing sufficient carbon monoxide vapor to entirely react with any potassium released from the GIC. They reasoned that if whole layers of intercalant are emptied during the stage transformation, the potassium vapor would react with the CO and a stage 3 KC_{36} compound would not form. Yet they produced a stage 3 compound without difficulty.

On the basis of their observations, Daumas and Hérold[49] proposed a GIC structure that is represented for a few stages in Fig. 7. The proposed structure has several intriguing properties. Every graphite interlayer space contains intercalant. The layer "kinks" exhibited in Fig. 7 have a lateral separation parallel to the layers that is several hundred angstroms. However in the region between kinks, the carbon layers are flat. Notice that the (00l) X-ray diffraction patterns from the structures shown in Fig. 7 would be es-

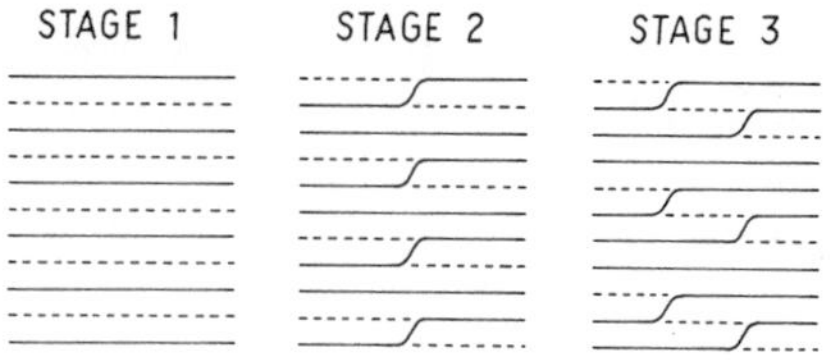

STAGE 1   STAGE 2   STAGE 3

Fig. 7. The Daumas-Hérold model of staging.[47] Carbon layers and intercalant layers are represented by solid and dashed lines, respectively. For $n > 1$ the domain sizes are macroscopic and may exceed 400 Å.

sentially indistinguishable from those of the corresponding structures exhibited in Fig. 2. Moreover, since every interlayer space contains intercalant, the staging transformation can result from lateral intercalant motion associated with kink propagation. This motion of the intercalant would be accompanied by the absorption or desorption of *some* intercalant according to whether the transformation is from higher to lower stage, or vice versa.

Recent X-ray experiments on stage 1 cesium-graphite[50] in which the melting of the cesium layers was shown to be closely coupled to the staging phenomena provided indirect microscopic evidence for the validity of the Daumas-Hérold model. These experiments are discussed in more detail in connection with structural phase transitions (Sections IV and V).

The most definitive evidence for the correctness of the Daumus-Hérold model[47] comes from recent electron microscope photographs recorded by Thomas and Millward,[51] who studied ferric chloride acceptor GICs. An example of their results appears in Fig. 8. The kinking of the graphite layers that is a key feature of the Daumas-Hérold model[47] is clearly evident in the remarkable photograph (i.e., Fig. 8). Also note that the localized stage of the

Fig. 8. High-resolution electron microscope image of $FeCl_3$-graphite exhibiting interpenetration of differently staged regions. From Ref. 51.

compound examined can be easily discerned. Moreover the carbon layer sequencing exhibits considerable variation indicating that the staging is impure.

Given the magnification used to record the photograph in Fig. 8 and the scale of that figure, slight distortions of the carbon layer in the regions between kinks would be difficult if not impossible to discern. However, extended X-ray absorption fine structure (EXAFS) measurements of potassium GICs have revealed that the carbon layers are distorted—perhaps puckered—in the neighborhood of the potassium intercalant ion.[52] Because these EXAFS measurements are novel to GICs and yield other important results indicating a departure from the classical model described by (1), I shall briefly summarize the EXAFS experiment here.

The EXAFS technique[53, 54] is now quite commonly employed to study the local environment (≈ 5 Å) of a specific atomic species in a disordered or ordered solid. The EXAFS is manifest as a series of oscillations $\Delta(k)$ (where k is the electron wave vector) in the cross-section for K-shell absorption, which are superposed on a smooth atomic background $\sigma_K^0(k)$. Thus,

$$\log(\text{absorptance})_{K\text{-shell}} = \sigma_K^0(k)[1+\Delta(k)] \tag{2}$$

An example of an EXAFS spectrum for stage 2 KC_{24} is shown in Fig. 9. The amplitude and phase of the oscillation $\Delta(k)$ are determined by the positions, types, and number of atoms in the neighborhood of the excited atom. The product of k and the oscillatory part of the absorption cross-section is shown in Fig. 10 for stage 2 KC_{24}. That product can be Fourier-transformed to yield the function $\Phi_\alpha(r)$ for a particular atomic species α, which in the case of KC_{24}, corresponds to the excited potassium atom.

$$\Phi_\alpha(r) = \int d^3k e^{i\mathbf{k}\cdot\mathbf{r}}[k\cdot\Delta(k)] \tag{3}$$

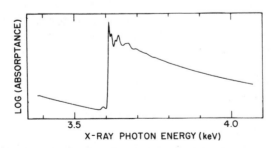

Fig. 9. The logarithm of the absorptance of a stage 2 intercalation compound of potassium-graphite as a function of X-ray photon energy including the onset of potassium K-shell absorption at 3.6 keV.

470   S. A. SOLIN

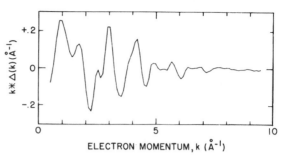

Fig. 10. The EXAFS oscillations $k*\Delta(k)$ extracted from the absorptance in Fig. 9 as a function of photoelectron momentum k.

But the function $\Phi_\alpha(r)$ can be computed theoretically and is given by the following expression, which yields a complex function of r:

$$\Phi_\alpha(r) = \sum_\beta \int_0^\infty \frac{dr'}{r'^2} P_{\alpha\beta}(r') \xi_{\alpha\beta}(r-r') \qquad (4)$$

where $r>0$ and $P_{\alpha\beta}$ is the radial distribution function of atom species β around the excited atom α. The peak function $\xi_{\alpha\beta}$, which is convolved with $P_{\alpha\beta}$ in (4), depends on the complex T-matrix of species β, the final state electron lifetime, and the phase shift due to the excited atom potential. Previous work[55] has shown that $\xi_{\alpha\beta}$ is insensitive to changes in crystal structure or local bonding. Therefore, it is common in EXAFS measurements to determine the peak function by examining a "standard" crystalline sample containing the species α and β separated by a known distance. Alternatively a theoretically calculated[56] peak function can be employed. Once $\xi_{\alpha\beta}$ is known, $P_{\alpha\beta}$ of the distorted sample can in principle be obtained.

There is considerable controversy about the validity of the procedures employed to obtain $\xi_{\alpha\beta}$ and thus about EXAFS results themselves.[57] To avoid that controversy, the EXAFS measurements of potassium GICs[52] described here were used to extract only relative distances referenced to stage 1 KC_8. The results were predicated on the valid assumption that only the $P_{\alpha\beta}(r)$ term in (4) is stage dependent.

Figure 11 shows the real parts (solid curves) and the magnitudes (dotted curves) of $\Phi_K^{(n)}(r)$ for stages 1 to 4 potassium-graphite at $\sim 80°K$. Here n is the stage number and the subscript K is the chemical symbol for potassium, the excited species.

Analysis of the entire structural region of the calculated functions $\Phi_K^{(n)}(r)$ indicates that the carbon layers do not remain flat on an atomic scale. Note that for 1.6 Å$<r<$5 Å there are three maxima in $|\Phi_K^{(n)}(r)|$ at 2.55, 3.42, and 4.14 Å (dotted curve Fig. 11a). These maxima are attributed to, but regularly displaced by 0.51 Å from the first, second, and possibly third nearest-carbon neighbor positions at 3.055, 3.935, and 4.65 Å, respectively. Those nearest-carbon-neighbor positions have been calculated by assuming that the potassium atoms sampled in the EXAFS experiment are located at sites over the centers of carbon hexagons. Note from Fig. 11 that in $\Phi_K^{(n)}(r)$ ($n = 2, 3, 4$), the second maximum at 3.42 Å is broadened and reduced in amplitude with respect to the corresponding feature in $\Phi_K^{(1)}(r)$, while the third maximum at 4.14 Å is not observed at all. Since the first-neighbor regions in $\Phi_K^{(n)}(r)$ are essentially stage independent, this broadening cannot be due to disorder in the potassium positions. Consequently, in stages 2, 3, and 4, the potassium atom's carbon neighbors more distant than the first occupy shells of increased width. Since one expects similar and only slight thermal effects in all stages at 80°K, this additional width is not vibrational in origin but indicates static disorder corresponding to carbon layer distortions that can be shown to be $\lesssim 0.2$ Å. The most likely form of distortion would be carbon layer puckering in the vicinity of K atoms.

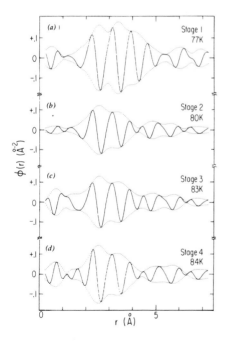

Fig. 11. The real part (solid curve) and the magnitude (dotted curve) of the Fourier transform of the EXAFS on the potassium K-shell absorption in potassium-graphite at $\approx$80°K. (a), (b), (c), and (d); stages 1, 2, 3, and 4, respectively. The data were transformed using a square window with k between 2.75 and 8.15 Å^{-1}, broadened by a Gaussian of width $\sigma_k = 0.5$ Å^{-1}.

3. Defects in the Classical Model of Staging

By assuming a Gaussian profile for $P_{\alpha\beta}(r)$ in the region of the first-neighbor carbon atoms surrounding the potassium atom, it is possible to fit the experimental value of $\Phi_\alpha(r)$ given by (3) with the theoretical expression by varying the Gaussian parameters of $P_{\alpha\beta}$. Such a procedure has been followed by Caswell et al,[52] and Table I lists their results for the stage dependence of d_{K-C}, the nearest-neighbor potassium—carbon distance, $T = 300°K$. If the carbon layers were flat, the package thickness d_2 (see Fig. 2) would be related to d_{K-C} by

$$d_2 = 2\left[d_{KC}^2 - d_{CC}^2\right]^{1/2} \tag{5}$$

where d_{CC} is the carbon—carbon intralayer nearest-neighbor distance. The values of d_2, deduced from (5) and the known stage dependence of d_{CC}[58] are also listed in Table I.

It is clear from Table I that d_2 is not stage independent, as has been assumed in the classical model of staging. The values of d_2 for stages, 2, 3, and 4 are identical to within experimental error but are significantly larger than that of stage 1. The EXAFS results for potassium provide the first evidence that the classical model is insufficient to account for staging in donor GICs. However, it has been known from X-ray diffraction studies that that model was not adequate for some acceptor GICs.[59]

In addition to the direct EXAFS evidence for carbon layer distortions, other more recent results provide confirmation that the carbon layers in potassium GICs are not truly planar. Moss and co-workers[60] have studied the temperature dependence of the X-ray diffraction from stage 2 potassium-graphite and have observed both ordered and disordered potassium layers. To explain the complex X-ray patterns of the ordered structure, they have

TABLE I

Nearest-Neighbor Potassium—Carbon Distances, d_{K-C}, and Carbon-Potassium-Carbon Package Thickness, d_2, as a Function of Stage for Potassium GICs at $T = 300°K$[a]

Stage	d_{K-C} (Å)	d_2 (Å)
1	3.055 ± 0.02	5.40 ± 0.02
2	3.138 ± 0.02	5.59 ± 0.02
3	3.144 ± 0.02	5.61 ± 0.02
4	3.140 ± 0.02	5.60 ± 0.02

[a] From Ref. 52.

proposed a model in which the ordering of the potassium layers is accompanied by a static distortion of the carbon layers. Leung et al.[61] have carried out careful studies of the (00l) X-ray intensities for various stages of potassium, rubidium, cosium, and bromine GICs. From a comparison of the observed (00l) intensities with those calculated as a function of the in-plane densities and c-axis charge distribution, they also concluded that the package thickness d_2 was not stage independent.

As an additional comment on intralayer distortions of the carbon planes, I note that Nixon and Parry[58] have made very careful X-ray measurements of the dependence of the in-plane carbon—carbon nearest-neighbor bond distance on stage number of potassium GICs. They found that D_{C-C} increased measurably with decreasing stage, as indicated in Fig. 12. The stage dependence of d_{CC} shown in Fig. 12 was deduced with the assumption that

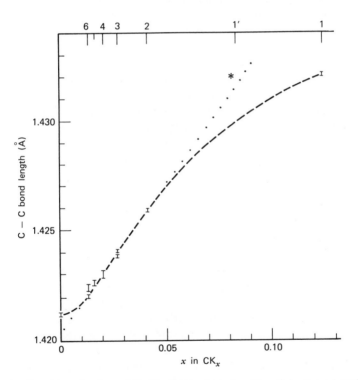

Fig. 12. Experimental values of the C—C intraplanar bond length in stages 1 to 6 potassium GICs as a function of the ideal composition KC. The indicated limits of error are $\pm 2\sigma$. The dashed curve indicates the sigmoidal dependence of bond length on x and has no other significance. The dashed curve represents the equation $d_x = (1.4203 + 0.1356x)$Å. The asterisk indicates the experimental value of d_x for stage 1 plotted against $x = \frac{1}{12}$ (stage 1') instead of $\frac{1}{8}$. From Ref. 18.

the carbon layers remained flat and merely expanded uniformly in two dimensions with decreasing stage. Nevertheless, it is likely that the layer distortions alluded to above are also stage dependent and that, therefore, the minimum value of d_{CC} will depend on stage.

Evidence has also been put forth that there is an increase in d_{CC} with decreasing stage in the case of acceptor compounds,[62] in particular $FeCl_3$ GICs. The changes in d_{CC} with stage will, of course, markedly affect many of the physical properties of GICs, particularly the intraplanar vibrations of the carbon layers.

4. Variable Stoichiometries

It has been almost an article of faith in GIC research that the layer stoichiometries in GICs were well defined with integer ratios for the number of carbon atoms per intercalant atom or molecule. This view permeates the interpretation of experimental results and is especially prevalent when applied to the widely studied alkali metal GICs. Thus, when bulk compositions intermediate between those corresponding to integer stage are observed, they are attributed to macroscopic or microscopic inhomogeneities. For example, Hérold's isobar measurements of weight uptake shown in Fig. 3 were compared with X-ray measurements by Nixon and Parry.[30] The X-ray results are indicated by the solid dots in Fig. 3. Vertical lines connecting the dots indicate mixed stages of macroscopic size; that is, inhomogeneities of the order of 500 Å or more exist in the sample and give rise to multiple sets of separately indexable X-ray reflections. Such mixed stages were also detected much earlier by Rudorff and co-workers.[63] Note, however, from Fig. 3 that it is extremely difficult to distinguish higher stages ($n > 3$) in the isobar because the plateau regions become vanishingly small.

As another example of the commitment to discrete layer stoichiometries in the case of microscopic inhomogeneities, consider again the work of Metz and co-workers discussed previously.[39-41] To account for the breadth of the $(00l)$ X-ray diffraction peaks of $FeCl_3$ GICs, they invoked a model in which packages containing an $FeCl_3$ layer together with one or more carbon layers were stacked regularly and/or irregularly along the c-axis. However, the layer stoichiometry was assumed to be the same for all packages and is fixed at six carbon atoms per $FeCl_3$ formula unit (i.e., C_6FeCl_3). Like Hérold,[29] Parry,[30] and others, Metz et al. base their analysis on a comparison of the measured weight uptake of a given GIC with its X-ray diffraction pattern. But weight uptake provides a very inaccurate measurement of the actual bulk stoichiometry and virtually no information on the layer stoichiometry. Weight uptake measurements are prone to error because they do not distinguish between truly intercalated material and material that is absorbed, pinned to defect sites in the graphite host, coalesced in microcracks, and so on. Such material may be present in quantities significant enough to affect the weight uptake measurements yet, as a result of small particle size and/or structural disorder, may defy detection in the usual Bragg diffraction discrete X-ray measurements. Other indirect methods for monitoring staging and stoichiometries such as measurements of sample dilation and electrical resistivity suffer from the above-mentioned deficiencies.

Fortunately, a clever microscopic method for measuring the concentration of intercalant or equivalently the bulk stoichiometry of a GIC has been developed by Caswell.[64] We sketch this procedure in the following discussion.

Suppose one has access to a GIC prepared from intercalant species S with a known bulk stoichiometry. The saturated stage 1 compound can usually serve this purpose. Let μ_S^0 be the linear X-ray absorption coefficient of the intercalant atoms or molecules at the wavelength of the X-ray photons used for diffraction studies. Also, we define μ_S to be the linear absorption coefficient for a sample whose concentration of S is less than that of the stage 1 standard sample. The concentration of the latter relative to that of the former is given by

$$C \equiv \frac{\mu_S}{\mu_S^0}, \qquad 0 \leq C \leq 1 \tag{6}$$

and $C = 0$ corresponds to pristine graphite. If ϕ is the angle between the c-axis of the GIC (prepared from HOPG) and the incident X-ray beam, then the transmitted X-ray intensity $I(C, \phi)$ corresponding to a source intensity I_S is given by

$$I(C, \phi) = I_S \exp\left\{-\left(\mu_S^0 C + \mu_g\right)(\sec \phi)t - \alpha\right\}. \tag{7}$$

Here μ_g is the linear absorption coefficient of the graphite, t is the sample thickness, and α contains factors corresponding to attenuation by other absorbers including the glass encapsulating container and air in the beam path. By measuring $I(1,0)$ it is possible to obtain a value for $(\mu_S^0 + \mu_g)t$ using the expression

$$\frac{I(1,\phi)}{I(1,0)} = \exp\left\{-\left(\mu_S^0 + \mu_g\right)t\left[\sec \phi - 1\right]\right\} \tag{8}$$

Since the mass absorption coefficients of the intercalant and of carbon can be obtained from tabulations of such data for all elements,[65] the quantities μ_S^0 and μ_g can also be deduced for the standard sample (usually stage 1). Thus

476 S. A. SOLIN

the value of t can be obtained from (8). The resultant t represents an optical thickness of the sample and is much more accurate than macroscopic values measured mechanically or with a telescope. For instance, Caswell[64] measured the optical thickness of a stage 1 cesium GIC to be 0.32 ± 0.005 mm. The corresponding thickness measured mechanically was 0.30 ± 0.05 mm and was considerably less accurate.

The concentration of intercalant in any GIC containing the same species S as the standard sample can be determined by measuring the X-ray transmission of that sample at $\phi=0$. From (8) one finds

$$C=1-\frac{1}{\mu_S^0 t}\ln\left(\frac{I(C,0)}{I(1,0)}\right) \qquad (9)$$

With the method described above, an accuracy in C of $\lesssim 4\%$ has been achieved.

Caswell has used the X-ray technique to study the concentration dependence of the inter- and intraplanar structure of stage 1 cesium-graphite.[64] Some of the results of this study are shown in Fig. 13 as solid dots. Caswell's results are relatively unique[66] because they involve *in situ* X-ray diffraction measurements with the GIC in equilibrium with the vapor of the intercalant species (in this case at a pressure P_{Cs} for Cs vapor). In contrast, most studies of GICs are carried out on metastable nonequilibrium samples that are

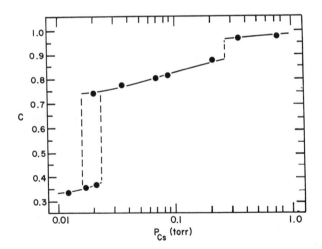

Fig. 13. Equilibrium cesium concentration for $T_S=748°K$ as a function of P_{Cs} [see text and (17)]. The solid circles are from Ref. 64 and the solid curve is from Ref. 67.

quenched at room temperature from the much higher temperature of preparation.

The most significant feature about the data of Fig. 13 is the continuous variation of concentration above $C \cong 0.7$. Each of the data points in this range corresponds to a *homogeneous* stage 1 compound as determined from ($00l$) X-ray diffraction measurements. Moreover the X-ray patterns indicated that the equilibrium stage 1 compound was relatively pure and exhibited c-axis correlations. However, ($h0l$) reflections also exhibited a considerable degree of stacking faults.

The discontinuity at $P_{Cs} = 0.27$ torr is associated with in-plane melting of the cesium layer and is discussed in more detail in Section IV.A.1. The discontinuity at $P_{Cs} = 0.27$ torr corresponds to a change in stage from stage 1 to stage 2 and exhibits considerable hysteresis. Notice also that within the concentration range of the stage 2 compound, the concentration is continuously variable. That concentration range also exhibits macroscopic regions of homogeneous stage 2 structure.

It is intriguing to note that the solid line of Fig. 13 is the data of Salzano and Aronson (SA),[67] whose measurements predate Caswell's[64] by some 16 years. They measured the concentration dependence of heavy alkali GICs using both the Knudsen effusion technique[67] and radioactive tracer techniques.[68] However SA, who lacked X-ray structural data, interpreted the continuous variation in C to be due to mixed macroscopic stages, although they did claim to observe a metastable MC_{10} (M = K, Rb, Cs phase). It is now clear of course that there is nothing magic about MC_{10}, since the concentration can vary continuously over a given range for each stage.

Other evidence for a continuous variation in layer and bulk stoichiometry with homogeneous staging has been presented by Underhill et al.[69] They pointed out that the structure factor for the ($00l$) X-ray diffraction pattern contains information on the in-plane charge density of the intercalant layers and, therefore, on the layer stoichiometry. By measuring the structure factor carefully and comparing their results with calculations, they deduced that stage 1 alkali metal GICs could possess a nonideal bulk stoichiometry such as $MC_{9.6}$. However, they also reported observation of compounds with increased density (e.g., $MC_{6.8}$ for the heavy alkali metal GICs). Such high densities are inconsistent with the ionic size for rubidium and cesium and with the available planar structures. The reason for this inconsistency may rest with the difficulty of making accurate integrated intensity measurements of the ($00l$) reflections.

Notwithstanding the historical bias toward ideal stoichiometries, recent theoretical results due to Safran[70, 71] predict the existence of homogeneously staged nonideal compounds and thus confirm the above-described experimental results. Safran focused on a phenomenological model, the primary emphasis of which is staging and associated stage-to-stage phase transitions. He argued that intralayer structural phase transitions involved energy differences that were small compared to the differences associated with staging and could, therefore, be ignored in the first approximation. Thus he adopted an inplane continuum model based on the following Hamiltonian

$$H = \int d\mathbf{r}\, \tilde{H}(\mathbf{r}) \tag{10}$$

where

$$\tilde{H}(\mathbf{r}) = -\mu \sum_i \sigma_i(\mathbf{r}) + \frac{1}{2} \sum_{ij} \int d\mathbf{r}\, W_{ij}(\mathbf{r},\mathbf{r}')\sigma_i(\mathbf{r})\sigma_j(r') \tag{11}$$

and

$$W_{ij}(\mathbf{r},\mathbf{r}') = V_{ij}(\mathbf{r},\mathbf{r}') - U(\mathbf{r},\mathbf{r}')\delta_{i,j} \tag{12}$$

In (10) to (12), i is a layer index corresponding to the interspace between two graphite layers, $\mathbf{r}$ is a vector in the plane perpendicular to the c-axis, $V(\mathbf{r},\mathbf{r}')$ is the intraplanar interaction, μ is the chemical potential, and $\sigma_i(\mathbf{r})$ is the local planar intercalant density normalized to that of a saturated layer. Thus, $0 \leq \sigma_i \leq 1$, and the upper limit reflects the assumption of hard-core repulsion between intercalant atoms or molecules.

When treated self-consistently in the mean field approximation, (10) to (12) yield

$$\sigma_i(r) = \left\{ \exp\left[\frac{1}{k_\beta T} \left(-\mu + \sum_j \int d\mathbf{r}\, W_{ij}(\mathbf{r},\mathbf{r}')\sigma_j(\mathbf{r}) \right) \right] + 1 \right\}^{-1} \tag{13}$$

The model above was further simplified by characterizing each plane by its average density σ_i and parameterizing the interactions by the planar averages of $U(\mathbf{r},\mathbf{r}')$ and $V_{ij}(\mathbf{r},\mathbf{r}')$, namely, U_0 and V_{ij}, respectively.

The staging phase diagram has been calculated for a power law behavior of V_{ij} given by

$$V_{ij} = \tfrac{1}{4} V_0 |Z_{ij}|^{-4} \tag{14}$$

where Z_{ij} is $d_{ij}/3.35$ Å and d_{ij} is the distance along the c-axis between intercalant layers i and j. The results are shown in Fig. 14, in which the ab-

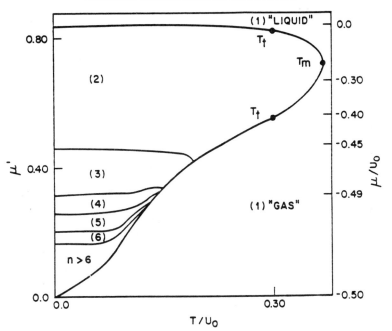

Fig. 14. Staging phase diagram for $V = \frac{1}{2} V_0$ and $\alpha = 4$ (see text) as a function of temperature T and chemical potential μ, both normalized to V_0. For clarity, the phase diagram is given as a *linear* function of $\mu' = (\mu/V_0 + \frac{1}{2})^{1/\alpha}$ leading to a nonlinear μ scale on the right-hand side of the figure. The integers in parentheses correspond to the stable, pure stage phases; T_m and T_t are the maximum and tricritical temperatures, respectively. From Ref. 71.

scissa is given in temperature units scaled to U_0 and

$$\mu' = \left(\frac{\mu}{U_0} + \frac{1}{2} \right)^{1/4} \tag{15}$$

linearizes the ordinate scale, which is nonlinear in μ. In Fig. 14, T_m is the maximum temperature at which a stage 2 compound could exist, while T_t is the temperature above which the transition from stage 1 to stage 2 is continuous. Transitions between all stages with $n > 2$ are discontinuous (i.e., first order). "Gas" and "liquid" refer to *homogeneous* stage 1 compounds of low and high intercalant concentration, respectively.

The key feature of Fig. 14 relevant to the present discussion is the existence for fixed T/U_0 of a range of chemical potentials, hence concentrations (or equivalently stoichiometries) that yield pure homogeneously staged

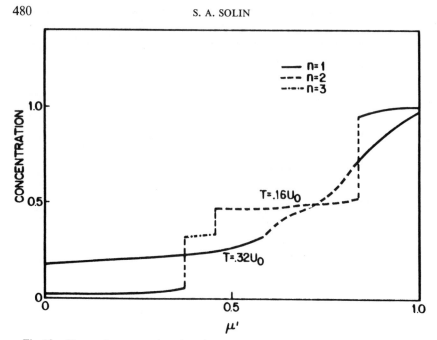

Fig. 15.   Stage and concentration plotted as function of chemical potential μ' (see legend of Fig. 14) at $T=0.16U_0$ and $T=0.32U_0$, where $T>T_t$, and transition from stage 2 to stage 1 is continuous. From Ref. 71.

compounds. For instance, at sufficiently high T/U_0 all stages should transform to a homogeneous stage 1 "gas" phase. These points are well illustrated in Fig. 15, which shows the stage and concentration dependence as a function of chemical potential for temperatures above and below T_t. The qualitative behavior exhibited in Fig. 15 is very similar to that observed by Caswell[64] for a stage 1 cesium GIC as noted in Fig. 13. In particular, it is clear that even when $T<T_m$ and the stage-to-stage phase transition is first order, there is within each stage a range of stoichiometries consistent with pure staging.

III.   SAMPLE PREPARATION AND INTERCALATION CHEMISTRY

For completeness I must comment on the chemical techniques and methods for preparing GICs. However, this subject warrants a complete treatise of its own, as evidenced by the excellent review articles[72, 73] addressing it. The reader is referred to those articles for detailed information. Since intercalation chemistry is peripheral to the focus of this chapter, I merely categorize below the general and most common methods used to prepare GICs.

A. Intercalation Techniques

1. The Hérold Method

In the Hérold method[29] the intercalant is loaded into a tube containing pristine graphite and is sealed under vacuum. The intercalant and the graphite are spatially separated and the former is heated to a temperature t_1 which, under proper conditions, produces a saturated vapor of intercalant at the graphite sample. The graphite is held at a temperature $t_2 > t_1$ and reacts with the intercalant vapor to form a GIC, the stage and/or stoichiometry of which depend on the temperature difference $t_2 - t_1$ and time of reaction. The Hérold method has been used successfully for both donors and acceptors. It is generally limited to use with relatively volatile intercalants.

2. The Solution-Reaction Method

In the solution reaction method of preparation the intercalant is dissolved (usually at room temperature) into a solvent.[74] Pristine graphite is then immersed into the solution and reacts selectively with the solute. Bromine GICs have been successfully prepared using this method with carbon tetrachloride as the solvent.[74]

3. Direct Liquid Immersion

Materials of low volatility and/or those that react with graphite to form compounds other than GICs at temperatures corresponding to very low intercalant vapor pressures can be prepared to direct immersion of the graphite host into the liquid intercalant. Thus lithium, which at 400°C forms LiC with graphite[75] and has a relatively low vapor pressure at that temperature, can be readily intercalated by directly immersing the graphite into pure liquid lithium or into a liquid lithium-sodium alloy.[77]

4. Electrolytic Reaction

Graphite serves as the anode or cathode constituent of an electrolytic cell, and the electrolyte contains the desired intercalant. Acid salts such as sulfuric acid GICs have been prepared by this method.[78] The primary limitation of the electrolytic reaction technique is related to the availability of a suitable electrolyte.

5. Other Techniques

Several unusual methods have been used to prepare GICs. For instance, ternary alkali GICs containing mercury have been prepared by heating pressed compacted constituents prepared from powdered graphite.[79, 80] Compounds have also been produced by chemical oxidation or reduction of other binary GICs. Thus $FeCl_2$ GICs can be prepared by reacting $FeCl_3$ with hydrogen.[81]

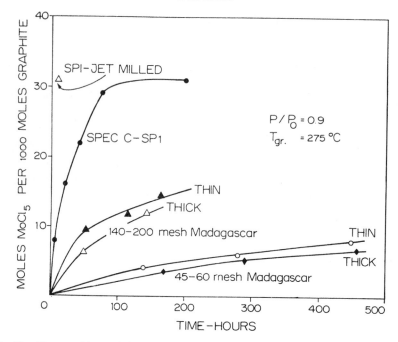

Fig. 16.   The rate of intercalation of various sizes of graphite by $MoCl_5$ vapor near saturation.
From Ref. 83.

B. Dynamics of Intercalation

The rate at which a given species of intercalant enters a graphite host depends on several factors.[82] These include the size, shape, and degree of graphitization of the graphite particles, as well as the mass and the packing fraction in the case of graphite powders. Naturally, the rate of intercalation is significantly affected by the vapor pressure of the intercalant and the temperature of the GIC when the Hérold method is used. Dramatic evidence of the dependence of the intercalation rate on crystal size is shown in Fig. 16, which shows data reported by Hooley[83] for the $MoCl_5$ intercalant. To date no quantitative models exist that can accurately account for the data of Fig. 16.

IV. STRUCTURAL PHASE TRANSITIONS IN DONOR GRAPHITE INTERCALATION COMPOUNDS

One of the most active areas of GIC research during the past several years has been the research on structural phase transitions. Most of the experimental attention has been focused on the heavy alkali metal intercalants potassium, rubidium, and cesium, perhaps because these species are known to form well-staged, relatively reproducible compounds. Though K, Rb, and Cs GICs have similar, if not identical, three-dimensional stage 1 structures, their behavior differs significantly vis-à-vis the structural phase transitions they appear to exhibit. The reader should note, however, that the phase transitions in alkali GICs are the subject of a great deal of controversy, and a coherent picture of the species dependence of such transitions is still lacking. Nevertheless, it is appropriate to review recent results, which I do by considering first temperature-induced transitions, then pressure-induced transitions in the alkali donor GICs. Section V presents a corresponding, yet more abbreviated discussion for acceptor intercalants.

A. Temperature-Induced Transitions

1. Stage 1 Alkali GICs

a. **Structures Under Ambient Conditions.** The room temperature, ambient pressure metastable structures of the heavy alkali GICs are very closely related and are discussed as a class. Figure 17 represents the structure of a planar graphitic layer. The carbon atoms lie on the hexagon corners and are threefold coordinated to form a hexagonal lattice. The layer primitive cell (dashed line in Fig. 17) contains two carbon atoms. In the 1930s Schleede and Wellmann[3] studied the structures of stage 1 GICs that have the idealized saturated stoichiometry of MC_8 (M = K, Rb, Cs). Subsequently Rudorff and co-workers proposed a structure in which the intercalate ions form ordered layers that are commensurate with the graphite layers and, with reference to the carbon layers, form a $(2 \times 2)R0°$ superlattice.[86] Thus, in the

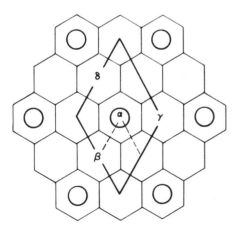

Fig. 17. *c*-Axis planar projection of the MC_8 commensurate structure. The corners of the hexagons represent the locations of carbon atoms, and the solid circles represent the locations of M atoms. The $(2 \times 2)R0°$ planar superlattice cell is depicted by the heavy lines. The two-atom planar cell of the graphite layer is depicted by the dashed lines and their solid line symmetric counterparts. The four equivalent hexagon center sites of the $(2 \times 2)R0°$ cell are labeled α, β, γ, and δ.

Rudorff structure, the M ions are located over the centers of carbon hexagons halfway between the adjacent carbon planes. This structure is one in which first and second nearest-neighbor hexagon sites are excluded, presumably as a result of the hard-core repulsion and ionic size.

The Rudorff structure accommodates four sets of specific sites in the expanded primitive cell (heavy lines in Fig. 17); these are labeled α, β, γ, and δ. Parry and co-workers[30, 87] confirmed the Rudorff structures using X-ray diffraction studies of stage 1 alkali GICs. They showed that for potassium[30] and rubidium,[87] the successive intercalant layers occupy α-, β-, γ-, and δ-sites, giving rise to a three-dimensional superlattice with a four-layer repeat distance. In contrast, CsC_8 exhibits[88] a three-layer $\alpha\beta\gamma$ repeat sequence in which the unoccupied δ-sites form tunnels along the c-axis. Notice that the $(2 \times 2)R0°$ planar superlattice produces the ideal MC_8 bulk and layer stoichiometries. The former is a consequence of the stage 1 character independent of the particular stacking sequence vis-à-vis $\alpha\beta\gamma\delta$ or $\alpha\beta\gamma$ arrangements.

In more recent X-ray single-crystal studies, the structures of the MC_8 compounds have been refined by Guérard and co-workers.[89–91] Their results, like Parry's,[30, 87] confirm the in-plane MC_8 structure, but they also find more complex stacking arrangements that preserve the $\alpha\beta\gamma\delta$ patterns. For instance, whereas Parry reports the space group of KC_8[30] to be $Fdd2$, Lagrange et al.[89] find that KC_8 is a mixture of three different crystalline phases and is orthorhombic with space group $Fddd$. The three $Fddd$ unit cells are built up from an ideal graphite plane but are rotated one from the other by 120° about the c-axis. This rotation gives rise to the apparent hexagonal symmetry of the ordered three-dimensional structures. It is important to note that such a structure is quite prone to stacking faults associated both with $\alpha\beta\gamma\delta$ sequence alterations and positive and negative 120° c-axis rotations. The energy differences associated with such stacking faults can be expected to be quite small.

b. Order-Disorder Transitions. Order-disorder phase transitions have been observed in some stage 1 heavy alkali GICs. Consider the case of CsC_8.[92] When this compound is heated in the presence of cesium vapor, the Cs layers undergo an order-disorder transition, the temperature and properties of which depend on the vapor pressure of the cesium surrounding the GIC.[64, 92] Naturally, the destruction of in-plane order is accompanied by a reduction in the correlation between the positions of Cs atoms in different intercalant layers. Also note that the cesium layers melt; that is, they disorder at a temperature much lower than the temperature at which the graphite layers themselves disorder.

Figure 18 shows the in-plane discrete X-ray diffraction ($\mathbf{k}_s - \mathbf{k}_i = \mathbf{q} \perp c$) from three-dimensionally ordered CsC_8 (curve c) and the in-plane diffuse

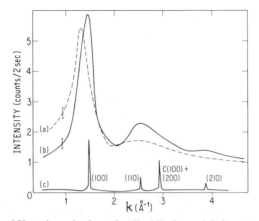

Fig. 18. Scattered X-ray intensity from the "liquid" phase of CsC_8 at $T=700°$K (curve a),
the lattice-gas disordered phase of $Cs_{1-x}C_8(x\approx0.2)$ at $T=700°$K (curve b), and CsC_8 at $300°$K
(curve c). The Bragg peaks in curve c are indexed on the $(2\times2)R0°$ structure shown in Fig. 17.
In curves a and b the statistical fluctuations (vertical error bars) have been averaged out and
the pure graphite (100) Bragg peak at 2.9 $\overset{\circ}{A}^{-1}$ has been subtracted for clarity.

scattering that occurs when CsC_8 is heated to $700°$K with and without Cs
vapor from excess cesium metal present in the sample tube (curves a and b,
respectively).[92] The in-plane room temperature pattern of Fig. 18 is
confirmation of and can be indexed on the MC_8 in-plane ordered structure
proposed by Rudorff[24, 31] and observed by others. The diffuse scattering
patterns are explained with reference to Fig. 17.

When excess cesium vapor is present in the sample tube, the concentration of the cesium intercalant layers will be very close to the saturation value
corresponding to a CsC_8 stoichiometry. In this case the Cs atoms cannot
disorder by hopping from site to site because there are no such available sites,
even for correlated hopping. Thus disorder develops by way of deregistry of
the cesium atoms, which move from hexagon center sites and acquire an
average interparticle distance that increases from 4.92 to 5.78 Å, the latter
being determined from application of the Ehrenfest relationship[93] to curve a
of Fig. 18. The shift to lower q of the first peak in curve a of Fig. 18 relative
to the (100) peak in curve c is consistent with the expansion of the Cs—Cs
intralayer distance upon melting. In addition, the diffraction pattern of curve
a is consistent with a liquidlike disorder of the cesium layers.

When no excess cesium is present during the heating process, cesium
is expelled from the GIC and is irreversibly absorbed in the walls of the
sample tube. In this case the first melting occurs at $608°$K, the same temperature at which melting with excess cesium present in the sample tube occurred. However, on subsequent cooling and reheating, melting of the cesium

layers occurs at considerably lower temperature, ($\approx 500°$K). In this case, the melting of the cesium layers corresponds to a lattice-gas type of disorder.[94] The vacancies created by the expulsion of cesium allow the remaining cesium ions to hop among hexagon sites, yet remain locally registered. The corresponding diffraction pattern of curve b in Fig. 18 yields, through the Ehrenfest relation, an average interparticle distance of 4.92 Å, which corresponds to registry, and the first peak in the diffraction pattern of curve b is coincident in q with the (100) peak of the ordered phase. Also, the higher degree of positional correlation associated with lattice-gas disorder is apparent in the enhanced oscillations present at $q > 2$ Å^{-1} in curve b relative to curve a.

Curve b in Fig. 18 was obtained from a sample with approximate stoichiometry $CsC_{9.6}$ corresponding to about 20% vacancies in the saturated CsC_8 compound. Note however, that at $T = 700°$K, well above the melting point of the Cs layers, the (00l) diffraction patterns indicate that the sample is a *homogeneous* stage 1 compound in excellent agreement with the model proposed by Safran[70, 71] and discussed in Section II.B.4. When the sample temperature is reduced below the melting temperature, macroscopic stage 2 regions develop and coexist with stage 1 regions in which the local stoichiometry is C_8Cs with the attendant $(2 \times 2)R0°$ superlattice structure.[92] The transition from homogeneous stage 1 to mixed stage can be easily understood in terms of the Daumas-Hérold model[47] of staging (Section II.B.2). Since, in that model, each carbon interlayer space contains intercalant, clusters of stage 1 regions can form lateral islands contiguous to stage 2 regions. On melting, the atoms in such islands migrate to uniformly fill each interlayer space (at a concentration less than that of CsC_8) so that a homogeneous stage 1 GIC results.

The measurements discussed above were refined by Caswell, who studied *in situ* the in-plane and c-axis structures of stage 1 cesium-graphite as a function of sample temperature and cesium vapor pressure.[64] Some of his results were discussed in Section II.B.4. Others are shown in Fig. 13, which corresponds to a sample temperature of 748°K. Recall that the discontinuity in concentration at $P_{Cs} = 0.27$ torr is associated with the simultaneous melting of the cesium layer and ejection of cesium atoms from the GIC. This ejection occurs when the temperature-dependent vapor pressure of the GIC, $P_{CsC_8}(T)$, exceeds P_{Cs}, the pressure of the surrounding cesium vapor, which is in turn determined by the temperature of the excess cesium metal in the sample tube. Sample decomposition occurs by way of the reaction

$$\frac{x}{x-8} CsC_8 \rightleftharpoons \frac{8}{x-8} CsC_x (gas) \qquad (16)$$

Moreover, Caswell showed that the temperature T_S at which the cesium layers in stage 1 CsC_x melt is related to the vapor pressure of the intercalate by the equation

$$-\ln\left[\frac{P_{CsC_8}(T_S)}{P_0}\right] = \frac{\Delta H}{RT_S} - \frac{\Delta S}{R} \qquad (17)$$

Here P_0 is a standard pressure (chosen to be 1 atm, ΔH and ΔS are the enthalpy and entropy of reaction for (16) and have been measured by Salzano and Aranson[95] to be 43.7 kcal/mole and 43.4 cal/(mole) (°K), respectively, and R is the gas constant. A plot of $\ln P_{CsC_8}(T_S)$ versus $1/T_S$ from (17) is shown as a solid line in Fig. 19. Also plotted in Fig. 19 is the measured dependence of $\ln P_{Cs}$ on $1/T_S$. Those data were obtained by fixing P_{Cs} and monitoring, with in-plane X-ray diffraction, the melting temperature T_S, as the temperature of the GIC was varied. This procedure was then repeated for several values of P_{Cs} as indicated in Fig. 19. The agreement between the calculated values of $P_{C_8Cs}(T_S)$ using (17) and the measured values of P_{Cs} is excellent and confirms the above-described analysis.

To date, there have been no thorough studies of structural alterations in KC_8 at elevated temperatures. Preliminary neutron diffraction studies have been reported for RbC_8 by Ellenson et al.[96] In those studies, no excess

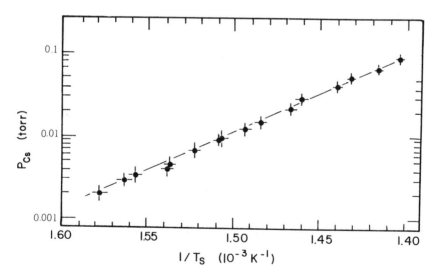

Fig. 19.   Inverse temperature ($1/T_S$) of the structural transition in CsC_8 as a function of P_{Cs} (see legend of Fig. 13). Data for solid circles from Ref. 64. The curve is the equilibrium vapor pressure of CsC_8 from Ref. 68.

rubidium metal was retained in the sample tube. However, the free volume not occupied by the sample was negligibly small. Therefore the equilibrium rubidium vapor pressure at a given temperature was approximately that of the sample itself.

The reciprocal lattice diagrams of RbC_8 measured at several temperatures by Ellenson et al.[96] are shown in Fig. 20. At room temperature (290°K) the patterns exhibit a high degree of in-plane order as evidenced by a set of sharp reflections for $q\|a^*$. However, the c^* streaks in the 290°K pattern, which were attributed to a "mixture of structures or a disordered state,"[96] are, in my opinion, most likely to be associated with stacking faults (see discussion below of stage 2 MC_{24} alkali GICs). The 290°K pattern at the bottom of Fig. 20 is that of the as-grown sample that, given its size (≈ 1 cm^3), may not

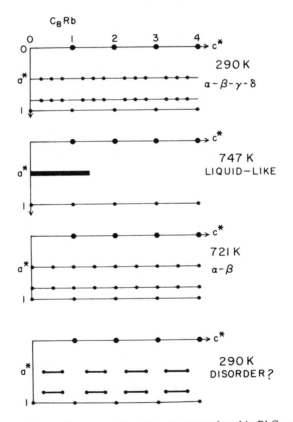

Fig. 20. Reciprocal lattice diagrams of the various structures found in RbC_8 as a function of temperature. The time sequence reads from bottom to top. From Ref. 96.

have achieved structural equilibrium at the time intercalation was terminated. Upon heating to $\approx 721°K$, a well-defined pattern indicative of an $\alpha\beta\gamma\delta$ intercalant layer stacking sequence develops. But this sequence may be a metastable and/or nonequilibrium structure of the type observed by Caswell.[64] Caswell monitored the onset and disappearance of several stacking sequences in CsC_8 including $\alpha\beta$, $\alpha\beta\gamma$, and $\alpha\beta\gamma\delta$ as equilibrium was approached.

At a temperature of $747°K$, c-axis correlations between the rubidium layers disappear and a "liquidlike" c^* scan at $0 < a^* < 1$ is observed. This pattern, if taken literally, indicates a degree of in-plane correlation of the Rb atoms that is much too high to be consistent with a flat c^* scan, that is, the c^* streak in the $747°K$ pattern is too narrow in the a^* direction. However, since no diffractometer c^* scans of the diffuse scattering at $747°K$ are presented, it is very difficult to pin down the actual Rb interlayer correlations. Also note that the onset temperatures for the appearance and disappearance of the middle two patterns of Fig. 20 were not reproducible with cycling. This is probably due to Rb vacancy formation, which is time dependent at elevated temperatures and is associated with expulsion of Rb from the GIC and irreversable absorption of Rb by the sample capsule. After being maintained for several days at $721°K$ and cooled to room temperature, the sample exhibited the pattern marked $290°K$ and shown in the upper part of Fig. 20. This pattern is well indexed on the expected $\alpha\beta\gamma\delta$ stacking sequence. It is further evidence that the as-grown sample was not fully annealed and possessed a metastable, highly faulted stacking arrangement of the Rb layers.

The temperature dependence of the electron diffraction from RbC_8 was measured by Kambe et al.[97] using a commercial electron microscope. These investigators reported the observation of $\alpha\beta\gamma\delta$ stacking for $110°K < T < 300°K$, but observed, in the range $300 \leq T \leq 330 \pm 108$, a reversible staging transition to a phase that exhibited other than $\alpha\beta\gamma\delta$ stacking (e.g., $\alpha\beta$, $\alpha\beta\gamma$, disordered, etc.).

When the incident electron beam propagates along the sample c-axis, the resulting in-plane diffraction pattern will be identical to the pattern of pristine graphite for the $\alpha\beta\gamma\delta$ stacking arrangement.[98] Kambe et al.[97] noted that "well-annealed" samples of RbC_8 confirmed this prediction, whereas as-grown samples always exhibited in-plane superlattice spots. They, therefore, emphasized the necessity of "annealing" the sample, by which they apparently mean maintaining the GIC at or near its preparation temperature for several days. Since the rate at which an equilibrium structure is achieved depends on many factors (see Section III.B), no general recipe for producing a "well-annealed" sample can be uniformly applicable, even for a given species of intercalant.

When the RbC_8 was heated in the vacuum of the electron microscope to temperatures well in excess of 330°K, desorption naturally occurred. Thus, Kambe et al.[97] could not confirm the staging phase transition which was observed at 721 and 747°K using neutron diffraction.[96] In addition, it is not technically feasible to use electron diffraction to obtain quantitative information on the in-plane diffuse scattering associated with a disordering, if any, of the Rb layers at elevated temperatures. However, it is disconcerting to note that in the temperature range $300 < T < 330$°K in which the electron diffraction and neutron diffraction measurement of RbC_8 overlap, there are conflicting results—no staging phase transition is observed with neutrons. This difference cannot be attributed to spatial inhomogeneities that are resolved by the focused electron beam, since the electron diffraction patterns were "independent of the region on the sample selected for observation"[97] at a given temperature. Indeed, similar disparities between electron and neutron diffraction results occur for LiC_6, which we now discuss.

The in-plane structure of a stage 1 lithium GIC is shown in Fig. 21.[75, 99] Because the lithium ion is much smaller than those of the heavy alkalis,[100] it can adopt the more dense MC_6 structure corresponding to a $(\sqrt{3} \times \sqrt{3})R30°$ superlattice. The primitive cell of that superlattice (and that of the graphite layer) is shown in Fig. 21 and can accommodate three sets of sites labeled α,

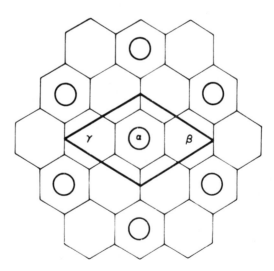

Fig. 21. c-Axis planar projection of the MC_6, M=Li, Eu etc. commensurate structure. The $(\sqrt{3} \times \sqrt{3})R30°$ planar superlattice cell is depicted by the heavy lines. The three equivalent hexagon center sites are labeled α, β, and γ. See also legend of Fig. 17.

β, and γ. Unlike the heavy alkalis, which adopt complex stacking arrangements of intercalate layers in their three-dimensionally ordered superlattice structures, LiC_6 exhibits the most primitive and basic stacking sequence, namely, $\alpha\alpha$.[75] Thus all lithium layers and all carbon layers are equivalent in stage 1 LiC_6 at room temperature.

The intraplanar and interplanar structures of LiC_6 have not yet been successfully studied at elevated temperatures. However, there exist both electron diffraction[101] and neutron diffraction[102] studies of LiC_6 at reduced temperatures. Using the former approach, Kambe et al.[101] reported a staging phase transition to an $\alpha\beta\gamma$ stacking sequence at $T < 200°K$ that was signaled by the disappearance of $(h00)$ diffraction spots. In contrast, Rossat-Mignod et al.[102] saw no evidence for any such transition in neutron diffraction studies of LiC_6 down to $10°K$, at which temperature the $\alpha\alpha$ stacking sequence still persisted.

c. Phase Diagrams and Theoretical Calculations. There have been two theoretical attempts to calculate detailed properties of the phase transitions at elevated temperatures in stage 1 alkali GICs. Bak and Domany[103] have addressed order-disorder transitions in CsC_8, RbC_8, and LiC_6. It is clear from the presence of (hkl) neutron or X-ray reflections at room temperature that the ordering in RbC_8 and CsC_8 must be three dimensional. Thus a four-state, two-dimensional Potts structural model[104] (corresponding to occupation of the α, β, γ, or δ in-plane sites of the MC_8 structure) that allows for a second-order transition is not appropriate to those GICs.

Bak and Domany[103] showed that according to the Landau-Lifshitz theory of phase transitions,[105] the CsC_8 three-dimensional structure is described by a six-component vector order parameter. When the free energy is expressed in terms of that order parameter, the third-order term is nonzero. Thus according to the Landau criteria the order-disorder transition in CsC_8 at $\approx 550°K$ must be first order. This result is consistent with the measurements of Clarke et al.,[92] who reported the transition to be nearly first order.

In considering the high-temperature transition at $721°K$ in which RbC_8 transformations from a "liquidlike" phase to an $\alpha\beta\alpha\beta$ stacked ordered arrangement, Bak and Domany[103] also find that the transition must be first order and it, too, is clearly three dimensional. However, RbC_8 is characterized by a three-component vector order parameter, and it is the sixth-order anisotropy terms in the Landau free-energy expansion that stabilize the ordered phase. There are no experimental data yet to confirm the predicted first-order nature of the $721°K$ transition in RbC_8.

A high-temperature order-disorder transition is predicted for LiC_6 to be also first order by Bak and Domany,[103] but this prediction, too, has not yet been confirmed experimentally.

Using a lattice-gas model, Lee et al.[106] have studied both order-disorder and order-order transitions in KC_8 and RbC_8, which have identical ground-state $\alpha\beta\gamma\delta$ four-layer, repeat stacking structures. In contrast to Bak and Domany,[103] they find that continuous second-order transitions can occur in RbC_8. A lattice-gas Hamiltonian was constructed using five interaction parameters: J_0 and J_0', which represent the first- and second-neighbor intralayer interactions, and J_1, J_2, and J_3, which are the first, second, and third nearest-neighbor interactions along the c-axis. Then

$$H = \left\{ J_0 \sum_{(0)} + J_0' \sum_{(0')} + J_1 \sum_{(1)} + J_2 \sum_{(2)} + J_3 \sum_{(3)} \right\} n(r_i) n(r_j) \qquad (18)$$

where $n(r_j) = 1$ or 0, according to whether the site at r_j is occupied, and the summations are over all pairs of sites corresponding to nonzero values of the interaction parameters. By subdividing the ground intralayer state into four interpenetrating $\alpha, \beta, \gamma, \delta$ $(2\times2)R0°$ sublattices, any planar state of the system can be constructed from the occupation probabilities of each of the sublattices. In the disordered state, all sublattices are occupied with equal

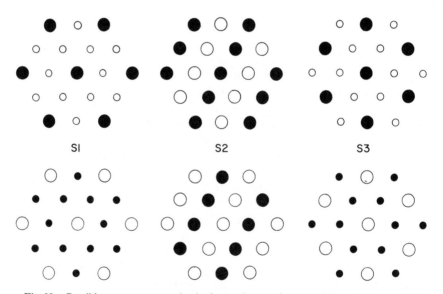

Fig. 22.   Possible structures appearing in the continuous phase transitions of KC_8 and RbC_8. The upper and lower diagrams show $\delta\rho = \rho - \rho_0$ in the ordered structures S_1, S_2, and S_3 in the $(2n+1)$th and $2n$th layers, respectively. Solid and open circles denote sites with positive and negative $\delta\rho$, respectively, and the relative magnitude $|\delta\rho|$ is represented by the circle size. In S_3 the small circles can vary in density, but they retain the same symmetry arrangement as the large circles. From Ref. 106.

probability. In the $T=0°$K ordered state three of the sublattices have zero occupation probability, while one has an occupation probability of 1.

Like Bak and Domany,[103] Lee et al.[103] applied the Landau-Lifshitz theory and in particular the Lifshitz condition to identify the ordered structures that could evolve from the lattice-gas disordered states by way of a continuous second-order phase transition. The possible structures identified are shown in Fig. 22. Note that this figure represents long-range order in the probability distribution of site occupancy. Therefore the occupancy of adjacent intralayer sites that would be precluded in a frozen ordered structure can occur nonsimultaneously in the structures of Fig. 22 as a consequence of correlated intersite hopping.

Since the Landau-Lifshitz theory[105] applies only to temperature regions near the phase transition, Lee et al.[106] used the Hamiltonian of (18) and the mean field approximation to compute the relevant phase diagrams for both order-disorder and order-order transitions. Their results are shown in Fig. 23, where $\tilde{J}_0 = J_0 + J_0'$, $\tilde{J}_1 = J_1 + J_3$, $K_i = \langle N_i \rangle / N$ $(i = 1 \cdots 4)$ is the probability of occupation of sublattice i, N_i is the occupation number of intercalant atoms in sublattice i, and N is the total number of intercalant atoms. As can be seen from Fig. 23a, for the appropriate interaction parameters the system undergoes a continuous transition into the S_2 ordered phase at T_2 and a second continuous order–order transition into an S_4 phase at lower temperature T_4. The S_4 phase is a four-layer repeat structure identical in form to

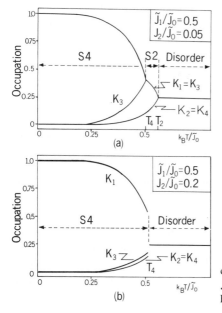

Fig. 23. Phase diagrams for sublattice occupation versus temperature for $\tilde{J}_1 / \tilde{J}_0 = 0.5$. (a) $\tilde{J}_2 / \tilde{J}_0 = 0.05$ and (b) $\tilde{J}_2 / \tilde{J}_0 = 0.2$ (see text). From Ref. 106.

the $T=0$ structure of MC_8 compounds but with site occupancies not fixed at zero or one. Lee et al. also found that if $J_2 \ll \tilde{J}_1$, the intermediate phase would be another two-layer repeat structure labeled S_2'. At sufficiently low temperatures, $K_1 = 1$ and $K_i = 0$ ($i \neq 1$), in agreement with observation. This is also true for the phase diagram of Fig. 23b, which was generated using a different set of interaction parameters. In that case there is a first-order transition to the S_4 ordered state at T_4 and no order-order transitions occur.

The publications of Lee et al.[106] and of Bak and Domany[103] represent important contributions and constitute a significant challenge to experimentalists. Inherent in the calculations of both sets of investigators is the assumption that the phase transitions predicted occur at constant intercalant density, that is, with a constant number of intercalant ions present in the GIC. Such a condition is trivial to achieve for the temperature region $T \lesssim 300°K$ but may be extremely difficult to achieve at the elevated temperatures corresponding to T_4 and T_2 of Fig. 14.

2. Stage 2 Alkali GICs

a. **Background.** The stage 2 heavy alkali GICs constitute the most thoroughly studied (with respect to structural phase transitions) of any of the graphite intercalates. As noted above, there are several conflicting analyses of the structural properties of these compounds, especially the stage 2 potassium and rubidium GICs. The following discussion briefly reviews the early (pre-1977) work on the structures of these compounds and then considers the post-1977 period, during which many new results were reported.

The stage 2 heavy alkali GICs were studied originally by Rudorff and Shulze.[107] They proposed an intralayer structure in which the M atoms formed a commensurate $(\sqrt{12} \times \sqrt{12})R30°$ superlattice (Fig. 24). It is important to emphasize that this proposal was motivated by the authors' belief that homogeneous compounds formed with only the idealized bulk stoichiometry MC_{24}. This then necessitates an MC_{12} layer stoichiometry that is consistent with the $(\sqrt{12} \times \sqrt{12})R30°$ superlattice structure. We now know that such a superlattice could not exist without creating well-defined (perhaps faulted) stacking sequences along the c-axis, the primary signature of which would be three-dimensionally derived (hkl) Bragg reflections associated with the ordered M ions. Yet no stacking correlation was observed by Rudorff and Shulze.[107]

In 1968–1969 Parry and coworkers[87, 88] reexamined the room temperature structures of stage 2 heavy alkali GICs. They concluded that the metal ions in KC_{24} exhibited a lattice-gas type of intralayer disorder in which these ions occupied sites over the centers of carbon hexagons.[30] This assertion was supported by structure factor calculations. However, in the case of RbC_{24} and

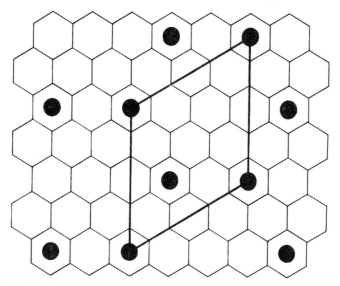

Fig. 24. c-Axis planar projection of the Rudorff MC_{12} commensurate structure. The metal atoms are represented by the solid circles. The $(\sqrt{12} \times \sqrt{12})R30°$ planar superlattice cell is represented by the heavy lines.

CsC_{24} they noted that such calculations resulted in major discrepancies between the calculated and observed structure factors.[108, 109] In the case of RbC_{24} it was suggested that only a fraction of the Rb atoms lie over hexagon center sites. This was the first suggestion of mixed-phase, disordered structures in the heavy alkali GICs, and it may have been remarkably anticipatory as we shall see.

Nixon and Parry also observed a stacking sequence for MC_{24} (M = K, Rb, Cs), which is denoted A/AB/BC/C and for KC_{24}, for example, results in a measured c-axis repeat distance of 26.25 ± 0.05 Å.[30] The capital letters A, B, and C refer to the stacking arrangement of the carbon layers. The diagonals correspond to disordered intercalant layers. Notice that even though the intercalant layers are disordered, they are bounded by carbon layers that form an eclipsed AA or BB configuration when viewed along the c-axis just as in the stage 1 MC_8 ordered compounds and in contrast to pristine graphite, which stacks in a staggered AB configuration.[16]

Nixon and Parry reasoned that the disordered layers in stage 2 MC_{24} GICs should "freeze" into ordered structures at sufficiently low temperatures. They therefore measured the temperature dependence of the X-ray diffraction patterns of these compounds and observed, as expected, a temperature-dependent order-disorder transition the temperatures of which depended on

496   S. A. SOLIN

TABLE II
Summary of Temperatures (°K) at Which Anomalies Occur in the Resistivity
$\rho(T)$, the Specific Heat $C(T)$, and the X-Ray Diffraction, Neutron
Diffraction, and Electron Diffraction from Stage 2 Heavy Alkali GICs.[a]

Material	$\rho(T)$	$C(T)$	X-ray scattering	Neutron scattering	Electron diffraction
KC$_{24}$					
	T_l 95 (100)		98 ± 3 (108)		86 (133)
	T_u 124 (100)		126 ± 0.5 (123)		130 (133)
RbC$_{24}$					
	60 (136)	48 (136)			
	T_l 106 (100)	106 (136)	110 (139)	110 (136)	
	110 (136)				
	T_u 165 (136)	165 (136)	159 (109)	165 (136)	170 (97)
	172 (110)				
	204 (136)	215 (136)			
CsC$_{24}$					
			50 (112)		
	T_l 180 (110)		163 (109)		
			165 (112)		
	T_u 230 (110)		228 (112)		
					305 ± 15 (155)[b]

[a]Relevant reference numbers are indicated in parentheses. The temperatures T_l and T_u are discussed in the text.
[b]Sample stoichiometry somewhat uncertain.

intercalant species and are listed in Table II.[108, 109] The planar ordering of the intercalant ions was accompanied by a staging transition to $A\alpha AB\beta BC\gamma C$ in the cases of RbC$_{24}$ and CsC$_{24}$ and to $A\alpha AB\beta BC\gamma CA\alpha'$ $AB\beta'BC\gamma'C$ in the case of KC$_{24}$. However, the specific in-plane structures to which α, β, and γ correspond were not identified. The primed and un-primed intercalant layers are located in identical carbon atom environments. The fact that the low-temperature stacking arrangements of RbC$_{24}$ and CsC$_{24}$ were identical but different from that of KC$_{24}$ may have been related to the differences in in-plane structures alluded to above.

b. New Results. In recent studies of the temperature dependence of the resistivity of MC$_{24}$ GICs Onn et al[110] observed discontinuities in $\rho(T)$ versus T, which they associated with structural phase transitions. Their results for KC$_{24}$ are shown in Fig. 25. The discontinuities in $\rho(T)$ versus T are clearly evident. Similar results have also been reported for RbC$_{24}$ and CsC$_{24}$. These together with the KC$_{24}$ data are also tabulated in Table II. In each case, two discontinuities occur at temperatures that have been labeled T_u and T_l. It is clear from Table II that the T_l discontinuities in ρ versus T for KC$_{24}$ and

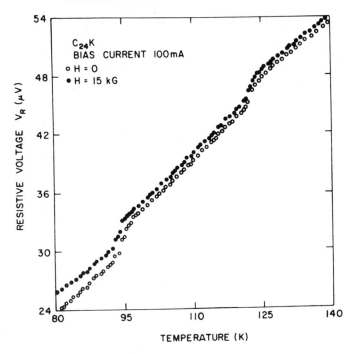

Fig. 25. The basal plane resistivity versus temperture of KC_{24} in zero field and in a field of 15 kG. From Ref. 110.

CsC_{24} are probably associated with the order-disorder transition observed in diffraction studies.

Data of the type shown in Fig. 25 are significant in that they provided the first evidence for multiple phase transitions in the MC_{24} GICs. While such measurements as ρ versus T clearly signal the occurrence of a phase transition, they are of little value in identifying the structural alterations involved. For this, thorough in-plane and c-axis diffraction studies are required. To date, only, in the case of CsC_{24} have such studies been carried out; and I now discuss the results of those studies.

CsC_{24}.   Figure 26a shows the in-plane ($q \perp c$) structure factor $S(q)$ of CsC_{24} prepared from HOPG.[111-114] The detailed planar positional correlations of the Cs atoms can be ascertained from the planar radial distribution function (RDF).[115] This can be obtained from the Fourier-Bessel inversion of $S(q)$. Thus,

$$\text{RDF} = 2\pi r \rho(r) = 2\pi r \rho_0 + r\int_{0.06\text{Å}^{-1}}^{15\text{Å}^{-1}} q[S(q)-1]J_0(qr)\,dq \qquad (19)$$

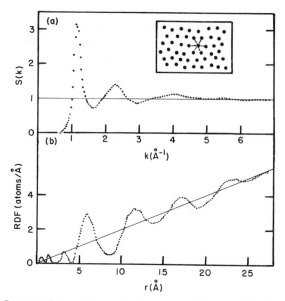

Fig. 26. (a) Structure factor $S(k)$ for the Cs layer in CsC_{24} at 300°K. The full data (not shown) extend to 15 Å^{-1}. Inset: schematic structure. (b) The radial distribution function (RDF) obtain from $S(k)$ in (a). The small ripples at $r<3.5$ Å are finite k artifacts.

Here, $\rho(r)$ is the areal density, so $2\pi r\rho(r)$ is the number of atoms in a planar ring of thickness dr with radius r, ρ_0 is the average macroscopic areal density, and J_0 is the zeroth-order Bessel function. The planar RDF that results when (19) is applied to the data of Fig. 26a is shown in Fig. 26b. Notice that oscillations indicative of short-range order occur about the line $2\pi r\rho_0$ the slope of which gives the average density. The oscillations at low $r(<4$ Å) are artifacts of the Fourier-Bessel transform that result from the finite upper q limit in (19). They correspond to an error of $\frac{1}{4}$ atom in the RDF.

Most significant in Fig. 26b is that the most probable nearest-neighbor distance as given by the position of the first peak in the RDF is 5.95 ± 0.1 Å. This distance is unrelated to any of the characteristic distances of the graphite layer such as the 4.92 Å hexagon center site distance of the $(\sqrt{12} \times \sqrt{12})R30°$ Rudorff structure. From the area under the 5.95 Å peak in the RDF, the nearest-neighbor coordination number has been determined and is found to be 6 ± 0.25. In contrast, the corresponding number for the Rudorff $(\sqrt{12} \times \sqrt{12})R30°$ structure is 3. The results of Fig. 26b are consistent with a six-fold coordinated, close-packed arrangement of Cs atoms, which can be characterized as a disordered triangular lattice and is shown schematically in the inset of Fig. 26a. Note that if one calculates the average interparticle distance for a uniform close-packed structure with a layer stoichiometry of CsC_{12}, a value of 6.02 Å is found. This value is consistent with the nearest-neighbor distance derived above.

A simple classical liquid theory has recently been applied by Plischke[116] to calculate the intraplanar structure of CsC_{24} at room temperature. He assumed that each cesium atom is singly ionized to Cs^+ upon intercalation and that the cesium ions interact by way of a screened Coulomb potential. That potential was determined from the self-consistent field approximation using a simplified model for the graphite band structure. The potential is constructed so as to be hard core like at small distances; that is, it is infinite for distances less than or equal to the ionic diameter of cesium. The potential $V(r)$ so derived was then used in the Percus-Yevick equation[117] to deduce the direct correlation function $C(r)$

$$C(r_{12}) = \left[1 - \exp\left(\frac{1}{k_B T} V(r_{12}) \right) \right] g(r_{12}) \qquad (20)$$

The pair correlation function $g(r)$ (or equivalently the planar RDF) was obtained from $C(r_{12})$ by solving the Ornstein-Zernike equation[118]

$$g(r_{12}) = 1 + C(r_{12}) + \rho_0 \int d^3 r_3 [g(r_{13}) - 1] C(r_{13}) \qquad (21)$$

Figure 27 compares the results of Plischke's calculation with the measured

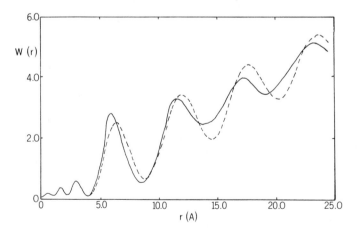

Fig. 27. Comparison of the data of Fig. 26b (solid curve) with the function $W(r) = 2\pi rng(r)$ (see text) calculated using the Percus-Yevick approximation (dashed curve). From Ref. 116.

500   S. A. SOLIN

RDF of Fig. 26b. Given the crudeness of the calculation, the agreement with experiment is quite reasonable. In particular, the difference in the positions of the theoretical and experimental first peaks is relatively small and might be improved by a readjustment of some of the parameters that determine the potential. Note also that Plischke assumed that the ionic diameter of Cs in CsC_8 was 3.39Å.[119] (This value is unreasonably small, as will be shown in Section IV.C)

Till now we have addressed only the intraplanar correlations in CsC_{24} at room temperature. The effect of reduced temperature on the in-plane correlations of the Cs atoms is illustrated in Fig. 28, which shows the temperature dependence of the first in-plane diffraction peak at $q = 1.158 \text{Å}^{-1}$.[112, 114] Notice that with decreasing temperature the peak narrows and grows in intensity. The peak has been fit with a Lorentzian given by

$$I(\varepsilon) \propto \left[1 + \xi^2(T)\varepsilon^2\right]^{-1} \tag{22}$$

where $\varepsilon = q_0 - k$, $q_0 = 1.158 \text{Å}^{-1}$ and $\xi(T)$ is the in-plane correlation length. However, the data are not sufficiently accurate to preclude other equally good fits, for example, those based on a power law.[120] Notice that the q_0 peak never actually becomes Bragg-like (see from Fig. 29), since ξ does not diverge to infinity. Thus, the transition at $T_l = 163°K$ reported as an order-disorder type by Parry and Nixon[108] is *not* confirmed by the data of Figs. 28 and 29, since the correlation range grows rapidly below that temperature but does not become infinite. The pinning of the in-plane correlation range may be associated with the planar defects contemplated in the Daumas-Hérold[47] structural model. Note that in the absence of careful diffractometer line shape measurements of the type shown in Figs. 28 and 29, mere growth of the correlation range can easily be erroneously misconstrued as evidence for long-range crystalline order. Only in the case of CsC_{24} reported here have sufficiently accurate analyses been carried out to make this distinction.

The in-plane correlations of Cs atoms in CsC_{24} cannot develop without attendant interplane correlations. Thus if at room temperature there is no

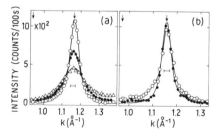

Fig. 28. Temperature dependence of the in-plane X-ray scattering intensity from stage 2 CsC_{24} in the vicinity of the first peak in $S(k)$ at 1.158 Å^{-1} (see Fig. 26a). (a) Triangles, 236°K, solid circles, 170°K; open circles, 157°K. (b) open circles, 90°K; solid circles 10°K. The solid curves are the Lorentzian fits discussed in the text. The asymmetry in some of the peaks is due to the presence of other features at $k \approx 1.3 \text{ Å}^{-1}$.

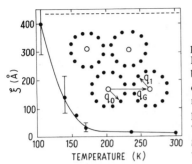

Fig. 29. The temperature dependence of the in-plane correlation length ξ detemined from the Lorentzian fits shown in Fig. 28. Values of ξ have been corrected for the instrumental contribution. The dashed line at the top represents the instrumental resolution limit. Inset: Schematic diagram of the in-plane reciprocal lattice of CsC_{24} at 150°K. The wave vectors of the primary cesium reflections and of the graphite reflections are q_0 and q_G, respectively. The wave vectors of satellite reflections are $q_1 = q_G - q_0$.

interplanar correlation, it can be expected to develop as the temperature is reduced and the in-plane correlation range grows. This crossover in dimensionality from two dimensions to three dimensions is beautifully displayed in Fig. 30, which shows (10l) X-ray scans along the c^*-axis in reciprocal space.[114]

Consider first the diffraction pattern of a two-dimensional crystal. This pattern would consist of rods of intensity in q-space perpendicular to the reciprocal plane and passing through it at the positions corresponding to Bragg spots.[121] For a polycrystalline collection of such two-dimensional crystals (e.g., HOPG) the pattern would consist of concentric, hollow, circular cylinders with a common axis. These cylinders would have a negligibly small wall thickness and their surfaces of revolution would be generated by spinning the above-mentioned rods of intensity about the c^*-axis. To the extent that the two-dimensional crystallites become disordered, the wall thickness of these cylinders will increase and ultimately merge. The resultant distribution of intensity is represented in the inset of Fig. 30.

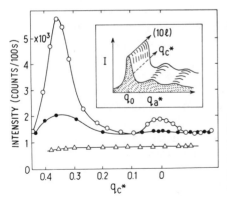

Fig. 30. Temperature dependence of the diffractometer scans along the (10l) direction in CsC_{24} (see inset). Note, q_c^* (in Å^{-1}) = $2\pi l/d_c$, where d_c is the stage 2 c-axis repeat distance. The peak at $q_c^* = 0.34$ Å^{-1} ($l=3$) gives $d_c = 55.4$ Å. Triangles, 300°K; solid circles, 170°K; open circles, 130°K.

When Cs atoms in CsC_{24} exhibit intralayer correlation but no interlayer correlation, an X-ray scan along $q \parallel q_{a*b*}$ will be independent of the constant q_{c*} value at which it is obtained. Therefore, for this case of two-dimensional correlations only, a scan with $q \parallel q_{c*}$ will yield a constant X-ray intensity for any fixed value of q_{a*b*}, but the magnitude of that intensity will depend on the choice of q_{a*b*}. Thus a flat $q \parallel q_{c*}$ scan is *direct* evidence for a two-dimensional system. When the two-dimensional Cs layers begin to develop three-dimensional correlations, the $q \parallel c*$ scans will develop peaks that for an ordered three-dimensional structure will be Bragg reflections. These points are illustrated in the inset of Fig. 30.

As can be seen from Fig. 30, the flat $(10l)$ scan $(q \parallel q_{c*})$ shows that at $300°K$ Cs atoms exhibit only in-plane two-dimensional correlations. However, at approximately $170°K$ there is a crossover to three-dimensional behavior as evidenced by the broad peak at the $l = 3$ $(q_c^* \sim 0.35)$ position. With further cooling to $100°K$, two peaks at $l = 3$ and $l = 0$ $(q_c^* = 0.0)$ are clearly discernible. These peaks are not of instrument-limited width, however, and indicate a c-axis correlation range of only ~ 40 Å. Therefore, at $T_l \cong 170°K$ the intraplanar correlation range $\xi(T)$ begins to grow rapidly with decreasing temperature and simultaneously there is a crossover from two- to three-dimensional correlations in the Cs ion positions.

When CsC_{24} is further cooled to temperatures below $140°K$, mixed-phase in-plane structures develop.[111, 112] These consist of macroscopic (>400 Å) regions of $(2 \times 2)R0°$ superlattice coexisting with the unregistered highly correlated yet noncrystalline phase discussed above. Evidence for the coexistence of these phases has been presented by Clarke and is shown in Fig. 31.[112] The insets indicate that the peaks corresponding to the $(2 \times 2)R0°$ superlattice are Bragg-like and instrument-resolution limited. In contrast, the peaks corresponding to the unregistered phase exhibit the characteristic tails associated with a lack of long-range order.[121] In the range $10°K < T < 140°K$, the relative concentration of the $(2 \times 2)R0°$ phase grows at the expense of the unregistered phase; however, the latter never disappears.

The structural studies of CsC_{24} discussed above were performed on samples prepared from HOPG. Thus a considerable amount of detail about angular correlations between Cs–Cs bonds in the Cs layers is lost because of the random orientation of the crystallites about the c-axis of HOPG. To overcome this difficulty, Clark et al.[111-114] carried out X-ray diffraction studies of CsC_{24} prepared from single-crystal graphite. They observed an in-plane diffraction pattern shown schematically in Fig. 29 (inset).

The pattern shown in Fig. 29 is well understood and results from two real-space, threefold, coordinated triangular arrangements of cesium atoms, each of which generates a hexagonal sextet of diffraction spots rotated by $\pm 14°$ about the c-axis with respect to the $\langle 100 \rangle$ graphite direction. These

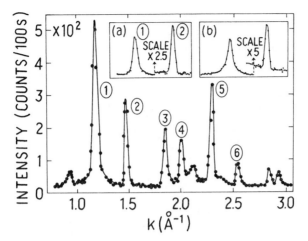

Fig. 31. In-plane diffraction profile of HOPG CsC_{24} at $10°K$. Peaks 1, 4, and 5 are from the unregistered Cs layer; peaks 2 and 6 are from registered $(2 \times 2)R0°$ regions. Peak 3 is a q_1 satellite (see inset of Fig. 29), and peak 7 is a combination graphite–$(2 \times 2)R0°$ cesium reflection. Insets: peaks 1 and 2: (a) at $10°K$ and (b) at $90°K$. From Ref. 112.

structures coexist in macroscopic domains. Note that the characteristic wavevector q_0 of the 12 satellite spots shown in Fig. 29 is equal in magnitude to the q vector of the first peak in the structure factor of the disordered Cs layers ($T \simeq 300°K$; see Fig. 26a). This together with the temperature dependence of the rotation angle (see discussion below), which is unrelated to any characteristic angle of the graphite lattice, reconfirms the nonregistry of the Cs atoms with the carbon layers. Patterns similar to that shown in Fig. 29 were first reported by Parry, who recently reinterpreted them as the results of mixtures of several commensurate structures.[122]

The temperature dependence of the average Cs—Cs bond direction ϕ_p relative to the graphite $\langle 100 \rangle$ direction is shown in Fig. 32. The data of that figure were obtained with fixed $q = q_0$ by rotating the CsC_{24} single crystal about its c-axis in the X-ray diffractometer while monitoring the diffracted intensity. A plot of the temperature dependence of ϕ_p is shown in the inset of Fig. 32 (solid circles). This plot is suggestive of an Ising-like orientational phase transition. The solid line fit to the data (inset) yields an exponent $\beta = 0.47$ and a transition temperature of $228°K$.

Table III summarizes the temperature dependence of both the positional and orientational correlations of Cs ions in CsC_{24}. The coupling between these two types of correlation is also temperature dependent, as the table indicates, but the independent observation of each with the detail described above is essentially unique to CsC_{24} and feasible only with GICs.

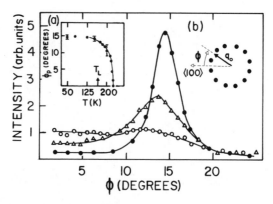

Fig. 32. Temperature dependence of the 12 primary q_0 reflections from single-crystal CsC_{24}. Solid circles, 148°K; triangles, 169°K; open circles; 185°K. Insets: (*a*) Temperature dependence of the angular position of the reflection relative to a graphite $\langle 100 \rangle$ direction. The solid curve is a power law fit to the data and is discussed in the text. (*b*) Geometry of χ-scans used to obtain the data in this figure.

KC_{24}. I now consider the temperature-dependent structural alterations exhibited by stage 2 potassium-graphite KC_{24}. This GIC was recently reexamined by Hastings et al.[123] who used X-ray diffraction techniques. They observed for the first time diffraction effects corresponding to both T_u and T_l (see Table II), which were identified as 126.0 ± 0.5°K and ≈ 93°K, respectively. They also reported that above T_u the potassium layers were two-dimensionally disordered and uncorrelated with one another.

At a temperature $T < T_l$, Hastings et al. found results in agreement with the earlier work of Parry and co-workers.[108, 109] In particular, they reported

TABLE III
Cesium Atom Correlations as a Function of Temperature[a]

Temperature	Positional	Orientational
$T \geq T_u = 228$°K	Two-dimensional unregistered liquid	[100] Preferred orientation
$T_l = 165$°K $\leq T \leq T_u$	Two-dimensional unregistered liquid	Orientational epitaxy with T-dependent rotation
50°K $\leq T \leq Tl$	Highly correlated unregistered liquid with 30 correlations	T-Dependent orientational epitaxy
$T \leq 50$°K	Incommensurate solid plus commensurate solid $(2 \times 2)R0°$ and $(3 \times 3)R0°$	

[a]From Refs. 111–114.

that the potassium layers are ordered and exhibit a stacking sequence $A\alpha AB\beta BC\gamma CA\alpha'AB\beta'BC\gamma'C$. Hastings et al.[123] also were unable to identify the intralayer ordered structure, a difficulty they attributed to the presence of multiple phases. For $T_l < T < T_u$ an intermediate-phase structure is observed and it has a stacking sequence $A\alpha AB\beta BC\gamma CA \cdots$ identical to the sequence found for RbC_{24} and CsC_{24} for $T = 159°K$ and $T = 163°K$, respectively. However, the intermediate phase also exhibited unusual diffraction profiles, which were convincingly interpreted as arising from stacking faults.

The $q \| c^*$ scans observed by Hastings et al[123] for $T = 120°K$ ($T_l < T < T_u$) are shown in Fig. 33. The arrows indicate the positions at which Bragg reflections would occur in the unfaulted structure. Note that the $(10l)$ scan (h

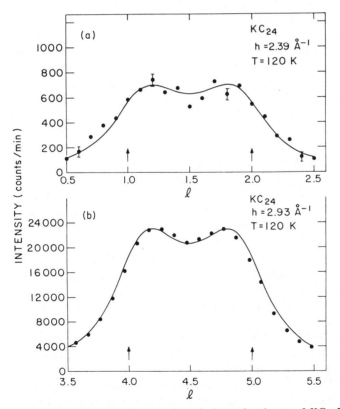

Fig. 33. (a) Scan along l of a powder peak from the intercalant in stage 2 KC_{24} HOPG at $T = 120°K$. (b) Scan along l of the $(10l)$ graphite peak at $T = 120°K$. Solid circles correspond to the measured data and the curves correspond to the stacking fault calculation (discussed in the text) with $\alpha = 0.38$. From Ref. 123.

$=2.935 \text{Å}^{-1}$ Fig. 33b), which is attributed solely to the graphite layers, is essentially temperature independent in the range $10°\text{K}<T<300°\text{K}$, whereas the broad doublet of Fig. 33a, which corresponds to the potassium layers, develops into the expected Bragg reflections at $T=T_l$.

To calculate the scattering profiles of Fig. 33, the following model was adopted.[123, 124] A layer sandwich consisting of an intercalant layer bounded by eclipsed carbon layers was constructed. If we characterize such a sandwich by a single capital letter A, B, C, $\cdots$, an unfaulted region of stage 2 GIC will be described by the stacking sequence ABCABC $\cdots$, which includes the staggering of carbon layers in adjacent sandwiches. This stacking sequence is identical to that of the face-centered-cubic crystal structure. Therefore, some of the diffraction patterns of face-centered cubic metals with stacking faults[125] will replicate those of faulted KC$_{24}$. Hastings et al.[123] employed an exact method for obtaining the relevant correlation functions of a disordered one-dimensional system. The method involves adjustment of a single parameter, α, to fit the observed diffraction patterns of Fig. 33. The parameter α is the probability of finding a fault between adjacent sandwiches. The solid curves of Fig. 33 are fits to the data with a value of $\alpha=0.38$. These fits provide convincing evidence for the stacking fault model.

The in-plane disordered structure of KC$_{24}$ as well as the structural transition at T_u were also studied with X-ray diffraction by Zabel et al.[126] These authors found a value of $T_u=122.9°\text{K}$, in good agreement with the value reported by Hastings et al.[123] They observed no evidence for the transition at T_l in the in-plane diffraction patterns, which supports the interpretation that that transition results from stacking rearrangements.

Zabel et al.[126] carefully studied dimensionality effects in KC$_{24}$ by examining the X-ray diffraction intensity along various lines in reciprocal space. Some of their results appear in Fig. 34. Figure 34a shows a $q \| c^*$ scan (($h0l$) $h=1.7\text{Å}^{-1}$) that probes interlayer correlations as discussed in the preceding subsection. Figure 34b shows a scan with $q \perp c^*$ and lying in a plane that intersects the c^*-axis at $q_{c*}=0.2\text{Å}^{-1}$. This scan distinguishes between two-dimensional, short-range order and the three-dimensional, long-range order (see inset in Fig. 30). Notice from Fig. 34a that above T_u (crosses) the c^* scan is flat, indicating that the potassium layers are uncorrelated. Below T_u (dots) that scan exhibits a "Bragg-like" peak indicative of three-dimensional ordering and interlayer correlations. The in-plane, short-range-order diffuse scattering is clearly evident from the data of Fig. 34b, which were recorded at $T_u+2°$ C. This scan depicts the ridgelike nature of the diffraction in a plane parallel to the a^*b^*-plane but well removed from the Bragg-like reflection shown in Fig. 34a.

Hastings et al.[123] and Zabel et al.[126] studied the temperature dependence of the intensity of the Bragg peak at 1.70Å^{-1} in KC$_{24}$. While both groups

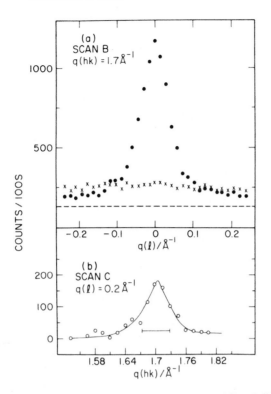

Fig. 34. (*a*) Scan along *l* from the intercalant in stage 2 KC$_{24}$. Solid circles → [T_u(= 122.9°K) − 24.5°K], x → T_u + 2°K.

concluded that the phase transition at T_u was higher order (i.e., not first order), only Zabel et al.[126] obtained a quantitative fit to the intensity data. If the long-range order vanishes as $|T - T_u|^{\beta}$, the X-ray intensity $I(T)$ is given by

$$I(T) \propto \left(\frac{T - T_u}{T_u} \right)^{2\beta} \tag{23}$$

Zabel et al. found an exponent $\beta = 0.18 \pm 0.01$ and $T_u = 122.9°$ as already noted. However great caution must be exercised in interpreting the exponent derived in such fitting procedures, since the presence of multiphase regions could "smear out" a first-order phase transition and make it appear to be of higher order.

The subject of the registry or lack of registry of the potassium atoms in KC$_{24}$ in particular and other stage 2 alkali GICs has received considerable attention recently. As noted above, EXAFS studies[52] of potassium GICs indicated that potassium atoms in KC_{24} were registered with the carbon layer and that the two-dimensional disorder at $T < T_u$ was of the lattice-gas type. Zabel et al.[127] have calculated the in-plane diffraction pattern of KC_{24} using a particular lattice-gas model. In that model two assumptions were adopted: (a) the occupation probability of a first- and second-neighbor hexagon from a filled hexagon was zero, and (b) all other hexagons have an equal occupation probability of two-thirds. Using these rules, sites were chosen at random but consecutively "by hand," the final decision of occupancy or nonoccupancy being based on dice tossing. The procedure above was employed to produce four independently constructed lattice-gas structures, each of which contained 141 atoms on a net of 800 hexagon centers. The scattered X-ray intensity in electron units was calculated from the expression

$$I_{e.u.} = \frac{f_K^2}{N} \sum_m \sum_n J_0(kr_{mn}) \tag{24}$$

where f_{K^+} is the ionic form factor for potassium and J_0 is the zeroth-order Bessel function that results from cylindrical averaging of the term $\exp\{i[\mathbf{k} \cdot (\mathbf{r}_m, \mathbf{r}_n)]\}$ in the Debye scattering equation.[128] The results obtained by averaging the diffuse scattering contributions to the intensities produced by the above-mentioned four independent structures are shown in Fig. 35. These are to be compared with the experimentally measured diffuse scattering contributions from KC_{24} shown in Fig. 36 (top). The calculated curve does not provide a reasonable fit to the observed diffraction, the most significant difference being the large discrepancy in the width of the first peak in $S(k)$. Such a discrepancy would not be removed by inclusion of the Debye-Waller factor in the lattice-gas calculation.

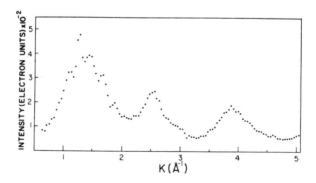

Fig. 35.   X-Ray intensity distribution calculated using a lattice-gas model for KC_{24}. From Ref. 127.

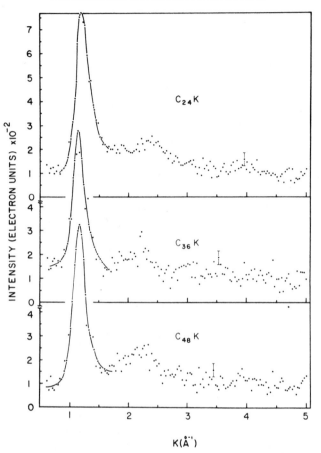

Fig. 36. X-Ray intensity distribution from the disordered potassium layers in stages 2 (KC$_{24}$), 3(KC$_{36}$), and 4 (KC$_{48}$) potassium GICs at room temperature. Solid curves are drawn as an aid to the eye only. From Ref. 127.

If the rules for occupancy and exclusion adopted by Zabel et al.[127] are retained and a Monte Carlo computer program technique is used together with periodic boundary conditions and with a net containing 10,000 hexagon sites, the lattice-gas structure produced cannot achieve the required density corresponding to KC$_{12}$ but falls short by some 13%.[129,130] Thus the "by-hand" method employed by Zabel et al.[127] apparently incorporates some bias toward increased intralayer packing. Nevertheless, a calculation of the intensity distribution for KC$_{24}$ using the Monte Carlo-generated lattice-gas yields results that are essentially indistinguishable from those of Fig. 35.

Moss and Zabel[131] have recorded optical transforms of the two-dimensional lattice gas prepared by the dice-throwing method described above and have compared these transforms to X-ray diffraction photographs of KC_{24}.[132] They note that as expected, the optical transforms exhibit both the periodicity and sixfold symmetry of the graphite lattice, whereas the diffraction photographs do not. This observation also supports a nonlattice-gas, liquidlike disorder in KC_{24}.

In further support of the liquidlike disorder in KC_{24} above T_u, Plischke[116] has applied to that GIC the same classical liquid theory used to calculate the RDF of CsC_{24} ($T>T_u$). He also claims to find reasonable agreement between theory and experiment, but does not present a comparison of the calculated $S(k)$ with the data of Fig. 36.

The EXAFS results and X-ray diffraction results for KC_{24} ($T>T_u$) yield apparently conflicting interpretations of the nature of in-plane disorder. Yet both these measurements show no evidence of any intercalant crystalline ordering at $T>T_u$. In contrast, Berker et al.[133] have interpreted their electron diffraction studies of KC_{21} at $T>T_u$ to indicate that expanded incommensurate-ordered structures coexist as islands in a sea of disorder. However, they do not specify the nature of the disordered state. The incommensurate structures, if unexpanded, would have been commensurate $(\sqrt{7} \times \sqrt{7})R19.1°$ and $(\sqrt{3} \times \sqrt{3})R30°$ superlattices, therefore they are labeled I_7 and I_3 by Berker et al.[133] The large islands ($\gg 50 Å$) in which the I_7 and I_3 structures are claimed to exist are orientationally locked to the graphite lattice (see discussion of CsC_{24} and inset, Fig. 32), but smaller islands ($\lesssim 50 Å$) of the I_7 and I_3 structures are also present above T_u, though these are orientationally unlocked from graphite.

The absence of any nongraphite-based Bragg reflections in the X-ray diffraction data of KC_{24} ($T>T_u$)[123, 126] is not compatible with the presence of I_7 and I_3 incommensurate structures. This absence cannot be attributed to the small crystallite size of the islands. Even for $50 Å$ islands, the breadth of the X-ray reflection at $2\theta = 20°$ that is produced with $MoK\alpha$ radiation will be only $\Delta(2\theta) \simeq 0.75°$. Thus, such reflections should be easily discerned in the X-ray diffraction pattern. When the crystallites of I_3 and I_7 are much greater than $50 Å$, the associated reflections, if present, will be Bragg-like and virtually impossible to miss.

Island size should have no effect whatsoever on the ability of the EXAFS technique to detect incommensurate structures, since that technique constitutes a local ($\lesssim 5 Å$) structural probe. In fact, the EXAFS probe tends to "see" selectively (i.e., to emphasize) locally ordered arrangements, such as hexagon center site occupancy. Therefore, the EXAFS and X-ray results would be consistent with a structure in which most but not all of the atoms were instantaneously unregistered with the graphite lattice. Such a structure is quite plausible, since center site occupancy is temporally consistent with the dynamics of a liquidlike potassium layer.

As a final comment on temperature-induced phase transitions in KC_{24}, we note that the results of Berker et al.[133] for $T > T_u$ constitute one aspect of a multiphase diagram they have presented for temperature versus in-plane density. This phase diagram is shown in Fig. 37, where I represents incommensurate and C corresponds to commensurate $(\sqrt{3} \times \sqrt{3})R30°$ and $(\sqrt{7} \times \sqrt{7})R19.1°$ superlattices of potassium ions, while D represents the unspecified disordered phase. The in-plane density of 0.2 corresponds to a stoichiometry of approximately KC_{21} measured using the relative integrated intensities of the $(00l)$ reflections. Such measurements of stoichiometry could contain significant inaccuracies because standard "Ω" scans[93] were not employed to take into account intensity contributions associated with the mosaic spread of HOPG-based GICs.

The phase diagram of Fig. 37 points out the multiphase nature of the in-plane structures of KC_{24}, previously noted by others.[123, 126] However, there is no supporting X-ray evidence for an I_7 structure at $T_u < T < T_l$. While Hastings et al.[123] inferred a contribution to the X-ray diffraction from an unspecified incommensurate phase in that temperature range, more recent and more careful measurements by Mori et al.[134] indicate that the observed diffraction patterns, including relative intensities, may be attributed solely to modulations of the graphite planes by the potassium. Also note that

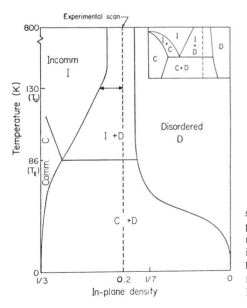

Fig. 37. Schematic phase diagram for stage 2 potassium-graphite based on temperature-dependent electron diffraction results (dashed line). The commensurate-incommensurate transition on the left can be a priori either second order (main frame) or first order (inset), respectively. From Ref. 133.

according to the phase diagram of Fig. 37, the ordered incommensurate structures should persist up to temperatures well above 800°K for a stage 2 GIC with a potassium in-plane density slightly higher than KC_{24}. This would require an unreasonably large intralayer potassium-potassium interaction. Suffice it to say that the phase diagram for KC_{24} proposed by Berker et al.[133] is interesting but awaits further verification from additional (X-ray, neutron, etc.) diffraction experiments.

RbC_{24}. I now address stage 2 RbC_{24}. The original measurements of the temperature dependence of the basal plane resistivity of RbC_{24} by Onn et al.[110] were repeated by Suematsu et al,[135] who found not two, but four discontinuities in ρ versus T. These anomalies occurred at 215, 165, 110, and 48°K and were assumed to be associated with order-disorder transitions. Other resistivity results of Seumatsu et al., together with anomalies found in the specific heat of RbC_{24} at similar temperatures, are tabulated in Table II.

The temperature dependence of the elastic neutron scattering from RbC_{24} was studied by Suzuki et al.[136] They reported a set of structures very much like those found in CsC_{24}.[111-114] Above T_u, the Rb atoms adopted a liquid-like structure and were unregistered with the graphite lattice. Below T_u they froze into a state of true long-range order (unlike Cs atoms in CsC_{24}, for which the planar correlation range was finite); the structure was that of a triangular lattice that was orientationally locked and rotated by 9.3° with respect to the graphite layers. Following an apparent pattern of discrepancy, electron diffraction results on KC_{24} were interpreted by Kambe et al.[97, 137] to indicate that the ordered state below T_u was a $(\sqrt{7} \times \sqrt{7})R19.1°$ superlattice. Also in disagreement with both the above-mentioned neutron diffraction results as well as X-ray diffraction results, Kambe et al.[97] found that at temperatures 20 to 30°K above T_u another ordered phase develops by way of an order-order transition. They suggest that the ordered phase is a superposition of several commensurate structures including the Rudorff $(\sqrt{12} \times \sqrt{12})R30°$ superlattice, but the diffraction pattern of the high-temperature ordered phase was too complex to index definitively. Notice that the temperature at which the claimed order-order transition occurs ($T \approx 200°K$) is close to the highest temperature ρ versus T discontinuity reported by Suematsu et al.[135] Nevertheless, no evidence of the disordered phase ($T > T_u$) was found from electron diffraction, even though it is the only phase found at these temperatures in other diffraction studies (e.g., X-ray[139] and neutron).[136] As with KC_{24}, these discrepancies cannot be attributed to the failure of X-ray and neutron probes to detect reflections from small crystallites of the unusual phases.

Laue and Burger precession X-ray photographs of RbC_{24} in the temperature range $110°K < T < 300°K$ were recorded by Yamada et al.[138, 139] At $T =$

$300°K$ they found an unregistered liquidlike structure akin to that of CsC_{24} at $300°K$. However, at $T=80°K$ a pattern clearly indicating a lattice-gas disorder was observed, superposed on the graphite reflections, each accompanied by weak satellite reflections. Second-order satellites were also observed at $180°K$. The satellite reflections were attributed to an $(8\times8)R0°$ superlattice. At a temperature of $110°K$ the diffuse rings transform into another set of satellite spots essentially identical to those observed for CsC_{24} at low temperature $(T<T_u)$. These spots appear to correspond to an incommensurate structure with wave vector $\mathbf{k}=0.335\mathbf{a}^*+0.105\mathbf{b}^*$, $|\mathbf{k}|=1.18\ \text{Å}^{-1}$ $\phi_p =13°$, where $\mathbf{a}^*$ and $\mathbf{b}^*$ are the basal plane primitive vectors of the carbon layer. Note, however, that $\mathbf{k}$ is very close to the wave vector $\mathbf{k}=(\frac{1}{3})\mathbf{a}^*+(\frac{1}{9})\mathbf{b}^*$ that corresponds to a $(9\times9)R0°$ commensurate structure. No evidence was found by Yamada et al.[139] for the $(\sqrt{7}\times\sqrt{7})R19.1°$ commensurate structure that was reported by Kambe et al.[97, 137]

Yamada et al.[139] have proposed model structures of RbC_{24} to explain the $T=180°K$ diffraction patterns. They suggested that those patterns arise from two types of wave: (1) longitudinal lattice distortion waves $(L-LDW)$ corresponding to deformations in the graphite layers due to coupling between graphite and rubidium, and (2) probability density waves (PDW), which correspond to the spatial dependence of the probability of hexagon center site occupancy by rubidium ions in the lattice-gas phase. Figure 38 shows the PDW patterns calculated by Yamada et al.[139] for both the $(8\times8)R0°$ and $(9\times9)R0°$ unit cells. Notice that there are clusters of sites with equal probability of occupation. Thus there is uncertainty in the rubidium ion locations above what is indicated in Fig. 38.

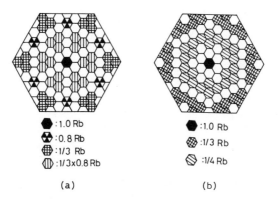

| :1.0 Rb
| :0.8 Rb
| :1/3 Rb
| :1/3x0.8 Rb

| :1.0 Rb
| :1/3 Rb
| :1/4 Rb

(a)   (b)

Fig. 38. Schematic description of probability density wave (PDW). Hexagons denote the statistical distribution of rubidium ions within the cluster. (a) $(8\times8)R0°$ unit cell. (b) $(9\times 9)R0°$ unit cell. From Ref. 138.

3. Higher Stage Alkali GICs

The data available from structural studies of higher stage ($n \geq 3$) alkali GICs is quite sparse compared to the data for stage 1 and stage 2 compounds. In general it appears that the higher stage compounds behave in a manner very similar but not necessarily identical to the manner in which the stage 2 compounds behave vis-à-vis temperature-induced phase transitions. Clarke et al. find that the orientational-phase transition seen in CsC_{24}^{111} also occurs in CsC_{36}^{112} and that the behavior of these two compounds is essentially the same. Similarly, Kambe et al.[97] report that stages 3, 4, and 5 rubidium GICs behave essentially the same; the stage 2 compound is also included in this group, the only significant difference being that the relevant ordering temperatures decrease slightly with increasing stage.

In contrast to CsC_{36} and CsC_{24}, which behave the same, KC_{36} and KC_{24} exhibit significant differences.[127] In the room temperature liquid phase, the first peaks in $S(k)$ occur at different positions. Similarly in the ordered phase at $T < T_u$ the wave vector for the potassium (100) reflection is smaller in stage 3 than in stage 2. The ordered structure has expanded in the latter relative to the former. Also note that the rotation angle of the orientational lock-in is about 12° in KC_{36} as compared to 7.5° in KC_{24}.

Mori et al.[134] have formulated a relationship between the rotation angle ϕ of the orientational lock-in and the lattice misfit between the carbon layer and the intercalate layer. They have applied this model to several alkali GICs. The lattice misfit Z is defined as $|G_{100}^I|/|G_{100}^G|$, that is, as the ratio of the

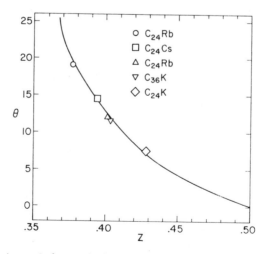

Fig. 39.   Rotation angle θ versus lattice misfit Z for alkali GICs (see text). The solid curve corresponds to the equation $\cos(30 - \theta) = (2Z^2 + 1)/2\sqrt{3}\,Z$. From Ref. 60.

magnitude of the intercalant lattice (100) wave vector to that of the graphite (100) wave vector. From the q-space geometry of the two superposed lattices Mori et al. find

$$\cos(30-\theta) = \frac{2Z^2 + 1}{2\sqrt{3}\,Z} \tag{25}$$

When this calculated result is compared with the experimentally determined values of θ and Z for several alkali GICs (Fig. 39), the agreement between the calculated and experimental curves is seen to be quite good.

B.   Pressure-Induced Transitions

1. KC_{24}

Till now we have considered only temperature-induced structural phase transitions in donor GICs. In very recent experiments Wada et al.[140, 141] have shown that structural transitions in alkali GICs can also be induced when hydrostatic pressure is applied. The most clear-cut and elegant example of such a pressure-induced transition involves KC_{24}.

The pressure dependence of the $(00l)$ X-ray diffraction patterns of KC_{24} is shown in Fig. 40. The ambient pressure diffraction pattern Fig. 40a, is clear

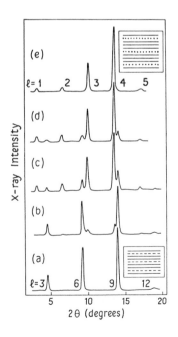

Fig. 40.   Pressure dependence of the room temperature $(00l)$ X-Ray diffraction profiles of KC_{24} prepared from HOPG. (a) Ambient pressure, (b) 2.5 kbar, (c) 4.0 kbar, (d) 5.5 kbar, (e) 6.5 kbar. Insets: (a) Stage 2 stacking (intercalant planes indicated by dashed lines), (e) Stage 3 stacking (ordered intercalant layers indicated by dotted lines).

evidence for a pure stage 2 KC_{24} GIC. However as the pressure is increased, notice that an additional set of (00l) reflections appears. These reflections gain intensity with increasing pressure at the expense of the stage 2 reflections until at $P \approx 7$ kbar, the pattern corresponds to a pure stage 3 compound. Note that the transition from stage 2 to stage 3 exhibited in Fig. 40 is completely reversible and that these two stages coexist in macroscopic ($\gtrsim$ 400 Å) domains at intermediate pressures.

The reversibility of the stacking transition from stage 2 to stage 3 is definitive evidence that this transition occurs at constant stoichiometry. For if potassium were ejected from the graphite layers when pressure was applied, it would have coalesced in the pressure-transmitting fluid (degassed paraffin oil) and would not have reentered the host lattice, thus rendering the transition irreversible. Therefore, the homogeneous stage 3 phase has a bulk stoichiometry close to KC_{24}, which necessarily requires a layer stoichiometry of KC_8. Such a stoichiometry immediately suggests that simultaneous

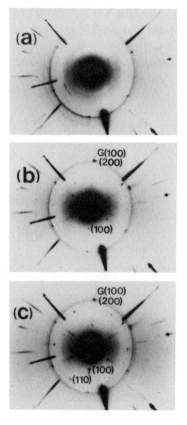

Fig. 41. X-Ray oscillation photographs of KC_{24} single crystals at (a) 1 kbar, (b) 4 kbar, and (c) 11 kbar. The intense streaks are X-ray reflections from the diamond anvils and the diffraction ring is from the beryllium-copper gasket of the pressure cell. Reflections $G(100)$ contain contributions from both host and potassium atoms. Other indices indicate potassium superlattice reflections.

with the above-described staging transition, the potassium atoms undergo an intralayer order-disorder transition to a $(2 \times 2)R0°$ superlattice identical to the structure in stage 1 heavy alkali GICs. Indeed such a structure is quite consistent with the observed pressure dependence of the in-plane X-ray diffraction patterns observed by Wada and co-workers[140, 141] (Fig. 41). Notice in particular that the $(2 \times 2)R0°$ spot patterns also evolve with pressure and saturate at a pressure of ≈ 7 kbar.

2. Other Donor GICs

A systematic study of the pressure dependence of the structure of cesium GICs and rubidium GICs has been carried out by Wada and Solin[142] and by Wada,[143] who also studied the pressure dependence of the Raman spectra. The effect of pressure on higher stage ($n > 2$) potassium GICs has also been examined.[144] Except for the stage 1 alkali GICs, to be discussed below, all alkali GICs studied to date exhibit staging transitions to pure or to mixed phases of one or more higher stages that coexist at elevated pressures. These staging transitions are accompanied by an intralayer ordering into a $(2 \times 2)R0°$ superlattice structure. Preliminary indications are that for $n \geq 2$, the *even-stage* heavy alkali GICs transform at sufficient pressure to higher pure stages compatible with the fixed stoichiometry of the ambient pressure compound. For example, stage 4 RbC_{48} transforms at ≈ 10 kbar to a pure stage 6 compound with bulk stoichiometry RbC_{48} and layer stoichiometry RbC_8.[143] In contrast, there is evidence that *odd*-stage GICs such as stage 3 RbC_{36} do not evolve to higher pure stages but always exhibit multiphase staging regardless of the applied pressure.[143] For the odd-stage alkali GICs, no pure higher stage is consistent with both the bulk stoichiometry and a layer stoichiometry of MC_8 ($M = K, Rb, Cs$).

Wada and co-workers have also studied the pressure dependence of the X-ray diffraction of stage 1 heavy alkali GICs.[141-145] For hydrostatic pressures up to 13 kbar, no intraplanar or interplanar structural transitions have been observed. The $(2 \times 2)R0°$ superlattice persists, albeit with a very slight in-plane contraction in its lattice parameters. There are, however, quite significant changes in the c-axis lattice parameters of stage 1 alkali GICs with pressure. In the pressure range 0–13 kbar, they exhibit a linear contraction of the c-axis lattice parameter. Given the assumption that the in-plane contractions are negligible compared to the c-axis contractions, the compressibility of the GIC can be expressed as

$$k_c = \frac{1}{c_0} \left(\frac{\partial c}{\partial P} \right)_{P=0} \cong \frac{1}{C_{33}} \qquad (26)$$

TABLE IV

Compressibilities, Package Thickness, and Coefficients of
Least-Squares Analysis (see text) of Stage 1 Heavy Alkali
GICs and of Pristine Graphite[a]

	Compressibility, $kc = \ln \partial c / \partial P$ $(\times 10^{-12} \text{ cm}^2/\text{dyne})$	Package thickness, c_0 (Å)
Graphite	2.73 ± 0.09	3.353 ± 0.003
KC_8	2.13 ± 0.09	5.355 ± 0.004
RbC_8	2.24 ± 0.12	5.657 ± 0.007
CsC_8	1.56 ± 0.05	5.943 ± 0.004

[a]From Refs. 142–145.

Here P_0 is the reference pressure and is equal to 1 bar, c is the pressure-dependent c-axis lattice parameter, and C_{33} is a component of the elastic stiffness tensor. The compressibilities of pristine graphite and stage 1 MC_8 GICs (M = K, Rb, Cs) measured by the X-ray methods described above are given in Table IV. Also shown in Table IV are the corresponding package thicknesses and the coefficient of least-squares analysis that indicates the accuracy of the straight-line fit to the data using (26). The compressibility given for RbC_8 is in reasonable agreement with the value determined from neutron scattering studies of its phonon dispersion curves.[96] The value given for pristine graphite is in excellent agreement with the results deduced using other techniques.[146, 147] These favorable comparisons confirm the validity of the X-ray technique for compressibility measurement and support the assumptions made in deriving (26). The compressibilities of various phases of higher stage heavy alkali GICs have also been measured by Wada.[143] These are tabulated in Ref. 143.

C.   Conflicts and Comments

In the foregoing discussion I have presented the facts as reported in the literature on phase transitions in donor GICs. By now the reader will no doubt have discerned the serious discrepancies and inconsistencies in the different experimental results on the structural properties of the alkali GICs. In no case are the observed discrepancies at the data level; that is, there are no examples of disagreements in which the same compound was measured with the same techniques in different laboratories. Thus the discrepancies involve interpretations of mutually exclusive measurements of what are assumed to be the same GICs.

Two major points of contention have emerged to date from the structural studies of alkali GICs. One involves the question of lattice-gas versus liquid-like behavior in the disordered phases of stage 2 MC_{24} compounds ($M =$ K, Rb), but no controversy exists for CsC_{24}. The basis for the disagreement rests, in my opinion, with the multiphase nature of the GICs in question. Thus the lattice-gas–liquid controversy may be simply a manifestation of the selectivity of a given probe to distinguish a particular structure; for example, EXAFS sees preferentially atoms located on hexagon centers.

The other major point of disagreement is related to the conflict between the results obtained using electron diffraction and other scattering techniques. At the heart of the disagreement is the observation with electron diffraction of certain ordered structures such as $(\sqrt{3} \times \sqrt{3})R30°$ superlattices, which have not been observed using X-ray or neutron diffraction. As noted above, this lack of confirmation cannot be attributed to finite crystal size effects. Therefore, I examine more carefully the viability of some of the structures reported in electron diffraction studies.

Consider the $(\sqrt{3} \times \sqrt{3})R30°$ structure, which is depicted in Fig. 21, and its possible observation in CsC_{24}. In that structure the cesium ions would be six fold coordinated and have a nearest-neighbor distance of 4.26 Å. If one naively assumed that cesium ions in CsC_{24} had an ionic radius equal to the quoted value 1.69 Å,[100] the ions could indeed occupy a structure as dense as $(\sqrt{3} \times \sqrt{3})R30°$. However, there is considerable evidence that the Cs—Cs pair potential is extremely repulsive at such small distances.

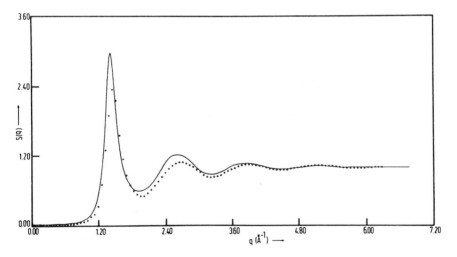

Fig. 42.   Structure factor $S(q)$ for bulk liquid cesium at 303°K. Solid curve, X-Ray, results; points, neutron results. From Ref. 148.

Note first that the in-plane liquid structure factor of CsC_{24} (Fig. 26a) is remarkably similar in quantitative detail to the structure factor of bulk liquid Cs,[148] which is shown in Fig. 42 for the corresponding temperature of 303°K. In particular the most probable near-neighbor separations between Cs ions is the same to within experimental error both in bulk liquid cesium and in CsC_{24}. There also is a zero probability of finding a pair of cesium ions at a separation as small as 4.26 Å in both cases, as can be deduced from the relevant RDFs. The pair potential for liquid cesium has been determined from the liquid structure factor using the Percus-Yevick[117] approximation (see Fig. 43). Notice that the potential diverges at an approximate separation r of 4.8 Å and that it is attractive at $r > 4.9$ Å. The rigidity of the electron gas in bulk

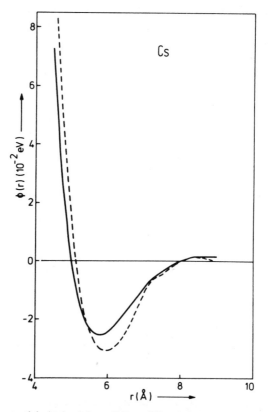

Fig. 43.   Pair potential obtained from X-Ray diffraction data from bulk liquid cesium at 373°K. Dashed curve hypernetted chain (HNC) approximation, solid curve, Heine-Abarenkov-type and Shaw-type model potential (see text). From Ref. 148.

liquid cesium contributes to the repulsive part of the potential. This contribution would be less significant in CsC_{24}, since part of the electron charge is delocalized in the bounding carbon layers. Nevertheless, the similarity of the liquid structure factors of CsC_{24} and bulk liquid cesium, together with the pair potential derived from the latter, suggests that a $(\sqrt{3} \times \sqrt{3})R30°$ structure cannot be sustained in graphite at room temperature and pressure regardless of the stage. The presence of that structure even at liquid helium temperatures is also very unlikely, since the pair potential is not expected to exhibit a dramatic temperature dependence.

As a final comment on this point, consider the pressure-induced order-disorder transitions discussed above. There is clear evidence that hydrostatic pressure produces a densification and $(2 \times 2)R0°$ ordering in MC_{24} compounds. Yet in no case was a more compact structure [e.g., $(\sqrt{3} \times \sqrt{3})R30°$] observed. Not even when stage 1 MC_8 compounds were compressed hydrostatically to 13 kbar did they exhibit any evidence in the X-ray diffraction patterns for an order-order transition to a $(\sqrt{3} \times \sqrt{3})R30°$ structure. Thus it appears that that particular structure may be spurious and an artifact of the electron diffraction experiments. For the reasons just expressed, I believe that $(\sqrt{3} \times \sqrt{3})R30°$ superlattices are unlikely in rubidium GICs and of marginal probability in potassium GICs. Otherwise those species would exhibit MC_6 stage 1 stoichiometries, which to my knowledge have never been observed during the 50 years in which the stage 1 alkali GICs have been studied with X-ray diffraction.

Other structures such as $(\sqrt{7} \times \sqrt{7})R19.1°$ and $(\sqrt{39} \times \sqrt{39})R0°$, which have been seen only with electron diffraction, are not precluded by ion size effects but must be verified by other techniques before they are seriously addressed in calculations of the relevant phase diagrams.

V. STRUCTURAL PHASE TRANSITIONS IN ACCEPTOR GRAPHITE INTERCALATION COMPOUNDS

A. Temperature-Induced Transitions

In contrast to the extensive studies of structural phase transitions in donor GICs there have been very few temperature-dependent studies of such transitions for acceptor GICs. To date only one acceptor system, nitric acid,[148–150] has been studied with anything like the detail of the donor studies described above. Besides nitric acid, temperature-induced order-disorder phase transitions have been reported for only a few other acceptor GICs. In these latter cases, the data and analyses are best described as fragmentary and preliminary.

522   S. A. SOLIN

1. Nitric Acid

Pure stages of nitric acid GICs ranging from stage 1 to stage 4 have been prepared and analyzed by Parry and co-workers,[149] Ubbelohde,[150] and Fuzellier.[151] There appears to be two idealized compositions depending on whether the HNO_3 molecules form a single or a double layer.[151] These compositions are respectively $C_{5n} HNO_3$ ($n=1,2,...$) with a sandwich thickness of $d_2 = 7.89 Å$ and $C_{8n} HNO_3$ ($n=1,2,...$) with a value $d_2 = 6.55 Å$. The nitric acid species readily form GICs with mixed stages,[149] but with care (e.g., slow intercalation) pure, well-staged compounds can be prepared.

The temperature dependence of the resistivity of stages 1 and 2 HNO_3 GICs[150] (Fig. 44) indicates a clear discontinuity at $T \approx -20°$ C: both the thermal expansion and the thermoelectric power of the stages 1 to 4 nitric acid GICs also exhibit anomalies at that temperature.[150] The most intriguing aspect of the anomalies in the transport properties and thermal expansion is

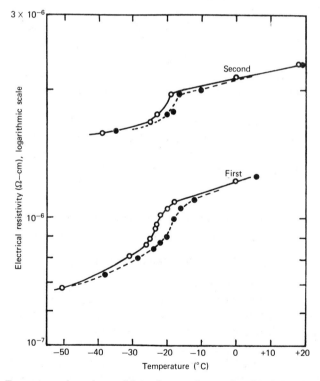

Fig. 44. Temperature dependence of the resistance of stages 1 and 2 nitric acid GICs. Original resistivity $\rho = 4.16 \times 10^{-5}$ Ω-cm. Open circles, temperature increasing; solid circles, temperature decreasing. From Ref. 150.

the stage independence of those features. The $-20°C$ features have been associated with an intralayer-interlayer, order-disorder transition in the HNO_3 layers.[149]

Nixon et al. found from X-ray measurements[149] that at room temperature the HNO_3 layers were disordered. Although this disorder was characterized as liquidlike, the assumption was based on a qualitative comparison of calculated and measured X-ray intensities. Nevertheless, data of Bottomley et al.[149] suggest a lattice-gas type of disorder to the HNO_3 GICs. The liquid-like disorder at room temperature is however clearly two dimensional, since X-ray photographs contain "streaks" of constant intensity parallel to the c^*-axis. The room temperature stacking sequences are given as A/A/A,..., A/AB/BC/CA/A,..., A/ABA/ACA/A,..., and A/ABAB/BCBC/CACA/A,..., for stages 1 through 4, respectively. Here the diagonals indicate a disordered HNO_3 liquid layer.

At $-20°$ C, the X-ray measurement indicates that each of the HNO_3 GICs, stages 1 through 4, undergoes an order-disorder transition. The HNO_3 molecules freeze into an incommensurate, ordered, intraplanar structure, the complete details of which are unknown. However, there is evidence that the HNO_3 molecules are not all coplanar in this ordered state.[149] As expected, the interlayer coupling, however weak, generates strong interlayer correlations between the ordered HNO_3 layers. Therefore, the resultant structures of HNO_3 GICs of stages 1 through 4 are clearly three dimensionally ordered at $T< -20°$ C.

2. Other Acceptor GICs

Order-disorder transitions have been reported for some metal fluoride GICs such as antimony pentafluoride and arsenic pentafluoride. In the case of AsF_5 the identification of an order-disorder transition appears to be based primarily on the observation of the temperature dependence of the motionally narrowed ^{19}F nuclear magnetic resonance (NMR) line.[152] Such an identification is highly speculative, since the intercalant species can certainly freeze into a statically disordered "glasslike" structure at reduced temperature. Thus broadening of the NMR line does not necessarily imply crystalline order.

In the case of AsF_5 there are also NMR data[153] indicating motional narrowing above the "melting" temperature. Hastings et al.[154] have reported X-ray studies of order-disorder transitions in low-stage AsF_5 GICs. However, details of these studies were not available at this writing.

Order-disorder transitions have been reported for Br_2, ICl, and IBR GICs.[155] These reports were based on electron diffraction studies of compounds from which much of the intercalant was desorbed in the microscope

S. A. SOLIN

vacuum. Therefore, there is considerable uncertainty about the stoichiometry as well as the ordered structures of those compounds.

Markiewicz[156] has recently employed a very simplified model to calculate
the critical temperature for order-disorder transitions in acceptor GICs. He
suggests that such transitions are driven by the ordering of the ionized fraction of molecules within a single layer. The model employed is applicable
only if the intercalant layer contains fully ionized species, which may also be
interspersed with neutrals. Thus it contains suppositions that are highly controversial and about which there is much debate.

The transition temperature for rational values of the charge exchange, f,
per intercalant molecule is deduced by computing the energy of a commensurate ionic lattice occupying part of the available carbon sites while
neutral molecules fill some of the remaining sites. An excited state of this
structure would consist of a Frenkel defect[157] in which a neutral and an
ionized molecule exchanged sites. The melting temperature T_c was estimated
from the energy E_D to create the Frenkel defect in analogy with AgBr[158] in
which the phase transition is, of course, first order and is created by an
avalanche of Frenkel defects. Thus $T_c \cong E_D/9K_B$. The values of T_c deduced

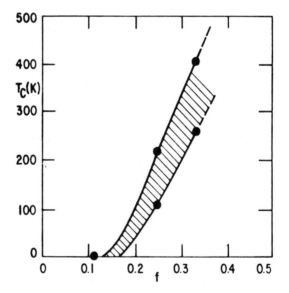

Fig. 45. Order-disorder temperature T_c as a function of charge transfer, f, for acceptor GICs.
Shaded region indicates allowed temperatures for varying degrees of hole screening; lower
bound = perfect screening (see text); upper bound = uniform hole density. Calculations were
made only at the points marked by solid circles. The curves are symmetric around $f = 0.5$. From
Ref. 156.

as a function of f and with E_D calculated both for perfect intralayer screening and for no screening are shown in Fig. 45. As can be seen, the deduced temperatures are at least of the right order of magnitude.

While the model employed by Markiewicz is interesting, a determination of its validity and/or success must await the acquisition of detailed experimental evidence (primarily X-ray and neutron diffraction) confirming and elucidating order-disorder transitions in all the acceptor GICs except those prepared with nitric acid. Note, however, that according to Fig. 45, GICs for which the charge exchange was <0.1 should exhibit no crystalline ordering of the intercalant layers even at $T=0$. Such a situation seems unlikely for massive molecular species whose zero-point motion should not be significant enough to induce a disordered state.

B.   Pressure-Induced Transitions

To date, there has been only one measurement of the effect of hydrostatic pressure on an acceptor GIC. Iye et al.[159] have studied the effect of hydrostatic pressure on the X-ray diffraction, the relative resistivity in the basal plane and along the c-axis, and the de Haas-van Alphen effect in several stages of HNO_3 GICs up to pressures of approximately 8 kbar. Their X-ray diffraction studies, which consisted of powder patterns, yielded no direct evidence of any intraplanar or interplanar phase transition. But these X-ray measurements were limited to 3.2 kbar and must therefore be considered to be preliminary. In contrast, the basal plane and c-axis pressure-dependent resistivities both indicated anomalies such as are shown for the latter in Fig. 46 for stages 3 and 4 HNO_3 GICs for pressures in the range of 5–6 kbar. Corresponding anomalies were also seen in the c-axis resistivity at the same pressures.[159]

Any attribution of the anomalies seen in the resistivity to order-disorder transitions would be speculative in the absence of confirmatory diffraction data. Nevertheless the data of Fig. 46 are highly suggestive of significant structural alterations in the pressure range 5–6 kbar. It is also interesting to note that the pressures at which the observed anomalies occurred were clearly stage dependent. If these anomalies do indeed correspond to structural phase changes, this result is in marked contrast to the temperature-induced transitions in HNO_3 GICs, which were found to occur at $-20°C$ independent of stage.

Iye et al.[159] also studied the compressibility of HNO_3 GICs by monitoring the changes in lattice parameters with pressure. Their results for pristine graphite and for a stage 3 HNO_3 GIC are summarized in Table V. Like Wada et al.,[140-145] who studied donor compounds, Iye et al.[159] found that the a-axis compressibility was negligible.

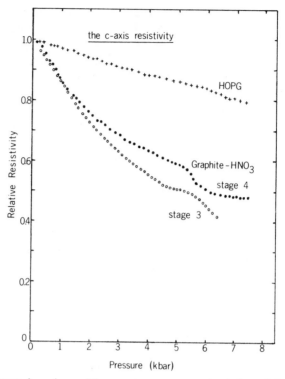

Fig. 46. Pressure dependence of the room temperature c-axis relative resistivity $\rho_c(P)/\rho_c(0)$ of pristine graphite (crosses), stage 4 HNO_3-graphite (solid circles), and stage 3 HNO_3-graphite (open circles). The resistivity values of the intercalates at zero pressure are 67 and 18 $\mu\Omega$-cm, respectively. From Ref. 159.

TABLE V
Lattice Parameters and Their Pressure Coefficients of Pristine Graphite and
Stage 3 Graphite-HNO_3^a

	Graphite	Stage 3 graphite-HNO_3
a_0 (Å)	2.468 ± 0.007	2.462 ± 0.008
$-\dfrac{d}{dp} \ln a (\times 10^{-12} \, cm^2/dyne)$	0.1 ± 0.3	0.0 ± 0.3
c_0 (Å)	6.722 ± 0.008	14.48 ± 0.01
$\dfrac{d}{dp} \ln c (\times 10^{-12} \, cm^2/dyne)$	2.9 ± 0.7	5.0 ± 1.0

^{a}From Ref. 159.

VI. CONCLUDING REMARKS

I have attempted to separate reality from myth, with the hope that by dealing with the former, no matter how complicated and difficult it is, we will enhance our ability to explain the intriguing physical phenomena exhibited by graphite intercalation compounds. Those phenomena are rooted in the many unusual structural properties and structural alterations such compounds exhibit. However, it should be clear from the preceding material that the structural studies of GICs carried out to date have raised many new questions while answering satisfactorily only a few of the old ones. For instance, although we may reasonably conclude that the room temperature structures of stage 1 heavy alkali GICs are well characterized and crystallographically accounted for, very little is known about the detailed structures of most acceptor compounds. Present work has shed light on some details of the short-range, two-dimensional disorder exhibited by higher stage ($n \geq 2$) heavy alkali GICs; yet the disordered structures of the rubidium and potassium compounds are still subject to considerable controversy at the most basic level—are they lattice-gas- or liquidlike or both.[2] Essentially, nothing is known about the nature of the short-range order in disordered acceptor GICs.

Although GICs are an ideal arena for the study of phase transitions, crossover phenomena, and dimensionality effects and their interrelationships, such studies are clearly in their infancy. Thus several clear-cut tasks lie ahead if the study of the structural properties of GICs is to progress in a serious way. We must:

1. Establish experimental phase diagrams that relate the structure of a GIC to temperature, externally applied pressure, and intercalant concetration.

2. Carry out detailed quantitative studies of the phase transitions already observed. Such studies will include the careful determination of critical exponents and the correlation of those exponents with the dimensionality of the GIC at temperatures and pressures spanning the phase transitions of interest.

3. Remove the discrepancies between electron diffraction results and results obtained using other diffraction techniques.

4. Carefully analyze X-ray and neutron diffraction line profiles and relate them to current theories of phase transitions in two dimensions.[160, 161]

If we can successfully accomplish the tasks, the renewed enthusiasm for GIC research will be entirely justified.

528   S. A. SOLIN

Acknowledgments

In preparing this chapter I drew heavily on knowledge gained from fruitful interactions with many collaborators and colleagues, who contributed immensely to the formulation of several of the ideas expressed here. Although the responsibility for the contents of this chapter rests solely with me, I take great pleasure in identifying those who were so generous with their intellectual contributions. I benefited greatly from several discussions with J. E. Fischer, D. Guérard, C. Horie, P. M. Horn, M. S. Isaacsson, T. Kaplan, S. Mahanti, S. C. Moss, R. J. Nemanich, M. F. Thorpe, M. Utlaüt, and H. Zabel. I am especially indebted to my close collaborators N. Caswell, Roy Clarke, D. M. Hwang, and N. Wada. Finally, like most experimentalists who study GICs, I owe considerable thanks to A. W. Moore, who has so generously supplied the pyrolytic graphite that is used for host material in much of the GIC research carried out all over the world.

The preparation of this chapter and the research described here in which I participated were supported by U.S. AROD under grant DAAG 29-80-K-0003 and by the National Science Foundation under grant DMR 80-10486.

References

1. A. Weiss, *Angew. Chem.*, **75**, 755 (1963)
2. W. Rüdorff, *Adv. Inorg. Chem. Radiochem.*, **1**, 223, 1959.
3. A. R. Ubbelohde and F. A. Lewis, *Graphite and Its Crystal Compounds*, Clarendon, Oxford, 1960.
4. G. R. Hennig, in *Progress in Inorganic Chemistry*, Vol. 1, F. A. Cotton, Ed., Interscience, New York, 1959, p. 125.
5. A. Hérold, N. Platzer, and R. Setton in *Les Carbones*, Vol. II, Masson, Paris, 1965, p. 462.
6. R. C. Croft, *Q. Rev. Chem. Soc.*, **14**, 19 (1960).
7. A. W. Moore, in *Chemistry and Physics of Carbon*, Dekker, New York, 1973 p. 69.
8. F. L. Vogel, *J. Mater. Sci.*, **12**, 982 (1977).
9. J. E. Fischer, in *Physics and Chemistry of Materials with Layered Structures*, F. Levy, Ed., Reidel, Dordrecht, 1979, p. 481.
10. J. E. Fischer, *Physica*, **99B**, 383 (1979).
11. L. B. Ebert, *Annu. Rev. Mater. Sci.*, **6**, 181 (1976).
12. M. S. Whittingham and L. B. Ebert, in *Physics and Chemistry of Materials with Layered Structures*, F. Levy, Ed., Reidel, Dordrecht, 1979, p. 533.
13. S. A. Solin, *Physica*, **99B**, 443 (1979).
14. M. S. Dresselhaus, G. Dresselhaus, P. C. Eklund, and D. D. L. Chung, *Mater. Sci. Eng.*, **31**, 141 (1977).
15. A. Hérold, in *Physics and Chemistry of Materials with Layered Structures*, F. Levy, Ed., Reidel, Dordrecht, 1979, p. 323.
16. R. W. G. Wycoff, *Crystal Structures*, Vol. I, Oxford University Press, London, 1962, p. 26.
17. S. A. Saffran, *Mater. Sci. Eng.*, **31** (1977), and references therein.
18. D. E. Nixon and G. S. Parry, *J. Phys. C* **2**, 1732 (1969).
19. W. Rudorff, *Z. Anorg. Allg. Chem.*, **254**, 319 (1947).
20. A. Hérold, *Mater. Sci. Eng.*, **31**, 1 (1977).
21. L. C. Hsu, H. Zelig, M. Rabinovitz, I. Argranat, and S. Sarig, *Inorg. Nuclear Chem. Lett.*, **11**, 601 (1975).
22. N. Caswell and S. A. Solin, *Solid State Commun.*, **27**, 961 (178).
23. Y. N. Novikov and M. E. Vol'pin, *Russ. Chem. Rev.*, **40** 733 (1971).
24. W. Rüdorff, *Chimia*, **19**, 489 (1965).

25. D. Billaud and A. Hérold, *Bull. Soc. Chim. Fr.*, **103**, 2042 (1974).
26. S. A. Solin, in *Proceedings of the Fourtheenth International Conference on the Physics of Semiconductors*, Kyoto, September 1979, H. Kamimura, Ed., in press.
27. S. A. Saffran and D. P. Hamann, *Phys. Rev. Lett.*, **42**, 1410 (1979).
28. S. A. Saffran, in *Proceedings of the Second International Conference on Intercalation Compounds of Graphite*, F. L. Vogel and A. Hérold, Eds., *Synthetic Metals*, in press.
29. A. Hérold, *Bull. Soc. Chim. Fr.*, 999 (1955).
30. D. E. Nixon and G. S. Parry, *J. Phys.* **D1**, 291 (1968).
31. W. Rüdorff and E. Shulze, *Z. Inorg. Chem.*, **277**, 156 (1954).
32. J. E. Fischer and T. E. Thompson, *Phys. Today*, **31**, 36 (1978).
33. A. R. Ubbelohde, *Proc. R. Soc. London, Ser. A*, **327**, 289 (1972).
34. A. R. Ubbelohde, *Mater. Sci. Eng.*, **31**, 341 (1980).
35. F. R. Gamble and T. H. Geballe, in *Treatise on Solid State Chemistry*, Vol. III, N. B. Hannay, Ed., Plenum, New York, 1976, p. 89.
36. M. L. Durus and G. Hennig, *J. Am. Chem. Soc.*, **79**, 5897 (1957).
37. W. Rüdorff and E. Shulze, *Z. Angew. Chem.*, **66**, 305 (1954).
38. N. Caswell, S. A. Solin, and W. Metz, *Bull. Am. Phys. Soc.*, **24**, 375 (1979).
39. W. Metz and D. Hohlwein, *Carbon*, **13**, 87 (1975).
40. W. Metz and E. J. Shulze, *Z. Kristallogr.*, **147**, 409 (1975).
41. D. Hohlwein, thesis, University of Hamburg, unpublished.
42. P. Oelhafen, E. Hauser, U. Gubler, J. Krieg, F. Greuter, P. Pfluger, V. Geiser, S. Stolz, and H. J. Guntherodt, in *Proceedings of the Second International Conference on Intercalation Compounds of Graphite*, F. L. Vogel and A. Herold, Eds., *Synthetic Metals*, in press.
43. P. Pfluger, P. Oelhafen, H. U. Kunzi, R. Jeher, E. Hause, K. P. Ackerman, M. Miller, and H. J. Guntherodt, *Physica*, **99B**, 395 (1980).
44. S. Y. Leung, C. Underhill, G. Dresselhaus, and M. S. Dresselhaus, *Solid State Commun.* **33**, 285 (1980).
45. S. A. Solin, *Mater. Sci. Eng.*, **31**, 153 (1977).
46. N. Caswell, S. A. Solin, and C. N. Papatheodorou, to be published.
47. N. Daumas and A. Hérold, *C.R. Acad. Sci., Ser. C*, **286**, 373 (1969).
48. D. E. Nixon, G. S. Parry, and A. R. Ubbelohde, *Proc. R. Soc. London, Ser. A*, **291**, 324 (1966).
49. N. Daumas and A. Hérold, *Bull. Soc. Chim. Fr.*, **5**, 1598 (1971).
50. Roy Clarke, N. Caswell, and S. A. Solin, *Phys. Rev. Lett.*, **42**, 61 (1979).
51. J. M. Thomas and G. R. Millward, *Mater. Res. Bull.*, **15**, 671 (1980).
52. N. Caswell, S. A. Solin, T. M. Hayes, and S. J. Hunter, *Physica*, **99B**, 463 (1980).
53. F. W. Lytle, D. E. Sayers, and E. A. Stern, *Phys. Rev. B* **11**, 4825 (1975).
54. E. A. Stern, D. E. Sayers, and F. W. Lytle, *Phys. Rev. B* **11**, 4836 (1975).
55. P. A. Lee and G. Beni, *Phys. Rev. B* **15**, 2862 (1977).
56. B. K. Teo and P. A. Lee, *J. Am. Chem. Soc.*, **101**, 2815 (1979).
57. T. M. Hayes, *J. Non-Cryst. Solids*, **31**, 57 (1978).
58. D. E. Nixon and G. S. Parry, *J. Phys. C* **2**, 1732 (1969).
59. A. Hérold, in *Physics and Chemistry of Materials with Layered Structures*, F. Levy, Ed., Reidel, Dordrecht, 1979, Table XVII, p. 412.
60. M. Mori, S. C. Moss, Y. M. Yan, and H. Zabel, to be published.
61. S. Y. Leung, C. Underhill, G. Dresselhaus, T. Krapcher, M. S. Dresselhaus, and B. Wuensch, *Bull. Am. Phys. Soc.*, **25**, 297 (1980).
62. S. Flandrois, J. Masson, J. Rouillon, J. Gaultier, and C. Hauw, in *Proceedings of the Second International Conference on Intercalation Compounds of Graphite*, F. L. Vogel and A. Hérold, Eds., *Synthetic Metals*, in press.

530   S. A. SOLIN

63. W. Rüdorff, *Adv. Inorg. Chem. Radiochem.*, **1**, 223, 1959.
64. N. Caswell, *Phys. Rev.*, in press.
65. *International Tables for X-Ray Crystallography*, Vol. III, Kynoch Press, Birmingham, England, 1972, p. 162.
66. B. Carton, thesis, University of Nancy, unpublished. See also A. Hérold, in *Physics and Chemistry of Materials with Layered Structures*, F. Levy, Ed., Reidel, Dordrecht, 1979 p. 323.
67. F. J. Salzano and S. Aronson, *J. Inorg. Nuclear Chem.*, **26**, 1456 (1964).
68. F. J. Salzano and S. Aronson, *J. Chem. Phys.*, **45**, 4551 (1966); **47**, 2978 (1967); **43**, 149 (1965); **42**, 1323 (1965).
69. C. Underhill, T. Krapcher, and M. S. Dresselhaus, in *Proceedings of the Second International Conference on Intercalation Compounds of Graphite*, F. L. Vogel and A. Hérold, Eds., *Synthetic Metals*, in press.
70. S. Safran, *Phys. Rev. Lett.*, **44**, 937 (1980).
71. S. A. Safran, in *Proceedings of the Second International Conference on Intercalation Compounds of Graphite*, F. L. Vogel and A. Hérold, Eds., *Synthetic Metals*, in press.
72. E. Stumpp, *Mater. Sci. Eng.*, **31**, 53 (1977).
73. M. C. Robert, M. Oberlin, and J. Mering, in *Chemistry and Physics of Carbon*, Vol. 10, P. Walker and P. A. Thrower, Eds., Pergamon, Oxford, 1973, p. 141. See also Refs. 2 to 6.
74. J. J. Song, D. D. L. Chung, P. C. Eklund, and M. S. Dresselhaus, *Solid State Commun.*, **20**, 1111 (1976).
75. D. Guérard and A. Hérold, *Carbon*, **13**, 337 (1975).
76. *Handbook of Chemistry and Physics*, C. D. Hodgman, Ed., 44th ed., Chemical Rubber Publishing Co., Cleveland, 1963, p. 2437.
77. S. Basu, C. Zeller, P. Flanders, C. D. Fuerst, W. D. Johnson, and J. E. Fischer, *Mater. Sci. Eng.*, **38**, 275 (1979).
78. D. P. Riley, *Proc. Phys. Soc. London*, **57**, 486 (1945).
79. P. Lagrange, M. El Makrini, D. Guérard, and A. Hérold, *Physica*, **99B**, 473 (1980).
80. P. Lagrange, M. El Makrini, D. Guérard, and A. Hérold, in *Proceedings of the Second International Conference on Intercalated Compounds of Graphite*, F. L. Vogel and A. Hérold, Eds. *Synthetic Metals*, in press.
81. For FeCl$_3$ see W. Rüdorff and H. Schulze, *Z. Anorg. Allg. Chem.*, **245**, 121 (1940). Also, G. Schoppen, H. Mayer-Spasche, L. Siemsgluss, and W. Metz, *Mater. Sci. Eng.*, **31**, 115 (1977); for FeCl$_2$ see Y. Novikov, M. E. Vol'pin, V. E. Prusakov, R. A. Stukan, V. I. Gol'danskii, V. A. Semiou, and Y. T. Struchkov, *Zh. Strukt. Khim.*, **11**, 880 (1970).
82. J. G. Hooley, *Carbon*, **13**, 469 (1975).
83. J. G. Hooley, *Mater. Sci. Eng.*, **31**, 17 (1977).
84. A. Schleede and M. Wellman, *Z. Phys. Chem.* B 18, 1 (1934).
85. See Refs. 24 and 31 and references therein.
86. The notation $(S \times T)R\theta°$ corresponds to a superlattice cell with primitive translations of length Sa and Ta, respectively, and noted by $\theta°$ with respect to the primitive cell of the carbon layers, the lattice constant of which is a.
87. G. S. Parry, *Mater. Sci. Eng.*, **31**, 99 (1977) and references therein.
88. G. S. Parry, in *Proceedings of the Third Conference on Industrial Carbons and Graphite*, Society of the Chemical Industry, London, 1971.
89. P. Lagrange, D. Guérard, and A. Hérold, *Ann. Chim. Fr.*, **3**, 143 (1978).
90. P. Lagrange, thesis, University of Nancy, 1977.
91. P. Lagrange, D. Guérard, M. El Makrini and A. Hérold, *C. R. Acad. Sci.*, Ser. C, **287**, 179 (1978).
92. R. Clarke, N. Caswell, and S. A. Solin, *Phys. Rev. Lett.*, **42**, 61 (1979).

93. H. P. Klug and L. E. Alexander, *X-Ray Diffraction Procedures for Polycrystalline and Amorphous Materials*, Wiley, New York, 1974, and references therein.
94. C. N. Yang and T. d. Lee, *Phys. Rev.*, **87**, 410 (1952).
95. F. J. Salzano and S. Aronson, *J. Chem. Phys.*, **42**, 1323 (1965).
96. W. D. Ellenson, O. Semmingsen, D. Guérard, D. G. Onn, and J. E. Fischer, *Mater. Sci. Eng.*, **31**, 137 (1977).
97. N. Kambe, G. Dresselhaus, and M. S. Dresselhaus, *Phys. Rev.* **B 8**, 3491 (1980).
98. E. L. Evans and J. M. Thomas, *J. Solid State Chem.*, **14**, 99 (1975).
99. R. Juza and V. Wehle, *Naturwissenschaften*, **52**, 20 (1965).
100. L. Pauling, *The Nature of the chemical Bond*, 3rd ed., Cornell University Press, Ithaca, N.Y., 1960.
101. N. Kambe, M. S. Dresselhaus, G. Dresselhaus, S. Basu, and J. E. Fischer, *Mater. Sci. Eng.*, **40**, 1 (1979).
102. J. Rossat-Mignod, D. Fruchart, M. J. Moran, J. W. Milliken, and J. E. Fischer, in *Proceedings of the Second International Conference, on Intercalation Compounds of Graphite*, F. L. Vogel and a. Hérold, Eds., *Synthetic Metals*, in press.
103. P. Bak and E. Domany, *Phys. Rev.*, **B 20**, 2818 (1979).
104. R. B. Potts, *Proc. Cambridge Phil. Soc.*, **48**, 106 (1952).
105. L. D. Landau and E. M. Lifshitz, *Statistical Physics*, Addison-Wesley, Reading, Mass., 1969, Chapter 13.
106. Chang-Rim Lee, H. Aoki, and H. Kamimura, *J. Phys. Soc. Japan*, **49**, 870 (1980).
107. See Refs. 2 and 37 and references therein.
108. G. S. Parry and D. E. Nixon, *Nature (London)*, **216**, 909 (1967).
109. G. S. Parry, D. E. Nixon, K. M. Lester, and B. C. Levene, *J. Phys. C 2*, 2156 (1969).
110. D. G. Onn, G. M. T. Foley, and J. E. Fischer, *Mater. Sci. Eng.*, **31**, 271 (1977).
111. R. Clark, N. Caswell, S. A. Solin, and P. M Horn, *Phys. Rev. Lett.*, **43**, 2018 (1979).
112. R. Clark, in *Proceedings of the International Conference on Ordering in Two Dimensions*, in press.
113. See Ref. 26 and references therein.
114. R. Clarke, N. Caswell, S. A. Solin, and P. M. Horn, *Physica*, **99B**, 457 (1980).
115. S. Ergun and M. Berman, *Acta Crystallogr.*, Sect. A, **29**, 12 (1973).
116. M. Plischke, to be published.
117. J. K. Percus and G. J. Yevick, *Phys. Rev.*, **110**, 1 (1958).
118. H. E. Stanely, *Introduction to Phase Transitions and Critical Phenomena*, Oxford University Press, New York, 1971.
119. S. Geller, *Z. Kristallogr.* **125**, 1 (1967).
120. Y. Imry and L. Gunther, *Phys. Rev. B.* **3**, 3939 (1971).
121. A. Gunier, *X-Ray Diffraction in Crystals, Imperfect Crystals and Amorphous Bodies*, Freeman, San Francisco, 1963.
122. G. S. Parry, in *Proceedings of the Second International Conference on Intercalation Compounds of Graphite*, F. L. Vogel and A. Hérold, Eds., *Synthetic Metals*, in press.
123. J. B. Hastings, W. B. Ellenson, and J. E. Fischer, *Phys. Rev. Lett.*, **42**, 1552 (1979).
124. D. E. Moncton, F. J. DiSalvo, J. D. Axe, L. J. Sham, and B. R. Patton, *Phys. Rev. B* **14**, 3432 (1976).
125. B. E. Warren, *X-Ray Diffraction*, Addison-Wesley, Reading, Mass., 1969, pp. 264–275.
126. H. Zabel, S. C. Moss, N. Caswell, and S. A. Solin, *Phys. Rev. Lett.*, **43**, 2022 (1979).
127. H. Zabel, Y. M. Yan, and S. C. Moss, *Physica*, **99b**, 453 (1980).
128. See Ref. 93, p. 793.
129. M. Winokur, J. H. Rose, and Roy Clarke, to be published.
130. E. Garboczi and M. F. Thorpe, unpublished results.

131. S. C. Moss and H. Zabel, to be published.
132. S. C. Moss, Y. M. Yan, and H. Zabel, *Bull. Am. Phys. Soc.*, **23**, 298 (1980).
133. A. N. Berker, N. Kambe, G. Dresselhaus, and M. S. Dresselhaus, *PRL* **45**, 1452 (1980).
134. See Ref. 60 and references therein.
135. H. Suematsu, M. Suzuki, and H. Ikeda, *J. Phys. Soc. Japan*, **49**, (1980) in press.
136. M. Suzuki, H. Ikeda, H. Suematsu, Y. Endoh, and H. Shiba, *Proceedings of the Fourth Yamada Conference on Physics and Chemistry of Layered Materials*, Sendai, 1980, *Physica B*, in press.
137. M. S. Dresselhaus, N. Kambe, T. Krapcher, and A. N. Berker, in *Proceedings of the Second International Conference on Intercalation Compounds of Graphite*, F. L. Vogel and A. Hérold, Eds., *Synthetic Metals*, in press.
138. Y. Yamada, T. Watanabe, T. Kiichi, and H. Suematsu, *Proceedings of the Fourth Yamada Conference on Physics and Chemistry of Layered Materials*, Sendai, 1980 *Physica B*, in press.
139. Y. Yamada, T. Watanabe, T. Kiichi, and H. Suematsu, to be published.
140. R. Clarke, N. Wada, and S. A. Solin, *Phys. Rev. Lett.*, **44**, 1616 (1980).
141. N. Wada, R. Clarke, and S. A. Solin, *Proceedings of the Second International Conference on Intercalation Compounds of Graphite*, F. L. Vogel and A. Hérold, Eds., *Synthetic Metals*, in press.
142. N. Wada and S. A. Solin, *Proceedings of the Fourth Yamada Conf. on Physics and Chemistry of Layered Materials*, Sendai, 1980 *Physica B*, in press.
143. N. Wada, thesis, University of Chicago, to be published.
144. N. Wada, to be published.
145. N. Wada, Roy Clarke, and S. A. Solin, *Solid State Commun.*, in press.
146. O. L. Blakslee, D. G. Proctor, E. J. Seldin, G. B. Spence, and T. Weng, *J. Appl. Phys.*, **41**, 3373 (1970).
147. R. Nicklow, N. Wakabayashi, and H. G. Smith, *Phys. Rev. B* **5**, 4951 (1972).
148. M. J. Huijben and W. van der Lugt, *Acta. Crystallogr.*, *Sect. A*, **35**, 431 (1979).
149. M. J. Bottomley, G. S. Parry, and A. R. Ubbelohde, *Proc. R. Soc. London*, *Ser. A*, **297**, 291 (1964).
150. A. R. Ubbelohde, *Proc. R. Soc. London*, *Ser. A*, **304**, 25, (1968).
151. H. Fuzellier, thesis, University of Nancy, 1974.
152. L. B. Ebert and H. Zelig, *Mater. Sci. Eng.*, **31**, 177 (1977).
153. L. B. Ebert, D. R. Mills, and J. C. Scanlon, *Mater. Res. Bull.* **14**, 1369 (1979).
154. J. B. Hastings, C. R. Fincher, Jr., M. Moran, and J. E. Fischer, to be published.
155. D. D. L. Chung, G. Dresselhaus, and M. S. Dresselhaus, *Mater. Sci. Eng.*, **31**, 107 (1977).
156. R. S. Markiewicz, in *Proceedings of the International Conference on Ordering in Two Dimensions*, in press.
157. N. B. Hannay, *Solid State Chemistry*, Prentice-Hall, Englewood Cliffs, N.J., 1967.
158. Z. Matyas, *Czech J. Phys.*, **4**, 14 (1954).
159. Y. Iye, O. Takahashi, S. Tanuma, K. Tsuji, and S. Minomura, in *Proceedings of the Second International Conference, on Intercalation Compounds of Graphite*, F. L. Vogel and A. Hérold, Eds., *Synthetic Metals*, in press.
160. B. I. Halperin and D. R. Nelson, *Phys. Rev. Lett.*, **41**, 121 (1978).
161. J. M. Kosterlitz and D. J. Thouless, *J. Phys. C.* **6**, 1181 (1973).

ANGLE-RESOLVED PHOTOEMISSION AS A TOOL FOR THE STUDY OF SURFACES

E. W. PLUMMER AND W. EBERHARDT

Department of Physics
University of Pennsylvania
Philadelphia, Pennsylvania

CONTENTS

I. INTRODUCTION

Angle-resolved photoelectron spectroscopy, as a tool for studying surfaces, has come a long way in the past 5 years. A 1975 review of the applications of photoemission spectroscopy to the study of metal surfaces clearly showed that the adaptation of angle-resolved detection was in its infancy.[1] In 1975 only four or five papers had been published, investigating the angular dependence of the photoemitted signal from intrinsic or extrinsic surface states.[2] For this chapter on angle-resolved photoemission from surfaces, however, more than 400 papers were compiled. The science always drives the advancement of a new technique. In this case it is the ability to measure the electronic properties of a surface in microscopic detail. But there have been two important developments that assured the short-term success and the long-range applicability of angle-resolved photoemission. First is the utilization, appreciation, and expansion of synchrotron radiation sources throughout the world.[3] The second development is the application of group theory to the measurement process.[4-6] The continuum of polarized intense radiation from the synchrotron makes it possible to do measurements that are not possible with conventional light sources available in the laboratory. The symmetry rules make interpretation of the data easy. This chapter stresses the importance of a variable-energy, polarized light source to achieve the maximum advantage from symmetry rules governing the excitation mechanism.

Photoelectron spectroscopy is an experimental technique that measures the kinetic energy of an emitted electron when a photon is absorbed. If an electron is to be photoemitted, the energy of the incident light must be larger than the ionization potential of an atom or molecule, or the work function of a solid. Therefore ultraviolet[7] to X-ray light sources[8] are commonly used. For a given energy of the incident light ($\hbar\omega$), the peaks in the energy distribution of the emitted electrons reflect the nature of the quantum states of the system being studied. The spectra of atoms or molecules have discrete lines from the different quantum energy levels[7, 8] The spectrum from a solid may show a broad continuum in the region of the occupied valence bands. Conventional angle-integrated photoelectron spectroscopy measures the energy and intensity of the peaks in the spectra as a function of the physical and chemical properties of the sample. For example, the photoelectron spectra of CO could be recorded as an isolated molecule, bound to a metal in the form of a carbonyl [e.g., $W(CO)_6$] or adsorbed onto a surface.[9]

Angle-resolved detection adds a new dimension to photoelectron spectroscopy. The intensity and energy dependence can be measured as a function of the collection angle with respect to some fixed axes. Angle-resolved photoemission becomes more powerful for systems with a high degree of symmetry. For example, the emission intensity from a gas-phase diatomic molecule is measured with respect to the direction of the polarization of the light, which is the only axis in the experiment. If the molecule is fixed in space, there are two important directions, the direction of the light polarization and the molecular axis. If the molecules are arranged in a two-dimensional ordered array, there may be as many as four symmetry axes to be considered. We illustrate by examples what can be learned from angle-resolved photoelectron spectroscopy of surfaces. It is impossible to discuss in this chapter all the systems that have been studied with angle-resolved photoemission, or to consider such important measurements as high-resolution, core-level spectroscopy, which do not require angle-resolved detection.[10]

II. THE PHOTOIONIZATION PROCESS

A. Theory

This section presents the theory of photoionization and illustrates its application to randomly oriented, molecular oriented, and two- and three-dimensional ordered systems. The transition probability per unit time between two eigenfunctions of the same Hamiltonian H_0 denoted by ψ_i and ψ_f is given by Fermi's golden rule, if the perturbation H' is small.

$$\frac{d\omega}{dt} = \frac{2\pi}{\hbar} |\langle \psi_f | H' | \psi_i \rangle|^2 \{\delta(E_f - E_i - \hbar\omega)\} \tag{1}$$

The perturbation to the system caused by the incident radiation is found by replacing the momentum operator $\mathbf{P}$ by $\mathbf{P} + (e/c)\mathbf{A}$, in the original Hamiltonian H_0. So that the full Hamiltonian is

$$\mathcal{K} = H_0 + \frac{e}{2mc}(\mathbf{A} \cdot \mathbf{P} + \mathbf{P} \cdot \mathbf{A}) - e\Phi + \frac{e^2}{2mc^2} |A|^2 \tag{2}$$

where $\mathbf{A}$ and ϕ are the vector and scalar potentials of the incident light field. It is always possible to choose a gauge where $\Phi = 0$. The differential photoionization cross-section can be written as:

$$\frac{d\sigma}{d\Omega} \propto |\langle \psi_f | \mathbf{A} \cdot \mathbf{P} + \mathbf{P} \cdot \mathbf{A} | \psi_i \rangle|^2 \{\delta(E_f - E_i - \hbar\omega)\} \tag{3a}$$

The diamagnetic term $|A|^2$ is always small and is neglected. If we use the commutator $[\mathbf{P}, \mathbf{A}] = -i\hbar \nabla \cdot \mathbf{A}$ we have

$$\frac{d\sigma}{d\Omega} \propto |\langle \psi_f | 2\mathbf{A} \cdot \mathbf{P} - i\hbar \nabla \cdot \mathbf{A} | \psi_i \rangle|^2 \{\delta(E_f - E_i - \hbar\omega)\} \tag{3b}$$

The term $\nabla \cdot \mathbf{A}$ is usually assumed to be small or zero. If the vector potential $\mathbf{A}$ is written as a plane wave in free space, then

$$\mathbf{A}(r, t) = \mathbf{A}_0 \exp(-i\omega t + i\mathbf{q} \cdot \mathbf{r}) \tag{4}$$

This is a transverse wave, so that $\mathbf{A}_0 \cdot \mathbf{q} = 0$ and $\nabla \cdot \mathbf{A} = 0$. At a surface $\nabla \cdot \mathbf{A}$ is usually not zero and may not be small compared to $\mathbf{A} \cdot \mathbf{P}$.[11-13] There is always a discontinuity in the component of $\mathbf{A}$ perpendicular to the surface whenever the dielectric constant is not equal to 1 inside the material. This causes a charge imbalance in the surface region, and longitudinal fields may result from the interaction of the electrons in the solid with the surface charge.[14] These induced electromagnetic fields differ significantly from the long-wavelength transverse applied field (4). We proceed in the next three sections ignoring the $\nabla \cdot \mathbf{A}$ term and return to discuss its importance in Section II.F.

If we assume that $\nabla \cdot \mathbf{A}$ term in (3b) is small and that the vector potential is represented by (4), the differential cross-section can be simplified. The momentum of the light is so small compared to the electron momentum or, alternatively, the wavelength of the light is so large compared to atomic dimensions, that we can treat $\mathbf{A}$ as a constant. The momentum of the photon is ignored, which is a good approximation below photon energies of ~ 10 keV. Equation 3 becomes

$$\frac{d\sigma}{d\Omega} \propto |\langle \psi_f | \mathbf{P} | \psi_i \rangle \cdot \mathbf{A}_0|^2 \delta(E_f - E_i - \hbar\omega) \tag{5a}$$

This "electric dipole" matrix element can be converted into two alternate forms by using the commutation relation of H_0 with $\mathbf{P}$ and $\mathbf{r}$, yielding[15]

$$\frac{d\sigma}{d\Omega} \propto |\langle \psi_f | \mathbf{r} | \psi_i \rangle \cdot \mathbf{A}_0|^2 \delta(E_f - E_i - \hbar\omega) \tag{5b}$$

$$\frac{d\sigma}{d\Omega} \propto |\langle \psi_f | \nabla V | \psi_i \rangle \cdot \mathbf{A}_0|^2 \delta(E_f - E_i - \hbar\omega). \tag{5c}$$

where V is the potential in H_0. Note that all forms of (5) are incorrect if $\nabla \cdot \mathbf{A} \neq 0$.

Until now we have said nothing about the character of the wave functions ψ_i and ψ_f. They could be the exact many-body eigenstates of the system, but it is preferable to work with single-particle or self-consistent field wave functions whenever possible. Since $\mathbf{P}$ is a single-particle operator, the differential cross-section in the simplest single-particle representation is the matrix element between a one-electron wave function ϕ_i for the electron in the ith orbital of the neutral and a wave function u_k representing the excited electron with momentum k. This single-particle model leads to a simple picture; there will be a peak in the photoelectron spectrum for each single-particle state of the system (energy must be conserved). It is a straightforward operation to illustrate the ramification of a breakdown in the single-particle picture.[15a] We write the ground-state N particle wave function ψ_0^N as

$$\psi_0^N = \phi_i \Psi_i^{N-1}$$

Where ϕ_i is the single-particle wave function of the electron to be removed and Ψ_i^{N-1} is the properly antisymmetrized determinant of the remaining $N-1$ electrons. Likewise the final state in the excitation process can be written as

$$\psi_f = u_k \Phi_{i,j}(N-1)$$

where the index i for the $N-1$ wave function $\Phi_{i,j}$ denotes that there is an electron missing from the ith orbital, which has been removed by the adsorption of the photon. The index j labels all excited states of the ion with a hole in the ith orbital. The photoionization cross-section becomes

$$\frac{d\sigma}{d\Omega} \propto |\langle u_k | \mathbf{A} \cdot \mathbf{P} | \phi_i \rangle \langle \Phi_{ij} | \psi_i \rangle|^2$$

If an independent-particle picture is correct, there will be no rearrangement of charge in the ionic state Φ_{ij} because of the missing electron. In this case $\langle \Phi_{ij} | \psi_i \rangle = \delta_{i,0}$ and we are back to the single-particle picture. But we know that there will be a rearrangement of the wave functions in the ion, which means that the function Ψ_i^{N-1} is not an eigenstate of the ion; consequently when it is projected onto the true eigenstates of the ion Φ_{ij}, it will have an overlap with states where $j \neq 0$ (i.e., excited states of the ion). The excited states have two holes and one excited electron, and are therefore commonly referred to as two-hole, single-particle states. Each ionic configuration (i, j) has its own energy. Consequently we would expect to see satellite lines accompanying the main single-particle peak in the photoelectron spectra when the single-particle picture breaks down. These satellite lines are referred to as shakeup.

The conservation of energy in (5) can be evaluated in more detail. The total energy before absorption of the photon must equal the total energy after absorption:

$$\hbar\omega + E_0^N = E^{N-1}(i) + E_{\text{kin}}(i)$$

or

$$E_{\text{kin}}(i) = \hbar\omega + E_0^N - E^{N-1}(i) \qquad (6)$$

where E_0^N is the total energy of the neutral particle system in the ground state,

$E^{N-1}(i)$ is the total energy of the $N-1$ ionic system in the ith state, and $E_{\text{kin}(i)}$ is the kinetic energy of the peak in the photoelectron spectrum corresponding to the ith state of the ion. Note here that we can easily change (6) to account for the excited states of the ion discussed in the preceding paragraph. We just add a second index to the ionic energy, $E^{N-1}(i, j)$.

Equation 6 shows that the kinetic energy of the photoemitted electron from the ith state increases linearly with increasing $\hbar\omega$. It is useful to define an energy that is characteristic of the state i and independent of the photon energy. This energy is called the binding energy $E_B(i)$, defined

$$E_B(i) = \hbar\omega - E_{\text{kin}}(i)$$
$$E_B(i) = E^{N-1}(i) - E_0^N. \tag{7}$$

Photoelectron spectroscopy is an ion spectroscopy. It measures the differences in the energies of the various ionic states i. For example, there is only one peak in the photoelectron spectrum of atomic hydrogen, at $E_B = 13.6$ eV. On the other hand, the photoelectron spectrum of neon shows three peaks. One at $E_B = 870$ eV for removal of a $1s$ electron, one at $E_B = 48.4$ eV for removal of a $2s$ electron, and a peak at $E_B = 25.6$ eV for the $2p$ levels.[8]

Figure 1 shows an example of the photon energy dependence of the spectra for a molecule, carbon monoxide, that is discussed in more detail later. The symmetry of a CO molecule is $C_{\infty v}$, so the individual molecular orbitals are classified according to σ or π symmetry with respect to the molecular axis. The most deeply bound level is the O_{1s} state denoted as the 1σ level, at a binding energy of 542.6 eV. The C_{1s} level (2σ) is next at $E_B = 296.2$ eV (not shown in Fig. 1). All the remaining outer energy levels are molecular. The C—O σ-bonding level (3σ) is at $E_B = 38.3$ eV, and the doubly degenerate bonding π levels (1π) are at $E_B = 16.8$ eV. There are two additional σ states that are like lone pair orbitals on the oxygen (4σ) and carbon (5σ) at $E_B = 19.7$ and 14.0 eV, respectively. All the photoelectron spectra shown in Fig. 1 are plotted as a function of binding energy (7). Curve a is for $\hbar\omega = 18$ eV, so photoionization can occur only from the two outer valence orbitals, 5σ and 1π. The 4σ is too deep, $E_b \gg \hbar\omega$. The kinetic energy of an electron photoexcited from the 5σ level is $\hbar\omega - E_B = 4$ eV. Curve b is for $\hbar\omega = 34$ eV, where the 5σ, 1π, and 4σ states can be seen. The kinetic energy of an electron excited from the 5σ level is now 20 ev. Curve c is for $\hbar\omega = 1254$ eV, where all the CO energy levels are excited.[8,17] At $\hbar\omega = 1254$ eV an electron excited from the CO 5σ level has a kinetic energy of 1240 eV.

Figure 1 shows that (7) works, that is, the binding energy is independent of $\hbar\omega$. The relative intensity is not independent of $\hbar\omega$. For example, notice how the σ states increase in intensity relative to the 1π as the photon energy is increased. These cross-sectional effects are due to the change in the matrix

PHOTOELECTRON SPECTRA OF CO

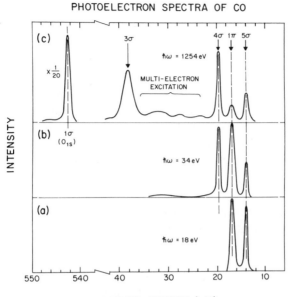

Fig. 1. Photoelectron spectra for CO as a function of photon energy.[8,16,17] These spectra are artistic reproductions of real spectra; thus the instrumental resolution is fixed. Curve c does not show the C_{1s} level.

element given in (5) as the final state changes due to increasing the photon energy. One other point illustrated in Fig. 1 is worth mentioning. The broad spectral features between the 3σ and 4σ lines are a consequence of the breakdown of the single-particle picture.[18] They correspond to excited states of the ion in which there are two holes and one excited electron. An example would be an ionic state in which the 4σ electron has been removed and an electron has been excited from the occupied 1π to the unoccupied 2π.

The next sections discuss the advantages of angle-resolved photoelectron spectroscopy. Section II.B describes what happens when a randomly oriented gas-phase molecule is fixed in space. Sections II.C to II.E will describe angle-resolved measurements on ordered two-dimensional and three-dimensional systems respectively. Finally Section II. F returns to the problem of $\nabla \cdot \mathbf{A}$ in (3b).

B. Excitation of Molecules

The angular dependence of the photoemission from a randomly oriented gas-phase molecule is not isotropic because the polarization vector fixes a direction. The simplest way to see this effect is to assume that the final state in the differential cross-section (5) is a plane wave; $\psi_f \propto e^{i\mathbf{k \cdot r}}$. The average overall molecular orientation gives a differential cross-section[1]

$$\frac{d\sigma}{d\Omega} = I\cos^2\gamma$$

where γ is the angle between the polarization vector $\mathbf{A}$ and the collection direction. The intensity is always zero when $\gamma = 90°$ for a free-electron final state. General theoretical considerations for gas-phase photoionization show that the differential cross-section is the form[19]

$$\frac{d\sigma}{d\Omega} = I\left[1 + \frac{\beta}{2}(3\cos^2\gamma - 1)\right] \tag{8}$$

where β is the asymmetry parameter, which depends on the initial and final wave functions; β can range from -1 to $+2$; $\beta = 2$ for a plane wave final state.

Figure 2 shows in column a the differential cross-section for the 1π level at $\hbar\omega = 21$ eV and 4σ at $\hbar\omega = 41$ eV of gas-phase CO^{20} These cross-sections were calculated by Davenport,[21] with the β's being 0.4 for the 1π and 1.35 for the 4σ. The low value of β for excitation from the 1π at $\hbar\omega = 21$ eV makes the angular distribution nearly isotropic [$\beta = 0$ in (8)]. The β-value for the 4σ approaches 2, so that the angular distribution approaches $\cos^2\gamma$. Columns b and c show the dramatic effects in the differential cross-section when the axis of the molecule is fixed in space. Column b is for the molecular axis perpendicular to the polarization direction, and column c is for the molecular axis parallel to the polarization. The variations in the differential cross-sections are much larger for the oriented molecule than for the randomly oriented molecule. The angular patterns are so unique that they can be used to determine the symmetry of an orbital if the molecular axis orientation were known, or vice versa.

Figure 2 allows us to illustrate for the first time the power of symmetry arguments applied to the differential cross-section in (5). If we place the detector along the molecular axis of CO, the final state ψ_f must have σ symmetry, because π symmetry states have zero amplitude along the axis. If $\mathbf{A}$ is aligned along the molecular axis, $\mathbf{A \cdot P}$ in (5a) is of σ symmetry. The matrix element is nonzero only if ψ_i has σ symmetry. The π states cannot be excited into the direction of the molecular axis with a polarization field parallel to the axis. The differential cross-sections in $3c$ and $4c$ of Fig. 2 show this behavior. If the detection is along the axis but the polarization is perpendicular to the molecular axis, then $\mathbf{A \cdot P}$ has π symmetry and only initial states of π symmetry can be excited. The matrix element must be symmetric with respect to all the symmetry operations of the molecule that leave the detector

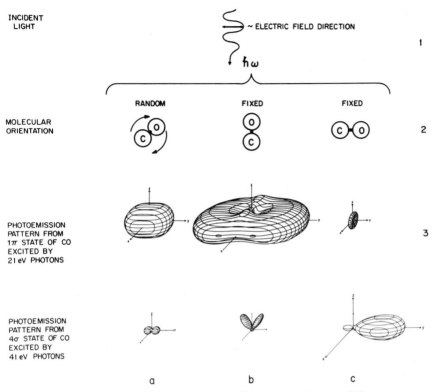

INCIDENT
LIGHT

~ ELECTRIC FIELD DIRECTION

1

$\hbar\omega$

RANDOM   FIXED   FIXED

MOLECULAR
ORIENTATION

2

PHOTOEMISSION
PATTERN FROM
1π STATE OF CO
EXCITED BY
21 eV PHOTONS

3

PHOTOEMISSION
PATTERN FROM
4σ STATE OF CO
EXCITED BY
41 eV PHOTONS

a   b   c

Fig. 2. Differential photoemission cross-sections for excitation from two different quantum states of the CO molecule for various orientations with respect to the exciting electric field. Row 1 shows the field orientation, row 2 the molecular orientation, rows 3 and 4 the differential cross-sections for the 1π and 4σ levels, respectively.[20,21]

position invariant, or it will vanish (see 3b and 4b). The differential cross-section shown in 4b of Fig. 2 indicates that there is a plane perpendicular to the polarization direction (xz-plane of Fig. 2) where the cross-section from a σ state is always zero when **A** is perpendicular to the molecular axis. If we define a plane by the molecular axis and the detection direction, the final state must always be even with respect to reflection about this plane, otherwise $\psi_f = 0$ in the plane. An odd final state has to have a node in the plane because it is a continuous function. When the polarization vector is perpendicular to this plane, it is odd with respect to reflection about the plane. Therefore only initial states that are odd with respect to this plane can be excited. The only odd states are π states.

If we knew the orientation of the CO molecules, we could identify σ and π states by placing the detector along the axis and flipping the polarization from parallel to perpendicular with respect to the molecular axis. If on the other hand, we know the symmetry of an orbital, we can determine the molecular axis. For example, if the detector were perpendicular to the polarization, we would see a null in the σ cross-section when the polarization was perpendicular to the molecular axis. If we aligned the polarization and the detector, we would see a null in the π cross-section when **A** was parallel to the molecular axis.

Let us jump ahead slightly to illustrate the use of symmetry rules to determine the orientation of the CO axis when CO is adsorbed on a surface. In 1976 Smith et al.[22] showed the first example of simple-symmetry selection rules to determine the geometrical configuration of an adsorbed molecule. Figure 3 shows two angle-resolved photoelectron spectra for CO adsorbed on Ni(100).[22] The vertical bars at the top are the energies of the gas-phase CO peaks. The connecting lines show how these levels shift when CO is bound to the surface. Section IV.A explains in detail why the levels move in the way they do. Here it is important to understand only that the peak in Fig. 3a that is about 11 eV below the Fermi energy is the CO 4σ level. The peak that is about 8 eV below E_f is the nearly degenerate CO 1π and the bonding 5σ level. The peak near the Fermi energy is due to the Ni $3d$ bands. Curve a is for normal emission with p-polarized light, which has components of **A** parallel and perpendicular to the surface normal. Curve b is normal collection for s-polarized light, which has **A** perpendicular to the surface

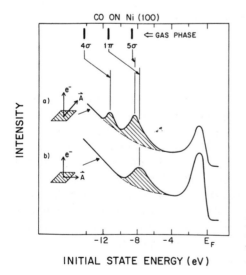

Fig. 3. Normal emission photoemission spectra for 3×10^{-6} torr-sec exposure of CO onto Ni(100) surface at room temperature ($\hbar\omega = 28$ eV for both curves). Curve a is for p-polarized light, and curve b is for s-polarized light.[22]

normal. The 4σ level is missing in Fig. 3*b*; therefore, **A** is perpendicular to the molecular axis and consequently the molecular axis is parallel to the surface normal. This also means that the peak in curve *b* of Fig. 3 is due to excitation from the 1π level. The 1π is 7.9 eV below E_f, while the 5σ and 4σ are 8.3 and 11.2 eV, respectively, below E_f. The 5σ and 1π levels of adsorbed CO have switched order compared to gas-phase CO.

The previous discussions have concentrated on the effects due to the symmetry of the initial state ψ_i. In many cases the effects due to the symmetry of the final state are just as dramatic. Such diatomic molecules as CO, N_2, and NO have resonances in the continuum.[16, 23] These resonances are due to virtual bound states trapped by an angular momentum barrier.[21, 24] For CO, N_2, and NO this virtual bound state has σ symmetry. Figure 4*a* shows the measured[16] and calculated[25] absolute cross-sections for the ionization from

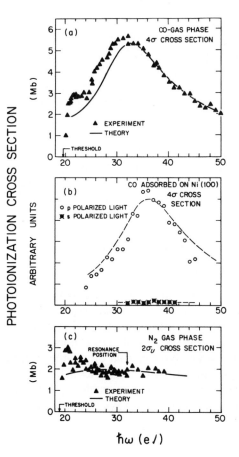

Fig. 4. Partial photoionization cross-sections from the 4σ level of CO and $N_2(2\sigma_u)$ as a function of photon energy. (*a*) Gas-phase CO: experiment[16] and theory.[25] (*b*) CO adsorbed on Ni(100),[26] showing normal collection with *p*-polarized light (open circles) and collection 7° off-normal and *s*-polarized light (solid crossed circles). (*c*) Gas phase N_2: experiment[16] and theory.[21] The dashed line through the open circles in (*b*) is the theoretical curve from (*a*) shifted by ~4 eV.

the CO 4σ level through the photon energy range where the resonance exists. The theoretical description of the resonance is excellent. The resonance in the gas phase occurs at a photon energy of 32 eV or a kinetic energy of about 12 eV. Figure 4b shows that the resonance still exists for CO adsorbed on a surface.[26] The resonant photon energy is shifted to about 36 eV and the kinetic energy to about 20 eV.

The final state in the resonance is σ symmetry. This means that only the component of the polarization along the axis of the molecule can excite a σ-symmetry initial state into the σ-symmetry final state. Therefore if the CO molecule is bound with its axis parallel to the surface normal as Fig. 3 indicated, the resonance cannot be observed with s-polarized light, because A is perpendicular to the molecular axis. The second curve in Fig. 4b shows the cross-section for 7° off-normal collection and s-polarized light.[26] There is no resonance present, that is, the molecular axis is perpendicular to the polarization vector A.

Figure 4c shows the photoionization cross-section from the N_2 $2\sigma_u$ level, which is the equivalent of the CO 4σ level. N_2 has higher symmetry because there is inversion symmetry about the plane perpendicular to the molecular axis, and equidistant from each N atom. This means that each wave function has odd (u, $ungerade$) or even (g, $gerade$) symmetry with respect to this mirror plane. The resonant state in N_2 has σ_u symmetry.[21] We have already presented the symmetry arguments to prove that a σ can be excited into a σ state only by the component of A parallel to the molecular axis. The polarization vector, which must be parallel to the molecular axis to excite σ into σ, is odd with respect to the new mirror plane in the N_2 molecule. Therefore, the matrix element is $\langle odd|odd|odd\rangle = 0$. The resonance is ruled out for the σ_u initial states, but exists for initial states of σ_g symmetry.[16, 21] The arrow in Fig. 4c shows where the resonant state would have been.

When N_2 is adsorbed on the surface, the symmetry will be broken and the resonance can be seen. This is one of two examples given in Section IV.

C. Two-Dimensional Systems

Surfaces are two-dimensional and surfaces are what we are writing about, so it may seem unnecessary to have sections for two-dimensional and three-dimensional systems. However a surface is a two-dimensional system that terminates a three-dimensional system, and we must illustrate how to recognize and analyze structure in a photoemission spectrum originating from a two- or three-dimensional system. Fortunately nature furnishes materials that are periodic in two directions. The layer compounds, such as, TaS_2 and $TaSe_2$, as well as graphite and intercalated graphite, have been used as prototype two-dimensional systems.

It is inherently easier to interpret angle-resolved photoemission data from an ordered two-dimensional system than from an ordered three-dimensional system. The excited electrons must escape into the vacuum to be measured. When they cross the "perfect" two-dimensional interface between the solid and vacuum, the parallel component of momentum ($k_{\parallel}$) is conserved,* but the perpendicular component ($k_{\perp}$) is changed. This means that when $k_{\parallel}$ is measured outside, we know $k_{\parallel}$ inside:

$$k_{\parallel} = \sqrt{\frac{2m}{\hbar^2} E_{\mathrm{kin}}} \sin\theta \qquad (9)$$

where θ is the angle between the surface normal and the detector. The $k_{\perp}$ is not conserved across the surface, and knowing $k_{\perp}$ (outside) does not tell you the value of $k_{\perp}$ (inside), unless you know the nature of $\psi_f(k)$ or $E(k)$ inside the solid (see next sections). This complication with three-dimensional systems drove such early investicators as Smith and Traum away from measurements of three-dimensional solids like GaAs.[27] They turned their attention instead to the investigation of two-dimensional materials.[28-30] The first demonstration that the band structure of layer compounds could be mapped out using angle-resolved photoemission measurements was the work of Smith, Traum, and DiSalvo.[28]

The layer compounds such as transition-metal dichalcogenides (e.g., TaS_2) consist of covalently bonded sandwiches (S—Ta—S) that are loosely coupled to one another by van der Waals interaction. They naturally cleave between the sandwiches, leaving a well-ordered surface exposed. Figure 5a shows angle-resolved photoelectron spectra from TaS_2 as a function of the polar angle of collection.[29] The peaks in the spectra move in energy as a function of the collection angle or using (9) as a function of $k_{\parallel}$. This is the band structure of the solid, $E(k_{\parallel})$. Figure 5b plots the peak positions versus the measured $k_{\parallel}$ (9) from data like that shown in Fig. 5a.[29] The solid lines are the calculated band structure of IT-TaS$_2$ by Mattheiss,[31] with the S $2p$ derived bands (below -1 eV) shifted upward by about ~ 0.8 eV to agree with the data.

The movement in energy of peaks in the photoelectron spectra with collection angle shown in Fig. 5a could not have been anticipated from our discussion of angular dependence from atoms or molecules Section II.B. The intensity of the emission from a specific quantum state of an atom or molecule changes with collection angle (Fig. 2), but the energy is independent of the geometry of the detector. Section II.B discussed atomic and molecular

---

*$k_{\parallel}$ is conserved $\pm \mathbf{G}_s$, where $\mathbf{G}_s$ is a surface reciprocal lattice vector.

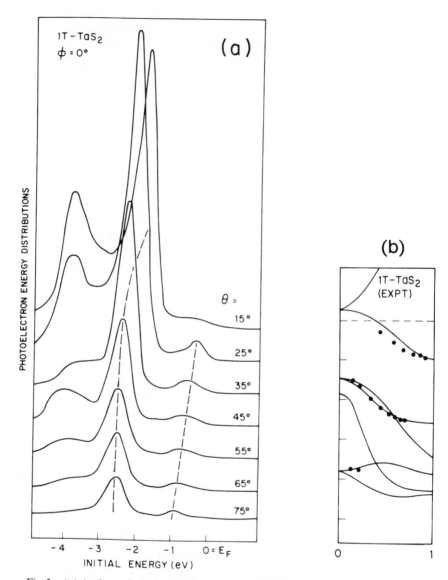

Fig. 5. (a) Angle-resolved photoelectron spectra of 1T-TaS$_2$ at $\hbar\omega = 10.2$ eV for a sequence of polar angles.[29] The dashed lines show the dispersion of the peaks. (b) Peak positions in energy versus parallel wave vector $\mathbf{k}_{\parallel}$. Solid curves are the theoretical band structure of TaS$_2$.[31]

physics, and the behavior shown in Fig. 5 is solid-state physics. The wave vector $k_\parallel$ is a good quantum number of the system, and the energy of the eigenstate depends on $k_\parallel$. This is the electronic band structure of a solid, which can be measured using angle-resolved photoemission.

The position in energy of the peaks in the angle-resolved photoelectron spectra from a two-dimensional material can be used directly to map out the band structure. The intensity variations as a function of collection angle can in principle tell us about the spatial character of the wave functions involved in the transition. For example, let us assume that the final state in the excitation is a plane wave $e^{i \mathbf{k}_f \cdot \mathbf{r}}$; then the matrix element in (5a) becomes[32]

$$M_{1f} \propto \mathbf{A} \cdot \mathbf{k}_f \phi(k_f)$$

where $\phi(k_f)$ is the Fourier transform of the initial state ψ_i. As we move the detector or the sample, we change k_f, and M^2 should change, reflecting the characteristics of $\Phi(k_f)$ modulated by $\mathbf{A} \cdot \mathbf{k}_f$.

The measurement of the angular dependence of the emission from TaS_2 was one of the first measurements to combine angle-resolved detection, polarized synchrotron radiation, and single-crystal samples.[30] These measurements were made by fixing $\mathbf{k}_f$ and $\mathbf{A}$ and rotating the sample about an axis normal to the surface (azimuthal pattern). In this experiment $|k_\parallel|$ stays fixed but the direction of $\mathbf{k}_f$ with respect to the two-dimensional crystallographic directions changes. Figure 6a shows the geometry of the IT-TaS_2 surface. The arrow and the surface normal define the plane of detection for Fig. 5. Figure 6b shows an "azimuthal plot" compared to the theoretical calculation of Liebsch.[33] These curves are the intensity of the Ta $5d$ band peak (peak near E_f in Fig. 5a) as a function of azimuthal direction of the detector. The polar collection angle is 42° from the normal, and the polarization vector is in the surface plane perpendicular to the detector direction.

The first general comment about Fig. 6b is that the free-electron final state model predicted zero amplitude in this geometry because $\mathbf{A} \cdot \mathbf{k}_f = 0$. The detector is located in a plane normal to the polarization vector of the light. Smith et al.[30] showed that this was an artifact of using a single plane wave. A final state that is composed of a linear combination of plane waves does not predict that the intensity is zero in the geometry of Fig. 6. This should not be surprising, since any state ψ_f can be written as a linear combination of plane waves. When ψ_f is written in this form, the angular dependence is not the Fourier transform of the initial state.[30]

The second observation concerns the symmetry of the azimuthal pattern shown in Fig. 6b. The Ta layer or the S layers have sixfold symmetry, but when they are stacked together the symmetry is only C_{3v}. Therefore, the

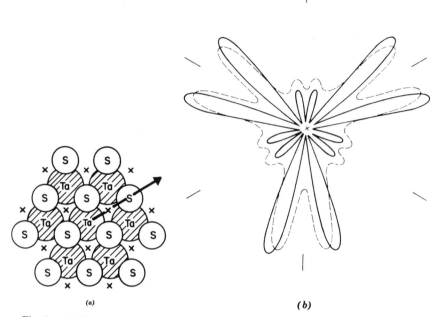

Fig. 6. (a) Geometrical structure of 1T-TaS$_2$. Crosses are the positions of the S atoms in the plane under the Ta. (b) Comparison of experimental azimuthal plot (dashed)[30] with corresponding theoretical results (solid)[33] ($\hbar\omega = 16$ eV).

azimuthal pattern should be three fold, not six fold symmetric, which it is. There are three mirror planes 120° apart for this structure. When the detector is in one of the mirror planes, the final state must be even with respect to the mirror plane. The amplitude of ψ_f would be zero in the mirror plane if it was odd. We denote the even final state symmetry by $\langle \oplus |$. In the experiment shown in Fig. 6b the polarization vector is perpendicular to the plane in which the detector is positioned. Therefore, when the detector is in one of the mirror planes of the crystal as indicated by the six lines in Fig. 6b, $\mathbf{A} \cdot \mathbf{P}$ is odd with respect to the mirror plane. Our photoexcitation matrix element is $\langle \oplus | \theta | \psi_i \rangle$. The ψ_i must be odd or the matrix element is zero. Since the filled portion of the Ta $5d$ band, which is measured in the experiment, is even with respect to the mirror plane,[31, 33] the intensity must be zero in the directions marked by the lines in Fig. 6b. The theoretical curves goes to zero in the mirror plane directions, but the experimental curve does not. This could be a consequence of not having 100% polarization, of the finite angular acceptance, or of an imperfect sample. If the light is not perfectly polarized, there will be a component of **A** in the mirror plane. This component is even so $\langle \oplus|\oplus|\psi_{5d}\rangle$ is allowed, since $|\psi_{5d}\rangle = |\oplus\rangle$.

Liebsch[33] calculated the angular dependence for the Ta $5d$ band using unpolarized light and a plane wave final state, in an attempt to evaluate the effects of the real final state. The plane wave final state azimuthal pattern did not look like the experimental or correct final state theoretical curve for the same geometry.[33] There appears to be no easy way to isolate initial and final state angular variations in the experimental azimuthal or polar angle plots. If the initial state is a core level, the angular dependence will be due to the final state[34] (see Section IV.B).

D. Photoexcitation in a Periodic Potential

In discussing the photoexcitation process in the presence of a two- or three-dimensional periodic potential from a conceptual point of view, we ignore practical problems in photoemission such as transport across the surface and strong inelastic scattering in the excited state.

The simplest of all periodic potentials is the constant potential. The wave functions are plane waves and the dispersion is given by $E = \hbar^2 k^2/2m$. Figure 7a shows the dispersion for the electron and photon. The difference in momentum between the initial and final states that conserves energy is

$$\Delta k = k_f \left[1 - \sqrt{1 - \frac{\hbar\omega}{E_f}} \right]$$

This momentum is much too large to be furnished by the momentum of the photon. Energy and momentum cannot be simultaneously conserved in

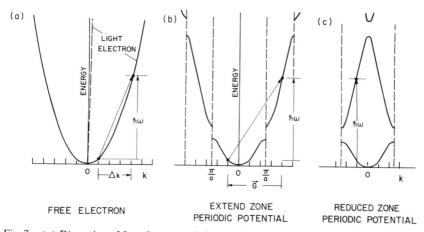

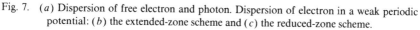

Fig. 7. (a) Dispersion of free electron and photon. Dispersion of electron in a weak periodic potential: (b) the extended-zone scheme and (c) the reduced-zone scheme.

photoexcitation from a translationally invariant system for these low-energy photons. The photoexcitation matrix element in the form of (5c) would have told us that the cross-section is zero, since $\nabla V = 0$.

A spatially varying potential must be introduced to photoexcite an electron. We want to consider a periodic potential where

$$V(\mathbf{r} + \mathbf{R}) = V(\mathbf{r})$$

and $\mathbf{R}$ defines the points in the Bravais lattice[35]

$$\mathbf{R} = N_1 \mathbf{a}_1 + N_2 \mathbf{a}_2 + N_3 \mathbf{a}_3 \tag{10}$$

with the N's being integers and $\mathbf{a}$'s the primitive lattice vectors. The Fourier transform of a periodic function like $V(r)$ can only have components with the periodicity of the real Bravais lattic,[35] so

$$V(r) = \sum_G V_G e^{i\mathbf{G} \cdot \mathbf{r}} \tag{11}$$

The $\mathbf{G}$'s are the reciprocal lattice vectors. If the $\mathbf{a}$'s defining the real-space primitive lattice are orthogonal, the $\mathbf{G}$'s are given by

$$\mathbf{G} = M_1 \frac{2\pi}{a_1^2} \mathbf{a}_1 + M_2 \frac{2\pi}{a_2^2} \mathbf{a}_2 + M_3 \frac{2\pi}{a_3^2} \mathbf{a}_3.$$

It is well known that for a weak periodic potential gaps are created when there are states in Fig. 7a with the same energy separated in $\mathbf{k}$ by a reciprocal lattice vector $\mathbf{G}$. Figure 7b shows the dispersion for an electron in a potential that is periodic, with a lattice spacing of a.

The photoionization matrix element in the form of (5c) applied to the potential given in (11) gives

$$\frac{d\sigma}{d\Omega} \propto \left| \sum_G \langle \psi_f | \mathbf{A} \cdot \mathbf{G} V_G e^{i\mathbf{G} \cdot \mathbf{r}} | \psi_i \rangle \right|^2 \delta(E_f - E_i - \hbar\omega) \tag{12}$$

The initial and final states must both be Bloch states, such that $\psi_\mathbf{k}(\mathbf{r} + \mathbf{R}) = e^{i\mathbf{k} \cdot \mathbf{R}} \psi_\mathbf{k}(\mathbf{r})$,[35] where $\mathbf{R}$ is given by (10). These states can be written as

$$\psi_{n\mathbf{k}}(\mathbf{r}) = e^{i\mathbf{k} \cdot \mathbf{r}} u_{n\mathbf{k}}(\mathbf{r})$$

with $u_{n\mathbf{k}}(\mathbf{r} + \mathbf{R}) = u(\mathbf{r})$, n is a band index. Putting this into (12) gives

$$\frac{d\sigma(k_f)}{d\Omega} \propto \left| \sum_G \langle u_{n\mathbf{k}_f} | \mathbf{A} \cdot \mathbf{G} V_G | u_{m(\mathbf{k}_f - \mathbf{G})} \rangle \right|^2 \tag{13}$$

where $E_f = E_i + \hbar\omega$ and $\mathbf{k}_f = \mathbf{k}_i + \mathbf{G}$. The crystals lattice furnishes the momentum in quantities of $\mathbf{G}$. Equation 13 brings out the fact that direct interband optical transitions are *Umklapp* processes because they involve not only the absorption of a photon $\hbar\omega$, but also diffraction against the crystal lattice through a reciprocal lattice vector $\mathbf{G}$.

Figure 7b shows an allowed transition conserving energy and momentum with the addition of a reciprocal lattice vector. This way of plotting the bands is called the "extended-zone scheme" in k-space. Since all functions in the presence of the periodic potential must be periodic in k-space, every band can be folded back into the first zone ($-\pi/a < k < \pi/a$ in Fig. 7). Figure 7c shows the reduced-zone scheme corresponding to the extended zone of Fig. 7b. The interband transition is vertical and has been called a "direct transition" because the value of reduced k is conserved. The reduced-zone scheme is convenient to determine which transitions are allowed for a given photon energy. Equation 13 must be used to decide whether the transition is forbidden by symmetry. The extended-zone scheme of Fig. 7b is more descriptive for photoemission.[36] The investigator decides what final state energy and direction are of interest, then connects all allowed initial states at the correct energy, displaced from the final state by any $\mathbf{G}$.

The foregoing discussion has implicitly assumed an infinite three-dimensional periodic potential. Let us now discuss briefly the effect of a surface. The easiest case to understand is the free-electron system bounded by a barrier. If z is defined as the direction normal to the surface, we have

$$\nabla V = \frac{\partial V}{\partial z} \hat{e}_z$$

where $\hat{e}_z$ is the unit vector in the z-direction. Now (5c) becomes

$$\frac{d\sigma}{d\Omega} \propto |\langle \psi_f | A_z \frac{\partial V}{\partial z} | \psi_i \rangle|^2 \delta(E_f - E_i - \hbar\omega)$$

This is what has been referred to in the literature as the surface photoelectric effect.[36, 37] Since it picks out only the A_z-component of the polarization field, it is often referred to as the vectorial photoeffect.[38] The Fourier transforms of $V(z)$ has every component allowed since it is not periodic. Therefore conservation of k_z is not a question. However $\mathbf{k}_\parallel$ must still be conserved in the excitation if there is a periodic potential in the parallel direction. Section II.F discusses the surface photoeffect in more detail.

E. Three-Dimensional Systems

The electronic structure of a surface is intimately related to the bulk electronic structure. Therefore, it is important in studies of surface properties to be able to measure the bulk band structure. Angle-resolved photoemission is the only experimental tool presently available that can measure the energy dispersion and symmetry of the bulk bands. Yet it is not a straightforward process to obtain the three-dimensional energy bands from the angle-resolved photoemission data. This is because the three-dimensional solid is viewed by the detector through a two-dimensional window—the surface. The kinetic energy and the momentum of the emitted electron are determined by the angle-resolved analyzer:

$$E_{kin} = \frac{\hbar^2}{2m}\left(k_\parallel^2 + k_\perp^2\right) \tag{14}$$

where $k_\parallel$ and $k_\perp$ are the components of momentum in the vacuum, parallel and perpendicular to the surface. Knowing the momentum and energy outside, we would like to determine the momentum and energy of the photoexcited electron inside the crystal and then E_i and k_i of the initial state from which the electron was excited.

Energy is conserved in the excitation process, so the total energy of the final state E_f is given by

$$E_f = E_i + \hbar\omega$$

If we arbitrarily define the zero of energy as the Fermi energy, the kinetic energy outside is

$$E_{kin} = E_f - e\phi \tag{15}$$

where ϕ is the work function. Using this definition of total energy, the initial state energy E_i is always negative, being zero at the Fermi energy. This notation was already used in Fig. 3 and 5.

The momentum is not conserved as the electron crosses the surface. First there is a potential energy step at the surface that decreases the component of the kinetic energy perpendicular to the surface as an electron emerges from the solid. Second, because of the periodic array of ion cores, the dispersion of an electron inside the crystal is not free electron-like. That is, E_f does not depend on the square of momentum as it does in free space [see (14)]. All that we know about the momentum of the final state inside the solid is that the component parallel to the surface will be conserved moduli, a surface reciprocal lattice vector G_S,

$$k_\parallel(\text{outside}) = k_\parallel(\text{inside}) + G_S \tag{16}$$

The perpendicular component of the photoexcited state inside the solid is undetermined experimentally. The value of $k_\perp$ could be anywhere on a rod in k-space with a fixed $k_\parallel$. This is illustrated in Fig. 8, where the bulk

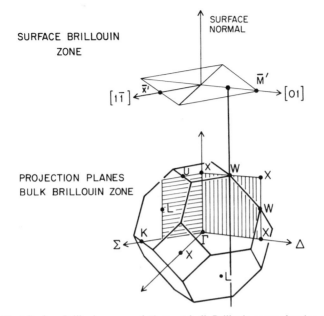

Fig. 8. (*Top*) Surface Brillouin zone and (*bottom*) bulk Brillouin zone; showing the projection of the bulk onto the surface. The solid line represents a rod connecting all points in the bulk zone with the one point in the surface zone with the same value of $k_\parallel$.

Brillouin zone is shown at the bottom for a face-centered cubic (fcc) solid and the surface Brillouin zone (SBZ) for the (100) surface is shown at the top. The determination of $k_\parallel$ (outside) fixes a point on the two-dimensional SBZ; $k_\perp$ can have a value anywhere along the rod extending into the three-dimensional Brillouin zone. One such rod is shown in Fig. 8. The Appendix presents the SBZ and bulk Brillouin zones for the low-index faces of body-centered cubic, face-centered cubic, and hexagonal close-packed crystals.

We view the excitation process in the direct transition model (see Fig. 7). The crystal lattice acts as the source of momentum to conserve momentum between the initial state and the final state. Thus,

$$\mathbf{k}_f = \mathbf{k}_i + \mathbf{G}_B \tag{17}$$

where $\mathbf{G}_B$ is the bulk reciprocal lattice vector.[35]

Equation 17, when combined with (15) and (16), gives the energy and momentum of the initial state

$$k_\parallel \text{ (inside)} = k_\parallel \text{ (outside)} - \mathbf{G}_S - (\mathbf{G}_B)_\parallel \tag{18a}$$

$$E_i = E_{\text{kin}} + e\phi - \hbar\omega \tag{18b}$$

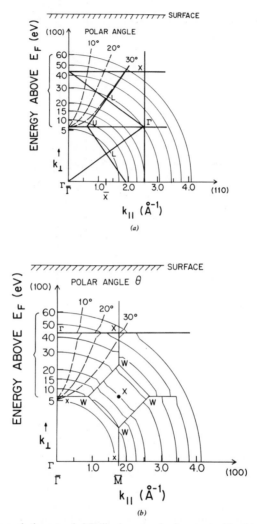

Fig. 9. Cuts through the extended Brillouin zone of a face-centered cubic crystal. The vertical axis is $k_\perp$ relative to a (100) surface. The horizontal axis is the momentum parallel to the surface. ($\mathbf{k}_\parallel$). (a) $\mathbf{k}_\parallel$ is in the (110) mirror plane. (b) $\mathbf{k}_\parallel$ is in the (100) mirror plane. The circles are contours of a constant energy inside the crystal, assuming a free-electron final state. The dashed lines show the $\mathbf{k}$ versus E location for transitions seen with a detector of fixed collection angle. The effects of band gaps in the final state are visible in (b).

Where $(G_B)_{\parallel}$ is the component of the bulk reciprocal lattice vector, parallel to the surface. $k_{\perp}(i)$ can be determined only by knowing $E_f(\mathbf{k})$. Figure 8 shows the cuts through the bulk Brillouin zone from the two mirror planes of the (100) face. Figures 9a and 9b plot these two slices in k-space in the extended zone scheme. The parallel component of momentum is plotted on the parallel axis, and the perpendicular component of momentum is displayed on the vertical axis.

Two examples illustrate how these diagrams are used. First, consider the case of a detector that is normal to the (100) surface (i.e., $\mathbf{k}_{\parallel}=0$). Then any direct transition occurs to a final state along the rod Γ to X to Γ, and so on, in the extended bulk zone. If the detector is set off of normal at $\mathbf{k}_{\parallel} = \bar{M}$ (Fig. 9b), then the final state of the excitation lies along the X-W-X symmetry line in the bulk. If E_f versus $\mathbf{k}_f$ is known, then $k_{\perp}$ (inside) can be calculated from the experimentally determined $\mathbf{k}_{\parallel}$ (outside) and E_{kin}. Once $\mathbf{k}$ (final) has been determined, the possible $\mathbf{k}_i$'s can be determined by subtracting bulk $\mathbf{G}$'s (Fig. 7b).

Normal emission simplifies this process of analysis. The component of momentum parallel to surface is zero, so (18a) becomes

$$\mathbf{k}_{\parallel}(i) = -\mathbf{G}_S - (\mathbf{G}_B)_{\parallel}. \tag{19}$$

We will neglect any Umklapp process and discuss its effects later; this assumption gives $\mathbf{k}_{\parallel}$ (final)=0 for normal emission. The final state lies in a high-symmetry direction of the bulk. This high symmetry rules out nearly all symmetry final states because they have no amplitude in the direction of the detector.[39] For example, for an fcc structure the $\Delta_1(\Sigma_1$ and $\Lambda_1)$ band is the only detectable final band for (100) [(110) and (111)] surfaces in the normal emission collection geometry. Figure 10 shows the calculated band structure for copper in the Γ-X direction, which is the normal direction to a (100) surface.[40] On the right we show a hypothetical normal emission photoemission spectrum taken at a photon energy of about 26 eV. There are two final state bands of symmetry Δ_1 and Δ_5 shown in the folded-back, reduced-zone scheme. We can immediately ignore the Δ_5 band because it has the wrong symmetry. Peak 1 in the hypothetical spectrum has an initial state energy of -3.6 eV and a final state energy with respect to the Fermi energy of 22.4 eV. Figure 10 shows that this energy corresponds to a point A_1 on the final band in the extended-zone scheme with $k_{\perp} = 1.64(2\pi/a)$, where a is the lattice constant. The $\mathbf{G}_B$ that folds this band back into the reduced-zone scheme is $G(200)=4\pi/a$. The bulk reciprocal lattice vector takes the point A_1 into B_1 in the reduced zone with $k_{\perp}$ (final)$=0.36(2\pi/a)$. The energy and momentum of the initial state have been determined, $E_i = -3.6$ and $k_{\perp}(i) = 0.36(2\pi/a)$. $(k_{\parallel}(i)=0)$. Following the same procedure for peak 2 in the spectrum gives $E_i = -7$ eV and $k_{\perp}(i)=0.48(2\pi/a)$.

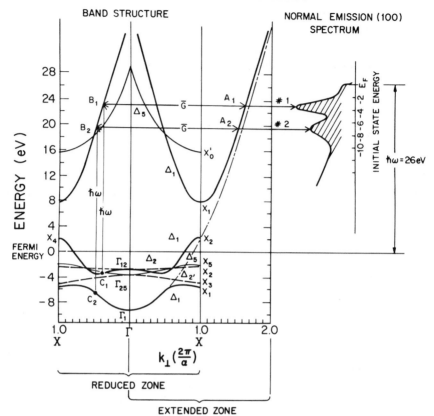

Fig. 10.   Reduced and extended zones of Cu in the Γ-X direction.[40] Dot-dashed curve is the free-electron band. Hypothetical photoemission for $\hbar\omega = 26$ eV is shown on the right.

Normal emission in the (100) direction limits the final state symmetry to Δ_1 bands. The dipole excitation restricts the possible initial state bands to Δ_1 and Δ_5.[39] Therefore, the Δ_2 and Δ_2' bands in Fig. 10 are dashed because they can never be observed in normal emission. The dipole selection rules are even more restrictive. The Δ_1 band can be excited only by a component of the electromagnetic field parallel to the Δ-axis, while the Δ_5 band is excited by a component perpendicular to the Δ-axis. This means that the Δ_1 initial state bands cannot be observed with s-polarized light.

Figure 10 illustrates clearly that if $\hbar\omega$ is changed, the peak or peaks in the spectrum will shift in initial state energy because the allowed transitions will shift in $k_\perp(i)$ and consequently in energy for any band which has dispersion. Figure 11 shows this effect for normal emission from Cu(111).[41] In

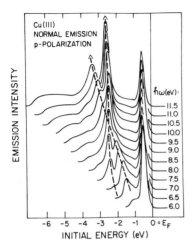

Fig. 11. Normal emission energy distributions obtained from Cu(111) using p-polarized light at photon energies $6.0 \leq \hbar\omega \leq 11.5$ eV.[41]

normal emission from this face only Λ_1 final bands are allowed by symmetry. The two peaks seen at approximately -2.5 and -3.2 eV in the $\hbar\omega = 11.5$ eV spectrum disperse as the photon energy is decreased, crossing over at $\hbar\omega$ ~ 8.5 eV. These peaks must be due to bulk bands that disperse as $k_\perp$ is changed. The peak in the spectra at $E_i = -0.4$ eV does not move with changing photon energy. It is due to excitation from a surface state, which is independent of $k_\perp$ and consequently $\hbar\omega$. To interpret the data in Fig. 11, we must know the final state bands in the (100) direction.

The simplest approximation to the final state band is the free-electron dispersion $E_f = \hbar^2(k_f^2/2m)$. Figure 10 shows this band by the dot-dash curve. The zero of kinetic energy is taken at the bottom of the s band (Γ_1). The free-electron band is a good approximation for the calculated band, except near the zone edge where a gap is formed. The initial state band dispersion has been measured for several faces of copper using the free-electron band assumption and normal emission collection.[42, 43] Figure 12 presents the data of Thiry et al.[43] for normal emission from the (110) face of copper, compared to the calculated bands of Burdick.[44].

The initial state dispersion is plotted as a function of the final state momentum. The magnitude of the $G(1,1,0)$ reciprocal lattice vector is $\sqrt{2}\,4\pi/a$. The Σ_2 band shown by the dashed curve in Fig. 12 cannot be excited into normal emission. The agreement between experiment and theory is excellent throughout the zone, except for the points indicated by the triangles. The authors indicate that this band is probably due to an Umklapp process involving the reciprocal lattice vector $G(111)$ instead of $G(110)$, which is needed to explain the other points. These few points occurring between $42 < \hbar\omega < 52$

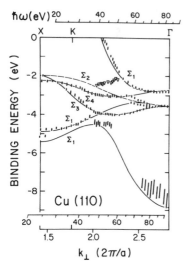

Fig. 12. Experimentally determined valence band for Cu(110) using a free-electron final band.[43] Solid and dashed lines from a calculation by Burdick.[44]

eV are the only indications that the free-electron final band is not a good approximation to the real final band.

Knapp et al.[41] improved on the free-electron approximation by measuring portions of the final band and fitting a final band calculated by Janak, et al.[45] to the data. Figure 10 shows that there is a large band gap at X above the Fermi energy. The upper portion of this gap at X_1 is above the vacuum level, and its position can be measured by the change in the secondary electron emission. The position in k-space where the Δ_1 band crosses the Fermi energy is known to be 0.82 $2\pi/a$ from deHaas-Van Alphen measurements. If the intensity at the Fermi energy is measured as a function of photon energy in normal emission from Cu(100), a peak in the intensity should be seen when a direct transition occurs from the Δ_1 band at the Fermi energy to the Δ_1 final band (Fig. 10). Figure 13 shows the results of this measurement, fix-

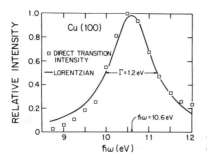

Fig. 13. Interband transition intensity versus $\hbar\omega$ for an initial energy of 0.13 eV for normal emission from Cu(100).[41]

ing the final Δ_1 band at 10.6 eV above the Fermi energy for $k_\perp = 0.82$ $2\pi/a$.[41] The width of 1.2 eV is due to final state lifetime, caused by inelastic scattering in the excited electron. The 1.2 eV width corresponds to a mean free path of approximately 14 Å.[41]

Figure 14 shows the results of the measurements of Knapp et al.[41] normal to the (100) and (111) faces of copper. These measurements show the symmetry restrictions discussed previously. The Δ_2 and Δ_2' bands cannot be seen in normal emission [(100) direction]. The Λ_1 band in the (111) direction (Λ-axis) cannot be excited by s-polarized light (Δ's in Fig. 14). The data in Figure 14 are reported to be accurate in k to 5% of the magnitude of the zone edge.[41]

The band gap above the vacuum level has also been observed for Ni(100).[46–48] In this case the position of the final band at Γ was determined by observing the photon energy at which the initial state energy of the s band was a maximum (farthest from the Fermi energy), which means that the transition originated from Γ.[47]

If the measurements are made at nonnormal collection, then the symmetry is lower and the class of final states with nonvanishing intensity at the detector is much larger. When the set of allowed final states is larger, there are fewer initial states forbidden by symmetry. Figure 9 showed that in principle we can reach any symmetry point or axis of the bulk Brillouin zone

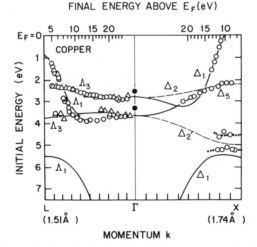

FINAL ENERGY ABOVE E_F(eV)

MOMENTUM k

Fig. 14. Experimentally determined initial band dispersions (E vs. $\mathbf{k}$) for copper along the Γ-Λ-L and Γ-Δ-X symmetry lines.[41] *Left*: Cu(111); *right*: Cu(110). Open circles correspond to s-p polarization, open triangles to s polarization. The solid and dashed curves are theoretical bands.[45]

using one crystal surface. This is especially important in the case of semi-conductors, where only a few surface orientations are available by cleaving or a sputter-annealing preparation.

The analysis of nonnormal collection data is simplified if the final bands are free electron-like. Then $k_{\parallel}$ is given by (9) and

$$k_{\perp} = \left(\frac{2m}{\hbar^2} \left\{ E_{kin} \cos^2\theta + U_0 \right\} \right)^{1/2}$$

where θ is the collection angle and U_0 is the inner potential. The inner potential in this model is the sum of the work function e_{ϕ} and the Fermi energy E_F. The application of these equations is illustrated in Fig. 9a for the Brillouin zone of Ni(100) in a cross-section along the (110) mirror plane.[47] The circles around the origin mark the location of the total **k** vector according to $\mathbf{k}^2 = 2m/\hbar^2(E_f + U_0 - e\phi)$ within the mirror plane for an electron with a certain E_f, with respect to the Fermi energy. Knowing the inner potential U_0 and the work function, we can calculate $|k_{\perp}|$ and $|\mathbf{k}_{\parallel}|$ and obtain the dashed lines on Fig. 9a, which locate the position of direct transitions as a function of energy for a fixed collector angle. The intersection of a dashed line and a circle gives the location in k-space of a transition to a specific final state energy. For example, in Fig. 9a, at a final state energy of about 28 eV and a collection angle of about 30°, we expect to see transitions from the L point in the bulk zone; for $\mathbf{k}_{\parallel} = \overline{X}$ in the SBZ (Fig. 9a).

This free-electron final state approximation and nonnormal collection has been used by Chiang et al.[49] to map out the bands of GaAs. They used a cleaved single crystal with a (110) surface orientation. The agreement between theory[50] and experiment is quite good, as shown in Fig. 15. Other surfaces of GaAs can be prepared only by molecular beam epitaxy.

In reality, the free-electron bands must have gaps at the zone boundaries. This means that if the detector is fixed in angle and the photon energy is increased, the zone boundary will be reached at some photon energy $\hbar\omega$. When the photon energy is increased, the observed interband transitions will suddenly change in intensity because the final state has a gap in energy, and only evanescent final states can contribute to the intensity. As the photon energy is increased, the top of the gap is reached and the intensity increases. This is illustrated schematically for the collection lines in Fig. 9b. At approximately 50 eV final state energy and 20° collection, the Δ-axis (Γ to X) will be crossed and observe rapid changes should be observed in the intensity of the direct transitions. Since $k_{\parallel}$ is known from the energy and angle, one can uniquely determine $k_{\perp}$. Dietz and Eastman[51] used this procedure on Cu(111) to identify points on the Σ-axis.

There is a more accurate method for determination of the initial and final state band structure at high-symmetry critical points, which is independent

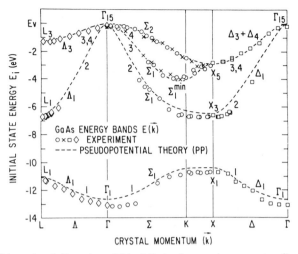

Fig. 15.   Valence band dispersions $E(\mathbf{k})$ of GaAs along major symmetry directions. Circles are experimental points obtained from normal emission spectra; crosses, squares, and diamonds are experimental points obtained from off-normal emission spectra.[49] The dashed curves are theoretical dispersion curves for valence bands.[50] The symmetry characters of the bands and the critical points are labeled.

of a free-electron final state model. Fig. 9*b* shows that if the detector is fixed at a value of $\mathbf{k}_{\parallel} = \overline{M}$ in the [100] direction of the (100) SBZ, then the rod extending back in $k_{\perp}$ goes through X, W, X, W, and so on. Therefore, as the photon energy is swept, keeping the $\mathbf{k}_{\parallel}$ value of momentum fixed, we measure only points in the band structure along the X-W line. Figure 16 shows a plot of this type of measurement for aluminum, compared to the calculated band structure in the extended zone as a function of $k_{\perp}$.[52] The data are collected as a function of photon energy (top scale) and converted to E versus $k_{\perp}$ (initial) by assuming a free-electron final band structure.[53] Qualitatively the free-electron final state works, but quantitatively discrepancies can be seen. The W point is observed at a photon energy of about 18 eV instead of 21 eV as predicted by the free-electron final state bands. Also there is a region around the X point at $k = 3.11$ Å^{-1} where there is no apparent dispersion ($27 \lesssim \hbar\omega \lesssim 43$). This shows the presence of a final state band gap at the X point. The data in Fig. 16 immediately give the initial state energies of the critical points (X and W). Final state bands can be fitted to the high-symmetry points and used to determine the dispersion of the bands between X and W. We will use these data in Section III to determine the band gap in aluminum in which a surface state exists.

The last method we discuss is based on a simple geometrical triangulation scheme: when the same interband transitions are observed for at least two

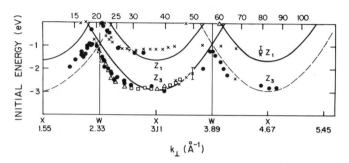

Fig. 16.   The curves are the calculated energy bands of Al along the XWX line.[52] Solid (dashed) curves are even (odd) with respect to the (100) mirror plane of a (100) Al crystal. The points are data from Levinson et al.[53]

different surface orientations of the crystals, the rods at fixed parallel momentum will intersect for these two orientations somewhere within the bulk Brillouin zone, thus giving the total electron momentum. The difficulty involved in this method is to make sure that one really observes the same bulk interband transitions. Because of these pitfalls this method has not been applied routinely; even though it was proposed in 1964.[54] One of the nicest examples was published just recently for gold.[55] A normal emission spectrum at $\hbar\omega = 16.8$ eV of Au(111) was compared to a series of spectra taken at the same photon energy of a Au(112) crystal in an off-normal direction in the (112) plane (Fig. 17). By variation of the polar angle θ, the same interband transition was found for $\theta = 25°$ (see Fig. 17). The crossing of the parallel momentum rods locates the total momentum at point a in Fig. 18. This result was confirmed by taking a similar set of spectra in an off-normal direction of Au(110).

In the discussion of the various methods of band structure determination with photoemission, we have employed the direct interband transition model but have avoided until now the inherent possibility of bulk and surface Umklapp processes. Bulk Umklapp processes are automatically included if we take into account all the existing final state bands in the reduced-zone scheme. A Fourier decomposition of the final state wave function gives the main direction, according to the final state momentum $\mathbf{k}_f$, and we are able to differentiate between primary cone emission, $\mathbf{k}_f \| \mathbf{k}_i$, or secondary cone emission; where $\mathbf{k}_f$ is not parallel to $\mathbf{k}_i$, according to Mahan.[36] In the free-electron final state approximation one can include *Umklapp* processes according to

$$E_f = \frac{\hbar^2 |\mathbf{k}_i + \mathbf{G}|^2}{2m}$$

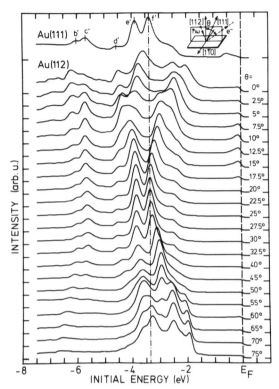

Fig. 17. Normal photoemission spectrum from Au(111) and family of spectra from Au(112) obtained at different collection angles $\hbar\omega = 16.85$ eV.[55]

where $\mathbf{k}_i$ is the momentum of the electron in the initial state, E_f is the final state energy, and $\mathbf{G}$ is a reciprocal lattice vector. Primary cones are characterized by $\mathbf{G} \| \mathbf{k}_i$, and for secondary cones $\mathbf{G}$ is not parallel to $\mathbf{k}_i$.

We want to point out here that the introduction of bulk Umklapp processes is somewhat artificial, because these Umklapp processes are automatically included if one takes the real final states into account in the one-electron direct interband transition picture. The question remains open whether there is a characteristic intensity difference for the observed interband transitions that can be related to the direction of the main Fourier component ($\mathbf{k}_f$) of the final state. There is little information available, but it seems that primary cone emission ($\mathbf{k}_i \| \mathbf{k}_f$) dominates the spectra by about one order of magnitude in intensity over secondary cone emission ($\mathbf{k}_f \| \mathbf{k}_i$).[47, 56]

In general surface Umklapp processes produce the same effect as bulk Umklapp processes unless the surface is reconstructed. In these cases the SBZ of the reconstructed surface is smaller than for the clean surface. Adding or

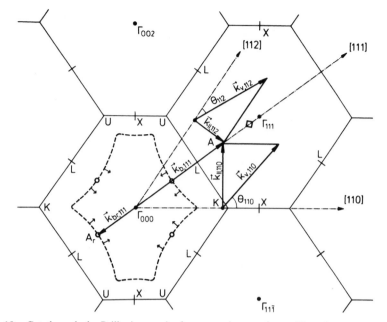

Fig. 18.   Cut through the Brillouin zone in the repeated zone scheme. The relevant wave vectors of the transition causing peak f in Fig. 17 are shown.[55]

subtracting a surface reciprocal lattice vector takes you to a different rod through the bulk Brillouin zone. The smaller the surface zone, the larger the number of rods to be considered in the three-dimensional zone. Van der Veen et al.[57] measured the normal emission spectra from unreconstructed (1×1) and reconstructed (1×5) Ir(100) and compared to the angle-resolved spectra to angle-integrated spectra. The angle-resolved spectra from the (1×5) structure look like the angle-integrated spectra because there are many more points in the three-dimensional Brillouin zone contributing to the signal.

We have tried to illustrate which transitions are allowed and where in energy and collection angle they will be observed. This can be accomplished by considering conservation of energy, and crystal momentum and the symmetry of the initial and final states. When it comes to determining the intensity of a peak in a spectrum, the matrix element must be evaluated. This requires a model of photoexcitation that can be evaluated numerically.

Historically the first model of solid-state photoemission was the three-step model of Berglund and Spicer.[58] Step 1 is the absorption of a photon, resulting in the excitation of an electron inside the crystal; step 2 is the propagation of the excited electron to the surface, and step 3 is the escape into the vacuum of the excited electron. The intensity of a direct transition in step 1 can be calculated in a band structure model.[59] The propagation to the surface can be evaluated by knowing the direction of propagation of the final band state and including an effective mean free path. The escape depth in most materials in the photon energy range being considered here is about 10 to 20 Å.[60] Finally the escape across the surface can be calculated by matching plane waves to bulk states conserving $\mathbf{k}_{\parallel}$.

This three-step model has obvious limitations. It cannot explain any phenomena in which the presence of the surface affects the wave functions of the initial or final states. For example, excitation from a bulk initial state into an evanescent final state[61] or from a surface initial state into a bulklike final state[62] cannot be included. A more severe problem is that the short mean path means that k is not a good quantum number for the final state, and consequently direct transitions should be smeared out in $\mathbf{k}$-space.[1] Schaich and Ashcroft realized all the shortcomings of a three-step model and developed a one-step model using quadratic response formalism.[63] Mahan[36] showed that conceptionally the final state in the photoemission matrix element is equivalent to the time-reversed state used in a low-energy electron diffraction (LEED) analysis. Liebsch developed the one-step LEED final state formulation of photoemission[64, 34] and applied it to emission from adsorbate levels[64] and layer compounds.[33] Pendry has applied this approach to both surface and bulk emissions.[65] At present this approach of using LEED scattering state for the final state in the photoemission matrix element is the most attractive.

F. Spatially Varying Photon Fields

We have assumed that the spatial variations in the vector potential $\mathbf{A}(r)$ are small on the scale of atomic dimensions. This leads to the three different forms of the differential cross-section given in (5), where $\mathbf{A}$ is treated as a constant. If we insist on using a single-particle representation for ψ_i and ψ_f, where the ψ's are the eigenfunctions of the Hamiltonian in the absence of the field, we are then forced to include a "screened" value for $\mathbf{A}$.[66] This dielectric response of the solid to the incident light is a manifestation of many-body effects, which could influence the photoemission properties considerably. The resultant screened field may differ considerably from the incident field in magnitude. But more significantly, if there are rapid variations in the screened field, the $\nabla \cdot \mathbf{A}$ term may become very important in the photoexcitation matrix element [see (3)]. We now discuss the nature and importance of these local field effects.

The simplest effect of the dielectric response of the sample is to create a different $\mathbf{A}$ deep inside the solid compared to the incident field $\mathbf{A}$. This can be evaluated using the macroscopic Fresnel theory.[67] The solid is represented by a frequency-dependent dielectric constant, $\varepsilon(\omega)$ and is separated from the vacuum by an infinitely sharp boundary. The boundary conditions at the solid-vacuum interface (Maxwell's equations) lead to the continuity of the component of the radiation field parallel to the surface, while the perpendicular component of the transverse field has a discontinuity at the surface when $\varepsilon(\omega) \neq 1$. Such a step function discontinuity in A_z would make the spatial variation of A_z very important in the matrix element, since the Fourier transform of a step function has nonzero amplitude at all frequencies. The screened field could produce the momentum required in the excitation process.

Any correct microscopic theory of the solid must produce a field that is continuous across the surface region. The physics of the situation is quite clear[68] even if the calculational procedure appears complicated.[11, 12, 69, 70] The discontinuity in the macroscopic component of the field normal to the surface means that $\nabla \cdot \mathbf{E} \neq 0$, so there will be a charge imbalance in the surface region. This charge imbalance can set up longitudinal fields caused by the interactions of the electrons in the metal with the surface charge.[71] Such a longitudinal field will differ significantly from the long-wave length transverse field. The nature of the longitudinal field depends on the frequency-dependent dielectric response of the system. For a nearly free-electron metal (e.g., Al or Na), the electron gas is incompressible below the plasma frequency $\omega_p = (4\pi n_0 e^2/m)^{1/2}$ (n_0 is electron density), which means that the longitudinal fields may exist only in the surface region and are evanescent into the bulk. This produces Friedel oscillations, with wavelengths much shorter than the wavelength of the incident light. Above the plasma energy, the longitudinal waves propagate into the solid (i.e., plasmons).

The behavior of the longitudinal waves can be calculated only in a nonlocal dielectric scheme.[11, 12, 69, 70] That is why the formulation in the theoretical papers is so formidable. But there is a separate effect at the surface that can be treated qualitatively in a local picture. Makinson[68] pointed out in 1937 that since the electron density $n(\mathbf{r})$ changes rapidly at the surface, a local dielectric constant will also be a rapidly varying function of z. If $\varepsilon(z, \omega) = 0$ at some point near the surface, the local field would become very large.[67] We can write a frequency-dependent local dielectric function starting from the Drude model[66]

$$\varepsilon(\mathbf{r}, \omega) = 1 - \frac{\omega_p^2}{\omega(\omega + i/\tau)} g(r) \tag{20}$$

where i/τ is the term representing absorption, and

$$g(r) = \frac{n(\mathbf{r})}{n_0} \tag{21}$$

is the ratio of the local charge density compared to the equilibrium bulk density; $g(r)=0$ in the vacuum and 1 in the solid. When $\omega<\omega_p$, there is a maximum in $\text{Re}(\varepsilon^{-1})$ at z_0 given by

$$g(z_0)=\left(\frac{\omega}{\omega_p}\right)^2 \tag{22}$$

Therefore when $\omega<\omega_p$ there is a peak in the local field at a position where the local charge density satisfies (22). The full width half-maximum (FWHM) of this peak is

$$\Delta g=\frac{g(z_0)}{\tau^2\omega_p} \tag{23}$$

Several calculations have been done using a nonlocal dielectric constant to estimate the longitudinal fields,[11, 12, 69, 70] but only one calculation exists that includes a realistic surface potential.[11] Feibelman[11] used a Lang-Kohn[72] self-consistent surface potential to calculate the fields at the surface of jellium. Several of his curves are shown in Fig. 19. Figure 19a is for $\hbar\omega<\hbar\omega_p$, and Fig. 19b shows the fields above $\hbar\omega_p$. All the curves show the longitudinal fields penetrating into the solid. For $\hbar\omega<\hbar\omega_p$ they are damped, while the curves for A_z^2 above $\hbar\omega_p$ (Fig. 19b) show larger oscillations and no damping. Figure 19a also shows the peak in A_z^2 near the surface caused by $\text{Re}(\varepsilon^{-1})$ reaching a maximum. This peak is shaded for convenience.

Identifying the contributions to the photoemission signal from the surface (surface photoeffect) is an old experimental and theoretical problem.[38, 66, 69] Recent experiments using all the advantages of angle-resolved photoemission and synchrotron radiation have been able to clearly identify the contributions to the photoemission matrix element from the spatially varying photon field at the surface of Al(100).[13, 73] The measured photoionization cross-section is in quantitative agreement with Feibelman's calculated cross-section, using the fields shown in Fig. 19.[13]

The measurements can best be understood if we return to (3) the original equation for photoionization. Let us rewrite this equation as:

$$\frac{d\sigma}{d\omega}=|\langle\psi_f|\overline{A}_0\cdot\overline{P}+\overline{P}\cdot\overline{A}_0|\psi_i\rangle+\langle\psi_f|\Delta\mathbf{A}(\mathbf{r})\cdot\mathbf{P}+\mathbf{P}\cdot\Delta\mathbf{A}(\mathbf{r})|\psi_i\rangle \tag{24}$$

where $\mathbf{A}_0$ is a macroscopic transverse field of the form of (4), inside the solid, $\mathbf{A}_0$ is the internal Fresnel field, and $\Delta\mathbf{A}(\mathbf{r})$ is the spatial variation of the local field,

$$\Delta\mathbf{A}(\mathbf{r})=\mathbf{A}(\mathbf{r})-\mathbf{A}_0 \tag{25}$$

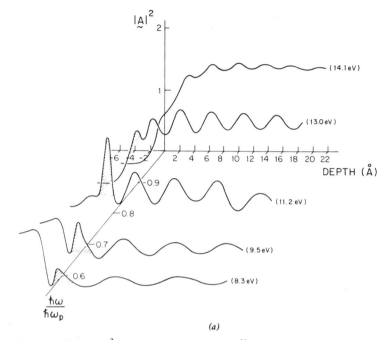

Fig. 19.   Local fields $|A_z|^2$ near the surface of jellium.[11] (a) For $\hbar\omega < \hbar\omega_p$. ($b$) $\hbar\omega > \hbar\omega_p$.

Written in this fashion, the first term in (24) is the "ordinary" single-particle matrix element, which can be written in any of the forms of (5). The second term in (24) is the correction due to the spatially varying field. The first term in (24) can be written in the form of (5c):

$$\langle \psi_f | \mathbf{A}_0 \cdot \mathbf{P} + \mathbf{P} \cdot \mathbf{A}_0 | \psi_i \rangle = 2 \langle \psi_f | \mathbf{A}_0 \cdot \mathbf{P} | \psi_i \rangle$$

$$= -2 \cdot \frac{i}{\omega} \langle \psi_f | \nabla V | \psi_i \rangle \cdot \mathbf{A}_0$$

$$= \frac{-2i}{\omega} \left\{ \mathbf{A}_0 \cdot \langle \psi_f | \nabla V_{\text{bulk}} | \psi_i \rangle + \mathbf{A}_0 \cdot \langle \psi_f | \nabla V_{\text{surface}} | \psi_i \rangle \right\}$$

$$(26)$$

The separation of the bulk and surface ∇V terms is a little artificial; what we mean is that the first term represents direct interband transitions (Sections II.D and II.E). This bulk term was eliminated in the measurements by measuring the photoionization cross-section of the intensity at the Fermi level edge and of the surface state[74, 75] in normal emission. There are not direct transitions from the Fermi energy for at least a 40 eV photon energy range for normal collection on Al(100). We know the symmetry of the two initial

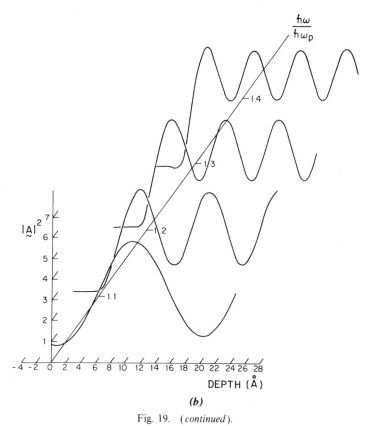

Fig. 19. (*continued*).

states investigated; they are both completely symmetric with respect to all symmetry operations of the surface.[13] The final state in normal emission is also completely symmetric; therefore only the z-component of **A** can be involved in the excitation. This reduces (24) to

$$\frac{d\sigma}{d\Omega} \propto \left| \frac{-2iA_0}{\omega} \left\langle \psi_f \left| \frac{\partial V_{\text{surface}}}{\partial z} \right| \psi_i \right\rangle + \left\langle \psi_f \left| \Delta A_z(z) P_z + P_z \Delta A_z(z) \right| \psi_i \right\rangle \right|^2 \quad (27)$$

Figure 20 shows the photoionization cross-section for the Fermi edge of Al(100) in normal emission for p-polarized light.[13] The solid curve is the theoretical calculation due to Feibelman.[11] The cross-section is much larger below $\hbar\omega_p$ than above, because the rapid variation in $\mathbf{A}(r)$ near the surface dominates the cross section (see Fig. 19a). The dashed curve in Fig. 20 is from Feibelman for the $\partial V_{\text{surface}}/\partial z$ term in (27) with a constant field A_0.

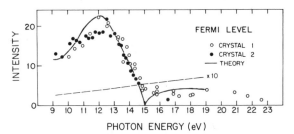

Fig. 20.   Photoionization cross-section for the Fermi energy electrons emitted normal to the Al(100) surface as a function of photon energy.[13] Curves are theoretical calculations by Feibelman[13] using a constant field (dashed) and a spatially varying field (solid).

An equivalent calculation for aluminum with ion cores also produced a structureless cross-section as a function of photon energy.[76] Experimentally this means that at $\hbar\omega \sim 12$ eV the cross-section is at least 25 times larger than the $\partial V/\partial z$ term would predict. The theory predicts that the correction term in (27) is six times larger than the single-particle term at $\hbar\omega = 12$ eV (cross-section is the square of the amplitude). At $\hbar\omega = 23$ eV the correction term in (27) is not important.

Levinson et al.[73] measured the absolute cross-section in Fig. 20 at $\hbar\omega = 13$ eV. It is 10^{-3} electron/(photons)(eV)(steradian) with a 50% experimental error. The value calculated by Feibelman[10] is 0.7×10^{-3} electron/photon-eV-steradian.

The cross-section is dominated by the peak in $A(\mathbf{r})$ near the surface. Its rapid variation produced a large ΔA_z term, which produced momentum for the transition. This peak can be understood qualitatively with a spatially dependent dielectric constant [see (20)].[68] Therefore we would expect to see this effect whenever $\varepsilon(\omega) < -1$ and ε_2 is relatively small, or τ large in (20). If the adsorption (ε_2) is too large, the peak will be smeared out in space and $\partial A/\partial z$ will be reduced.

III. CLEAN SURFACES

The introduction of a surface to terminate a periodic solid changes the wave functions, especially in the surface region. The presence of the surface means that the "running" Bloch waves must now be combined in "standing" waves to match continuously to the wave functions in the vacuum.[78] These new "standing" waves will have different amplitudes at the surface and in the bulk, producing a local density of states at the surface different from that in the bulk.[79] There also exist completely new states at the surface, called surface states. These are electronic states at the surface that exist in a forbidden band gap of the bulk bands and decay exponentially into the bulk (i.e., evanescent) states. For semiconductors a basic understanding of surface states was developed in the 1930s.[80, 81] Much indirect evidence exists to prove the existence of surface states on semiconductor surfaces.[82] The development of angle-resolved photoemission has made it possible to directly observe and characterize surface states.

Ironically the greatest progress has been made on metal, not semiconductor surfaces. It wasn't until the work of Forstmann and Pendry[83] in 1970, showing that surface states could exist in a s-d hybridization gap, that it became apparent that surface states could be found on metal surfaces. The first experimental documentation of a surface state on a metal was in 1970,[84] yet at the present time surface states have been observed on more than 25 single-crystal metallic surfaces.

We first describe characteristic measurements for surface states on metals, then discuss semiconductor surfaces, and finally briefly address the issue of local density of states at a surface. Before we begin, it is appropriate to define more precisely the terms "surface state," "surface resonance," and "local density of states." The latter term is nearly self-explanatory: the local density of states at the surface $\rho(r, E)$ is the sum over all the states at r with energy E

$$\rho(r, E) = \sum_\alpha |\psi_\alpha(r)|^2 \delta(E - E_\alpha) \qquad (28)$$

Where to evaluate $\rho(r, E)$ is another question: $\rho(r, E)$ could be integrated over the unit cell, over some radius around each surface atom, or down to a depth z_0 below the surface. Surface states or resonances can be defined more precisely. An electronic state at the two-dimensional surface can be characterized by its energy E and wave vector parallel to the surface $k_\parallel$. Every bulk state is characterized by E and $\mathbf{k} = (\mathbf{k}_\parallel, \mathbf{k}_\perp)$, where $\mathbf{k}_\perp$ is the component of $\mathbf{k}$ perpendicular to the surface. Surface states are states at the surface with an E and $\mathbf{k}_\parallel$ such that there are no states in the bulk with the same values of E and $\mathbf{k}_\parallel$, independent of $\mathbf{k}_\perp$. This type of electronic state cannot mix with any bulk state (i.e., a surface state).

Surface states must lie in a band gap in the projection of the bulk states onto the surface. This projection is accomplished by plotting E versus $\mathbf{k}_\perp$ for the bulk bands at a fixed $\mathbf{k}_\parallel$ and projecting these states onto the SBZ at the specific value of $\mathbf{k}_\parallel$ (see Fig. 8). This procedure is carried out for all $\mathbf{k}_\parallel$ in the SBZ. For example, if we wanted to find the projection of the copper bulk bands onto the Cu(111) surface, we could start at $\mathbf{k}_\parallel = 0$. The rod extending into the bulk with $\mathbf{k}_\parallel = 0$ is the Λ-axis (see Appendix) or the Γ-L direction shown on the left of Fig. 14. When we project the bands shown in Fig. 14 onto the $k_\parallel = 0$ point of the SBZ, we find that there are two regions of energy that have no bulk states. The first is the "neck" of Cu, which extends from the Fermi energy to about 0.8 eV. The second gap is a hybridi-zation gap extending from 4 to about 5.5 eV. A peak seen in a normal emis-sion photoelectron spectra from Cu(111) with an initial state energy from 0 to -0.8 eV from -4 to -5.5 eV cannot be a bulk state. For example, the peaks at -0.4 eV in the spectra shown in Fig. 11 are due to a surface state.[41]

We can write down the first two rules for identifying a peak in an angle-resolved photoemission spectrum as a surface state.

1. The energy E and parallel momentum $k_\parallel$ of the state must lie in a gap in the projection of the bulk band structure onto the SBZ. These projec-tions must consider the symmetry of the bulk states, an issue that is dis-cussed later.
2. The initial state energy of the peak must be independent of the exciting photon energy for a fixed $k_\parallel$. If a peak moves as $\hbar\omega$ is changed, then it is due to an initial state whose energy depends on $k_\perp$ (bulk state) or to a final state (fixed kinetic energy).

Figure 11 shows that the surface state at -0.4 eV on Cu(111) has an ini-tial state energy independent of $\hbar\omega$, while the transitions from the bulk bands move with changing photon energy.

A surface resonance is an electronic state at the surface that can mix with bulk states. Therefore a surface resonance will not satisfy Rule 1; that is, it does not lie in a gap in the projected bulk bands. If the surface resonance mixes strongly with the bulk band, it will become smeared out in energy, which will make it difficult to observe experimentally.

A. Surface States on Metals

Metals do not have total band gaps at the Fermi energy between the con-duction and valence bands like the gaps that exist in semiconductors or in-sulators. Therefore surface states will exist in "partial gaps" in metals. The term "partial gap" indicates that there are restricted regions of the SBZ that have no bulk states. These regions are confined to specific energies and val-ues of $k_\parallel$.

The most famous partial gap in a metal is the "neck" in the Fermi surface of Cu in the (111) direction. For $k_\parallel = 0$ on the Cu(111) surface the gap ex-tends about 0.8 eV below the Fermi surface (Fig. 14). As $k_\parallel$ increases, the gap decreases in size until the neck closes at $k_\parallel \sim 0.25$ Å^{-1}. Figure 21 shows the projection of the bulk bands onto the SBZ and the measured dispersion of the surface state.[85] The surface state disperses up toward the Fermi en-ergy as $k_\parallel$ increases, but its dispersion is not as rapid as the dispersion of the bulk band edge. The surface state runs into the projection of the bulk bands at $k_\parallel \sim 0.25$ Å^{-1}. This means that the one data point on the left that is outside the gap in the projected bulk bands is probably a surface reso-nance.

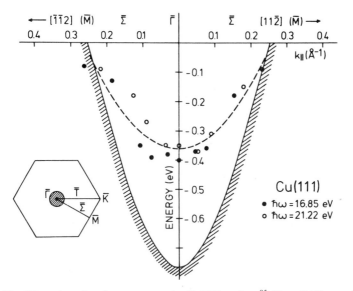

Fig. 21.   Dispersion of surface states on the Cu(111) surface.[85] The solid line and shaded area enclose the partial gap in the projected bulk band structure.[86] The figure at the left shows the SBZ and the size of the neck at the Fermi energy.

The neck in Cu can be represented by an effective mass perpendicular to the Γ-L direction, where $E = \hbar^2 k^2 / 2m^*$. The effective mass determined from cyclotron resonance[87] is $0.46m$, and it is $0.34m$ from band structure calculations.[44] The effective mass of the surface state is larger than the band edge, as can be seen from Fig. 21. The dashed line through the surface state data points has an effective mass of about $0.7m$.[85] Other measurements on this surface state give an effective mass smaller than $0.7m$. The original measurements on Cu(111) by Gartland and Slagsvold[88] gave an effective mass of $(0.42 \pm 0.05)m$, and more recent measurements by Knapp et al.[41] give an effective mass of about $0.4m$. This surface state is in an s-p gap and is primarily s-p in character.[86, 89]

The surface states that are calculated[90] or observed[13, 74, 75] on single-crystal faces of aluminum are s-p in character just like the Cu(111) surface state. These states are slightly different from the Cu(111) state in that they exist in partial band gaps where both the bottom and top band edges are below the Fermi energy. For example, the rod extending into the bulk for $k_\parallel = 0$ on the Al(100) surface passes along the Δ-axis (i.e., from Γ to X in the bulk zone: Fig. 8). The experimental data for Al shown in Fig. 16 are for the bulk direction perpendicular to the Γ-$X(\Delta)$ axis. There is a gap at the X point between -1.25 and -2.9 eV. Figure 9b shows that the bulk bands between X and W (plotted in Fig. 16), project out along the $\overline{\Gamma}$-$\overline{M}$ direction in the SBZ

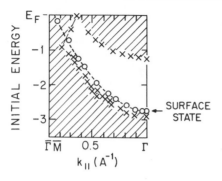

Fig. 22.   Measured bulk bands at Al projected onto the $\overline{\Gamma}$-$\overline{M}$ direction of the SBZ of Al(100)[53] and the experimentally determined dispersion of the surface state.[75]

(Fig. 8). Figure 22 shows the measured bulk bands (crosses) after the data in Fig. 16 have been corrected for the shape of the final bands,[53] projected onto the SBZ, and the data of Hansson and Flodstrom[75] for the dispersion of a surface state found in this gap. In this case the effective mass of the lower band edge and the surface state is about $0.96m$, while the effective mass of the upper band edge is about $1.7m$ near $\overline{\Gamma}$.

The surface state band lies very close to the bottom band edge, so it penetrates very far into the crystal, probably farther than calculated because most of the calculations place the surface state in the middle of the gap.[90] As the surface state band approaches the Fermi surface, it crosses into a region that has no gap in the projection of the bulk states; thus it becomes a resonance. The mixing with the bulk states cannot be too large in this region, since the surface resonance is still observed in the experiment.[74, 75]

Figure 23 shows two energy distributions taken at $\hbar\omega = 12$ eV normal to an Al(100) crystal.[13] The surface state is the big peak 2.75 eV below the Fermi energy. This surface state at $\overline{\Gamma}$ in the SBZ is in the gap of the Δ_1 bulk band. This band and consequently the surface state are completely symmetric with respect to all symmetry operations of the (100) surface. The final state in the photoexcitation matrix element must also be completely symmetric with respect to all symmetry operation if it is to have amplitude at the detector. This means that only the perpendicular (to the surface) component of $\mathbf{A}$ can be involved in the excitation because the parallel component is odd and would make the matrix element zero. When the light is s-polarized or normally incident to the surface, the surface state as well as all bulk transitions should be invisible in normal emission. Figure 23 shows that the surface state intensity as well as the Fermi step height decreased dramatically as the angle of incidence was changed from 45° to 15°. Macroscopic calculations of the electromagnetic field using measured optical constants predict that $|A_z|^2$ at $\theta_I = 15°$ should be about 0.3 of $|A_z|^2$ at $\theta_I = 45°$. The intensity in the $\theta_I = 15°$ curve is approximately 25% of the $\theta_I = 45°$ curve, which shows that the symmetry of the surface state is correct.

NORMAL EMISSION Aℓ (100)

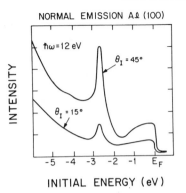

INITIAL ENERGY (eV)

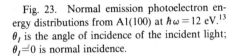

Fig. 23. Normal emission photoelectron energy distributions from Al(100) at $\hbar\omega = 12$ eV.[13] θ_I is the angle of incidence of the incident light; $\theta_I = 0$ is normal incidence.

We can illustrate another symmetry property of any truly two-dimensional system using the surface state on Al(100). Figure 24a shows the extended SBZ for Al(100). All the data shown in Figs. 22 or 23 have been at $\overline{\Gamma}(1)$ ($k_{\parallel} = 0$) or along the line from $\overline{\Gamma}(1)$ to $\overline{M}$. If the electronic states at the surface are two dimensional, the energy of the surface state at $k_{\parallel} + G_{\parallel}$ must be the same as at $k_{\parallel}$, where $G_{\parallel}$ is a surface reciprocal lattice vector. This is just Bloch's theorem in two dimensions. Therefore we should be able to observe the surface state with an energy of -2.75 eV at $\overline{\Gamma}(1)(k_{\parallel} = 0)$ or $\overline{\Gamma}$ in the second zone ($\overline{\Gamma}(2)$) or the third zone $\overline{\Gamma}(3)$ (Fig. 24a). The photoelectron spectra a, c, and d shown in Fig. 24b are for $k_{\parallel} = (0,0)[\overline{\Gamma}(1)]$, $k_{\parallel} = (2.19\ \text{Å}^{-1}, 0)[\overline{\Gamma}(2)]$, and $k_{\parallel} = (2.19\ \text{Å}^{-1}, 2.19\ \text{Å}^{-1})[\overline{\Gamma}(3)]$. The surface state appears in all three spectra, as it should. Curve b is taken at the $\overline{X}$ point [$k_{\parallel} = (1.1\ \text{Å}^{-1}, 0)$] to show that the surface state does not exist at this point in the SBZ. The data in Fig. 24b prove that the peak at -2.75 eV in normal emission is due to a two-dimensional electronic state with the periodicity of the Al(100) face.

There are several other interesting features in Fig. 24b. First the peaks A and B, and probably C are due to bulk direct transitions. It was generally believed[75] that bulk transitions could not be observed in Al because it was a nearly free-electron metal. The complete band structure of Al can be mapped out if the correct polarization and photon energies are used.[53] The second point of interest is that at $\hbar\omega = 50$ eV the intensity of the surface state at $\overline{\Gamma}$ in the second Brillouin zone is about 10 times larger than the intensity in the normal emission spectrum ($\overline{\Gamma}(1)$). If the photon energy is changed to 70 eV, the intensity of the surface state in the first zone is about 10 times larger than the intensity in the second zone. Figure 25 plots the intensity of the surface state at $\overline{\Gamma}$ in the first and second zones as a function of photon energy.[53] The intensity of the surface state at $\overline{\Gamma}$ in the first zone peaks at about 70 eV, while the intensity in the second zone peaks at about 48 eV. Bloch's theorem applied to the surface electronic states says that the energy of the states at two

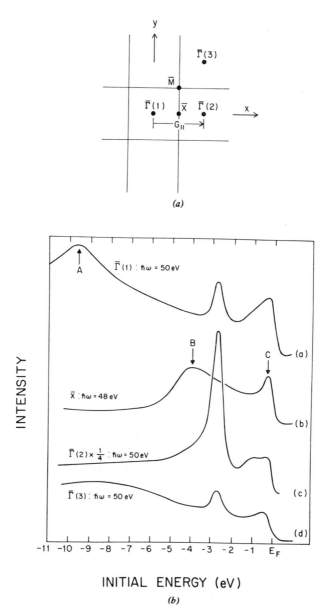

Fig. 24. (a) Extended-surface Brillouin zone of Al(100). (b) Energy distributions at four values of $k_{\parallel}$ in the SBZ.[53] The light is p-polarized, the plane of polarization is the xz-plane in (a). The polar collection angles for the four curves are 0° (a), 20° (b), 41° (c), and 68° (d).

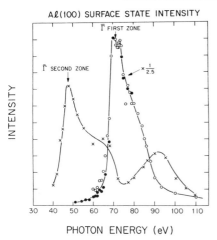

Fig. 25. Relative photoionization cross-sections for the surface state measured at $\bar{\Gamma}$ in the first and second zones (Fig. 24a).[53] The open and solid circles for the first zone represent data on two different crystals.

values of $k_{\parallel}$ separated by $G_{\parallel}$ must be the same. The intensity can be quite different.

This rapid variation in the cross-section of a surface state at high photon energy was explained by Louie et al.[62] for the surface state on Cu(111). The surface state at $\bar{\Gamma}$ on Al(100) occurs in the band gap formed at the bulk Brillouin zone edge at the symmetry point X. The wave function of the surface state is composed of combinations of the bulk states at the gap, since Γ-X is the direction in the bulk normal to the (100) surface. Louie[62] showed theoretically that if the surface state is primarily derived from a single bulk band (as it is in Al), the photoionization intensity should peak when the photon energy was correct to excite the surface state into a final state bulk band with a value of $k_{\perp}$ corresponding to a band extremum. This implies that a substantial contribution to the photoemission intensity results from transitions from the surface state into bulk final states. Specifically for the surface state on Al(100), we would expect to see enhancement in the cross-section of the surface state at kinetic energies of the final state corresponding to extremum points in the bulk bands at X in the Γ-X direction.

A free-electron final band picture for Al predicts that there are bands crossing the X point in the Γ-X direction at about 72 eV and 45 eV above the Fermi energy. The band crossing at 72 eV is the one with final state momentum normal to the surface. In the reduced-zone scheme it is folded back with a reciprocal lattice vector of the form $G(0,0,2)$, where the last index is the z-direction. Therefore when the surface state is excited into this final state it will come out normal to the surface and peak at a photon energy $72 + 2.75$ eV $= 75$ eV. This is very close to the peak in the cross section of the surface state seen in normal emission (Fig. 25). The final band crossing X at 45 eV has been folded back into the reduced, zone scheme with a reciprocal lattice vector of the type $\bar{G}(1,1,1)$, which can be visualized in Fig. 9a. The final direction of an electron excited into this band is $(k_\perp, k_\parallel = 2.19$ Å$)$, that is, it comes out at $\bar{\Gamma}(2)$ in Fig. 24a. Therefore the intensity of the surface state peak seen in the second zone should peak at a photon energy of $45 + 2.75$ eV $= 48$ eV, which is very close to the observed peak in the $\bar{\Gamma}(2)$ cross-section. The curves in Fig. 25 can be used to determine the critical points in the final band structure.

Surface states have been reported or calculated on aluminum,[13, 74, 75, 90] copper,[41, 85, 86, 88, 89, 91, 92] silver,[92] gold,[93] nickel,[94–98] iron,[99] tungsten,[100–103] molybdenum,[100, 104, 105] niobium,[106] cobalt,[107] titanium,[108] and lead.[109, 110] The $W(100)$ face is probably the most studied.[2, 84, 100–103] There are three different bands of surface states or resonances extending throughout most of the SBZ. The symmetry and dispersion of these bands have been measured.[100–102] We do not discuss this system here because it is not clear at present what effect the reconstruction of this face has on the surface states, or vice versa.[102]

The final example involves magnetic surface states on Ni(100).[95] We use this example because it illustrates several important procedures associated with identifying surface states. Figure 26 shows four sets of angle-resolved energy distributions from a clean and a contaminated Ni(100) surface. Two geometries are shown for each column. The left-hand column is for two measurements in the $\bar{\Gamma}$-$\bar{X}$ direction in the Ni(100) SBZ (see Fig. 8). The value of $k_\parallel$ is 1.30 Å^{-1} (with $E_{\text{initial}} = 0$) in both measurements (the zone edge at $\bar{X}$ is 1.26 Å^{-1}). In the top set of curves on the left the light is s-polarized with A along the $\bar{\Gamma}$-$\bar{X}$ direction in the surface and the detector is in the plane defined by the surface normal and the $\bar{A}$ vector ($\bar{\Gamma}$-$\bar{X}$ direction; see Fig. 8). The curves in the bottom set are the same except that A is perpendicular to the plane of collection. In both cases the detector is in a mirror plane of the (100) crystal, so it must have even symmetry with respect to reflection about this mirror plane. If the final state had odd symmetry, $\psi_f = 0$ at the detector. This means we can denote the final state as even $\langle\psi_f| = \langle\oplus|$. The polarization vector is in the plane of collection for the top curves and perpendicular to the mirror plane for the bottom, so $\mathbf{A}\cdot\mathbf{P}$ is even (odd) for the top (bottom) curves. This means that the matrix element [see (5)] is $\langle\oplus|\oplus|\psi_i\rangle$ for the top curves and $\langle\oplus|\ominus|\psi_i\rangle$ for the bottom. Therefore the only initial states that can be excited in the top (bottom) curves are even (odd) with respect to reflection about the $\bar{\Gamma}$-$\bar{X}$ mirror plane. The curves on the right are for $k_\parallel = 0.89$ Å^{-1} in the $\bar{\Gamma}$-$\bar{M}$ direction of the Ni(100) SBZ (Fig. 8); $\bar{M}$ is at $k_\parallel = 1.78$ Å^{-1}. These curves are for p-polarized light with A in the plane formed by the crystal normal and the $\bar{\Gamma}$-$\bar{M}$ direction. In the top curves the detector is in this plane, while it is perpendicular to the plane of incidence of the light in the bottom. It is easy to show that the symmetry of the matrix element

Ni (100)

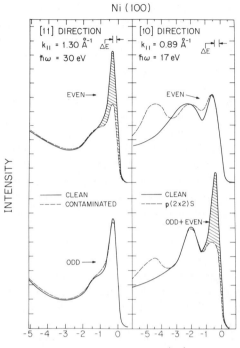

Fig. 26.   Angle-resolved photoelectron spectra from Ni(100). *Left*: spectra taken at $\hbar\omega = 30$ eV using *s*-polarized light; *right*: spectra taken with *p*-polarized light at $h\omega = 17$ eV.[95] The shaded areas indicate the disappearance of surface states after the crystal has been contaminated.

$\langle \psi_f | A \cdot P | \psi_i \rangle$ is $\langle \oplus | \oplus | \psi_i \rangle$ for the top curves and $\langle \oplus | \oplus + \ominus | \psi_i \rangle$ for the bottom. The latter matrix element has both odd and even $\overline{A} \cdot \overline{P}$ symmetry components because $A_\perp$ is even and $A_\parallel$ is odd (perpendicular and parallel with respect to surface).

Figure 26 shows that surface states can be destroyed by contaminating the surface and that they may have specific symmetry with respect to a mirror plane of the surface. The sensitivity to contamination is generally used as one test for surface states or resonances. We list this as our third criterion for surface states.

3.   A peak in a photoelectron spectrum that is caused by a surface state or resonance should be sensitive to the surface conditions (perfection and cleanliness). This "crud" test should not be used by itself, especially in angle-resolved photoelectron spectroscopy, because the intensity of a

bulk transition can be affected by the surface order. If the effect of contamination is the same for a wide variety of photon energies and a fixed $k_\parallel$, it is fairly safe to assume that the changes are caused by the destruction of an intrinsic surface state or resonance. We return to this point later with examples from Pd(111).

The data in Fig. 26, coupled with data at different photon energies, indicate the existence of surface states or resonances in both the $\overline{\Gamma}\text{-}\overline{M}$ and $\overline{\Gamma}\text{-}\overline{X}$ directions of the SBZ of Ni(100).[95] The surface peak in the $\overline{\Gamma}\text{-}\overline{X}$ direction (Fig. 26, left) has even symmetry with respect to the $\overline{\Gamma}\text{-}\overline{X}$ mirror plane, while the surface state (or resonance) in the $\overline{\Gamma}\text{-}\overline{M}$ direction is odd with respect to the $\overline{\Gamma}\text{-}\overline{M}$ mirror plane. Plummer and Eberhardt[95] showed that criteria 2 and 3 for the existence of a surface state could be satisfied in the range $0.9 < k_\parallel < 1.6$ Å^{-1} for the even surface peak in the $\overline{\Gamma}\text{-}\overline{X}$ direction and in the approximate range 0.7 Å$^{-1} < k_\parallel$ for the odd state in the $\overline{\Gamma}\text{-}\overline{M}$ direction. In both directions the surface state or resonance appeared within -0.1 eV of the Fermi energy. The only remaining test for a surface state is criterion 1, that is, the state must lie in a gap in the projection of the bulk bands.

If the calculated bulk bands are projected onto the Ni(100) SBZ (Fig. 8), there are no gaps near the Fermi energy in either direction. This does not mean that the peaks in Fig. 26 correspond to surface resonances. The surface electronic states shown in Fig. 26 have even or odd symmetry with respect to the two mirror planes of a (100) surface, so we need to project out separately the even and odd bulk states along these two directions. Since nickel ferromagnetic, there is a band structure for both the minority and majority spin states. Figure 27 shows the projection of the odd (even) minority and majority bands onto the $\overline{\Gamma}\text{-}\overline{M}$ ($\overline{\Gamma}\text{-}\overline{X}$) direction of the Ni(100) surface.[95] Now it is obvious that the surface peaks in Fig. 26 are surface states; in fact they are magnetic surface states. The odd symmetry surface state in the $\overline{\Gamma}\text{-}\overline{M}$ direction is in a gap in the projection of the odd majority bands, while the even surface state in the $\overline{\Gamma}\text{-}\overline{X}$ direction is in a gap of the minority bands.

These surface states on Ni(100) can be used to demonstrate two problems frequently encountered in applying criteria 1 through 3 to determine that a peak in a spectrum is a surface state. The first criterion says that the surface peak must fall in a gap in the projected bulk band structure. For the examples used in this chapter the Cu(111) face is the only one for which the calculated bulk band structure gave the same projection as the measured band structure. The measured gap at X in the Al(100) face differs in magnitude and position from the calculated gap with respect to the Fermi energy.[53] The measured band structure of Ni[47, 111, 112] is considerably different from the equivalent calculations.[113] For example, the d bands are about 30% nar-

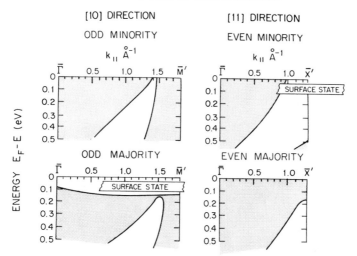

Fig. 27.   Projection of the bulk bands on Ni onto the two symmetry directions of the SBZ.[95]

rower than calculated,[111] and the measured exchange splitting[112] is 0.3 eV instead of the calculated value of 0.6 eV.[113] The projections shown in Fig. 27, which were obtained from experimentally determined energy bands,[95, 111] would look quite different if the calculated band structure of Ni were used. Therefore the experimentalist must be cautious when inferring, based only on a calculated band structure, that a surface-sensitive peak is or is not a surface state.

The second problem associated with the identification of a surface state is the separation and identification of a surface peak in a set of spectra in which many bulk transitions are present. Plummer and Eberhardt[95] could not track the odd surface state in the $\bar{\Gamma}$-$\bar{M}$ direction (Fig. 27) back to $\bar{\Gamma}$, in the region of $k_{\parallel} \lesssim 0.7$ Å^{-1}. Bulk transitions mask the surface state if it exists in this region of the SBZ. There were peaks in the spectra in this region of $k_{\parallel}$ that satisfied criterion 3. They were surface sensitive, but they did not satisfy criterion 2. The peaks appeared at different initial state energies at different photon energies. Erskine[96] has attempted to track this odd state back to $\bar{\Gamma}$. There are many cases like this one, in which the bulk transitions make it difficult to identify surface states. Figure 28a presents an especially difficult case, Pd(111). The spectra in Fig. 28a are for clean and low-temperature H adsorption at three different photon energies. There are many changes induced by H adsorption and unfortunately many of the changes are in the d-band region of the clean Pd spectra. Figure 28b plots the difference curves

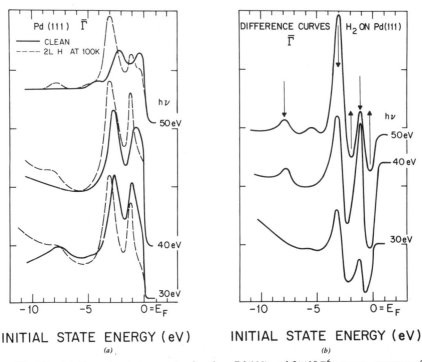

Fig. 28.   (a) Normal emission spectra for clean Pd(111) and 2×10^{-6} torr-sec exposure of H_2 at 100°K.[110] (1L = 10^{-6} torr·sec). (b) Difference curves for the three set of spectra in (a), where the clean curves have been subtracted from the H adsorption. Arrows pointing down (up) are extrinsic (intrinsic) surface states or resonances.

for the three photon energies shown in Fig. 28a. Comparing the actual energy distributions and the difference curves for several photon energies is one procedure for identifying extrinsic and intrinsic surface states. For example, if there is a positive peak in the difference curves at the same initial state energy for all photon energies and a clear peak in the energy distributions for at least one H adsorption spectrum, the peak is most likely an H-induced state. Likewise if there is a negative dip in all the difference curves and a peak in the clean spectra, there should be a clean surface state or resonance at this energy. Comparing Fig. 28a and Fig. 28b reveals extrinsic H-induced states at approximately -1, -3, and -8 eV with intrinsic states at about -0.2 and 2.0 eV. In these situations it is important to have several photon energies available. In general a reader should not trust difference curves for angle-resolved photoelectron spectroscopy unless data for several photon energies are presented.

B. Semiconductor Surface States

The presence of surface states on semiconductor surfaces was verified early using angle-integrated photoemission.[114, 115] Moreover, the existence of such states is obvious through the "pinning" of the Fermi level at the surface. In bulk semiconductors the Fermi level shifts, depending on the doping level, from the top of the valence band to the bottom of the conduction band. In contrast, early angle-integrated photoemission and work function measurements[116, 117] showed that the Fermi level is pinned at the surface, almost independent of the doping level. Even though the existence of surface states on semiconductors was confirmed relatively early, little is known about these states compared to surface states on metals.

Semiconductor surfaces are more complex than metal surfaces, and harder to prepare and characterize. Only a few low-index surface orientations are natural cleavage planes of single-crystal semiconductors [e.g., the (111) surfaces of Si and Ge and the (110) surface of GaAs]. Other surface orientations are accessible only by molecular beam epitaxy or sputter-annealing. In the case of semiconductors sputter-annealing is known to introduce surface defects and roughness.[118] The reproducibility of photoemission data from cleaved surfaces is not very high. The number of steps varies from cleave to cleave. In many cases this is obvious by visual inspection. Another problem is the density and distribution of domains of reconstruction on the surface.

Most semiconductor surfaces reconstruct. The best known example is the Si(111) surface. A freshly cleaved surface shows a (2×1) superstructure. Upon heating to 400°C this irreversibly transforms into a 7×7 reconstructed surface. A Si(111) (1×1) surface structure can be prepared by rapid quenching from high temperatures or by 1 to 5% of a monolayer of an impurity such as tellurium or aluminum, which stabilizes the surface.[119] The Si(111) surface is only one example of a variety of reconstructed semiconductor surfaces. Models for these surfaces have been developed, mostly on the basis of dynamic LEED calculations. They clearly show the fundamental difference between a metal surface and a semiconductor surface. For metals, reconstruction is the exception, not the rule. The general behavior is a few percent contraction of the surface plane toward the second plane, with no lateral reconstruction of atoms within the surface plane. In contrast, semiconductors with their covalent (Si, Ge, diamond) or heteropolar bonding (GaAs or III–Vs or VIs) nearly always exhibit reconstructed surfaces, and the perturbation can extend into the second, third, fourth, or deeper layer.[120]

The presence of these reconstructions and atomic displacements on semiconductor surfaces should make the study of their electronic structures very exciting. Surface states per se are a unique detector for these rearrangements, provided we understand the connection between the electronic

structure and the atomic structure of the surface. The main objective of angle-resolved photoemission studies has been the determination of the structure of semiconductor surfaces by calculating the position of surface states
as a function of the surface structure, and comparing to data. Surface states
on semiconductors are in general s-p-like, but unlike the case of metals this
does not imply that they are delocalized. Calculations show that semiconductor surface states are rather localized within certain bonds. Examples are
the "dangling-bond" and "back-bond" surface states. Figure 29 shows the
charge density plot of the empty Ga surface state that is located in energy
just at the bottom of the conduction band.[121] Even though this state is s-like,
it clearly is very localized within the surface layer. This indicates a rather
unique situation, where certain surface states might be correlated with very
specific bond length or bond angle changes.

In reality, however, the situation is less simple because of the deep penetration of the reconstruction. At present band structure calculations con-

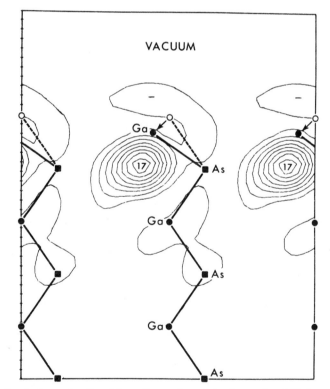

Fig. 29.   Pseudocharge density of the cation surface states for a relaxed GaAs surface. The
charge-density contour map is normalized to 1 electron per unit cell: $\Omega_c = 812$ Å.[121]

sider only rearrangements of the atoms within the first layer and at best an additional contraction of the surface layer. On the other hand, dynamic LEED calculations have included reconstruction within the second and deeper crystal layers. For Si, deviations from bulk atomic position up to the fourth layer have been postulated.[120] This clearly indicates the qualitative difference of the structural models derived from state-of-the-art LEED and photoemission studies.

Chadi recognized early that band structure calculations compared to angle-resolved photoemission data may lead to structural models that are not realistic.[122] He therefore introduced the total energy as an additional functional that must be minimized for a correct geometric structure. This is certainly an important improvement on the calculational scheme and, as we will see later, it has important consequences in the case of the structural determination of H atoms on transition metal surfaces.

At the present time, angle-resolved photoemission as a structural tool for semiconductor surfaces, is certainly inferior to dynamic LEED. Therefore we regard a determination of the atomic structure of the surface that is based solely on the determination of the electronic structure as ambiguous, even if the calculations include the total energy. If angle-resolved photoemission and LEED or other methods go hand in hand in a structural determination, then certainly both methods reinforce each other and the final results appear to be more reliable. In the following we show how a structural model was developed for GaAs(110) that satisfies both LEED and photoemission results.

The natural cleavage plane of GaAs is the (110) surface, on which equal numbers of Ga and As atoms are present. The crystallographic structure of this surface is illustrated in Fig. 30. The LEED pattern of the GaAs(110) surface shows a 1×1 pattern with no indication of a superstructure. This however does not imply that there is no reconstruction present at the surface. It indicates only that equivalent atoms of the surface lattice must deviate in the same way from the ideal bulk lattice positions. Indications for such a behavior were derived from a comparison of (angle-integrated) photoemission results with band structure calculations for the ideal truncated bulk lattice. These calculations[124, 125] predicted surface states within the band gap of GaAs, whereas the photoemission data showed that this was not the case. This result has been confirmed by angle-resolved photoemission (occupied states).[125] The unoccupied states can be detected only by absorption or yield spectroscopy, and the excitonic shifts present in absorption increase the uncertainty of the energy determination of the empty surface states. If we assume a 0.5 eV excitonic shift, these unoccupied states are located just at the bottom of the conduction band.

The first detailed LEED analysis of the reconstruction was presented by Lubinski et al.[126] They followed an idea proposed earlier[127] and calculated

Ga As (110)

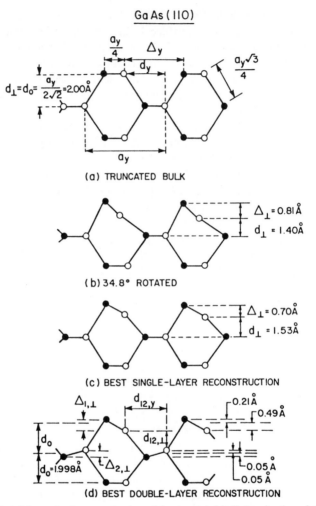

Fig. 30. (a) Schematic planar projection of the truncated bulk termination of GaAs(110), (b) bond-rotated geometry, (c) best single-layer reconstruction, and (d) best two-layer reconstruction.[123]

the effects on the dynamic LEED I–V or I vs. V curves of a rotation of the topmost GaAs pairs normal to the bond. This reconstruction is illustrated in Fig. 30b. The agreement between their calculation and their data was best for a rotation angle of $\omega = 34.8°$. Later the methods of analysis were refined, and now the rotation angle is close to 27.3°. More important, however, deviations from the ideal lattice position are present down to the third layer. For a summary of the LEED results on GaAs(110), see Duke et al.[128] and references therein.

How do the photoemission results and the electronic structure calculations agree with the structural model established by LEED? First of all, the band structure calculations do not take into account displacements of atoms in the second and third layers. The rotation angle as determined by LEED is $\omega = 27.3 \pm 2°$, whereas the photoemission results converged to 27° with an uncertainty of around 5°. Table I lists the experimental and theoretical re-

TABLE I
Surface States on GaAs(110)[a]

	Experimental energy (eV)[b]			Theoretical energy (eV)[b]							
Symmetry point	Ref. 129	Ref. 130	Ref. 131	Ref. 131	Ref. 131	Ref. 132	Ref. 132	Ref. 132	Ref. 133	Ref. 134	Ref. 135
$\bar{\Gamma}$	0.3	0.5	0.5	0.	0.4		0.5	0.1	0.5		0.3
	0.7	0.85					0.5		0.7	0.7	1.0
		(1.9)							2.2		
	3.4	(3.0)					3.35		3.0		
		(4.6)									
	6.0										
$\bar{X}'$	0.9	0.9		0.5	0.9		1.05	0.9	0.9	1.0	0.7
	1.4	1.4					1.4	1.2	1.6	1.2	0.9
	1.6	1.9	2.0		2.0	2.0	2.0	1.8	1.85	1.7	1.2
									2.1		
	4.0	4.1		3.8	3.7		3.45		3.6	3.2	3.5
	5.0								5.5		
$\bar{X}$		1.3	1.2	0.5	1.2	0.1	1.35	0.9	1.1	1.1	0.9
	1.8	2.1	2.2	2.1		1.9	2.2			1.4	1.6
									2.5		
	2.8	3.0							(2.7)		
		3.7		3.5		3.3	4.3	4.5	4.1	3.7	3.6
		(5.5)		5.3	5.1	6.0	5.8	5.6	5.8		5.1
									6.2		5.3
$\bar{M}$	1.3	1.0		0.4	1.2	0.2	1.1	0.9	1.05	0.8	0.7
	1.5	(1.5)					1.7	1.6	1.75		1.2
		3.3				2.8			3.3		2.7
	3.8	3.9					3.9	4.1	3.7		
			5.7	5.8	5.6				6.3	5.7	5.6
						6.4	6.4	6.5	6.5		5.8
Angle											
ω				0°	19°	0°	34.8°	BR[c]	25.2°	34.8°	27°
$\Delta\omega$					19°≤ω ≤25°						20°≤ω ≤28°

[a] Deeper lying states ($E_B \geq 7.0$ eV) omitted.
[b] Features reported as weak given in parentheses.
[c] Bond relaxation model after Ref. 136, which corresponds to $\omega = 27°$ for a single-layer reconstruction.

sults for surface states on GaAs(110). Since most of the surface states in GaAs(110) are degenerate with underlying bulk bands, it is experimentally and theoretically difficult to define a surface state. As the tabulated results show, there is disagreement not only about the energy position of the surface states but also about the number of surface states at the critical points in both experimental and theoretical studies.

First we discuss the experimental results. Here the major disagreement is in the number of states found for the critical points. Once the surface states are identified, the binding energies agree reasonably well. Therefore the experimental results seem to depend mostly on the method of identification of surface states, when they lie on top of bulk interband transitions. Since relatively few primary spectra have been published, we cannot judge which interpretation is correct. Only one of the groups[129] has used the full power of angle-resolved photoemission in determining the symmetry of the surface states after excitation with polarized radiation. This symmetry information is essential in the comparison with calculations if structural information is to be extracted.

Before we make a comparison with the theory, a few comments about the computational methods are appropriate. All but one of the calculations summarized in Table I uses the tight-binding formalism. This procedure contains as parameters certain overlap integrals that are fitted to give agreement with the known bulk band structure. Nevertheless there are for all the calculations appreciable differences in the calculated bulk density of states (cf. Fig. 7 of Ref. 134, Fig. 3 of Ref. 137, and Fig. 2 of Ref. 132). The major problem, however, is the identification of surface states. This is illustrated by a comparison of two calculations by the same author for only slightly different geometries.[132, 133] In the second calculation the criterion for surface states is relaxed somewhat and consequently the number of calculated surface states rises considerably[133] (from 13 to 33; see Table I).

The experimental and theoretical uncertainty concerning the nature of a surface state on a semiconductor surface undoubtedly is the major reason for the qualitative differences found in different columns of Table I. Given this problem, coupled with the nonreproducibility of data due to sample preparation, the photoemission comparison of theory and experiment can at best be said to confirm the LEED analysis. The experimental situation could be partially clarified by combining angle-resolved photoemission and LEED current and voltage measurements. This would serve two purposes. First it would allow a comparison of the quality of the samples through the LEED data, Second, the combined data could be used to evaluate the relative sensitivity to imperfections of both techniques. It would also be informative to have theoretical calculations of the photoemission spectra from surface defects.

Results on surface states on the GaAs(100) and GaAs(111) surfaces have been published.[138, 139] These surfaces must be prepared by molecular beam epitaxy, and they have complicated superstructures; therefore it is not appropriate to discuss the results until they are more complete and have been confirmed by independent groups.

Probably the most widely investigated semiconductor is silicon, either with its natural cleavage face (111) or in the (100) orientation. In general the GaAs(110) results suggest that it is very difficult to obtain reproducible photoemission results. The problems that occur with Si are even larger than for GaAs. The Si(111) surface, as cleaved, exhibits a (2×1) reconstruction that transforms into a (7×7) reconstructed surface upon heating to about 400°C. A (1×1) structure might also be present during the transition from (2×1) to (7×7).[140] The (2×1) reconstruction consists, in its simplest form, of a raising and lowering of adjacent rows of surface atoms. This model was proposed by Haneman[141] and seems to be valid with some modifications regarding the positions of atoms in subsurface layers.[142] Figure 31 shows a model of this surface and surface Brillouin zone.

The unreconstructed Si(111) surface should have a half-filled, dangling-bond surface state band, which would make this surface metallic. The reconstruction causes this band to split into two bands, one of them filled and the other empty, with an optical gap of 0.26 eV as determined by infrared absorption.[143] This obviously lowers the total energy of the surface and makes

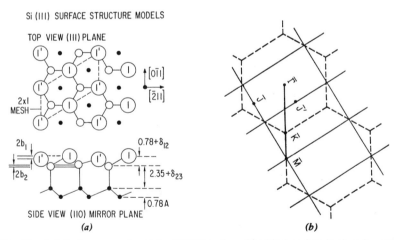

Si (III) SURFACE STRUCTURE MODELS

TOP VIEW (III) PLANE

2xI MESH

SIDE VIEW (IIO) MIRROR PLANE

(a) *(b)*

Fig. 31. (a) Surface geometry of the Si(111) surface. The ideal surface is given by setting $\delta_{12} = b_1 = \delta_{23} = d_2^e = 0$ (i.e., atom 1 = atom 1'). The (2×1) surface is given by displacing alternate rows of surface atoms 1 and 1' outward and inward, respectively. The outer double layer spacing is 0.78 Å and the second to third layer spacing is 2.35 Å (Si—Si bond length).[149] (b) The SBZ for Si(111).

TABLE II
Experimental Surface States on Si(111) (2×1)[a]

	Ref. 144		Ref. 145		Ref. 146		Ref. 147		Ref. 148	
	Band edge at $\bar{\Gamma}'$[b]	Dispersion	Band edge at $\bar{\Gamma}'$	Dispersion	Band edge at $\bar{\Gamma}'$	Dispersion	Band edge at $\bar{\Gamma}'$	Dispersion	Band edge at $\bar{\Gamma}'$	Dispersion
	0.75 eV	Down 850 meV towards $\bar{K}$ up 650 meV (weak feature) toward $\bar{K}$	0.25 eV	Up 500 meV toward $\bar{J}'$	0.7 eV	600 meV upward toward $\bar{K}$	0.45 eV	Flat ≤100 meV toward $\bar{J}'$	0.7 eV	Flat ≤100 meV toward $\bar{J}$; down ≤200 meV toward $\bar{J}'$
	1.25 eV	Flat ≤100 meV toward $\bar{J}'$			1.5 eV	Flat (≤200 meV) toward $\bar{K}$	The authors measure a value of 0.85 eV relative to E_F and at $n=2\times10^{19}/cm^3$; doping E_f is about 400 meV above the valence band edge.[116]			
	1.7 eV	Down toward $\bar{J}'$			2.6 eV					
	3.0 eV	Up toward $\bar{J}'$								
	The dispersion is measured in the direction toward the corner $\bar{K}$ of the unreconstructed SBZ. Depending on the orientation of domains, this measures either toward $\bar{J}$ or to a point close to $\bar{J}$ (see Fig. 31).		The authors do not give the sample doping; we have assumed the samples to be intrinsic Si, which places E_f at 340 meV above the valence band edge.[116]		The dispersion is measured in the direction toward the corner $\bar{K}$ of the unreconstructed SBZ. Depending on the orientation of domains, this measures either toward $\bar{J}$ or to a point close to $\bar{J}'$.					

[a] All binding energies (BE) measured relative to the valence band edge.
[b] Measured at $K_{\parallel} = 0.25\,\bar{\Gamma}\bar{J}$.

it semiconducting. These dangling-bond surface states were among the first surface states detected by (angle-integrated) photoemission.[114, 115] These states are centered about 0.5 eV below the valence band maximum, but extend into the gap, which makes detection easy. The surface states of Si(111) (2×1) were studied relatively early by angle-resolved photoemission, but we still do not have a consistent picture of the dispersion and location of these bands. This is demonstrated in Table II, which gives the experimental measurements. One of the reasons the data do not agree might be the formation of domains on the (2×1) reconstructed surface. Most of the authors are aware of this problem and claim to have selected only single-domain samples, but this selection seems to be done on a rather arbitrary basis. These photoemission studies should be accompanied by recording LEED I vs. V curves, so one sample could be compared to another. This might indicate when domain structure was important. Domains cannot be the only source of error. Table II gives measured values for the binding energy of the surface states at $\bar{\Gamma}$ that should be independent of the presence of domains. Parke et al.[145] reported a spread in the positions of the surface state bands of up to 0.4 eV even for samples that initially had been characterized as good cleavage faces. The other references listed in Table II give no quantitative information about the reproducibility of their results.

Table II indicates that the measured position of the topmost surface states, which is the famous dangling-bond state, varies at $\bar{\Gamma}$ between 0.1 and 0.75 eV. Some of the authors observe no dispersion for this state, whereas others claim to have observed 0.6 eV upward or 0.8 eV downward dispersion for this state. These measurements are carried out in different directions, or in no specified direction, with respect to the (2×1) domain SBZ (see Fig. 31), but these directions either pass through $\bar{J}$ or pass through a point close to $\bar{J}'$ in the SBZ, so that a large dispersion in the $\bar{\Gamma}$-$\bar{K}$ unreconstructed zone direction is inconsistent with no dispersion along $\bar{\Gamma}$-$\bar{J}$ or $\bar{\Gamma}$-$\bar{J}'$ in the (2×1) domain SBZ.

This section has illustrated the state of confusion that exists both in the experimental data and in the theoretical calculations of surface states on semiconductors. This emphasis represents the opinion of two outsiders to the field. We refer the reader to a review article of this subject by Eastman,[149] who is much more knowledgable about the subject and almost as pessimistic as we are. Eastman's review article[149] discusses structural determinations of Si(001) and Si(111) surfaces including the (7×7) Si(111) structure.[150]

C. Surface Local Density of States

Thus far we have emphasized the capabilities of angle-resolved photoemission when all the components of the matrix element (5) have specified symmetries. For example, we use single-crystal surfaces so that the detector can be placed in a high-symmetry direction, polarized light so that the symmetry of the polarization with respect to the final state can be varied to pick out specific symmetry initial states, and variable photon energy to "tune in" specific symmetry final states. In the example of angle-resolved photoemission presented in this section, we want to remove all the angular anisotropic effects caused by the terms in the photoexcitation matrix element.[151]

The objective is to measure the change in the local density of states (28) at the surface compared to the bulk. In general, there should be a narrowing of the bandwidth at the surface compared to the bulk bandwidth.[78] This can be easily understood in terms of the lowered coordination of the atoms at the surface. For example, a nearest-neighbor, tight-binding calculation of the surface and bulk density of states predicts that the second moment μ of the bandwidth at the surface is proportional to the number of nearest neighbors Z.[152] The second moment is

$$\mu = \frac{\int \rho(E)(E-\bar{E})^2 \, dE}{\int \rho(E) \, dE}$$

with

$$\bar{E} = \frac{\int \rho(E)E \, dE}{\int \rho(E) \, dE}$$

This implies that the full width at half-maximum of the d bandwidth at the surface should be approximately proportional to $Z^{1/2}$. For fcc metals, $Z=8$, 7, and 9 for the (100), (110), and (111) surfaces, respectively. This predicts reductions in the d-band FWHM at the surface of 18.4, 23.6, and 13.4%, respectively.

The density of states could be measured in photoemission if all the energy and angular dependence in the matrix element (5) were eliminated. Then the intensity would be a sum of the initial and final density of states. Mehta and Fadley[151] eliminated most of the matrix element effects by using X-ray photons and polycrystalline samples. The monochromatized X-ray source produced photons of 1487 eV energy, so that electrons excited from the valence band of the sample have a kinetic energy of about 1480 eV. There are so many final state bands at these energies that a reduced-zone scheme picture (like Fig. 10) has been described as "spaghetti." This implies that there will be no effects due to final state bands. The polycrystalline samples help to scramble any angular effects due to initial states. If it is true that there are no angular effects caused by the matrix element and the cross-section is constant over the energy range of the band, the density of states can be measured.

The surface density of states can be observed by using angle-resolved detection.[151] The mean free path for electrons of about 1500 eV kinetic energy in copper or nickel is approximately 15 Å.[60] If the photoemitted electrons are collected normal to the surface, the measurement samples down to a depth given by the inelastic scattering length Λ_e. When the collection is near grazing, the emitted electrons come from a much smaller depth because they are traveling nearly parallel to the surface. The signal at the analyzer can be written as[151]

$$I(E) = \int^\infty \rho(z, E) \exp - \left(\frac{z}{\Lambda_e \cos \theta'} \right) \tag{29}$$

where $\rho(z, E)$ is the local density of states defined by (28) and θ' is the internal angle of the escaping electron. The internal angle is related to the external collection angle θ by $\sin \theta' / \sin \theta = (E_{kin} / E_{kin} + U_0)^{1/2}$, where U_0 is the inner potential discussed in Section II.E. If the collection is normal to the surface ($\theta' = 0$) then $\Lambda_e \cos \theta' = 15$ Å, while grazing collection at $\theta = 85°$ reduces the z escape depth ($\Lambda_e \cos \theta'$) to 1.9 Å; $\theta' = 82.6°$ when $\theta = 85°$ if the inner potential U_0 is 14 eV.[151]

Mehta and Fadley[151] measured a reduction in the FWHM of the d bands of Cu (Ni) of 11.8% (10.7%) as the collection angle was changed from 50° to 85°. The theoretical prediction were 17.5% for Cu[151, 153] and 30% for Ni.[98, 151] The measured second moment μ decreased by 19% between 50° and 85° collection for Cu and 21% for Ni. The theoretical calculations for Cu[153] are in excellent agreement with the measurements at all angles of collection. The calculations for Ni[98, 151] disagree in that they predict a μ for the bulk of about 2 (eV)^2 and a 10.8% reduction between 50° and 85°, whereas the data show that $\mu = 0.9$ (eV)^2 for small angles (bulk) and a 21% reduction in μ between 50° and 85°.

These measurements show that the picture of band narrowing at the surface is qualitatively correct. For a copper surface the observed band narrowing is quantitatively explained by theoretical calculations of the surface and bulk density of states.[153] The theoretical calculations do not explain the Ni data because the calculated bandwidth is too large. This is consistent with the angle-resolved band structure measurements on nickel.[95, 111, 112]

Fuggle and Menzel have also reported a slight d-band narrowing at grazing emission from silver,[154] and Citrin and Wertheim have observed the effect on gold.[155]

IV. ADSORPTION

The three preceding sections have discussed angle-resolved photoemission and its application to the measurement of electronic properties of the bulk and clean surface. Now we consider adsorption of foreign atoms or molecules onto a clean, well-characterized surface. Angle-resolved photoemission allows us to obtain a detailed picture of the electronic properties of the surface-adsorbate complex. The electronic properties of the system are used to determine the microscopic nature of the adsorbate-substrate and adsorbate-adsorbate interactions. A key to understanding the surface bond is the surface geometry. The geometry can in principle be determined from the electronic energy levels, since the geometric and electronic configurations are intimately related.

There are three different approaches to determine the geometry of an adsorbate using angle-resolved photoemission. The orientation of the bond axis of a molecular adsorbate such as carbon monoxide can be determined from simple symmetry considerations, using the excitation from a molecular orbital of the molecule not involved in the bonding to the surface. The experimentally determined energy positions and dispersion of the energy levels involved in the bond to the surface can be compared to theoretical calculations for bonding in different sites. Finally, the diffraction of the outgoing electron by the substrate atoms can be measured and calculated to determine the site. The last procedure is easier to interpret when the excitation is from a core level of the adsorbate, so that there is no angular dependence caused by the initial state. This section is separated into two parts, dealing with valence levels and core levels. The core-level section deals solely with structural determinations via scattering of the photoexcited electron.

A. Valence Levels

To illustrate both the capabilities and the limitations of angle-resolved photoemission as a tool capable of determining the surface bond of an adsorbate with the substrate and the adsorbate-adsorbate interaction, we use several selected examples. In considering the first, chalcogen adsorption on nickel and aluminum, we demonstrate the ability of this technique to determine the symmetry of adsorbate levels, the dispersion of bands, and the symmetry-induced effects caused by the substrate. The second set of examples, hydrogen adsorption on titanium and palladium, demonstrates the capability of determining the bonding geometry from the comparison of theoretical calculations and data and the effect of bonding on intrinsic d surface states. Chlorine adsorption on silicon also is used to illustrate the capability of determining the bonding site. The adsorption of carbon monoxide, nitrogen, and ammonia serves to illustrate, both the effects on the molecular orbits due to coordination to the metal surface and the determination of bonding orientation of the molecule. Finally the case of xenon on palladium provides an example of the effects observed in a physisorbed layer.

1.   Chalcogen-Ordered Overlayers

Sulfur adsorbs into the fourfold site of an Ni(100) surface as shown in the real-space representation of Fig. 32. It forms two ordered overlayers, a $p(2$

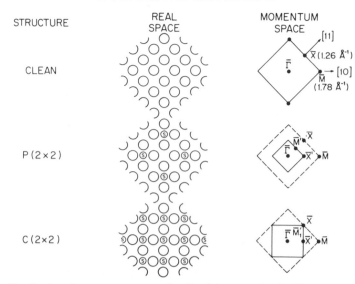

Fig. 32. Real- and momentum-space unit cells of the two ordered sulfur structures $p(2\times2)$ and $c(2\times2)$ on Ni(001). The small circles in the real-space representation are the surface Ni atoms.

$\times 2$) structure at one-fourth of a monolayer and a $c(2\times2)$ structure at one-half of a monolayer.[156] The Ni—S bond length is 2.19 Å. An isolated S atom has an electronic configuration $1s^22s^22p^63s^23p^4$ with an ionization potential of 10.4 eV. When this atom is adsorbed onto a surface, the bonding will be primarily through the $3p$ electrons. Since the surface reduces the symmetry, the Pz (z normal to surface) orbital will not be degenerate with the $PxPy$ orbitals. If the surface bond to the $PxPy$ orbitals is stronger than the bond to the Pz orbital, we should find the $PxPy$ electronic levels lower in energy. When the S concentration on the surface is increased to form an ordered overlayer, there will be interaction between the S atoms. This can occur if there is direct overlap of the S wave functions, or by way of indirect interaction through the substrate. The discrete energy levels of the single adsorbed atom become energy bands for the ordered overlayer. That is, the energy position of a specific level depends on the value of $k_\parallel$ being measured.

We can predict in a qualitative way what will occur when the S atoms are close enough to form bands, by considering a two-dimensional isolated S layer, assuming only nearest-neighbor interactions. Figure 33 shows the Pz and $PxPy$ wave functions for the $c(2\times2)$ configuration at the three high-symmetry points in the surface Brillouin zone ($\overline{\Gamma}$, $\overline{X}'$, $\overline{M}'$). Since $\overline{\Gamma}$-$\overline{X}'$ and $\overline{\Gamma}$-$\overline{M}'$ are mirror planes of the SBZ, all functions are odd or even. The Pz state is even in both mirror planes, but there is one even and one odd band derived from the Px and Py functions. At $k_{\|} = 0(\bar{\Gamma})$, Fig. 33 shows that the Pz-derived state is bonding because all the nearest-neighbor S atoms have the same sign. The $PxPy$ S states are degenerate and antibonding. Therefore, as we make the S—S spacing smaller, the Pz level at $\bar{\Gamma}$ should drop in energy and the degenerate Px, Py level should rise. At $\bar{M}'$ the Pz-derived band is antibonding because all nearest-neighbor atoms are of opposite sign. The even and odd $PxPy$ states ($Px+Py$ and $Px-Py$) are degenerate and of a bonding character. Therefore, the Pz and Px, Py levels at $\bar{M}$ should behave in an opposite way from the corresponding states at $\bar{\Gamma}$ as the S—S spacing is decreased. The Pz level should rise in energy, and $PxPy$ should drop. This suggests that the Pz band may cross over the $PxPy$ bands.[158] The dispersion curves at the bottom of Fig. 33 show what the bands from $\bar{\Gamma}$ to $\bar{M}'$ should look like for the two-dimensional S layer. The $Px+Py$ (even) and $Px-Py$ (odd) bands are degenerate only at $\bar{\Gamma}$ and $\bar{M}'$. The behavior at $\bar{X}'$ is quite different. The Pz level is "nonbonding," since it has two nearest neighbors of each sign. The next nearest neighbors are anti-bonding. The odd and even $PxPy$ levels are not degenerate. The even Px state is strongly bonding, but the odd Py state is still antibonding.

The two even bands in either the $\bar{\Gamma}$-$\bar{M}$ or the $\bar{\Gamma}$-$\bar{X}'$ direction do not mix with each other because of the inversion symmetry about the center of the two-dimensional S layer. The Pz band is odd with respect to reflection about the plane and the $PxPy$ band is even. When the S layer is placed on the surface, this symmetry is destroyed and the two bands will hybridize, creating a hydridization gap.[158] This is shown schematically by the dotted lines in the $\bar{\Gamma}$-$\bar{X}'$ direction at the bottom of Fig. 33. The Pz band at $\bar{\Gamma}$ is pure Pz, but in the middle of the zone it is mixed with Px. The odd state cannot mix with the other bands because the mirror plane symmetry still exists.

The S—S spacing in the $p(2\times2)$ structure is about 5 Å, so we would expect little dispersion due to direct S—S overlap (the covalent radius of S is 1.04 Å). The photoelectron spectra should reveal the splitting of the Pz and $PxPy$ orbitals due to the interaction of the S with the Ni substrate, but the energy of these levels should be independent of $k_{\|}$. The spacing in the $c(2\times2)$ structure is 3.5 Å, which should result in dispersion, and the angle-resolved photoelectron spectra should exhibit the qualitative features discussed above and shown in Fig. 33 (bottom).

Normal emission collection ($k_{\|} = 0$ or $\bar{\Gamma}$) can be used to identify the Pz and $PxPy$ energy levels. The states at $\bar{\Gamma}$ have C_{4v} symmetry, with the Pz state having a_1 and the $PxPy$ state having e symmetry. The final state, which is an outgoing wave to the detector, has a_1 symmetry. Therefore, as explained in Section II, the symmetry of the polarization vector can be used to pick out specific symmetry initial states. If the polarization vector is in the plane of the surface, it can excite only e-symmetry initial states. If it is normal to the

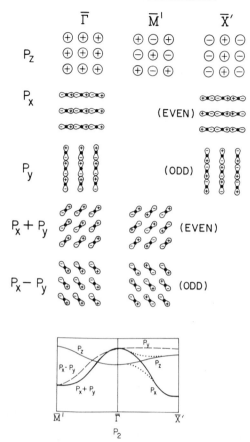

Fig. 33.   Schematic representation of the Pz and the odd and even Px, Py sulfur orbitals at the three high-symmetry points in the SBZ of the $c(2 \times 2)$ structure of Fig. 32. The bottom figure shows the dispersion expected in a tight-binding model.

surface, only a_1-symmetry initial states can be excited. Figure 34 shows the measurements that were used to determine the energy position of the a_1 and e states at $\bar{\Gamma}$ for both the $p(2 \times 2)$ and $c(2 \times 2)$ structures.[159] The peaks in the spectra within a couple of electron volts of the Fermi energy are due to excitations from the d bands of Ni, while the peaks between 4 and 6 eV below the Fermi energy are the S $3p$-derived states. The spectrum at the bottom left is for nearly s-polarized light where only the e state can be excited. It is at a position of 4.9 ± 0.15 eV below the Fermi energy. The top spectrum on the left is for p-polarized light where both the a_1 and e states can be excited. The shaded region shows the contribution to this spectrum from the e state

NORMAL EMISSION

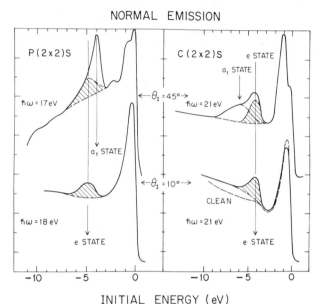

INITIAL ENERGY (eV)

Fig. 34.   Normal emission spectra for $p(2 \times 2)$S and $c(2 \times 2)$S on Ni(100) for p-polarized light at two angles of incidence.[159]

as determined from the bottom spectrum. The a_1 state is at an energy of 4.24 ± 0.05 eV below the Fermi energy. The $PxPy$ (e) energy level is below the Pz (a_1) energy level, indicating that the $PxPy$ orbitals are more bonding to the substrate than the Pz orbitals. Measurements of the dispersion of these levels with $k_\parallel$ showed less than 0.15 eV dispersion.[159]

The same measurements as described above for $p(2 \times 2)$S are shown for $c(2 \times 2)$S in Fig. 34 (right). The a_1 state drops in energy to 5.9 eV below the Fermi energy, while the e state rises to 4.44 eV.[159] The a_1 and e states invert their order from $p(2 \times 2)$ to $c(2 \times 2)$. This is exactly what would have been anticipated from the discussion about the effects of dispersion. The e state is antibonding, while the a_1 state is bonding at $\bar{\Gamma}$ with respect to the sulfur overlayer. Figure 35 shows the measured dispersion of the two even bands and one odd band along each of the two mirror planes. The shaded regions are the projected bulk bands of the appropriate symmetry.

The simplest measurement to make and to understand is to map the dispersion of the odd band. To do this, the detector is placed in a mirror plane and the polarization vector is set perpendicular to this mirror plane. The collection angle or the photon energy is varied so that E can be recorded versus $k_\parallel$. The odd symmetry bulk bands in Ni all lie above the S state, so

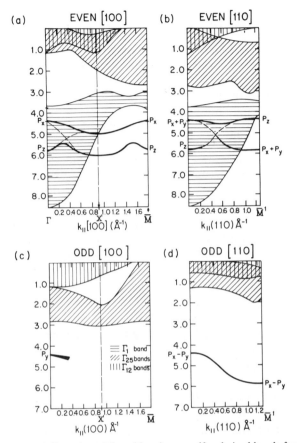

Fig. 35.   Measured dispersion of the odd and even sulfur-derived bands for a c(2×2) over-layer on Ni(100).[159] The shaded areas are the projection of the odd and even bulk bands onto the Ni(100) surface.

there is no interference from bulk transitions. Figure 35d shows the measured $Px - Py$ band in the $\overline{\Gamma}$-$\overline{M}$ direction. It starts at 4.4 eV below the Fermi surface and disperses downward to 5.9 eV. At $\overline{M}$ the odd $Px - Py$ and even $Px + Py$ bands are degenerate. Figure 35b shows the measurements for the even bands, which were made with the light polarization vector in the plane of detection. At $\overline{M}$ there are two bands, 4.3 and 6 eV. The 6 eV band is the $Px + Py$ band because it is at the same position as the $Px - Py$ level at $\overline{M}$. Therefore, the 4.3 eV band at $\overline{M}$ must be the Pz band. At $\overline{\Gamma}$ the order is reversed, Pz at 5.9 eV and $PxPy$ at 4.4 eV. The bands will cross in the middle of the zone. When they cross they will hybridize because of interaction with the substrate. Figure 35b shows that this happens in the middle of the zone. The bottom band is not shown all the way across the zone because the data could not be fitted with Gaussian or Lorentzian peaks. Liebsch has shown that the line shape is very complicated in this region because of hybridization and mixing with the substrate Ni 4s band.[158, 159]

The measurements in the $\bar{\Gamma}$-$\bar{X}'$ direction are shown in Fig. 35a and Fig. 35c. The odd band could not be tracked all the way across the zone because of intensity problems.[159] Figure 32 showed that the surface Brillouin zone for the c(2×2)S overlayer is only half as long in this direction as the clean zone. This means the S-derived bands should be symmetric about a value of $k_\parallel$ equal to one-half the clean zone edge. Figure 35a shows that this happens, assuring the existence of a c(2×2) structure. Because of the differences in the c(2×2) and clean surface zone, the Pz and PxPy states are not pure at $\bar{X}'$, but it is obvious from Fig. 35a that the even bands are crossing and that hybridization has occurred in this direction. The dashed curves indicate what the bands could have looked like before hybridization.

The major difference between c(2×2) and p(2×2)S overlayers is the amount of dispersion of the 3p levels. The p(2×2) spectra show very little dispersion with $k_\parallel$, while the c(2×2) structure exhibits a bandwidth of about 1.5 eV. The magnitude of this dispersion is approximately what is calculated for a free c(2×2)S layer,[159] indicating that the dispersion is predominantly a consequence of the direct S—S interaction. The difference between the p(2×2) and c(2×2) spectra cannot be totally explained in terms of S—S-induced dispersion. The average energy of the c(2×2) Pz band is 5.35 eV, which is 1 eV larger than the Pz level in p(2×2)S (4.2 eV). The average energy for the Px, Py states is, on the other hand, about the same for p(2×2)S and c(2×2)S. This means that the centroid for the density of states has shifted by about 0.4 eV to lower energy for c(2×2)S compared to p(2×2)S. A simple analysis would predict that the S in the c(2×2) configuration is bound more tightly to the surface than in the p(2×2) configuration because the energy levels have dropped (higher binding energy). This is in contradiction to the fact that the p(2×2) structure forms before the c(2×2) structure.

Two theoretical calculations have been performed for c(2×2)S on Ni(100). Liebsch carried out a KKR non-self-consistent calculation[159] and Richter et al.[160] used a self-consistent linear-augmented plane wave scheme adapted to thin films. Both theoretical techniques give excellent agreement with the ordering of the levels, the dispersion, and the hybridization. The non-self-consistent calculation does not give quantitative agreement with the data for the magnitude of the binding energies, presumably because of the non-self-consistency. The self-consistent calculation of Richter et al.[160] produces binding energies at the symmetry points of the SBZ that are in good agreement with the data: −4.4 (−5.6) eV and −5.6 (−3.9) eV for the Px, y (Pz) band at $\bar{\Gamma}$ and $\bar{M}$, respectively. The equivalent experimental values are -4.4 (-5.9) eV and -5.9 (-4.3) eV at $\bar{\Gamma}$ and $\bar{M}$, respectively. The general picture of the bonding of S to the Ni(001) surface is the same in both calculations. The S $3p$–Ni $3d$ levels form bonding-antibonding orbitals as the sulfur approaches the surface. The antibonding levels are all above the Fermi energy at the S—Ni equilibrium spacing.[158] The dispersion in the S-derived bonding bands is primarily a result of the S—S direct interaction. The hybridization of the two even S bands is caused by interaction with the Ni $4s$ band.

The S-derived bands can be discussed in terms of surface states or surface resonances, except now they are extrinsic states. The bands in Fig. 35 that do not overlap a bulk band (shaded region) are extrinsic surface states. The direct width of these states is zero because they cannot mix with any bulk states. On the other hand, the S bands that overlap bulk bands will be broadened. The line shape can in principle be quite complicated because the S state can mix differently with various regions of the bulk bands. In the extreme case the interaction of the adsorbate levels with the substrate S band would broaden the adsorption band to such an extent that it could not be seen experimentally. As the adsorbate band emerges from the bulk bands at the SBZ zone edge (Fig. 4a), it would become narrow and visible. Liebsch calculated that the maximum width of the s bands occurred at $\bar{\Gamma}$ for the Pz level. The inherent width was about 1 eV.[159] Richter et al.[160] calculated a width of about 0.6 eV for the Pz level. The broadening is small because the strongest interaction is between the S $3p$ and Ni $3d$ levels and the S bands are split off from the Ni $3d$ bands. Experimentally the broadening of the S levels due to interaction with the substrate bands cannot be observed.[159] The lifetime effects of the hole state dominate the spectral width.

Figure 36 shows a charge-density difference plot for $c(2 \times 2)$S on Ni(100).[160] The difference between the $c(2 \times 2)$S/Ni and the isolated S layer and Ni surface is shown. Several features of this plot are easy to understand in terms of our previous discussions. The excess charge between the S and Ni atoms is the bond between the S Px, y and Ni $3d$. The S Pz level is not symmetric about the plane of S atoms. There is excess charge outside and a depletion of charge between the S atom and the surface. This is what causes the hybridization of the even Pz and Px, y bands. The surface has destroyed the inversion symmetry about the plane containing the S atoms.

There are new features revealed by Fig. 36, as well. The first observation is that the charge disturbance is localized to the first Ni plane, because the bonding is primarily with the Ni $3d$ states. The second important observation is that the bonding charge between the S and Ni is counteracted by a charge depletion under the surface Ni atom. This charge depletion will undoubtedly cause a change in the Ni—Ni spacing between the surface plane

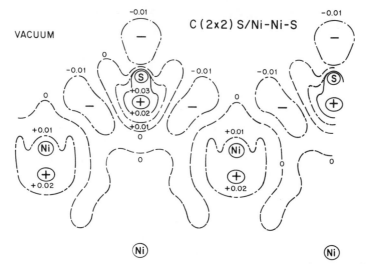

Fig. 36. Charge-density difference plot for $[c(2\times2)S$ on Ni(100)]–[S layer]–[Ni layer].[160] The units are in terms of e/a_0^3 where a_0 is the Bohr radius.

and the second layer, probably resulting in lattice expansion. The surprising result shown in Fig. 36 is the large charge depletion in the Pz orbital below the S atom. The origin of this effect is not clear at present. It could be due to the hybridization of the Pz band with the $PxPy$ band or to the $S(Pz)$—Ni(4s) interaction, which "stretches" the Pz.[160]

The bonding of chalcogens to Ni involves the d electrons of the transition metal substrate. Aluminum, on the other hand, is a substrate for which the bonding is due to the s and p electrons of the substrate. The energy dispersion of (1×1) overlayers of oxygen, sulfur, selenium, and tellurium on Al(111) have been measured using angle-resolved photoemission. The oxygen-adsorbed layer forms a (1×1) structure,[161] bonding in the threefold site with a Al—O bond length of 2.12 Å as determined by LEED[162] and 1.79 Å from SEXAFS (surface extended X-ray absorption fine structure) measurements.[163] Sulfur, selenium, and tellurium also form hexagonal superstructures aligned to, but not in registry with, the Al surface.[164] The S—S, Se—Se, and Te—Te spacings are 1.22, 1.29, and 1.40 times larger, respectively, than the Al—Al surface spacing.[164] This unique situation allows us to evaluate the effect of bonding of a commensurate and an incommensurate superstructure to a nearly free-electron metal.

Again, as we did in Fig. 33 for a square surface lattice, we can predict qualitatively the dispersion for this hexagonal surface lattice. Figure 37 shows the real lattice and the reciprocal lattice of the two-dimensional hexagonal

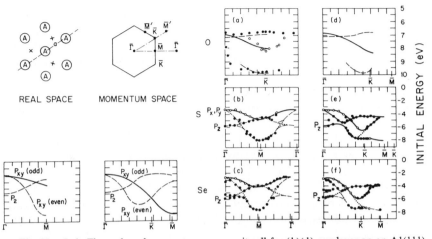

Fig. 37. *Left*: The real- and momentum-space unit cell for (1×1) overlayer on an Al(111) surface, showing the dispersion expected from this isolated (1×1) chalcogen layer (bottom); *right*: the experimental dispersion curves for O, S, and Se,[157,164,165] where odd bands are solid lines; open circles are data points of odd symmetry. The lines in the oxygen data are from Hofmann et al.[164] and the data points from Eberhardt and Himpsel.[165]

unit cell. The two symmetry points in the SBZ are $\overline{M} = \pi/(a \cos 30)$ and $\overline{K} = \pi/[a(\cos 30)]^2$. The reader can construct the equivalent picture of the Pz and Px, Py wave functions for this hexagonal structure, as shown in Fig. 33, for the square array. The complication is that $\overline{K}/\pi$ is not a multiple of $1/a$ ($\overline{K}/\pi = 1.33/a$). This problem can be circumvented by plotting the wave functions at $\mathbf{k} = 1.5K$ ($\overline{M}$ point in second zone with $\pi/k = a/2$) and at $k = 0.75\overline{K}$ ($\pi/k = a$), along the lines in k-space from $\overline{\Gamma}$ to $\overline{K}$. Along this direction in k-space the even Px, Py orbital is antibonding at $\overline{\Gamma}$, becomes strongly bonding at $k = .75\overline{K}$, and again becomes antibonding at $\overline{M}$ in the second zone. The odd $PxPy$ orbital is degenerate with the even one at $\overline{\Gamma}$ and becomes more bonding as k increases, reaching its maximum bonding configuration at $\overline{M}$ in the second zone. The two $PxPy$ orbits are degenerate at $\overline{K}$, but do not mix because they have different symmetries with respect to the mirror plane. The Pz orbital is bonding at $\overline{\Gamma}$ and antibonding at $\overline{K}$. This tight-binding two-dimensional dispersion is shown on the bottom left of Fig. 37. In the $\overline{\Gamma}$-$\overline{M}$ direction the even Px, Py orbital becomes strongly bonding at $\overline{M}$, while the odd component at $\overline{M}$ is just slightly more bonding than at $\overline{\Gamma}$.

When this hexagonal two-dimensional layer is placed in contact with the substrate, we expect to see shifts in the positions of the Pz and Px, Py bands relative to each other and mixing of the bands due to the reduced symmetry. The two even bands will hybridize where they cross, forming a gap just as they did for S on Ni(100). A new effect should appear on Al(111) because the two-dimensional symmetry of the surface layer is reduced when it is placed in contact with the substrate. The two broken lines shown in the momentum space of Fig. 37 show that both the $\bar{\Gamma}$-$\bar{M}$ and $\bar{\Gamma}$-$\bar{K}$ directions are mirror planes for the two-dimensional overlayer. The X's in the real lattice are the positions of the Al surface atoms. The $\bar{\Gamma}$-$\bar{K}$ direction is not a mirror plane for the adsorbed layer-substrate system. This means that the interaction with the substrate will allow the even and odd bands of the two-dimensional overlayer in the $\bar{\Gamma}$-$\bar{K}$ direction to hybridize. Specifically, we should see a gap open up a $\bar{K}$ between the even and odd Px, Py bands.

Figure 37 shows the measured dispersion for oxygen, sulfur, and selenium on Al(111): (a), (b), and (c) are for the $\bar{\Gamma}$-$\bar{M}$-$\bar{\Gamma}$ direction, which is a mirror plane. The chalcogen bands behave exactly as expected, except the Pz and Px, Py bands are nearly degenerate at $\bar{\Gamma}$ for oxygen. The odd and even Px, Py bands disperse downward from $\bar{\Gamma}$ to $\bar{M}$, with the even band being the most bonding at $\bar{M}$. The even Px, Py and Pz bands hybridize when they cross, forming a hybridization gap. The dispersion of the overlayer bands repeat with the appropriate symmetry; that is, $\bar{\Gamma}$ in the second zone is the same as $\bar{\Gamma}$ in the first zone. Figures 37d, e, and f show the dispersion along the $\bar{\Gamma}$-$\bar{K}$-$\bar{M}$ direction. To understand what is occurring in this direction, we track the bands of Se on Al (Fig. 37f). At $\bar{\Gamma}$ the Pz band is below the degenerate $PxPy$ bands. At $\bar{M}$ we know from Fig. 37c that the $PxPy$ bands have binding energies of -7.6 and -4.1 eV. The even $PxPy$ band at $\bar{M}$ along the $\bar{\Gamma}$-$\bar{M}$ direction is odd at $\bar{M}$ in the $\bar{\Gamma}$-$\bar{K}$-$\bar{M}$ direction. Therefore, the band at about -7.5 eV binding energy in Fig. 37f at $\bar{M}$ should be the odd $PxPy$ band. Figure 37f shows that the -7.5 eV band is observed in the odd geometry. But for $k_{\parallel} < \bar{K}$ the same band is seen in the even geometry. The reduced symmetry of the overlayer-substrate system compared to the isolated overlayer has removed the $\bar{\Gamma}$-$\bar{K}$ direction as a mirror plane. This allows the two-dimensional odd and even $PxPy$ states to mix, and consequently form a gap near $\bar{K}$. There are two hybridization gaps in the $\bar{\Gamma}$-$\bar{K}$-$\bar{M}$ direction; one where even Pz and $PxPy$ cross and one where the even and odd $PxPy$ bands cross. The magnitudes of the latter gaps are 1.2, 1.3, 1.6, and 2.0 eV for O, S, Se, and Te, respectively.[157, 164] The magnitude of the gap at $\bar{K}$ does not seem to depend on whether the overlayer is commensurate (O) or incommensurate (S, Se, or Te) with the substrate.

The reduction in the symmetry of the overlayer-surface complex compared to just the overlayer removes $\bar{\Gamma}$-$\bar{K}$ as a mirror plane, and consequently there are two $\bar{M}$ points. They have been labeled $\bar{M}$ and $\bar{M}'$ in Fig. 37. Hofmann et al.[164] measured the oxygen-induced bands on Al(111) in both directions and found identical dispersion. In a two-dimensional system in contact with a three-dimensional substrate the energy of the states at $\bar{M}$ and $\bar{M}'$ are

TABLE III

Chalcogen Adsorption

A. Oxygen: Ionization Potential 13.6 eV, Covalent Radius 0.74 Å

Substrate	Structure	Nearest-neighbor spacing (Å)	$\bar{\Gamma}$ energies (eV)		Dispersion (eV)		Ref.
			Pz	$PxPy$	Pz	$PxPy$	
Al(111)	(1×1)	2.86	7.1	6.7	0.3	~3	164
			6.7±0.2	7.3±0.3	~0.2	~2.4	165
Ni(100)	$p(2\times2)$	4.98	8	6		—	166
	$c(2\times2)$	3.52	6	5.5		~1	167
Cu(100)	$c(2\times2)$	3.61		5.5		~1.3	168
Ni(111)	($\sqrt{3}\times\sqrt{3}$)	$R30°*$	5.4	5.6			169

B. Sulfur: Ionization Potential 10.4 eV, Covalent Radius 1.04 Å

Substrate	Structure	Nearest-neighbor spacing (Å)	$\bar{\Gamma}$ energies (eV)		Dispersion (max) (eV)		Ref.
			Pz	$PxPy$	Pz	$PxPy$	
Al(111)	$\sqrt{1.5}(1\times1)$	3.5	5.9	3.4	1.5	4.6	157
Ni(100)	$p(2\times2)$	4.98	4.24	4.9		<0.15	
	$c(2\times2)$	3.52	5.86	4.44	1.5	1.4	159
Ni(111)	$p(2\times2)$	4.98	4.2	5.5			169
Cu(100)	$p(2\times2)$	5.1	4.9	5.3	~0		170

C. Selenium: Ionization Potential 9.75 eV, Covalent Radius 1.17 Å

Substrate	Structure	Nearest-neighbor spacing (Å)	$\bar{\Gamma}$ energies (eV)		Dispersion (eV)		Ref.
			Pz	Pxy	Pz	Pxy	
Ni(100)	$p(2\times2)$	4.98	~4	~4		~0	171
Ni(100)	$c(2\times2)$	3.52	4.4	3.6	1.2	1.8	171
Ni(111)	$p(4\times4)$		4.0	6.2	0	1.2	169
Ni(111)	$\sqrt{3}\times\sqrt{3}R30°$	4.31	~4.2	~5.8		<1.0	169
Al(111)	(1×1)	3.7	5.6	2.8	1.7	4.9	157

D. Tellurium: Ionization Potential 9.01 eV, Covalent Radius 1.37 Å

Substrate	Structure	Nearest-neighbor spacing (Å)	$\bar{\Gamma}$ energies (eV)		Dispersion (eV)		Ref.
			Pz	$PzPy$	Pz	$PxPy$	
Al(111)	(1×1)	4.0	4.2	2.0	1.2	4.8	152
Ni(111)	$\sqrt{3}\times\sqrt{3}R30°$	4.31	4.4	5.3			164

always the same. This can be proven by using time reversal. The intensity in a photoemission experiment can be different from $\overline{M}$ and $\overline{M}'$ because the final state is a three-dimensional state.

Table III summarizes the data for chalcogen adsorption obtained using angle-resolved photoemission. There are several general conclusions:

1. The Pz level is above the $PxPy$ level for bonding chalcogens at low density on transition metals. As the adsorbate-adsorbate spacing decreases and dispersion becomes important, the Pz drops below the $PxPy$ levels at $\overline{\Gamma}$. The density of states for the Pz band is generally 1.5 to 2.0 eV above the $PxPy$ density of states for chalcogens on Al.

2. The magnitude of the dispersion of the $PxPy$ bands for the chalcogens on Al is abnormally large, compared either to the equivalent dispersion on Ni or to the Pz band on Al. The reason is that the wave functions at $\overline{M}$ or $\overline{X}'$ for chalcogens on Ni are more localized[5] than the equivalent functions on Al. The bonding on Ni is through the $3d$'s, while it is with delocalized s and p functions on Al.[172]

We can restate everything presented in this section on chalcogen adsorption in terms of group theoretical nomenclature. The isolated S overlayers have D_{4h} (Ni structure) and D_{6h} (Al structure). The notation (D_{nh}) means that there is an n-fold rotation axis, a mirror plane, perpendicular to the n-fold axis, and a two-fold axis perpendicular to the n-fold axis. When the overlayer is placed on the surface, the last two symmetry operations described above are removed, so the symmetry classification becomes C_{nv}. The S overlayer on Ni has fourfold symmetry if it is bound to the surface in a four fold site (Fig. 32). The resultant symmetry is C_{4v}. If the S atoms were bonded to the Ni surface in the bridge site, the symmetry of the surface complex would be C_{2v}. However C_{2v} symmetry is so low that there can be no degenerate energy levels (i.e., all S bands would mix when they crossed). The data in Fig. 35d prove that the symmetry of the $c(2\times2)$S on Ni(100) system is higher than C_{2v}, because the odd band does not mix with the other bands.

The D_{6h} symmetry of the isolated chalcogen layer is lowered to C_{nv} when it is placed in contact with the Al substrate. The two rotated hexagonal structures (chalcogen and Al surface) form a C_{3v} symmetry complex. This results in the odd and even chalcogen bands of the isolated layer (D_{6h} symmetry) mixing in the overlayer in the $\overline{\Gamma}$-$\overline{K}$ direction. The failure of the odd band in the $\overline{\Gamma}$-$\overline{K}$ direction to mix means that the symmetry is higher than C_{2v}, that is, it is C_{3v}.

2. Hydrogen

The adsorption of hydrogen onto a surface has been the prototype system for many surface techniques and theoretical models.[173] The reason is that a proton should be the simplest of chemisorbed species. Fortunately these studies of H adsorption are also relevant to technologically important phenomena such as embrittlement of steels, heterogeneous catalysis, and hydrogen storage. We present two examples of coupling angle-resolved photoemission with theoretical calculations to determine the nature of the bond between hydrogen and metal bond: hydrogen adsorbed on titanium and palladium. Feibelman and Hamann have carried out a detailed and critical evaluation of the accuracy with which the geometrical structure can be determined from the angle-resolved photoemission spectra,[108, 174, 175] using data for H on Ti. The case of H adsorbed on Pd(111) is used to present a detailed picture of the H-metal states involved in the surface bond.[110, 177]

The determination of the surface geometry has always been one of the major problems in surface science. The analysis of angle-resolved valence spectra in principle can yield the structure, because electronic and geometric properties are interrelated. This is not a direct structural technique, since it requires comparing calculated surface band structures for various proposed geometries with the measured data. We use the H on Ti(0001) example to illustrate the sensitivity (or lack of sensitivity) of this approach to the bonding geometry. The historical development of this system clearly demonstrates the pitfalls and uncertainties that are probably inherent in this procedure, even for a system as simple as an ordered monolayer of hydrogen on an unreconstructed metal surface.

The first paper by Feibelman and Hamann[175] adapted a self-consistent linear-combination-of-atomic-orbitals method[176] to calculate the local density of states at a Ti surface with H present. The only spectroscopic information was an angle-integrated photoemission spectrum of Eastman[178] on evaporated polycrystalline films. The experimental data showed that the presence of hydrogen on or near the Ti surface created a peak 5 eV below the Fermi energy,[178] and the work function increased by about 0.3 eV. The theoretical calculations for clean Ti(0001) showed the presence of a d surface state at the Fermi energy that was not present in the data of Eastman for H on Ti. Feibelman and Hamann[175] set out theoretically to find the geometry for H bound to a Ti(0001) surface such that (1) the H peak was 5 eV below the Fermi energy, (2) the clean surface state disappeared, and (3) the work function increased by a few tenths of an electron-volt. They assumed for computational simplicity that the H formed a (1×1) structure, that is, it was ordered with one monolayer coverage.

H atoms placed under the first surface plane in the tetrahedral or octahedral sites did not remove the d-like surface state at the Fermi energy, and the H-induced density of states peak was at about 6.30 eV below the Fermi energy. The work function increased slightly, which was satisfactory. When the H atoms were placed atop a Ti surface atom, there was no H-induced density of states below the d bands of the Ti. The only acceptable geometries seemed to be the surface site, which would be the natural extension of the hexagonal close-packed or face-centered-cubic structure. Feibelman and Hamann[175] investigated two possibilities for this site, the first with the Ti—H bond length given by the Ti metallic radius and the H covalent radius used by Louie.[177] This site was about 0.4 Å above the surface with a Ti—H bond length of about 1.78 Å. The second spacing was given by the Ti—H_2 bond length of 1.9 Å, giving a vertical spacing from the surface of 0.9 Å. H in either of the spacings destroyed the clean intrinsic surface state, and the work function increased by 0.3 to 0.5 eV. The peak in the density of states from the H-induced states occurred 4.4 and 5.1 eV below the Fermi energy for the Ti—H_2 and H covalent radius spacings, respectively. Therefore the first paper concluded that H was bound on the surface in the threefold site, which is the natural extension of the Ti(0001) surface. The bond length is given by the metallic radius plus the H covalent radius.

After the publication of the first theoretical paper on H on Ti(0001), Himpsel produced angle-resolved spectra for H adsorbed on Ti(0001).[108] The normal emission spectra are shown in Fig. 38. The angle-resolved data showed more detailed structure than the angle-integrated data on polycrystalline films. The normal emission clean spectrum shows a peak at the Fermi energy that is quenched by H adsorption. This is the surface state predicted by Feibelman and Hamann.[175] The adsorption of H produced two peaks in normal emission, at 1.3 and 6.9 eV below the Fermi energy. The 6.9 eV state

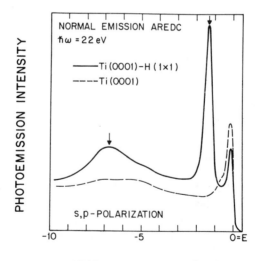

Fig. 38.   Normal emission photoelectron spectra for Ti(0001) and Ti(0001)—H(1×1).[108] The large increase in intensity near −7 eV is partially due to a final state effect.[108]

disperses as $k_\parallel$ is changed, with a maximum dispersion of 2.4 eV. The 1.3 eV H-induced peak does not disperse with $k_\parallel$ and exists only half-way out to the zone edge. The 1.3 eV H feature exists in the s-d hybridization gap in clean Ti.[175] The new data produced additional requirements for a theoretically acceptable geometry. The geometry chosen from the fit to the angle-integrated spectrum was unacceptable because there was no H-induced feature near -1.3 eV, in the hybridization gap. Feibelman and Hamann found that there was not a position on or under the surface where the theoretical surface band structure fit both the features exactly. They concluded that within the limitations of a local-exchange-correlation model band structure calculation to explain a measured excitation spectrum, there were several H-bonding sites that were spectroscopically "acceptable."[108]

We illustrate their calculations in Fig. 39 and Table IV. Feibelman and Hamann tried 25 different sites, of which we will discuss three. Figure 39a shows the projection of the bulk band structure onto the high-symmetry lines in the SBZ of a clean Ti(0001) surface. The calculation is for an 11-layer Ti film, and only the states that are even under deflection in the central plane are shown. The gap in the projected band structure near the center of the SBZ (near $\overline{\Gamma}$) extending from E_F to about -2.7 eV is a hybridization gap between the s and d bands of Ti. The heavy lines show states with large amplitude on the Ti surface layer. A d surface state exists at $\overline{\Gamma}$ (peak in clean spectrum of Fig. 38). This surface state exists throughout most of the SBZ. It mixes with the d bands (dispersing downward near $\overline{\Gamma}$) but reemerges from the bands in a large gap about $\overline{M}$. It is in a symmetry gap at $\overline{K}$ even though it appears to overlap the projected bulk bands. Figures 39b and 39c show the two-dimensional band structure for the two sites (out of 25) favored by calculations. The Δ's are the measured binding energies for the two H-induced peaks as a function of $k_\parallel$. Figure 39b is an adsorption site labeled "fcc surface site" because it is the site for continuing an fcc lattice instead of an hcp lattice. The intrinsic surface state at $E_F(\overline{\Gamma})$ is removed, and both H-induced features are present. The upper and lower H bands are too high in energy. The bond length used in the calculation is the Ti—H_2 bond length. Figure 39c is for a tetrahedral site under the first plane. The calculated position of the 1.3 eV H feature is reproduced exactly, while the lower state is too low in the calculations. The major problem in this last structure is that the intrinsic surface state is still present. This is a general result for sites below the surface when the surface states are d-like and very localized to the surface.[108, 179] The Ti—H bond length in the tetrahedral site is the Ti—H_2 length, which requires a 30% expansion at the surface of the plane spacing.

The origin of the problem of fitting both peaks in the spectrum can be understood. At the zone center the two peaks correspond to the bonding and antibonding combination of the H_{1s} with the Ti $3d_{3z^2-r^2}$ and $4s$. As the

(a)   CLEAN Ti SURFACE

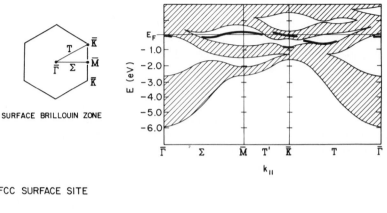

(b)   FCC SURFACE SITE

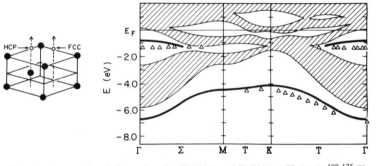

Fig. 39.   Calculated energy band dispersions for Ti(0001) and Ti(0001)—H (1×1).[108,175] The heavy lines are states with high wave function amplitude in the surface layers. The shaded regions are the projections of the bulk bands onto the surface using an 11-layer Ti film. Only states of even reflection symmetry with respect to the central plane are shown. (a) Clean Ti(0001), (b) H in the fcc surface site at left, (c) H adsorption in tetrahedral site with 30% surface plane expansion, and (d) H adsorbed in an atop position; experimental data shown by traingles.[108]

overlap of the H_{1s} and Ti $3d_{3z^2-r^2}$ wave functions increases, the splitting of the bonding-antibonding level increases. If the overlap is too large, the antibonding state is driven out of the hybridization gap into the bulk bands, where it disappears as an observable surface state. This is what happens in the fcc surface site shown in Fig. 39b when the spacing of the H from the surface is decreased. The bottom band can be made to agree with the data by decreasing the Ti—H bond length from the value of 1.90 Å used in Fig. 39b. When the calculated bonding H band agrees with the data, the top antibonding band is shoved up out of the gap, because the H_{1s} and Ti $3d_{3z^2-r^2}$

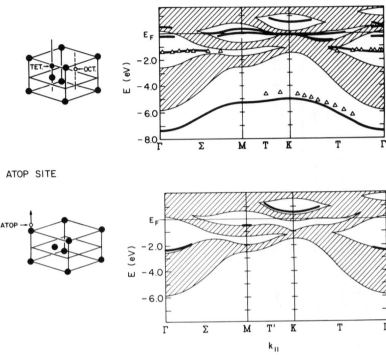

Fig. 39. (*continued*).

overlap has increased. The atop position in Fig. 39*d* is the opposite case. The overlap is so small that neither the bonding nor the antibonding state is split off from the bulk bands. Sites where either or both of the H-induced states are missing can be ruled out. Since both states cannot be fit with any geometry, Feibelman and Hamann[108] concluded that there were several geometries that must be considered to be acceptable.

Seven of the most acceptable sites are listed in Table IV, where the calculated work function change, the energy of the bonding H state at $\bar{\Gamma}$, the bonding-antibonding splitting, and the dispersion of the bonding state are compared to experimental data. The work function change is qualitatively correct for all sites except the fcc site with the short bond length. The energy of the bonding state at $\bar{\Gamma}$ is too large for the octahedral and tetrahedral sites unless the surface lattice is expanded. The antibonding H level is present in

TABLE IV
Potentially Acceptable Sites for the Bonding of Hydrogen on Titanium

Feature in spectra	Experimental results[a]	Theoretical calculations for specific sites with Ti—H bond length and lattice expansion given						
		Octahedral	Tetrahedral	Tetrahedral	Face-centered cubic	Face-centered cubic	Hexagonal close packed	Face-centered cubic
Ti—H bond length (Å)		2.07	1.78	1.90	1.78	1.90	1.90	1.90
Lattice expansion (%)	0.21–0.3	0	0	30	0	0	0	20
Change in work function (eV)		0.1	0.4	0.2	−0.4	0.3	0.5	0.2
Bonding H level at $\bar{\Gamma}$ (eV)	6.9	8.2	8.2	7.4	7.2	6.7	6.7	6.6
Dispersion in bonding level $\bar{\Gamma}$ to $\bar{K}$ (eV)	2.4	2.1	2.0	2.5	2.4	2.5	2.6	2.4
Energy splitting of bonding-antibonding level at $\bar{\Gamma}$ (eV)	5.6	7.5	7.8	6.0	>7.2	5.9	5.9	5.5
Calculated heat of adsorption (eV)		2.6	2.3	2.3	3.0	3.3	3.1	2.9
Charge on hydrogen atom (eV)		1.4	1.5		1.3[b]		1.2	

[a]All results from Refs. 108 and 175.
[b]This charge is calculated for hcp site at same spacing.

all the sites except the short fcc Ti—H bond length and the hcp site (not listed). If the square root of the sum of the squares of the percentage error (with respect to the experimental data) is calculated, the two fcc sites with bond lengths of 1.90 Å have the lowest root-mean-square (RMS) error. The tetrahedral site with the 30% lattice expansion has an RMS error larger by a factor of 2 than the two fcc sites. The RMS errors for all the other sites in Table IV are considerably larger.

Feibelman and Hamann[108] argue that three or four of the sites listed in Table IV must be considered to be spectroscopically "acceptable." This implies that even an extensive set of angle-resolved spectra is not sufficient to determine the bonding geometry unambiguously. The main reasons are associated with the approximations involved in the calculational procedure. The local-exchange-correlation calculation should not accurately fit the positions of flat bands such as the antibonding H band, because the state is very localized. Total energy calculations were performed as a means of deciding between spectroscopically "acceptable" sites.[108] The calculated heats of adsorption are listed in Table IV. The fcc site with the long Ti—H bond length is favored over the hcp site or any underlayer configuration. We believe it would have been justifiable to rule out the tetrahedral and octahedral sites because they did not destroy the intrinsic surface state, and the short bond length fcc and hcp sites because the antibonding H state did not exist. The remaining three sites, long bond length fcc and hcp, and the expanded lattice fcc site, cannot be separated spectroscopically.

The calculation procedure utilized by Feibelman and Hamann[176, 177] is one of the most sophisticated schemes applied to angle-resolved photoemission. It is self-consistent and there are no "muffin tin" or pseudopotentials. The only problem originates with the basis set required to obtain the rapid charge-density change at the surface. When the calculation scheme is less sophisticated or the surface more complicated (e.g., as in reconstructed semiconductors), it becomes more unlikely that the geometric structure can be determined by comparing theoretical and experimental valence band spectra. It will always be possible to rule out configurations that are qualitatively wrong, such as the top position for H on Ti, but there will be several spectroscopically "acceptable" geometric configurations.

Louie has employed a self-consistent pseudopotential theory using a local-density description of exchange and correlation to calculate the electronic states at a Pd(111) surface, with and without an ordered monolayer of H.[109, 177] His theoretical scheme utilizes a mixed-basis set (plane waves plus d-like Gaussians), which allows him to represent the surface region more easily than the straight Gaussian basis set technique employed by Feibelman and Hamann.[175] The only photoemission data available with which Louie could compare his calculation were angle integrated. The main features of these data were: (1) upon adsorption of a foreign atom or molecule, a reduction in the photoemission intensity occurs in the energy range from 0 to 2 eV below E_F,[180, 181] and (2) H adsorption induces a new peak in the spectrum 6.5 eV below the Fermi energy.[181] Louie compared the surface density of states for three different adsorption sites with the experimental angle-integrated data.[177] The two triagonal sites shown in Fig. 39b with a Pd —H bond length of 1.69 Å produced very acceptable agreement with the data. The on-top position (Fig. 39b) was qualitatively different from the data. The H-induced level was only 4.5 eV below the Fermi energy.

Figure 40 shows the calculated results of Louie[177] compared to new angle-resolved data for clean Pd(111) and a (1×1)H overlayer.[110] The heavy lines are surface states or resonances and the shaded region is the projected bulk bands. The calculation shows nine different surface bands on the clean surface. Six of these bands are observed experimentally with approximately a 0.5 eV error in the surface states at $\bar{\Gamma}$. At $\bar{M}$ in the SBZ there is a calculated surface band crossing the Fermi energy that is not observed, and the data show a surface-sensitive peak at about −1.2 eV that does not appear in the calculation.

When H is adsorbed, the surface states at $\bar{\Gamma}$ shift in energy and a new H-induced band is present below the bulk bands. The agreement between theory and experiment is very good except at $\bar{M}$. At $\bar{M}$ the dispersion of the low-lying H band is noticeably different in the data compared to the calculation, and the calculated band at about −0.6 eV is not observed experimentally. The calculations were carried out for the hcp surface site shown in Fig. 39b. The bonding to the substrate is primarily H_{1s} to Pd_{4d}. This can be seen in the charge-density contour plots shown in Figs. 41 and 42.[177] Figure 41 shows the charge density of the clean surface state 4.1 eV below the Fermi energy and the H-bonding level at −7.5 eV, both at $\bar{\Gamma}$. It is obvious that the H has interacted with the clean surface state to form the bonding level and the bonding is H_{1s} to the $Pd4d_{3z^2-r^2}$. Both the intrinsic surface state (Fig. 41a) and the extrinsic H-induced surface state are very localized, within the first Pd layer. There are two electrons contained in the bonding band, but the adsorbed H is nearly neutral. The additional charge in this band is located on the first Pd layer and is primarily d. Figure 42 shows the change in the −2.0 eV intrinsic surface state at $\bar{\Gamma}$ (a) when H is adsorbed (b). In both cases the charge density is localized on the first layer of Pd atoms, but the presence of the hydrogen has dropped the energy to −3.2 eV and a small charge density has appeared near the H atom. These charge-density plots show how localized the H bonding is and the d character of this bond.

The two examples, H on Ti(0001) and H on Pd(111), presented in this section clearly illustrate the sophistication of the measurement procedure and

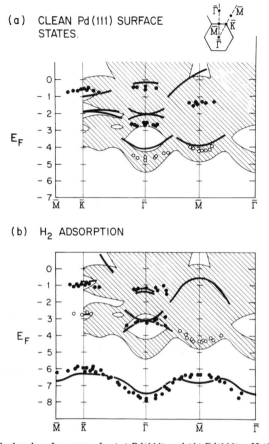

Fig. 40. Calculated surface states for (a) Pd(111) and (b) Pd(111)—H (1×1) compared experimental data.[110,177] The shaded regions are the calculated projection of the bulk bands onto the (111) surface 177. The heavy lines are calculated surface states and the circles are experimental points.[110] The open circles are uncertain.

the theoretical analysis. When good angle-resolved data exist, computer programs can be used to determine the most probable binding sites for ordered overlayers and to produce a detailed picture of the bonding (Figs. 41 and 42). This approach is obviously most powerful when applied to an ordered overlayer where the full capability of angle-resolved photoemission can be utilized.

The success of the procedure of comparing angle-resolved spectra with theoretical calculations for an adsorbate such as hydrogen also in large part

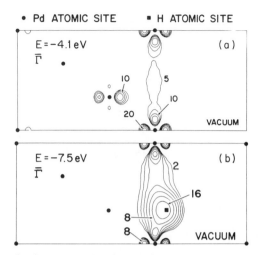

Fig. 41. Charge-density contour plots for (a) clean and (b) adsorbate surface states at $\overline{\Gamma}$. The charge densities are normalized to one electron per unit cell and plotted for a (110) plane cutting the (111) surface.[177] The position of the H atom is indicated by a solid square.

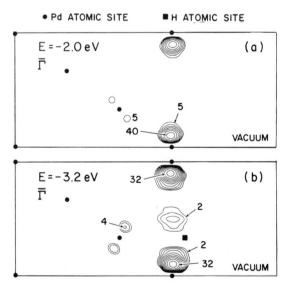

Fig. 42. Charge-density contour plots for (a) clean and (b) adsorbate surface states at $\overline{\Gamma}$. The charge densities are normalized to one electron per unit cell and plotted for a (110) plane cutting the Pd(111) surface.[177] The position of the H atom is indicated by a solid square.

depends on being able to see some adsorbate-induced feature in the spectra. We end this section with an example in which "nothing" is seen. When H is adsorbed on Pd(111) at room temperature the spectrum looks like clean Pd(111).[110] A figure equivalent to Fig. 28 for room temperature adsorption would show no changes. There would be no measurable split-off state, and the d surface states would be unaffected. Thermal desorption spectra and the work function change would show that H remained in the surface region. That is, the H would not have diffused into the bulk.

The most reasonable explanation is that an irreversible transition occurs as the substrate is heated, with the H going into or under the first plane of metal atoms. This site would account for the fact that the intrinsic d surface states are unperturbed. Why the split-off state is absent is still a mystery, but it could indicate that there is more overlap of the H_{1s} level with the metallic s band in the subsurface site than in the adsorption site. This irreversible transition has also been observed on Ni and Pt.[110]

3. Semiconductor Surfaces

In comparison to metal surfaces, there is very little experimental information available on angle-resolved photoemission spectra from adsorbates on semiconductor surfaces. There are several reasons for this lack of data. First, it is not easy to prepare reproducible surfaces. The clean surfaces generally reconstruct—for example, the (2×1) reconstruction of the Si(111) surface. This reconstruction may lead to the formation of domains on the surface. A small number of defects or impurities on the surface can stabilize a specific surface structure. For example, 1 to 5% of a monolayer of Te or Cl stablizes the Si(111) (1×1) surface structure.[119] In contrast to a metal surface, where reconstruction is uncommon and localized to the surface plane when it occurs, reconstruction in semiconductors is common and the distortions penetrate many layers into the solid.

The second experimental problem is that the sticking coefficient for adsorption from the gas phase is much lower for semiconductors than for metals. For example, an exposure of about 5×10^{-6} torr-sec of oxygen will form a monolayer on Ni(100), but it requires about 100 torr-sec exposure to form a monolayer on Si(111) or GaAs(110). This means that extreme care must be taken in handling the gas. Excited molecules produced in the chamber can have a large effect on the experiment.[182]

Because of the aforementioned experimental problem, the theory of the electronic structure of adsorbates on semiconductor surfaces has developed at a faster rate than experimental data. A specific case is H adsorption on Si. Several theoretical papers have been published,[183,184] but experimental data are available only in the form of a conference report[185] and are not in agreement with what is expected from theory. The observed H-induced structure shows no dispersion, while theory predicts about 1 eV dispersion.[184]

The one system that has been investigated theoretically and experimentally is Cl adsorption on Si. The atomic chemisorption of Cl on the Si(111) surface is one of the first adsorption systems to be investigated using angle-resolved photoemission.[186] In addition surface band structure calculations for this adsorption system have been carried out for two different geometric arrangements of the Cl overlayers.[183] In one position the Cl atoms sit atop the Si atoms with a distance given by the covalent radii of both atoms. In the other position the Cl is located closer to the surface plane in the three-fold hollow position, with the bond length between Cl and Si taken to be the sum of the ionic radii.[183] The experimental results for Cl on Si(111) indicate a behavior of the adsorbate bands very similar to what has been observed for metal substrates. Two bands are observed at $\bar{\Gamma}$, located at -7.5 and -4.5 eV relative to the top of the valence band. The adsorbate band at -4.5 eV displays a dispersion with $k_\parallel$ that has the symmetry of the clean surface Brillouin zone, proving that Cl is adsorbed in a 1×1 overlayer. The total bandwidth is about 1.5 ev. The high binding energy band shows no dispersion. Close to the zone boundaries there are also structures in the spectra due to intrinsic Si features that are shifted upon adsorption. In this study the authors try to identify the symmetry of the adsorbate bands by measuring the emission intensity versus exit angle. Analogous to the "oriented-molecule" picture, they conclude that the lower band (-4.5 eV) has π symmetry and the high-energy band has σ symmetry as derived from the P_z atomic orbitals. The best site as determined by a comparison of the experimental data with the theoretical calculation was the top position.[186]

The system of Cl on Si(111) is the only atomic adsorption system on a semiconductor surface that has been studied extensively in angle resolved photoemission. There are at present no angle-resolved studies of the oxygen adsorption on semiconductors, even though many angle-integrated studies have been carried out. Overlayers of metal atoms (Schottlky barriers) have been studied in angle-integrated emission. Recently metal overlayers on GaAs[187] and Si[188] have been studied in angle-resolved photoemission research. Band structure calculations for metallic overlayers on GaAs also exist.[189] In view of the problems associated with these studies, which were discussed in the section about clean semiconductor surfaces, we decided not to present these results in detail. In particular, the metallic overlayers, at least in the case of Si, cause rather "exotic" surface reconstruction. A $\sqrt{19} \times \sqrt{19}$ reconstruction has been observed after deposition of Ni on Si(111).[188]

It is apparent from this short discussion that much more experimental work is required in the area of adsorption on semiconductors. The theory for many systems already exists, but it must be checked experimentally before proceeding to new systems. The ability to grow ordered layers with molecular beam epitaxy opens a new area of research in which angle-resolved photoemission can be a very useful technique.

4. Molecular Adsorption

Any discussion of photoemission from molecular adsorbates must start with carbon monoxide. CO has been to angle resolved photoemission what jellium has been to the surface theorist. Molecularly adsorbed CO has furnished a test system for the new technique of angle-resolved photoemission. There are two reasons for the great usefulness of CO: (1) a lot is known about the photoemission from gas-phase $CO^{16,21}$ and transition metal complexes,[9] and (2) the bonding of CO to a transition metal is primarily through the highest occupied molecular orbital of CO. The lower lying molecular orbitals are not grossly perturbed by bonding. Therefore the gas-phase properties of the latter orbitals can be utilized to predict theoretically the angular distribution from a molecule of fixed orientation.[21]

We present and discuss the photoelectron spectra obtained when such simple molecules as CO and N_2 are adsorbed onto a surface. Again, as in the preceding sections, the examples are meant to illustrate what can be done, not what has been done. Figure 43 shows the calculated wave functions of these two isoelectronic molecules. CO has $C_{\infty v}$ symmetry while N_2 has $D_{\infty h}$ symmetry. We will use the $C_{\infty v}$ notation for both N_2 and CO; for example, the $N_2 - 2\sigma_u$ level ($D_{\infty h}$ notation) is the 4σ level in $C_{\infty v}$ notation. The 2π orbital is unoccupied in both molecules. Even though these two molecules are isoelectronic, their wave functions are quite different. The highest occupied orbital in CO (the 5σ) is a lone pair type of orbital on the C end. This is why the CO bonds through the C end to a transition metal. The CO 4σ orbital should be nearly unaffected by bonding. In contrast, the spatial extents of the N_2 4σ and 5σ orbitals are nearly identical, so both will be involved in the bond to the substrate. It is not obvious from Fig. 43 how N_2 will bond to a surface. It could bond in a linear configuration, through the 5σ and 4σ levels, or it could lie down to form a π bond. Almost any bonding configuration will destroy the $D_{\infty h}$ symmetry of the N_2 molecule allowing the $2\sigma_u(4\sigma)$ and $3\sigma_g(5\sigma)$ states to mix.

By now the reader should be convinced of the power of symmetry rules when applied to angle-resolved photoemission. Given any model for the adsorption of a molecule on a surface the predictions for the symmetry of the orbitals can be checked with this technique. All that is required is polarized light and an angle-resolved detector. In the early stages of the development of this field the applicability of symmetry operations was not appreciated, which led to studies in which angular profiles were fit by theory or interpreted by "intuition." It is still important to understand the variations in the

CO   N₂

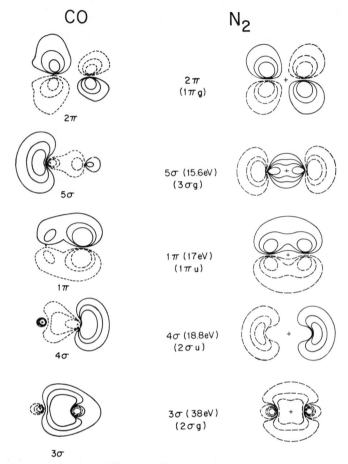

Fig. 43.   Wave functions for CO[190] and N₂.[191] The binding energies are given in parentheses. The contours have values of 0.3, 0.2, and 0.1. The carbon end of the CO molecule is to the left.

angular intensity, but this understanding is not necessary to determine the bonding orientation or to identify the symmetry of individual orbitals. We discuss these symmetry-derived observations for N_2 and CO and then compare a numerical calculation and experimental data for the angular profile from adsorbed NH_3.

The model for the bonding of CO to a transition metal is σ donation from the CO 5σ level to the metal and metal to CO back-donation into the CO 2π unoccupied orbital.[192] This can be restated in terms appropriate to photoelectron spectroscopy. The CO 5σ orbital will form a bonding orbital with the metal electrons, most probably split off from the d bands of the metal.

The d electrons will mix with the CO unoccupied 2π orbital, forming occupied bonding combinations. These metal-d-CO(2π) states will have a small amount of 2π character and will be spread out over the d bandwidth of the solid. The metal-d-CO (2π) interaction is bonding with respect to the substrate but the 2π is antibonding with respect to CO (Fig. 43). Therefore the bonding to the substrate weakens the C—O bond, which results (in the case of carbonyls) in a slight expansion of the CO bond length.[9]

The first published example of data using polarized light and angle-resolved detection, where the orbital symmetry or molecular orientation could be determined from symmetry arguments,[22] was shown in Fig. 3. Since the measurements were made before the importance of symmetry rules was understood,[4-6] the authors[22] interpreted their data using Davenport's[21] calculations for CO. We now know how to interpret or use the data without the aid of any calculation. If the molecular axis is parallel to the surface normal, then a σ state cannot be excited into normal emission (σ final state) when A is perpendicular to the molecular axis. Therefore the peak at about 11 eV in Fig. 3, which disappears in curve b for s-polarized light, is a σ state of a CO molecule standing upright on the surface. The only CO peak remaining in curve b is the 1π. Figure 44 shows these types of measurement for CO adsorbed on Ni(100), Ni(111), and Cu(100).[193] In these spectra the light is

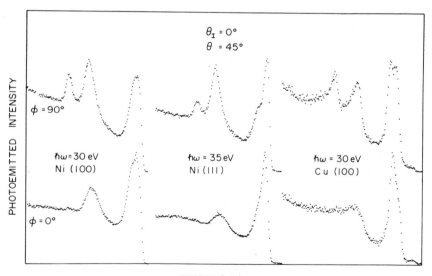

Fig. 44. Forbidden ($\phi=0$) and allowed ($\phi=90$) geometries for CO adsorbed on Ni(100), Ni(111), and Cu(100) surfaces.[193]

always in the plane of the surface (s-polarized); $\phi = 90°$ means that the detector is parallel to **A**, $\phi = 0$ is for the detector perpendicular to **A**. We refer to these two geometries as "allowed" and "forbidden." "Allowed" means that only initial states that are even with respect to the plane defined by the polarization vector and the detector are excited into the direction of the detector. In the "forbidden" geometry only initial states that are odd with respect to the plane perpendicular to **A** and parallel to the detector can be excited. Each spectrum at the bottom shows the position of the odd 1π orbital, while the top curves have three levels present: two σ states and one even π state. Table V lists all measurements of the positions of the three CO orbitals using symmetry selection rules. The data describe condensed films on CO and Cu,[194] physisorption on Al, weak chemisorption on Cu,[195] and chemisorption on Ni and Pt. One word of caution is needed. Most of these measurements were made about 30° off normal with approximately 34 eV photon energy. This corresponds to a value of $k_\parallel \sim 1 \text{ Å}^{-1}$. If there is measurable dispersion, the values listed in the top portion of Table V correspond to the binding energies of the two even (σ's) and one odd (π) bands out in the SBZ. We return later to discuss dispersion. The value of the σ states for the high-coverage CO on Pd(111) at 100°K given in the parentheses are for normal collection and differ from the values obtained at 30° collection because of dispersion. The bottom section of Table V lists several systems for which the dispersion has been measured.

Figure 3 showed what is listed in Table V; that is, all three CO levels of absorbed CO have a smaller binding energy than their counterpart in free CO. For example, the condensed film of CO on Cu has a uniform shift of approximately 1 eV in all the binding energies. The column labeled Δ_{avg} is the average shift of the 4σ, 1π, and 5σ levels of adsorbed CO compared to gas-phase CO. The binding energy of adsorbed CO is referenced to the Fermi energy, while the gas-phase values are referenced to the vacuum. Some appropriate work function needs to be added to the surface binding energies before the magnitude is compared to the gas-phase values. When the work function is included, there is still an average upward shift of 2 to 3 eV for the valence orbital of CO adsorbed on Ni compared to free CO. The 5σ moves up much less than the 1π and 4σ. A simple chemical bond model would predict that the bonding 5σ level should drop in energy compared to its gas-phase value. This is an initial state argument, and we must remember that photoelectron spectroscopy is an ion spectroscopy.

The shift in binding energy of a given level of a molecule between the gas phase and adsorbed on a surface can be written as

$$\Delta E = +\Delta E_{initial} + \Delta E_{relaxation} \qquad (30)$$

Where $\Delta E_{initial}$ is the shift in the initial state one-electron orbital, which can

CO Binding Energies (eV)[a]

CO environment	Δ_{avg} ($\Delta_{avg} - \phi$)	$E(5\sigma)$	$\Delta E(5\sigma) - \Delta_{avg}$	$E(1\pi)$	$\Delta E(1\pi) - \Delta_{avg}$	$E(4\sigma)$	$\Delta E(4\sigma) - \Delta_{avg}$	$E(3\sigma) -$	$\Delta E(3\sigma)\ \Delta_{avg}$	Heat of adsorption (eV)
Gas phase	0	14.0	0	16.9	0	19.7	0	38.9		
Condensed on Cu[194] at 10°K	5.2 (0.98)	8.8	0	11.7	0	14.5	0			
Physisorbed on Al at 20°K[b]	5.1 (~0.7)	8.9	0	11.9	−0.1	14.6	0			
Weak chemisorption[195] on Cu(100)	7.3 (3.0)	8.2	−2.0	8.65	1.0	11.8	0.6			0.63[196]
Ni(100) at 300°K[22]	7.8	8.2	−2.0	7.9	1.2	11.2	0.7			1.3[197]
Ni(100) at 300°K[26]	7.9	8.3	−2.2	7.8	1.2	10.8	1.0			
c(2×2)CO on Ni(100)[198]	8.2 (2.9)	8.0	−2.2	7.5	1.4	10.6	1.1			1.3[197]
Ni(111) at 300°K[193]	8.3	8.0	−2.3	7.1	1.5	10.7	0.7			1.21[199]
Ni(110) at 300°K[c]	8.0 (3.0)	8.0	−2.0	7.7	1.2	10.8	0.9			1.42[200]
Pd(111) at 300°K[c]		8.25				10.9				1.48[201]
Pd(111) at 100°K[c]	7.1 (1.5)	8.2 (8.6)	−1.3	8.2	1.6	11.3 (11.4)	1.3	30.6	1.2	0.87[201]
Pt(111) at 300°K[202]	6.5 (0.8)	9.8	−2.3	8.9	1.5	12.3	0.9			1.30–1.4[201]
Co(0001)($\sqrt{3}\times\sqrt{3}$)$R30°$[204]	8.1	7.9±0.05	−2.0	7.7±0.1	1.1	10.7±0.05	0.9			
Theory										
NiCO (HF)[205]	−1.68		−1.2		0.6		0.6		1.5	
NiCO (SFH)[206]	3.9		−2.9		0.5		1.0			
(Ni)$_5$ (CO)[207] (HFS-SCF)	3.0		−2.1		0.7		1.4		1.6	
NiCO (HFS)[208]	2.9		−1.5		1.2		.37		2.3	

CO Adsorption Systems Where Dispersion Has Been Reported

System	$E(5\sigma)$ at $\bar{\Gamma}$	Dispersion	$E(1\pi)$ at $\bar{\Gamma}$	Dispersion	$E(4\sigma)$ at $\bar{\Gamma}$	Dispersion
c(2×2)CO on Ni(100) at 300°K[221]	8.0				11.0	~0.15
"Hexagonal" CO on Ni(100) at 80°K[221]	8.6				11.3	0.25
"Hexagonal" CO on Pd(100)[223]						0.4
p(1×2)CO on Fe(110)[224]	8.0	~0.8	6.5		11	0.3
CO on Co(0001) at 100°K[204]	8.20±0.05	0.8±0.1	7.3±0.4		10.75±0.05	0.45±0.07

[a] This table lists only systems for which the 1π binding energy has been determined by "forbidden geometry" measurements (see Fig. 44). Binding energies are relative to Fermi energy for adsorption systems.
[b] Unpublished work by H. Levinson.
[c] Work by Plummer and Eiserhardt.

be defined only theoretically. A convenient definition is in terms of the Hartree-Fock orbital energies ε_i

$$\Delta E_{\text{initial}} = -\varepsilon_i(\text{coordinated CO}) - \varepsilon_i(\text{CO}) \qquad (31)$$

Where ε_i is always negative. If the ith CO orbital is bonding in the complex, then $\Delta E_{\text{initial}}$ will be positive. A positive $\Delta E_{\text{initial}}$ means that the binding energy increases or the initial energy becomes a larger negative number. The differential relaxation energy of the electrons around the hole created in the coordinated CO compared to the isolated CO, $\Delta E_{\text{relaxation}}$, is negative as defined in (31) because the hole is more effectively screened in the condensed or coordinated CO phase compared to the isolated CO,[209] lowering the binding energy of the coordinated complex. For example, the initial state shift ($\Delta E_{\text{initial}}$) in the 5σ level when it bonds to a cluster of five Ni atoms has been calculated to be 1.8 eV,[207] while the relaxation shift was $\Delta E_{\text{relaxation}} = -2.7$ eV.[207] The 5σ orbital of coordinated CO has a lower binding energy than free CO despite the bonding shift of nearly 2 eV. The same calculation predicted an initial state shift of -0.9 (-0.8) eV for the 1π (4σ) level with a relaxation shift of -2.8 (-3.6) eV. Two of the unfortunate truths of photoelectron spectroscopy are illustrated by this calculation. First, bonding orbitals do not necessarily drop in energy (higher binding energies). Second, relaxation shifts are not constant. For the $(\text{Ni})_5\text{CO}$ calculation there was a difference of nearly 1 eV in relaxation energy between the 5σ and 4σ levels.[207]

There is a general trend in the binding energies that would have been expected from our discussion of the bonding of CO to a transition metal. The bonding interaction of the 5σ should decrease the difference in binding energy between the 4σ and 5σ. The back-bonding into the 2π should lengthen the C—O bond, which will increase the energy spacing between the 4σ and 1π.[205] If values of $[E(4\sigma) - E(1\pi)]/[E(4\sigma) - E(5\sigma)]$ from Table V are compared with the heat of adsorption, there is a general correlation. When this ratio is large, the heat of adsorption is large. For weakly bound CO on Cu, the ratio of binding energies is 0.87 and the bond energy is 0.63 eV; for the low-temperature CO state on Pd(111) the ratio is 1.0 and the bond energy is about 0.9 eV. For room temperature adsorption of CO on Pt and Ni, the ratio of the binding energies is 1.2 to 1.4, and the heat of adsorption ranges from 1.3 to 1.5 eV.

There have been many attempts to identify a "2π" level in adsorbed CO. There is no 2π level—only 2π character mixed into surface metal d states. Therefore we should look for changes induced in the d bands in the "forbidden" or odd geometry. In this measurement the symmetry of the substrate must be considered. For example, for $c(2\times2)\text{CO}$ on Ni(100) the surface has C_{4v} symmetry. There are two mirror planes. Smith et al.[210] found a CO-induced peak +2.2 eV below the Fermi energy in the (100) mirror plane in the odd geometry and a peak at 1.3 eV below the Fermi energy in normal emission. The 2.2 eV peak must have been created by the interaction of the metal electrons with the odd π levels of CO. Studies of these types need to be expanded to include photon sweeps at fixed $k_\parallel$ to assure that the peak has not resulted from enhancement of a bulk transition.

We pointed out in the introduction that the symmetry measurement could be used to determine the geometry if the symmetry of the orbital were known. The intensity of the 4σ in the forbidden geometry relative to its intensity in the allowed geometry can be used to determine the possible degree of bend of the molecular axis with respect to the surface normal.[198] This requires a knowledge of the degree of polarization and of the angular profile of emission of the 4σ.[198] We do not discuss this approach, but say only that CO is standing straight up within experimental error in all the systems described here.

The final state resonance shown in Fig. 4 can also be used as a symmetry test for the orientation. When the light is incident normal to the surface, $\mathbf{A}$ is in the plane of the surface and the 4σ state cannot be excited into the resonance, if the molecule is standing straight up. Figure 45 shows the intensity of the two CO peaks in Figs. 3 and 44 as a function of photon energy for various geometries. Column a is for p-polarized light and normal emission. The resonance is observed in the 4σ cross-section for CO on all three crystals, because there is a component of $\mathbf{A}$ along the molecular axis. Columns b, c, and d are configurations in which the resonance should not be observed if CO is standing straight up. The light is p-polarized in column b, but the detector is 63° from the normal. Calculations show that the resonant intensity is emitted in a narrow angular cone along the molecular axis.[21] If the molecule is standing upright on the surface, 63° is such a large angle that no sign of the resonance should be seen.[21] Columns c and d are for s-polarized light and 7° and 45° collection, respectively. A σ initial state cannot be excited into the resonance with s-polarized light. The resonance is not observed in the 4σ cross-section in columns b, c, and d. Again we conclude from symmetry arguments that CO stands upright on the surface. How upright is upright must be determined by numerical analysis.[193,198] Allyn et al.[198] concluded that the experimental uncertainties limited the determination to about 15°.

Figure 45 shows an interesting feature in the cross-section from the composite 5σ and 1π peak (crosses). It shows a resonant behavior in column d for both Ni surfaces and in column c for Ni(111). Both these geometries are such that the excitation of a σ initial state into the σ resonant state is forbidden by symmetry. Obviously one of the two states involved in this transition does not have σ symmetry. The fact that the resonant behavior was not seen

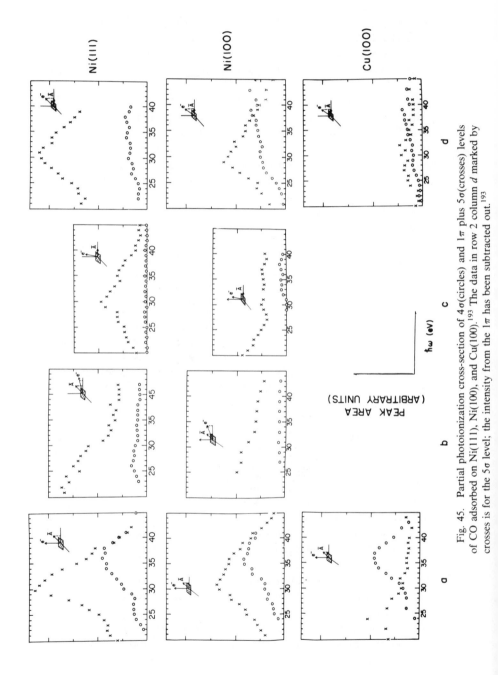

Fig. 45. Partial photoionization cross-section of 4σ(circles) and 1π plus 5σ(crosses) levels of CO adsorbed on Ni(111), Ni(100), and Cu(100).[193] The data in row 2 column d marked by crosses is for the 5σ level; the intensity from the 1π has been subtracted out.[193]

626

for the 4σ level in the geometries shown in columns c and d indicates that the final state still has σ symmetry. Therefore the initial state must have a reduced symmetry. An apparent explanation for this behavior must be checked in more detail experimentally. The CO—CO interaction within the overlayer will form bands if the overlayer is ordered, and the band formed from the 5σ will disperse upward toward the Fermi energy with increasing $\mathbf{k}_\parallel$, while the 1π band will disperse downward. The bands will hybridize, mixing even π and σ character into each one. This hybridization occurs in an isolated CO layer without the substrate present because there is no inversion symmetry about the plane of the CO layer. The π character mixed into the states at $k_\parallel \neq 0$ can be excited into the final state σ resonance with s-polarized light. Even though the isolated gas-phase π state does not couple to the resonance, it is not forbidden by any symmetry rule.[16,21] Therefore there is nothing to prevent the hybridized σ-π state from coupling to the resonance.

We turn now to N_2 adsorption. Figure 46 shows three spectra for N_2 adsorbed at 90°K onto Ni(110).[211] The spectra look similar to the CO adsorption spectra. There are two peaks instead of the three observed in the gas phase,[7,16] and the 5σ and 1π appear to be nearly degenerate. But there are obvious differences in the adsorbed N_2 spectra compared to the CO spectra (Figs. 3 and 44). The peak near -12 eV is very broad, and its shape changes with collection geometry and photon energy.[211] For example, in Fig. 44 the line shape is different for the three geometries shown. If the 4σ state has the same linewidth as the 5σ state (seen at -8.1 eV) we need more than three peaks to fit the broad peak near -12 eV. Curve a in Fig. 46 shows a three-peak decomposition of the -12 eV peak. The energies of these peaks have been taken from other spectra (e.g., curve b, where a clear peak is seen). These three peak positions are at -12.7, -11.8, and -10.7 eV. These multiple peaks are a result of a breakdown of the simple single-particle picture. They are two-hole–one-particle states discussed in Section II.[206]

Let us discuss the one-electron properties of N_2 adsorption before returning to the multielectron effects. We assume that the narrow peak seen at about -11.8 eV in curve b of Fig. 46 is the 4σ state. Comparison of curves b and c shows that the N_2 molecule is bound to the surface standing up, like CO. Umbach et al.[212] came to the same conclusion for N_2 adsorption on W(110). With the molecule standing upright on the surface, we can use symmetry rules to measure the binding energy of the 1π level. Curve c of Fig. 46 shows that the 1π level is at -7.8 eV. Curve b places the 4σ at -11.8 eV and the 5σ at -7.9 eV. The position of the 5σ in normal emission is -8.1 eV. Either there is dispersion or the 1π level in curve b is intense enough to shift the apparent peak position. Table VI lists the binding energies of the few N_2 adsorption systems that have been studied.

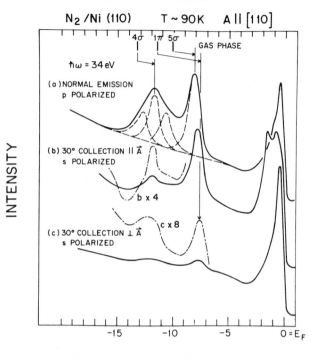

Fig. 46.   Angle-resolved photoelectron spectra of N_2 adsorbed on Ni(110) at 90°K.[211]

Table V and VI illustrate the differences between CO and N_2 adsorption. The bonding of CO is primarily through the 5σ level, leaving the 1π and 4σ levels relatively unchanged. The 5σ drops, relative to the 4σ, approximately 2.6 eV for strong chemisorption and 2.1 eV for weak chemisorption (CO on Cu). The 4σ-to-5σ spacing for adsorbed N_2 increases by 0.6 eV for Ni(110)[211] and over 1 eV for adsorption on Ni(100)[213] and W(110).[212] The 4σ and 1π energy spacing increases in adsorbed CO compared to gas-phase CO. This increase is 0.45 eV for strong chemisorption and 0.35 eV for weak chemisorption (Table V). In comparison, the 4σ-to-1π spacing for N_2 adsorbed on Ni(110) increases by 2.2 eV.

The explanation of these different shifts of N_2 compared to CO is quite straightforward.[206, 211–213] When the N_2 molecule interacts with the surface (end on), both the 4σ and 5σ levels overlap the metallic wave functions. The N_2 inversion symmetry is broken and the $2\sigma_u(4\sigma)$ and $3\sigma_g(5\sigma)$ functions can mix, creating new σ states of the bound complex that are much more like the

TABLE VI

N$_2$ Binding Energies (eV)

N$_2$ environment	$5\sigma(3\sigma_g)$ Binding energy	$\Delta E - \Delta E_{avg}$	$1\pi(1\pi_u)$ Binding energy	$\Delta E(1\pi) - \Delta E_{avg}$	$4\sigma(2\sigma_u)$ Binding energy	$\Delta E(4\sigma) - \Delta E_{avg}$	$3\sigma(2\sigma_g)$ Binding energy	$\Delta E(3\sigma) - \Delta E_{avg}$	$(\Delta_{avg} - \phi) - \Delta E_{avg}$
Gas phase	15.6	0	17	0	18.8	0	38	0	0
Condensed on Cu at 10°K[194]	11.1	0	12.5	0	14.3	0			4.5 (0.28)
Physisorbed on Pd(111)[211]	8.7	+0.1	10.3	-0.1	11.9	+0.1			6.8 (1.3)
Chemisorbed on Ni(110)[211]	8.1	-0.5	7.7	+1.2	11.7	-0.9			8.0 (2.5)
Chemisorbed on W(110)[212]	7.4				12±0.2				
Chemisorbed on Ni(100)[213]	7.6				12.4				
			Theory						
Ni—N$_2$[214]		-0.76		+0.18		+0.58			-0.8
Ni—N$_2$[206]		-0.3		0.9		-0.6			2.5

629

levels of CO. For example, compare the binding energies for CO adsorbed on Cu(100) (Table V) and N_2 adsorbed on Ni(110) (Table VI). These are both weak chemisorption situations. The binding energy of the 4σ state is 11.7 (11.8) eV; the 1π binding energy is 7.7 (8.6) eV and; the 5σ binding energy is 8.1 (8.2) eV, for N_2 on Ni(110) [CO on Cu(100)]. The σ states of N_2 are able to readjust themselves to maximize the bonding, forming orbitals very similar to CO, but the 1π level of N_2 is still quite different from the corresponding level of CO (see Fig. 43).

If our explanation is correct, the 4σ level in chemisorbed N_2 can be excited into the resonance. It was shown in Fig. 4 that this transition was forbidden in the gas phase, as a result of the inversion symmetry in N_2. The photoionization cross-section for normal emission and p-polarized light is shown in Fig. 47.[211] Both N_2 peaks in Fig. 46 exhibit the resonance. The resonance occurs at a kinetic energy of about 15 eV for both states, compared to 12 eV for gas-phase N_2.[16] Comparison of Fig. 47 for N_2 to Fig. 45 for CO reveals only one difference: namely, the 4σ intensity for N_2 is a little weaker than the equivalent CO intensity, compared to the 5σ level.

The calculated values listed at the bottom of Tables V and VI show only the theoretical shift of the levels between the isolated and coordinated molecule. This seems to be the only fair comparison, since the calculated binding energies of the isolated molecules do not agree with experimental data in many cases. The Hartree-Fock-Slater[207, 208] and many-body CNDO[206] (complete neglect of differential overlaps) formalisms reproduce qualitatively the average shifts (Δ_{avg}) and the shifts of the individual CO valence orbitals. They predict an average drop of the CO—5σ of 2 eV in agreement with experimental values obtained on Ni surfaces. Two of the schemes do not predict an increase in the 4σ to 1π binding energy.[206, 207] Both these calculations predict a larger relaxation energy (30) for the 4σ than for the 1π. The calculation by Rosen et al.[208] reproduces both the downward shift

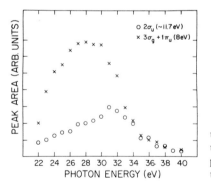

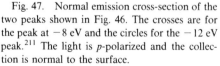

Fig. 47. Normal emission cross-section of the two peaks shown in Fig. 46. The crosses are for the peak at -8 eV and the circles for the -12 eV peak.[211] The light is p-polarized and the collection is normal to the surface.

of 5σ and the upward shift of the 1π. The many-body CNDO calculation is in excellent agreement with the N_2 data[206] (Table VI).

The straight ΔSCF Hartree-Fock calculation for Ni—CO[205] and Ni—N_2[214] does not work. The Δ_{aVG} is the wrong sign, and individual level shifts are conspicuously different from the experimental values. It is claimed that this procedure works better for N_2 adsorption than for CO adsorption,[214] but Tables V and VI show that the shifts calculated for Ni—CO are better than those for Ni—N_2. It is not understood why a Hartree-Fock calculation produces such poor results, compared to the other schemes, but perhaps all the other theoretical techniques include correlation effects to some degree.

Figure 46 for N_2 adsorption exhibited a mulipeak spectrum in the region of the 4σ. This behavior has been observed for CO adsorbed on Cu(100),[195] NO adsorbed on Ni(100)[215] and Ir(111),[216] as well as N_2 adsorbed on Ni and W(110).[212] These peaks are a result of multielectron excitations, producing two-hole, one-particle ionic states. These states (called shakeup states) seem to be most pronounced in such weakly chemisorbed species as CO on Cu or N_2 on Ni or W.[206, 217-220] When the coupling of the molecule to the surface is very small (physisorption), there is little charge transfer when the molecule is ionized. If the coupling to the surface is strong, the flow of charge is easy by redistributing charge in occupied orbitals with significant weight on both the metal and ligand. In the weak chemisorption case the orbitals of the ion can be considerably different from the equivalent neutral orbits yet have enough overlap with the substrate to result in the excitation of two-hole, single-particle ionic states upon ionization.[217-219] The angular and photon energy dependence of these multielectron states will be one of the areas of future experimental and theoretical work.

All the angle-resolved spectra of adsorbed CO and N_2 presented in this chapter look qualitatively similar: two peaks in the allowed geometry and one in the forbidden geometry. Figure 48 shows that this is not always the case, by displaying the angle-resolved spectra equivalent to those for N_2 on Ni(110) shown in Fig. 46; this example, however, shows Pd(111).[211] There are three peaks of relative intensity: 15, 55, and 30% at binding energies of 11.9, 10.3, and 8.7 eV, respectively. These three peaks are the N_2, 4σ, 1π, and 5σ levels, and they appear in all measurement geometries with approximately equal relative intensities. Their relative intensities are in excellent agreement with the gas-phase branching ratios.[16] The molecules are physisorbed in a random orientation at 45 °K on Pd(111). The same observation has been recorded for CO on Al(100) at 20 °K (see Table VI).

Let us return to the issue of dispersion in the energy levels of molecules adsorbed on surfaces. Horn et al.[221] measured the dispersion of the CO 4σ level for a $c(2\times 2)$ and compressed hexagonal overlayer structure on Ni(100).

E. W. PLUMMER AND W. EBERHARDT

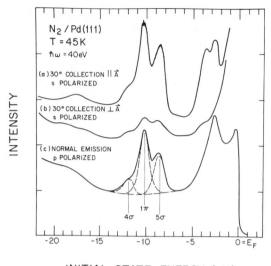

INITIAL STATE ENERGY (eV)

Fig. 48. Photoelectron spectra for N_2 adsorbed on Pd(111) at 45°K.[211]

The $c(2 \times 2)$ structure is the same as shown in Fig. 32 for S adsorption except the CO is bound in the atop site, with the CO—CO spacing being about 3.5 Å.[222] The measured dispersion of the 4σ level for the 0.5 monolayer $c(2 \times 2)$ structure was about 0.15 eV. The closer packed 0.8 of a monolayer coverage exhibited a 0.25 eV dispersion. Subsequently Horn et al.[223] performed the same measurements on a compressed-incommensurate ordered overlayer of CO on Pd(100). This structure corresponded to 0.8 of a monolayer and produced approximately 0.4 eV dispersion in the 4σ band, with the correct symmetry in the SBZ. The measured dispersions for CO on Pd(100) compared very well with the tight-binding calculations by Hermann and Batra for an isolated CO layer. This calculation predicted a dispersion of about 1 eV for the 5σ and 1π bands, which was not measured in the experiment.[223]

Jensen and Rhodin[224] have attempted to measure the dispersion of all three CO bands for a $p(1 \times 2)$CO layer on Fe(110). Their data are shown in Fig. 49. The LEED pattern is reproduced in Fig. 49a, and a geometry consistent with the LEED pattern is shown in Fig. 49b. The shaded regions of Fig. 49c show the measured dispersion, compared to a tight-binding calculation using parameters from the CO on Pd(100) work.[223] The dispersion in the (001) azimuth is very similar to the tight-binding calculation and is what "intuition" would predict. The 4σ dispersion is a few tenths of an electron volt, as it was for CO on Ni(100) or Pd(100).[221, 223] The dispersion of the 5σ is larger

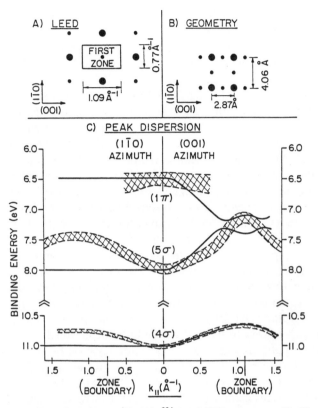

Fig. 49. $p(1 \times 2)$CO adsorption on Fe(110).[224] (a) LEED pattern at 75 eV: small dots, primary spots; large dots, CO-induced spots. (b) A geometry consistent with the LEED pattern. (c) Dispersion of CO levels in the two high-symmetry directions. Solid lines are tight-binding calculations.

because the wave function is spatially more extended (Fig. 43). There is about 0.8 eV dispersion in the 5σ. The 1π could not be tracked because of its weak intensity compared to the 5σ. All these measurements were performed in the plane of incidence with p-polarized light, so that all three bands are always excited.

The dispersion in the $(1\bar{1}0)$ azimuth is not what would have been expected. First the σ states do not have their minimum binding energy at the zone edge predicted from the LEED structure. Second, there is a related problem: namely, the dispersion is much larger than calculated. The dispersion reaches its extreme value at $\mathbf{k}_{\parallel} = 1.1$ Å^{-1} in the (001) azimuth as expected from the geometry shown in Fig. 49b, but in the $(1\bar{1}0)$ azimuth the extreme value of the dispersion is about 1.2 Å^{-1} which is not a multiple of the SBZ spacing shown in Fig. 49a (0.77 Å^{-1}). Obviously the geometries deduced from LEED and angle-resolved photoemission are different.

Future work will undoubtedly concentrate on separating out the π and σ bands and determining their hybridization. In cases such as the compressed layer on Pd(100), the symmetry of the overlayer is C_{2v}. In C_{2v} symmetry there cannot be degenerate levels, so the 1π level is split at $\bar{\Gamma}$ in the SBZ. This would result in different binding energies for the "odd" and "even" π bands at $\bar{\Gamma}$.[223] These effects are caused by the CO—CO lateral interaction.

We end this section with a brief discussion of azimuthal patterns from the valence orbitals of adsorbed molecules. Section II.C discussed azimuthal patterns from such layer compounds TaS$_2$ (see Fig. 6). An azimuthal pattern is obtained by fixing the polarization vector $\mathbf{A}$ and the detector direction, and rotating the sample about an axis normal to the surface. The variations in intensity of the photoemitted signal from a given molecular orbital as the crystal is rotated depend on the anisotropic nature of both the initial and the final states (5). Section IV.B discusses the simpler case of an initial state that is a core level of an adsorbate and consequently is spatially isotropic.

The naive expectation is that the symmetry of an azimuthal pattern will yield the symmetry of the adsorbate substrate complex. If the asymmetries are dominated by local molecular properties, long-range order is not necessary; that is, the bonding site symmetry can be determined. Likewise if we knew the angular behavior of the isolated molecule, the orientation of the molecule with respect to the surface could be determined.

Azimuthal patterns from adsorbed diatomics have been singularly uninformative. No one has reported any variation in the signal from a CO, N$_2$, or NO level with azimuthal angle.[22, 193, 215, 221] This means that scattering from the substrate atoms cannot be a major factor in these cases. NO adsorbed on Ni(100)[215] is believed to bond in a bent configuration with the NO molecular axis as much as 30° from the surface normal. It is likely that this bent molecule would align itself with the crystallographic directions of the surface, thus producing a four fold azimuthal pattern for adsorption on Ni(100). Loubriel et al.[215] saw no anisotropies in the experimental azimuthal patterns. The NO molecule must be rotating freely around the surface normal.

Purtell et al.[225] observed a very dramatic azimuthal pattern from ammonia (NH$_3$) adsorbed at about 200 °K on Ir(111). These azimuthal patterns from the NH$_3$ $1e$ orbital are shown in Fig. 50 for two photon energies. The assignment of the peak in the photoelectron spectra 11.3 eV below the Fermi energy to the N—H bounding $1e$ orbital of NH$_3$ was accomplished by measuring the dependence of the signal strength on the polarization direction of the light. These measurements indicated that the NH$_3$ bonds nitrogen end down, with the three H atoms symmetric about the surface normal. Figure 50 immediately shows that the NH$_3$ molecule is not rotating freely like the NO molecule on Ni(100). It is locked into a fixed orientation.

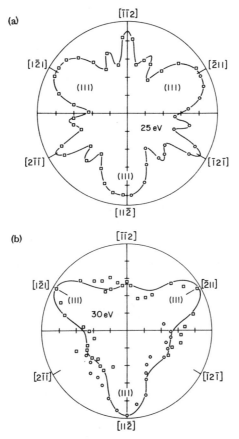

Fig. 50. Azimuthal plot of photoemission from the $1e$ orbital of NH_3 adsorbed on Ir(111) at 200°K.[225] Circles represent data and squares 120° rotation of data. (a) $\hbar\omega = 25$ eV; angle of incidence 45°, collection 45°. (b) $\hbar\omega = 30$ eV, the angle of incidence 45°, collection 42.3°.

This probably rules out the atop site. One would naively expect that the intense lobes in Fig. 50 would be along the direction of the N—H bonds. This assumption lead Purtell et al.[225] to propose a bonding site for NH_3 on Ir(111). This is the threefold site, which is the natural continuation of the fcc lattice. The three H atoms are in the [112] directions, that is, between the nearest-neighbor Ir atoms in the surface plane.

Unfortunately "intuition" and angle-resolved photoelectron spectroscopy are not always compatible. A calculation by Kang et al.[226] for free NH_3 shows that below $\hbar\omega = 120$ eV the three lobes come out between the N—H bond directions. These authors could not decide on the best site for NH_3 on Ni(111) by comparing calculated and measured azimuthal patterns. The important experimental observation is that the NH_3 is locked into the substrate orientation.

5. Physisorption

We have already mentioned the work of Norton et al.[194] on condensed and physisorbed layers of N_2 and CO on Cu and the angle-resolved measurements for physisorbed N_2 on Pd(111)[211] (see Fig. 48). This section describes the results of Scheffler et al.[227] on a hexagonal incommensurate layer of xenon on Pd(100). The spin-orbit splitting of the Xe $5p$ levels introduces a new phenonmenon that was not observed in the chalcogen overlayers discussed in Section IV.A.1.

Photoemission from the $5p$ shell of Xe leads to either a $^2P_{3/2}$ or $^2P_{1/2}$ ionic state in the L—S coupling limit. The binding energies of these two states in Xe are 12.12 and 13.43 eV, respectively. This 1.3 eV splitting is probably larger than the dispersion in the $5p$ bands of an ordered overlayer of physisorbed Xe, so we cannot use the arguments presented in Fig. 37 for a hexagonal overlayer to determine the dispersion. The spin-orbit coupling mixes the Pz and Px, Py states, so that, in general we will not have bands of pure spatial symmetry.

Consider what happens at $\overline{\Gamma}$ in the two-dimensional Xe layer. If the spin-orbit coupling were zero, we would have three bands at $\overline{\Gamma}$; one Pz (even) and two degenerate Px, Py (odd and even) as shown in Fig. 37. The Pz band is bonding and the degenerate $PxPy$ bands are antibonding. In contrast when the spin-orbit coupling is large, the crystal field produced in the Xe layer will split the gas-phase degenerate $P_{3/2}$ level into two levels, $m_j = \pm \frac{3}{2}$ and $m_j = \pm \frac{1}{2}$. The $P_{3/2}$, $m_j = \pm \frac{3}{2}$ level is pure $PxPy$ symmetry, while the $P_{3/2}$, $m_j = \pm 12$ and $P_{1/2}$, $m_j = \pm \frac{1}{2}$ levels are mixed Pz and Px, Py.[227] Since the Px, Py levels are antibonding at $\overline{\Gamma}$, we would expect to see the $P_{3/2}$, $m_j = \pm \frac{3}{2}$ level at a smaller binding energy than the $P_{3/2}$, $m_j = \pm \frac{1}{2}$ level. Notice that the symmetries of the three bands at $\overline{\Gamma}$ in the presence of spin-orbit coupling are altogether different from what was described in Section IV.A.1, where there was no spin-orbit coupling. For example, in the absence of spin-orbit effects a measurement made with normal collection and s-polarized light shows only one of the three p bands (Fig. 3). When the spin-orbit coupling is appreciable, three bands are visible in this measurement geometry.

Figure 51 shows the angular resolved measurements of Horn et al.[227] for an incommensurate ordered hexagonal overlayer of Xe on Pd(100). The density of the Xe is 5.75×10^{14} atoms/cm^2. There are five peaks in the spectrum for normal emission ($\theta = 0°$). Two of these peaks at 5.5 and 6.7 eV were shown to be a result of scattering from the out of registry substrate without conserving $k_\parallel$, so the very intense peaks at $20 < \theta < 40$ are partially scattered into the normal collection direction.[227] Measurements of the intensity of the

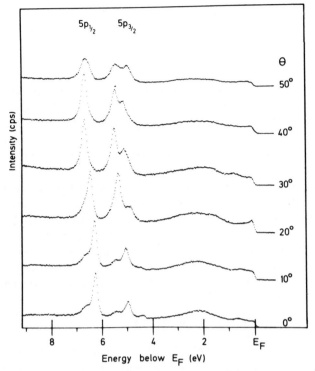

Fig. 51. Angle-resolved spectra for hexagonal layer of Xe on Pd(100) as a normal function of the collection angle θ; $\theta = 0°$ is normal.[227]

two peaks at 4.4 and 5.0 eV binding energy in normal collection as a function of the angle of incidence of the light proved that the 4.4 eV state was $PxPy$ in character and the 5.0 eV peak was dominated by Pz.[227] Therefore the following assignment is possible. The $P_{3/2}$, $m_j = \pm \frac{3}{2}$ level is at 4.4 eV; the $P_{3/2}$, $m_j = \pm \frac{1}{2}$ level is the 5.0 eV peak, and the $P_{1/2}$, $m_j = \pm \frac{1}{2}$ state has a binding energy of about 6.2 eV. The splitting at $\bar{\Gamma}$ between the $P_{3/2}$, $m_j = \pm \frac{1}{2}$ and $P_{1/2}$, $m_j = \pm \frac{1}{2}$ states is nearly the same as the gas-phase spin-orbit splitting, but the $P_{3/2}$, $m_j = \pm \frac{3}{2}$ state has about 0.5 increased splitting. By decreasing the coverage and watching the splitting go to zero, Scheffler et al.[227] showed that the splitting at $\bar{\Gamma}$ of the $m_j = \pm \frac{3}{2}$ and $m_j = \pm \frac{1}{2}$ states was due to the Xe—Xe crystal field.

The spin-orbit coupling removes much of the spatial symmetry of the wave function. The previous discussion illustrated the effects in normal emission. The effects of spin-orbit coupling should also be observed in the mirror plane measurements. In principle all three bands could have both odd and even character. Measurements on a physisorbed system where the dispersion and spin-orbit splitting have approximately the same magnitude should be very interesting. The odd geometry could be used to determine the odd character of each band as a function of $k_{\|}$.

B. Core Levels

Angle- and energy-resolved core-level photoemission is a tool capable of determining the geometrical structure of a surface.[34, 64] The use of a core level of the surface atom as the initial state considerably simplifies the photoemission matrix element (5), since a core level has no inherent angular anisotropy. Therefore all angular effects must be due to the final state.[64] In an independent-particle picture, the final state is represented by the wave function of the photoexcited electron. In 1974 Liebsch[64] showed that the existing formalism developed for dynamic LEED calculations,[228] could be adapted to the angle-resolved photoemission problem. The physics of this process is shown in Fig. 52. A core level of a surface atom is excited by absorption of a photon ($\hbar\omega$); ψ^0 represents a wave propagating directly to the detector while ψ' represents the composite of all the waves produced by scattering processes from neighboring atoms.[34] The wave function of the emitted photoelectron is represented by

$$\psi = \psi^0 + \psi' \tag{32}$$

In general the amplitude of ψ' will be smaller than ψ^0 because every scattered wave is backscattered at least once, and the backscattering cross-section is small. Therefore it is the interference between ψ^0 and ψ' that will

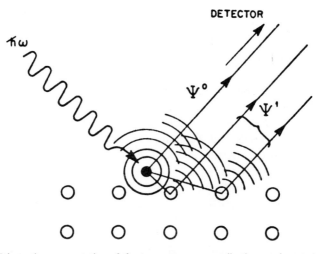

Fig. 52. Schematic representation of the two processes contributing to photoemission from a localized adsorbate atom core level.[244] The dark atom has been ionized.

cause variations in the intensity at the detector due to scattering from the substrate.

Liebsch illustrated both the applicability of the LEED formalism to the photoemission problem[64] and the potential of angle-resolved core-level photoelectron spectroscopy as a structural tool with a calculation for excitation from the $2s$ level of a sulfur monolayer on a nickel substrate.[34] This calculation showed that the interference between the direct wave ψ^0 and the diffracted wave ψ' resulted in variations of about $\pm 40\%$ in the cross-section as the measurement parameters were changed. These variations were very sensitive to the position of the adsorbed atom (the source). This calculation also showed that multiple scattering from the neighboring atoms was very important and could not be ignored. Finally Liebsch pointed out one real difference between the LEED-type final state and the real photoemission final state for an ordered overlayer. The ionized atom has a potential different from all the other adsorbate atoms. We return to this point later.

There are several experiments that will vary the outgoing wave vector $\mathbf{k}$ and consequently produce diffraction effects.[34] For example, the following three experiments suggest themselves.

1. Fix the direction of the detector and sweep the photon energy. As the photon energy increases, the wavelength of the excited electron will decrease; consequently the interference between the direct wave ψ^0 and the diffracted wave ψ' will go from constructive to destructive. The most common geometry is normal emission.

2. Fix the detector and rotate the sample about an axis normal to the surface. This changes the path length of the scattered waves and consequently the phase and amplitude of ψ' compared to ψ^0. These plots are called azimuthal patterns.

3. Fix the crystal and the incident light and change either the polar angle of collection or the azimuthal angle of collection. These experiments are usually dominated by the effects due to the direction of $\mathbf{A}$. Therefore they have not proved to be very useful.

We describe one measurement of type 1 and one of type 2. Each is compared to theoretical calculations using a multiple scattering LEED formalism.[229] Our objective is to illustrate the structural sensitivity of this technique and the reliability (or complications) of data analysis.

Normal emission collection is the simplest geometry of type 1 to analyze. The most extensively studied system is Se adsorption on Ni(100).[235-238] The early LEED work determined that the Se was bound in the fourfold site on the Ni(100) surface with a Se—Ni spacing of 2.34 Å. This corresponds to a vertical spacing ($d_\perp$) of 1.55 Å of the S above the center of the first Ni plane.[156] Figure 53 shows the intensity for the Se $3d$ emission for $p(2\times2)$ Se on Ni(100),[235] as well as several calculated curves for various bonding sites.[238]

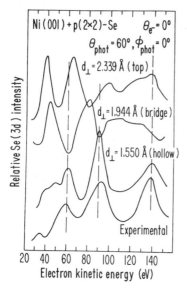

Fig. 53. Normal emission photoemission intensity versus kinetic energy for excitation of the Se(3d) level in $p(2\times2)$Se on Ni(100). The bottom curve is the experimental curve[235] and the top curves are calculated.[236,237] The binding energy of the Se(3d) level is 57 eV.[235]

All three sites have Se—Ni spacing of 2.34 Å. It is apparent from Fig. 53 that the bonding site can easily be determined from normal emission photon sweeps.

Li and Tong[236] showed that the theoretical curves were very sensitive to $d_\perp$. For example, the 90 eV (138 eV) peak in the hollow site curve moves to 72 eV (103 eV) as $d_\perp$ is changed from 1.55 to 1.94 Å. This is a shift of -45 eV/Å (-88 eV/Å). This type of change is qualitatively what should be expected from the change in the path length of the scattered wave compared to the primary wave. The important shift is the relative spacing between the two peaks, since the LEED calculation has an adjustable parameter called the inner potential. The inner potential can be used to adjust experimental and theoretical peak positions. The spacing between the two peaks is calculated to change by -42.5 eV/Å as the adsorbate substrate spacing is increased. The theoretical spacing can be made to fit the experimental spacing when $d_\perp \sim 1.63$ Å. This number compared to the 1.55 Å LEED value gives an indication that for most systems we could expect about 0.1 Å accuracy, which is comparable to results from LEED.

This procedure has been applied to $c(2\times2)$Se,[235, 236] $c(2\times2)$S [S(2p) emission],[235, 238] $c(2\times2)$Na [Na (2p) emission],[239] and $c(2\times2)$Te[239] on Ni(100). For all these systems except Te, the theoretical and experimental relative cross-section for normal emission agreed very well when the LEED-determined geometry was used in the calculation. The experimental and theoretical curves for $c(2\times2)$Te were qualitatively different.[240] The atomic cross-section for Te(4d) levels has pronounced structure in the first 100 eV of kinetic energy. The 4d wave functions have nodes that result in Cooper minima in the cross-section,[241] and the occupied 4d states couple strongly to the continuum f states.[242] These atomic effects make it hard to see the solid-state diffraction effects and require a more sophisticated theoretical scheme than is used in a LEED-type state.[242] In these cases a different type of photoelectron diffraction experiment is more appropriate, such as the azimuthal pattern described as type 2.

The "azimuthal pattern" measures the intensity from a core level at a fixed photon energy and polar collection angle as the crystal is rotated about an axis normal to the surface. These azimuthal patterns have been measured and calculated for $c(2 \times 2)$Te and Na on Ni(100).[243, 244] Figure 54 compares theory and experiment for three photon energies and 30° collection.[244] Although the Te and Na are believed to occupy the same fourfold hollow site on Ni(100),[156, 245] the azimuthal anisotropies are quite different, even for the same kinetic energy [2p binding energy is 31 eV for Na and 41 eV for Te(4d)]. For comparison, the 80 (90) eV Na azimuthal pattern is the same kinetic energy as the 90 (100) eV Te pattern. These differences arise from

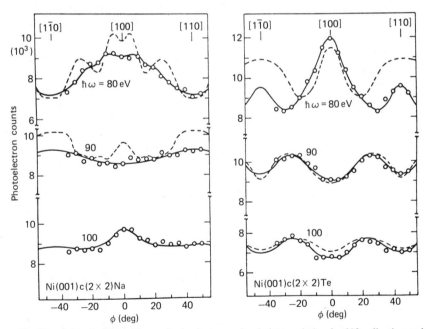

Fig. 54. Azimuthal dependence of adsorbate core-level photoemission for 30° collection and 80, 90, and 100 eV photon energies.[244] *Left*: 2p emission from $c(2 \times 2)$Na on Ni(100); *right*: 4d emission from $c(2 \times 2)$Te on Ni(100). Dashed curves are theoretical.[244]

two effects. First the Na—Ni bond length (2.84 Å[245]) is longer than the Te—Ni bond length (2.6 Å[156]), resulting in different phases between ψ^0 and ψ' in (32). Second, the Na($2p$) core level is excited into s or d final atomic states, while the Te($4d$) goes into p or f states.

The theoretical curves shown in Fig. 54 were normalized at $\phi=20°$. The agreement between experiment and theory, for the correct site, is good, reproducing the amplitude of the modulations and their shapes.[244] In this study an extensive series of calculations was performed, in which the polar angle and photon energy were varied, to see if any consistent fit with the experimental data could be achieved.[244] The only reasonable fit to the data was achieved when the adatom was placed in the fourfold hollow site. The vertical spacing was for Na 2.23 Å and 1.9 Å for Te, in agreement with LEED.[156, 245] The one problem was that the kinetic energy of the emitted electron from Te had to be increased by 8 eV compared to experiment, assuming a 11 eV inner potential.[244] A possible explanation is that adsorbed Na forms a metallic layer and the $2p$ core hole is screened during diffraction; but in the case of Te the surface layer is like an insulator, and the diffraction process occurs in the presence of the Coulomb field of the core hole.[244] This has the effect of increasing the kinetic energy of the electron near the core hole. We alluded to this problem at the beginning of the section. The Te on Ni(100) azimuthal patterns furnish the first hint that hole localization could be important.

Photoelectron diffraction as a structural tool is in its infancy. Only a few systems have been investigated, and consequently little correlation exists between normal emission photon energy sweeps and azimuthal patterns. Obviously the structural sensitivity of the two measurement geometries needs to be investigated on a common system. For example, Li and Tong[236] suggest that the structure in an azimuthal pattern might be totally a result of scattering in the overlayer. Smith et al.[244] show theoretical results for Na that disagree with this conclusion. Likewise, if the theoretical analysis of $c(2\times 2)$Te must include the effects of a localized core hole, then the effect should be more important in normal emission. Wendin[242] claims that the solid-state diffraction modulation calculated by Li and Tong[236] for Te on Ni will explain the experimental data[239] when the proper atomic cross-section is included. There is an inconsistency in these conclusions.

The most relevant question that should be answered is: Why should you do photoelectron diffraction experiments instead of LEED? The photoemission calculations appear to be more complicated than LEED because a photoemission matrix element must be combined with a time-reversed LEED calculation.[64] The photoemission experiments are undoubtedly more complicated than LEED. There are at least three distinct advantages of photoelectron diffraction over LEED.

1. A photon beam is not as destructive as a electron beam.
2. Core-level diffraction is atom specific. For example, Li and Tong[237] have shown theoretically that the C_{1s} normal emission intensity versus $\hbar\omega$ for adsorbed CO is sensitive to the carbon-metal spacing. The O_{1s} emission is very sensitive to the C—O spacing.
3. The interference process described by Fig. 52 is different from that due to a LEED beam. In photoelectron diffraction the oscillations in the signal are due to the interference of the diffracted beams with the primary beam, whereas a LEED state is just the sum of the diffracted beams. This might result in simplification of the theory when applied to photoelectron diffraction. Tong and Li[238] have shown that for kinetic energies larger than about 100 eV a calculation with a single backscattering reproduces the full multiple scattering calculation. This quasidynamic theory reduces computational time considerably.

The disadvantage of photoelectron diffraction is that the kinetic energy range is limited by existing monochromators. This limited photon energy range also limits surface EXAFS (extended X-ray absorption fine structure) measurements.[163, 246, 247] The interpretation of angle-resolved photoelectron diffraction measurements will always be more difficult than EXAFS. In EXAFS the total absorption cross-section is measured versus $\hbar\omega$. The effective 4π solid angle collection of all emitted electrons in an absorption experiment reduces the multiple scattering effects.[246] Consequently the analysis is much simpler and the oscillations are much smaller than in photoelectron diffraction. For example, the oscillation in the normal emission Se spectra shown in Fig. 53 was about 40% while one would expect $\lesssim 5\%$ oscillation in the equivalent EXAFS signal.

Fadley's[248–250] group has demonstrated the feasibility of a different experimental approach to structural determination using azimuthal patterns. Their approach is to use X-ray radiation ($\hbar\omega = 1487$ eV) for exciting core levels of less than about 600 eV binding energy, which gives the final electron a kinetic energy of about 1000 eV. They realized that if the measurements were made very close to grazing collection, the photoexcited electron would be forward scattered from the adjacent atoms and therefore the theoretical analysis could be simplified. The scattering could be treated in a single-scattering model.

The single-scattering model has proved to be successful in explaining their data. It is based on the following assumptions.

1. Excitation matrix elements are assumed to be given by plane wave final states. This is justified by the kinetic energy of the outgoing electron (~ 1000 eV).

2. A single-scattering or kinematical approximation is used, where the scattering from the j atom is given by a complex scattering factor

$$F_j(\theta_j) = |f_j(\theta_j)| \exp\{i\psi_j(\theta_j)\}$$

The scattering factor is determined from 21 partial-wave phase shifts calculated for purely atomic scattering or including the effects of valence electron redistribution due to bond formation.[248]

3. The inelastic scattering is included by way of an exponential damping.
4. Vibrational motion is incorporated by way of a Debye-Waller factor.
5. Scattering from a finite cluster of atoms is included. Typical sizes are in the range of 100 to 300 atoms.

It is appropriate that the last figure of this chapter be a set of Fadley "flower patterns" (Fig. 55). The data for the O_{1s} intensity from $c(2\times2)O$ on Cu(100) is shown on the top half of each drawing and the theory at the bottom. In processing, first the raw data were fourfold averaged; then the minimum intensity was subtracted from all data.[248]

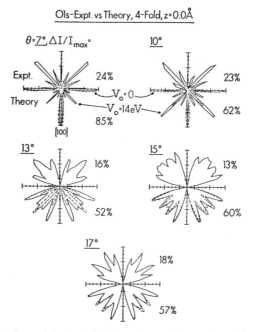

Fig. 55. A comparison of the fourfold-minus-I_{min} experimental and theoretical azimuthal distributions of the O_{1s} intensity from $c(2\times2)O$ on Cu(001).[250] The polar angles are measured from grazing. The anisotropies (%) are indicated for each item, with the experimental value on top and the theoretical value on the bottom.

Two theoretical calculations are shown in Fig. 55. The solid curve corresponds to a 14.1 eV inner potential with refraction included, and the dashed curve corresponds to no refraction (inner potential zero). The agreement between experiment and theory is very good. The theoretical curves have larger oscillations than the data that should be expected.[250]

The data were compared to theoretical calculations for various possible bonding sites and vertical displacements to determine the geometry of $c(2 \times 2)O$ on Cu(001). The best agreement between experiment and theory was found for the geometry of in-plane adsorption ($d_\perp = 0$) in fourfold holes (Fig. 32). This site has five equal O—Cu bonds of length 1.81 Å, which is reasonable because the bond length in Cu_2O is 1.84 Å. It is difficult at this stage of development to evaluate the accuracy of this technique, but it surely gives bonding site geometry and, presumably, a fairly accurate description of the vertical spacing.

APPENDIX
SURFACE AND BULK BRILLOUIN ZONES

The surface Brillouin zones of the three low-index faces of a face-centered cubic (fcc) and a body-centered cubic (bcc) crystal and the (0001) face of a hexagonal close-packed (hcp) crystal are shown in perspective to the bulk Brillouin zone.

A.I. Face-Centered Cubic

Figure 8 compared the (100) surface Brillouin zone to the bulk zone, and Fig. 9 plotted the cuts through the extended bulk Brillouin zone in the two symmetry directions for the (100) surface, $\overline{\Gamma}\text{-}\overline{X}$ and $\overline{\Gamma}\text{-}\overline{M}$. For an fcc crystal the bulk vectors shown in Fig. 8 have the form

$$\Gamma K = \left[\tfrac{3}{4}, \tfrac{3}{4}, 0\right] \frac{2\pi}{a}$$

$$\Gamma L = \left[\tfrac{1}{2}, \tfrac{1}{2}, \tfrac{1}{2}\right] \frac{2\pi}{a}$$

$$\Gamma W = \left[1, \tfrac{1}{2}, 0\right] \frac{2\pi}{a}$$

$$\Gamma X = \left[1, 0, 0\right] \frac{2\pi}{a}$$

where a is the bulk lattice constant. The two high-symmetry points in the (100) surface are $\overline{M}$ and $\overline{X}$ where,

$$\overline{\Gamma}\overline{M} = \frac{2\pi}{a} \qquad \text{and} \qquad \overline{\Gamma}\overline{X} = \sqrt{2}\,\frac{\pi}{a}$$

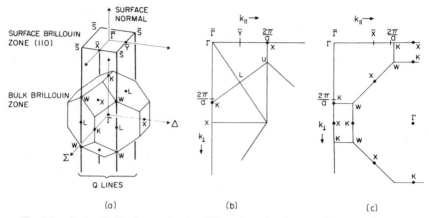

Fig. A.1.   Surface Brillouin zone for the (110) surface of an fcc crystal compared to the bulk Brillouin zone. (*a*) Projection of bulk zone onto SBZ. (*b*) and (*c*) Cuts through the bulk Brillouin zone in two directions of the SBZ; $k_{\parallel}$ is horizontal, $k_{\perp}$ is vertical.

Figure A.1 shows the surface Brillouin zone for the (110) surface and the cuts through the bulk extended zone in two directions for a (110) surface, where

$$\overline{\Gamma Y}=\frac{\pi}{a}, \qquad \overline{\Gamma X}=\frac{\sqrt{2}\,\pi}{a}, \qquad \overline{\Gamma S}=\frac{\sqrt{3}\,\pi}{a}, \qquad (110)$$

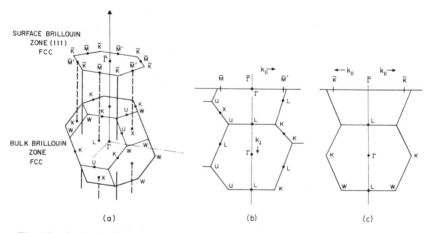

Fig. A.2.   Surface Brillouin zone for the (111) surface of an fcc crystal compared to the bulk Brillouin zone. (*a*) Projection of bulk zone onto SBZ. (*b*) and (*c*) Cuts through the bulk Brillouin zone in two directions of the SBZ; $k_{\parallel}$ is horizontal, $k_{\perp}$ is vertical.

Figure A.2 is for the (111) face of an fcc crystal, with

$$\overline{\Gamma K}= \frac{\sqrt{2}\,\pi}{a\cos^{2}30°} \qquad \text{and} \qquad \overline{\Gamma M}= \frac{\sqrt{2}\,\pi}{a\cos 30°}$$

A.II. Body-Centered Cubic

Figure A.3 shows the bulk Brillouin zone of a bcc crystal. The high-symmetry points are vectors of the form

$$\Gamma H=(0,0,2)\frac{\pi}{a}$$

$$\Gamma N=(1,1,0)\frac{\pi}{a}$$

$$\Gamma P=(1,1,1)\frac{\pi}{a}$$

The (100) SBZ is shown in Fig. A.3, where

$$\overline{\Gamma X}=\frac{\pi}{a} \qquad \text{and} \qquad \overline{\Gamma M}=\frac{\sqrt{2}\,\pi}{a}$$

bcc (100) SURFACE

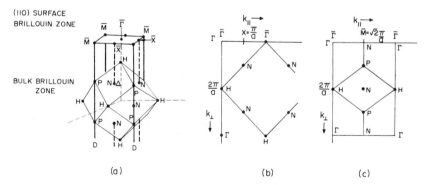

(a)   (b)   (c)

Fig. A.3. Surface Brillouin zone for the (100) surface of a bcc crystal compared to the bulk Brillouin zone. (a) Projection of bulk zone onto SBZ. (b) and (c) Cuts through the bulk Brillouin zone in two directions of the SBZ; $k_{\parallel}$ is horizontal, $k_{\perp}$ is vertical.

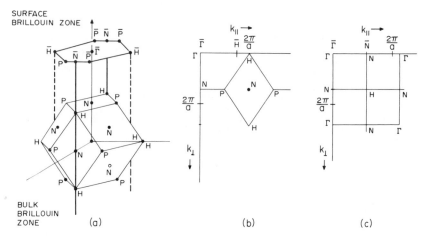

Fig. A.4. Surface Brillouin zone for the (110) surface of a bcc crystal compared to the bulk Brillouin zone. (*a*) Projection of bulk zone onto SBZ. (*b*) and (*c*) Cuts through the bulk Brillouin zone in two directions of the SBZ; $k_\parallel$ is horizontal, $k_\perp$ is vertical.

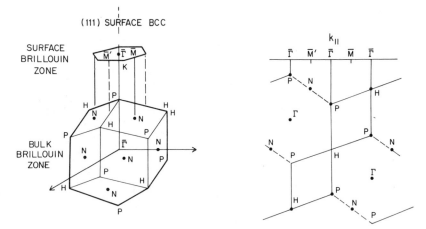

Fig. A.5. Surface Brillouin zone for the (111) surface of a bcc crystal compared to the bulk Brillouin zone. (*a*) Projection of bulk zone onto SBZ. (*b*) Cuts through the bulk Brillouin zone in two directions of the SBZ; $k_\parallel$ is horizontal, $k_\perp$ is vertical.

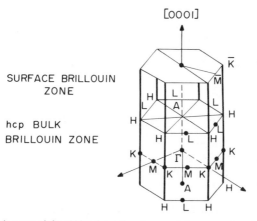

A.6. Surface Brillouin zone of the (0001) face of an hcp crystal compared to the bulk Brillouin zone.

Figure A.4 shows the (110) surface projections. The symmetry points of the surface are

$$\overline{\Gamma N} = \frac{\sqrt{2}\,\pi}{a} \qquad \text{and} \qquad \overline{\Gamma H} = \frac{3}{2}\frac{\pi}{a}$$

Figure A.5 is the (111) bcc surface, with

$$\overline{\Gamma M} = \frac{\pi}{\sqrt{2a}\,\cos 30°} \qquad \text{and} \qquad \overline{\Gamma K} = \frac{\pi}{\sqrt{2a}\,\cos^2 30°}$$

A.III. Hexagonal Close Packed

Figure A.6 shows the (0001) SBZ of an hcp crystal compared to the bulk Brillouin zone.

Acknowledgments

Most of the work was supported by the Materials Research Laboratory program under grant DMR-7923647 and ONR. We thank Roy Richter and John Wilkins for sending us unpublished results.

References

1. E. W. Plummer, in *Topics in Applied Physics*, Vol. 4, R. Gomer, Ed., Springer-Verlag, New York, 1975.
2. B. Feuerbacher and B. Fitton, *Phys. Rev. Lett.*, **30**, 923 (1970); **29**, 786 (1972); *Phys. Rev. B*, **8**, 4890 (1973); B. J. Waclawski, T. V. Vorburger, and J. R. Stern, *J. Vac. Sci. Technol.*, **12**, 301 (1975); W. E. Egelhoff and D. L. Perry, *Phys. Rev. Lett.*, **34**, 93 (1975).

3.   C. Kunz, in *Topics in Current Physics*, Vol. 10, Springer-Verlag, New York, 1979; *An Assessment of the National Need for Facilities Dedicated to the Production of Synchrotron Radiation*, National Academy of Sciences, Washington, D.C., 1976.

4.   J. Hermanson, *Solid State Commun.* **22**, 9 (1977).

5.   M. Scheffler, K. Kambe, and F. Forstmann, *Solid State Commun.*, **23** 7896 (1977).

6.   J. W. Davenport, *J. Vac. Sci. Technol.*, **15**, 433 (1978).

7.   D. W. Turner, C. Baker, A. D. Baker, and C. R. Brundle, *Molecular Photospectroscopy*, Wiley, New York, 1970.

8.   K. Siegbahn, C. Nordling, G. Johansson, J. Hedman, P. F. Hadin, K. Hamrin, U. Gelius, T. Bergmark, L. O. Wermo, R. Manne, and Y. Baer, *ESCA Applied to Free Molecules*, North Holland, Amsterdam, 1969.

9.   E. W. Plummer, W. R. Salaneck, and J. S. Miller, *Phys. Rev. B*, **18**, 1673 (1978).

10.   P. H. Citrin, G. K. Wertheim, and Y. Baer, *Phys. Rev. Lett.*, **41**, 1425 (1978). J. M. Duc, C. Guillot, Y. Lassailly, J. Lecnate, Y. Jugnet, and J. C. Verdrine, *Phys. Rev. Lett.*, **43**, 789 (1979).

11.   P. J. Feibelman, *Phys. Rev. Lett.*, **34**, 1092 (1975); *Phys. Rev. B*, **12**, 1319 (1975).

12.   K. L. Kliewer, *Phys. Rev. B*, **14**, 1412 (1976); *Phys. Rev. B*, **12**, 1319 (1975).

13.   H. J. Levinson, E. W. Plummer, and P. J. Feibelman, *Phys. Rev. Lett.*, **43**, 952 (1979); *J. Vac. Sci. Technol.*, **17**, 216 (1980).

14.   K. L. Kliewer and R. Fuchs, *Phys. Rev.*, **172**, 607 (1968).

15.   H. Bethe and E. E. Salpeter, in *Encyclopedia of Physics*, Vol. XXXV, Springer, Berlin-Göttingen-Heidelberg, 1957.

15a.   R. Manne and T. Aberg, *Chem. Phys. Lett.* **7**, 282 (1970).

16.   E. W. Plummer, T. Gustafsson, W. Gudat, and D. E. Eastman, *Phys. Rev. A*, **15**, 2339 (1977).

17.   U. Gelius, E. Basilier, S. Svensson, T. Bergmark, and K. Siegbahn, *J. Electron Spectrosc. Relat. Phenom.*, **2**, 405 (1974).

18.   J. Schirmer, L. S. Cederbaum, and W. Dumcke, *Chem. Phys.*, **26**, 149 (1977).

19.   J. Cooper and R. N. Zare, *Lectures in Theoretical Physics*, Vol. XI C, Gordon Breach, New York, 1969; also *J. Chem. Phys.*, **48**, 942 (1968).

20.   E. W. Plummer and T. Gustafsson, *Science*, **198**, 165 (1977).

21.   J. W. Davenport, Ph.D. thesis, University of Pennsylvania, 1976; *Phys. Rev. Lett.*, **36**, 945 (1976).

22.   R. J. Smith, J. Anderson, and G. J. Lapeyre, *Phys. Rev. Lett.*, **37**, 1081 (1976).

23.   T. Gustafsson and H. J. Levinson, *Chem. Phys. Lett.*

24.   J. L. Dehmer and D. Dill, *Phys. Rev. Lett.*, **35**, 213 (1975).

25.   N. Padial, G. Csanak, B. V. McKoy, and P. W. Langhoff, *J. Chem. Phys.*, **69**, 2992 (1978).

26.   C. L. Allyn, T. Gustafsson, and E. W. Plummer, *Chem. Phys. Lett.*, **47**, 127 (1977).

27.   N. V. Smith and M. M. Traum, *Phys. Rev. Lett.*, **31**, 1247 (1973).

28.   N. V. Smith, M. M. Traum, and F. J. DiSalvo, *Solid State Commun.*, **15**, 211 (1974).

29.   N. V. Smith and M. M. Traum, *Phys. Rev. B*, **11**, 2087 (1975).

30.   N. V. Smith, M. M. Traum, J. A. Knapp, J. Anderson, and G. J. Lapeyre, *Phys. Rev. B*, **13**, 4462 (1976).

31.   L. F. Mattheiss, *Phys. Rev. B*, **8**, 3719 (1973).

32.   J. W. Gadzuk, *Phys. Rev. B*, **10**, 5030 (1974).

33.   A. Liebsch, *Solid State Commun.*, **19**, 1193 (1976).

34.   A. Liebsch, *Phys. Rev. B*, **13**, 544 (1976).

35.   N. W. Ashcroft and N. D. Mermin, *Solid State Physics*, Holt, Rinehart & Winston, New York, 1976.

36.   G. D. Mahan, *Phys. Rev. B*, **2**, 4334 (1970).

37. K. Mitchell, *Proc. R. Soc., London, Ser. A*, **146**, 442 (1934).
38. For a clear discussion of this problem, see J. K. Sass and H. Gerischer, *Photoemission and the Electronic Properties of Surfaces*, B. Feuerbacher, B. Fitton, and R. F. Willis, Eds., Wiley, New York, 1978, Chapter 16.
39. W. Eberhardt and F. J. Himpsel, *Phys. Rev. B*, **21**, 5572 (1980).
40. N. V. Smith and L. F. Mattheiss, *Phys. Rev. B*, 1341 (1974).
41. J. A. Knapp, F. J. Himpsel, and D. E. Eastman, *Phys. Rev. B*, **19**, 4952 (1979).
42. J. Stohr, P. S. Wehner, R. S. Williams, G. Apai, and D. A. Shirley, *Phys. Rev. B*, **17**, 587 (1978); P. A. Wehmer, G. Apai, R. S. Williams, J. Stohr, and D. A. Shirley, in *Proceedings of the Fifth International Conference on Vacuum Ultraviolet Radiation Physics, Montpellier, France, 1977*, M. C. Castex, M. Pouey, and N. Pouey, Eds., p. 67, extended abstract.
43. P. Thiry, D. Chandesris, J. Lecante, C. Guillot, R. Pinchaux, and Y. Petroff, *Phys. Rev. Lett.*, **43**, 82 (1979).
44. G. A. Burdick, *Phys. Rev.*, **129**, 138 (1963).
45. J. F. Janak, A. R. Williams, and V. L. Moruzzi, *Phys. Rev. B*, **11**, 1522 (1975).
46. R. J. Smith, J. Anderson, J. Hermanson, and G. J. Lapeyre, *Solid State Commun.*, **21**, 459 (1977).
47. W. Eberhardt and E. W. Plummer, *Phys. Rev. B*, **21**, 3245 (1980).
48. F. J. Himpsel, J. A. Knapp, and D. E. Eastman, *Phys. Rev. B*, **19**, 2919 (1979).
49. T. C. Chang, J. A. Knapp, M. Aono, and D. E. Eastman, *Phys. Rev. B*, **21**, 3513 (1980).
50. K. A. Pandey and J. C. Phillips, *Phys. Rev. B*, **9**, 1552 (1974); K. C. Pandey, unpublished.
51. E. Dietz and D. E. Eastman, *Phys. Rev. Lett.*, **41**, 1674 (1978).
52. S. P. Singhal and J. Callaway, *Phys. Rev. B*, **16**, 1744 (1977).
53. H. Levinson, F. Greuter, and E. W. Plummer, to be published.
54. E. O. Kane, *Phys. Rev. Lett.*, **12**, 97 (1964).
55. P. Heimann, H. Moisga, and H. Neddermeyer, *Solid State Commun.*, **29**, 463 (1979).
56. F. J. Himpsel and W. Eberhardt, *Solid State Commun.*, **31**, 747 (1979).
57. J. F. Van der Veen, F. J. Himpsel, and D. E. Eastman, *Solid State Commun.*, **34**, 33 (1980).
58. C. N. Berglund and W. E. Spicer, *Phys. Rev.*, **136**, A1044 (1964).
59. N. V. Smith, *Phys. Rev. B*, **19**, 5019 (1979).
60. C. J. Powell, *Surf. Sci.*, **44**, 29 (1974).
61. P. J. Feibelman and D. E. Eastman, *Phys. Rev. B*, **10**, 4937 (1974).
62. S. G. Louie, P. Thiry, R. Pinchaux, Y. Petroff, D. Chandesris, and J. Lecante, *Phys. Rev. Lett.*, **44**, 549 (1980).
63. W. Schaich and N. W. Ashcroft, *Phys. Rev. B*, **8**, 2452 (1971).
64. A. Liebsch, *Phys. Rev. Lett.*, **32**, 1203 (1974).
65. J. B. Pendry, in *Photoemission and the Electronic Properties of Surfaces*, B. Feuerbacher, B. Fitton, and R. F. Willis, Eds., New York, 1978, Chapter 4, p. 87.
66. N. W. Ashcroft, in *Photoemission and the Electronic Properties of Surfaces*, B. Feuerbacher, B. Fitton, and R. F. Willis, Eds. Wiley, New York, 1978, Chapter 2, p. 21.
67. L. D. Landau and E. M. Lifshitz, *Electrodynamics of Continuous Media*, Pergamon, Oxford, 1960.
68. R. E. Makinson, *Proc. R. Soc. London, Ser. A*, **162**, 367 (1937).
69. K. L. Kliewer, in *Photoemission and the Electronic Properties of Surfaces*, B. Feuerbacher, B. Fitton, and R. F. Willis, Eds., Wiley, New York, 1978; Chapter 3, p. 45.
70. G. Mukhopadhyay and S. Lundqvist, *Solid State Commun.*, **21**, 629 (1977).
71. K. L. Kliewer and R. Fuchs, *Phys. Rev.*, **172**, 607 (1968).
72. N. D. Lang and W. Kohn, *Phys. Rev.*, 4555 (1970).

73. H. J. Levinson and E. W. Plummer, to be submitted.
74. P. O. Gartland and B. J. Slagsvold, *Solid State Commun.*, **25**, 489 (1978).
75. G. V. Hansson and S. A. Flodstrom, *Phys. Rev. B*, **18**, 1562 (1978).
76. J. B. Pendry, private communication; theory described in Ref. 77.
77. J. B. Pendry and J. F. L. Hopkinson, *J. Phys.* (*Paris*), *Colloq*, **7**, C4-142 (1978).
78. T. Forstmann, in *Photoemission and the Electronic Properties of Surfaces*, B. Feuerbacher, B. Fitton, and R. F. Willis, Eds., Wiley, New York, 1978, Chapter 8, p. 193.
79. D. Kalkstein and P. Soven, *Surf. Sci.*, **26**, 85 (1971).
80. I. Tamm, *Phys. Z. Soviet Union*, **1**, 733 (1932).
81. W. Shockley, *Phys. Rev.*, **56**, 317 (1939).
82. A. Many, Y. Goldstein, and N. B. Grover, *Semiconductor Surfaces*, North Holland, Amsterdam, 1965.
83. F. Forstmann and J. B. Pendry, *Phys. Rev. Lett.*, **24**, 1419 (1970); *Z. Phys.*, **235**, 75 (1970).
84. E. W. Plummer and J. W. Gadzuk, *Phys. Rev. Lett.*, **25**, 1493 (1970).
85. P. Heimann, J. Hermanson, H. Moisga, and H. Neddermeyer, *Surf. Sci.*, **85**, 263 (1979).
86. D. G. Dempsey and L. Kleinman, *Phys. Rev. B*, **16**, 5356 (1979).
87. J. F. Koch, R. A. Stradling, and A. F. Kip, *Phys. Rev.*, **133**, A240 (1964).
88. P. O. Gartland and B. J. Slagsvold, *Phys. Rev. B*, **12**, 4047 (1975).
89. J. B. Danese and P. Soven, *Phys. Rev. B*, **16**, 706 (1977); J. G. Gay, J. R. Smith, and F. J. Arlinghaus, *Phys. Rev. Lett.*, **42**, 332 (1979); J. A. Appelbaum and D. R. Hamann, *Solid State Commun.*, **27**, 881 (1978); S. Louie, P. Thiry, R. Pinchaux, Y. Petroff, D. Chandesris, and J. Lecante, *Phys. Rev. Lett.*, **44**, 549 (1980).
90. J. B. Pendry and J. F. L. Hopkinson, *J. Phys.*, **C4**, 142 (1978); E. B. Caruthers, L. Kleinman, and G. P. Alldredge, *Phys. Rev. B*, **8**, 4570 (1973); **9**, 3325 (1974); **9**, 3330 (1974); J. R. Chelikowsky, M. Schlüter, S. G. Louie, and M. L. Cohen, *Solid State Commun.* **17**, 1103 (1975); D. S. Boudreaux, *Surf. Sci.*, **28**, 344 (1971); D. Spanjaard, D. W. Jepsen, and P. M. Marcus, *Phys. Rev. B*, **19**, 642 (1979); H. Krakauer, M. Posternak, and A. J. Freeman, *Phys. Rev. Lett.*, **41**, 1072 (1978).
91. P. Heimann, J. Hermanson, M. Moisga, and H. Neddermeyer, *Phys. Rev. Lett.*, **42**, 1782 (1979); P. Heimann, H. Neddermeyer, and H. F. Roloff, *J. Phys. C*, **10**, L17 (1977); D. R. Lloyd, C. M. Quinn, and N. V. Richardson, *J. Phys. C*, **8**, L371 (1975).
92. H. F. Roloff and H. Neddermeyer, *Solid State Commun.*, **21**, 561 (1977); G. V. Hansson and S. A. Flodstrom, *Phys. Rev. B*, **17**, 473 (1978); P. Heimann, H. Neddermeyer, and H. F. Roloff, *J. Phys. C*, **10**, L17 (1977).
93. Z. Hussain and N. V. Smith, *Phys. Lett.*, **66A**, 492 (1978); G. Hansson and S. Flodstrom, *Phys. Rev. B*, **18**, 1572 (1978); P. Heimann, M. Moisga, and H. Neddermeyer, *Phys. Rev. Lett.*, **42**, 801 (1979).
94. F. J. Himpsel and D. E. Eastman, *Phys. Rev. Lett.*, **141**, 507 (1978).
95. E. W. Plummer and W. Eberhardt, *Phys. Rev. B*, **20**, 1444 (1979).
96. J. L. Erskine, *Phys. Rev. Lett.*, **45**, 1446 (1980).
97. P. Heimann and H. Neddermeyer, *J. Phys. F*, **6**, L257 (1976); P. J. Page, D. L. Trimm, and P. M. Williams, *J. Chem. Soc., Faraday Trans. I*, **70**, 1769 (1974); W. Eberhardt, E. W. Plummer, K. Horn, and J. Erskine, *Phys. Rev. Lett.*, **45**, 275 (1980).
98. D. Dempsey, W. Grise, and L. Kleinman, *Phys. Rev. B*, **18**, 1270 (1978); D. Dempsey and L. Kleinman, *Phys. Rev. Lett.*, **39**, 1297 (1977); S. C. Wang and A. J. Freeman, *Phys. Rev. B*, **21**, 4585 (1980).
99. P. Heimann and H. Neddermeyer, *Phys. Rev. B*, **18**, 3537 (1978); M. Tomasck and P. Mikusk, *Czech. J. Phys. B*, **28**, 904 (1978); B. Laks and C. E. T. Gonclaves Da Silva, *Solid State Commun.*, **25**, 561 (1978); D. G. Dempsey, L. Kleinman, and E. Caruthers, *Phys. Rev. B*, **12**, 2932 (1975); **13**, 1489 (1976); **14**, 279 (1976).

100. S. -L. Weng, T. Gustafsson, and E. W. Plummer, *Phys. Rev. Lett.*, **39**, 822 (1977); *Phys. Rev. B*, **18**, 1718 (1978).

101. B. Feuerbacher and R. F. Willis, *J. Phys. C*, **9**, 169 (1976); R. F. Willis, B. Feuerbacher, and H. E. Christenson, *Phys. Rev. Lett.*, **38**, 1087 (1979); B. Feuerbacher and B. Fitton, *Phys. Rev. Lett.*, **29**, 786 (1972); *Solid State Commun.*, **15**, 295 (1974); M. W. Holms, D. A. King, and J. E. Inglesfield, *Phys. Rev. Lett.*, **42**, 394 (1979); W. F. Egelhoff, J. W. Linnett, and D. L. Perry, *Phys. Rev. Lett.*, **36**, 98 (1976).

102. J. C. Campuzano, D. King, C. Somerton, and J. E. Inglesfield, *Phys. Rev. Lett.*, **45**, 1649 (1980); M. Holms and T. Gustafsson, to be published.

103. N. Nicolaou and A. Modinos, *Phys. Rev. B11*, 3687 (1975); *Surf. Sci.*, **60**, 527 (1976); *Phys. Rev. B*, **13**, 1536 (1975); N. Kar and P. Soven, *Solid State Commun.*, **20**, 977 (1976); R. V. Kasowski, *Solid State Commun.*, **17**, 179 (1975); N. V. Smith and L. F. Mattheiss, *Phys. Rev. Lett.*, **37**, 1494 (1976); M. Posternak, H. Krakauer, A. J. Freeman, and D. D. Koelling, *Phys. Rev. B*, **21**, 5601 (1980); W. R. Grise, D. G. Dempsey, L. Kleinman, and K. Mednick, *Phys. Rev. B*, **20**, 3045 (1979); H. Krakauer, M. Posternak, and A. J. Freeman, *Phys. Rev. Lett.*, **43**, 1885 (1979); J. E. Inglesfield, *Surf. Sci.*, **76**, 379 (1978).

104. G. Kerker, K. Ho, and M. Cohen, *Phys. Rev. B*, **18**, 5473 (1978); *Phys. Rev. Lett.*, **40**, 1593 (1978).

105. C. Noguera, D. Spanjaard, D. Jepsen, Y. Ballu, C. Guillot, J. Lecante, J. Paigne, Y. Petroff, R. Pinchaux, P. Thiry, and R. Cinti, *Phys. Rev. Lett.*, **38**, 1171 (1977).

106. S. G. Louie, K. Ho, J. R. Chelikowsky, and M. L. Cohen, *Phys. Rev. Lett.*, **37**, 1289 (1976).

107. F. J. Himpsel and D. E. Eastman, *Phys. Rev. B*, **20**, 3217 (1979).

108. P. J. Feibelman, D. R. Hamann, and F. J. Himpsel, *Phys. Rev. B*, **22**, 1734 (1980).

109. S. G. Louie, *Phys. Rev. Lett.*, **40**, 1525 (1978).

110. F. Greuter, W. Eberhardt, J. DiNardo and E. W. Plummer, *J. Vac. Sci. Technol.* (Detroit Meeting, 1980); W. Eberhardt, F. Greuter, and E. W. Plummer, *Phys. Rev. Lett.*, **46**, 1085 (1981).

111. F. J. Himpsel, J. A. Knapp, and D. E. Eastman, *Phys. Rev. B*, **20**, 624 (1979).

112. D. E. Eastman, F. J. Himpsel, and J. A. Knapp, *Phys. Rev. Lett.*, **40**, 1514 (1978).

113. C. S. Wang and J. Callaway, *Phys. Rev. B*, **15**, 298 (1977).

114. L. F. Wagner and W. E. Spicer, *Phys. Rev. Lett.*, **28**, 1381 (1972).

115. D. E. Eastman and W. D. Grobman, *Phys. Rev. Lett.*, **28**, 1378 (1972).

116. F. G. Allen and G. W. Gobeli, *Phys. Rev.*, **127**, 150 (1962).

117. J. van Laar and J. J. Scheer, *Surf. Sci.*, **8**, 342 (1967).

118. W. Mönch, *Festkörperprobleme*, Vol. XIII, Vieweg, 1973 p. 241.

119. H. D. Shih, F. Jona, D. W. Jepsen, and P. M. Marcus, *Phys. Rev. Lett.*, **37**, 1622 (1976).

120. T. D. Poppendieck, T. C. Ngoc, and M. B. Webb, *Surf. Sci.*, **74**, 34 (1978).

121. J. R. Chelikowsky, S. G. Louie, and M. L. Cohen, *Phys. Rev. B*, **14**, 4724 (1976).

122. D. J. Chadi, *Phys. Rev. Lett.*, **41**, 1602 (1978).

123. C. B. Duke, *Crit. Rev. Solid State Mater. Sci.*, **8**, 69 (1978).

124. J. D. Joannopoules and M. L. Cohen, *Phys. Rev. B*, **10**, 5075 (1974).

125. J. R. Chelikowski and M. L. Cohen, *Phys. Rev. B*, **13**, 826 (1976).

126. A. R. Lubinsky, C. B. Duke, B. W. Lee, and P. Mark, *Phys. Rev. Lett.*, **36**, 1058 (1976).

127. J. D. Levin and S. Freeman, *Phys. Rev. B*, **2**, 3255 (1970).

128. C. B. Duke, R. J. Meyer, A. Paton, P. Mark, A. Kahn, E. So, and J. L. Yeh, *J. Vac. Sci. Technol.*, **16**, 1252 (1979).

129. G. P. Williams, R. J. Smith, and G. J. Lapeyre, *J. Vac. Sci. Technol.*, **15**, 1249 (1978).

130. A. Huijser, J. van Laar, and T. L. van Roy, *Phys. Lett.*, **65A**, 337 (1978).

131. J. A. Knapp, D. E. Eastman, K. C. Pandey, and F. Patella, *J. Vac. Sci. Technol.*, **15**, 1252 (1978).

132. D. J. Chadi, *Phys. Rev. B*, **18**, 1800 (1978).
133. D. J. Chadi, *J. Vac. Sci. Technol.*, **15**, 1244 (1978).
134. C. Calandra, F. Manghi, and C. M. Bertoni, *J. Phys. C*, **10**, 1911 (1977).
135. J. R. Chelikowsky and M. L. Cohen, *Solid State Commun.*, **29**, 267 (1979).
136. A. Kahn, G. Cisneros, M. Boan, P. Mark, and C. B. Duke, *Surf. Sci.*, **71**, 387 (1978).
137. K. C. Pandey, J. L. Freeouf, and D. E. Eastman, *J. Vac. Sci. Technol.*, **14**, 904 (1977).
138. P. K. Larsen, J. H. Neave, and B. A. Joyce, *J. Phys. C*, **12**, L869 (1979).
139. K. Jacobi, C. V. Muschwitz, and W. Ranke, *Surf. Sci.*, **82**, 270 (1979).
140. W. Moch and P. Auer, *J. Vac. Sci. Technol.*, **15**, 1230 (1978).
141. D. Hanemann, *Phys. Rev.*, **121**, 1093 (1961).
142. R. Feder, W. Mönch, and P. O. Auer, *J. Phys. C*, **12**, L179 (1979).
143. G. Chiarotti, S. Nannarone, R. Pastore, and P. Chiaradia, *Phys. Rev. B*, **4**, 3398 (1971).
144. M. M. Traum, J. E. Rowe, and N. V. Smith, *J. Vac. Sci. Technol.*, **12**, 298 (1975).
145. A. W. Parke, A. McKinley, and R. H. Williams *J. Phys. C*, **11**, L993 (1978).
146. G. V. Hansson, R. Z. Bachrach, R. S. Bauer, D. J. Chadi, W. Gopel, *Surf. Sci.*, **99**, 13 (1980).
147. G. M. Guichar, F. Houzay, R. Pinchaux, and Y. Petroff, unpublished.
148. F. J. Himpsel, P. Heimann, and D. E. Eastman, submitted to *Phys. Rev.*
149. D. E. Eastman, *J. Vac. Sci. Technol.*, **17**, 492 (1980).
150. D. J. Chadi, R. S. Bauer, R. H. Williams, G. V. Hansson, R. Z. Bachrach, J. C. Mikkelsen, F. Houzay, G. M. Guichar, R. Pinchaux, and Y. Petroff, *Phys. Rev. Lett.*, **44**, 799 (1980).
151. M. Mehta and C. S. Fadley, *Phys. Rev. Lett.*, **39**, 1569 (1977); *Phys. Rev. B*, **20**, 2280 (1979).
152. M. C. Desjongueres and F. Cyrot-Lackmann, *Surf. Sci.*, **53**, 429 (1975); F. Cyrot-Lackmann, *J. Phys. Chem. Solids*, **29**, 1235 (1968).
153. K. S. Sohn, D. G. Dempsey, L. Kleinman, and E. Caruthers, *Phys. Rev. B*, **13**, 1515 (1976); **14**, 3185 (1976); **14**, 3193 (1976).
154. J. C. Fuggle and D. Menzel, *Surf. Sci.*, **53**, 21 (1975).
155. P. H. Citrin and G. K. Wertheim, *Phys. Rev. Lett.*, **41**, 1425 (1978).
156. J. E. Demuth, D. W. Jepsen, and P. M. Marcus, *Phys. Rev. Lett.*, **31**, 540 (1973).
157. K. Jacobi, C. V. Muschwitz, and K. Kambe, *Surf. Sci.*, **93**, 310 (1980).
158. A. Liebsch, *Phys. Rev. B*, **17**, 1653 (1978).
159. E. W. Plummer, B. Tonner, N. Holzwarth, and A. Liebsch, *Phys. Rev. B*, **21**, 4306 (1980).
160. R. Richter, O. Jepsen, and J. Wilkins, *Bull. Am. Phys. Soc.*, **25**, 193 (1980), and Private Communication.
161. S. A. Flodstrom, C. W. B. Martinson, R. Z. Bachrach, S. B. M. Hagstrom, and R. S. Bauer, *Phys. Rev. Lett.*, **40**, 907 (1978).
162. C. W. B. Martinson, S. A. Flodstrom, J. Rundgren, and P. Westrin, *Surf. Sci.*, **89**, 102 (1979).
163. L. I. Johansson and J. Stohr, *Phys. Rev. Lett.*, **43**, 1882 (1979).
164. P. Hofmann, C. V. Muschwitz, K. Horn, K. Jacobi, A. M. Bradshaw, K. Kambe, and M. Scheffler, *Surf. Sci.*, **89**, 327 (1979).
165. W. Eberhardt and F. Himpsel, *Phys. Rev. Lett.*, **42**, 1375 (1979).
166. K. Jacobi, M. Scheffler, K. Kambe, and F. Forstmann, *Solid State Commun.*, **22**, 17 (1977).
167. G. J. Lapeyre, J. Anderson, and R. J. Smith, *Surf. Sci.*, **89**, 304 (1979).
168. D. T. Ling, J. N. Miller, D. L. Weissman, P. Pianetta, P. M. Stefan, I. Landau, and W. E. Spicer, *Surf. Sci.*, **95**, 89 (1980).
169. T. W. Capehart and T. N. Rhodin, *J. Vac. Sci. Technol.*, **16**, 594 (1979).
170. D. T. Ling, J. N. Miller, D. L. Weissman, P. Pianetta, P. M. Stefan, I. Lindau, and W. E. Spicer, *Surf. Sci.*

171. K. Jacobi and C. V. Muschwitz, *Solid State Commun.*, **26**, 477 (1978).
172. D. R. Salahub, M. Roche, and R. P. Messmer, *Phys. Rev. B*, **18**, 6495 (1978).
173. R. Gomer, Ed., *Interactions on Metal Surfaces*, Vol. 4 in *Topics in Applied Physics*, Springer-Verlag, New York, 1975.
174. P. J. Feibelman and D. R. Hamann, *Solid State Commun.* **34**, 215 (1980).
175. P. J. Feibelman and D. R. Hamann, *Phys. Rev. B*, **21**, 1385 (1980).
176. P. J. Feibelman, J. A. Appelbaum, and D. R. Hamann, *Phys. Rev. B*, **20**, 1433 (1979).
177. S. G. Louie, *Phys. Rev. Lett.*, **42**, 476 (1979).
178. D. E. Eastman, *Solid State Commun.*, **10**, 933 (1972).
179. P. J. Feibelman and F. J. Himpsel, *Phys. Rev. B*, **21**, 1394 (1980).
180. H. Conrad, G. Ertl, J. Kuppers, and E. E. Latta, *Surf. Sci.*, **58**, 578 (1976).
181. J. E. Demuth, *Surf. Sci.*, **65**, 369 (1977).
182. W. Ranke and K. Jacobi, *Surf. Sci.*, **81**, 504 (1979).
183. M. Schlüter and M. L. Cohen, *Phys. Rev. B*, **17**, 716 (1978).
184. K. C. Pandey, *IBM J. Res. Dev.*, **22**, 250 (1978).
185. D. E. Eastman, F. J. Himpsel, J. A. Knapp, and K. C. Pandey, *Inst. Phys. Conf.*, **43**, 1059 (1979).
186. P. K. Larsen, N. V. Smith, M. Schlüter, H. H. Farell, K. M. Ho, and M. L. Cohen, *Phys. Rev. B*, **17**, 2612 (1978).
187. D. Denley, K. A. Millis, P. Perfetti, and D. A. Shirley, *J. Vac. Sci. Technol.*, **16**, 1501 (1979).
188. G. V. Hansson, R. Z. Bachrach, R. S. Bauer, and P. Chiaradia, *J. Vac. Sci. Technol.*,
189. W. E. Pickett, S. G. Louie, and M. L. Cohen, *Phys. Rev. B*, **17**, 815 (1978).
190. J. B. Johnson and W. G. Klemperer, *J. Am. Chem. Soc.*, **99**, 7132 (1977).
191. D. R. Salahub, private communication.
192. F. A. Cotton and G. Wilkinson, *Advanced Inorganic Chemistry*, 3rd ed., Interscience, New York, 1972.
193. C. L. Allyn, Ph.D. thesis, University of Pennsylvania, 1978.
194. P. R. Norton, R. L. Tapping, H. P. Brodia, J. W. Gadzuk, and B. J. Waclawski, *Chem. Phys. Lett.*, **53**, 465 (1978).
195. C. L. Allyn, T. Gustafsson, and E. W. Plummer, *Solid State Commun.*, **24**, 531 (1977).
196. J. Pritchard, *J. Vac. Sci. Technol.*, **9**, 895 (1972).
197. J. C. Tracy, *J. Chem. Phys.*, **56**, 2736 (1972).
198. C. L. Allyn, T. Gustafsson, and E. W. Plummer, *Solid State. Commun.*, **28**, 85 (1978).
199. G. Doyen and G. Ertl, *Surf. Sci.*, **43**, 197 (1974).
200. J. L. Falconer and R. J. Madix, *Surf. Sci.*, **48**, 393 (1975).
201. H. Conrad, G. Ertl, J. Koch, and E. E. Latta, *Surf. Sci.*, **43**, 462 (1974).
202. T. Gustafsson, private communication.
203. R. W. McCabe and L. D. Schmidt, *Surf. Sci.*, **65**, 189 (1977).
204. H. J. Freund, D. Heskett, and F. Greuter, private communication.
205. K. Hermann and P. S. Bagus, *Phys. Rev.*, **16**, 4195 (1977).
206. D. Saddei, H. J. Freund, and G. Hohlneicher, *Surf. Sci.*, **95**, 527 (1980).
207. D. E. Ellis, E. J. Baerends, H. Adachi, and F. W. Averill, *Surf. Sci.*, **64**, 649 (1977).
208. A. Rosen, P. Grundevik, and T. Morovic, *Surf. Sci.*, **95**, 477 (1980).
209. There is a possibility that correlation effects could make $\Delta E_{relaxation}$ positive.
210. R. J. Smith, J. Anderson, and C. J. Lapeyre, *Phys. Rev. B*, **22**, 632 (1980).
211. K. Horn, J. DiNardo, W. Eberhardt, H. J. Freund, and E. W. Plummer, to be published.
212. E. Umbach, A. Schichl, and D. Menzel, *Solid State Commun.*, **36**, 93 (1980).
213. P. Bagus, C. R. Brundly, K. Hermann, and D. Menzel, *J. Electron Spectrosc. Related Phenom.*, **20**, 253 (1980).
214. K. Hermann, P. S. Bagus, C. R. Brundle, and D. Menzel, submitted to *Phys. Rev. B*.
215. G. Loubriel, C. L. Allyn, T. Gustafsson, and E. W. Plummer, to be published.

216. P. A. Zhdan, G. K. Boreskov, A. I. Boronin, A. P. Schepelin, W. F. Egelhoff, and W. H. Weinberg, *Appl. Surf. Sci.*, **1**, 25 (1977).

217. K. Schönhammer and O. Gunnarsson, *Solid State Commun.*, **23**, 691 (1977); **26**, 399 (1978).

218. P. S. Bagus and K. Herman, *Surf. Sci.*, **89**, 588 (1979).

219. H. J. Freund and E. W. Plummer, submitted to *Phys. Rev.*

220. R. P. Messmer and S. H. Lamson, *Chem. Phys. Lett.*, **65**, 465 (1979); R. P. Messmer, S. H. Lamson, and D. R. Salahub, *Solid State Commun.*, **36**, 265 (1980).

221. K. Horn, A. M. Bradshaw, and K. Jacobi, *Surf. Sci.*, **72**, 719 (1978).

222. S. Anderson, *Solid State Commun.*, **21**, 75 (1977).

223. K. Horn, A. M. Bradshaw, K. Hermann, and I. P. Batra, *Solid State Commun.*, **31**, 257 (1979).

224. E. Jensen and T. Rhodin, *J. Vac. Sci. Technol.*,

225. R. J. Purtell, R. P. Merrill, C. W. Seabury, and T. N. Rhodin, *Phys. Rev. Lett.*, **144**, 1279 (1980).

226. M. W. Kang, C. H. Li, S. Y. Tong, C. W. Seabury, T. N. Rhodin, R. J. Purtell, and R. P. Merrill, to be published.

227. M. Scheffler, K. Horn, A. M. Bradshaw, and K. Kambe, *Surf. Sci.*, **80**, 69 (1970).

228. J. B. Pendry, *Low Energy Electron Diffraction*, Academic, New York, 1974.

229. The adaptation of Leed scattering formalism to photoemission has developed rapidly; see Refs. 47–52.

230. J. B. Pendry, *J. Phys. C*, **8**, 2413 (1975); *Surf. Sci.*, **57**, 679 (1976).

231. B. W. Holland, *J. Phys. C*, **8**, 2679 (1975); *Surf. Sci.*, **68**, 490 (1977).

232. S. Y. Tong and M. A. Van Hove, *Solid State Commun.*, **19**, 543 (1976).

233. C. H. Li, A. R. Lubinsky, and S. Y. Tong, *Phys. Rev. B*, **17**, 3128 (1978).

234. M. Scheffler, K. Kambe, K. Jacobi, and F. Forstmann, *Solid State Commun.*, **22**, 17 (1977); M. Scheffler, K. Kambe, and F. Forstmann, *Solid State Commun.*, **25**, 93 (1978).

235. S. D. Kevan, D. H. Rosenblatt, D. Denley, B. -C. Lu, and D. A. Shirley, *Phys. Rev. Lett.*, **41**, 1565 (1978); *Phys. Rev. B*, **20**, 4133 (1979).

236. C. H. Li and S. Y. Tong, *Phys. Rev. B*, **19**, 1769 (1979); *Phys. Rev. Lett.*, **42**, 901 (1979).

237. C. H. Li and S. Y. Tong, *Phys. Rev. Lett.*, **43**, 526 (1979).

238. S. Y. Tong and C. H. Li, *Crit. Rev. Solid State Sci.*, (1980).

239. G. P. Williams, I. T. McGovern, F. Cerrina, and G. J. Lapeyre, *Solid State Commun.*, **31**, 15 (1979).

240. I. T. McGovern, W. Eberhardt, and E. W. Plummer, *Solid State Commun.*, **32**, 963 (1979).

241. J. W. Cooper, *Phys. Rev.*, **128**, 681 (1962).

242. G. Wendin, *Solid State Commun.*, to be published.

243. D. P. Woodruff, D. Norman, B. W. Holland, N. V. Smith, H. H. Farrell, and M. M. Traum, *Phys. Rev. Lett.*, **41**, 1130 (1978).

244. N. V. Smith, H. H. Farrell, M. M. Traum, D. P. Woodruff, D. Norman, M. S. Woolfson, and B. W. Holland, *Phys. Rev. B*, **21**, 3119 (1980).

245. S. Andersson and J. B. Pendry, *Solid State Commun.*, **16**, 563 (1975); J. E. Demuth, D. W. Jepsen, and P. M. Marcus, *J. Phys. C*, **8**, L25 (1975).

246. P. A. Lee and J. B. Pendry, *Phys. Rev. B*, **11**, 2795 (1975).

247. P. H. Citrin, P. Eisenberger, and R. C. Hewitt, *Phys. Rev. Lett.*, **41**, 309 (1978).

248. S. Kono, C. S. Fadley, H. F. T. Hall, and Z. Hussain, *Phys. Rev. Lett.*, **41**, 117 (1978); S. Kono, S. M. Goldberg, H. F. T. Hall, and C. S. Fadley, *Phys. Rev. Lett.*, **41**, 1831 (1978); *Phys. Rev. B*, **22**, 6085 (1980).

249. L. -G. Petersson, S. Kono, H. F. T. Hall, C. S. Fadley, and J. B. Pendry, *Phys. Rev. Lett.*, **42**, 1545 (1979).

250. C. S. Fadley, S. Kono, L. -G. Petersson, S. M. Goldberg, H. F. T. Hall, J. T. Lloyd, and Z. Hussain, *Surf. Sci.*, **89**, 52 (1979).

AUTHOR INDEX

Tittel, F. K., 14(28), *43*
Tobias, R. S., 21(65), *44*
Tobocman, W., 264(11, 12), 266(11), 268(11), 269(36), 270(36), *272, 273*
Toennies, J. P., 208(46), *235,* 239(16), 259(71), *262, 263,* 281(45), 282(45), 297(45), *307*
Tomasck, M., 578(99), *652*
Tong, S. Y., 635(226), 639(236, 237, 238), 640(236, 237, 238), 642(236), 643(237, 238), *656*
Tonner, B., 597(159), 598(159), 599(159), 600(159), 601(159), 605(159), *654*
Top, Z. H., 205(21), 206(21), 235, 278(42), 279(43), 281(42), 288(62, 63), 289(43), 291(43), 292(43), 293(43), 306(62, 63), *307, 308*
Tossell, J. A., 75(73), *112*
Toxvaerd, S., 373(38, 39), 385(38), 386(38), 389(38, 39, 52), 413(52), *452, 453*
Tracy, J. C., 623(197), *655*
Traum, M. M., 545(27, 28, 29, 30), 546(29), 547(30), 548(30), 590(144), 638(244), 641(243, 244), 642(244), *650, 654, 656*
Tredwell, C. J., 28(75), *44*
Trickey, S. B., 205(26), *235*
Tricomi, F. G., 126(18), *188*
Triezenberg, D. G., 358(24), 409(24), *452*
Trimm, D. L., 578(97), *652*
Truhlar, D. G., 198(11), 199(11), 216(59), 223(86), 226(105, 106), *236, 237,* 289(67), *308*
Tsien, T. P., 239(26, 27), *262*
Tsuji, K., 525(159), 526(159), *532*
Tsuzuki, T., 347(45), *355*
Tully, J. C., 274(28, 29, 35), 283(48, 50), 288(27), 289(48), *307*
Turner, D. W., 534(7), 627(7), *650*
Turner, J. W., 311(3), 316(18), 320(18), 348(48), *354, 355*

Uang, Yea H., 373(34), 390(34), *452*
Ubbelohde, A. R., 456(3), 463(33, 34), 467(48), 521(149, 150), 522(149, 150), 523(149), *528, 529, 532*
Uhlenbeck, G. E., 186(80), *189*
Umanski, S. Ya., 274(24), *307*
Umbach, E., 627(212), 628(212), 629(212), 631(212), *655*
Underhill, C., 466(44), 473(61), 477(69), *529, 530*

Une, T., 185(76, 77), *189*

van der Lugt, W., 519(148), 520(148), 521(148), *532*
Van der Veen, J. F., 564(57), *651*
van der Waals, J. D., 358(2), 373(2), 383(2), 413(2), *451*
Van Hove, M. A., *656*
van Kampen, N. G., 358(6), 439(6), *452*
van Laar, J., 583(117), 587(130), *653*
Van Leuven, P., 115, 116(4), 117(4), 158(48), *188, 189*
van Roy, T. L, 587(130), *653*
Van Stryland, E. W., 10(22), *43*
Van Vechten, J. A., 63(48), *112*
Van Vleck, J. H., 140(29), 154(38), 175(67), 183(29), *188, 189,* 223(81), *236*
Velarde, M., 335(36), *355*
Venzl, G., 227(109), *237,* 257(66), 259(73), 260(73), *263*
Verdrine, J. C., 535(10), *650*
Verhaar, B. J., 174(65), *189*
Vezzetti, D. J., 285(55), *307*
Vidal, C. R., 116(3), 171(3), *188*
Vieceli, Jr., J. J., 358(25), 409(25), 440(25), 445(25), 448(25), 451(25), *452*
Vinatieri, J. E., 439(84), 443(91), *453, 454*
Vinitsky, S. I., 157(45), *188*
Vogel, F. L., 456(8), *528*
Vol'pin, M. E., 460(23), 481(81), *528, 530*
von der Linde, D., 9(18), 14(18, 27), 25(67), 26(67), *43, 44, 45*
von Jena, A., 8(16, 17), 15(16, 17), 17(16, 17), 28(72, 79), 30(79), *43, 44*
Von R. Schleyer, P., 48(6), *111*
Vorburger, T. V., 534(2), 578(2), *649*

Waclawski, B. J., 534(2), 578(2), 622(194), 623(194), 629(194), 636(194), *649, 655*
Wada, N., 515(140, 141), 517(140, 141, 142, 143, 144, 145), 518(142, 143, 144, 145), 525(140, 141, 142, 143, 144, 145), *532*
Wade, K., 51(26), 92(26), *111*
Wagner, L. F., 583(114), 591(114), *653*
Wakabayashi, N., 518(147), *532*
Waldeck, D., 7(14), 8(14), 9(14), 10(14), 11(14), 32(91), 33(91, 92), *43, 45*
Walgraef, D., 311, 312(5), 318(20), 320(26), *354, 355*
Walker, R. B., 206(34), 207(41), 226(99), 238(11), 253(57), 257(64), *261, 262, 263,* 289(69), *308*

SUBJECT INDEX

679